W0037151

EXPERIMENTAL STRESS ANALYSIS

Experimental Stress Analysis

Proceedings of the VIIIth International Conference on Experimental Stress Analysis, Amsterdam, The Netherlands, May 12–16, 1986
Organized by: Netherlands Organization for Applied Scientific Research (TNO) on behalf of The Permanent Committee for Stress Analysis

edited by

H. WIERINGA

TNO-IWECO
Delft, The Netherlands

1986 **MARTINUS NIJHOFF PUBLISHERS**
a member of the KLUWER ACADEMIC PUBLISHERS GROUP
DORDRECHT / BOSTON / LANCASTER

Distributors

for the United States and Canada: Kluwer Academic Publishers, 190 Old Derby Street, Hingham, MA 02043, USA
for the UK and Ireland: Kluwer Academic Publishers, MTP Press Limited, Falcon House, Queen Square, Lancaster LA1 1RN, UK
for all other countries: Kluwer Academic Publishers Group, Distribution Center, P.O. Box 322, 3300 AH Dordrecht, The Netherlands

Library of Congress Cataloging in Publication Data

ISBN-13:978-94-010-8465-9   e-ISBN-13:978-94-009-4416-9
DOI: 10.1007/978-94-009-4416-9

Preface

Designing and manufacturing structures of all kinds in an economic and a
safe way is not possible without doing experimental stress analysis. The
modernity of structures, with their higher reliability demands, as well
as today's more stringent safety rules and extreme environmental
conditions necessitate the improvement of the measuring technique and the
introduction of new ones.
Although theoretical/mathematical analysis is improving enormously, an
example of which is the finite element model, it cannot replace
experimental analysis and vice versa. Moreover, the mathematical analysis
needs more and more accurate parameter data which in turn need improved
experimental investigations. No one can do all those investigations on
his own. Exchange of knowledge and experience in experimental stress
analysis is a necessity, a thing acknowledged by every research worker.
Therefore, the objective of the Permanent Committee for Stress Analysis
(PCSA) is to promote the organization of conferences with the purpose
disseminating new research and new measuring techniques as well as
improvements in existing techniques, and furthermore, to promote the
exchange of experiences of practical applications with techniques.
This VIIIth International Conference on Experimental Stress Analysis on
behalf of the PCSA is one in a series which started in 1959 at Delft
(NL), and was followed by conferences at Paris (F), Berlin-W, Cambridge
(UK), Udine (I), Munich (FRG) and Haifa (Isr.). Such a Conference will be
held in Europe every fourth year, half-way bewteen the IUTAM Congresses.
Close co-operation exists between PCSA on the one hand and the
Gemeinschaft für Experimentelle Spannungsanalyse (GESA) of the VDI/VDE
Gesellschaft Mess- und Regelungstechnik (FRG) and with the Society of
Experimental Mechanics (SEM) on the other.
In this book the papers have been collected that will be discussed at the
VIIIth Conference on Experimental Stress Analysis.
The topic of the Conference is "Current Developments and Perspectives in
Experimental Mechanics". The PCSA expresses its thanks to the authors who
took the trouble to forward papers comprising their experiences in the
field of experimental mechanics and hopes that these contributions will
fulfill your expectations and help you in solving your stress analysis
problems. Moreover, PCSA expects the Conference to contribute to the
development and deepening of personal contacts between the participants
from all over the world.
It also likes to express its thanks to the Paper Selection Committee
members for their evaluation of the papers presented and for their advice
and support in drafting the programme. Acknowledgement of thanks is also
expressed to the Netherlands Organization for Applied Scientific Research
(TNO) and in particular to the fellow-workers of the TNO Corporate
Communication Department and those of TNO-IWECO; Institute of Mechanical
Engineering. Without their efforts this Conference would not have been
possible.

May 1986
H. Wieringa
Chairman Permanent Committee
for Stress Analysis

Organizing Committee

H. Wieringa, Chairman
H. van den Berg, Secretary

Permanenent Committee for Stress Analysis

H. Wieringa (The Netherlands)-Chairman
S.I. Andersen (Denmark)
V. Askegaard (Denmark)
M.C. Azevedo (Portugal)
A. Berkovits (Israel)
A.A. Betser (Israel)
R. Bourgois (Belgium)
A. Bray (Italy)
R. Fidler (United Kingdom)
K. Fristedt (Sweden)
J.J. Geerlings (The Netherlands)
M. Gibstein (Norway)
K.H. Laermann (Federal Republic of Germany)
A. Lagarde (France)
A.R.G. Lamas (Portugal)
A. Martin (France)
Chr. Rohrbach (Federal Republic of Germany)
W. Schumann (Switzerland)
P. Stanley (United Kingdom)
D.S. Theocaris (Greece)
M.E. Tschinke (Italy)

Paper Selection Committee for the VIIIth International Conference on Experimental Stress Analysis

H. Wieringa, Chairman (The Netherlands)
K.H. Laermann (Federal Republic of Germany)
P. Stanley (United Kingdom)
J.J.P. Geerlings (The Netherlands)
A.U. de Koning (The Netherlands)

Scientific Secretariat / Editor

H. Wieringa
TNO-IWECO
P.O. Box 29
2600 AA Delft
The Netherlands
Phone: +31/15 608608
Telex: 38192 iweco nl

TABLE OF CONTENTS

MEASURE AND DESIGN OF STRESSES IN ADHESIVE BONDED TRUSSES : EXTENSOMETRI-
CAL AND LASER-ELASTICIMETRICAL METHODS

L. BEN AICHA[*], Y. GILIBERT[**], A. RIGOLOT[***]

E.N.S.T.A. - L.M.E.
Groupe Composites et Collage
Centre de l'Yvette, Chemin de la Hunière
91120 PALAISEAU

1. INTRODUCTION
    Used since the remotest antiquity, the bonded assemblage has been deve-
loped in the technological field since the beginning of the 20th century.
    However, the fondamental researches about this way of bonding dates
back to, at the very most, forty years. Coupled with important progresses
in joint design and improved knowledge of bonding processes, the result is
an ever growing variety of adhesive bonded assemblages, depicted for instance in references [1].
    One of the aims of this paper is to show with the help of examples,
that the extensometrical method with electrical gauges is suitable for the
study of the mechanical behaviour of bonded structures.
    This method, elaborated by GILIBERT [2] in the years 1971-73, allowed
to validate a finite elements method [5].
    In 1977-78, GILIBERT and RIGOLOT have developed an analytical theory
of double lap bonded joints by using the method of matched asymptotic ex-
pansions [3] [4] ; since then, the comparison between measurements and
computations allowed conclusive progress with regard to this complex domain of solid mechanics.
    In 1985, with the collaboration of FERRE [6], BEN AICHA and GILIBERT
used a laser-photoelasticimetry method to measure the stresses in double
lap bonded joints with the aim of confirming theoretical results.

2. PREPARATION OF THE SURFACE AND MEASUREMENT OF THE GEOMETRY
2.1. Properties of the Adherends
    A low carbon steel (0.18 % carbon ; XC 18 French standard equivalent
to SAE-AISI 1017) was used for the adherends of the specimens studied. The
properties of the steel were controlled using mechanical measurements and
microscopic observations. The Young's modulus (E_T = 207 700 MPa) and the
Poisson's ratio were determined using tension test with strain gauges at
a very low strain rate ($\dot{\varepsilon}$ about 10^{-6} s^{-1}) on steel specimens with 100 mm2
square cross section and 250 mm length. The Brinnell hardness (HB = 210)
test showed the uniform quality of the material, whereas microscopic observations indicated a ferritic microstructure.

[*]Thésard ENSTA/LME/GCC

[**]Chef du Groupe Composites et Collage, Maître de Conférences,
    Docteur ès Sciences

[***]Conseiller Scientifique du G.C.C., Professeur des Universités à Reims

Wieringa, H (ed), Experimental Stress Analysis.
© *1986. Martinus Nijhoff Publishers, Dordrecht.*

2.2. Preparation of the Rough Adherends, Finishing and Blasting

A shaping machine with a traversing tool was used in the preparation of the rough adherends.

This was followed by milling and after which a grinding was effected with a horizontal surface grinding machine under different conditions : coarse or fine grain grinding wheels. For all these operations, machining conditions (speed, feed, depth of cut, lubrication, sharpening of the tools) were strictly controlled and are described in Reference 2,Chapter 4

The finishing procedure for some of the specimens was completed with sand blasting under four conditions using different particle diameters (115, 169, 282 and 423 μm) or with shot blasting. The particles with a 115 μm diameter were made of high purity white alumina ($Al_2O_3 > 99,5$ %), whereas the particles with diameters of 423, 282 and 169 μm were of brown corundum, i.e. less pure alumina (composition : $Al_2O_3 > 94$ % ; $TiO_2 < 2.5$ % ; $SiO_2 < .5$ %). The blasting pressure used was 0.4 to 0.5 MPa. The sand jet was inclined at 60° from the specimen surface and displaced at a speed of 24 mm s^{-1}. The shot blasting was conducted with iron shots 1 mm in diameter.

In this paper, the various surface states are designated as follows :
RF - fine ground state ;   RFS - fine ground sans blasted state ;
RG - coarse ground state ; RGS - coarse ground sans blasted state ;
RGG - coarse ground shot blasted state.
The numbers following RFS and RGS designate the particle mesh size.

2.3. Measurement of the Surface Roughness Parameters

The surface profiles were determined using probes with an air bearing of 25 μm and 750 μm travel distances. The specimens used had the same section as the adherends and were machined and treated at the same time under the same conditions. The surface profiles were measured, before and after blasting, along reference paths, the ends of which are marked with two notches remaining after blasting.

Data were recorded and processed by a computer to determine the various parameters of the surface defects describing the total profile, roughness and waviness. The calculations were done using a method developed by B. Scheffer of RENAULT (FRANCE).

We observed that the surface roughness parameters alter very quickly during the first few seconds of blasting and do not change after a while. We selected a time lapse of 30 seconds which ensured a stable state of these parameters.

Measurements showed that the surface roughness parameters are not dependent on the direction of measurement for blasted states, whereas condition as ground showed some anisotropy.

Reference 2, chapter IV, table IV.1 presents values of the surface roughness parameters for the various surface states studied. These are the total depth of roughness, R_t ; the average depth of roughness, R ; the maximum depth of roughness, R_{max} : the roughness levelling depth, R_p ; the arithmetical mean deviation from mean line of roughness, R_a ; and average spacing of roughness, A_R.

From these data, it can be seen that the surface roughness parameters for sand blasted states depend on both the initial grinding and the mean diameter of the particles used for blasting. For all surface preparations, the maximum depth of roughness R_{max} is very close to the total depth R_t, which indicates that surfaces have no aberrant defects of roughness and are consequently fairly homogeneous ; for the fine ground state, the roughness parameters increase steadily with the particle diameter whereas for

the coarse ground state these increase notably only for a diameter superior to 200 µm.
Other surface defect parameters were computed but proved of little importance as far as the mechanical properties of the joints are concerned. However, they confirmed that the prepared surfaces had well defined and reproducible parameters.

2.4. Properties of the adhesive
We used the commercial adhesive, Eponal 317, of Sonal, France, which is a two component system (an epoxy resin containing mineral fillers and a hardening agent). It was processed at room temperature ; at such temperature, about 20 minutes are available for use and polymerization rate reaches 90 % within 2 hours.
We determined its mechanical, physical, structural and chemical properties The elastic constants were measured at room temperature (20°C) using two methods. Tensile test using strain gauges and a very slow loading rate (ε about 10^{-6} s^{-1}) showed an elastic brittle behavior. Using traction test the Young's modulus was found to be 5800 MPa, the Poisson's ratio was 0.327 and the strength was 30 MPa ; whereas ultrasonic measurements gave 5940 MPa for the Young's modulus and 0.323 for the Poisson's ratio. The measured density was 1.433 kg dm^{-3}.

3. DESIGN AND MANUFACTURE OF THE TEST-SPECIMENS-LOADING CONDITIONS
We used specimens with two double-lap joints. One of these double-lap joints was used for strain measurements. The other joint, which was tightened transversely by a binding clip, ensured axial and uniform load in the adherends, while the clip eliminated any local bending. The extremities of the test-specimen were fixed on a universal testing machine by knee joints. In fact, this design of the specimen ensured symmetrical fractures in the tested part, and the clip decreased considerably the statistical deviation in the measurements, as shown in Reference 2, Chapter 2. Figure 1 shows a schematic view of the specimen.
The parts (10 mm x 4.5 mm cross-section for parts 2' ; 10 mm cross-section for parts 1') were produced at a ± 0.01 mm tolerance. The length of the overlap in the tested part of the specimen was 88 mm. A thickness of 0.5 mm was selected for the adhesive joint and controlled by gauged spacers. Preliminary tests, described in Reference 2, Chapter 11, showed that this thickness gives the best mechanical behavior for these dimensions of the metallic parts and this length of the joint.
The test-specimens were loaded in tension at a slow rate (1000 N per minute) at room temperature (20°C).

4. PRINCIPLE OF THE MECHANICAL MEASUREMENTS
Since mechanical quantities inside the joint cannot be obtained directly, we used electric strain-gauges placed at different points on external surfaces of the metallic adherends. Figure 1 shows the position of the gauges ; their distances from the middle of the joint are as follows :
- gauge 1 : - 37   mm
- gauge 2 : - 18.5 mm
- gauge 3 :   0
- gauge 4 : + 18.5 mm
- gauge 5 : + 37   mm
The positioning of the gauges is described in detail in Reference 2, (Chapter 5).

A sensitive and precise (± 0.25 %) bridge (VE 20 Micromeasurement) was used.
Figure 2 shows typical strain gauge recordings for a ground sans blasted state. Other recordings are given in Reference 2.
Principle of the mechanical Measurements is described by references 2 and 7

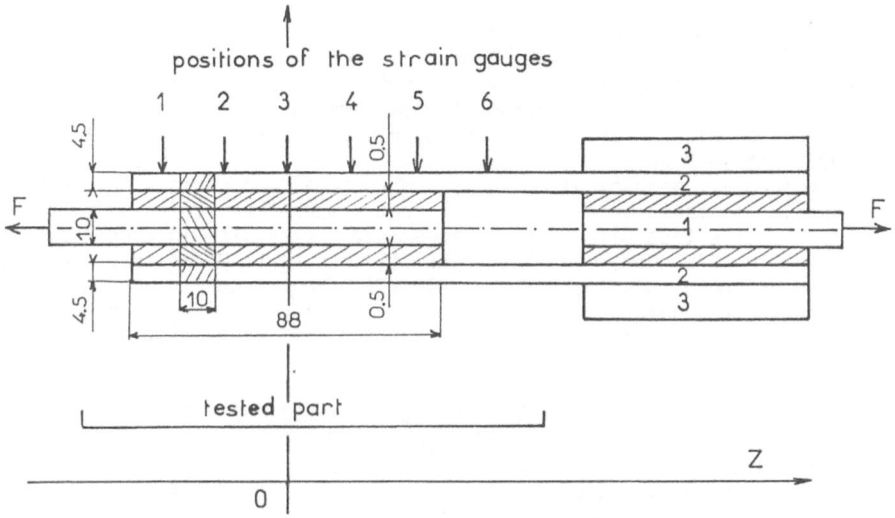

Figure 1. Schematic view of the test-specimen (dimensions in mm). A view of the cross-section is superimposed between gauges 1 and 2. 1' and 2' : adherends ; 3' : binding clip.

Visual and microscopic observations of the fractured specimens have confirmed that in the central part of the joint a progressive shear fracture occurs inside the adhesive ; at the ends of the joint one can see either a shear fracture inside the adhesive or an adhesive failure between the adhesive and the metallic adherend along the upper or lower face, induced during the final phase of the fracture.
These facts differentiate the ground sand blasted state from the ground state, which showed predominantly metal-resin adhesive fractures.
The various stages for the ground sand blasted states, (i) elastic behavior, (ii) initiation of micro-cracks, (iii) coalescence of the micro-cracks, (iv) stable propagation of the cracks, and finally, (v) unstable propagation up to final fracture, proved to be absolutely reproducible (Reference 2, Chapter 8).
These experimental results can be partly explained by the various theoretical analyses [3, 4, 8], which have analyzed stress fields in adhesive joints. However, these theoretical analyses, which are based on continuum mechanics and use idealization of interfaces, cannot explain the significant differences observed using the above measurement procedure, and are presented below.

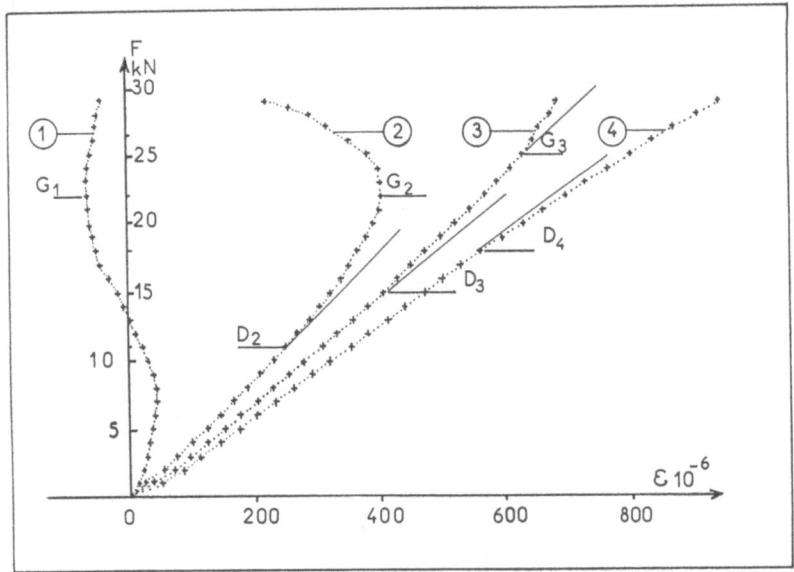

Figure 2. Strain versus tensile load for a ground sans blasted state
(outputs of gauges 1-4). Non-linearity occurs over point D_2.

5. EFFECTS OF ROUGHNESS ON MECHANICAL RESULTS

Here are presented the effects of roughness on the three typical mechanical
parameters : elastic limit (crack initiation threshold), crack propagation
threshold, and ultimate strength.
We found that roughness can be adequately represented by using the total
depth of roughness R_t.
We observed a very small scattering in the measured values, which gives
proof of the importance of our careful and well-defined process in preparation and testing. Deviation in the results is generally higher for the
propagation threshold and the ultimate strength than for the elastic limit.
Further, deviation in the ultimate strength is higher for non-blasted surfaces than for blasted surfaces (Reference , Chapter 8). Typical values,
corresponding to the results presented in Figure 3, are (i) 0.25 kN for the
crack initiation threshold (about 12 kN), (ii) 0.5 kN for the crack propagation threshold (about 24 kN), and (iii) 0.5 kN for the ultimate strength
(about 30 kN).
For the sand blasted states, we found that the mechanical parameters were
influenced by the initial grinding and the roughness, which is related to
the sand particle size.
The fine grinding is equivalent or better than the coarse grinding for the
three mechanical parameters ; it gave better values for the elastic limit
with all diameters.
For both types of grinding, we observed that the strain measurements given
by the gauges at the external surface of the adherends were significantly
smaller (for any position) and more uniform along the specimen, when the
surfaces were prepared using particles with a diameter of 169 μm. Further-

more, this preparation shows the best values for the three mechanical parameters.

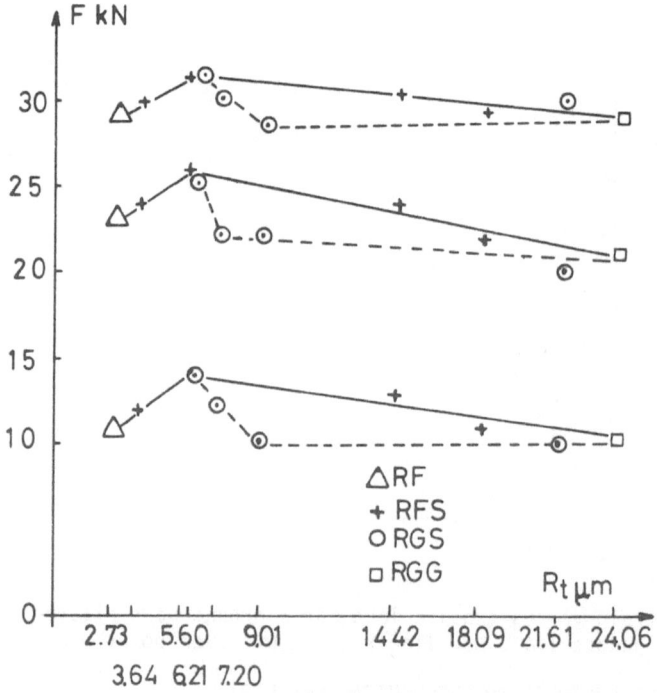

Figure 3. Mechanical parameters versus the total depth of roughness : from
the bottom, (i) crack initiation threshold, (ii) crack propagation threshold, and (iii) ultimate strength. For fine grinding without blasting (RF)
the values are in accordance with the extrapolation (with respect to the
roughness) of those obtained for fine grinding with blasting (RFS).

Figure 3 summarize the results. Almost all of the plotted points represent
the mean value of four tests. In Reference 2, Chapter 9, one can find the
curves representing F versus the other roughness parameters, R_a, R_p, R ;
they have the same behavior as the curves in Figure 6.
Figure 3 shows that for fine grinding, with or without blasting, the three
mechanical parameters reach a maximum value when the total depth of roughness equals 5.6 μm. It seems possible to associate this property with the
fact that the total depth of roughness is approximately the size of the
fillers in the resin.
No precise trends can be found in the case of coarse grinding.

6. COMPARISON BETWEEN THEORETICAL AND EXPERIMENTAL RESULTS
For an optimal state (e_j = 0,5 mm), we see on figure 4 a good correlation
in the elastic domain between theoretical results [3] [4] [8] and experimental measures. Some difference appears in the neighbourhood of the free
end of the upper adherend : it was impossible to introduce in computations
the effects of the angular singularities in these regions.

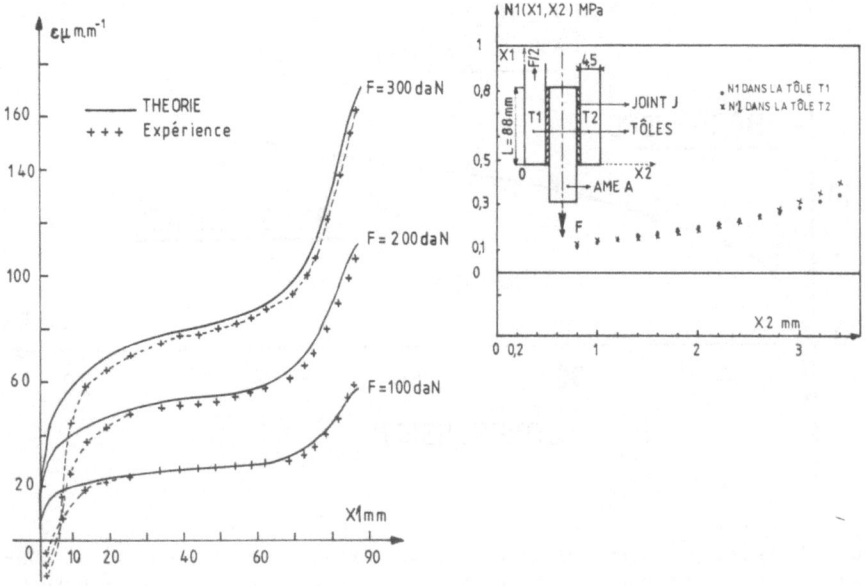

Figure 4. Longitudinal deformations versus distance of the measurement. Localization from the free ends of the adherends (T_1 and T_2, figure 5)

Figure 5. Principal stresses in adherends T_1 and T_2 at abscissa $X_1 = 10,5$ mm versus the thickness coordinate X_2.

7. STRESS MEASUREMENTS BY A LASER PHOTOELASTICIMETRICAL METHOD

The truss is made in the same way that the one depicted in Figure 1, whose adherends in polycarbonate (PSM-1) are joined by PC1 adhesive (Vishay Micro-mesures, Young's modulus : $E_{PSM.1}$ = 2400 MPa ; E_{PC1} = 3150 MPa. POISSON's ratio : $\nu_{PSM.1}$ = 0.38 ; ν_{PC1} = 0.36 ; Figure 5.)

Principal stresses, measured by an automatic laser-photoelasticimeter, are presented in Figure 6. It's to be noted that the shape of the curve in figure 6 is similar to the experimental one depicted in figure 4.,especially in the neighbourhood of the free edge of the adherend.

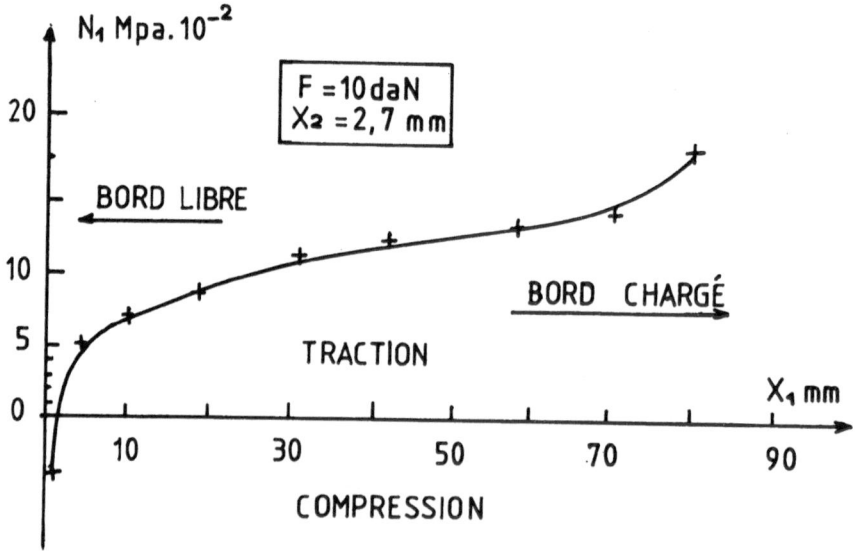

Figure 6. Principal stresses along the overlap in the adherend T1,at abscissa : X_2 = 2,7 mm,versus the length-coordinate X_1 (related to figure 5).

CONCLUSION
Our experimental results show that the surface roughness has a significant
influence on the mechanical properties (crack initiation threshold, crack
propagation threshold, and ultimate strength) of joints. We have found the
best conditions for joint design as follows :
- fine grinding is better than coarse grinding ;
- sand blasting improves mechanical properties ;
- sand particle diameter must be selected so as to give a total depth of
roughness close to the size of the adhesive fillers.
However, we are quite satisfied of reaching one of the aim of this work
which is the reproducibility of optimal specimens.

REFERENCES

1. BLOMQUIST E.M. Adhesives-past, present and future. American Society for
   Testing and Materials, 1963.
2. GILIBERT Y. Contribution à l'étude de l'adhésivité de matériaux collés
   par l'intermédiaire de résines époxydiques. Thèse de doctorat d'état de
   physique, Université de Reims, 1978.
3. a) GILIBERT Y., RIGOLOT A. Analyse asymptotique des assemblages-collés
   à double recouvrement sollicités au cisaillement en traction. J. Mech.
   Appl. Vol. 3, n° 3, pp. 341-372, 1978.
   b) Analyse asymptotique des assemblages collés à double recouvrement
   sollicités au cisaillement en traction. C.R. Acad. des Sc. Paris, Mécanique des Solides, t. 288, Série B, pp. 387-390 (25 Juin 1979).

4. RIGOLOT A. Régularisation de l'effet Saint-Venant dans une structure plane composite collée - Actes du Congrès Euromech 184 - Cachan - Septembre 1984.

5. AIVAZ-ZADEH M.S., VERCHERY G., GILIBERT Y. Eléments finis d'interface. Application aux assemblages collés et structures stratifiées. Actes du colloque : Tendances actuelles en calcul des structures, 6-7-8 Novembre 1985, Pluralis, Paris.

6. FERRE M. Ecole Nationale Supérieure de Techniques Avancées, 32 Boulevard Victor, 75015 PARIS.

7. GILIBERT Y., VERCHERY G. Influence of surface roughness on mechanical properties of joints. Title of book : ADHESIVE JOINTS : formation characteristics and testing, pp. 69-84, 1984, PLENUM PUBLISHING CORPORATION.

8. GILIBERT Y., RIGOLOT A. Théorie élastique de l'assemblage collé à double recouvrement : utilisation de la méthode des développements asymptotiques raccordés au voisinage des extrêmités. R.I.L.E.M., Matériaux et Constructions, N° 107, pp. 363-387, DUNOD, Paris.

AUTOMATED TESTING OF CONCRETE COMPRESSIVE PROPERTIES

V. Cervenka, P. Bouska

Building Research Institute, Czech Technical University
Solinova 7, 166 08 Praha 6, CSSR

INTRODUCTION

Material properties of concrete are normally determined by standard testing methods according to the code procedures. These testing methods were developed namely for the requirements of quality control. The test for each material parameter is typically performed on a special test specimen.For example, the concrete strength is tested on cubes or cylinders, the modulus of elasticity on prismatic specimens, etc. The measuring devices are often not quite efficient and do not allow automation of the testing process. The use of resistance wire strain gagues glued on the specimen surface has also many disadvantages, such as short measuring base, limited range, etc. Therefore a question is often raised about the reproducibility of such tests.

At the Building Research Institute of the Czech Technical University in Prague a research is conducted concerning the process of damage of the concrete under variable loading history. A method of automated testing of concrete was developed within this project. Its original purpose was to provide accurate measurements of stress-strain diagrams in the whole range of compressive loading and to provide data on the damage of the material by microcracks by means of volumetric strain. The method is suitable for both, scientific research of material properties, and due to its efficiency, also for large scale quality control purposes. It is restricted to the uniaxial compressive loading with variable loading history. However, this paper deals only with the loading under monotonously increased strain.

The research was performed in the laboratories of the institute. The used electronic measuring devices and systems originate from the company Peekel Instruments, who supplied the equipment and provided valuable consulting.

DESCRIPTION OF THE SYSTEM

The scheme of the system is shown in Fig. 1. It consists of three parts (two hardware parts and one software part):

1. Hydraulic testing machine with programable loading.
2. Computer-controlled measuring system.
3. Software for load control, measuring and data processing.

Before going to the detail description, the overall function of the system is discussed.

A special cage with strain transducers is attached to a concrete specimen of cylindric or prismatic shape.The specimen is placed in a loading machine. The deformation gauges and the electro-hydraulic system of the testing machine are permanently connected to a computer-controlled data acquisition system. The loading of specimens and measurements are done by means of computer programs, which are operated from a computer terminal placed near the testing machine. The uniaxial compressive test is performed under a constant rate of the longitudinal strain. Measured data are saved in disk files. In this way a series of specimens can be tested. After the testing is completed the data are reduced by means of another set of programs. Several material parameters, such as compressive strength or modulus of elasticity, are evaluated and stress-strain diagrams are plotted. The parameter evaluation is performed for each specimen individually and statistical analysis for a series of specimens can be also performed.

The results are produced in the form of printed tables and plotted diagrams. The text processor is used for the final report writing.The whole system can be characterized as a computer aided system for experimental investigation.

TEST SPECIMENS

The test specimens can be cylinders with the diameter either 100 mm or 150 mm, prisms with the cross section either 100x100 mm or 150x150mm. The standard cylinders 150 mm in diameter and 300 mm in length are most commonly used. However, the authors also used prismatic specimens 100x100x400 mm, and 150x150x450 mm. These specimen shapes guarantee that the failure volume is located within the measured one. The contact planes of the heads of a specimen are provided with sulphur caping, in order to assure regular contact conditions in loading machine.

TESTING MACHINE

The loading of specimen is done in the AMSLER 5000 kN press, Fig. 6. It was an old testing machine which was rebuilt and upgrated by adding the electronically controlled servovalv. A new electro-hydraulic system with the feed-back loop, as illustrated in Fig. 2., was designed and installed by Peekel. Its function can be briefly described as follows:

The desired value (D.V., Fig. 2) of controlling parameter is generated by a computer program and converted by Dynalog to the analog signal. The loop of the servo-system then adjusts the hydraulic pressure until the chosen controlling parameter, which can be either the measured force or strain (M.V.,Fig.2), is equal to the desired value.In this way the hydraulic servo-system always follows the loading path generated by computer.

Here, the loading was performed under the strain control, with a constant rate of strains. Due to the capacity limitations in the hydraulic oil flow the machine can be used only for moderate rates of loadings.

The purpose of this paper does not allow to discuss all interesting aspects of this venture. However in the end we have the machine with required parameters at a fraction of the cost of a new one.

MEASURING EQUIPMENT

A special strain gauge was designed for this purpose by Dr. Petrik (Ref. 1,2). It is a contact gauge whose scheme is shown in Fig. 4. It contains two linear variable displacement transducers from Peekel. They are parallel-mounted and measure an identical displacement over the same base. The first transducer serves for strain measurement and the second one for the test machine control. (In a case of the force control only one transducer is needed.) The length of the measuring base can be 70 or 150 mm.

Two such gauges are mounted into a special cage as shown in Fig.3. This cage can be mounted on cylindrical (or prismatic) specimens. The two longitudinal strains are measured by these gauges on the opposite sides of a specimen. The transverse strain is measured by a transducer (No.3,Fig.3) mounted in the cage. This gauge measures the change of diameter of the specimen.

An analog signal for the test machine control is received from the full bridge connection of the two transducers on the opposite sides of a specimen, in order to receive an average value. The other two longitudinal transducers measure the strains on the sides of a specimen. The averaging is done by software.

The cage can be quickly mounted on a specimen (Fig.5). The range of strain is approximately +-0.01, depending on a base length, which is more than sufficient for this purpose. The cracking and spalling of the concrete does not effect the function of the gauges.

The loading force is measured by a load cell built in the testing machine.

The measurements are done by Autolog with following parameters: carrier frequency 1 kHz, rate 22 ch./s, resolution 1 bit = $1\mu\mathcal{E}$, range +- 20.000 bits.

Both data acquisition systems, Dynalog and Autolog, are controlled by the computer PDP 11/40 with 64 kw memory and with the necessary pripheral units (see Fig. 1). Disk drives with removable cartridges RK05 and floppy disk drives RX02 are used for the data file storage. The system is operated from the VT100 terminal. The LA34 printer is used for the printed output. The graphs can be inspected on the graphic display and plotted on the X-Y plotter.

MEASURED VALUES AND THEIR EVALUATION

The measuring is done at the prescribed increments of longitudinal strain $\Delta\mathcal{E}$. The following values are recorded at each point: longitudinal strain $\mathcal{E}_1$, transverse strain $\mathcal{E}_2$, stress σ , time. The number of recorded points per test can range from tens to several hunderts. From these data a variety of useful relationships can be derived.

The uniaxial stress-strain diagram for the compressive test is constructed.From this diagram following material parameters are derived: σ_u -compressive strength, $\mathcal{E}_u$ - strain at the peak of the diagram, $E_{0,2}$ - secant modulus of elasticity and $\nu_{0,2}$ - Poisson ratio at $0,2\sigma_u$ (see Fig. 7).

The stress-volumetric strain diagram is constructed, where volumetric strain is defined as $\mathcal{E}_v = \mathcal{E}_1 + 2\mathcal{E}_2$. The volumetric strain reflects the damage of concrete by microcracks prior the failure. The onset of the microcracking is described by the discontinuity limit (Ref. 3,4). This limit can serve as an estimate of the long term strength and of the low cycle fatique limit. The following material parameters are derived from the volumetric diagram: σ_c - stress and ε_c - volumetric strain at the discontinuity limit, and the relative discontinuity limit

The seven above defined material parameters are derived for each specimen individually. Then, statistical analysis is made for a series of specimens.

The method for deriving the average stress-strain diagram for a series of tests is based on normalized diagrams. Each diagram is normalized with respect to the peak coordinates by the transformation $F= \sigma / \sigma_u$, $S= \varepsilon_1 / \varepsilon_u$, where F and S are the normalized stress and normalized strain, respectively. All normalized diagrams are passing through the same peak point, as illustrated on Fig.9.a. Now, an average F is found in a vertical section through the family of normalized curves for chosen S. In this way an average normalized diagram is calculated. The average diagram is found by transforming the average normalized diagram to the average peak coordinates $\bar{\sigma}_u$, $\bar{\varepsilon}_u$. This procedure must be employed, if the peak of the average diagram shall be located at the average strength level of the series.

Similar procedure is used for the evaluation of the average stress-volumetric strain diagram. The stress normalization is the same as in the previous diagram. The volumetric strain is normalized by the transformation $V = \varepsilon_v / \varepsilon_c$, where V is the normalized volumetric strain.

Further, the diagrams for the transverse strain ε_2 or for the Poisson ratio $\nu = - \varepsilon_2 / \varepsilon_1$ can be plotted.

SOFTWARE

A special software was developed for this measuring system. It was originally written in BAISIC from Peekel, which is a BASIC compiler with additional commands for measuring. Most of the programs are now also written in FORTRAN. The software has three parts.

The first program controls the loading at constant strain rate. It also takes readings of the gagues at required intervals and stores data in the disk files. The second set of programs performs data processing including the evaluation of the material parameters, statistical analysis and deriving of the average diagrams. The third set of programs is for printing of the results and plotting of diagrams.

The software runs under the operating system RT11 from DEC.

APPLICATION

The system was succesfully used for testing of hunderts of specimens from variety of types of concrete. It was applied to a research of the analytical form of stress-strain diagram of concrete and to several industrial development projects concerning fiber reinforced concrete, atmospheric corrosion, and other topics.

An example of a typical testing series is shown here. It consisted of 12 specimens from normal structural concrete. The specimens were tested under the strain rate $5\mu s$/s. The resulting diagrams are shown in Fig. 8.a,b,c. On the left side are plotted the families of curves for all specimens. On the right side are plotted the average diagrams.

The average diagrams were obtained using normalized transformation as described earlier. The normalized diagrams for this series are plotted on Fig.9.a,b. The shape of the average diagram, Fig. 8.a-right, was further analyzed.

A first derivative of the curve was found numerically, Fig. 10. It has a meaning of tangent modulus of elasticity. As the relationship for the derivative is almost linear in the ascending branch of the diagram, this part of the diagram can be described by a second degree parabola. This has an important consequence for the analytical form of the stress-strain diagram.

The strain control has a primary effect on the descending branch of the stress-strain diagram. This will be demonstrated on comparison with another series, made from similar concrete, tested by the same hardware, but with the manual control of the testing machine. The resulting diagrams are shown on Fig. 11. The strain-time relationships are plotted in Fig. 12. The variation in the strain rates of these tests can be compared with the constant strain rate of the computer-controlled tests of the previous series, Fig.13.

Under the manual control the loading path is uncertain and depends on the stiffness of the specimen, the stiffness of the machine, and on the other effects. In the post-peak state the specimens are often loaded under high strain rates and the descending branch is not measured accurately. In many cases it is not recorded at all.

The complete stress-strain curve, such as the one shown in Fig. 8.a, can be reliably obtained only from strain-controlled tests.

CONCLUSIONS

The system for automated testing of concrete compressive properties can be succesfully used for the experimental determination of the material parameters of concrete, which are needed for either scientific or industrial purposes. The stress-strain diagram icluding the descending branch is obtained under strain-controlled loading. The stress-volumetric strain diagram is also obtained and can serve as a measure of damage of concrete by microcracks.

REFEREENCES
1. CERVENKA, V.: A System for Automated Testing of Concrete. (In Czech), Stavebnicky casopis, 32 c.4, VEDA, Bratislava 1984, p. 287.
2. CERVENKA, V., PETRIK, M.: Experimental Investigation of Low Cycle Fatique of Concrete Bazed on Volumetric Strain, In: 15th Yugoslav Congress Theor. App. Mech., Kupari, 1981.
3. CERVENKA, V.: Behavior of Concrete under Low Cycle Repeated Loadings, CEB Bulletin d'Information., No.132, Rome, May 1979., p. 15-20.
4. RUSCH, R.: Physikalische Fragen von Betonpruefung Zement-Kalk-Gips, H.1, 1959, p. 1-9.

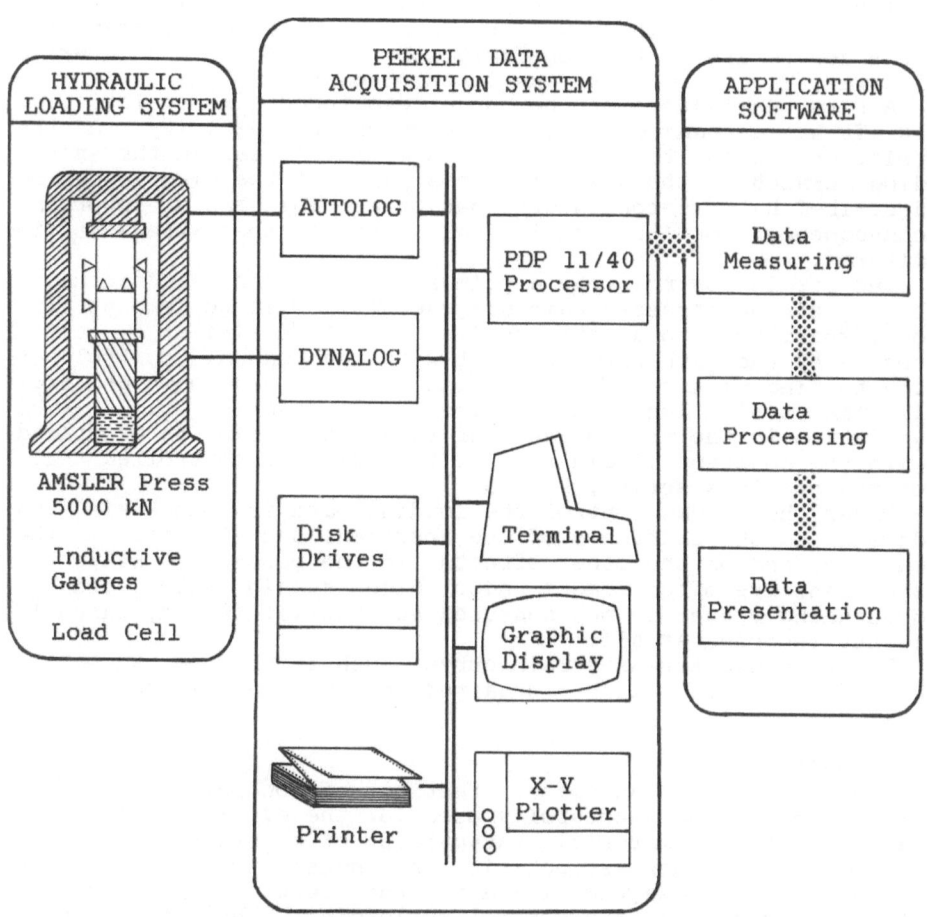

Fig. 1. Scheme of the system for automated testing of concrete.

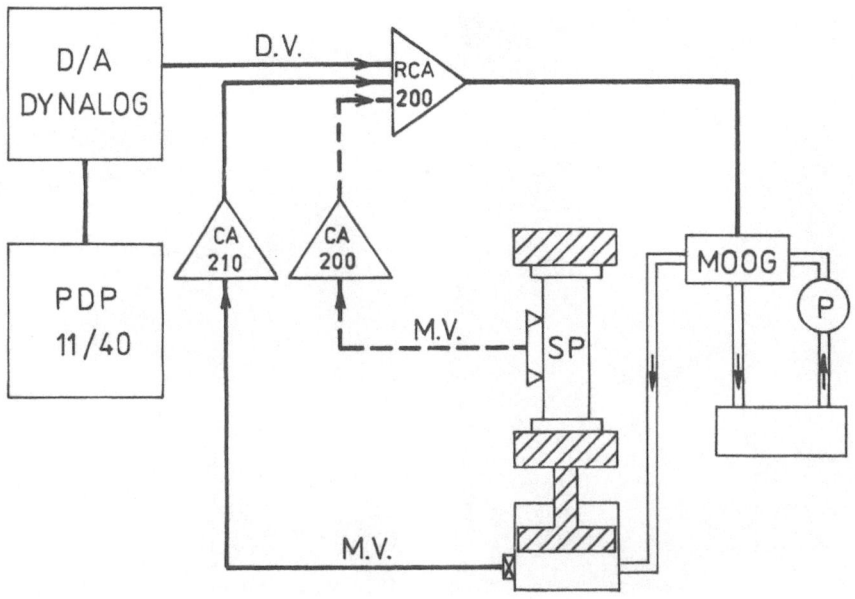

Fig. 2. Flow chart of the elektro-hydraulic loading system

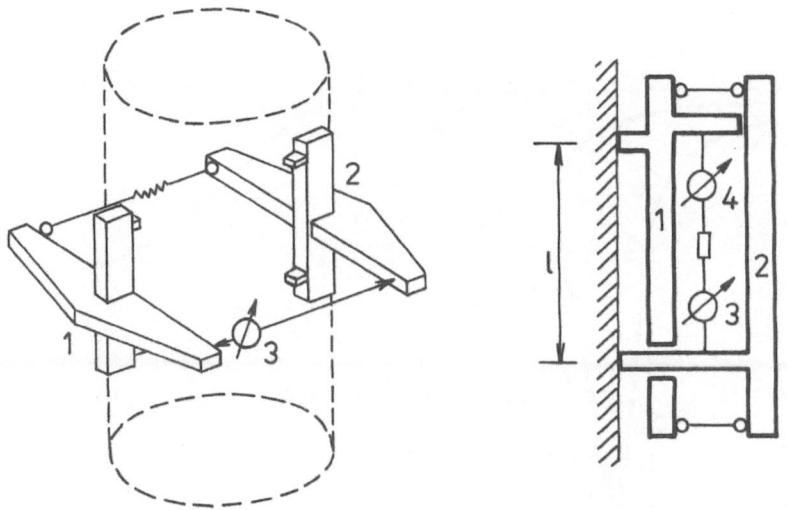

Fig.3. Scheme of the gauge cage.   Fig.4. Scheme of the con-
                                          tact   strain gauge.

Fig.5. View of the cage
with specimen.

Fig.6. View of the testing
machine AMSLER 5000 kN.

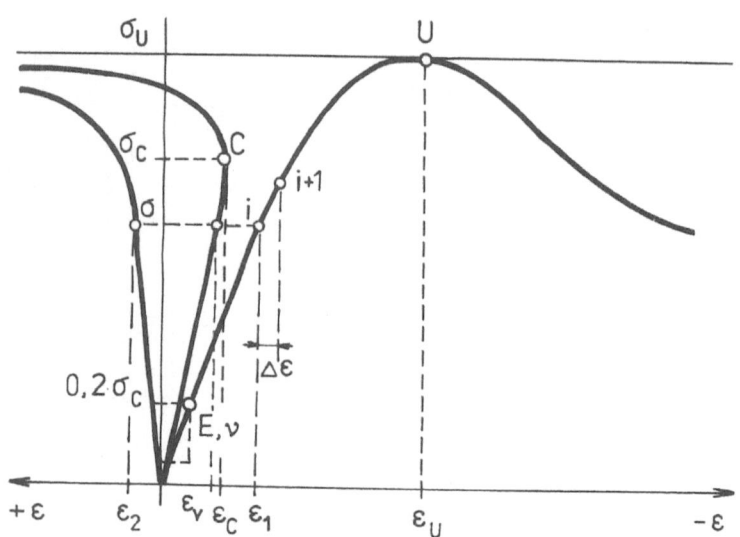

Fig.7. Measured and evaluated values.

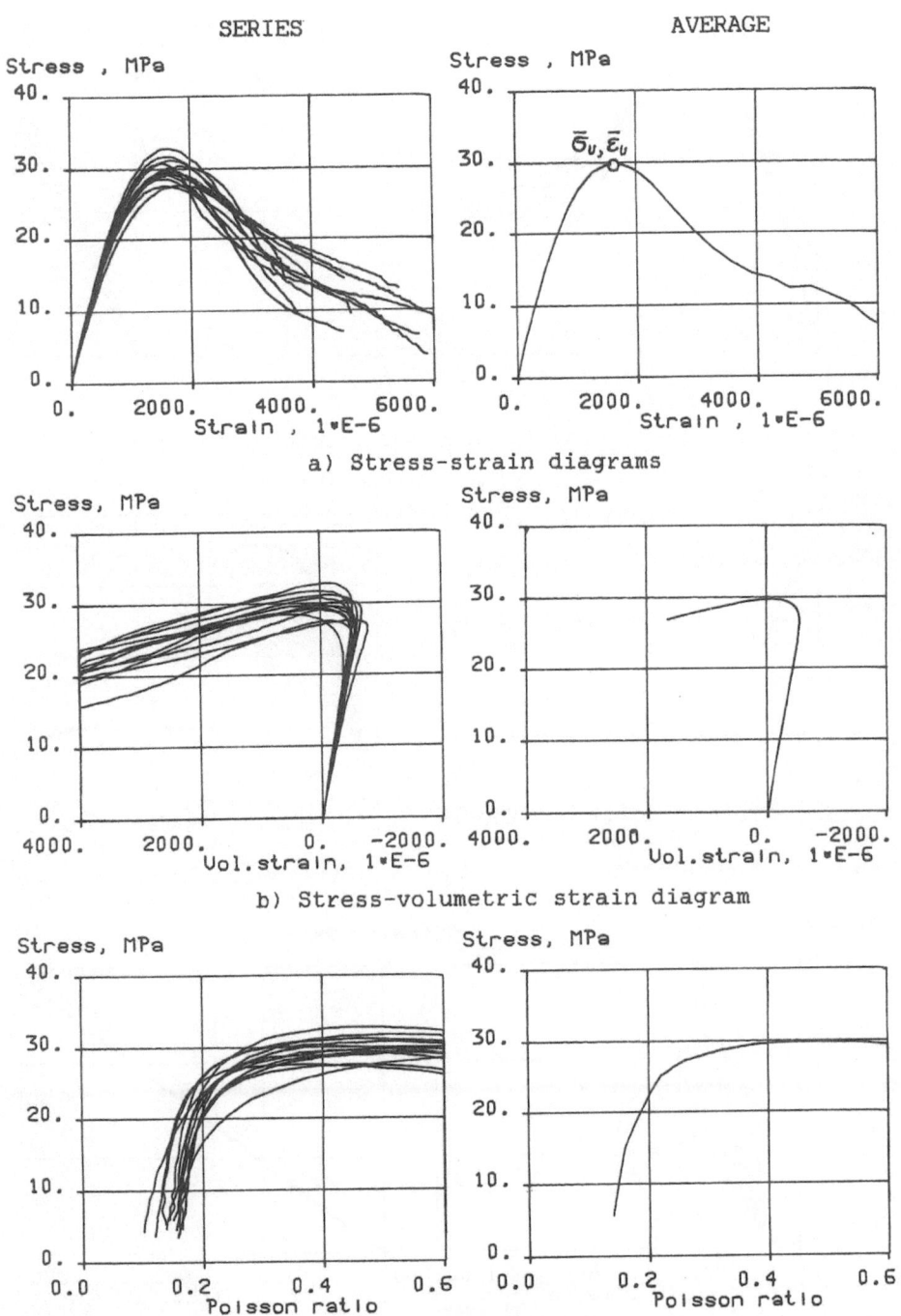

a) Stress-strain diagrams

b) Stress-volumetric strain diagram

c) Stress-Poisson ratio diagram

Fig.8. Processing of the diagrams for a series of specimens.

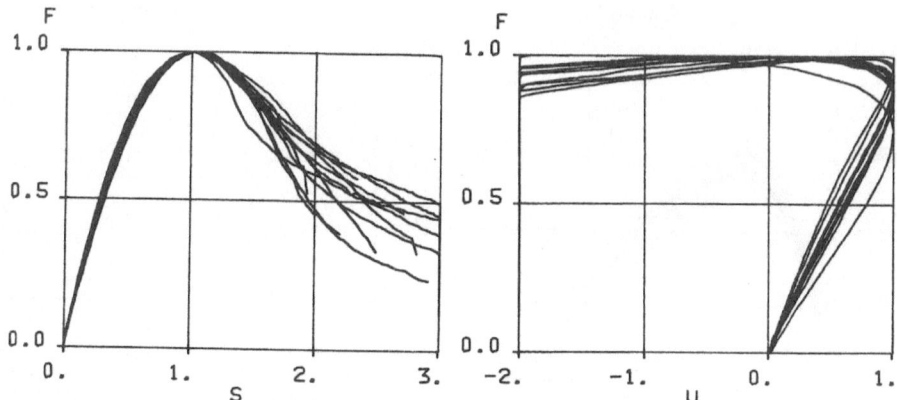

Fig.9. Normalized stress-strain and volumetric diagrams.

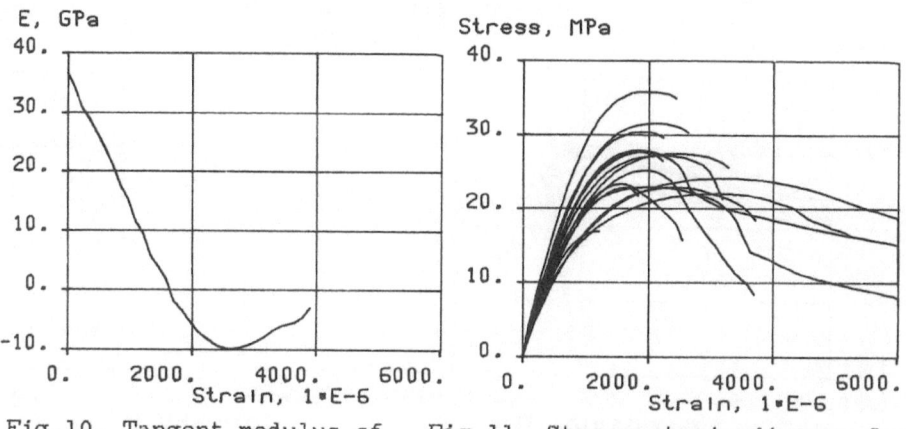

Fig.10. Tangent modulus of elasticity (Fig.8.a).

Fig.11. Stress-strain diagram for manually controlled tests.

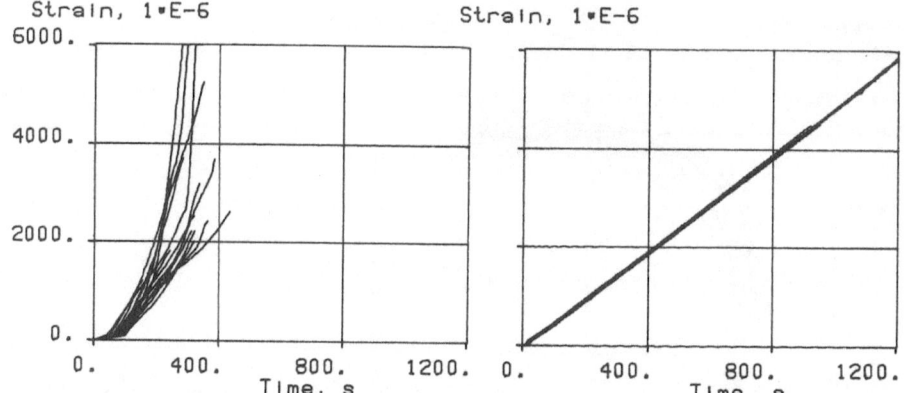

Fig.12. Strain-time relation (manual control).

Fig.13. Strain-time relation (computer control).

METALLIC STRUCTURES UNDER MECHANICAL AND CYCLICAL THERMAL LOADING

M. COUSIN, J.F. JULLIEN
Laboratoire BETONS ET STRUCTURES - I.N.S.A. de LYON -
69621 VILLEURBANNE CEDEX FRANCE

ABSTRACT

This study concerns the behaviour of metallic structures submitted to the combination of a primary loading and a secondary cyclic thermal sollicitation.

An original experimental device was conceived and carried through in order to apply to thin stainless steel cylinders a tension and an axial cyclic gradient of temperature of variable intensity.

Numerical calculations are made by taking into account several models of the material behaviour laws. The identification of the different models is made from monotonous and cyclic characterisation tests according to temperature.

The comparison of the test results with those stemming from calculation allows a better comprehension of the thermoelastoplastic behaviour of the structure.

I - INTRODUCTION -

More and more complex models allow the production of the cyclic-plastic behaviour of materials. Nevertheless, many difficulties still remain in analysing the behaviour of structures.

The very conservative nature of the rules of construction shows the uncertainties within this field. A better comprehension of the phenomena is likely to be achieved by experimental studies. Most of the experimentations that have been developed these last few years concern the structure of "type BREE" essentially. As a matter of fact, concerning the question of the three bars as well as for the test of tensile -twisting or tensile-bending, such is the force applied that the sollicitations that they produce are a state of stresses which is the same on the whole structure. Therefore, such studies bring out behaviour of materials more than behaviour of structures. Tests on structures are still very rare and tend especially to observe the global behaviour.

It has become necessary obviously to increase experimentations where the combination of primary and secondary sollicitations touches a restricted volume of the structure and where an adapted device of measures allows the study of the local and global behaviour of structures.

This is why we have developed this experimental study the results of which are compared with those stemming from numerical calculation. The study concerns the behaviour of large size, thin, stainless steel cylinders submitted to a tensile and to a cyclic thermal axial step.

II - <u>EXPERIMENTAL DEVICE</u> (1) -

The size of the specimens are :
- diameter : 400 mm
- length : 1 000 mm
- thickness : 2 mm

The cylindrical shell is obtained by rolling and welding

II.1 - <u>Loading</u> (figure 1) :

The primary loading is a tensile whose intensity is constant in time. This loading is applied by means of an hydraulic jack and bars of transmissions. The secondary loading is a cyclic thermal axial step. The intensity of this gradient can reach 350°C/cm. This value corresponds to an intensity of thermal stress three times the yield stress of the material. The axial step in temperature is obtained by forced internal convection. The escape gases of a burner are aired into the cylinder where a deflector forces them to hit the internal side very locally at high speed. A watering by means of jets of water focused on the same level allows cooling and obtaining the desired temperature. A little rubber joint separates the water and the hot gases.

The radial thermal step is insignificant, considering the small thickness of the cylinder and the mode of heating used here.

II.2 - <u>Device of measurements</u> :

The interest in the chosen method for creating thermal sollicitation is to keep the external side of the cylinder quite inoccupied for the measurements. The device of measurements concerns :

- the knowledge of loading by two load cells for the tensile load and by thermocouples for the axial thermal step. These termocouples (nickel chrome - alloying nickel) of 0.2 mm of diameter are microwelded on the outside of cylinders.

- the global and local behaviours : it is necessary to have the knowledge of field of displacements.

The gradient of strain is very significant and the temperature is continually variable. For these reasons, strain gauges are used only in the zones where the temperature stays fixed in the cold zone and where the gradient of strain is small.

The measurements are done by special axial and radial transducers.

The instrumentation must allow the measurements of small residual displacements around important displacements caused by the primary and secondary sollicitations.

. Axial displacement (figure 2) : the sensitive part of the axial transducers is an LVDT. They are put at the same level in front of the cold zone and fixed on the cylinder. The transmission of displacement is carried out by a thin bar in invar steel, microwelded at the other extremity.

. Radial displacement : the radial displacement in a point is transmitted perpendicular to the cylinder by a thread in invar steel of 0.2 mm of diameter at a distance of one meter. The threat is set by a spring. The sensitive part is an LVDT. The important length allows us to neglect the axial displacements. Computer equipement is used for the data acquisition and the automatic piloting of loading (figure 3).

III - <u>EXPERIMENTAL AND NUMERICAL RESULTS</u> -

III.1 - <u>Applied loadings</u> :

The exact definition of the secondary loading is given by the temperature cards along a generator at different steps of the rising and falling temperature.

III.2 - <u>Modelization of the material</u> :

Characterisation tests under monotonous tensile and cyclic strains imposed upon at different temperatures allow us to point out the material behaviour law according to different types of modelization.

III.2.1 - <u>Kinematic hardening model</u> :

As for this type of modelization, we have chosen a bilinear hardening. Figure 4 shows the model built from the monotonous traction curves.

III.2.2 - <u>CHABOCHE model</u> (2) :

This five parameters model is identified from cyclic tests done on adapted test specimen. The chosen model does not take any isotrope hardening into account and proves the kinematic hardening by summing up two exponentials.

III.3 - <u>Numerical modelization</u> :

The numerical calculations are carried out with the INCA part of CASTEM code developed by the French Atomic Energy Commission (3). This code allows us to take account of different models of behaviour of material, variable with the temperature.

The structure is axisymetric and two types of elements are considered : the linear element at two nodes or the linear element at two nodes with integration in the thickness in plasticity.

The cylinder is meshed on 52 cm in 280 elements. The gap of temperature between two nodes does not exceed 10°C (figure 5).

III.4 - <u>Comparison of experimental and numerical results</u> :

The comparison concerns the results of test 6 ($G_{max} = 300°C/cm$; P = 470 kN).

III.4.1 - <u>Displacements</u> (Figures 6, 7) :

After 3 cycles the chaboche model and the kinematic model describe the behaviour of the structure perfectly as it is proved by the different curves to the displacements.

If the ultimate state is the same : the stabilization of the structure, there are differences after a larger number of cycles. Results of calculation with the numerical models bring out a different behaviour on a level with the warm and cold hinges.

III.4.2 - <u>Strains</u> :

The stabilization of plastic axial strains is visualized by strain gauge measurements in the cold part of the cylinder.

Figure 8 shows comparison between hoop strains obtained by numerical calculation and by experimental tests from radial displacements. Figure 9 shows that near the temperature step we have accomodation after 10 cycles and in the other part after 3 or 4 cycles only.

IV - CONCLUSION -

The comparison of the results bring out an imperfect determination of displacement values for the ultimate state by means of numerical modelization.

However the correct description of the behaviour in the course of the first cycles is very promising for the capacities of the calculation codes in order to find the solution for such problems.

An improvement is possible in the continuation of the development of models which should take into account complex states of stresses.

In the part of a Benchmark's calculation, the cold creep effects have been studied. The results are better but are still different to experimental results.

The main purpose of this work is to bring out points of reference not only useful for the validation of the codes, but also for the simplified methods which remain a finality within this field.

V - REFERENCES -

(1) - COUSIN M. - "Contribution à l'étude des structures métalliques sous chargements mécanique et thermique cyclique" - Thèse d'Etat - Mars 1984 - INSA de LYON

(2) - CHABOCHE J.L., DANG VAN K., CORDIER G. - "Modelization on the strain memory effect on the cyclic hardening of 316 stainless steel" SMIRT 1979

(3) - HOFFMANN A., COMBESCURE A., CHAVANT C. - "Calcul linéaire et non linéaire des coques" IPSI Janvier 1977

This research work was conducted in Concretes and Structures Laboratory of INSA LYON with the backing of "Ministère de la Recherche et de l'Industrie" and EDF/SEPTEN.

VI - KEYWORDS -

SHAKEDOWN - RATCHETING - CYCLICAL LOADING - METALLIC STRUCTURES - THERMAL LOADING - PLASTICITY

FIGURES -

1. Applied loadings
2. Axial displacement transducer
3. Data recording equipment
4. Kinematic hardening model
5. Mesh of the structure
6. Experimental radial displacement
7. Radial displacement - comparison
8. Hoop strain - comparison
9. Evolution of radial displacement during some cycles

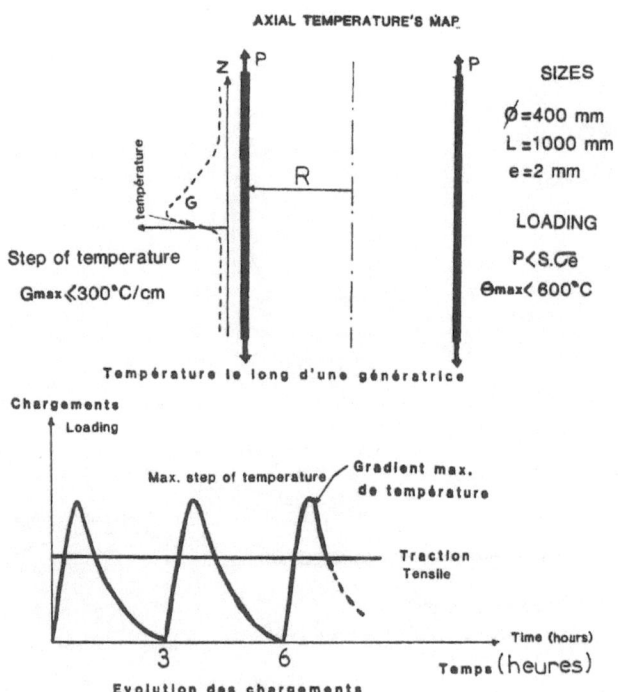

AXIAL TEMPERATURE'S MAP

Z

P

P

SIZES

ϕ =400 mm
L =1000 mm
e =2 mm

température

G

R

LOADING

Step of temperature

P<S.Ge

Gmax ≤300°C/cm

Θmax< 600°C

Température le long d'une génératrice

Chargements
Loading

Max. step of temperature

Gradient max.
de température

Traction
Tensile

3   6

Time (hours)

Temps (heures)

Evolution des chargements
LOADING VARIATION

Figure 1

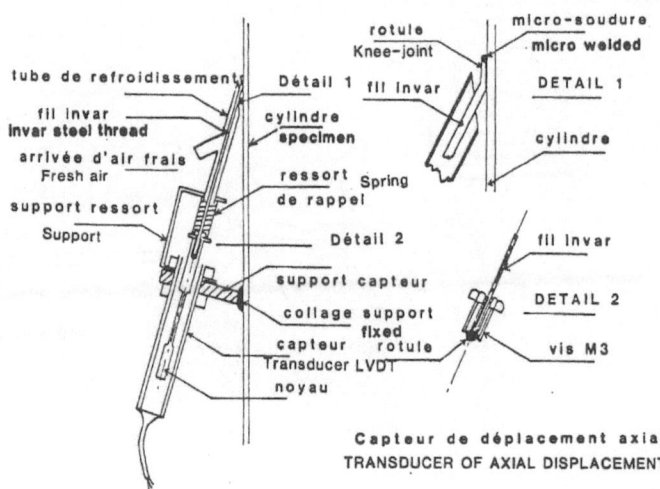

tube de refroidissement

rotule
Knee-joint

micro-soudure
micro welded

Détail 1   fil invar

DETAIL 1

fil invar
invar steel thread

cylindre
specimen

cylindre

arrivée d'air frais
Fresh air

ressort   Spring
de rappel

support ressort
Support

Détail 2

fil invar

support capteur

DETAIL 2

collage support
fixed

capteur   rotule
Transducer LVDT

vis M3

noyau

Capteur de déplacement axial
TRANSDUCER OF AXIAL DISPLACEMENT

Figure 2

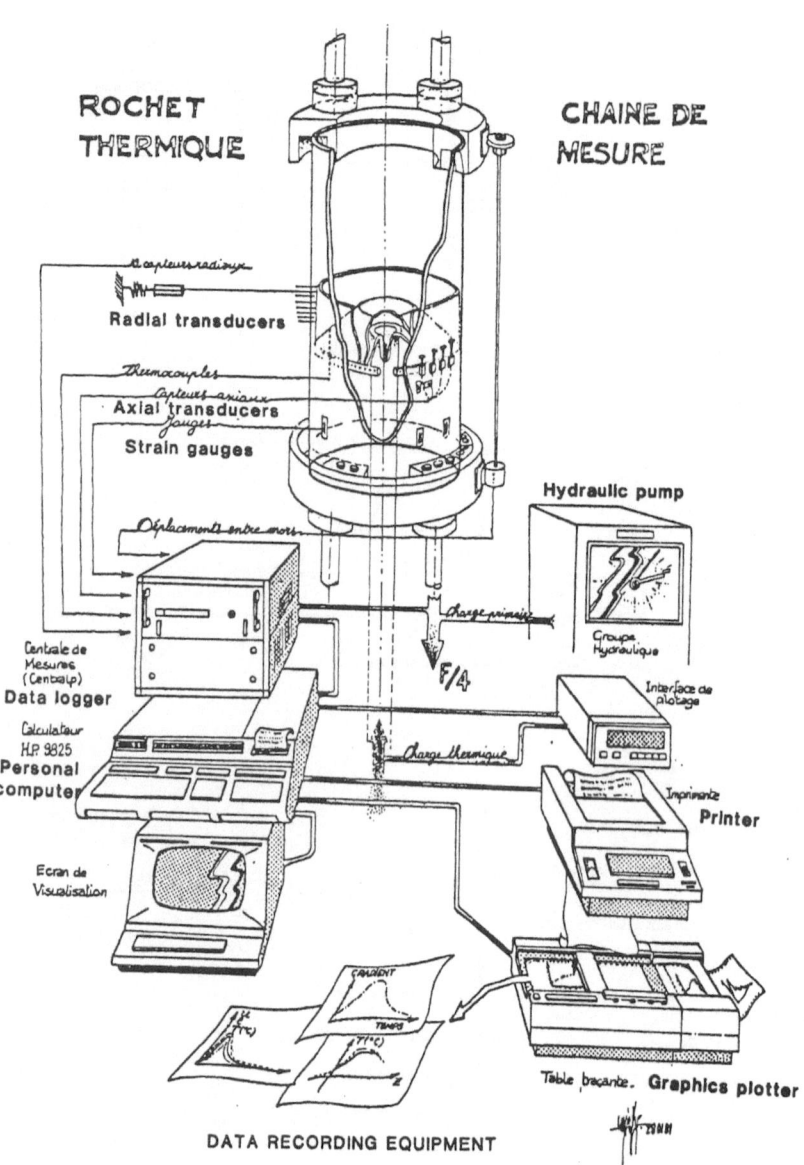

Figure 3

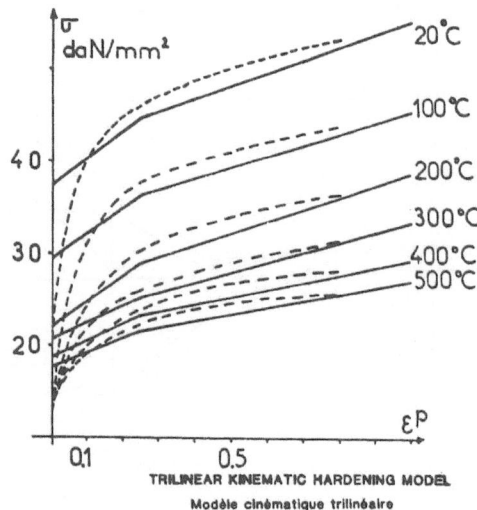

TRILINEAR KINEMATIC HARDENING MODEL

Modèle cinématique trilinéaire

Figure 4

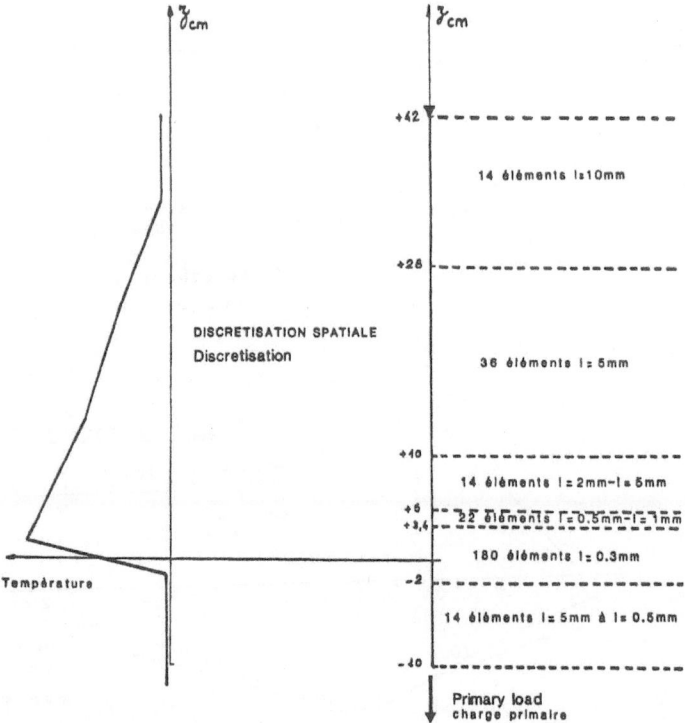

Figure 5

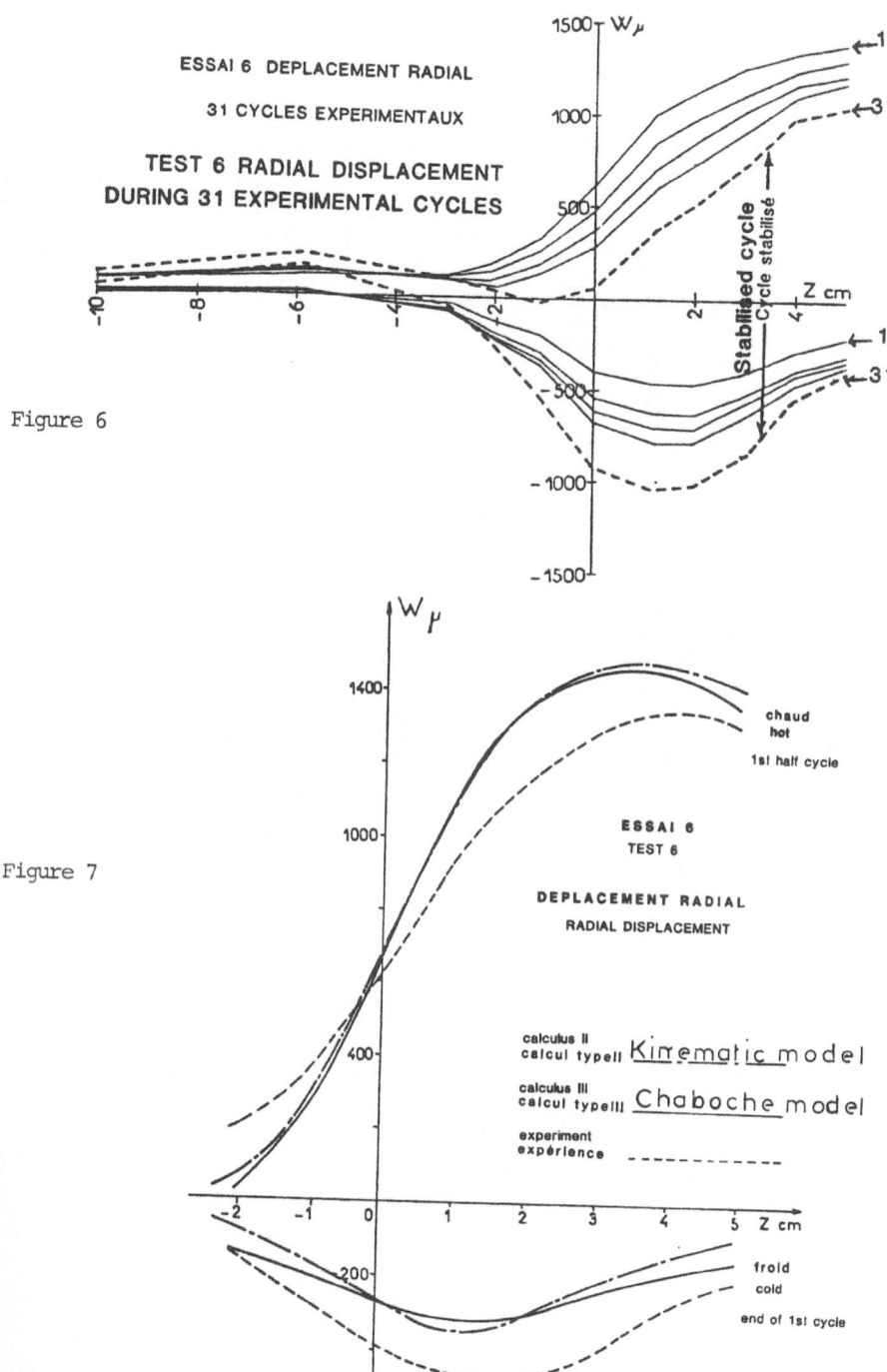

Figure 6

ESSAI 6 DEPLACEMENT RADIAL

31 CYCLES EXPERIMENTAUX

**TEST 6 RADIAL DISPLACEMENT
DURING 31 EXPERIMENTAL CYCLES**

Figure 7

ESSAI 6
TEST 6

DEPLACEMENT RADIAL
RADIAL DISPLACEMENT

calculus II
calcul typeII Kinematic model

calculus III
calcul typeIII Chaboche model

experiment
expérience - - - - - - - -

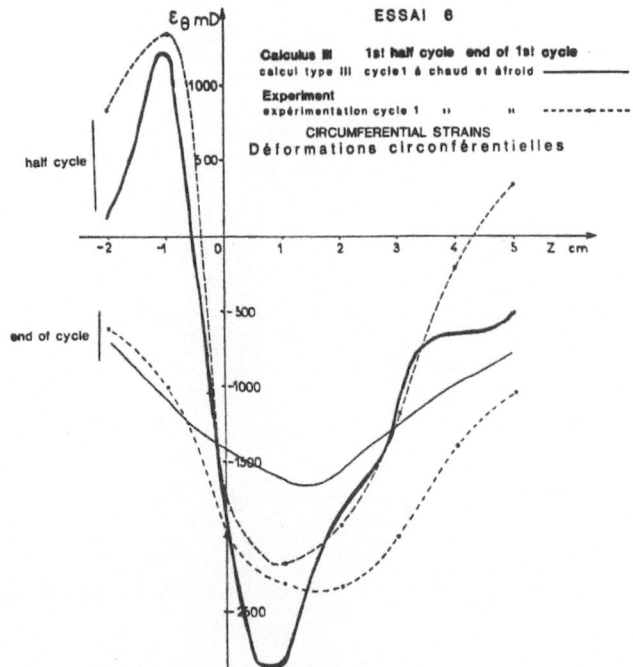

ESSAI 6

Calculus III   1st half cycle   end of 1st cycle
calcul type III   cycle1 à chaud et à froid ————————

Experiment
expérimentation cycle 1   "   "   ---·---·---

CIRCUMFERENTIAL STRAINS
Déformations circonférentielles

Figure 8

EVOLUTION OF RADIAL DISPLACEMENT DURING SOME CYCLES

EVOLUTION DU DEPLACEMENT RADIAL   AU COURS DES CYCLES

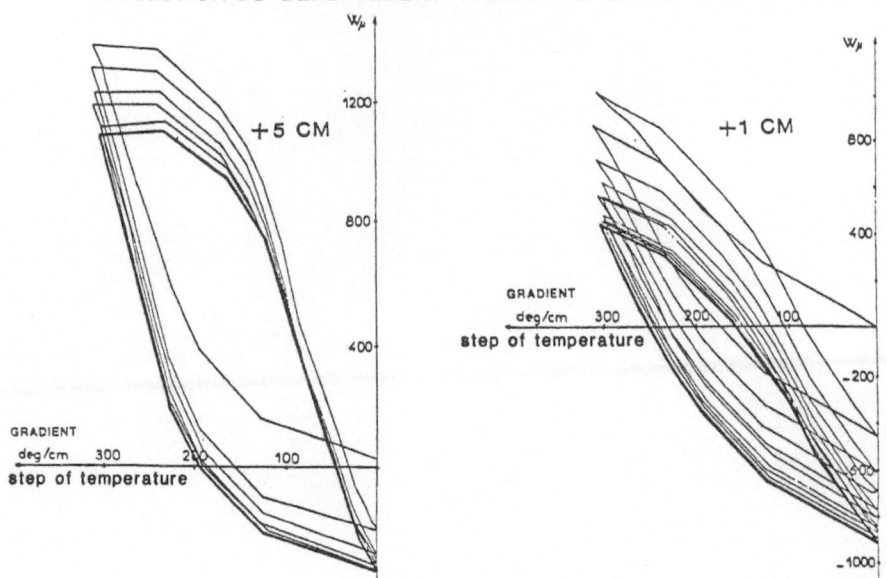

Figure 9

RESPONSE OF SPHERICAL SHELLS UNDER APEX LOAD TO VARYING PROPORTIONS

Sameh S. Issa
Experimental Stress Analysis Laboratory, Department of Civil Engineering,
Kuwait University, P.O. Box 5969 Al-Safat, Kuwait.

1. ABSTRACT

Two thin spherical closed segments of equal base-diameters, with a low
and a high rises, were selected as testing objects. A uniform distributed
load over a small circle at the apex and pinned-edge-conditions have been
considered. Spherical segments are investigated analytically by the force
method, numerically by implementing the finite element method and experimentally by means of electrical resistance strain gages. Apex loads inside
the shallow zone seem to require special considerations.

2. INTRODUCTION

The characteristics of spherical caps depend, among other factors on the
ratio of rise to diameter of the outer-most parallel circle. This ratio
will be referred to hereafter as the arching parameter (AP). The performance of spherical shells as structural elements varies, in terms of their
AP, from that of a circular plate at one extreme to that of a spherical
shell at the other. Shallow spherical shells whose geometrical properties
eliminate some terms of the general solution of spherical shells, represent
an intermediate stage.

Theoretical analyses of spherical shells cited in literature handle usually either the case of clamped edges or rolling supports along the outer-
most parallel circle. A uniform distributed load over the entire surface-
area is a common case of loading. Herein, spherical shells with hinged boundary conditions and apex load within the shallow zone are considered. These
are of practical importance in the area of civil engineering. Besies, the
validity of existing approaches under the above mentioned case of loading
seems to require further study.

The present work investigates the surface stresses of two spherical
shells whose AP are selected so apart to provide a shallow and a nonshallow
spherical segments. Surface-stress-distribution is presented as a criterion for assessing the characteristics of both shells as well as the approaches employed.

Spherical shells are defined by the radius of the sphere, R, the rise,
h, and the diameter of the outer-most parallel circle, $2a_0$ (Fig. 1), known
also as the base diameter. In general, theoretical solutions of spherical
shells involve analytic functions, power series or periodic functions [1-9].
Several theoretical approaches are based on establishing differential equations of the second order whose solutions result in the so-called stress
and displacement functions [1-3]. These functions contain arbitrary constants that could be determined from the boundary and transition conditions.
The "force method" is adopted herein due to its flexibility in modifying
the solution to match with the present boundary conditions.

The scope of analysis is widened to include a numerical approach, namely
the finite element method. A simple computer program is prepared to

Wieringa, H (ed), Experimental Stress Analysis.
© *1986. Martinus Nijhoff Publishers, Dordrecht.*

generate the required data for establishing the finite element grid of
any spherical shell. This makes finite element analysis convenient, if
proved to reflect the true response
of spherical segments under the present conditions. Based on a slice of
the spherical cap that is limited by
two meridional planes which form between them a five-degrees-dihedral
angle, and the outer-most parallel
circle, as shown in Fig. 2, the finite element grid (a gore) is defined.
The package SAP V was invoked in this
analysis. The possible rotation of
the finite element about the normal
to its surface is taken into account.
In the case of shallow shells, this
rotation becomes a significant problem. The selected gore along with
the boundary elements, described below, provide a finite element grid that can be implemented for spherical segments of different
APs.

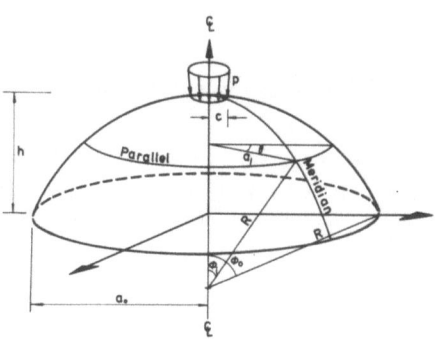

The analytical and numerical results are
checked experimentally. For this purpose,
spherical segments supplemented with a
sufficient number of measuring stations were
considered. Components of each measuring station and their arrangement within the station
as well as the selection of their locations
on both surfaces of the spherical segment were
designed to provide a thorough scanning of
surface stresses.

The applicability and particularities of
each approach are pointed out. The results
obtained from the three approaches are compared and discussed.

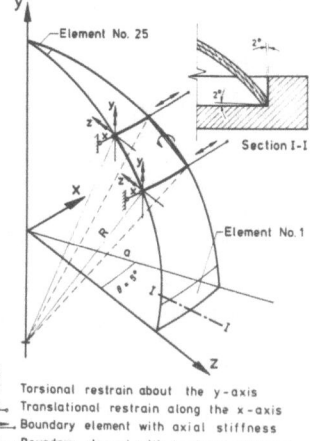

Torsional restrain about the y-axis
Translational restrain along the x-axis
Boundary element with axial stiffness
Boundary element with torsional stiffness

FIGURE 2. Details of the selected gore

3. MATHEMATICAL FORMULATIONS

In general, exact theoretical solutions
for thin shells can be achieved by means of superimposing the results obtained from the membrane and bending theories. The final results are worked out in two steps. First the membrane forces are determined for the
boundary conditions shown in Fig. 3a. This solution assumes unrestrained
deformation along the edge of the outer-most circle of the spherical shell
under consideration. In a further step, the flexibility coefficients at
the edge of the spherical shell are determined. These are the torsional
flexibility about the tangent to the outer-most parallel circle and the
translational flexibilities in both horizontal and vertical directions.
This leads to the formulation of the edge deformation due to unit edge actions, M, H and V as shown in Fig. 3b. The actual values of M, H and V are
obtained by satisfying the existing boundary conditions. This approach is
known as the force method [7].

Spherical segments are shells of revolution subjected, in the present
case, to an axisymmetrical load. It follows that the internal forces

$$N_{\phi\theta} = N_{\theta\phi} = M_{\phi\theta} = M_{\theta\phi} = 0, \tag{1}$$

in which $N_{\phi\theta}$ and $N_{\theta\phi}$ are in plane shear forces; $M_{\phi\theta}$ and $M_{\theta\phi}$ are twisting moments. All internal forces are given in terms of force or moment per unit length. The membrane theory assumes zero moment and zero shear at the outer-most circle, Fig. 3a, it follows that the transverse shear forces vanish:

$$Q_\phi = Q_\theta = 0. \tag{2}$$

A uniform circular load, p, covering an area of a small radius, c, is considered as acting radially at the apex. Significant meridional and circumferential membrane forces, outside the loaded area, are given by:

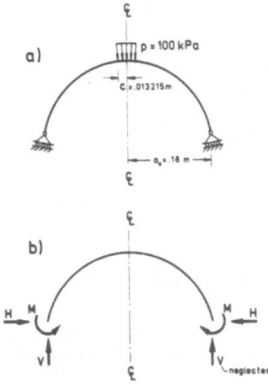

a) p = 100 kPa
c = .013215 m
a_o = .18 m

b) H M M H
V neglected

FIGURE 3. Boundary conditions and edge unit loads

$$N_{\phi_i}^m = \frac{-\text{ total load}}{2\,\pi\,a_i\,\sin\,\phi_i} = \frac{-\Sigma V}{2\pi R \sin^2 \phi_i} = \frac{-P}{2R\,\pi\sin^2 \phi_i},$$

$$N_{\theta_i}^m = \frac{P}{2R\,\pi\,\sin^2 \phi_i} \tag{3}$$

in which $P = \Sigma V = p\pi c^2$.

The edge deformations, resulting from the membrane analysis, namely the horizontal translation Δ_e and the meridional rotation β_e can be determined from

$$\Delta_e = \frac{R \sin \phi_o}{Et} (N_\theta^m - \mu N_\phi^m), \tag{4}$$

$$\beta_e = \frac{\cot \phi_o}{Et} (1+\mu)(N_\phi^m - N_\theta^m) - \frac{d}{Rd\phi}\left(\frac{\Delta_e}{\sin\phi_o}\right),$$

herein E is the elastic modulus; t is the thickness and μ is Poisson's ratio [7]. For the present case, eqs. (4) can be written as:

$$\Delta_e = \frac{P(1+\mu)}{2\,\pi\,E\,t\,\sin\,\phi_o} \qquad ; \qquad \beta_e = 0. \tag{5}$$

Referring to Fig. 3b and the boundary conditions of the present case, the meridional moment at the edge, M, is equal to zero and the vertical shear, V, is neglected. Under the latter condition, the present approach becomes less suitable for shallow spherical shells. However, in the case of non-shallow spherical shells or shallow spherical shells with small characteristic length, $\ell = \sqrt[4]{(Rt)^2/12\,(1-\mu^2)}$, when compared with the base radius, a_o, the deflections and internal forces due to edge loadings are limited to a narrow strip next to the edge [1,7].

The horizontal deflection due to unit horizontal force, H, can be obtained from the relation:

$$\Delta_{e_1} = \frac{1}{Et}[R \sin \phi_o\,(2K \sin \phi_o - \mu \cos \phi_o)]. \tag{6}$$

Thus, the condition of consistent deformation along the horizontal direction is

$$\Delta_e + \Delta_{e_1} H = 0. \tag{7}$$

The actual value of H can be determined based on eqs. (5,6,7). At this stage the internal forces N_ϕ, N_θ, M_ϕ and M_θ can be obtained as functions of H. These internal efforts can be superimposed on those obtained from the membrane analysis to achieve the final solution.

Based on the classical linear theory, the above mentioned two steps can be combined [10] to yield a set of formulas for the internal forces of closed spherical shells with any boundaries at lower edge. Under the present conditions of loading and boundaries, the above mentioned formulas have been modified to take the following form

$$N_{\phi_i} = \frac{-P}{2R \pi \sin^2 \phi_i} + \cot \phi_i \ [S_1 \ F_1 \ (\alpha_i) + S_2 \ F_2 \ (\alpha_i)]$$

$$N_{\theta_i} = \frac{P}{2R \pi \sin^2 \phi_i} + K[S_1 \ F_3 \ (\alpha_i) - S_2 \ F_4 \ (\alpha_i)]$$

$$M_{\phi_i} = \frac{-R}{2K} \ [S_1 \ F_4 \ (\alpha_i) + S_2 \ F_3 \ (\alpha_i)]$$

(8)

$$M_{\theta_i} = \frac{-R}{2K} \ \{-S_1 [\frac{\cot \phi_i}{K} \ F_2 \ (\alpha_i) - \mu \ F_4 \ (\alpha_i)] +$$

$$S_2 [\frac{\cot \phi_i}{K} \ F_1 \ (\alpha_i) + \mu \ F_3 \ (\alpha_i)]\}$$

in which

$$\alpha_i = \phi_i/\phi_o \qquad\qquad ; \quad K = \sqrt[4]{3(1-\mu^2)} \ (\frac{R}{t})^2 \ ;$$

$$S_1 = - \ \frac{(N_\theta^m - \mu \ N_\phi^m)}{2K} \qquad ; \quad S_2 = \frac{N_\theta^m - \mu \ N_\phi^m}{2K} \qquad ;$$

$$F_1 (\alpha_i) = e^{-\alpha_i} \cos (\alpha_i) \qquad ; \quad F_3 (\alpha_i) = e^{-\alpha_i} [\cos (\alpha_i) + \sin (\alpha_i)];$$

$$F_2 (\alpha_i) = e^{-\alpha_i} \sin (\alpha_i) \qquad ; \quad F_4 (\alpha_i) = e^{-\alpha_i} [\cos (\alpha_i) - \sin (\alpha_i)].$$

It is noted that S_1 and S_2 are factors which depend on the nature of boundaries and type of loading. The F_i-factors are trigonometric and exponential functions of geometry which are obtained from the general solution of the differential equation:

$$W^{IV} + 4 \ K^4 \ W = 0 \quad ,$$

(9)

where W is the normal deflection of the middle surface. For further details the reader is referred to Hampe [10].

A computer program was developed for the chosen approach and utilized in determining the surface stress on both sides of the two shells under consideration.

4. FINITE ELEMENT MODELLING AND ANALYSIS

A finite element discretization that consists of a gore and appropriate boundary elements as shown in Fig. 2 is proposed. The gore which has been described above, is subdivided into 25 finite elements of the quadrilateral type for thin plates and shells. This type of finite element can be modified, if necessary, to a triangular element as it is the case at the apex.

Due to rotational symmetry of spherical shells, material particles are restrained to circumferential translation and meridional rotation. All nodal points of the finite elements that are lying in the Y-Z-plane, see Fig. 2, can be controlled through the X-translation, Y-rotation and Z-rotation boundary conditions of each. The circumferential tangents to the other meridional plane of the gore are not coinciding either with X-axis or the Z-axis. Consequently, additional circumferential boundary elements of axial stiffness are required. Also, boundary elements of torsional stiffness are arranged between each two successive nodes along the meridian of the latter meridional plane. Edge-nodes are restrained against translation in the

three principal directions. Rotation at the outer-most parallel circle is
only permitted about its circumferential tangents.

In general, for the analysis of shallow shells, finite elements of the
type selected herein tend to rotate about their normals; this can be allevi-
ated by additional boundary elements at the nodes. However, the present
arrangement of finite-element-model and boundary conditions provide all necessary constrains to hold the element from the said rotation. A typical
quadrilateral element that is used in the present finite element model,
along with the proposed boundary conditions, is illustrated in Fig. 2.
Several runs of SAP V, having additional boundary elements at the nodes for
a large number of different AP were carried out. Comparing the obtained
surface stresses with those resulting from the proposed finite-element-discretization, shown in Fig. 2, no difference was recognized. This confirms
that the suggested finite-element-discretization may be used for different
arching parameters (AP), which also includes with slight modification,
circular plates.

5. EXPERIMENTAL WORK AND SETUP

The response to varying geometrical proportions should be assessed separately for shallow and nonshallow spherical shells. Therefore, the APs of
the two models used are selected far apart. This would enhance the assessment resulting from comparing the output of the suggested approaches to include spherical shells under apex loads in general. One model with a very
low AP, $h/2a_o = 1/24.27$, (shell 1, of radius 1.0997 m), and another model
with a high AP, $h/2a_o = 1/3.883$, (shell 2, of radius 0.2211 m) are tested.

The models were formed by heating and vaccuming process. For this purpose negative aluminium molds with tiny hole at the inversed crown, were
machined. The outer end of this tiny hole was connected through a rubber
house to a vacuum pump. This arrangement, with the exception of the vacuum
pump, was stationed in a programable oven. Circular plates with appropriate diameters were cut out from acrylic sheets of approximately 2 mm thickness. Spherical shells were produced by placing a circular acrylic plate
on the top, see Fig. 4, of the negative mold and heated gradually up to its
softening point. The vacuum pump was then started and its operation continued under gradually decreasing temperature. The circular plate was sucked
down until it touched the negative mold. The
deformed plate remained under suction until
ambient temperature was reached. The edge of
the obtained spherical segment was machined to
provide first a parallel circle of 36 cm diameter and then an edge form to satisfy the mechanism of a "pinned" edge as shown in Fig. 2
Section I-I.

Spherical shells are surfaces of double curvature whose surface-area is less than that of
the circular plates they are formed from. In
the case of small curvatures, (Shell 1) practically no difficulties were encountered in forming the spherical segment. On the other hand,
forming shell 2 was faced with the problem of
producing spherical segment of 0.12878 m^2 surface area from a circular
plate of o.13896 m^2 surface area. Also, the diameter of this circular plate was equal to the total meridional length of the spherical segment as
given by $2R[\text{arc sin } (a_o/R)]$.

Thus, the circumference of the circular plate required for shell 2 was

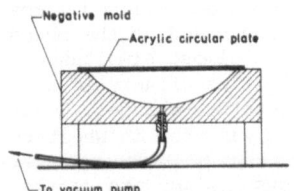

FIGURE 4. Negative mold

1.3214 m while that of the outer-most-parallel circle was 1.1310 m. This difficulty was circumvented
by unfolding the spherical segment according to the
principles of descriptive geometry as shown in Fig.
5. The resulting shape, with slight magnification
of diameter to prepare edge-conditions in line with
Fig. 2 section I-I, was sawed out from an acrylic
sheet and formed as described above. The resulting
meridional edges were glued using a mixture of
grinded acrylic and chloroform.

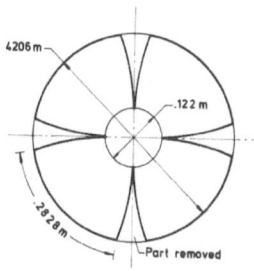

FIGURE 5. Acrylic sheet used to form shell 2

Due to axisymmetrical loading at the apex, meridians and parallels of the spherical shells are
principal directions. As a result, electrical resistance strain gauge were fixed in pairs so that
the longitudinal axis of the parallel gauge coin-
cides with the transverse axis of the meridional
one. It follows that each two gauges represent a measuring station that
yields sufficient data for complete definition of the surface-state of
stresses. Short strain gauges of 0.6 mm grid were used to measure peak
strains. Measuring stations on the upper surface at the apex had to be
omitted due to the presence of apex loads. Because of symmetry, the measuring station on the other side of the apex consists of one strain gauge
that was fixed in an arbitrary direction. A total of nine measuring stations were located along the meridian. Each two stations, apart from the
one at the apex, were arranged at two opposite sides of the shell-surface
with a common normal to the middle-plane.

Specially prepared cylinders out of polyurethane,
which were cast directly on the unloaded spherical
segments, were used to transmit loads to small circular areas at the apex as shown in Fig. 6. These
cylinders provide better contact between load and
shell to guarantee uniformity; further due to large
Poisson's ratio of the polyurethane, it could be
assumed that load pressure was acting normally to
the surface.

Fig. 7 shows the experimental setup that consists
of the spherical shell loading frame, specially designed load cell and a scanning system. A high recording speed of the scanning device was chosen to
obtain almost simultaneous measurements. Physical
properties of acrylic are greatly affected by any
temperature-changes. Therefore, a low voltage, 1
volt, was used in Wheatstone's bridge.

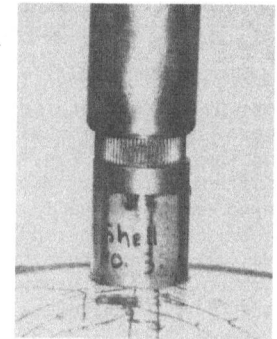

FIGURE 6. Polyurethane cylinder

6. RESULTS AND DISCUSSION

The results obtained in the present work were conducted for the two spherical segments described above when acted upon by a normal load uniformly
distributed over a small circle of radius c at the apex. The intensity of
the considered load, p, is 100 kPa and its radius, c, is 0.013215 m. It
should be noted that the arching parameter of a spherical segment which is
given by the expression $0.5 \tan (\phi_i/2)$, is independent of the radius of the
sphere. Also AP is not a constant parameter for a given spherical segment,
but it varies as a function of ϕ_i. It then follows that, even in the case
of spherical segments with high arching parameters, there is always a small
segment at the apex that possesses the proportions and characteristics of a
shallow spherical shell. This fact has significant influence on the local

stresses next to the acting load, if this apex load is either concentrated or uniformly distributed covering a circular area within the proportions referred to above. This phenomenon along with the fact that the edge horizontal force, H, causes greater bending moment next to the edge as PA increases, are the reasons for a wavy shape of the experimental curves representing the surface stresses in shell 2.

FIGURE 7. Experimental setup

Figure 8 to 15 present a comparison between the output of the three chosen approaches. They illustrate the normal stresses, of shells one and two, on both the upper and lower surfaces as a function of the normal distance from the axis of revolution. The results obtained from the finite element solution, when compared with experimental ones outside the apex region, show fair agreement. However, this agreement is improved for higher arching parameter as shown in Figs. 12 to 14. Within the apex-region, the finite element approach yields remarkable high stresses. The discrepancy between the experimental results at the apex and those obtained by the finite element and force methods is due to the linear elasticity assumed in the two later methods, while it is known that acrylic material is nonlinearly elastic (similar to concrete). It seems that, for practical design, the numerically determined stresses at the apex can be reduced by 40 to 50%. Parallel stresses that are conducted experimentally on the lower surface of both shells are higher than those resulting from the finite element and force methods.

The force method and the resulting special formulation of internal forces yield surface stresses at the apex-region that are hardly influenced by the boundary conditions. It results that the membrane part of the solution dominates. This explains the unique sense of stresses obtained on both sides of shells in the meridional as well as in the parallel directions. Consequently, this method is not suitable for the analysis of shallow spherical shells. Furthermore, it does not consider the characteristics of the shallow part, at the apex-region of the non-shallow spherical shells when they are subjected to apex load within their shallow segment.

7. CONCLUSIONS

The force method is usually recommended for the analysis of non-shallow shells. However, it gives unsatisfactory results for nonshallow spherical shells when acting upon by apex loads within its shallow region. A finite element discretization, in the form of a gore, along with appropriate boundary elements is suggested. This finite element discretization, in spite of the high stresses it yields at the apex, proved to be applicable for spherical segments of different arching parameters. The problem of high stresses at this crown can be amended, for practical design, by reducing them with a factor of 40%. It is noted that the finite element results get relatively closer to the experimental ones for higher arching parameter.

The experimental output of this work serves not only the goals of the present study but also provide a tool for checking on new theories, assumptions or hypotheses that might be considered by others in the area of spherical segments with hinged edge.

38

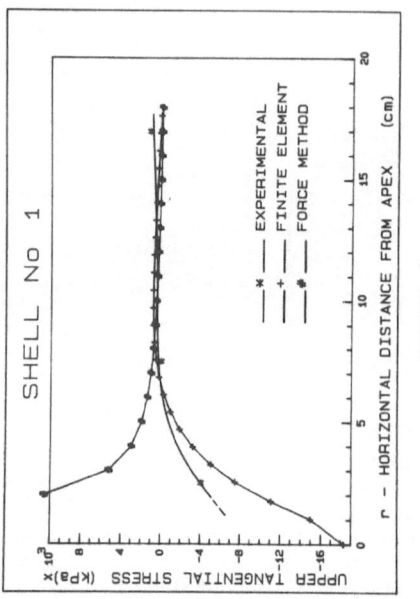

FIGURE 8

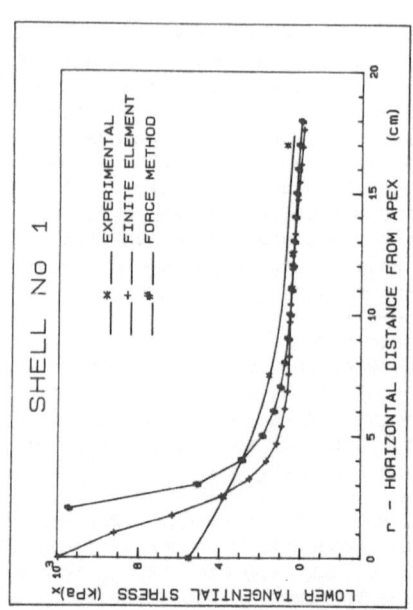

FIGURE 9

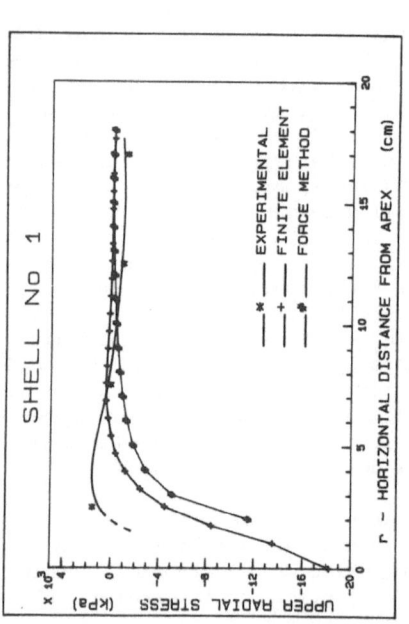

FIGURE 10

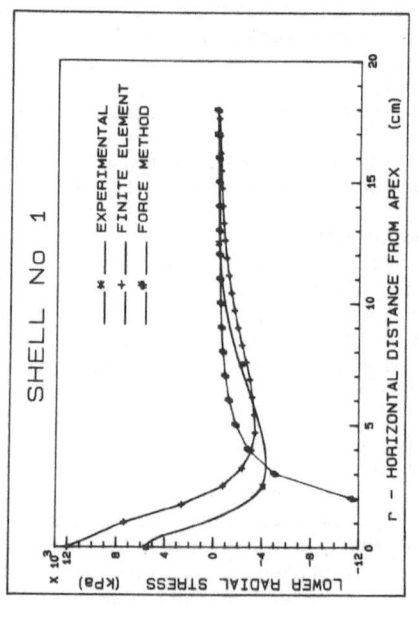

FIGURE 11

Meridional and Parallel Surface Stresses

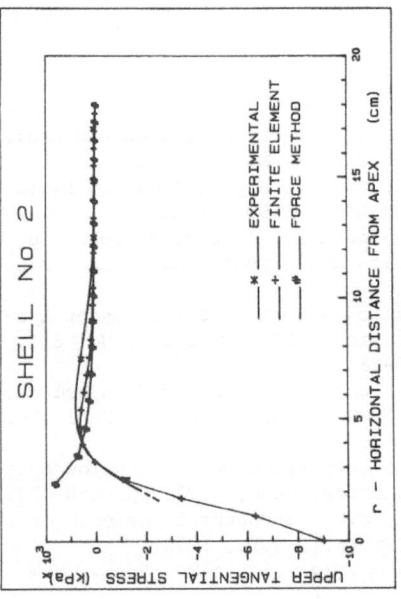

FIGURE 13

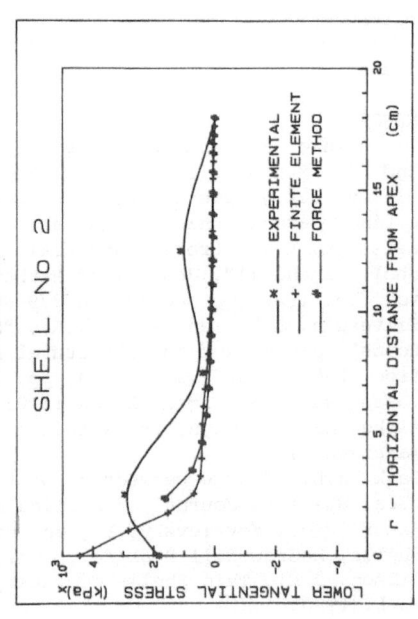

FIGURE 15

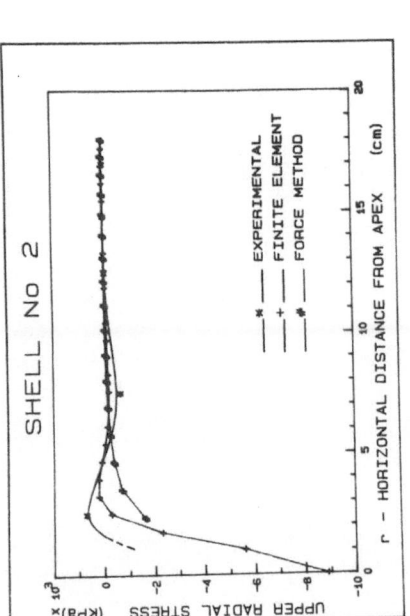

FIGURE 12

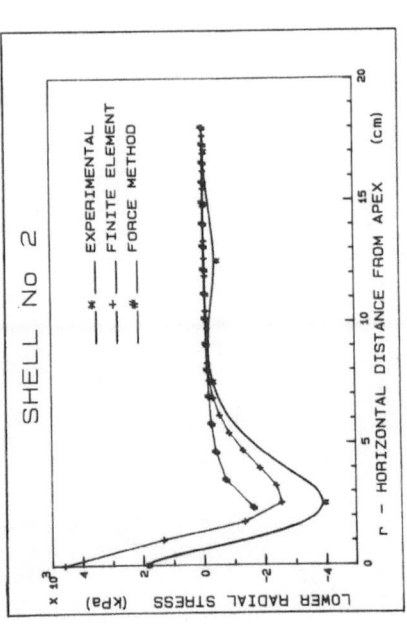

FIGURE 14

Meridional and Parallel Surface Stresses

ACKNOWLEDGEMENT

The financial support through Kuwait University Grant No. EV011 is acknowledged.

REFERENCES
1. Timoshenko, S., and Woinowsky-Krieger, S., Theory of Plates and Shells, 2nd Ed., McGraw-Hill Book Co., Inc., New York, N.Y., 1959.
2. Vlasov, V.Z., General Theory of Shells and Its Applications in Engineering, Nasa Technical Translation, NASA T T F-99, 1964.
3. Reissner, E., "Stresses and Small Displacements of Shallow Spherical Shells I and II", Journal of Mathematics and Physics, Cambridge, Mass., Vol. 25, 1946, pp. 80-85 and 279-300.
4. Silverman, I.K., and Mays, J.R., "Shallow Spherical Shells Under Apex Loads", Journal of the Structural Division, ASCE, Vol. 100, No. ST1, Proc. Paper 10262, Jan. 1974, pp. 249-264.
5. Łukasiewiez, S., Local Loads in Plates and Shells, Sijthoff & Noordhoff, Alphen aan den Rijn, the Netherlands, PWN-Polish Scientific Publisher, Warszawa, 1979.
6. Deak, A.L., "Large Deflection of a Linearly Viscoelastic Shallow Spherical Shells", Journal of Applied Mechanics, June, 1972, pp. 469-474.
7. Baker, E.H., Kovalevsky, L., and Rish, F.L., Structural Analysis of Shells, McGraw-Hill Book Co., Inc., New York, N.Y., 1972.
8. Gibson, J.E., Thin Shells Computing and Theory, Pergamon Press, New York, 1980.
9. Reissner, H., "Spannungen in Kugelschalen (Kuppeln)", Festschrift Heinrich Mueller-Breslau, Alfred Kroener Verlag, Leipzig, 1912.
10. Hampe, E., Statik Rotationssymmetrischer Flaechentragwerke, VEB Verlag fuer Bauwesen, Berlin, 1964.

STUDY OF STRESS DISTRIBUTION IN THE WALLS OF VESSEL CONTAINERS FOR HIGH - PRESSURE GAS

D.R.MOCANU, N.IONESCU, E.SPIREA

INCERTRANS - Bucharest and I.P.I.C.C.F.- Buzău , ROMANIA

1. INTRODUCTION

The implicated revolution vessels,named according to directions in force-vessel-containers-are used for depositing fluids in the form of under pressure compressed,liquefied or dissolved gas (oxygen,nitrogen,argon,carbon dioxide,various combustible gas etc.).They belong to the category of curved plates with reduced thickness,that's why,when dwelling them on,it is considered that the container walls behave like a membrane,i.e.they can take over only stretching stresses.

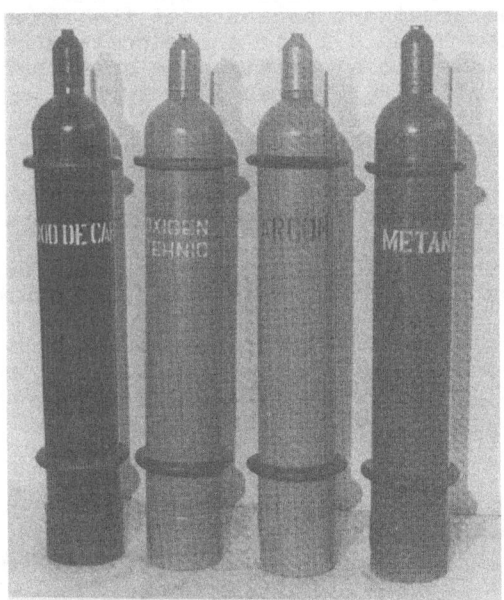

FIGURE 1. Vessel - containers

2. MANUFACTURE TECHNOLOGY

Besides the economical reasons,the pipe manufacture of vessel - containers has the advantage of manufacture rigorous inspection in comparison with the use of the metal block plastically deformed by matrix forging in depth at a 1 - d rate,i.e. (6...8):1,referring the reported height to diameter (fig.2).

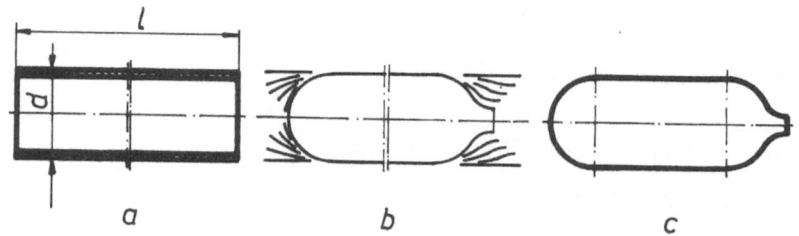

a b c

FIGURE 2. Pipe plastic deformation (ogivation) for obtaining vessel-containers. a)- Pipe cut at the technological length l. b) - Neck and bottom progressive deformation (ogivation). c) - The vessel-container in final shape.

For making the final shape of fig.2c,having as initial stage the shape of fig.2a,there is produced a controlled plastic deformation,according to an own technology.Fig.2b shows,schematically,the sequence of plastic deformation,so as from the cylindrical shape to be obtained the ogive shape at the neck and bottom of the vessel - container.
Looking the deformations sequence at the vessel bottom and neck,there can be said it is a dislocation,because at microstructural level,in a certain pipe area there are produced,network localized (limited) perturbations,that separate the zones of a crystal where a sliding took place form the zones where this one didn't took place.Investigations in the two ogivated regions show that as the dislocation shifts itself,the sliding appears in the surface area where it moves itself.It is observed the importance of this aspect(point), not only to account for crystals sliding but,in the same time, for correlation and interdependence of plastic deformation from yield stress,flow,fatique and brittle facture.

Ogive configuration of vessel - container bottom and neck, confirms the hypothesis of the applied manufacture technology, that is to say a proportional simultaneous combination, resulting a helical - marginal dislocation (fig. 3 and fig.4)

Fig.3.

FIGURE 3. Vessel - container bottom and neck. a) - ground - view. b) - cross - section.

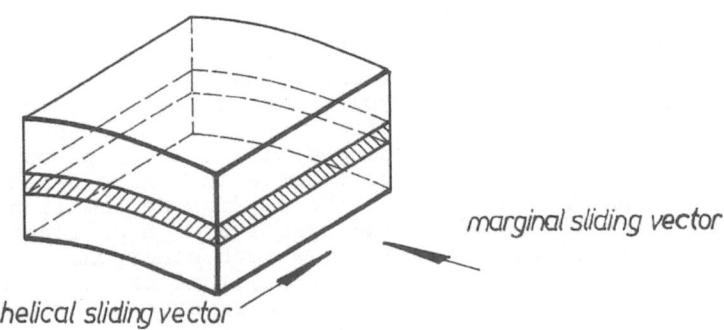

marginal sliding vector

helical sliding vector

FIGURE 4. Helical - marginal dislocation at vessel bottom and neck ogivation.

From fig.4 there can be seen that the plastic defor - mation by ogivation represents a sliding of some regions of the crystal,one over the other,along certain crystallogra- phic planes,whem the tangent stress reaches a certain cri- tical value dependent on the heating temperature.The fine structure of sliding lines was studied at very great magni- fications,with the help of electronic microscopes.The limi- ted number of sliding systems (sliding plane + sliding direction) is the reason of the great dependence given the helical - marginal deformability positioning at vessel - container ogivation (fig.5).

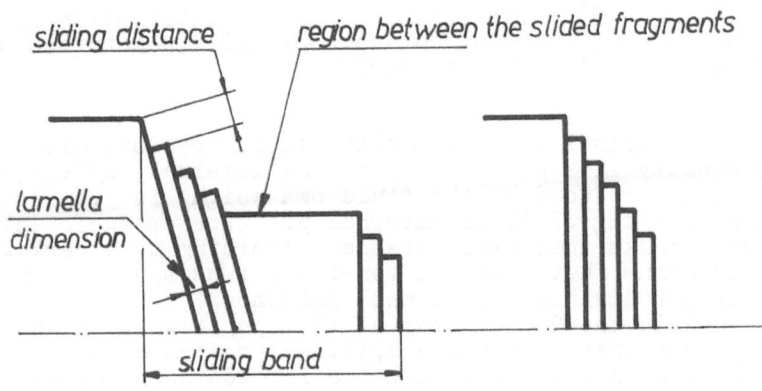

sliding distance   *region between the slided fragments*

lamella dimension

sliding band

fig.5

44

FIGURE 5. Scheme of a helical – marginal deformation (cross-section). a) – Correct deformation, with minimum residual stresses. b) – Inadequate deformation with great residual stresses.

This is due to the fact that the inverse stress, developed as result of dislocations heaping up at barriers during the first loading cycle, helps the dislocations movement when the sliding direction changes. Moreover, when the sliding direction is changed dislocations of opposite sign are created by the same sources that hase produced the dislocations that determined the sliding deformation in the initial direction. Because the opposite sign dislocations attract themselves and annihilate each other, the total effect is a network " softening" (fig.2b); that explains the fact that the shape of the flow curve in inverse direction is under the flow curve of the initial direction (fig. 6).

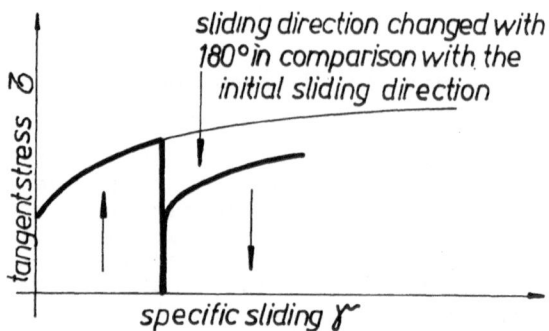

sliding direction changed with 180° in comparison with the initial sliding direction

FIGURE 6. Effect of changing the force applying direction on – curve at helical – marginal deformation.

That δ – γ curve explains stress distribution into the vessel – container bottom and neck walls, determined by the ogivation applied technology.

3. STRENGTH CALCULUS
Into the vessel – container, the fluid pressure exerts itself uswally in a uniform way on the internal surface, and the pressure produced by the fluid own weight is proportional to height L (fig .7). The internal pressure produces into the vessel walls scattered stresses, according to a certain stress spectrum that develops tensile stresses, uniformly distributed (divided) onto wall thickness.
Because in the 2nd chapter of this paper we said the vessel – container is monoblock, without bottom welding (fig. 7), walls thickness calculus is made so as at the value of the hydraulic test pressure, the material stress in any point would not go beyond 75 % the yield stress guaranteed for the quality type of the used steel.

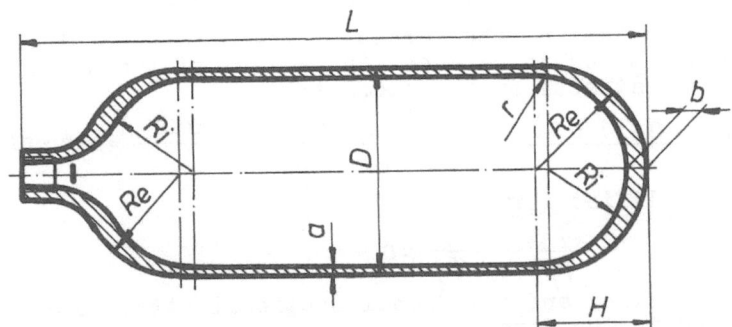

FIGURE 7. Seamless monoblock vessel - container (cross - section)

Assuming the vessel is subjected to an internal pressure p, there is separated an abcd infinitezimal element, of dS1 and dS2 sides, wraped up by the meridian curves M_1 and M_2 and the parallel circles φ_1 and φ_2 (fig. 8).

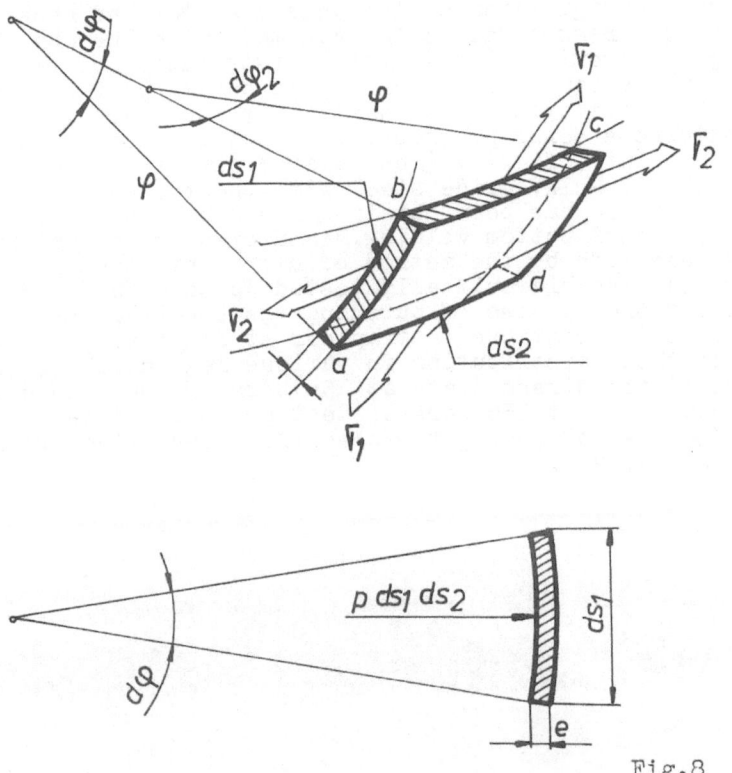

Fig.8

FIGURE 8. Vessel – container bottom cross – section

The rezultant of the forces on the ad and bc sides is normal and tends to balance the effect of the normal pressure that acts on the element:

$$\frac{\vec{V_1} \cdot e \cdot ds_1 \cdot ds_2}{} + \frac{\vec{V_2} \cdot e \cdot ds_1 \cdot ds_2}{} = p \cdot ds_1 \cdot ds_2 \qquad (1)$$

or, in simplified form:

$$\frac{\vec{V_1}}{\varphi_1} + \frac{\vec{V_2}}{\varphi_2} = p \qquad (2)$$

The parallel and meridional sectional stresses are expressed in the form:

$$M_1 = \vec{V_1} \cdot e \qquad M_2 = \vec{V_2} \cdot e \qquad (3)$$

4. STRESS DISTRIBUTION

On the basis of these reasons, using the methods for the stresses experimental determination, there has been established stating from known data (hydraulic test pressure, pipe minimum thickness, internal and external radius, Ri and Re), the optimum configuration of the vessel bottom and neck.

Bottom thickness establishing was made with the object of obtaining a convenient (covering) distribution of the stresses in the connection area of the convex part (neck) (bottom) to the cylindrical part ; any necessary increase of the thickness being progressive proportional from the connection point (fig.7) (correlated with fig.5 and 6.)

Using the 31 Mn12V grade steel with the technology in chapter 2 , there has been obtained the monoblock vessel-container, without bottom welding. Measurement of actual walls thickness was made by the method of ultrasonic transparency.

The vessel was hidraulically tested up to a 5oo bars pressure for the purpose of pursuing the behaviour, determining the specific deformations that appear at increasing pressures, as well as their distribution in various regions. There has been pursued the stress state at 15o bars pressure as nominal work pressure and at 25o bars as test pressure, at temperatures of + 6o°C and – 4o°C, using the electrical resistive stress analysis (fig. 9).

Fig. 9

FIGURE 9. Electroresistive transducers location on the vessel
container.
The accomplished measurement show that the stress distribution on the vessel bottom and neck determines a security
(rate) of 1 . 97 al 15o bars and 1. 53 at 225 bars.
Stress determination on the vessel bottom and neck was
doubled by using the reflexion photoelasticity method, that
emphasized the adequate (suitable) behaviour of the chosen
steel grade, technological manufacture method and relaxing
heat treatment (fig.1o)

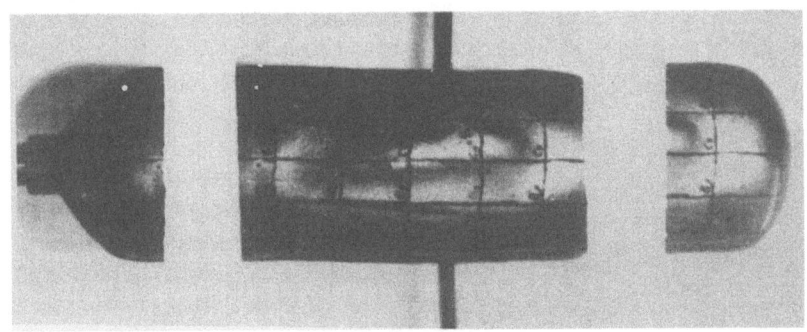

FIGURE 1o.Stress distribution measurement into the vessel-
container body by the reflexion photoelasticity method.
By repeated measurements of stress distribution, using the
electrical resistive stress- analysis and reflexion photoelasticity, there has been determined the optimum shape of the
vessel bottom and neck, thus determining a rational use of the
material, in full security conditions when operated.
From the accomplished tests it results that the b thickness
(fig.7),in the middle of the bottom,must satisfy the following conditions when the connection radius r is greater or
equal whith o.o75 D,i.e.

$$b \geqslant 1.5a \qquad \text{for } H/D \geqslant 0.2 \qquad (4)$$
$$b \geqslant a \qquad \text{for } H/D \geqslant 0.4 \qquad (5)$$

5. CONCLUSIONS
On the basis of these tests there has been improved the
shape coefficient of the vessel bottom and neck ,as parameter
dependent on the pipe H/D rate.
Under that form, the dimensioning of vessel-containers
will be made in practical trust conditions given the theoretical reasons and the laboratory tests,modifying the present
official regulations for the design,manufacture and control
of the vessel - containers,on the basis of this contribution.
We must add that beginning with 1985 there are manufactured,
in serial production,vessel - containers with the optimised
shape resulting from this paper.

REFERENCES

1.Mocanu D.R.: Rezistenţa materialelor,Editura Tehnică,
  România,198o.
2.Mocanu D.R.: Analaiza experimentală a tensiunilor,
  Editura Tehnică,România 1976.
3.Iosipescu N.: Introducere în fotoelasticitate, vol. I şi II
  Editura Tehnică,România 196o.
4.Vishay : Bulletin SFC - 3oo.

THE EXPERIMENTAL ANALYSIS OF THE STRESS DISTRIBUTION IN A 1000m³ SPHERICAL STEEL TANK FROM THE PETRO-CHEMICAL INDUSTRY

V. RADIANOV

Scientific Researcher - Building Research Institute - INCERC -   Bucharest
ROMANIA

1. INTRODUCTION

The growth of the chemical industry in Romania during the last two deca - des has led to the development of large petrochemical plants. These plants have been provided with storage tanks such as the 1000 m³ spherical   steel tanks for propylene.

The paper describes the experimental analysis on the stresses distribu - tion in a 1000 m³ spherical steel tank made of 34 steel segments (RV - 52 steel), with a shell thickness of 45 mm used to store propylene and a work- ing pressure of 21 bars.

This experimental analysis was initiated by the designer of this project in order to homologate the tank prototype /1/. The experimental analysis on the stresses distribution in the 1000 m³ spherical steel tank was performed during the water pressure test and after the thermal treatment   for   the stress relieving after mounting.

This experimental test aimed the followings:
- to check the calculation hypotheses and methods;
- to check the new execution technology of tank segments in the plant;
- to check the stresses in the support areas where the possibilities for a theoretical predictions are limited;
- to check the stresses in zones with deviations from the sphere geometry above those admitted for a correct execution, but approved by the designer;
- to check the tank strength capacity.

Fig. 1 shows   the   1000 m³   spherical steel tank under the   experimental   test of stress distribution.

To perform such an analysis the   resis- tive electric tensometry method has been used /2/, that established the strains of the tank shell and the stresses by means of strength theories. This   method   has been chosen both due to   the relatively short time for the analysis and to   the large number of measuring points. At the same time, the use of this method   made possible the centralization of measure- ments in the automated measuring   equip- ment AUTOLOG-PEEKEL INSTRUMENTS B.V. /3/.

Fig. 1

2. LOADING INSTALLATION

The most unfavourable loading schemes and the size of loadings led to the design of the 1000 m^3 spherical steel tank, with the 45 mm shell thickness and a calculation pressure equal to 21 bars. The tank loading was done according to a loading scheme suggested by the designer of the tank and to the Romanian code /3/;the reference loading level (L-r.t.l.) is equal to the calculation pressure (cp=21 bars) and the maximum loading level(L-m.t.l) is equal to the test pressure ($p_p = 1.47 \times p_c = 31$ bars), all these pressures being measured in the lower side of the tank.

Thus, two installations have been built, the former for tank water filling and the latter for increasing the hydraulic pressure inside the tank up to the maximum loading level (L-m.t.l.= 31 bars) in 15 minutes

Fig.2 shows the schemes of the two installations.

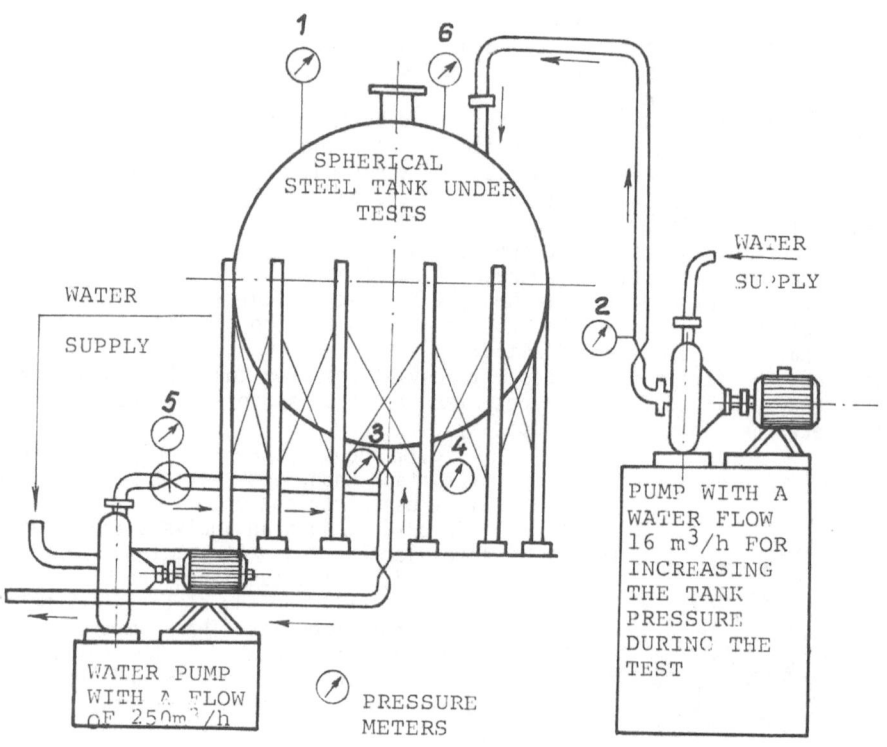

Fig. 2

Several manometers of minimum 0.5 bars sensitivity and the   measuring range up to 45 bars were flexible mounted inside the tank in order to mea - sure the hydraulic pressure as follows:
- a manometer on the pipe that increases the hydraulic pressure near   the aggregate that increases the hydraulic pressure;
- two manometers at the top part of the tank;
- two manometers at the lower part of the tank.
During the test the hydraulic pressure inside the 1000 m^3 spherical steel tank was measured both at the top and lower part of the tank, but the pressure that was predominant during the test was that measured at' the low- er part of the tank.

3. INSTALLATION FOR MEASURING THE SPECIFIC STRAINS

The installation for measuring the specific strains during the experimen tal analysis performed on a 1000 m^3 spheric steel tank used in the   petro- chemical industry consists of a tensometer installation with   resistive straingauges and a measuring station placed near the tank under   testing.
3.1. Strain gauge installation. The strain gauge installation was   mounted on the external surface of the tank shell on the following places: on   the top calotte, on the supports no.4 and no.5, on the meridiam that passes bet ween supports no.4 and no.5, on the top and lower parts of all supports and on the lower calotte and on the internal surface of the tank shell the stra in gauges have been mounted near the inlet opening, at the upper side   of all supports and on certain areas of the lower calotte.

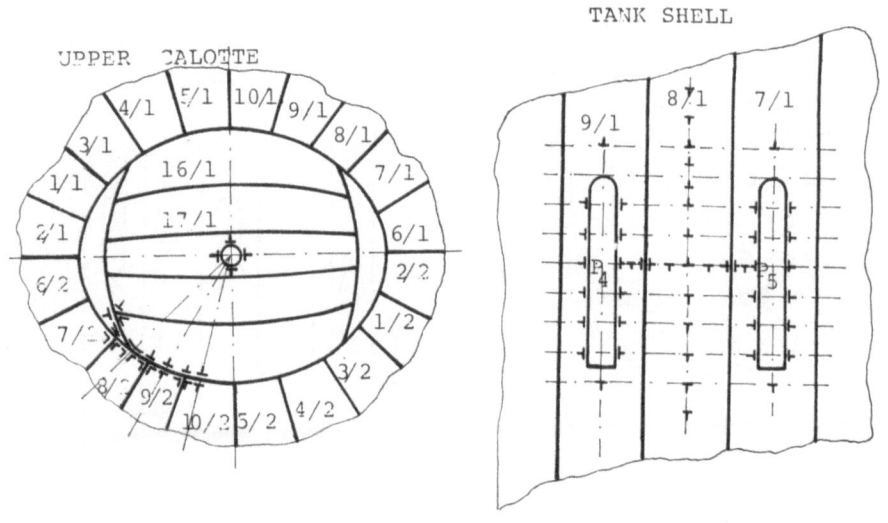

Fig. 3   Fig. 4

Fig. 3,4,5,6 show the positions, measuring points on various zones   of the tested 1000 m^3 spherical steel tank. The distribution of these measur - ing points has been done according to the project of this test, drawn up by our specialists and according to the test specification established by the tank designer.

There have been 128 measuring points in all, provided with 256 resistive strain gauges disposed on the tank as follows:
- 48 pcs. on the upper calotte;
- 64 pcs. on the tank shell;
- 74 pcs. on the lower calotte;
- 40 pcs. on the supports;
- 30 pcs. on the internal surface of the tank shell.

Each measuring point was made of two resistive strain gauges perpendicu- larly on each other , namely, the former along the direction of the meridi- an and the latter along the parallel lines. Each measuring point was thermo compensated with a resistive strain gauge stuck on a steel plate fixed near the respective measuring point. Romanian resistive strain gauges have been stuck with an adhesive of the type M-Bond 200/5/ and protected with a speci al putty.

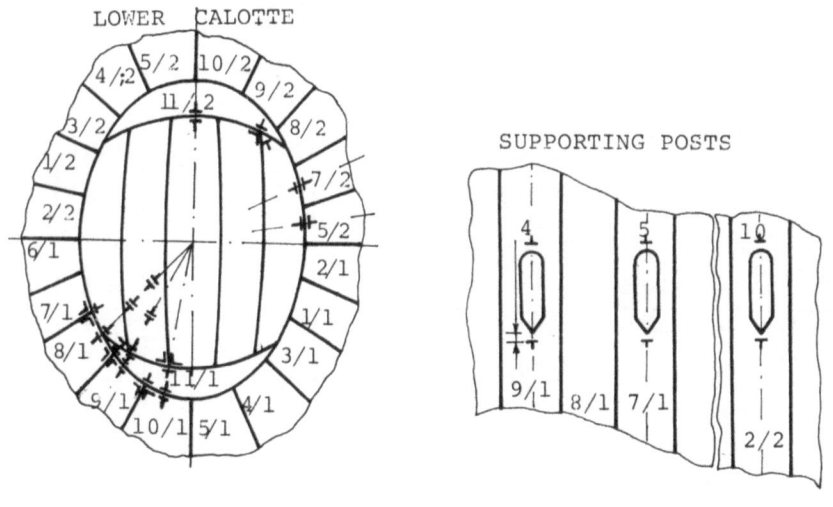

Fig. 5   Fig. 6

Generally, all resistive strain gauges have been placed near the   weld- ing seams where the quality of the base material is supposed to be changed.

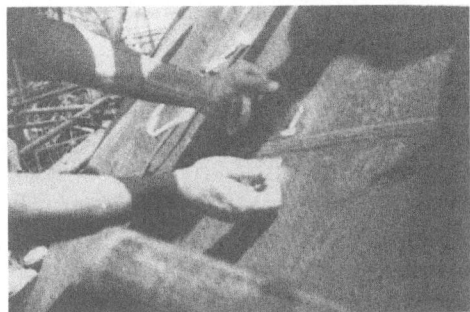

Fig. 7

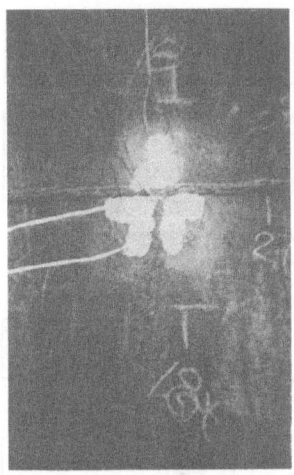

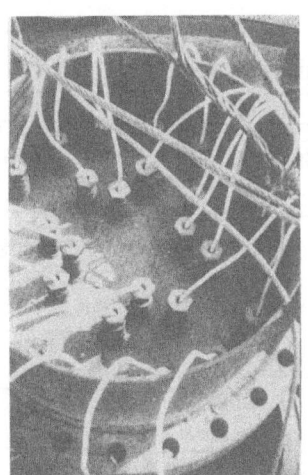

Fig. 8   Fig. 9

Fig. 7 shows how the resistive strain gauges are on the external   surface
of the tank shell;Fig.8 shows three measuring points at the crossing of the
welding seams on the external tank shell; in Fig. 9 the cables for the   resistive strain gauges disposed on the internal surface of the tank   shell
have been pulled out using the blind flange provided with inlets.

3.2. Measuring apparata
The measuring apparata consists of an AUTOLOG PEEKEL INSTRUMENTS B.V. pro
vided with 260 measuring channels, central unit, PDP 11/04   microprocessor
and a deck printer of the type LA-180.

4. THE STAGES OF EXPERIMENTAL ANALYSIS

This experimental analysis has been performed according to working diagram by loading-unloading cycles of the tank /4/. These loading - unloading
cycles had the following characteristic loading levels:
- zero level of the test load (L-zero.t.l.) - empty tank;
- zero level of the overpressure of test load(L- zero overpr. t.l.) - tank
filled with water;
- the reference level of the test load (L- r.t.l.= cp)- tank filled with
water and an overpressure of 21 bars;
- maximum level of the test load (L-m.t.l. = 1.47 x cp) - tank   filled
with water and an overpressure of 31 bars.

There were intermediate loading stages between these characteristic ones
of the experimental analysis that provided a better understanding of   the
stresses in the tank shell and avoids the possible accidents. After the value of the maximum test load level was reached the overpressure was reduced
to the reference level of the test load and the tank was kept under   this
load for 18 hours.

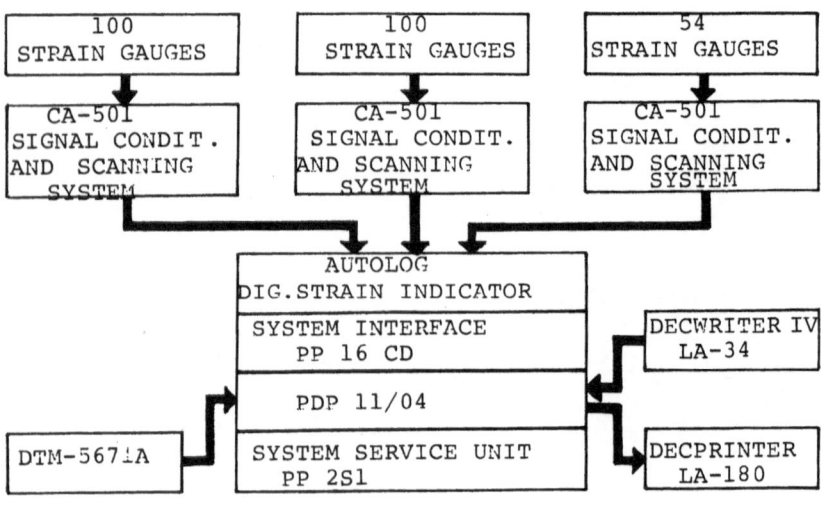

Fig. 10

Fig. 10 shows the automated strain gauge installation scheme with 260 AUTO-
LOG measuring points.

5. THE RESULTS OF THE EXPERIMENTAL ANALYSIS

Using the measuring programme adapted to PDP 11/04 microprocessor   and
knowing the elestic constants of the materials the tank was made of(E= elas
ticity module and μ = Poisson coefficient), tables have been obtained   on
the deck printer that contained the average strains for each loading level
and specific to all these the equivalent stresses, equivalent strains, equi
valent stresses and safety coefficients.

The calculation formulae used in the measuring programme were as follows:

$$\sigma_p = \frac{E}{1 - \mu^2}(\varepsilon_p + \mu \varepsilon_\theta)$$
in daN sq.cm.

$$\sigma_\theta = \frac{E}{1 - \mu^2}(\varepsilon_\theta + \mu \varepsilon_p)$$
in daN sq.cm.

$$\varepsilon_{ech.} = \sqrt{\varepsilon_\gamma^2 + \varepsilon^2 - \varepsilon_p \cdot \varepsilon_\theta}$$
in $\mu m \ m^{-1}$

$$\sigma_{ech.} = \sqrt{\sigma_\gamma^2 + \sigma_\theta^2 - \sigma_p \cdot \sigma_\theta}$$
in daN sq.cm.

$$C_1 = \frac{R_m}{\sigma_{ech}} \qquad\qquad C_2 = \frac{R_{p\ 0.2}}{\sigma_{ech}}$$

in which:

ε_p = real mean strain along the meridians;

ε_θ = real mean strain along the parallels;

σ_p = real stress along the meridians;

σ_θ = real stress along the parallels;

E = elasticity module of the material

μ = Poisson coefficient

$C_{1,2}$ = safety coefficients

R_m = material breaking strength;

$R_{p\ 0.2}$ = material yielding strength.

The stress caused by the water filling of the tank, namely, the changing of level (L- zero t.l.) with the level (L- zero overpr. t.l.) gave reduced values of the real reduced stresses in the tested points.
    The experimental analysis on the distribution of pressures inside the spheric steel tank proved the followings:
- the real equivalent stresses increase almost proportional to the hydrau - lic pressure inside the tank;
- real equivalent stresses on the lower side of the shell are smaller with 5% than those on the external surface of the shell;
- at the reference level of the testing load (L - r.t.l. = cp = 21 bars)the highest values of the equivalent stresses were included between the values 1600 daN sq. cm. and 1760 daN sq.cm. along the meridians that passes through the supports no.4 and no.5, respectively 0.338 R_m and 0.533 R_p 0.2, and for the other tested points the values of the real equivalent stresses were smaller;
- at the maximum level of the testing load (L- maximum t.l. = 1.47 x cp = = 31 bars) the highest values of the equivalent stresses were between 2575 daN sq.cm. and 2680 daN sq.cm and found on the meridian that passes through the supports no.4 and no.5 that represents 0.515 R and 0.812 R_p 0.2; in the

other tested points the value of the equivalent stresses do not exceed the value of 2180 daN sq.cm that represents 0.419 R_m and 0.661 R_p 0.2;
- the pressure was kept constant for 18 hours   at the reference level of the testing level (L-r.t.l.= cp= 21 bars) and led to small increases   of the strains;
- the remanent stresses in the measuring points with maximum   equivalent stresses had small insignificant values;
- the maximum values of the real equivalent stresses were noticed at the up-per sides of the supports and at the welded corner joints of the upper and lower calotte and the tank shell.

6. CONCLUSIONS

The experimental analysis on the distribution of stresses in the 1000 m^3 spherical steel tank with 45 mm shell thickness proved that the tank had an elastic behaviour at the reference level of the testing load (L-r.t.l. =cp= =21 bars) and at the maximum level of the testing load (L-m.t.l.=1.47 x cp= = 31 bars).

At both testing load levels, the value of 0.9 x R_p 0.2 was not   exceeded in no measuring point, and the remanent strains did not exceed the   values recommended by the Romanian codes.

Strains increase almost proportional to the hydraulic pressure inside the tank. At the same time, it has been established that zones with deviations from the sphere geometry lead to stress concentrations.

If at the maximum testing load level (L-m.t.l.= 1.47 x cp = 31 bars), in certain measuring points on the tested tank, the equivalent stresses had va-lues close to 0.815 Rp 0.2, in the other points the equivalent stresses did not exceed the value of 0.665 Rp 0.2.

It has been recommended that the 21 bars pressure measured at the   lower side should be the limit pressure in the case of the 1000 m^3 spherical steel tank with the tank shell of 45 mm thickness.

This experimental analysis on the distribution of stresses in a   1000 m^3 spherical steel tank from the petrochemical industry provided data for the design of these tanks and for the improvement of these experimental   ana-lyses using the resistive strain gauge method for the in - situ tests on structures.

REFERENCES

1. BALAN,St.   - Incercarea Construcţiilor. Editura Tehnică,
   et.al.   Bucureşti, 1965.

2. MOCANU,D.R.   - Analiza experimentală a tensiunilor. Editura Tehnică
   Coordinator   Bucuresti,1977.

3. * * *   - Instrucţiuni AUTOLOG-PEEKEL INSTRUMENTS B.V.,Holland
                     1979.

4. * * *   - Incercarea in-situ a construcţiilor prin   încărcări
                     statice, STAS - 1336-80.

5. AVRIL, J.   - Encyclopedie Vishay d'analyse des contraintes,Vishay
   Coordinator   Micromesures, Paris.

SMALL-SCALE TESTING OF END-RESTRAINED TUBULAR BEAM-COLUMNS

R. C. BARROS

Department of Civil Engineering, Faculdade de Engenharia da Universidade do Porto, Porto 4099, Portugal.

1. INTRODUCTION

It has been common practice in the design of tubular columns to assume an "effective lenght factor" which is intended to include the effects of end restraints. Usually this factor is in the range of 0.8 to 1.0 but there is little experimental evidence to support this.

More recently it has become important to be able to estimate the effect of semi-rigid end constraints on the buckling behaviour of beam-columns. To this end the Structural Stability Research Council established its Task Group 23 "Effect of End Restraint on Initially Crooked Columns" (1). The work repor ted herein outlines research fostered by this emphasis, and is an attempt to provide an experimental data base from which the buckling behaviour of tu bular columns with semi-rigid end restraints may be estimated.

The test sequence reported in this paper involves the testing of four small scale tubular column specimens, tested under semi-rigid end conditions. The specimens were hot-formed, eliminating the residual stress patterns inherent in large-diameter cold-formed columns. The test sequence does allow study of the interaction of such effects as column out-of-straightness, load eccentri city and, primarily, end restraints. A relatively small diameter-to-wall thickness ratio ensured that general yielding would be in the instability mo de.

The problem of buckling of tubular columns is complicated by the fact that the buckling direction of such columns is not readily predetermined, and it is thus important to permit buckling in any direction. The end restraint sys tems developed for this purpose adequately met this criterion. The achieve- ment in designing the end-restraint systems allowed total monitoring of the spatial buckling of the beam-columns tested.

2. EXPERIMENTAL PRELIMINARIES

2.1. Development and description of the end-restraints

Not only did the end restraints have the requirements of being reliable and behaving predictably for a column under a given axial load, but they needed to provide uncoupled end restraint in two perpendicular directions at each end; also preferably, they should provide rotation restraint which could be varied between experiments. The detailed development and calibration of these restraints led to the end restraints system shown diagrammatically in Figure 1. In this figure it can be seen that the rotational restraint is being provided by perpendicular torsion bars which are essentially behaving in an uncoupled manner in perpendicular directions.

Calibration experiments, described by Barros (2, 3), produced functions which related the end restraint moment in the torsion bar, M_{TB}, and the end rotation about that axis θ, in the form

$$M_{TB} = A \left[\frac{\theta}{(\frac{L}{L_{REF}})} \right]^B$$

where L is the lenght of the torsion bar during the experiment, and L_{REF} is some reference lenght of the bar. The parameters A and B are specific to a particular torsion bar. A typical end restraint function is shown in Figure 2.

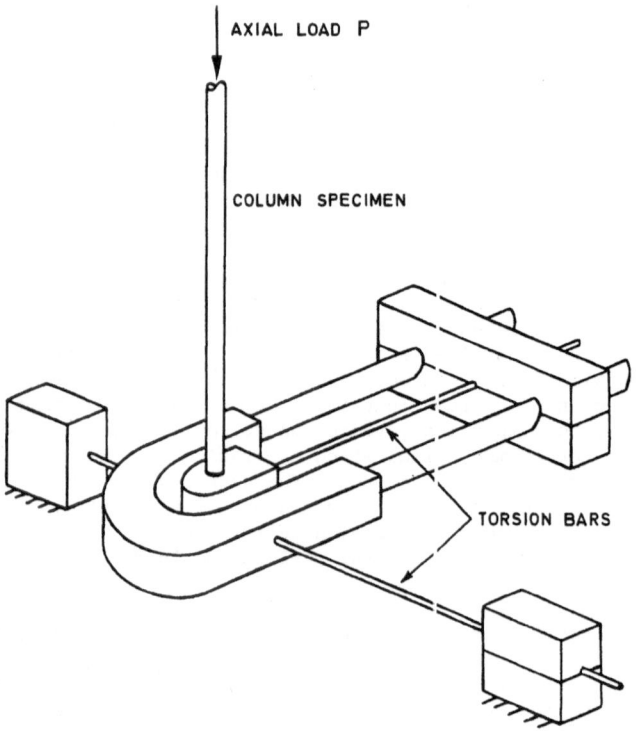

Figure 1. Diagram of an End-Restraint System

2.2. Characteristics of the column specimens

In the column testing sequence four hot-formed steel tubular columns each 24 ins. (0.610 m) long were tested in compression. The column specimens had an outside diameter of 0.5 ins and an average thickness of 0.09 ins. Preliminary tensile coupon tests indicated that the material had a yield stress of 63.4 psi (444 MPa), and an elongation at failure of 14.5%.

The out-of-straightness patterns were carefully measured on each specimen along longitudinal perpendicular planes. Table 1 gives the magnitude of out of-straightness and also the form of the out-of-straightness pattern, as mea sured prior to testing using an alignment jig with a dial gage with 1/1 000 in. precision.

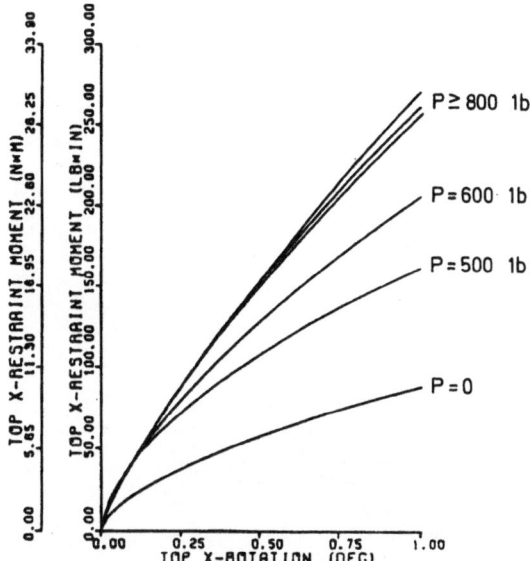

Figure 2. Top Restraint Moment Function
(for positive x rotation)

Table 1. Maximum Column Out-of-Straightness

Specimen Number	X-Direction		Y-Direction	
	Inches (mm)	Form of Curvature	Inches (mm)	Form of Curvature
1	0.0033 (0.085)	Single	0.0068 (0.17)	Single
2	0.0016 (0.041)	Double	0.0072 (0.18)	Single
3	0.014 (0.36)	Single	0.0082 (0.21)	Single
4	0.0043 (0.11)	Single	0.0028 (0.07)	Single

3. TESTING PROGRAM

The four column specimens were tested in a CGS electro-hydraulic universal testing machine. Each column was supported at each end by the complex
end-restraint systems already described. In the process of column testing
the rotation of each end restraint (in each direction) was observed, together with axial deformation, and mid-height lateral deflection of each specimen in perpendicular directions. A photograph of a column just at the end of
testing is shown in Figure 3.

The relative end eccentricities were measured during testing between the
center of tubes and the center of end blocks, but the absolute magnitudes at
each end were established by subsequent theoretical analyses (2) to be those

60

shown in Table 2.

Figure 3. Column Specimen at End of Test

Table 2. End Eccentricities

Specimen Number	Bottom Head		Top Head	
	X-Direction (inches)	Y-Direction (inches)	X-Direction (inches)	Y-Direction (inches)
1	-0.12	0.15	-0.01	0.04
2	-0.12	0.15	-0.01	0.04
3	-0.12	0.08	-0.04	0.0
4	-0.12	0.15	-0.01	0.04

The major differences between specimens tested was in the rotational resistance offered to the ends. These differences are summarized in Figure 4, where the ratio of end stiffness to reference stiffness is shown for each specimen.

SPECIMENS

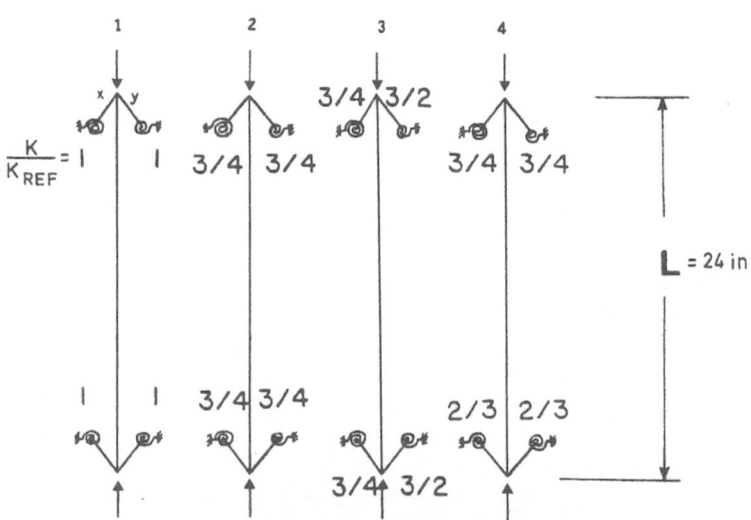

Figure 4. End Restraint Differences in Specimens

Typical results, plotting axial load vs. mid-height lateral deflection and mid-height displacement path, are shown in Figures 5 and 6. The solid triangles represent experimental values obtained, and the solid lines repre‐ sent subsequent theoretical analyses.

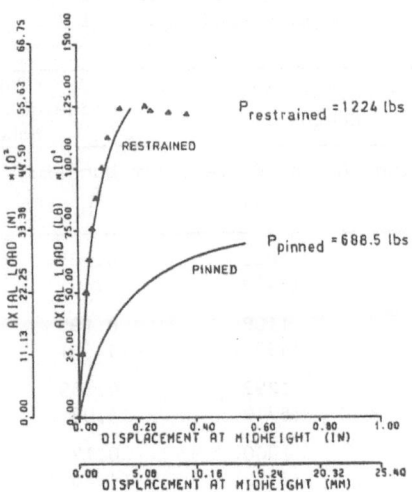

Figure 5. Comparison of Load-Lateral Displacement at
Midheight Curves (for Specimen 1)

62

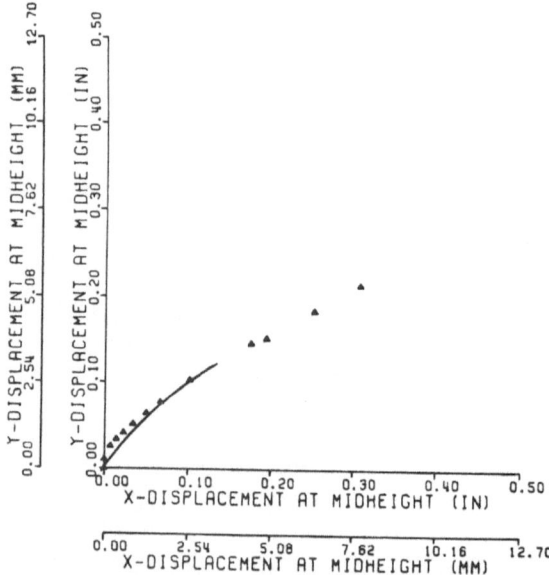

Figure 6. Comparison of Midheight Displacement Path
(for Specimen 1)

A tabular summary of the results of all four column tests is given in Table 3.

Table 3. Static Column Loads and Mid-height Deflection
at Buckling and at End of Tests

Specimen Number	Static Column Loads lbs (Newtons)		Total Midheight Deflection Inches (mm)	
	At buckling P_b	At end of test P_{end}	At buckling δ_b	At end of test δ_{end}
1	1260 (5607)	1221 (5433)	0.185 (4.70)	0.374 (9.51)
2	1179 (5247)	1108 (4931)	0.193 (4.90)	0.538 (13.7)
3	1454 (6470)	1292 (5749)	0.159 (4.03)	0.373 (9.47)
4	1333 (5932)	1300 (5785)	0.173 (4.40)	0.258 (6.55)

4. COMPARISON OF RESULTS AND DISCUSSION
    The theoretical analyses, modelling beam-column behaviour only up to buckling by the Influence Coefficient Method (4), employ a second-order Newton-

-Raphson iterative technique with a modified Crisfield Method near buckling (2, 5).

In Figure 5, the solid line most closely modelling the experimental results includes the effect of end restraints, while the other curve is an iden tical analysis with the effects of end restraints removed. It is clear that the end restraints have a dramatic effect both in increasing the column buck ling load and in decreasing the lateral deformation at which buckling takes place. This observation was confirmed on all four column specimens.

Examination of the load-deflection curves for pinned and restrained beam columns leads to the conclusion that inclusion of the developed end-restraint systems produces an increase in the maximum strength of the order of 65%-100%, and a decrease in the total buckling deflection at midheight of the or der of 63%-70%.

Tables 4 and 5 compare characteristic results of the four beam-column ex periments.

Table 4. Comparison of Buckling Loads

Specimen Number	Buckling Load lbs (Newtons)		% Error Experiment vs. Theory
	Theory	Experiment	
1	1224 (5536)	1260 (5607)	1.29
2	1150 (5118)	1179 (5247)	2.52
3	1408 (6266)	1454 (6470)	3.27
4	1293 (5754)	1333 (5932)	3,09

Table 5. Comparison of Total Buckling Deflections at Midheight

Specimen Number	Buckling Deflection in (mm)		% Error Experiment vs. Theory
	Theory	Experiment	
1	0.180 (4.57)	0.185 (4,70)	2.78
2	0.189 (4.80)	0.193 (4.90)	2.12
3	0.159 (4.03)	0.159 (4.03)	0.
4	0.171 (4.35)	0.173 (4.40)	1.17

Comparison of theoretical and experimental results showed that the buck-ling loads were estimated with errors less than 3.5%; the local buckling de flections at mid-height were estimated with errors less than 3%; and the buckling directions were estimated with errors of the order of 5%.

The effective lenght is defined as that length which gives on the critical Euler load (for pinned ends) the same load carrying capacity as the failure load for the actual restrained column. Using this approach, the effective lenght factors, of the four tubular beam-columns tested, were found to be 0.750, 0.780, 0.707 and 0.735, for specimens 1 to 4 respectively.

5. CONCLUSIONS

Four small-scale tubular steel column specimens have been tested under conditions of semi-rigid end connections, using specially constructed complex end restraint systems which allowed rotational resistance in perpendicular directions to be uncoupled. Parameters affecting column buckling behaviour which received special attention were end rotation restraint, column initial out-of-straightness and load eccentricity.

Theoretical analysis modelling the effect of spatial end-restraint by the Influence Coefficient Method, established convergent solutions up to the limit buckling load with high degree of accuracy. It was shown that provision of moderate end restraint, dramatically affected both the buckling load and the mid-height lateral deflection of the column at buckling.

REFERENCES

1. Chen WF: End Restraint and Column Stability. Journal of the Structural Division, ASCE, 106, No. ST 11, Proc. Paper 15796, 1980.
2. Barros RC: Buckling Analysis of End Restrained Imperfect Tubular Beam Columns. Ph.D. Dissertation, The University of Akron, Akron, Ohio, 1983.
3. Barros RC: Review of Elasto-Plastic Buckling of Tubular Beam Columns. Engenharia, Ano 1, nº 4, F.E.U.P., Porto, Janeiro-Abril 1984.
4. Chen WF and Atsuta T: Theory of Beam-Columns, Vol. 2, Space Behavior and Design. New York: Mc Graw-Hill Book Company, Inc, 1978.
5. Crisfield MA: A Fast Incremental/Iterative Solution Procedure that handles Snap-through. Computers and Structures, 13, 1981.

EXPERIMENTAL STRESS ANALYSIS WITH AUTOMATION OF MEASUREMENTS AND DATA PROCESSING USING VIBRO-WIRE GAUGES DURING THE CONSTRUCTION OF ONE SEGMENTAL BOX-GIRDER PRESTRESSED CONCRETE BRIDGE

T.JÁVOR, J.TRENČINA

Research Institute of Civil Engineering,Bratislava,ČSSR

1.INTRODUCTION

The technical development of the last decade led in all countries to a large scale building programme of concrete roads,bridges as well as public and industrial buildings. This development imposed high requirements upon the designer, as well as on the construction firms.Irrespective of recent methods of computer aided design,it is still very necessary to check the quality of structures,verify their static and dynamic behaviour,carry out long term observations of strains in use,investigate various effects of loading,etc.,and compare these experimental results with the assumption made by the designer.Introduction of new materials and structural form led to the development of in situ measurement techniques.The vibrating-wire method is a very progressive way of testing civil engineering structures in situ.

2.AUTOMATION OF VIBRATING-WIRE,INDUCTIVE AND RESISTANCE GAUGES

The experimental analysis of prestressed box-girder bridges during constructions as well as the long-term one,is made by the Czechoslovak embedded vibro-wire gauges.For long-term temperature measurements we used embedded vibro-wire thermometers.Regarding the requirements of maximum automation of our measurements we applied the impulse wire gauges with only two-wire connection to the measurement equipments.In this case the gauge wire is set in damped oscillation by an 0,4 ms impulse duration.After a delay of T= 10ms there are measured 100 oscillation periods.Then we can determine the wire oscillation frequency from the relation $f = 1/T$ .From this value,using the calibrated gauge constant we determine the concrete strain ε depending on frequency, i.e.

$$\varepsilon = A.f^2 + B.f + C ,$$

where A,B,C,are constants determined separately for each vibro-wire gauge,already during calibration.The transformation of values to mechanical stresses with relevant correlation effects of temperature changes is made by a computer programme for transformation of strain to stresses with regard to the creep,shrinkage and the Young-modul of elasticity,too.

For measuring purposes we use a measuring bus with installed data logger,type HP 3050 with relevant digital voltmeter frequency reader,programmed switch unit,tuner and XY-recorder.The system is oriented to the internationnally codified IEC-BUS and controlled by controller HP 9826 S.The measuring equipment enables measurements of gauge wire oscillation frequency by a velocity of 3 channels in one second.The measured values are stored in a magnetic casette unit securing a long-term storage and measuremnet evaluation.The measurement results are registrated in real time,in table form or graphically, by line recorder.The scheme of this system is given in Fig.1.

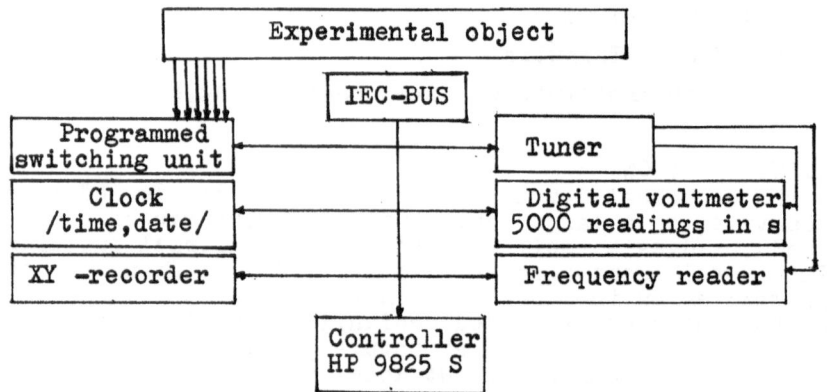

FIGURE 1. Block scheme of measurement equipment for long-term deformation observation by means of embedded vibro-wire gauges,

The inductive and resistance gauge form the further group and they are applied for various short-term measurements of deformations as:checking the strain in contact joints,strain evaluation of stiffening bars in the middle of bridge spans, etc.Naturally,besides electrical gauges there were applied mechanical gauges as Huggenberger gauges,surface straimeters or optical geodetic instruments,etc.The automation of measuring processes is advantageous also for the inductive and resistance gauges by means of various digital loggers.The main components of a such data logger are illustrated in Fig.2.

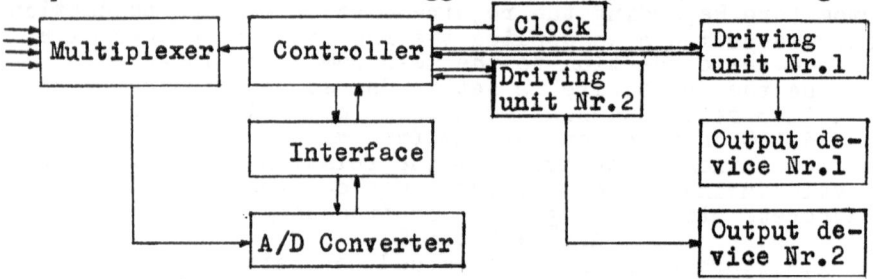

FIGURE 2. General block scheme of a digital data logger.

The multiplexer forms the input part of the data logger and the number of its switch elements gives the data logger capacity,i.e.the maximum recorder number,the signals of which the data logger is able to measure.The reed relays are mostly used as switching units.The digital integrate voltmeter with ranges from 10mV to 1000 V is very often employed as a measuring equipment.The measurements of particular channels can be

made at one common range or the ranges are switched automatically,either according to the given programme or according to
the value of just switched voltage.The automatic switching of
ranges affords a great variability of readers connection with
various levels of specific signals and guarantees measurements
at a maximum voltmeter sensitivity.A disadvantage is a more-
times greater delay in the measurement process and an unneg-
ligible wear of range-relay-contacts.A suitable compromise is
the application of range switching according to a programme.
The programme is set on so-called programme panels of the data logger,namely,as a rule,for whole input decades.For connection of the data logger with a PC /Personal Computer/,the
necessary range is set according to the PC programme.There
are various complements to the digital voltmeters enabling
for example a.c.voltage-measurements,resistance measurements,
etc.The computer internal clock gives the time indication in
the heading of measured data readings and generates the start
signals in set-up-time intervals.So the measuring cycles of
the data logger are secured automatically in intervals from
some decades of seconds up to some hours.A suitable comple-
ment is an equipment enabling the timing also during the fall
out of electrical energy.The input equipment is able to store
the measured data.Most often there are required two output
forms:print and paper tape.It is possible to realise it either by means of separate equipments,i.e.by a 32 characters
printer and a punch or by one equipment- a teleprinter with
a punch unit.
　　Already some years the data loggers MBH 5000 Metal Ltd,So-
lartron DTU Schlumberger,HBM US 100 or ADU 100,as well as UM
10 Metra Blansko are used for gauge measurements.The are some
systems composed of a PC and measuring equipment for observation and processing the experiments in line,as the computer
Logger 3366 Schlumberger with PDP 11/05,further the HP 3050 B
with a calculator.For their usage at greater experiments it is
necessary to complete them by periphery devices.
　　A completely automated data logger is able to assume by a
minicomputer the following functions: to start the measuring
cycle,to set up an arbitrary channel,to adjust the voltmeter
range and to stop the measuring cycle.The data from the data
logger,i.e.the measurement date and time,the number of the
measurement channel,the voltmeter data and the range are taken over in parallel code.The connection enabling these functions request relatively complex additional electronic circuits on the side of data logger and usually some arrangements
in its electronics,too.On the side of PC a usual parallel interface is sufficient.A greater number of selectors are necessary and if low-level logic circuits are used,the data logger
must be placed as close as possible to the PC.
　　The observation and evaluation programmes of experiments in
line are substantially the same ones as for off-line.So the
proper useful programmes can be elaborated in some of algo-
rithmic language /FORTRAN,ALGOL/ or,perhaps,more advantageously in conversation languages /BASIC/ providing that subrou-
tines programmes for the control of measuring equipments are
at disposal.The given complete systems use mostly various mo-

difications of the language BASIC.The on-line programme have
primarily to secure the experiment observation in such a way
that a wrong measured value is indicated,the relevant measurements automaticaly repeated,the conservation with operators
secured furnishing them the maximum information necessary for
evaluation of the experimental course and for eventual inter-
vention in this one.The final evaluation of the experiment has
to be involved in the programme.It is suitable to store the
partial data in an external memory and process them into the
final form only after having finished the experiment in the
off-line regime by a further evaluating programme.The on-line
regime is applicable for a observation of every experiments,
however,it is necessary to consider its effectivity for particular experiment types.This is important from the point of
view of the effective usage of the computing time.An example
of the effective on-line application at greater experiments
is the model-loading test or a loading test of construction
elements,where about 50 up to 200 values are measured and
where maximum of information and modification possibility of
the test course is necessary.

3.PHASES OF AUTOMATION PROCESS
   For ensuring a flexible operating system there were designed following experimental operation phases conformable to
the general experimental work:
- elaboration of the measurement programme,
- preparation of the experimental structure and of the loading equipment,
- calibration of the sensors,
- embedding the sensors into the measured structure,
- declaration of the parameters of the sensors,
- verification of the correct connection of sensors as well
  as of the function of the whole measuring and computation
  system,
- checking the long-term stability of the prepared measuring
  system including the embedded sensors,vibro-wire gauges,etc,
- starting the measurement by realization of zero-reading,
- measurements at various loading states of the observed
  structure,
- evaluation and graphical representation of measured results,
- storing the results in computer memory,
- elaboration of results in table form,
- end of the experiment.
The programming process we used is suitable for structure or
model tests.It is given in Fig.3.
   In our institute we used very oft the modified Hewlett-
Packard cybernetic system for model as well as structural analysis in situ.Its block scheme is in Fig.4.According to it the
transducers of strains,deflections and temperatures are built
in the tested structure.The sensors outputs are connected to
switch unit.By means of this switch unit the electric signals
from particular sensor are lead to the tuner and relevant mul-
timeter,where the analog signal is transforme to a digital va-
lue.This value is lead by a switching box into the calculator
memory for storing or further processing.The results we recei-

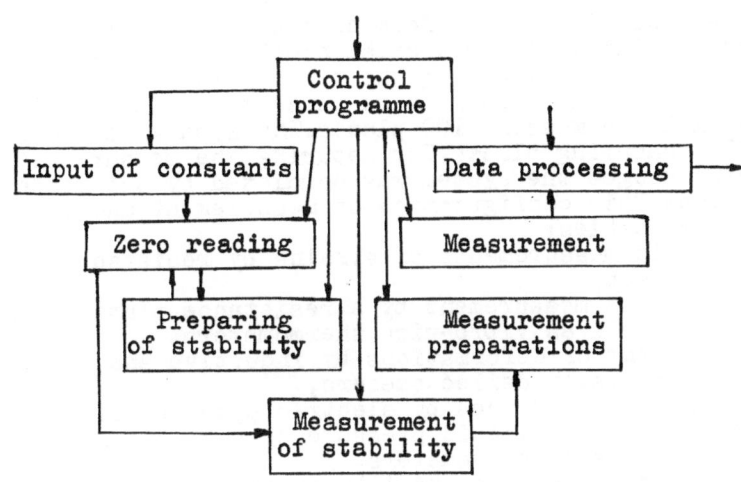

FIGURE 3. Structural scheme of the programme equipment for automated experimental tests.

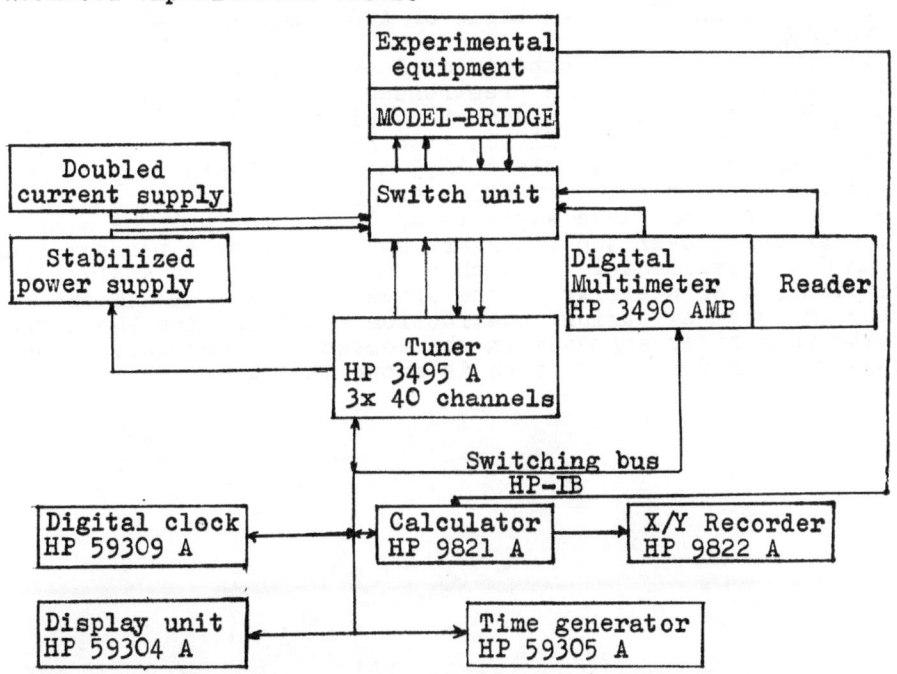

FIGURE 4. Block scheme of cybernetic system for model and in-situ research system Hewlett-Packard used at the Research institute of Civil Engineering in Bratislava for measurement of bridges.

ve in graphical or table form on the adjusted X/Y recorder or on the display unit.The digital clock and the time generator determined for programming the time course of measurements. Our system has at present time three tuners by means of which is possible to measure 100 places.The system can be enlarged for 400 measuring places.For securing the measurements by vibro-wire gauges the system was completed by the reader of frequences.This configuration of measurement and computer technique enables:

- automatic measurements of strains by resistance and vibro-wire gauges,
- temperature measurement by a resistance platinum thermometers as well as vibro-wire thermometers,
- measurement of deformations by inductive and resistance transmitters,or deflectometers,
- measurement of forces by electric dynamometers,
- measurements of liquid pressure by electric gauges,
- measurements by clinometers,
- immediate evaluation and illustration of values on X/Y recorder.

4.EXPERIMENTAL STRESS ANALYSIS OF A SEGMENTAL BOX-GIRDER BRIDGE

An especially great experimental observation was made during construction of the prestressed concrete box-girder bridge with precast segments near the village Podturen in Slovakia.The viaduct has 17 fields with spans from 30 m up to 70 m with piers of maximum height 32,9 m.The segmental superstructure was erected by an erection bridge with the cantilever method,which can place segments up to 80 t weight.The starting segments over the piers are stifftened by a precast cross beam.Our experimental research on the site should analyse the cooperation effectivity of the precast cross beam and the starting segment during construction and determine the stress state in further segments in the course of construction.The construction during erection is shown in Fig.5.

FIGURE 5.The segmental prestressed concrete precasted box-girder bridge during the erection and observation.

There were embedded about 330 vibro-wire gauges and vibro-wire thermometers.The deflection measurements were made by inductive transducers and by geodetic methods,the inclinations by inclinometers MAIHAK and Huggenberger levels,some short measurements by resistance strain gauges and the local stresses by photostress method.The course of strains during the erection of segmental precast elements measured by vibro-wire gauges is given in Fig.6.

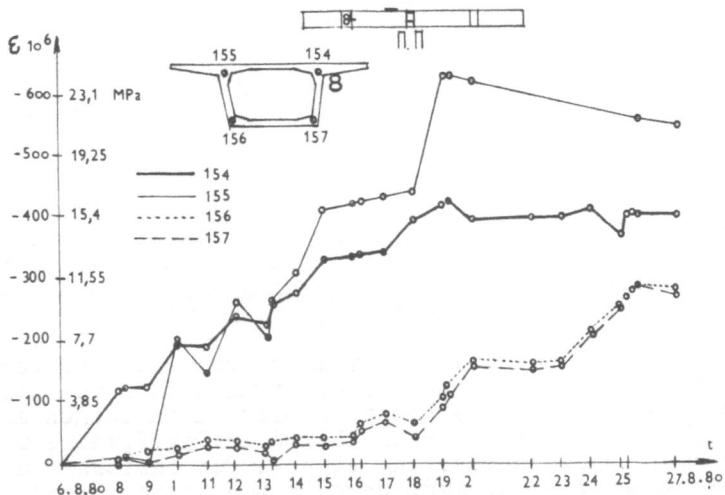

FIGURE 6.The course of strain at the corners of the second segmental box-girder element during the erection of the symetrical cantilever structure measured by vibro-wire gauges.

Very interesting is the analysis of the cooperation effectivity of the precast cross beam and the first box-girder segment over the bridge pier during the bridge erection.After prestressing low pressure stresses arose in the cross beam,in certain stages also tensile values due to cross contraction of the prstressed box-element acting by flexure on the cross beam-wall.A certain assymetry of stresses is caused by ground plan curvature of the bridge.The state of stress of the starting segment and the built-in cross beam was checked immediately after prestressing,than after the carriage transfer,during the further segment assembling and in one month intervals during the long-term observation.The total stresses assuming the Young-modul E= 38 500 MPa for longitudinal and vertical direction in the segment over the bridge pier are in Fig.7.

During the long-term observation the creep and the shrinkage are recommended to be checked on concrete samples situated in box-girder hollows and compared with samples stored in laboratory conditions where the changes of temperature and moisture are excluded.The long-term strain measurements of the bridge were continued after finishing the construction by automated equipments.The readings are recorded directly and the results are processed by a Orion data log-

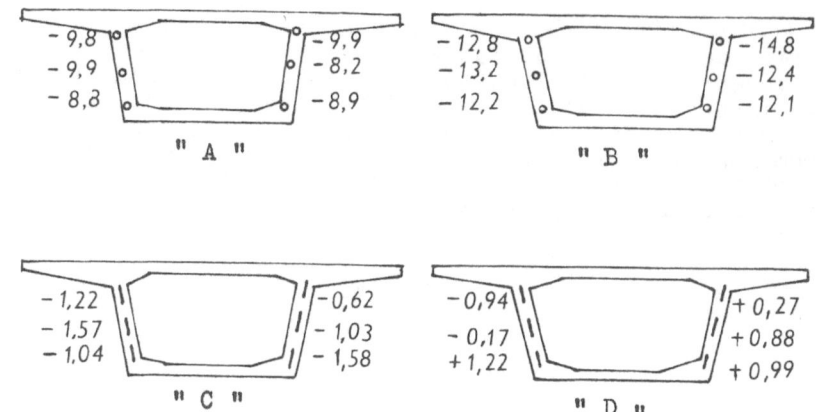

FIGURE 7. The total stresses in the first segment over the bridge pier in longitudinal direction after the prestressing /A/ and after the carriage transfer /B/ as well as in vertical direction after the prestressing /C/ and after the carriage transfer /D/. /The values are in MPa./

ger + IBM PC/XT Controller,by means of which the corresponding transformation of measured strains into the stresses is also made.These values are than plotted in tables and graphically.Such an example is showed in Fig.8,where T being thermometers,TS compesating samples for shrinkage and creep. The measurements are made once in two months during 5 years. The double diagrams are strains in symetrical corners of the same segmental element.The results obtained up to now show that the creep is stabilized after three yeras and the influence of the summer and winter seasons upon the deformations of the structures are very expressive.

5.CONCLUSION

The results from long-term observation of box-girder bridges contributed to the analysis of their durability and can be used as basic material for designing similar bridges.The analysis of thermal effects,shrinkage,creep as well as losses of prestressing by appropriate transformation methods of strains into the stresses have been verified by new theories of box-girder bridges and provide information about their real stress state.It is always necessary to have sufficient data from comparison samples concerning the creep and shrinkage as well as the E-modulus.

The described automated method of measurements by vibro-wire gauges and the appropriate method of data processing proved themselves very good.There is presented also an appropriate way of results interpretation as well as of their processing,the accordance degree between complex deformations obtained by measurements and forseen deformations being analysed.In accordance with present experience we have processed also instructions for long-term observations of concrete bridges and they have been accepted as RILEM Recommendations.

FIGURE 8. The automatic registered course of strain of various segmental precasted elements as well as compensating samples for creep and shrinkage and the temperatures measured by embedded vibro-wire gauges and thermometers during 5 years after the prestressing of the box-girder bridge.

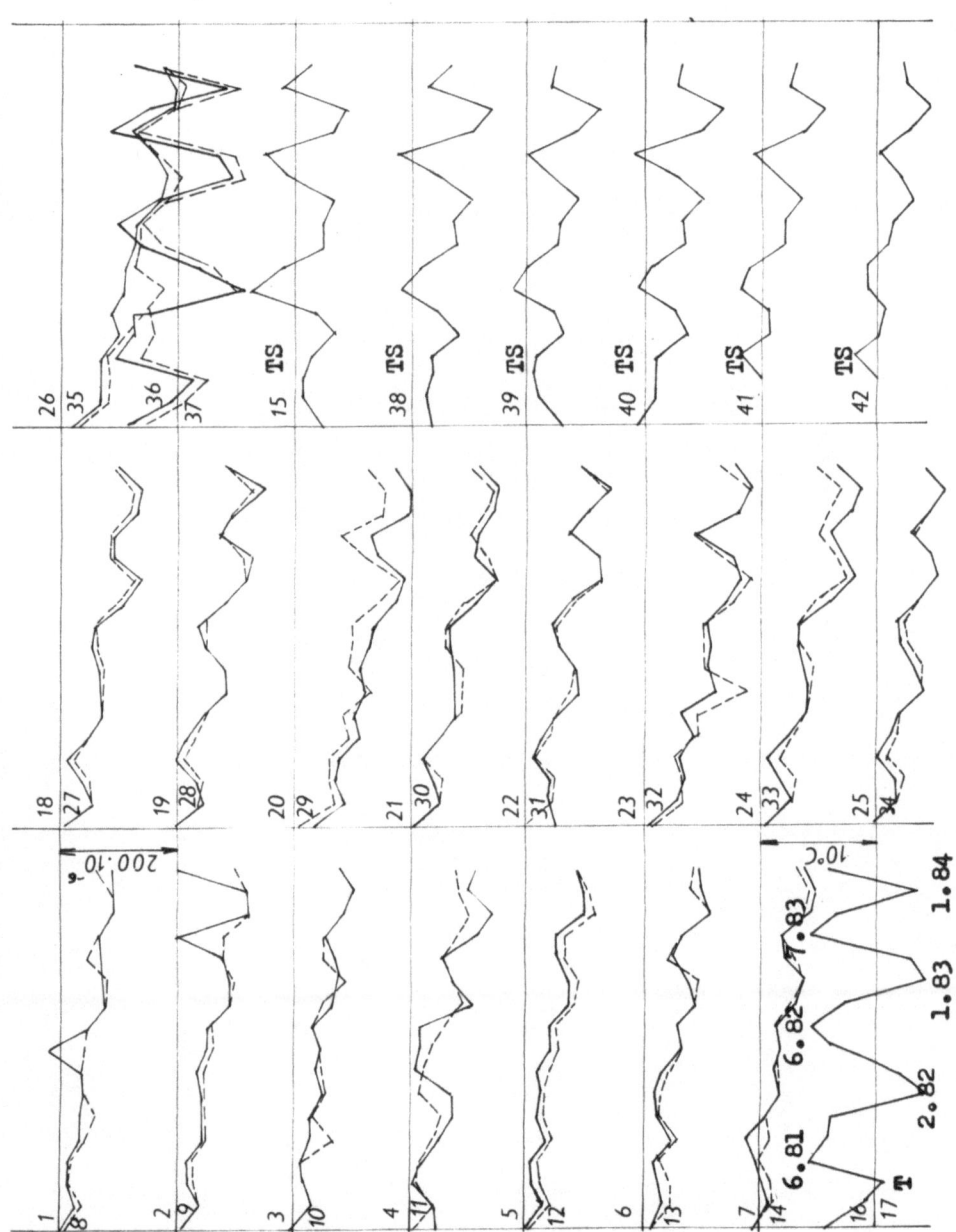

AN EXPERIMENTAL INVESTIGATION OF THE BEHAVIOUR OF FLAT RECTANGULAR PLATES
UNDER THE ACTION OF COMBINED IN-PLANE AND LATERAL LOADING

J.S.W. TAYLOR

Department of Mechanical Engineering, University of Surrey, Guildford, U.K.

1. INTRODUCTION

Experimental work has always played a key role in structural research and
continues to supply vital information to workers developing computer analyses
and design codes. This paper describes a series of small-scale model tests
conducted to establish the effect of uniform lateral loading (i.e.
hydrostatic pressure) upon the buckling behaviour and ultimate strength of
flat rectangular plates subjected to in-plane uniaxial compressive loading.
In the offshore environment engineering structures such as semi-submersibles
and TLPs make extensive use of stiffened and unstiffened panels that can
frequently be analysed as assemblages of flat plate elements - the in-plane
loading arising in the compression flanges of deck and pontoon structures
subjected to bending. Whilst some work has been conducted in this area (1,2)
it has been restricted to cases involving relatively low lateral pressures
(e.g. ship plating) and there is a paucity of information regarding the
integrity of plated structures operating at depths in excess of 50 metres.
The validity and economy of using small-scale model testing in this field has
been well-established (3), provided that sufficient care is taken in the
design and use of the test equipment and that the experimental data obtained
is analysed appropriately. This paper concentrates upon the development of
the test facility and the processing of the experimental data obtained. Test
results are also presented and compared to existing theoretical solutions for
thin rectangular plates.

2. TEST PROGRAMME
2.1 Description of Models

A total of eighteen rectangular plates have been tested under differing
combinations of axial compression (in the longitudinal direction) and lateral
pressure loading. Each plate model was fabricated from cold-rolled steel
sheet specially prepared to be equivalent to Grade 50 structural steel with a
nominal yield stress of 330 MPa in the annealed condition. Comprehensive
mechanical testing has been conducted on the sheets in order to ensure full
knowledge of the material properties for each individual model. The models
used in the test programme had nominal thicknesses of 0.8, 1.0 and 1.6 mm and
were cropped from their respective sheets prior to heat treatment. All of
the models tested had a nominal width (b) of 150mm and aspect ratios (a/b) of
1.5, 1.9 or 2.0.

2.2. Design of the Test Facility

The test facility allows a wide range of plate geometries to be tested
under the action of combined loading and with differing boundary conditions.
The test rig consists of a rectangular frame having two vertical guide
channels that ultimately locate the plates and support the scanning equipment
(Fig. 1). A pair of rigid transverse carriages with roller bearings at their
ends are fitted into the channels such that they can move freely in the
vertical plane. The carriages are attached to the plates by means of rows of
clamping screws. Use of a single pair of bearings on each carriage provides

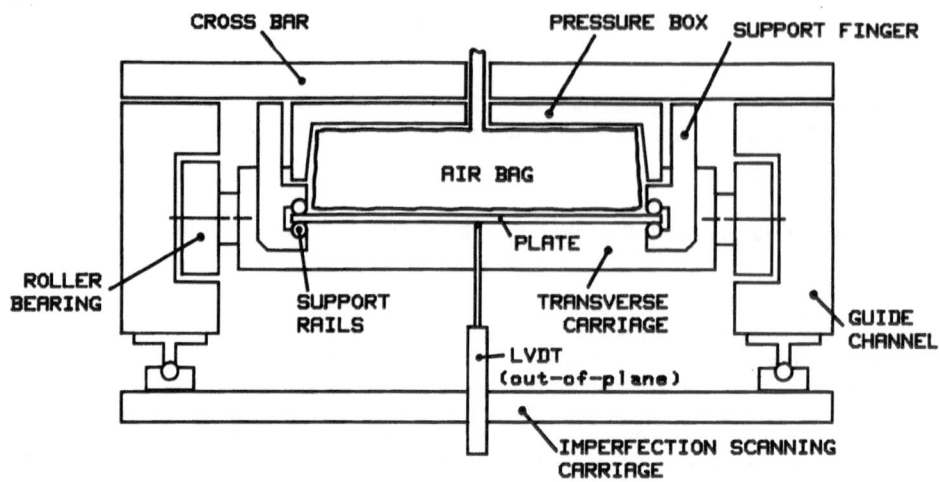

FIGURE 1. Sectional Plan of Combined Loading Test Rig

effective simple support conditions at the loaded edges of the plate, whereas installation of a second bearing set provides rotational fixity of these edges. The unloaded edges of the plates are constrained by means of a pair of small diameter support rails held in a row of 'support fingers'. The test rig is fitted into a servo-hydraulic test machine and the axial loading is applied through a self-aligning seating assembly attached to the fixed cross-head of the machine. Lateral pressure loading is applied by means of a neoprene air bag fitted in a pressure box retained by a series of cross-bars. The cross-bars also locate the support fingers thus forming a self-reacting sub-assembly that is fitted to the vertical guide channels of the frame. The plate support system and associated components are fully adjustable so as to allow control over the initial out-of-plane alignment of the plate. An important feature of the test rig is the imperfection/deformation scanning system - a carriage (mounted on linear bearings) that can be driven along the entire length of the vertical channels by a stepper motor. LVDTs mounted on the carriage can be used to map the initial imperfections of a plate prior to test and can subsequently monitor the deformation behaviour at selected positions without any loss of positional datums. The test rig can accomodate plates up to 2mm thick with lengths and widths of up to 1000mm and 250mm respectively.

2.3 Experimental Boundary Conditions

There are sixteen unique types of idealised boundary condition combinations possible for symmetric plate problems. The difficulties associated with the provision of 'ideal' experimental boundary conditions in terms of the flexural and membrane stresses developed at the plate edges in model tests are well known (4,5). Some excellent attempts have been made to simulate simply-supported boundary conditions (1,6), but the resulting physical arrangements are complex, expensive and not suitable for testing plates subjected to combined axial and lateral loading. The approach used by the author is to provide simple boundary conditions that are related to those encountered in practice and to quantify them by means of experimental measurement (e.g. strain gauging in the proximity of the supports). The test rig provides the loaded edges with rigidly supported, rotationally fixed or free flexural boundary conditions that approach the ideal. However, the

corresponding membrane conditions are less certain since the edges are
constrained to remain straight but it is unlikely that the clamping
arrangement produces zero shear stress (as is often assumed analytically) -
zero tangential movement is a more realistic assumption. The unloaded edges
of the test plates are nominally simply-supported but it is inevitable that
as loading progresses into the large deflection plastic regime there will be
some rotational clamping present. However under pressure-dominated combined
loading the amount of flexural restraint has only a small effect upon the
load-deflection behaviour and the resultant critical stress distribution.
The effective membrane boundary conditions for the unloaded edges are
difficult to ascertain since the frictional effects due to the lateral
loading and local plate deformations increase the amount of in-plane
constraint. These 'elastic' restraining forces generated in the experimental
arrangement are often found in plate elements of real structures and it is
possible to incorporate such effects into post-buckling analyses (7). It has
been established that the test rig provides experimental boundary conditions
that are well-behaved and repeatable.

2.4 Control of Test Equipment and Data Logging

A typical test requires a large amount of data to be logged including
end-shortening, out-of-plane deformations and surface strains and at the same
time involves precise control and monitoring of the loading and transducer
positioning systems. A microcomputer is used to control the test machine,
the transducer carriage stepper motor and a multi-channel data logger that
form the basis of the test instrumentation system. In this way it is
possible to acquire data at prescribed programmable intervals and positions
throughout a continuous loading sequence without incurring any errors due to
finite scanning times - loading pauses for data collection are avoided
completely. The microcomputer is also used to carry out some on-line
processing such as averaging and peak value detection. This leads to an
ability to display scaled and zero-corrected values of key quantities on a
CRT or plot the progress of derived parameter curves during the test. The
microcomputer also compacts the test data into an efficient coded format
prior to disc storage. This not only reduces the on-line data storage
requirements but also enhances the speed of subsequent data transmission and
post-processing. The servo-hydraulic test machine is interfaced to the
microcomputer so that Load, Position or Strain Control ramps can be
implemented under software control. The pressurising system used for the
tests had no feedback loop from the pressure transducer and required manual
adjustment.

2.5 Imperfection Scanning

Initial geometric imperfections have a significant effect upon the buckling
behaviour of plates (8,9) and it is important to be able to determine the
plate imperfections prior to testing. The scanning carriage can be fitted
with a cross-slide assembly that allows an array of transducers to be
positioned at predetermined stations across a plate. In combination with an
automated routine that uses the stepper motor to drive the carriage in small
increments along the plate, this enables a complete matrix of imperfection
data to be built up. Typically a 300mm x 150mm plate can be scanned in 40
minutes using a pitch of 2mm in both the longitudinal and transverse
directions. A carpet plot of the plate surface can be useful for the
identification of anomalies in the boundary conditions and areas of local
'damage' and can be used for the rational selection of positions for strain
gauging. It is also possible to identify cases in which the initial
imperfection profile could generate an out-of-plane deformation due to axial
load alone which would be in opposition to that induced by pressure alone

thus leading to the possibility of 'snap-through' during combined loading. Acquisition of the imperfection data in digitised form enables Fourier Transform techniques to be used for efficient information storage and for display of the imperfection amplitudes in the 'frequency domain' (i.e. imperfection wavelength expressed in terms of the leading dimensions of the model). Inspection of the Fourier coefficients of the measured imperfections can aid the interpretation of the buckling behaviour of a plate since a dominant imperfection wave number may provoke the model plate to buckle in a higher mode than that predicted by idealised analysis.

2.6 Test Procedure

A potential problem with the testing of thin plate models is the difficulty of alignment. It is important to ensure that the distance between the line of action of the compressive load and the mid-plane of the plate is minimised. This is particularly important when the simple support boundary condition is used at the loaded edges of the plate, but is also a factor in reducing the stressing of the test rig in the fixed edge tests. Simple determination of the position of the fitted plate by dimensional measurement is unreliable - relatively small misalignments can have a marked effect upon the plate performance since they act as initial imperfections. The Southwell plot technique (10) has long been used for the experimental determination of the Critical Load (Pc) associated with the buckling of columns. The technique can also be used to determine the overall effective imperfection (misalignment + initial imperfections) in a test model by extrapolating the critical load slope back to the lateral displacement axis. However, there are considerable problems associated with the adaptation of the technique to fully supported plates (11) and the use of simple Southwell plots for plate testing is discouraged. The approach adopted by the author has enabled the advantages of the Southwell plot to be utilised for setting up test plates whilst not depending upon the technique for subsequent determination of the buckling behaviour. The plate is installed in the transverse carriages (configured as simple supports) and is initially aligned by simple measurement of position. A small axial pre-load is applied followed by small compressive load increments within the elastic buckling range until reliable values of mid-span out-of-plane displacement are obtained. The resulting load/deflection behaviour is analysed using a Southwell plot and adjustments are made to the model's position (when unloaded) to compensate for the predicted effective imperfection. This process is repeated until an acceptably small overall imperfection is achieved. Following the imperfection adjustment the side rails for supporting the unloaded edges of the plate are fitted and the transverse carriages can have their secondary bearings installed for cases in which the loaded edges are to be rotationally constrained. When the side supports are fitted the clearance between the unloaded edges of the plate and the rails is adjusted using shims to provide continuous support whilst retaining freedom for the edges to slide in-plane between the rails. This is important if local 'patch loading' effects are to be avoided and the shims are smeared with molybdenum grease to minimise contact friction. The transducer scanning system was used to obtain complete out-of-plane deformation information at a number of positions along the length of the model during test so as to log the development of particular buckling/failure modes. In cases where only axial loading was applied the test machine was operated in position control with an effective 'strain' rate of 15 microstrain per minute. In cases where only lateral pressure was used a small compressive axial pre-load (typically 5% of Pc) was first applied in order to eliminate slack in rig and to help promote unambiguous continuous collapse. In each test the loading was applied gradually and removed at judiciously chosen stages so as to detect the onset of plasticity and permanent set in the plates. Once clear evidence of post-ultimate behaviour or fully developed collapse mechanisms was obtained the loading rates were usually increased so as to set clearly visible failure modes into the plates.

3. EXPERIMENTAL RESULTS
3.1 Axial Loading Behaviour

There are problems associated with the comparison of experimental plate behaviour data to the available theoretical analyses. The critical load Pc obtained from theoretical analysis of an axially loaded plate is a quantity that proves elusive in the experimental situation. The ultimate load carrying capacity ($Pmax$) of a plate is quite readily determined by experiment, but few theoretical analyses extend well into the post-buckling range and those that do are approximate and do not allow for buckle pattern 'snap-through'. Consequently, there is considerable interest in the comparison of experimental plate behaviour to the theoretical value of Pc. Several methods are available for the evaluation of Pc from experimental data – the standard Southwell plot technique is discounted due to the significant nonlinearities present when it is applied to plate tests – and each has been applied to the six plates that were subjected to axial compressive loading only. The 'Point of Inflection' method (12) takes the point of contraflexure Pcf on a graph of axial load vs. mid-span lateral deflection as being an indicator of Pc. The 'Strain Reversal' method (12) considers the load Psr at which the mid-span axial strain on the convex side of the plate begins to decrease as being indicative of Pc. The 'Maximum Membrane Strain' method (5) suggests that the averaged axial strain at the centre of the plate achieves a maximum value at a load Pm that is approximately equal to Pc. The 'Pivotal Concept' (13) can be applied to the determination of Pc by fitting the experimental data to the relationships derived from a perturbation solution of the Karman equations for elastic large-deflection behaviour. Figure 2 shows the results obtained for a 1.65mm thick plate with $(a/b)=1.90$ and typifies the buckling behaviour of the plates tested with their loaded edges 'built in'. The experimentally determined values of Pcf and Pm consistently exceed the critical load by a small margin (typically +5% of Pc) and the magnitude of this discrepancy appears to be controlled by the magnitude of the initial imperfections in the plate – acceptable results are obtained for plates in which the effective imperfection amplitude does not exceed $0.25t$ (where t is the plate thickness). The values of Psr were significantly lower than Pc for every test, discrepancies of more than 20% being common. Application of pivotal techniques to the test data yielded critical load estimates within 5% of the corresponding theoretical values and good agreement was maintained for initial imperfections as large as $0.50t$ – it was also possible to obtain acceptable correlation for larger imperfections by applying a least-squares fit technique (13). Successul use of the pivotal techniques required careful choice of data points and numerical instability occurred in several cases. In all six tests the point of contraflexure Pce on the characteristic Axial Load vs. End-shortening plot agreed very closely with the theoretical critical load – typical discrepancies being -4% of Pc. It was found that in most cases the load Psd at the point of contraflexure of an 'Axial Strain Difference' plot agreed more closely with Pc than did the corresponding value of Pm – in every test the value of Psd was within 3% of Pc. Photograph A shows the failure mode associated with these tests and is typical of the permanent buckles found in wide compression flanges taken beyond their ultimate load condition.

3.2 Lateral Loading Behaviour

Six plates were subjected to lateral uniform pressure loading alone so as to investigate deviations of behaviour from the available theoretical large-deflection and elasto-plastic analyses. Current design practice for laterally loaded plates involves a number of qualitative criteria since at present nonlinear analyses will not reliably deal with all aspects of the complex plastic response to lateral loading (14). The six plates tested had an aspect ratio of (a/b)=1.9 and thicknesses ranging from 0.6mm to 1.8mm. In each test the plate was loaded in stages so that the onset of appreciable permanent deformation (nominally 0.5t) could be detected – the pressure was subsequently ramped up until a full hinge mechanism had developed. Figure 3 shows the lateral loading response of a 1.64mm thick plate with its shorter edges rotationally fixed. Even at relatively low pressures there is significant nonlinearity in the pressure vs. lateral deflection curve (linear theory is about 6% in error when the ratio of central deflection to thickness (w/t) is of the order of unity). Figure 3a shows that for deflections in excess of (w/t)=1 linear theory is inadequate and that initial tensile yielding of the plate surfaces has induced significant permanent set in the plate. Considering this early onset of plasticity, Figure 3b shows a surprisingly good correlation between the experimentally recorded behaviour and that predicted by a large-deflection finite difference analysis for non-dimensionalised pressures up to Q=70. The deviation of the experimental curve from the large deflection analysis prediction becomes significant as the plastic hinges spread along the edges of the plate and as seen in Figure 3 and Photographs B and C, a fully plastic hinge mechanism is formed which controls the final 'collapse' of the plate. The yielding of the plate is a continuous process which although initiated at relatively low pressures does not have a significantly deleterious effect upon its stiffness – this is largely due to the increasing dominance of the membrane stresses at higher pressures. These tests confirm that, due to the membrane action, pressure loaded plates may sustain extremely high pressures if no limitation is placed upon the magnitude of the lateral deformation and permanent set. Practical design criteria usually specify a small permissable permanent set which limits the allowable pressure loading to a level well below the collapse pressure. The strain gauges installed on the test plates detected initial yielding of the material at lateral load levels which did not appear to induce any permanent set in the plate. The fact that this localised yielding produces little permanent set explains the apparent increased strength of plates loaded by lateral pressures in excess of the theoretical load necessary to induce initial plate yielding.

3.3 Combined Loading Behaviour

The six models subjected to combined axial and lateral loading had nominal aspect ratios of (a/b)=2 and a plate thickness of 1.6mm. Each plate was strain gauged, the installations consisting of back-to-back pairs of orthogonally aligned rosettes at mid-span and critical edge locations. Two models were loaded axially until unambiguous mid-span displacement was detected, at which point the load was held constant whilst the pressure was ramped up to cause fully plastic membrane conditions – in these cases it is not possible to define a 'collapse' load for the plate since transition from the fully elastic condition to the plastic 'membrane' condition is continuous. Two plates were first subjected to lateral pressures of sufficient size to cause the onset of plasticity and permanent set prior to the application of an axial load – in these cases post-buckling collapse could be detected, although in the case with relatively high pressure the ultimate load was not clearly defined. The last two plates were used to determine the load path dependency of the failure mode and to investigate the effect of snap-through – the latter situation being provoked by the introduction of an 'unnatural' imperfection mode by adjustment of the initial geometric imperfections. It was found difficult to interpret the strain gauge data - the strain readings tended to be irregular and were largely influenced by surface yielding at an early stage of plate loading. Nonetheless, the installations using gauges with a post-yield specification did provide useful information with regard to the local strain fields and stress distributions existing during snap-through phenomena. The strain gauge data has indicated that the important pressure-induced membrane effect is dependent upon both the membrane and flexural conditions operating at the plate edges. Figure 4a compares the behaviour of a plate subjected to a lateral pressure of 1 Bar (equivalent to 10 metres depth of water) prior to axial loading with that of a similar plate subjected to axial load alone. The pressure produces a 21% reduction in the ultimate load carrying capacity, whilst the membrane effects lead to a post-buckling path with a higher load level than the simple axial load test. In this case the lateral load induced a mid-span deflection greater than twice the plate thickness prior to the application of the axial load (Figure 4b) and yet the resulting reduction in ultimate capacity is far less than would have been the case for an initial geometric imperfection of that magnitude. Apart from cases in which the initially applied mode of loading took the plate very close to its failure condition, there was no significant load path dependency detected and the final plastic hinge failure mechanisms due to combined loading showed elements of the profiles associated with both modes of loading (see Photograph D).

4. CONCLUSIONS

Carefully designed small scale model tests can be used to provide experimental data of direct use and relevance to the design of plated structures. Use of carefully measured 'non-ideal' boundary conditions leads to a simple test arrangement that closely models practical plate edge conditions. Efficient data collection and processing techniques can enhance the interpretation of plate test data and reduce storage requirements. Some of the established techniques for the determination of experimental buckling loads must be used with caution and should be cross-verified by an alternative technique. Strain gauge installations on thin plate models often provide erratic information due to their sensitivity to local imperfections and large-deflection yield phenomena. The membrane stresses produced by the lateral pressure in combined loading cases reduce the deleterious effect of the additional mode of loading and it has been found that at design load levels there is no significant load path dependency due to the combined loading.

REFERENCES

1. Brown JC and Harvey JM : J. Mech. Engng. Sci., Vol.11., 1969, pp. 305-17.
2. Aalami B & Chapman JC : Trans.Roy.Inst.Nav.Arch., 114, 1972, pp.155-181.
3. Taylor JSW & Der Avanessian H : Proc. Ann. Conf. BSSM., Camb., Sep. 1985.
4. Clarkson J : Trans. Inst Nav. Archit., Vol.98., 1956, pp. 443-63.
5. Edlund BLO : Int.Skr.S 73:26, Chalmers Tek. Hogskola, Goteborg, 1973.
6. Moxham KE : Rep.CUED/C-Struct/TR.3, Cambridge Univ., 1971.
7. Rhodes J and Harvey JM : J. Mech. Engng. Sci., Vol.13., 1971, pp. 82-91.
8. Williams DG & Aalami B : Constrado Monograph, 1979.
9. Elishakoff : Proc. I.U.T.A.M. Symposium, U.C.L., Aug. 1982.
10. Southwell RV : Proc. Royal Soc. London, 135, 1932, pp. 601-616.
11. Spencer HH and Walker AC : Exp. Mech., Aug. 1975, pp. 303-310.
12. Hu PC, Lundquist EE and Batdorf SB : NACA Tech. Note No. 1124, 1946.
13. Souza MA, Fok WC and Walker AC : Exp. Techniques, Oct. 1983, pp. 36-39.
14. Hughes OF : Jnl. Ship Research, June 1981, pp.77-89.

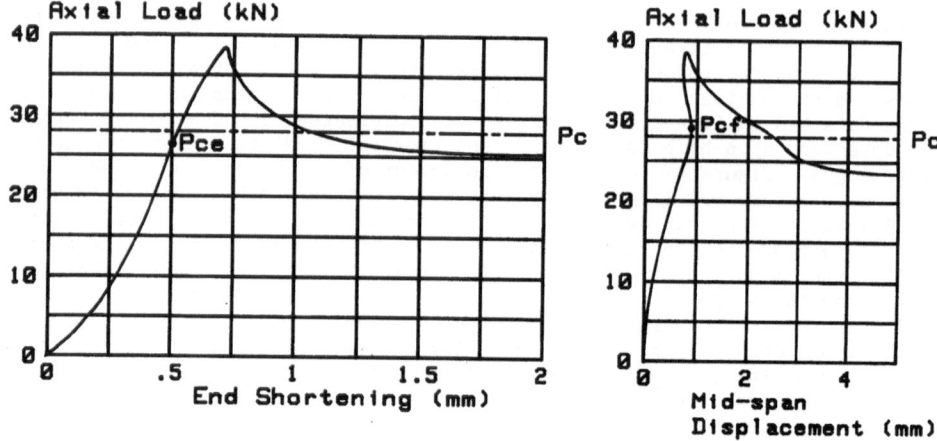

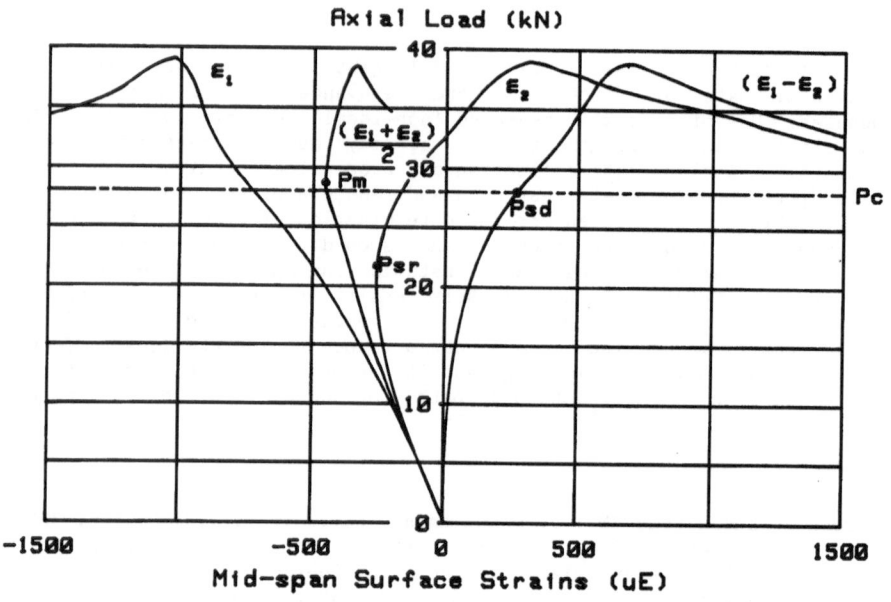

FIGURE 2. Buckling Behaviour due to Axial Loading only.

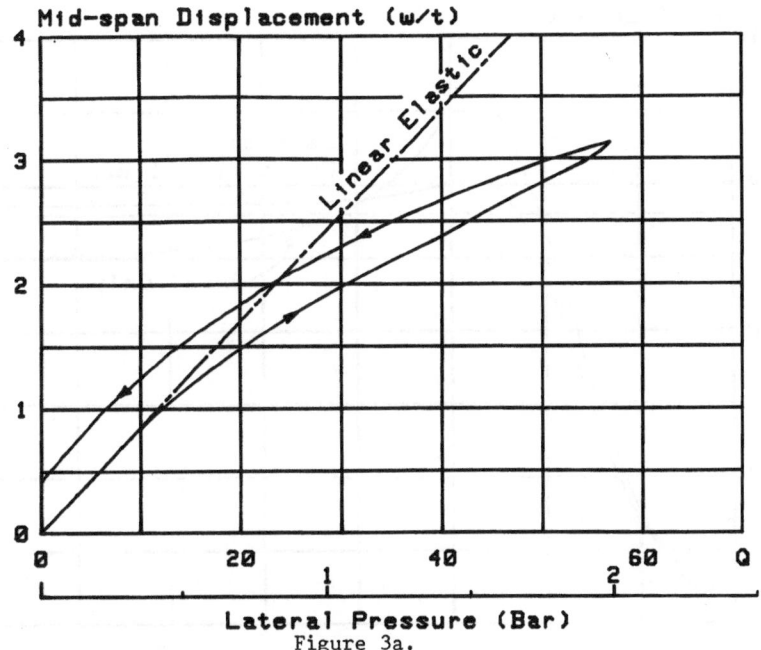

Figure 3a.

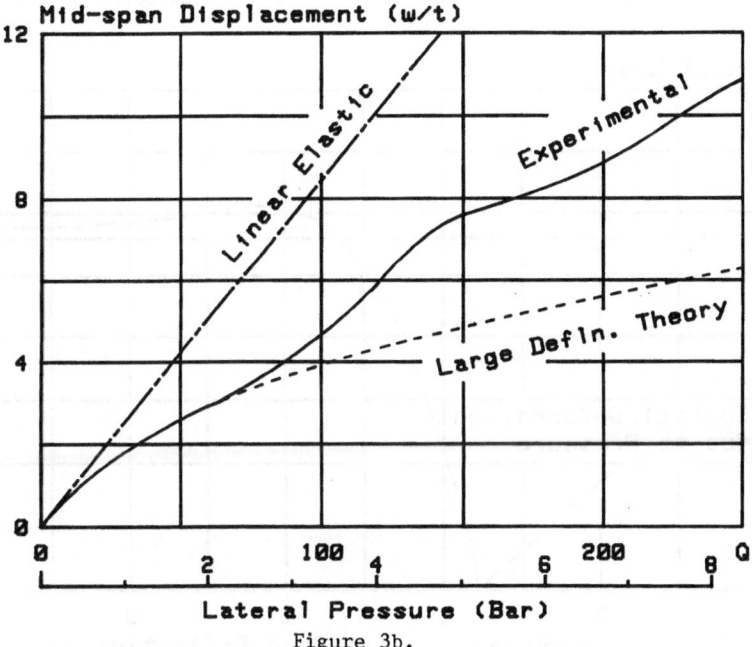

Figure 3b.

FIGURE 3. Plate Response to Lateral Pressure alone.

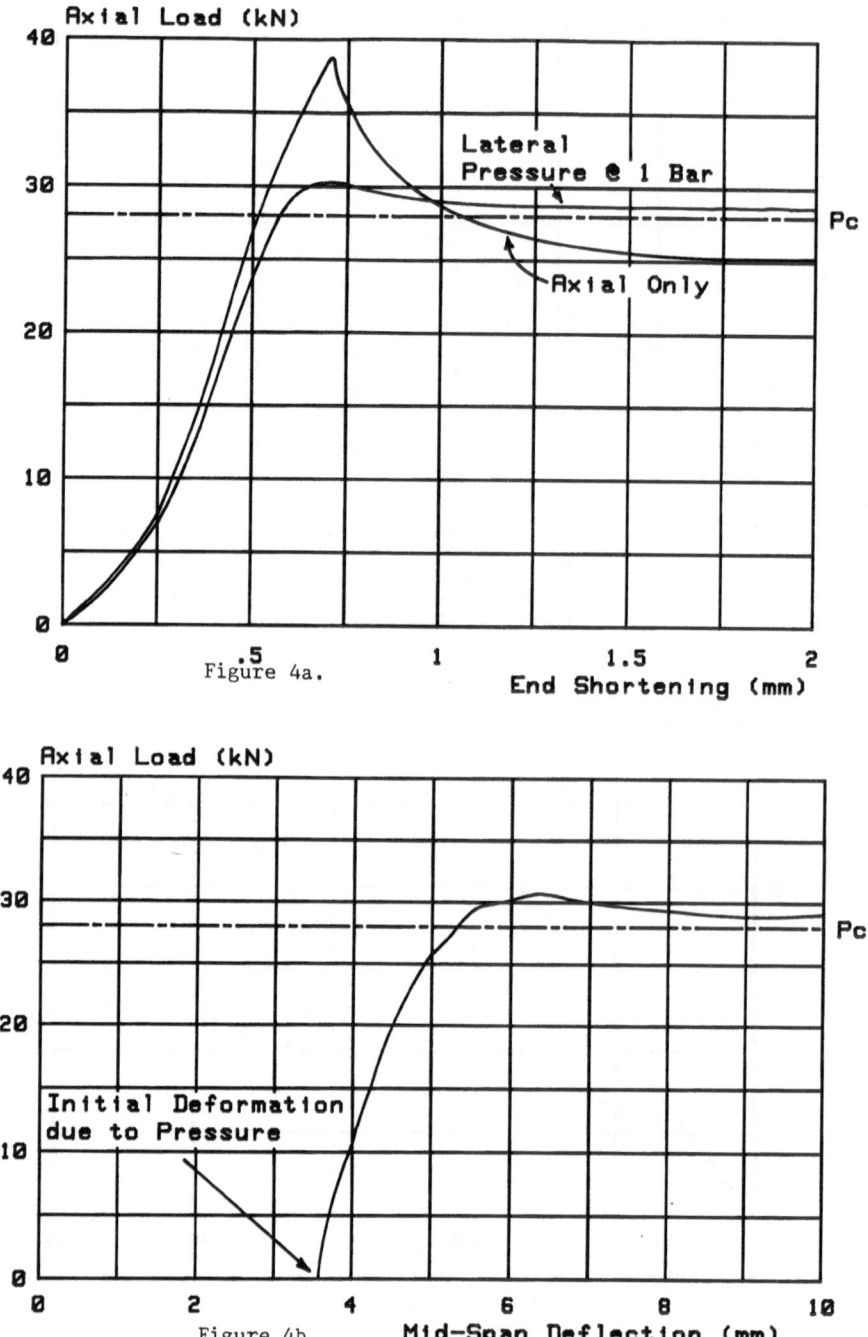

Figure 4a.

Figure 4b.

FIGURE 4. Response to Combined Axial and Lateral Loading.

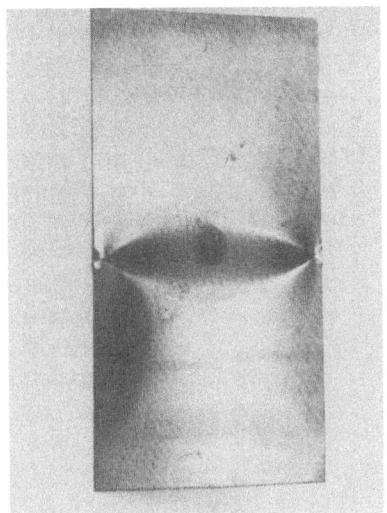

Photograph A.

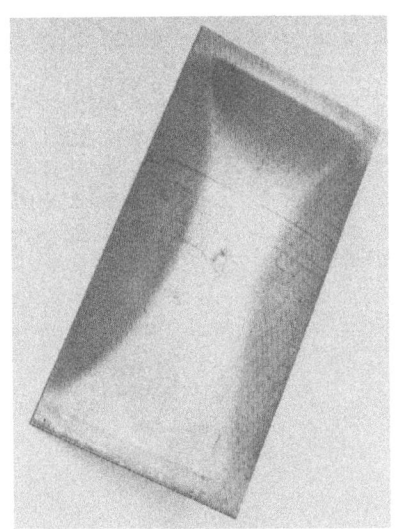

Photograph B.

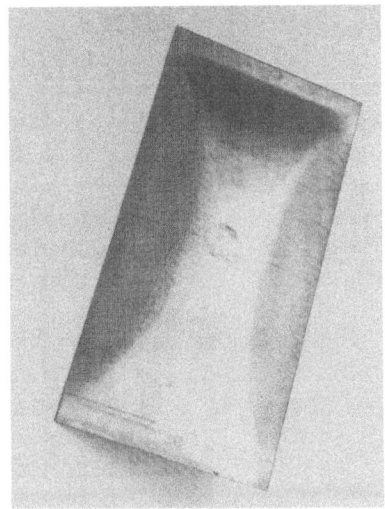

Photograph C.

Photograph D.

PHOTOGRAPHS OF TEST PLATES

A : Buckling due to Axial Load only.
B : Large Deflection Profile due to Lateral Pressure only.
C : Fully Developed Plasticity due to Lateral Pressure only.
D : Concave Face of Plate subjected to Combined Axial and Lateral Loading.

TECHNIQUES IN EXPERIMENTAL STRESS ANALYSIS FOR REINFORCED
CONCRETE STRUCTURES

R.H. SCOTT & P.A.T. GILL

DEPARTMENT OF ENGINEERING, UNIVERSITY OF DURHAM, ENGLAND.

SUMMARY
   Procedures for the measurement of reinforcement strains in
reinforced concrete structures are discussed. A technique is
described which uses electric resistance strain gauges
installed in a milled duct within the reinforcement to avoid
degredation of the steel/concrete interface. Details of
experimental work using this technique are given and typical
results presented to demonstrate its successful performance.
The use of embedment strain gauges to measure strains in the
concrete is also described. Future developments are
indicated.

1. INTRODUCTION
   A number of procedures have been developed over the years
for measuring longitudinal reinforcement strains. An indirect
approach is to interpolate from surface strain measurements
made either with a Demec gauge or with surface mounted strain
gauges, but this is obviously approximate since it is
difficult to perform the interpolation with any real degree of
confidence or accuracy. However, the Demec gauge can yield
very useful data when used carefully and for this reason it
has been widely adopted for a whole range of strain measuring
applications. A drawback is that since a gauge length of
200mm is typically used, the readings give average, rather
than local, strain values. Gauges with shorter gauge lengths
are available but these have reduced sensitivity and accuracy.
An alternative approach is to mount electric resistance strain
gauges directly onto a concrete surface, but this is well
known for its awkwardness even when big gauges are used.
Consequently the use of surface mounted strain gauges is
comparatively rare.
   Another method is to fix pins to the reinforcement which
then project through to the surface of the concrete in
specially formed ducts. Strain measurements are taken at the
surface using a Demec gauge. This is an inconvenient method
to perform in practice and the pins have to be kept short to
avoid flexing and consequent loss of accuracy. In addition,
it too yields average, not localised, strain values.
   The use of electric resistance strain gauges to measure
reinforcement strains is obviously attractive since the data
they yield are both more localised and an order of magnitude
more sensitive than Demec readings. However, bonding strain
gauges to the surface of the reinforcement degrades the bond

Wieringa, H (ed), Experimental Stress Analysis.
© *1986. Martinus Nijhoff Publishers, Dordrecht.*

characteristic between the rod and the surrounding concrete, and the lead wires are also a disturbance as they have to be taken out through the concrete to the sides or ends of the specimen. A partial solution to these problems is to install both the gauges and the wiring in a groove milled in the surface of the reinforcement. This works reasonably satisfactorily if only a few gauges are used but it still results in a rod surface which is different from the prototype. The problem worsens as the number of strain gauges to be installed increases and for this reason the procedure is unsuitable for large scale gauging operations.

The authors wished to obtain detailed measurements of the longitudinal reinforcement strain distributions in reinforced concrete members in order to investigate the effects of cracking in the concrete. When reinforced concrete members are subjected to tensile stresses, cracks develop in the concrete which profoundly effect the behaviour of the members. The reinforcement strains peak at the cracks, since here the reinforcement is carrying most of the load, and decline away from the cracks as load is shared with the surrounding concrete. When measuring these strains it was essential that there should be no degradation of the steel/concrete interface, and very detailed and localised data was required. This ruled out using all the methods outlined above and so a solution was sought which would enable the gauges to be installed internally and so leave the interface undisturbed. The procedure developed involved mounting the gauges in a duct milled longitudinally through the centre of the reinforcement. By using modern strain measurement technology and data acquisition systems, this permitted measurements to be undertaken in a more detailed manner than had been attempted previously (1,2). This procedure will now be described in more detail.

2. STRAIN GAUGING DETAILS

The technique involved milling two reinforcing rods down to a half round and then machining a 5mm wide by 2.5mm deep longitudinal groove in each to accommodate the strain gauges (gauge length 3mm) and their wiring (Fig. 1). After installation of the gauges, the two halves were glued together so that outwardly they had the appearance of a normal reinforcing rod, but with the lead wires coming out at the ends.

The strain gauges were installed using a cyanoacrylic adhesive and protected with a polyurethane varnish. Considerable care was needed in organising and successively bonding down the lead wires as these were added, starting at each end and working towards the middle. Finally the two halves of the rod were bonded together with an epoxy resin which also filled any remaining spaces in the duct.

The space available in the duct was severely limited which necessitated using very fine lead wires. Because of the space limitations a two wire, common dummy, installation was tried at first but gave problems with stability since the small lead wires, including those to the dummy gauges, were necessarily about four metres long. This was overcome by changing to a three wire, common dummy situation (despite requiring even finer lead wires) and by replacing the dummy strain gauges with precision resistors mounted within the data logging hardware. Up to 84 strain gauges were installed within each rod.

Before being cast into a reinforced concrete specimen, the rods were load cycled in order to minimise hysteresis and to check the performance of the strain gauge installation.

3. DATA COLLECTION

An automated data collection system was needed to handle the considerable quantities of data that the tests would generate. The chosen system consisted of two units; a data logger and a supervising microcomputer.

The logger handled 208 input channels using thirteen cards of sixteen channels each. It was constructed in modular form with individual modules for the microprocessor unit, analogue to digital converter, instrumentation amplifier and power supply units as well as the thirteen input cards. Twin constant current energisation was used which was switched to each channel in turn by reed relay scanners. A scanning speed of 8 channels/second was selected which was entirely satisfactory for the quasi-static nature of the data being recorded.

The logging hardware was controlled by the logger's own microprocessor which communicated with the operator via a RS232 serial line link connected either to a standard terminal (which permitted only a cryptic interaction) or to a supervising microcomputer. This latter option was selected by the authors as it permitted very sophisticated control of the logger to be achieved. The complex FORTRAN program which was written to interface the two units included extended file handling routines to facilitate data storage. Full details of both the hardware and the software have been reported elsewhere (3,4).

4. TESTING

The first programme of tests modelled the behaviour of the tension zones of reinforced concrete beams by testing simple tension specimens in the laboratory. Both short term and long term tests were undertaken.

Specimens for the short-term tests were 1500mm long with cross-sections ranging from 70x70mm up to 200x200mm. They were reinforced with either 12mm or 20mm diameter strain gauged rods positioned centrally. Both plain mild steel and ribbed high yield steel reinforcement was used. After some experimentation, a standard gauging layout was adopted which had 80 strain gauges spaced at 12.5mm centres along alternate halves of each rod over the central 1m of each specimen (Fig. 2a). A pair of gauges was also installed at each end of each rod, outside the zone of the concrete, to measure rod strains and so provide a correlation with the load measuring instrumentation. The specimens were loaded incrementally, usually up to yield of the reinforcement, in a purpose-built rig with each test being completed within one day. Strain gauge readings were recorded at every load stage and in particular very detailed information was sought before and after the formation of cracks.

Specimens for the long-term tests were 885mm long with a standardised cross-section of 100 x 100mm. They were tested in a servo-hydraulic testing machine and their reduced length was a constraint imposed by the machine dimensions. Only 12mm diameter reinforcement was used. The gauging layout adopted had 60 strain gauges spaced at 15mm centres over the full length of each specimen (Fig. 2b) and once again, additional gauges in the rod, external to the concrete, provided a correlation with the load measuring instrumentation.

Two types of long-term tests were undertaken: sustained load tests lasting up to 3 months and cyclic load tests lasting 3 weeks which, at a frequency of 20 cycles/hour gave some 10 000 cycles in total The initial procedure with both types of tests was to load up the specimen until a crack formed and with the sustained load tests, the load would then be maintained at this level. With the cyclic load specimens the cracking load would become the peak of the load cycle and generally the trough of the load cycle would be set at 50% of the cracking load, but with one test a value of 90% was used.

The second series of tests investigated the strain distributions in lapped tensile reinforcement using specimens having dimensions similar to those for the short-term tension specimens described above (Fig. 2c). The test procedure was also similar to that for the short-term specimens. Early results suggested the presence of steep strain gradients over a distance of approximately one rod diameter at the extreme ends of the lapped rods. To investigate these further, strain concentration gauges having five elements each of 1mm gauge length at 2mm centres were installed at each rod end. These gauges were particularly difficult to install in the duct because of their side lead attachments..

Some specimens also contained embedment strain gauges in the concrete (12mm gauge length, overall size 30mm x 9mm x 2.4mm). These were always restricted to one half of the specimens as it was considered that they might act as crack inducers. Depending on the specimen cross-section, one, two, three or five rows of embedment gauges were used to investigate the strain gradients from the reinforcing rod to the surface of the concrete. The gauges were positioned in the mould using a grillage of fine wires. Inevitably these were a potential source of disturbance in the concrete.

Surface strain measurements were taken on all specimens using a Demec gauge.

5. RESULTS

Space constraints preclude a full discussion of the very large quantity of data generated by these tests. Instead typical results will be presented to demonstrate the performance and flexibility of the strain measuring techniques.

Typical longitudinal reinforcement strain distributions for a short-term tension specimen are shown in Fig. 3. The detailed way in which the strain distributions were measured is clearly shown, with the strains peaking at each crack position. Strain distributions each side of a crack were essentially linear, indicating that zones of constant bond stress were present. Very detailed information concerning bond stress distributions was obtainable from these results.

Fig. 4 shows typical strain distributions for a specimen subjected to long-term sustained loads. This specimen had one crack only positioned approximately at a third point. Consequently on one side of the crack (Zone A) there was considerable debonding after the crack formed, whilst on the other there was a region where the bond was largely undisturbed (Zone B). Fig. 4 indicates that time-dependent changes in Zone A were more marked than those in Zone B due, it was believed, to the higher degree of initial debonding in the former region.

For the lapped specimen the results indicated that strain distributions along the rods were largely constant prior to cracking, with peaks similar to those in the tension specimens forming after cracking. Early results from the strain concentration gauges supported the existence of steep strain gradients at the extreme rod ends, but this investigation is still continuing.

Figure 5 uses the embedment gauge results to plot transverse strain distributions across the section. It shows the large degree of debonding which occurred between the reinforcement and the surrounding concrete when the crack formed, and how further increase in load was mainly carried by the reinforcement with little load being transferred into the concrete. The suspicions regarding the crack-inducing tendencies of the embedment gauges were largely confirmed as the first crack in the specimens which contained these gauges always occurred within the gauged zone and sometimes at strain levels which were untypically low. However, the effect was probably marginal since the second crack usually occurred at only a slightly higher load in the half of the specimen without the embedment gauges.

6. DEVELOPMENTS

The strain measuring technique has proved to be extremely successful and further developments are planned. The work on lapped reinforcement is being developed to include an investigation of laps in compression, and a comprehensive investigation of the reinforcement strain distributions in reinforced concrete beam/column connections has just commenced. This will involve gauging bent reinforcement, an obvious complication, and it is intended that the reinforcement will be bent prior to the duct being milled. Preliminary investigations indicate that the interface between the two rod halves should be in the plane of bending. Longer term plans are to move into the monitoring of full-size structures in the field.

7. CONCLUSIONS

The technique of internally strain gauging reinforcing rods has been developed to the stage where reliable detailed strain distributions can be obtained even over long time periods. Results to date are most encouraging and further developments of the technique are in hand.

ACKNOWLEDGEMENTS

The financial support of the Science and Engineering Research Council and the assistance of the authors' colleagues are both gratefully acknowledged.

REFERENCES

1. Mains RM: 'Measurement of the Distribution of Tensile and Bond Stresses along Reinforcement Bars'. Journal of American Concrete Institute. Vol. 3, pp. 225-252, 1951.
2. Nilson AH: 'Bond Stress-slip Relations in Reinforced Concrete'. Report No. 345. Department of Structural Engineering, Cornell University, New York, 1971.
3. Scott RH, Gill PAT and Munro M: 'A Modern Data Collection System and its Interfacing Requirements'. Proceedings of Civil Comp.83, London, 1983.
4. Scott RH and Gill PAT: 'A Modern Data Collection System'. Strain, Vol. 20. pp. 63-68, 1984.

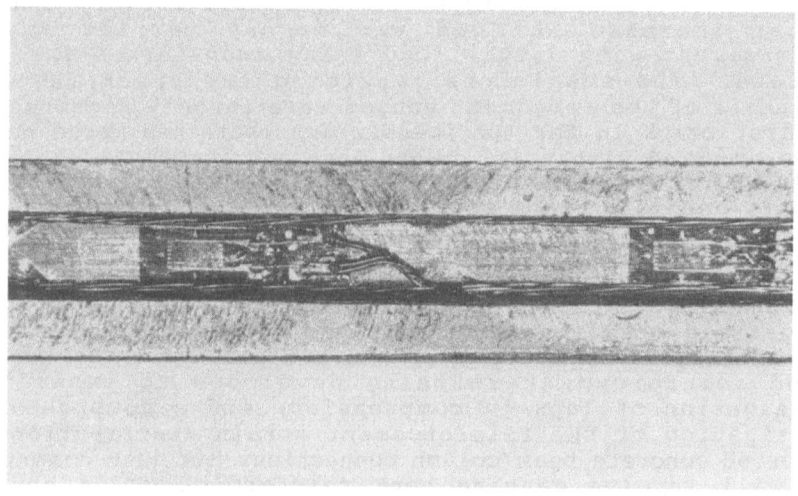

Fig 1: STRAIN GAUGE LAYOUT

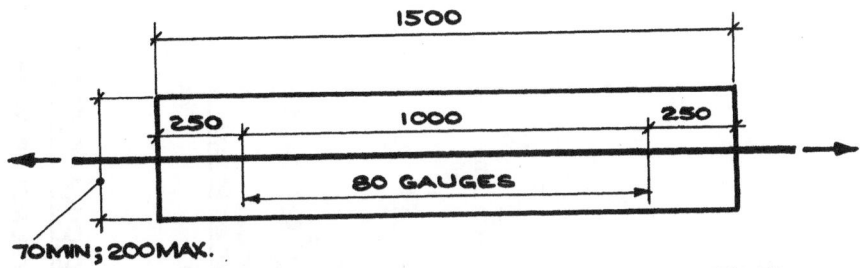

(a) <u>SHORT-TERM TENSION SPECIMENS</u>

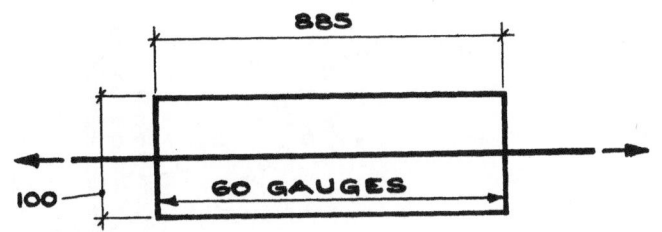

(b) <u>LONG-TERM TENSION SPECIMENS</u>

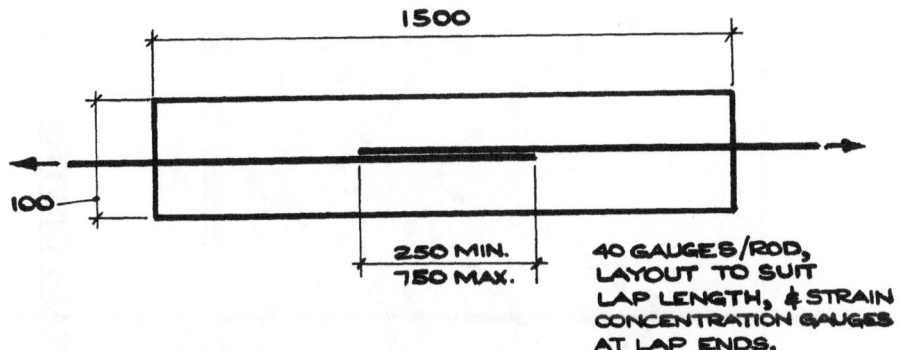

(c) <u>TENSION LAP SPECIMENS</u>

ALL DIMENSIONS IN <u>mm.</u>

Fig 2 : <u>SPECIMEN LAYOUTS</u>

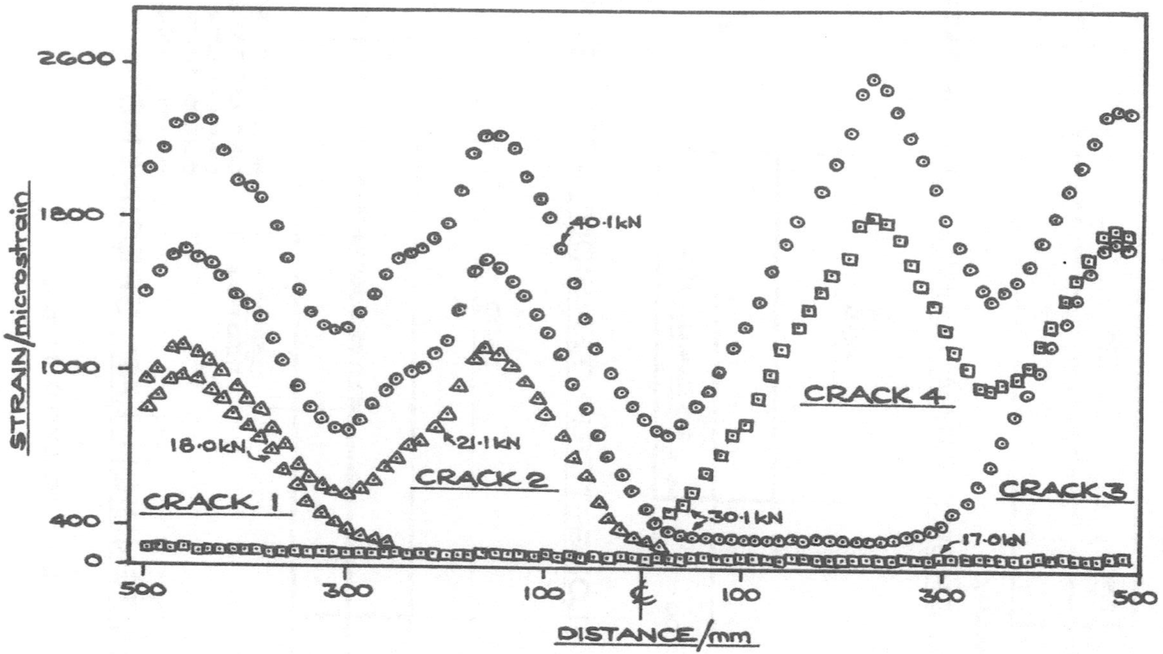

94

Fig 3: STRAIN DISTRIBUTIONS: TYPICAL SHORT-TERM SPECIMEN

(100x100 CROSS-SECTION; T12 REINFORCEMENT)

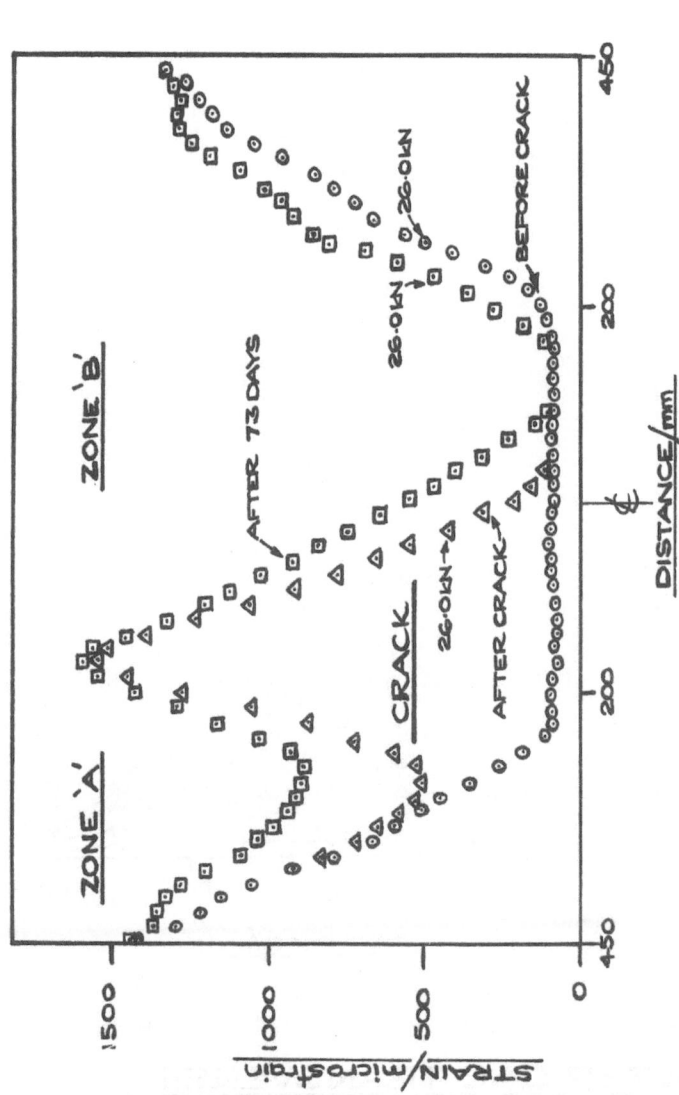

Fig4: STRAIN DISTRIBUTIONS:TYPICAL LONG-TERM SPECIMEN

(T12 REINFORCEMENT)

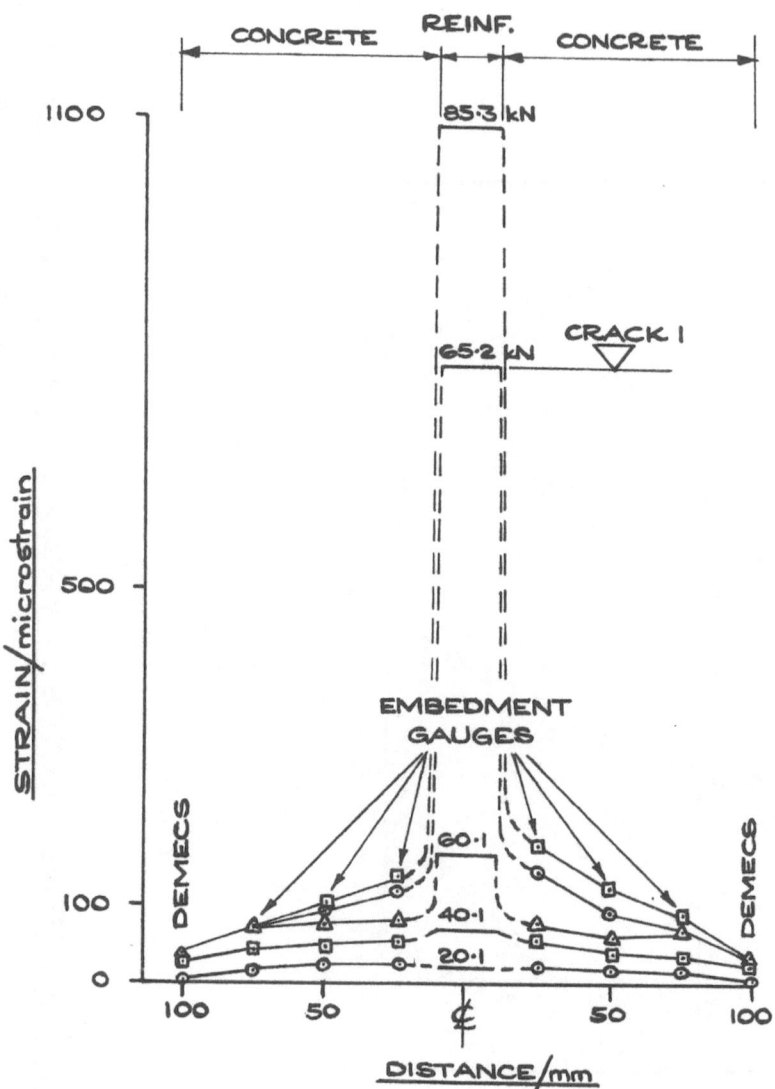

Fig5: **TYPICAL TRANSVERSE
STRAIN DISTRIBUTIONS**

EXPERIMENTAL AND THEORETICAL RESEARCH INTO THE COLLISION RESISTANCE OF
INLAND LPG-CARRIERS
G.T.M. Janssen, J.M.J. Oostvogels
TNO-IWECO, P.O. Box 29, 2600 AA Delft, the Netherlands

1. INTRODUCTION
When LPG is transported by inland-waterway carriers, the collision
resistance of these vessels is crucial for the determination of risk factors of this type of transport. Essential as to this determination is to
establish which penetration depths give rise to leakage of the gas tanks in
these vessels.
The most simple way to prevent leakage of the tank is the requirement that,
when a LPG-carrier collides, only the vessel's hull will be deformed, and
not the tanks themselves. The classification bureau "Germanischer Lloyd"
calculated (with Miniorsky-Woisin method) that this requirement would lead
to very low cruising speeds (2-8 km/h) [1,2,3].
So as to attain solutions that could be applied to practice, deformation of
the LPG-tanks themselves had to be allowed. The Bureau Veritas calculated
that this will lead to a considerable increase in the critical impact speed
(24-27 km/h) [1]. The calculation was carried out by means of the Rosenblatt
method [4,5,6,]. This method raised questions, however, as to the validity
of the boundary conditions applied. The cylindrical tank was, for instance,
thought to be clamped at the saddles. Moreover, it was found impossible to
show whether these results are conservative or not.
It was necessary to obtain more certainty as to the high critical impact
speeds and to gather more information about the depth of penetration at
which leakage occurs.
Therefore, TNO-IWECO was instructed to carry out two collision experiments
on scale 1:7, as well as collision calculations by means of the finite element method. This contribution starts with a brief description of the
Rosenblatt method and next gives a description of the experimental and
theoretical research at TNO-IWECO.
As starting point for the calculations and experiments an existing
LPG-carrier was taken (Pampero, Samoen and Passaat) (Fig. 1).

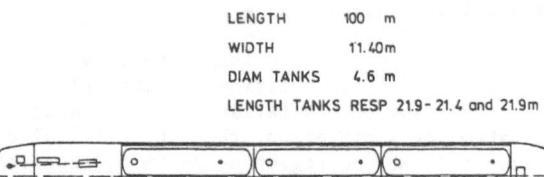

LENGTH	100 m
WIDTH	11.40m
DIAM TANKS	4.6 m
LENGTH TANKS RESP	21.9- 21.4 and 21.9m

Fig. 1: Division of the cargo tanks of an inland LPG-carrier

2. THE ROSENBLATT METHOD

The Rosenblatt method can be applied for the calculation of energy absorption when the extent of penetration is slight and it may be expected that membrane deformation (stretching) is the principal thing. Bureau Veritas applied this method not only when ships' hulls were slightly penetrated, but also for the calculation of the energy absorption when a cylindrical LPG-pressure tank [5,6] was penetrated. For this calculation the pressure tank was thought to be clamped over the saddles. It is assumed that when a bow is penetrated, the strips will be deformed into the triangular configuration as is shown in Fig. 2. Taking this situation as starting point, for each strip deformation energy can be calculated.

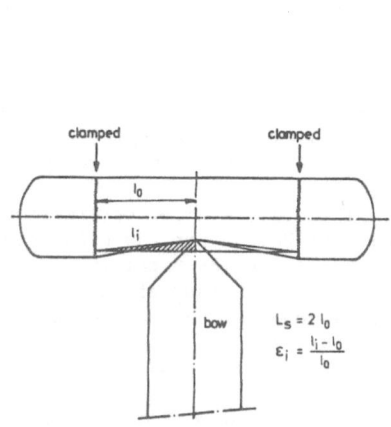

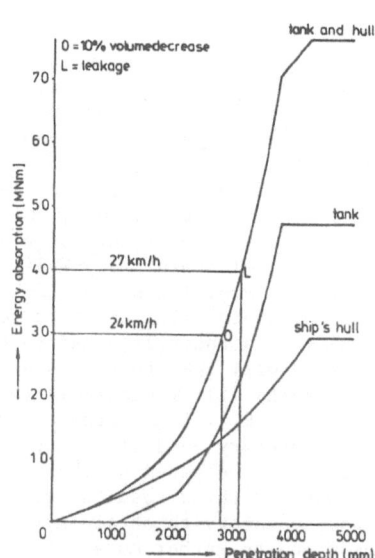

Fig. 2: Deformation mechanism according to the Rosenblatt method

Fig. 3: The energy absorption by ship's hull and tank (Rosenblatt method)

It was assumed that the tank would fail when a certain average elongation was exceeded or when liquid would flow from the safety valves of the tank. The latter situation is created when the tank decreases in volume by 10%, for the maximum degree of filling of the tank is 90%. The average critical elongation was defined as the elongation at which the tension became the average of the yield and ultimate stresses.

The results in Fig. 3 show that a pressure tank can absorb much energy in comparison with the amount of energy the ship's hull can absorb before a critical elongation is exceeded, or overflowing takes place. Moreover, it was predicted that an overflow occurs earlier than a leakage. If an overflow of LPG is taken for a criterion, then the critical collision speed of free-floating vessels is in the region of 24 km/h, assuming that the mass of both ships is 2,300 tons. If the results of this method could be proved to hold in practice, this would mean that at the customary cruising speeds in the Netherlands inland waterways (for big ships 20 km/h) there is no risk of tank leakage at collisions.

When the above method is applied, however, questions are arising as to the boundary conditions used. The cylindrical tank was, for instance, thought to be clamped at the saddles. Moreover, it has been assumed that the membrane deformation is prevailing when the tank is penetrated. Deformation of the saddles, the bending of the entire tank (banana shape) and the local bending in the tank wall are not included in the consideration. To gain a better insight into these problems, TNO-IWECO carried out collision experiments on scale 1:7 and finite-element calculations.

3. COLLISION EXPERIMENTS AND CALCULATIONS
3.1 The collision experiments
The collision experiments were carried out at a test tank on scale 1:7 [7]. As the wall-thickness of the real LPG pressurized tanks is about 21 mm, this means that the wall-thickness of the test tank became 3 mm. The diameter of the test tank was 657 mm and its length 3,057 mm. Because for the test tank the same material stresses were aimed at as in practice, its internal pressure is also the same as in practice. The penetration force becomes 1/49 of that in practice and the energy absorbed 1/343 of that in practice. Naturally, this only applies, if the test tank's geometry is exactly scaled down from the real tank and the material properties of both tanks are the same. As is shown in Fig. 4, this has been succeeded very well for the

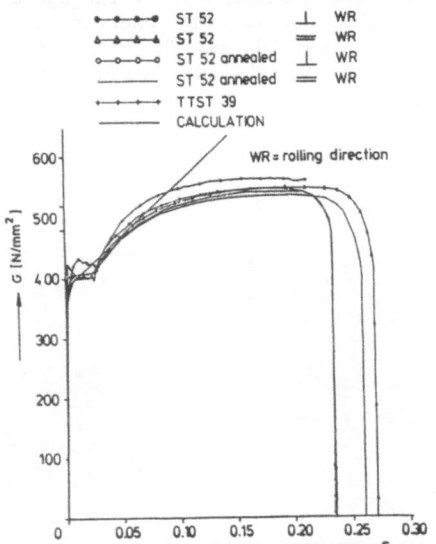

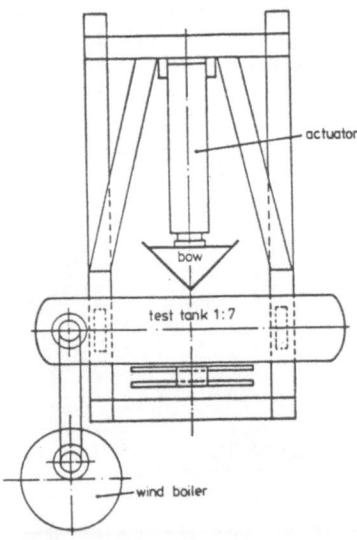

Fig. 4 : Tension curves of test-
tank and LPG-tank material

Fig. 5: Top view of the test
set-up

material properties. As regards the scaling down of the geometry, this has been done as well as possible by the factor 7.
In those parts that are deformed during the penetration, minor concessions have been made only in the saddle feet. So as to prevent the tank from collapsing at the bolted joint between the sole plates and the foundation, the sole plates under the tank feet, as well as the bolts have been made disproportionally thick. A top view of the test set-up is given in Fig. 5.

The bow, mounted on a hydraulic actuator of 1.400 kN at an oil pressure of 100 bar and a maximum stroke of 1,200 mm, consists of a V-shaped, stiffened structure having a top angle of 90° and a straight, vertical front. The air vessel, separated by a non-return valve, serves to simulate the gas, which at the real LPG-tanks comprises 10% of the tank volume.

For reasons of safety it was decided to entirely fill the tank with water. So as not to prevent the movement of the tank, the air receiver was resiliently mounted.

During the test the following quantities were recorded:
- The oil pressure in the hydraulic cylinder. From this, the penetration force was derived;
- The displacement of the bow. From this, the penetration depth in respect of the tank bases was derived;
- The pressure in the test tank. From this, the decrease in volume of the test tank is derived and the energy absorbed by the gas can be calculated.

Two tests were carried out. The first test was done with an internal gauge pressure of 1 bar in the tank, and the second test with 6 bar. The first tank collided with a constant bow speed of 87 mm/s up to a tank penetration of 460 mm. After the test, the tank proved to be leaky in the contact zone. Therefore, at the second test colliding was done intermittently so as to determine that penetration depth at which the leakage starts. The speed was chosen a factor 8 slower so as to enable the intermitting penetration. At the moment of leakage the bow penetration was 260 mm. The tank deformed after the test with 6 bar as shown in Fig. 6. As is shown in Fig. 7 the

Fig. 7: The deformed left tank-foot seen at the bow side. The double plate has been torn loose from the tank wall (6 bar)

Fig. 6: The deformation pattern after the test (6 bar)

doubler plate is torn loose at the saddle (front side), and at the back the
tank base is collapsed by the torsion buckle.
The 1 bar test yields simular results.

3.2 The collision calculations
Three calculations were carried out by means of the finite element
method. [8,9]:
* A bow penetration in the middle of the tank at an internal gauge
  pressure of 1 bar;
* A bow penetration in the middle at an internal gauge pressure of 6 bar;
* A bow penetration at a quarter of the distance between the saddles at an
  internal gauge pressure of 6 bar.

In Fig. 8 an example is given of a finite-element mesh (dotted lines) for a
central-bow-penetration case. For reasons of symmetry only half of the tank

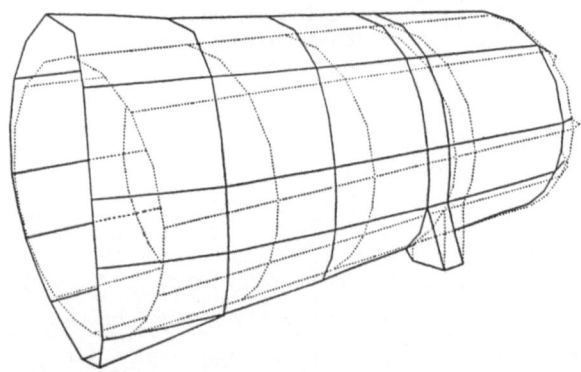

Fig. 8: The calculated deformations of the LPG-tank
(Finite element analysis)

considered. So as to avoid too long calculation times, a coarse element
model was chosen without any refinements. The calculations were carried out
with the TNO-IBBC finite element program DIANA. Eight-node, thick-walled
shell elements were chosen according to AHMAD [10]. In the calculation 5
layers per element were chosen over the thickness of the shell so as to
describe the elastic-plastic behaviour. Each layer has 4 integration
points. In each node the scale element has 3 displacements and 2 rotations
for degrees of freedom. Upon formulating the element, large displacements
have been taken into account (geometrically non-linear). At the saddles the
tank wall is thickened, because in the actual structure here a doubler
plate is mounted.
For a material model was chosen an elastic-plastic model having kinematic
hardening according to PRAGER (bilinear tensile curve) and the VON MISES
flow criterion [11].
The material data required were determined from a tensile test carried out

at a test piece of the tank material (Fig. 4). The tension curve to 10% has been approximated as well as possible.

The internal gauge pressure is composed of the pressure above the tank and the self-weight of the liquid. As the tank is being compressed, its volume will decrease. As a result the pressure in the tank will increase. After each displacement step (45 mm) the increase in the pressure was calculated, an adiabatic change of state being considered.

For practical reasons, the adaptation of the pressure load was each time carried out only after some steps.

Furthermore, the saddles are fully clamped at their lower sides. The problem that an increasing number of nodes will run against the bow and thereafter have to be prescribed, while displacing, was solved by means of springs having a non-linear spring characteristic. These springs are mounted between the "bow" (straight vertical line with nodes) and the tank wall. Now the characteristic of these springs is such that their stiffness is zero at a positive length and that they are very stiff at a negative length. The calculations now comprise the displacement by small steps of the nodes of the imaginary bow. An example of a deformed tank (solid lines) is shown in Fig. 8 at a penetration depth of 2,000 mm.

3.3 Results of the experiments and calculations
In Fig. 9 the experimental and calculated energy curves are shown.

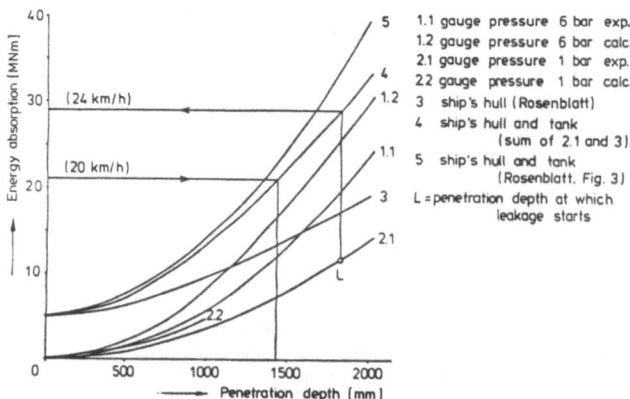

Fig. 9: Energy penetration diagram of central collision; the results are based on experiments, finite-element and Rosenblatt methods

So as to restrict computer costs, the calculation for the case with 1 bar of internal gauge pressure has only been continued to 1,000 mm. The experimental curves were obtained by integrating the experimental displacement curves, the displacement being scaled up by a factor 7 and the force by a factor 49. The curves calculated appear to be 30% higher than the experimental ones. The difference between experiment and calculation is probably

to a high extent caused by the coarse-element distribution. This results in
the model not being very well able to follow the complex deformation behaviour (i.a. buckling and wrinkling).
It is interesting to state that the increase in pressure results in a stiffening effect. The internal gauge pressure in the tank causes membrane
stresses. They increase the stiffness perpendicular to the tank wall.
At a penetration of 2,000 mm the decrease in volume appears to be maximally
5.5%. The critical volume decrease of 10% at which LPG would escape from
the tank, is not reached. Before this can occur, there is already leakage.
The saddles are heavily loaded and, therefore, also limit the energy to be
maximally taken up. It can be derived from the experiment that the point of
collapse occurs at 2,600 mm (scale 1:1).
When the excentric calculation was carried out at an internal gauge
pressure of 6 bar, it appeared that the energy-absorption-penetration curve
hardly differed from the centric case. The local behaviour of the saddles
differs greatly, however, from that at the centric calculation (chapter 4).

4. DETAIL CALCULATIONS
So as to further support the validity of the above experimental and
theoretical results, a number of details have been subjected to an additional consideration [12]. These details are:
● The behaviour of the LPG-tank under the influence of a local load
  resulting from a bow that is not uniform in shape. In this case the
  model situation is based on a web frame penetrating an LPG-tank. It has
  been shown that because of the local plastic deformation the unevenness
  of the front of the web frame will decrease so that the risk of local
  penetration, resulting in leakage, will be considerably reduced;
● The local behaviour at the tank feet. It has been concluded from the
  experiments and calculations that upon the centric collision collapse
  occurs at 2.600 mm penetration, and at the excentric collision at
  1,950 mm;
● The influence of the collision speeds and the mass forces. The possibility has been investigated of leakage in the first phase of the collision. By solving a equation of motion, both the notch depth in the wall
  resulting from an actually non-realistic sharp bow, and the pertinent
  depth appeared to be abt. 10 mm (sheet thickness is 20.8 mm). The energy
  dissipated appeared to be negligibly small in respect of the total
  energy absorption.
  Omitting the mass force does not reduce the effect of either the test
  or the calculations;
● The influence of the stiffening rings. The stiffening rings are meant to
  prevent buckling of the cylindrical-tank wall in case of vacuum. On
  account of a collapse analysis it was concluded that the energy dissipated when the stiffening rings collapse can be neglected in respect of
  the energy absorbed totally;
● The fastening of the tank at the ship's bottom. Both at the experiments
  and at the calculations it was assumed that the tank would not collapse
  at the bolted joint. So, it is important that this also holds true for a
  real vessel. In that case it is known that the displacements of the tank
  in respect of its surroundings will be limited. This is in particular
  significance for the protuding parts of the pressurized cylindrical
  tank. These are i.a. the dome, the suction well and the piping. It
  appears from the calculations, however, that the existing bolted joint
  will already collapse at a penetration of about 300 mm.
It can be shown that by combining some measures it can reasonably

ensured that the bolts do not collapse.
These measures are:
- Raising the quality of the bolts;
- Increasing the number of bolts;
- Applying larger bolt diameters;
- The adaptation of the safety stops.
- The behaviour of the dome, manhole and suction well at a collision. The displacements of the above components will run against frames, beams and plates in the surroundings. As concerns the dome, it was shown that probably first the cover plates collapse and then the dome.
The same applies to the manhole. The suction well will collide with the bottom frames. It is possible, however, to move the bottom frame over a short distance.
- The effect that the adjacent tank takes over a part of the collision load. It has been calculated that the tanks will run agains each other at a penetration depth beyond 2,000 mm. This means that in general we cannot count on the adjacent tank taking over a part of the collision load. At a penetration depth in the region of 1,700 mm the back of the tank will contact the struts. These will collapse very quickly and not play any role of significance in the total energy consideration.

5. CONCLUSIONS

On the basis of the results (Fig. 9) it can be concluded that the inland LPG carrier can stand a collision when the ramming ship of a comparable tonnage does not sail faster than 34 km/hr. The total energy-absorption is composed of the energy absorption of the ship's hull and the experimentally determined energy-absorption of the tank having 1 bar of internal gauge pressure (the most conservative combination). Because in the Netherlands inland waterways the sailing speed of these big ships is lower than 24 km/hr (20 km/hr), it may be concluded that upon collisions no leakage of the LPG-tank will occur.

The membrane-stretching deformation energy determines only partially the energy absorption. The bending deformation and also the deformation of the saddles considerably contribute to the amount of deformation energy produced. Therefore, it is recommended to modify the Rosenblatt method at this point.

ACKNOWLEDGEMENT

TNO-IWECO expresses its acknowledgement to the Ministry of Health, Physical Planning and the Environment for its permission to publish this article.

LITERATUUR

[1] VAN DER HORST, J. en VAN DER SCHAAF, J.
     Nautical risk of liquified gas carriers.
     Proceedings of the Fourth Industrial Loss Prevention Symposium,
     Harrogate, UK, September, (1983).
[2] MINORSKY, V.U.
     An analysis of ship collision with reference to protection of nuclear power plants.
     Journal of Ship Research, 3 October (1959).
[3] WOISIN, G.
     Die Kollisionsversuche der GKSS.
     Jahrbuch der Schiffbautechnischen Gesellschaft, 70, (1979).
[4] ROSENBLATT, M.

Tanker structural analysis for minor collisions.
U.S. Coast Guard, December (1975).

[5] EDINBERG, D.E.
The collision resistance of two inland LPG-carriers;
"Pampero" and "Chemgas 17" barge.
Bureau Veritas, Report no. 81-125, Paris (1981).

[6] EDINBERG, D.E.
An extension to the collision resistance of two inland LPG-carriers;
"Pampero" and "Chemgas 17" barge.
Bureau Veritas, Report no. 81-110, Paris, (1981).

[7] OOSTVOGELS, J.M.J.
Aanvaringsexperimenten uitgevoerd op een schaalmodel van een gastank
(only in Dutch).
TNO-IWECO, Report no. 5072080-82-1, (1982).

[8] HULSEN, M.A.
Aanvaringsberekeningen van een LPG-tank (only in Dutch).
TNO-IWECO, Report no 5071039-82-1, (1982).

[9] HULSEN, M.A.
Excentrische aanvaringsberekeningen van een LPG-gastank (only in
Dutch).
TNO-IWECO, Report no. 5071042-82-1, (1982).

[10] PUTHLI, R.S.
Geometrical non-linearity in collapse analysis of thick shells, with
application to tubular steeljoints.
Heron, 26, no. 2, (1981).

[11] PRAGER, W.
The theory of plasticity; a survey of recent achievements.
Proc. Inst. Mech. Engrs., 169, 41-57, (1955).

[12] JANSSEN, G.T.M. en HULSEN, M.A.
Detailberekeningen met betrekking tot de sterktebepaling van
LPG-scheepstanks bij een aanvaring (only in Dutch).
TNO-IWECO, Report no. 5071043-82-1, (1982).

EXPERIMENTAL STRESS ANALYSIS AT PLASTIC DEFORMATIONS

P. MACURA

THE IRON AND STEEL RESEARCH INSTITUTE, DOBRA, CSSR

1. INTRODUCTION

The technical progress in continuous casting of steel and in application of hardly-deformable materials is intimately associated with new requirements for analysing the state of stress at plastic deformation during the forming process. Little success was reached hitherto with exact analytical solution of that problem; for this reason, solution by means of the methods of experimental analysis of stress presents a valuable contribution in this field. To deal in detail with the problem of state of stress at forming The Iron and Steel Research Institute has available the visioplastic method, the photoplastic method for the surface foils with application of reflected light and the photoplastic method with light passing through the models prepared of amorphous mass.

2. THE VISIOPLASTIC METHOD

This method was used for determination of the state of strain and stress at the lateral surface of a rolling stock when rolling on smooth rolls. A rectangular network of coordinates was applied at such surface and the rolling process was screened by means of a high-frequency camera. The components of the vector of speed v_i /Fig.1/ and of the tensor of strain rate $\dot{\varepsilon}_i$ and $\dot{\gamma}_{ij}$ of the nodal points of a coordinate grid /Fig. 2/ were evaluated from a film record by applying the following relationships

$$\dot{\varepsilon}_i = \frac{\partial v_i}{\partial i} \quad ; \quad \dot{\gamma}_{ij} = \dot{\gamma}_{ji} = \frac{\partial v_i}{\partial j} + \frac{\partial v_j}{\partial i} \tag{1}$$

To determine the components of the tensor of stress one has to apply the theories of plasticity; here, the theory of plastic creep was applied. This is based upon precondition that the stress deviator D_6 and the deviator of increment of plastic strain $D_{d\varepsilon}^P$ are similar and coaxial

$$D_{d\varepsilon}^P = \lambda D_6 \tag{2}$$

Accordingly, among the components of the tensor of stress T_6 and the components of tensor of strain rate $T_{\dot{\varepsilon}}$ the following equations can be derived /1/ :

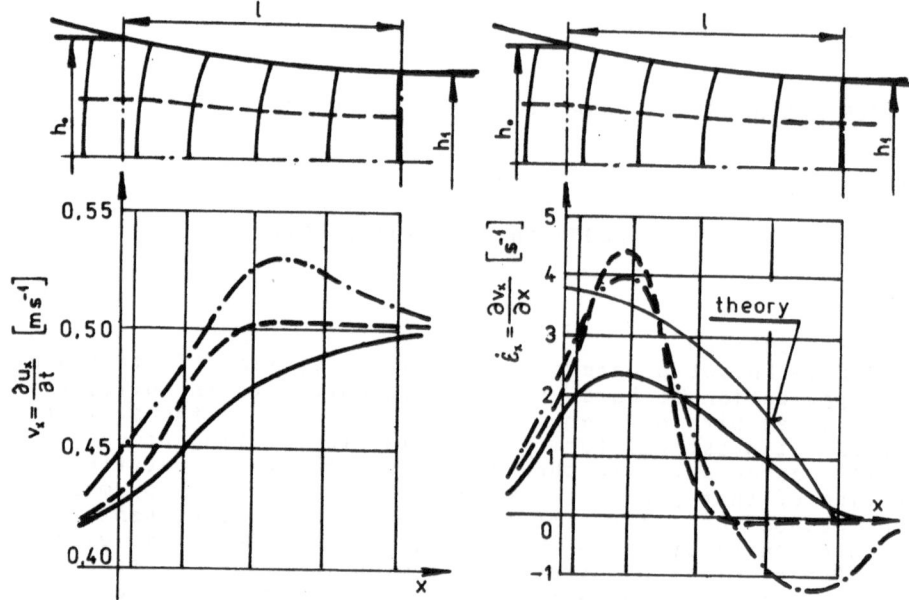

FIGURE 1. Courses of velocity v_x in three lines.

FIGURE 2. Courses of strain rates $\dot{\varepsilon}_x$.

$$\left.\begin{array}{l} \dot{\varepsilon}_i = \dfrac{S_{\dot{\gamma}}}{2\,S_{\tau}}\,(\sigma_i - \sigma_s) \\[2mm] \dot{\gamma}_{ij} = \dfrac{S_{\dot{\gamma}}}{S_{\tau}}\,\tau_{ij} \end{array}\right\} \qquad \sigma_s = \dfrac{\sigma_x + \sigma_y + \sigma_z}{3} \qquad (3)$$

In the system of equations (3) S_{τ} represents the intensity of shear stress defined by means of the second invariant of the deviator of stress $I_2\,(D_\sigma)$:

$$S_{\tau} = \sqrt{-I_2(D_\sigma)} = \sqrt{\frac{1}{6}\left[(\sigma_x - \sigma_y)^2 + (\sigma_y - \sigma_z)^2 + (\sigma_z - \sigma_x)^2 + 6\left(\tau_{xy}^2 + \tau_{yz}^2 + \tau_{zx}^2\right)\right]} \qquad (4)$$

and $S_{\dot{\gamma}}$ is the intensity of shear-stress rate defined in a similar manner as follows:

$$S_{\dot{\gamma}} = 2\sqrt{-I_2(D_{\dot{\varepsilon}})} = \sqrt{\frac{2}{3}\left[(\dot{\varepsilon}_x - \dot{\varepsilon}_y)^2 + (\dot{\varepsilon}_y - \dot{\varepsilon}_z)^2 + (\dot{\varepsilon}_z - \dot{\varepsilon}_x)^2 + \frac{3}{2}\left(\dot{\gamma}_{xy}^2 + \dot{\gamma}_{yz}^2 + \dot{\gamma}_{xz}^2\right)\right]} \qquad (5)$$

For solution of stress one has to determine experimentally the response of the formed material to the mechanical loading under various thermodynamical conditions of the forming process. Figure 3 shows the dependences of the intensity of shear stress S_{τ} on the intensity of shear strain $S_{\dot{\gamma}}$ at constant

intensity of shear-strain rate $S_{\dot{\gamma}}$, determined at the upsett-
ing tests run on a cam plastometer with lead at 20°C.

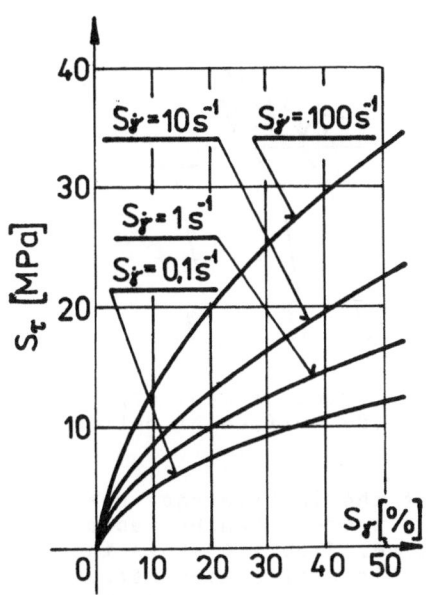

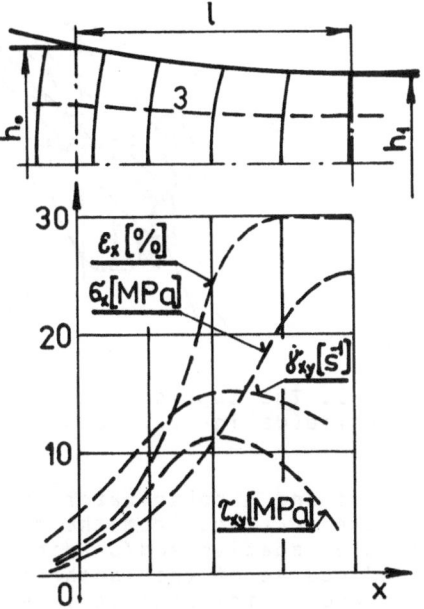

FIGURE 3. Strengthening
curves of lead.

FIGURE 4. Courses of parame-
ters $\varepsilon_x, \dot{\gamma}_{xy}, \sigma_x, \tau_{xy}$.

On the basis of the set of aquations (3), of static conditions of equilibrium and of the components of strain-rate te-
nsor $T_{\dot{\varepsilon}}$ as determined experimentally, one can then formulate
the components of stress tensor T_{σ} . In this respect Fig.4
shows some components of tensors of stress, of strain and of
strain rate along the fibre 3. This method is described in
detail elsewhere /2/, /3/.

3. THE PHOTOPLASTIC METHOD OF SURFACE FOILS

This method was used for solving the same problems as the
previous method; when combined with the orthodox photoelasti-
cimetry one can analyse the state of stress simultaneously at
the lateral side of rolling product and in the working rolls
and even at the contact area of rolls with the rolling product as well. The principle of method and the applied experimental equipment is illustrated in Fig.5.

The lateral side of the experimental lead sample is provided with an optionally active foil, stuck-on the side and the
sample is rolled on a laboratory rolling mill the rolls of
which are made of optically active CR 39 material. The Laboratory rolling mill is installed in a site lying between the
pollarizing filters of a transmission polariscope, thus providing the experimental analysis in rolls and at the contact
area of the rolls with the rolling product. To acquire a pattern of isoclines and isochromates in the optically active
foil the measuring instrument is equipped with another source

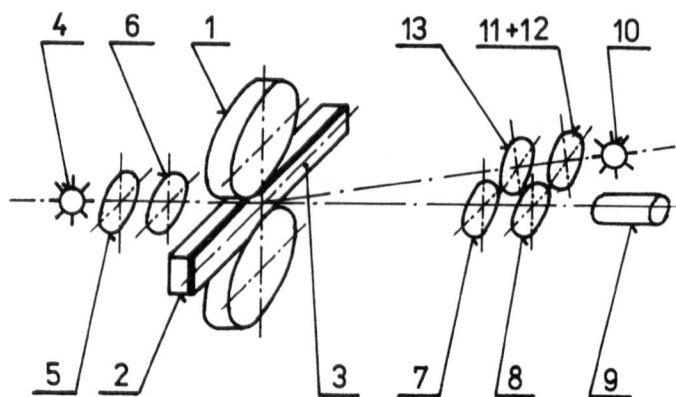

FIGURE 5. The applied optical set for measurement of the stress fields in rolls and at the side surface of rolling stock.

of light and a polarizator so that the interference phenomena from the roll and from the rolling product can be recorded at assingle negative and/or diapositive. Figure 6 shows the acquired course of the isochromatic lines of semi-orders in the rolls and in the optically-sensitive foil, stuck-on the lateral area of a rolling product.

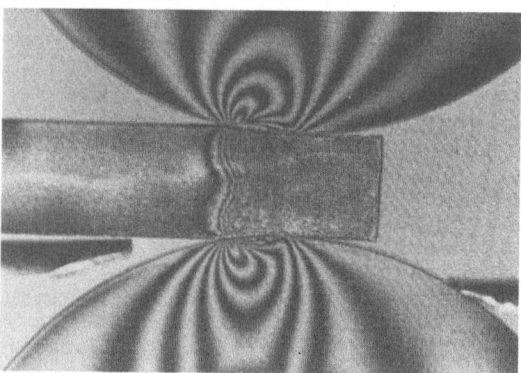

FIGURE 6. The course of the isochromatic lines of semi-orders obtained in the rolls and at the side surface of the rolling stock.

The analysis of stress in rolls and at the contact area of rolls with the rolling product has been run by the method of shearing stress along three horizontal and twenty vertical sections in the vicinity of the roll contact with the rolling product, see Fig.7.

To analyze the stress in rolling material one has to apply

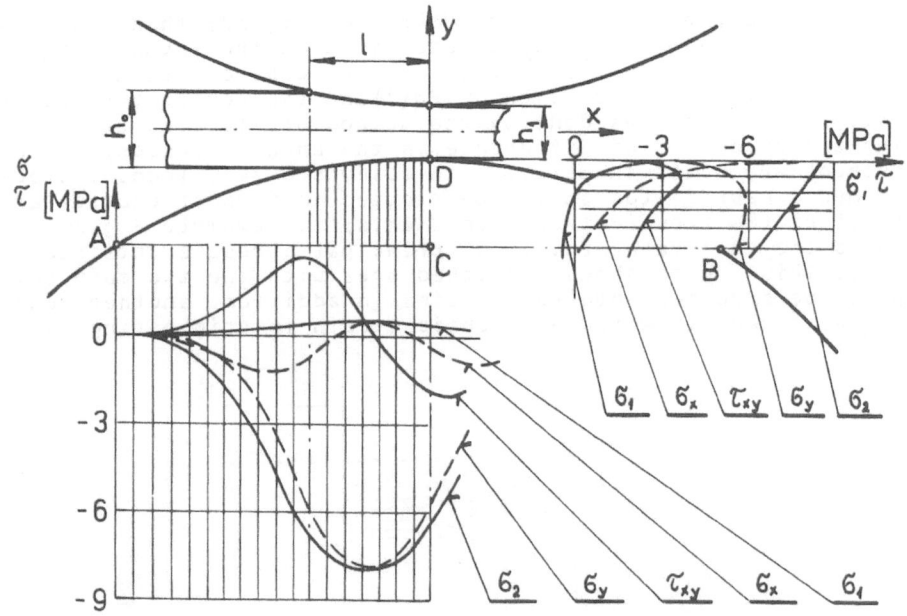

FIGURE 7. Courses of components of stress tensor

again the theory of plasticity such as the theory of elastic-plastic strains based upon the fundamental precondition, namely, the deviators of stress and strain are similar and coaxial :

$$D_\varepsilon = \psi \, D_6 \tag{6}$$

By setting into Eq (6) and by modifying one can derive the relationship among the components of the tensor of stress and strain as follows /4/ :

$$(6_1 - 6_2) = \frac{2 S_\tau}{S_\gamma} (\varepsilon_1 - \varepsilon_2) \tag{7}$$

In Eq (7) the term S_γ represents the intensity of shear strain defined by means of the second invariant of strain deviator $I_2(D_\varepsilon)$:

$$S_\gamma = 2\sqrt{-I_2(D_\varepsilon)} = \sqrt{\frac{2}{3}\left[(\varepsilon_1-\varepsilon_2)^2 + (\varepsilon_2-\varepsilon_3)^2 + (\varepsilon_3-\varepsilon_1)^2\right]} \tag{8}$$

To determine the difference of the principal stresses (6_1-6_2) by Eq(7) one has to assess the difference between the main relative strains $(\varepsilon_1-\varepsilon_2)$, which can be experimentally determined from the pattern of isochromates; moreover, one has to determine the functional relationship among the shear-stress intensity S_τ and the shear strain S_γ, which can be measured with the help of a cam plastometer /see Fig.3/; then, the intensity of shear strain S_{γ} is determined in the investigated point, which presents a problem for the case of planar stress.

For this reason solution of Eq(7) and the consecutive separation of stress was done by the method of stepwise approaching and by method of difference in the shear stresses; these methods are dealt in detail with elsewhere /4/. Then, the obtained tensor fields should be plotted in form of equiscalar level of the tensor-stress components; for example Fig.8 shows the equiscalar levels of principal stresses obtained here. This figure shows a hatched area with the two main compressive stresses involved herein; in addition, another zone is shown here in which no plastic strain is originating yet.

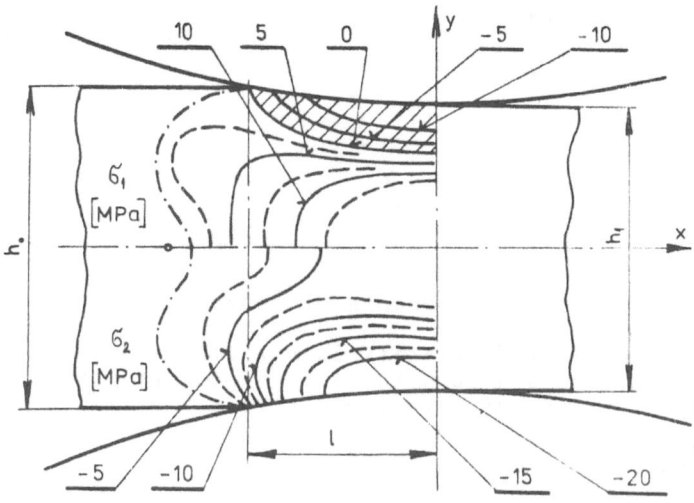

FIGURE 8. Courses of the main stresses in the effective zone of strain at rolling.

4. THE PHOTOPLASTIC METHOD WITH MODELS OF AMORPHOUS MASS

This method was used for analysing the state of stress and strain in cross-sectional area of the rolled products when rolling in shaped grooves. Application of this method is given in the case of rolling a square-area rod in oval groove. The geometrical conditions of rolling are illustrated in Fig. 9. The test sample is prepared of a foil of low-modulus epoxide resin in sizes identical with the input sectional area of the rolling stock. The sample is compressed between two transparent plates, namely, by means of templates with a shape corresponding to the grooving rolls. Various positions of approach of templates refer to the relevant cross-sectional area of the geometrical zone of deformation, see Fig.9. The deformed sample is put between the polarizing filters of a transmission polariscope and its interference phenomena in form of isoclinic and isochromatic lines are investigated as given in Fig.10; it shows, moreover, the obtained course of

the isochromatic lines of semi-orders for the I section illustrated in Fig.9.

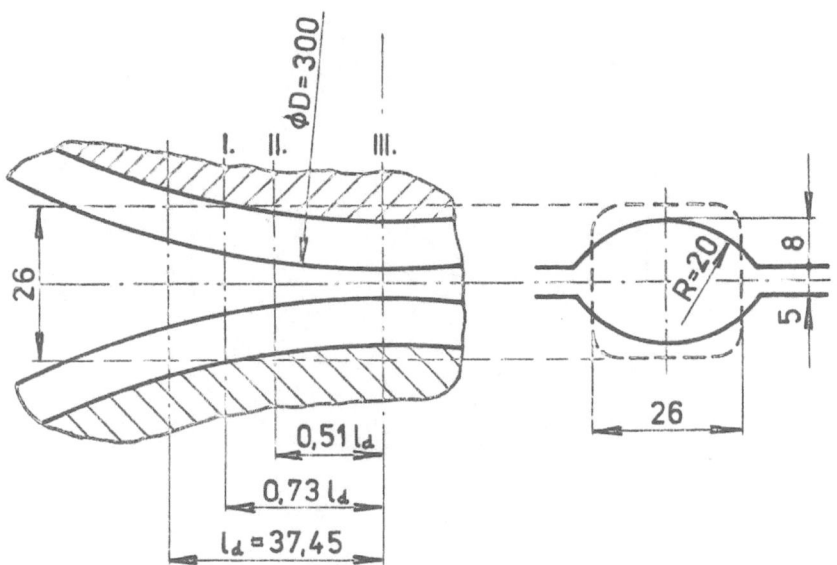

FIGURE 9. Geometrical conditions of rolling in square-oval
groove.

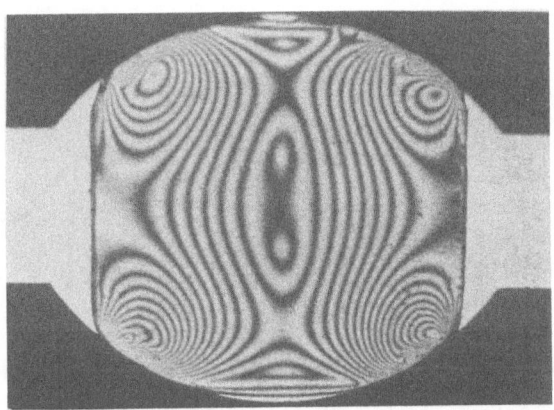

FIGURE 10. The obtained course of the isochromatic lines.

From the pattern of isoclines one can plot the isostate and
slip lines, see Fig.11. To analyze the stress one has to use
again the theory of plasticity; evaluation was made in a manner similar to the method cited above with application of
theory of elastic-plastic deformation and the method of consecutive approaching with the method of shear stresses. Figure 12 shows the obtained patterns of the equiscalar levels of

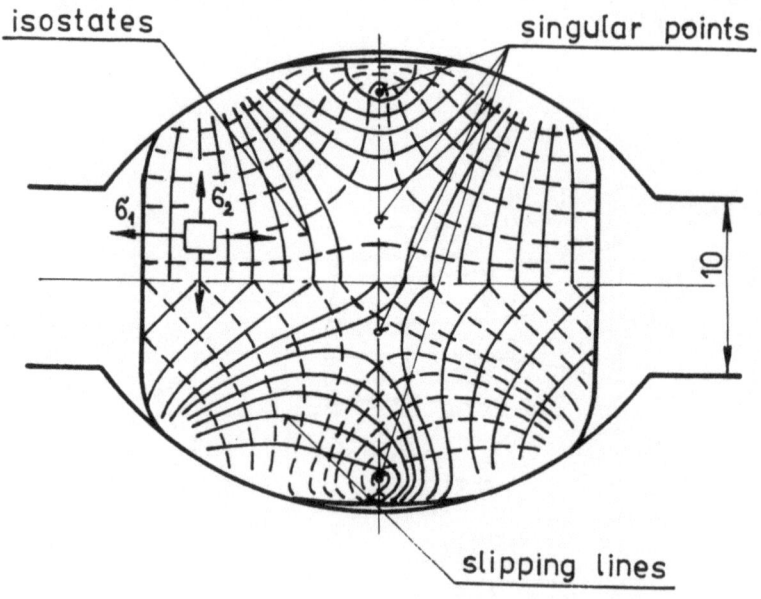

FIGURE 11. The courses of the isostates and slip lines.

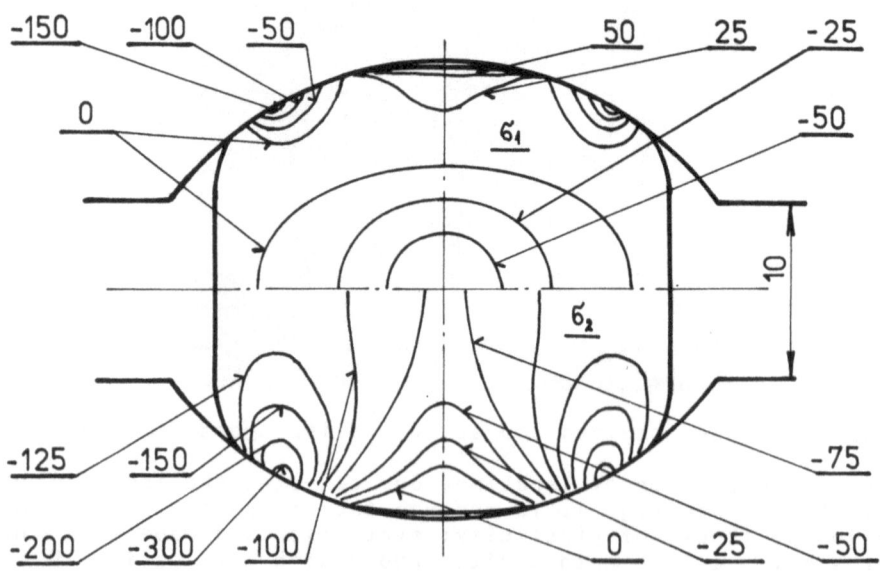

FIGURE 12. The equiscalar levels of main stresses σ_1 and σ_2 [MPA].

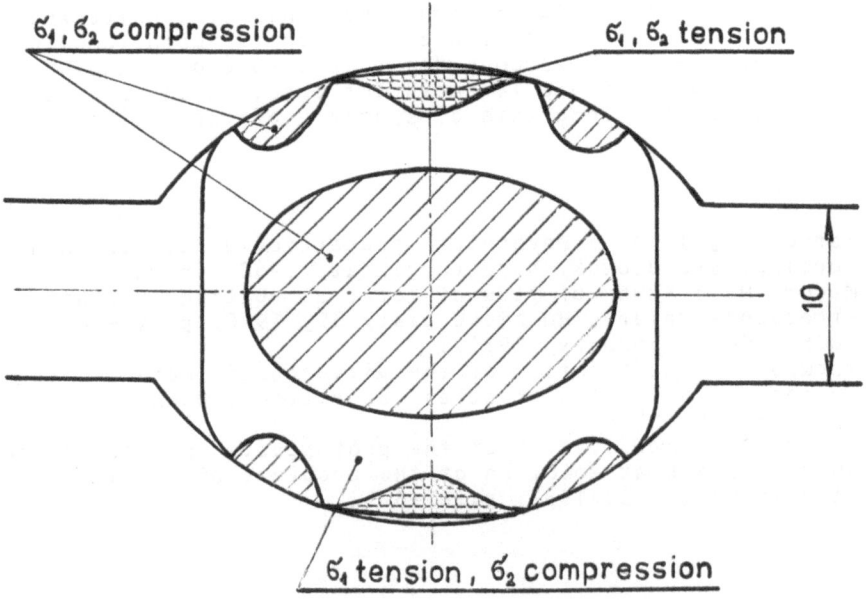

FIGURE 13. The zones of tensile and compressive main stresses.

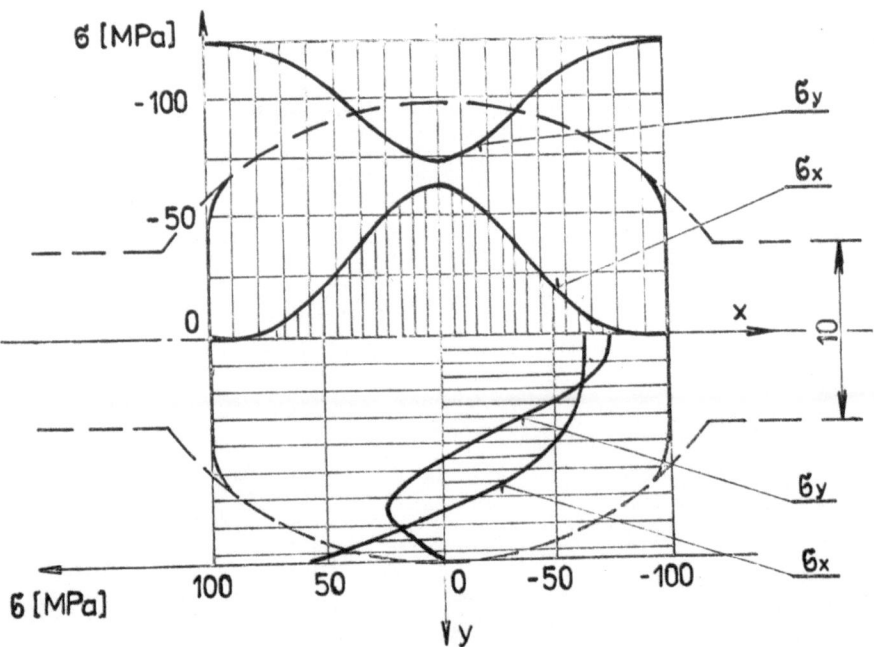

FIGURE 14. The courses of the main stresses along the axis of symmetry of the rolling product.

main stresses, while Fig.13 shows the zones of tensile and
compressive main stresses. The zones with two main tensile
stresses are rather hazardous due to possible origination of
material failure. Figure 14 shows the obtained patterns of
the main stresses along axis of symmetry of the rolling product.

REFERENCES

1. Tarnovskij I. J. : Mechanical properties of steels in hot
   forming. Sverdlovsk, Metallurgizdat, 1960 /in russian/.
2. Macura P. : Study of flow of metal at rolling by a high-
   -frequency camera. Hutnické Listy 35, 1980, p.244-247
   /BISI 19463, January, 1981/.
3. Macura P. : Application of the visioplastic method at solution of the states of stress at rolling. Hutnické Listy
   39, 1984, p.851-855.
4. Macura P. : Application of the photoplastic method of the
   surface foils at solution of the problems of forming.
   Strojírenství 32, 1982, p.665-671.

MICRO-COMPUTER BASED MEASUREMENT AND PROCESSING METHOD FOR
HUMAN LOCOMOTOR SYSTEM´S TAKE-OFF PROPERTIES EVALUATION

V. MEDVED

FACULTY OF PHYSICAL CULTURE
ZAGREB UNIVERSITY
Horvaćanski zavoj 15
41000 ZAGREB
YUGOSLAVIA

1. INTRODUCTION

Viewed as a mechanical system human body possesses interesting properties. Study of movement and locomotion of the body, as a whole, has revealed many features that can adequately be modeled, on the macroscopic level, by mechanical analogs. Biomechanical functioning of the body cannot eventualy be separated from the physiological one - these two approaches being essentially integrated (1,2). In spite of this, the study of mechanical (kinematic and kinetic) variables of movement only has nevertheless led to many important findings regarding the interplay between active (muscle generated), passive and inertial forces and moments of forces the body is constrantly being subjected to in the course of various movement patterns.

In particular, among locomotor system´s properties being important for successfull realization of many movement structures, the so-colled take-off propertie deserves some attention. Being mainly the manifestation of the so-colled explosive power of the organism, it is generally being estimated via standard jumping type tests like Sargent, Abalak and jump in the horizontal direction, all these giving the end result expressed as some height or length achieved, respectively. It has been found that the dynamic strength of muscles responsible for realization of the above mentioned movement structures lies in good correlatiœ with static strength, measured in isometric conditions. It has also been found that this property is best enhanced through dynamic exercises, often taking the advantage of the muscle prestretching mechanism (jump from some height).

Biomechanical approach to the study of take-offs can throw further light on basic physiological mechanisms functioning in these cases. Research is currently in progress, in many laboratories in the world, toward finding multiple temporal and other relations between various biomechanical variables (including myoelectric signals) with the prospective goal to better explain the functionality principles and energetics of the neuromuscular system.

There are movement classes posing higher demands on the locomotor system in terms of higher speeds achieved, impact forces exerted and, consequently, higher stresses imposed on the biological tissue.Often, they are characterized by better skill and coordination. An example of such movement class can be encountered in sport gymnastics, embodying literally hundereds of various elements of technique. Gymnastic authorities stress the importance of proper take-off in the execution of complete performance (3, citation): "The effectiveness of the take-off sets the uppermost limits of what the gymnast can hope to attain during the airborne phase of any skill. It is during this time that the path (trajectory) followed by the performer´s center of gravity and the amount of rotary motion (angular momentum) available for skill are irrevocably established."

Having the main goal to search for a simple yet still powerfull enough method of take-off capabilities evaluation one gets attracted by one specific kinetic variable; ground reaction force - easily prone to measurements in situational conditions by the force platform. Taking this rationale, an experimental method has been developed based on measurement and processing of ground reaction force signals recorded during take-off phase preceding certain movement pattern. Take-off´s quality, presumably, is a strong determinant of success of performance of that pattern. The method as such is particularly suited to the study of skilled movements; like those occuring in sport gymnastics - supposing that skill level can be ascribed adequate criterion.

2. DESCRIPTION OF THE METHOD - AN EXAMPLE OF SPORT GYMNASTIC MOVEMENTS

The method can proceduraly be devided into three sequential steps: measurement and data acquisition, data processing and analysis and result interpretation. Each of these steps will be described and illustrated with some examples taken from a recent take-off study in the field of sport gymnastics, results of wich will be presented elsewhere (4).

2.1. Measurement and data acquisition

The hardware part of the instrumentation used includes one force platform, dimensions 40 x 40 cm, multichannel amplifier system RM Dynograph Recorder, Beckmann Inc., coupled via analog to digital (A/D) converter to the Apple II micro-computer equipped with standard peripherals: video display, diskette unit und printer Epson RX-80 FT. The system is housed in the comfortable facility of biomechanics laboratory.

Force platform is of the strain gage type, having three separate Wheatstone type measuring bridges with force transducers, each measuring one component of the total applied force. Platform channels are connected to amplifiers via couplers (one pair coupler-amplifier per channel), all being part of the RM Dynograph Recorder. Coupler-amplifier subsystem provides proper balancing and signal conditioning and amplification, resulting with signals in the conventional Volt range on the output of the RM Dynograph Recorder. Regarding the frequency content of measured force signals, they have, for the present purpose, been analog lowpass-prefiltered with a cut-off frequency of 150 Hz.

A/D converter card is built arround the Analog Devices chip AD 574A. It is of the successive approximation type, has the precision of 12 bits and has 16 channels (of wich only three are used in the present application).

The software developed for this purpose includes several routines written in BASIC and Assembler, enabling easy definition, control and realization of the measurement and acquisition procedure.

A program UPIS A/D,written in BASIC, provides definition and control of the A/D conversion process of multiple channels of information. The number of channels has been for the present purpose fixed to three, while other conditions have been determined resulting in basically two fixed standardized versions, suited to corresponding applications. For the purpose of recording force signals manifested in gymnastic elements of technique performed from the standing position (backward somersault and straddle jump) the version with sampling frequency of 128 Hz and duration of time sequence of 2 seconds has been used. Alternatively, in elements where take-off has been preceded by short run (running forward somersault and layout forward roll) the version with sampling frequency of 512 Hz and record length of 500 milliseconds has been used.

One Assembler routine, called TIBET A, forming the nucleus of the conversion method, provides multiple A/D conversion of one or more channels of analog information. Parameters of conversion (number of channels, starting channel, sampling frequency, number of samples) are defined via above mentioned programs in BASIC. Sampling time equals among all channels. One elementary conversion cycle takes 46 μseconds (determining maximum possible sampling frequency). Determination of actual sampling time is done by software delay loop.

The system has been used in the investigation on the sample of 16 women gymnasts, age ranging from 8-17, skill level ranging from regional to international. Each of the above mentioned elements of technique has been performed six times, while take-offs have been realized from the force platform. Each trial was graded by two competent gymnastic judges. Signals have been stored on diskettes.

As an illustration, one can show typical vertical component of force signals of one of participants in the study (age 12, weight 332 N).

abscissa 1 cm = 185 msec
ordinate 1 cm = 460 N

backward somersault   straddle jump

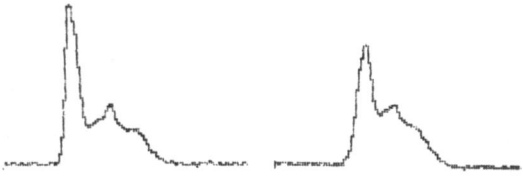

abscissa 1 cm =   46 msec
ordinate 1 cm = 2300 N

forward somersault   layout forward roll

2.2. Data processing and analysis

Data are processed off-line using Apple II micro-computer.
Considering only the vertical component of force, out of all
signals recorded basic parameters are being extracted and/or
numerically computed. These are: absolute maximum or maxima
(two, in forward somersault and layout forward roll), duration
of final take-off force segment, impulse (time integral of
force) during the final take-off period and gradient (maximal
positive time derivative).

For the purpose of force signal parameter extraction, a
BASIC program called MONITOR has been developed, providing display, in graphical and numerical forms, of time function of
signals recorded. Special adaptation has been provided enhan-
cing visual tracking of signal's morphology, so the user has
the possibility to scan the entire record and to extract time
and amplitude values of interest.

General statistics is then computed of all extracted and
computed parameters, of their normalized (with respect to body
weight) values and of grades, in the whole trial space, per
each element.

While force parameters cannot be considered predictors of
success of performance in the strict causal sense, they, as an
objectively measurable physical quantity, can be treated as
such in the formal sense. Therefore, linear regression equations are computed between each individual parameter (raw and
normalized) and grades, in the whole trial space, for each
element of technique.

2.3. Results interpretation and take-off properties evaluation

Results obtained in this way can be interpreted following
two distinct, but at the same time strongly interrelated, phi-
losophies; the main goal can be either to broaden the basic
knowledge in the area by incorporation new experimental observations, or the primary objective might be to interpret each
subject's results comparing it against normative data. At the
present time, the first one of these objectives is being pursued, in order to form an initial data base and calculate
normative values of parameters for the population defined by
this sample. Later, data obtained in further measurements - in
standardized conditions - shall be compared against normative
values, so in this way it will be possible to evaluate individual take-off properties. Simultaneously, testing protocol

will be reduced and optimized, without, hopefully, loosing useful informations.

Statisctics over all subjects gives the range of normal values of force parameters, in each element performed. Here, the investigation (4) has resulted with some data which could have been expected and can serve as a basis for establishment of normative (reference) values for the population defined by the sample having been tested. On the other side, some extremely high values of force have been measured and have indicated that human locomotor system might indeed be exposed to high impulsive loading during impacts and take-offs. These findings serve as a means of prediction of the nature and amount of mechanical stress experienced by the body. Such findings are used in the process of estimation and, subsequently, prevention of injuries of the musculo -scelatal system.

Correlation analysis serves to detect good formal predictors of the skill of performance, as reflected in take-off force signals. The investigation has shown that, generally, "running" elements of technique display higher correlations then elements performed from the standing position. Efforts are in progress now (5) to reduce and condense the testing protocol making it more practical, fast and user oriented. Similar procedures, however, can be conceptualized for testing of other locomotor activities including take-off, as well.

In addition, a further improvement in data processing is planned toward the so-called vectordiagram signal representation, similarly like in (6 and 7), where it has been successfully used in the field of rehabilitive medicine to detect certain "force signatures" of pathological walking.

3. CONCLUSION

It is possible to summarize crucial features of the presented method:

- accuracy, precision, repeatability and noninvasiveness of the method are generally satisfying.
- the method is relatively easy to use supposed the operator is familiar with elementary micro-computer handling, including running of application programs. This makes it suitable for providing fast measurements on larger samples of subjects (the question of sample size severly hampers the progress in the relatively still young science of modern biomechanics because of complexity and cost of measurement equipment, processing methods and modeling methodology).
- efforts to make this method less expensive in terms of both man-time and instrumentation will probably continue in the direction of designing a simpler version, giving fast feedback.
- the method is open to inclusion of new variables (myoelectric signals).
- results are obtained under standard conditions and can serve as a suitable starting point for building a data base compatible with future expert systems. (Similar approach can be pursued in other kinesiological activities, too.)

Tne method as such is applicable in the follow-up of training process in sport gymnastics, in estimation of stress and injurie prevention. Besides, it enables a better insight in fundamental mechanisms of visco-elastic behaviour of human locomotor system.

Acknowledgement

This investigation has been supported by the grant from the Republic´s Selfmenagement Community of Interest of Croatia in 1985.

REFERENCES

1. Komi, P.V.: Biomechanics and neuromuscular performance, Medicine and science in sports and exercise, 16, 1, 26-28, 1984.
2. Nashner, L.M.: Towards compatible physiological and bio-mechanical approaches to the study of movement, IX International Congress of Biomechanics, Waterloo 1983.
3. George, G.S.: Biomechanics of women gymnastics, Prentice Hall, Inc. Englewood Cliffs, N.J., 1980.
4. Medved, V., Wagner, I. and Živčić, K.: Statistical analysis and predictive value of take-off force vector in selected gymnastic skills (in preparation).
5. Medved, V., Wagner, I. and Živčić, K.: Towards objective skill level estimation in sport gymnastics (in preparation).
6. Crenna, P., Frigo, C.: Monitoring gait by a vector diagram technique in spastic patients, from: Clinical Neurophysiology in Spasticity, ed. P.J. Delwaide, R.R. Young, Elsevier Publ. BV 1985, 109-124.
7. Pedotti, A.: Functional evaluation and recovery in patients with motor disabilities, in: Use of computers in Aiding the Disabled, ed. J. Raviv, North-Holland Publ. Comp. IFIP-IMIA, 1982, 53-71.

BUCKLING OF AXIALLY COMPRESSED IMPERFECT CYLINDERS AND RING STIFFENED CYLINDERS UNDER EXTERNAL PRESSURE

Nicolas WAECKEL   EDF/SEPTEN - 12-14 av. Dutrievoz
                                    69628 VILLEURBANNE CEDEX
Jean-François JULLIEN   ) INSA de LYON - Bâtiment 304 - 20 av.
Alain KABORE   ) Albert Einstein - 69621 VILLEURBANNE
) CEDEX

ABSTRACT -

   The main purpose of this paper is to present the experimental device which has been developed at INSA LYON (Ref. 1), in order to study the buckling behaviour of cylindrical shells under axial loading or lateral pressure.

   Specific measurement systems are necessary for these types of tests. In particular, we describe an original application of the projected fringes method to the case of a complete cylindrical shell. Also experimental results are presented.

I - INTRODUCTION -

   Buckling of thin shells is a complicated phenomenon which has instigated many theoretical and experimental studies over the last forty years. Buckling concerns aeronautical industry, offshore and nuclear structures and civil engineering construction (tank, silos, etc ...).

   Our paper is limited to the cases of thin cylindrical shells under axial compression or lateral external pressure. The axial compression buckling is difficult to study because of the sensitivity of the phenomenon to several parameters such as axisymetry of the loading or initial imperfections. Contrary to the lateral pressure, axial compression leads to a violent and fast buckling which occurs in a few milliseconds.

   If we compare the many experiments which have been performed on axial compressed cylinders, we notice that the results are very scattered and that experimental loads are often very much inferior to the classical buckling load value given by :

$$P_{cl} = \frac{2\pi Ee^2}{\sqrt{3}(1-\nu^2)}$$

where E is the young modulus and e the thickness of the shell. The average of the results of the best experiments of the literature is equal to 0.63 Pce. The scatter of these results and the gap with theoretical values are due to a bad mastery of the initial imperfections map.

   In order to study the influence of intentional initial geometric imperfections on buckling behaviour we have developed a specific experimental device which is described in chapter II.

   In chapter III of this paper, we relate experimental results obtained on this device with thin cylindrical shells under axial compression or lateral external pressure.

II - EXPERIMENTAL SET-UP -

The testing device which has been specially conceived and set up for buckling experiments is described in figure 1 and 2. Its characteristics allow axial compression up to 10 tons,

In order to increase the frequency and the number of buckltests, we have chosen technical solutions which gain time without harming the quality mechanical clamps, automatic loading process and scanning system, etc...

II.1 - Specimens :

Cylindrical specimens (length L = 150 mm, internal radius R = 75 mm, thickness 0.120 < e < 0.200 mm) are obtained by electrodeposition of nickel on an aluminium or a centrifugated cast carbon-epoxy mandrel. The forms of the imperfections are previously, carefully machined on this mandrel by a numerically controlled milling machine. High quality specimens were obtained by this process : the shells present no lap joint and thickness variations are less than 3 %. The mechanical properties are homogeneous and similar to those of stainless steel at 400°C and are measured for each specimen tested.

The ring stiffened cylinders are also obtained by electrodeposition of nickel. However, the aluminium mandrel consists of N thin steel rings inserted between N + 1 coaxial aluminium ferrules. The cristalline structure of the electrodeposited nickel on steel being near epitaxic, the joint between the shells and the steel stiffeners is near-perfect.

Two types of internal ring stiffeners have been studied according to the width l of the stiffener which is equal to 2,5 mm (type I) or 1 mm (type II).

II.2 - Measurement of axial loading and displacements :

The special testing device consists of an hydraulic jack which imposes an axial displacement with a precision of ± 1 micron.

The axiality of the loading is maintened by means of an adjustable ball-joint screws system and three load cells B (NOVATECH UK). The central load cell A (STRAINSERT) gives the value of the axial loading effectively applied to the shell.

The axial displacement is measured by means of central LVDT C (SCHAE-VITZ ± 2,5 mm) and three peripherical LVDT'S D (SCHAEVITZ ± 2,5 mm).

The automatic control system of the testing apparatus uses all these data for piloting. Its time of response must be very short (less than one millisecond) in order to "follow" the evolution of the shell when buckling occurs.

Recording and processing of experimental data is made on an HP 9826 computer. The external lateral pressure is obtained by the means of a vaccum pump ; a force regulation system cancels the axial compression.

II.3 - Measurement of radial displacement :

The knowledge of the initial geometry of the test specimens is primordial for 2 buckling experiment and its numerical interpretation.

The automatic rotative scanning system described in figure 2 is connected to a computer and allows the recording and the transformation of the complete surface of the shell in Fourier's series.

The scan of the structure can be made at different phases of the experiment, during pre-or post-buckling stage.

The radial displacements are measured by means of a contact less capacitive transducer (see figure 2).

II.4 - Measurement of the lateral rigidity :

The contactless transducer for radial displacement measurement can be replaced by a special transducer conceived for measuring the évolution of the lateral rigidity of the shell interms of axial compression (see figure 2). Exploiting the measurements of this transducer allows us to anticipate good accuracy (about 5 %) of the critical buckling load of the shell without destroying it.

II.5 - Optical measurement of the geometry :

The traditional measurements (axial load, end shortening, local radial displacement, ...) which are made during the test are completed by a new optical measurement system based on the principle of projected fringes (réf. 1 to 3). This system allows the vizualisation of the surface of the entire specimen (see figure 3).

The optical device presents 8 fringes projectors which project vertical gratings on 8 eighths of the cylinder. The density of the projected gratings is 40 lines/mm. On the white painted shell, the pitch of each grating varies from 0.45 mm to 0.97 mm. The projected image of these 8 gratings is them reflected in four 45° inclined mirrors which allow the focusing of the virtual image in one point of the complete surface of the cylinder. The focus point is in the image plan of a camera whose role is to take a negative of the initial state of the shell. This negative called reference negative, is replaced in the image plan of the camera. When the surface of the cylinder deforms the projected gratings deform too, and their images are going to interfere with the initial grating image recorded on the reference negative.

The interfering fringes thus produced are representative of the radial displacements of the shell (see figure 4). It's possible to film the evolution of the geometry continuously. Neverthless during the few milliseconds that buckling is occuring, a fast-movie camera would be necessary.

The optical method is well suited to follow the pre-buckling growth of localized initial geometrical imperfections.

The simplified value of the radial deflection w at a point M (α) can be expressed by the classical formula :

$$w = \frac{np(\alpha)}{tg\theta(\alpha) + tg\beta(\alpha)}$$

where n is the fring order at point M (α)
        p is the pitch of the projected grating on the cylinder (if the pitch of the grating projected by the projector is constant, the pitch of the cylinder grating is variable with the angular position α of point M)
        θ and β are the angles of projection and observation of point M

The fringe order n is determined from the zones where radial deflection is zero (at the clamps for instance) or by comparison to LVDTS measurements.

The pitch p(α) is measured on a negative KODALITH roled up around the cylinder and then exposed to the projected grating. So p(α) is known with a precision fo a few microns. The projection angle $\theta(\alpha)$ is precisely measured for each projector by vizualising their pencils of rays using a smoke producing device.

The observation angle $\beta(\alpha)$ is neglected because the distance between the cylinder and image plan where fringes are produced is sufficiently big.

This optical method which gives information on radial deflection all around the shell at all the time presents a precision of about $1/10°$ of mm on a range of 7 or 8 mm. The processing of the interference fringes can be manual or automatic.

III - EXPERIMENTAL AND NUMERICAL RESULTS -

III.1 - Cylinders under axial compression :

The main purposes of the experimental program concerning cylinder under axial compression are :
- to produce thin cylindrical specimens with purely geometric imperfections,
- to study the critical role played by well defined imperfections of modal shapes developed from theoretical and experimental analysis of near-perfect cylinders.

A few local imperfections whose forms are representative of industrial defects observed on the thin structures of LMFBR, are also studied.
- to give complete and precise experimental data to validate recent elements calculations.
- to understand the buckling behaviour better just prior to instability.

Several types of imperfections with different amplitudes have been experimentally studied : a complete modal imperfection corresponding to the pattern of an Euler bifurcation mode, a local modal imperfection, an axisymetric modal imperfection, and industrial imperfections.

III.1.1 - Cylinders without intentional imperfections :

Experimental results from more than fifty perfect cylindrical specimens show that critical values (buckling stress σ_{cr}, circumferential mode n, buckling endshortening $\Delta\ell$) are very reproductible. Moreover, experimental critical stress values range from 84 to 88 % of the classical theoretical buckling stress values ($\sigma_{c\ell} = Ee/R\sqrt{3}(1-\nu^2)$. Subsequently, the tests on the near perfect specimens will be considered as reference-tests with respect to the specimens with intentional geometric initial imperfection.

The experimental buckling stress is slightly superior ot the limit of proportionality of the material ; subsequently, buckling occurs in elasto-plastic domain. Calculations show that elasto-plastic analysis leads to a good approximation of the experimental critical load (within 10 %). As the amplitude of the non intentional random imperfections are small (less than 20 % of thickness in average) an elasto-plastic calculation on the perfect geometry is sufficient to determine the critical load. Numerical circumferential buckling modes are always superior to experimental buckling modes which are between 10 and 12. This is due to the fact that during the very short period of time when buckling is occuring, the experimental circumferential wave number decreases.

III.1.2 - Cylinder with a complete periodic imperfection :

For 3 values of the amplitude of the imperfection we have compared the experimental results with those obtained on the reference near perfect cylinders : the reduction of the critical stress and the reduction of the initial stiffness are about 50 % (see figure 5).

The post buckling behaviour is stable which is amazing for an axial compressed cylinder. The buckling stress is lower than proportional limit of the material, so the buckling can be considered, in the aggregate, as elastic buckling. Neverthless during the buckling phase, successive local plastic hinges appear. These hinges are on a linte whose slope is about 45°.

One complete modal imperfect cylinder has been computed with diffe-rent 2D or 3D finite elements codes. The computations give good results of critical values providing that, in addition to the main imperfection cente-red on circumferential mode n = 16, another imperfection centered either buckling mode n = 8, or on mode n = 4.

III.1.3 - Cylinder with a localized axisymetric imperfection :

The experimental critical stress values are reduced from 40 to 50 % according to the amplitude of the imperfection (figure 5). The post-buck-ling behaviour is stable or unstable according to the amplitude. The buck-ling appears near the convex axisymmetric imperfection under the form of 11 or 12 circumferential waves.

III.1.4 - Conclusion :

The many tests performed in this study show that it's possible, on the one hand, to improve the quality of the tests and to reduce, in the case of near perfect cylinders, the gap between experimental and classical critical stresses values and, on the other hand, to reduce the scatter of experimental results and subsequently, to control the parameters influencing the critical behaviour of the structure more closely. Figure 5 shows that the very critical role played by the complete modale imperfection, the type of imperfection often arbitrarily taken into account in design calculations is experimentally confirmed.

The experiments validate present numerical methods, but also show the sensitivity of the results to various parameters.

III.2 - Cylinders under lateral pressure (4) -

The main purposes of the experimental program concerning ring stiffe-ned cylinders are :
- to produce near perfect thin ring stiffened cylinders (i.e without geome-trical imperfections and without residual stresses due to the joints between the ring stiffeners and the shell)
- to study the buckling behaviour to the shells according to the spacing or the inertia of the stiffeners
- to understand the pre and the post-buckling behaviour of the ring stiffe-ned shells better and to give complete and precise experimental data to validate recent finite element calculations.

III.2.1 - Experimental and numerical results :

The test results are presented on table 1. More than fifteen stiffe-ned or unstiffened cylinders have been tested. For each type of specimen, experimental results show that critical values (buckling stress σ_{cr}, cir-cumferential mode n), are very reproductible which confirms the validity of the experimental methodology.

The experimental critical stress values range from 90 to 100 % of the analytical or numerical buckling stress value (see table 1).

The analytical results are issues from the Von Mises theory by taking the distance between two stiffeners (75 mm or 50 mm or 50 mm) instead of the length of the cylinder from the discreet stiffened theory. The numerical results are obtained by using the finite elements method with the CASTEM system codes. As the buckling occurs in the linear elastic domain of the material, Euler buckling load calculations are sufficient.

III.2.2 - Conclusion :

The bucling of the cylindrical shells under lateral pressure leads to a stable post-buckling behaviour with a post-buckling bifurcation load 2 or 3 times higher than the buckling bifurcation load. The load supported by the structure increases until the implosion occurs.

Experimental and numerical buckling patterns are either localized or generated according to the inertia of the stiffeners (see figure 6).

Experiments show that the circumferential buckling mode number n is generated by the half wave length $\lambda = \pi/n$ of the initial local geometrical imperfection which has the largest relative amplitude. Consequently, in usual design calculation methods, taking into account, the modal defect which is characterised (amplitude and period) by the largest measured local imperfection, seems reasonable.

ACKNOWLEDGEMENTS -

The studies presented in this paper were financed by the "Bureau de Contrôle des Constructions Nucléaires (BCCN)" and by "Electricité de France DER". The work was carried out in the laboratory of "Bétons et Structures" of the "Institut National des Sciences Appliquées de LYON" France.

REFERENCES -

(1) - N. WAECKEL - "Instabilité de structures minces comportant des imperfections géométriques initiales" - Thèse de Doctorat d'Etat ès Sciences - Université LYON 1 / INSA - mars 1984

(2) - F.K. LIGTENBERG - "The moiré method : a new experimental method for the determination of moments in small slab models" - Proc. SESA Vol. WII n° 2 pp. 83-98 - 1954

(3) - B. DESSUS, M. LEBLANC - "The fringe method and its application to the measurement of deformations, vibrations, contour lines and differences of objects" - Opto-electronics 5 pp. 369-391

(4) - A. KABORE - "Flambage de coques raidies sous pression externe" Thèse de Docteur-Ingénieur, Université LYON 1 / INSA - Avril 1985

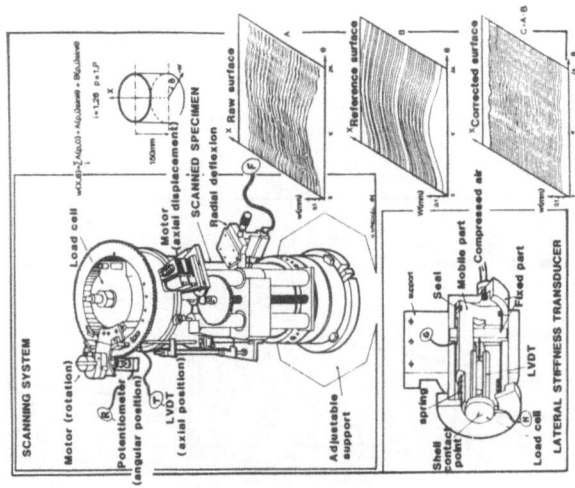

FIG 2

SCANNING SYSTEM

Motor (rotation)

Load cell

Motor (axial displacement)

Radial deflexion

SCANNED SPECIMEN

X Raw surface

X Reference surface

X Corrected surface

Potentiometer (angular position)

LVDT (axial position)

Adjustable support

LATERAL STIFFNESS TRANSDUCER

Mobile part

Compressed air

Fixed part

LVDT

Seal

support

spring

Shell contact point

Load cell

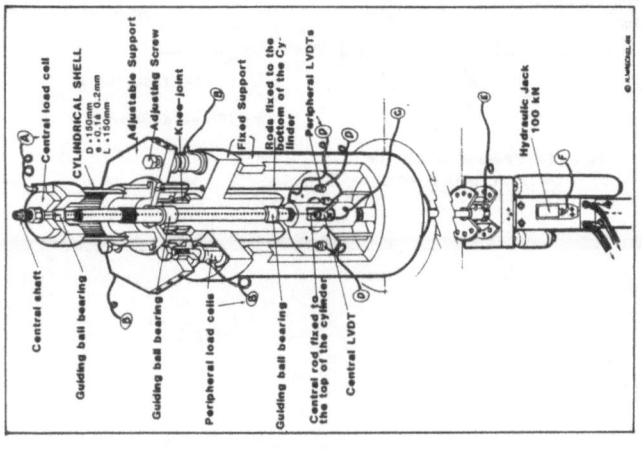

FIG 1   EXPERIMENTAL SET-UP

Central load cell

CYLINDRICAL SHELL
D = 150mm
e = 0.1 è 0.2mm
L = 150mm

Adjustable Support

Adjusting Screw

Knee-joint

Fixed Support

Rods fixed to the bottom of the Cy-linder

Peripheral LVDTs

Central shaft

Guiding ball bearing

Guiding ball bearing

Peripheral load cells

Guiding ball bearing

Central rod fixed to the top of the cylinder

Central LVDT

Hydraulic Jack
100 kN

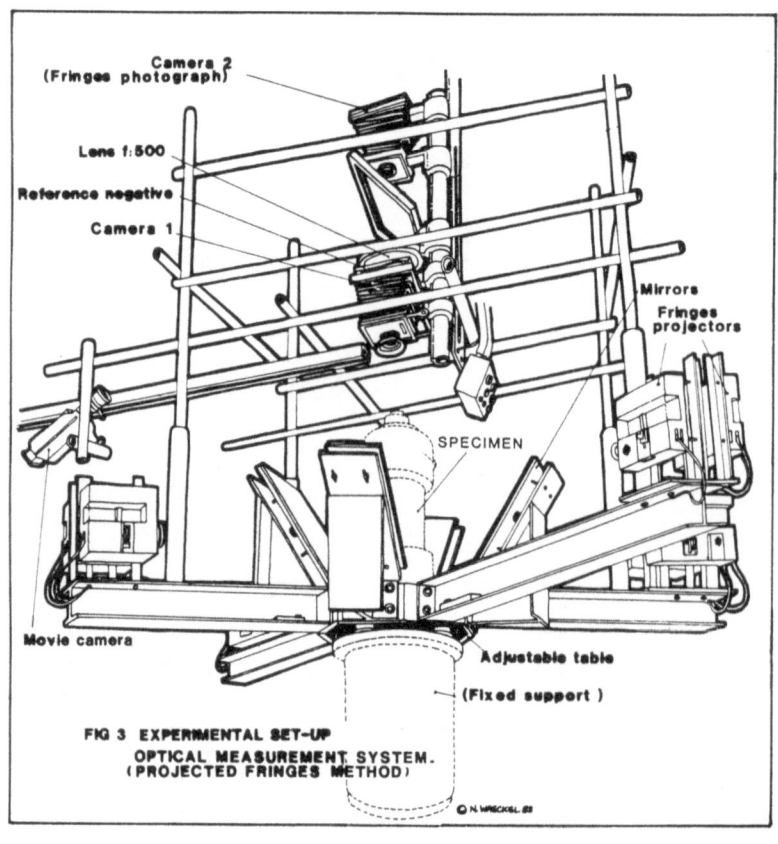

Camera 2
(Fringes photograph)

Lens f:500

Reference negative

Camera 1

Mirrors

Fringes projectors

SPECIMEN

Movie camera

Adjustable table

(Fixed support)

FIG 3 EXPERIMENTAL SET-UP
OPTICAL MEASUREMENT SYSTEM.
(PROJECTED FRINGES METHOD)

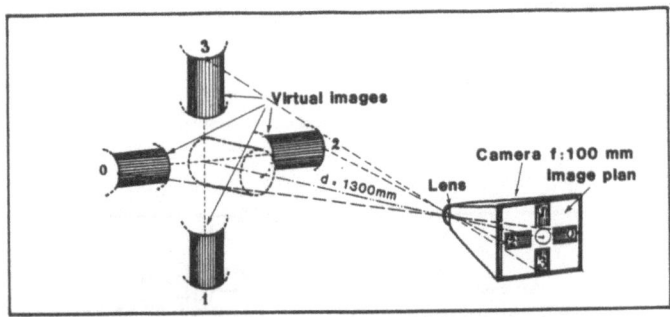

Virtual images

Camera f:100 mm
Image plan

d: 1300mm   Lens

Elastic fem calculation

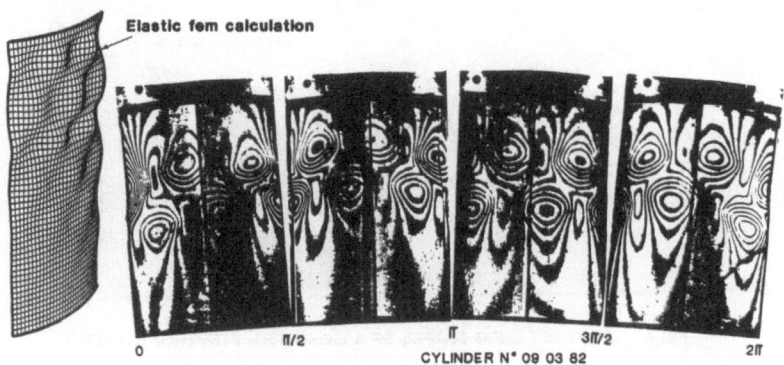

0   Π/2   Π   3Π/2   2Π

CYLINDER N° 09 03 82

FIG 4   EXPERIMENTAL RESULTS - BUCKLING SHAPE

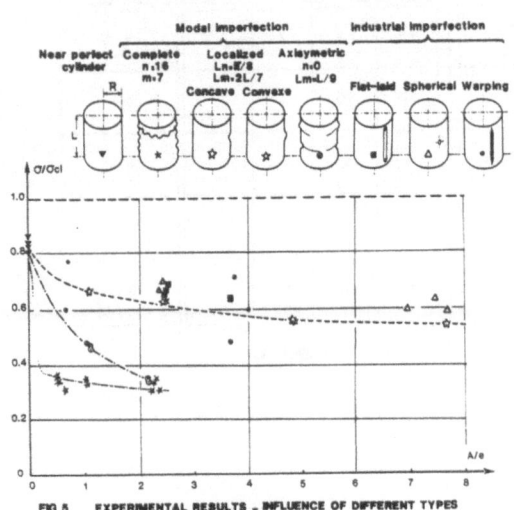

FIG 5   EXPERIMENTAL RESULTS - INFLUENCE OF DIFFERENT TYPES
OF INITIAL IMPERFECTIONS ON THE BUCKLING LOAD
OF AXIALLY COMPRESSED CYLINDERS.

405 < R/e < 450

132

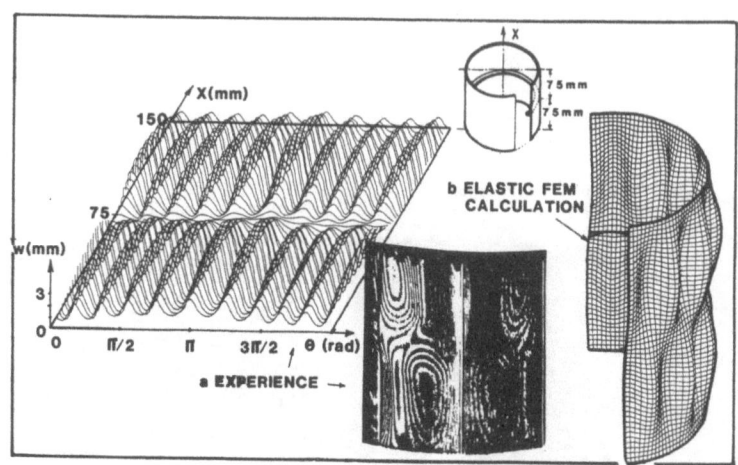

FIG 6   POST BUCKLING SHAPES OF A MONO-RING STIFFENED CYLINDER

	n_{exp}	$n_{fem\ calc}$	m_{exp}	$m_{fem\ calc}$	$\dfrac{\sigma_{euler} - \sigma_{exp}}{\sigma_{exp}}$
TYPE I	12	13	2	2	5 %
TYPE I	15	15	3	3	4 %
TYPE II	8	9	1	1	8 %
TYPE II	8	9	1	1	8 %
	8	10	1	1	5 %

TABLE 1

STRESS ANALYSIS OF THREADED CONNECTIONS SECURED BY LOCKNUTS

I.M. ALLISON

Mechanical Engineering Department,
University of Surrey, United Kingdom.

1. INTRODUCTION

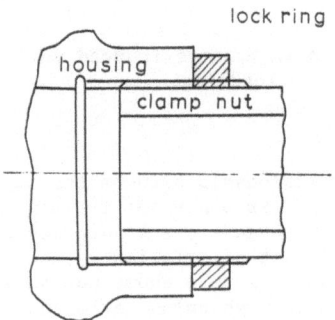

Figure 1   Threaded connection secured by locknut.

The primary cause of a major structural failure was attributed to the loosening of a threaded connection secured by a lockring.   Investigation of this problem has involved both experimental and analytical studies of the way in which external load is transferred through a thread previously preloaded by tightening a lockring with a prescribed torque.

Details of a typical joint are shown in figure 1. Tightening the locking ring produces a compressive force at the interface with the housing, a corresponding tension in the clamp nut, and a contact load distribution along the thread flank which is a function of the flexibility of the mating components. When additional load is applied the preloaded assembly acts as a monolithic structure and additional contact loads are developed on the thread flanks. Depending upon the direction of the external load this will be opposite in sign to the initial flank load induced in either the housing or the lock ring. The purpose of the present study was first, to develop a procedure for determining the thread flank contact load and secondly, to relate the local contact load to the peak stress in the thread root fillet radius.

134

2. INTERFACE LOAD – TIGHTENING TORQUE RELATIONSHIP

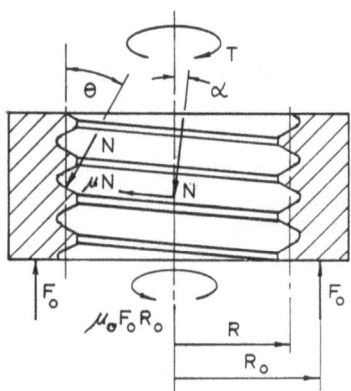

Figure 2   Forces and torsional moments induced during tightening of
lockring.

The generally accepted relationship between the applied torque and
resultant preload can be employed in analysing the behaviour of a wide
variety of joint arrangements with sufficient accuracy for design purposes,
provided an appropriate coefficient of friction is introduced.   Treating
the thread as an inclined plane, figure 2 shows how an analysis involving
simple statics gives the required tightening torque as:

$$T = F_o R \left\{ \mu_o \frac{R_o}{R} + \mu \sec\theta + \tan\alpha \right\} \qquad (1)$$

where

F_o load applied to housing

R thread contact radius

R_o effective radius of lockring – housing interface

μ coefficient of friction at thread contact

μ_o coefficient of friction at lockring – housng interface

α thread helix angle

θ thread flank angle

3. LOAD DISTRIBUTION IN A TYPICAL THREAD

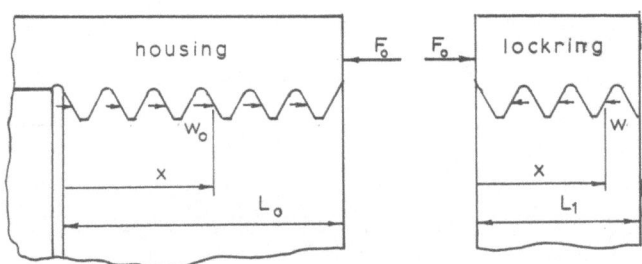

Figure 3 Notation for thread loads induced during tightening of lockring.

Conventional methods of analysis, which take no account of thread flexibility are unsuitable for calculating the flank load distribution and critical stresses in threaded components. However, an elastic analysis which includes the effects of both the thread geometry and the overall component stiffness predicts ω_o , the axial load per unit length along the contact length of the housing thread, as illustrated in figure 3, by;

$$\omega_o = F_o \sqrt{\frac{A+A_o}{2kEAA_o} \tan\alpha} \ \frac{\cosh mx}{\sinh mL_o} \qquad (2)$$

where:

A_o cross sectional area of housing

A cross sectional area of clamp nut

α thread helix angle

L_o thread engagement length

k transverse stiffness of thread form

$$m^2 = \frac{A+A_o}{2kEAA_o \tan\alpha}$$

E elastic modulus of thread material

It is clear that a similar expression may be derived for the locknut in which case, for preloading the lockring against the housing, the contact load ω_1 will be in the opposite direction to ω_o and will be given by:

$$\omega_1 = -F_o \sqrt{\frac{A+A_1}{2kEAA_1} \tan\alpha} \left(\frac{\cosh m_1 x}{\tanh m_1 L_1} - \sinh m_1 x\right) \quad (3)$$

136

where

A_1 cross sectional area of lockring

$$m_1^2 = \frac{A + A_1}{2kEAA_1 \tan \alpha}$$

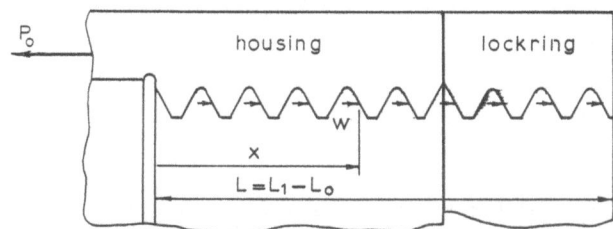

Figure 4   Notation for thread loads induced during external loading of threaded assembly.

Once the initial assembly has been made the combination of the locknut and housing shown in Figure 4 can be treated as an integral unit, and the thread contact load distribution due to an additional external load evaluated by simply substituting the appropriate geometric parameters into equation (5) which becomes

$$\omega = P_0 \sqrt{\frac{A + A_0}{2kEAA_0} \tan \alpha} \left(\frac{\cosh mx}{\tanh mL} - \sinh mx \right) \quad (4)$$

where

P_0 external load applied to clamp nut

$L = L_0 + L_1$ engaged length of clamp nut thread.

It follows that distributions of the contact load can be plotted along the engaged length of the thread for both the initial assembly condition and the addition of external load.

4. THREAD LOAD DISTRIBUTION IN TYPICAL JOINT.

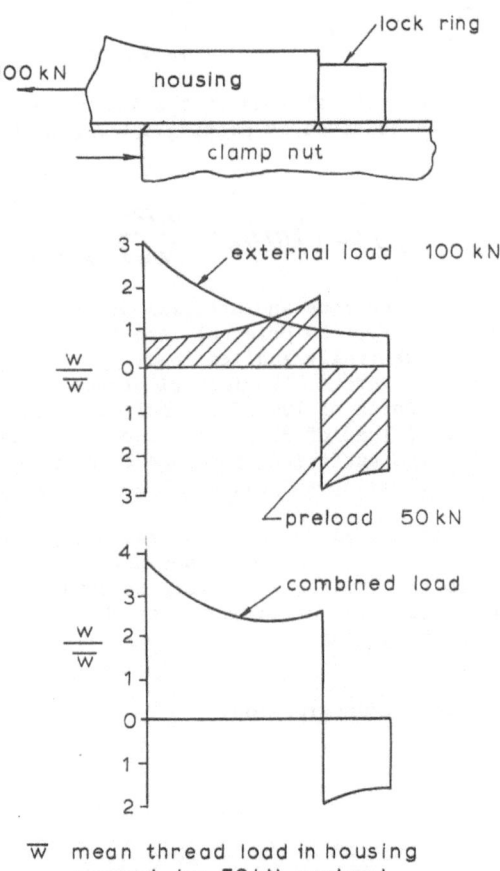

Figure 5   Contact load distribution along thread flanks.

Details of the thread load distributions for a particular example are
given in figure 5.   It should be noted that in this particular example the
cross sections of the housing and lockring are similar, and a sufficiently
accurate estimate of the thread contact load distribution due to the
external force can be obtained by assuming the housing cross section
applies over the full engaged length of the clamp nut thread.   At the
expense of deriving a more complicated relationship the analysis can be
extended to include variations in cross section when this is required.

138

As a direct consequence of this analysis the axial load at a distance from the loaded end of the clamp nut is given by the expression

$$P = P_o \left\{ \cosh mx - \frac{\sinh mx}{\tanh mL} \right\} \tag{5}$$

and it follows that the residual load at the interface between the housing and the locknut will be obtained by putting $x = L_o$ in the above expression and writing

$$P_I = F_o - P_o \left\{ \cosh mL_o - \frac{\sinh mL_o}{\tanh mL} \right\} \tag{6}$$

Substituting the dimensions and material properties used in this example it is found that F_o, the original interface compression, will be reduced by 15% of the externally applied load, P_o. For the joint configuration shown in figure 4, the initial preload of 50 kN will be reduced to 35kN when a force of 100 kN is applied to the clamp nut. It follows that an external load of 330kN will need to be applied before the joint deformation becomes sufficient to cause loss of contact at the interface between the lockring and the housing.

For a correctly assembled joint it is reassuring to note that the addition of a substantial external load does not cause a material reduction in the preload across the housing – lockring interface. Thus this simple analysis explains the security of threaded assemblies which include a properly tightened locking ring.

5. THREAD FILLET STRESS CONCENTRATIONS

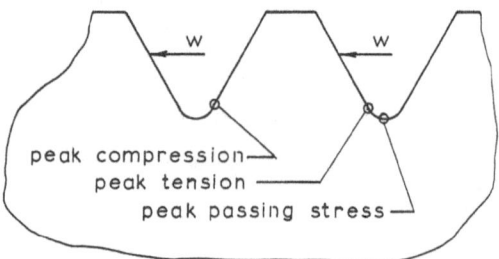

Figure 6   Stress concentrations in thread fillet radius.

Examination of the isochromatic patterns in slices cut from a stress frozen three dimensional photoelastic model of a preloaded threaded connection indicates that the stress distribution in the thread fillets is both complex and very sensitive to the details of the contact loading condition on the thread flank. Figure 6 illustrates how independent local stress concentrations are induced in the thread fillet by both thread bending and the passing load. For preloading of the lockring configuration used in this example, it is found that the effect of thread bending predominates, and it follows that the critical stresses in the thread fillet are likely to be related to the local flank load. Assuming this acts at the effective diameter of the thread it is possible to define a nominal bending stress at the root fillet by:

$$\sigma = \frac{3\sqrt{3}\ \omega_o}{2\rho} \qquad (7)$$

Since the analytical solution expresses ω_o in terms of the interface compression between the lockring and housing it is convenient to non-dimensionalise the fillet bending stress in terms of τ , the mean shear stress on the clamp nut thread so that

$$\tau = \frac{F_o}{\pi D_c L_I}$$

where D_c core diameter of the clamp nut.

Figure 7 shows a comparison between the analytical values of σ/τ and the corresponding non-dimensional fillet stresses measured in the photoelastic model clamp nut. In general the fillet stresses predicted by the simple theory compare favourably with the measurements made on the photoelastic model. However the experimental observations exhibit substantial scatter, and this appears to be associated with the nature of the local thread contact conditions which vary along the thread contact length. More importantly significant differences are associated with the thread run out which causes increased fillet stresses to be induced away from the end faces of the component. It follows that the location of the critical fillet stress will not be predicted correctly although the magnitude is sufficiently close for practical design purposes.

Clearly when a substantial external load is involved, the effects of thread bending and the local passing load at the loaded end of the assembly could interact, and the effect of thread run out could become of prime importance in preducting the fatigue life of the joint. Further tests are planned to investigate this aspect of preloaded threads.

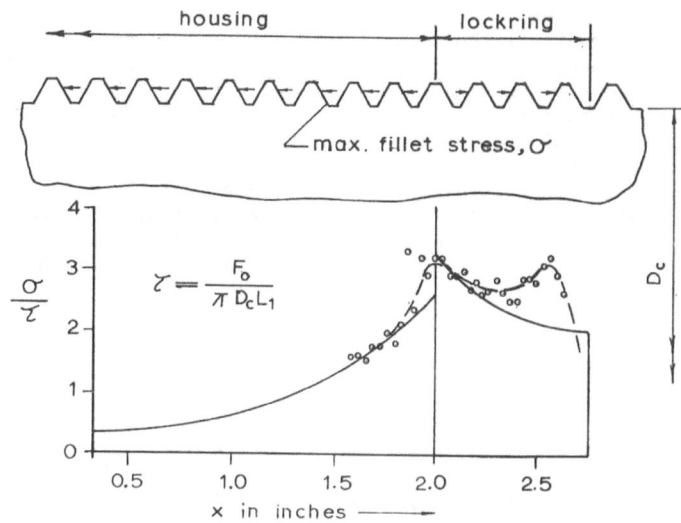

calculated nominal value
photoelastic data

Figure 7   Fillet stress distribution in clamp nut.

6. CONCLUSIONS

1. In a threaded connection the variation in thread flank load is highly dependent upon the joint geometry, particularly the relative stiffness of the component body and the thread form.

2. For a given tightening torque the preload and hence maximum stress are a function of the friction in the thread and at the nut interface.

3. The maximum thread fillet stresses, and hence mode of failure, are a complex function of the shank torque and tension, and of the local bending of the thread form.

4. The way in which the flank load is distributed along the thread length makes the locknut an effective form of fastener, which is insensitive to the application of external load.

5. A simple method has been proposed for calculating the thread flank load distribution and hence the critical fillet stresses.

6. Further studies will be required to confirm the effectiveness of this solution although it has been shown to provide an acceptable estimate of the maximum stresses in a particular example.

7. The results of tests undertaken on a three dimensional photoelastic model suggest that further investigation of the effect of thread run-out is required.

REDESIGN AND DEVELOPMENT OF A FORCE TRANSDUCER OF THE BUCKLE TYPE

G.W.M. Peters[*], A.A.H.J Sauren, H.v. Mameren[*]

  Eindhoven University of Technology,
  P.O. Box 513, 5600 MB Eindhoven, The Netherlands.
* Rijks-University Limburg, Maastricht, The Netherlands

ABSTRACT

A redesign of the so-called buckle transducer is presented that is used for
the local measurement of static tensile forces in membrane-like connective
tissue structures around the elbow joint. This new design involves size
reduction and elimination of false output sources as well as the development
of a special tool for the proper installation of the transducer.
A mechanical model of the buckle transducer has been developed to
investigate the importance of various design parameters. This model has
proven to be an important tool for optimizing the design. Test measurements
show the experimentally determined transducer characteristics to be in good
agreement with model predictions and the transducer to come up to
expectations.

INTRODUCTION

The so-called buckle transducer is a suitable and, to the best of our knowledge, only means for direct measurement of static tensile forces in rope or
membrane-like structures. Direct measurement of forces is an important advantage because the mechanical properties of the material under consideration need not to be known. The buckle transducer was introduced by Salmons[1] and used among others by Barnes and Pinder[2], Walmsley et al.[3],
and Lewis and Frasier[4]. Lewis et al.[5] made design improvements and used
the buckle transducer for in vitro force measurements in knee ligaments.

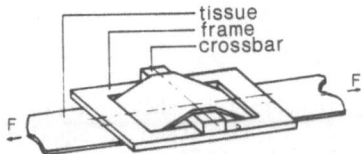

Fig. 1. Basic principle of the buckle transducer

The buckle transducer basically consists of a frame and a removable
crossbar, the crossbar being much stiffer than the frame. As shown in Fig. 1
a strip of tissue is led through the frame and over the crossbar. Exertion
of a tensile force (F) on the strip will cause the frame to deform. This
deformation results in a straingage voltage output that is directly related
to the tensile force, thus providing a means for direct force measurement.
From some pilot studies we concluded that none of the earlier designs of the
buckle transducer were suitable for our purpose: the in vitro measurement of

Wieringa, H (ed), Experimental Stress Analysis.
© *1986. Martinus Nijhoff Publishers, Dordrecht.*

tensile force distributions in membrane-like connective tissue structures
around the elbow joint. Apart from their relatively large size these designs
are prevented from a successful application towards our objective by three
major disadvantages, affecting measurement accuracy and reproducibility.
First, as the frame is considerably wider than the tissue strip led through
(Fig. 2a), deformation of the buckle will result not only from interaction
with this strip but also from an unknown interaction with the surrounding
material

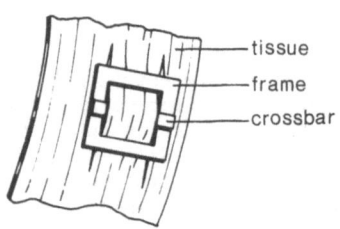
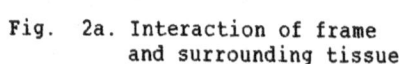
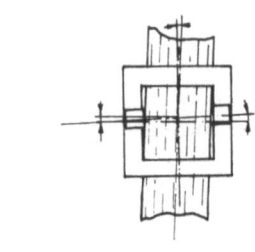

Fig. 2a. Interaction of frame   Fig. 2b. Alignment errors
          and surrounding tissue

Secondly, the actual mutual alignment of crossbar, frame, and tissue strip
may deviate considerably from the desired configuration because their relative positions are not uniquely defined (Fig.2b). Finally, the transducer's
characteristics appeared to show significant hysteresis.
Starting from these experiences and with the aid of a specially developed
mechanical model we arrived at a new transducer design. Because of both the
small size of the new transducer and some stringent prescriptions for its
installation a special installation tool was designed as well. The
transducer design and the installation tool are presented in the next
section. In the subsequent two sections attention is paid to the formulation
of the mechanical model of the buckle transducer and its application for
design optimization, especially with regard to the reduction of hysteresis
phenomena. Finally some test results are presented.

REDESIGN OF THE BUCKLE TRANSDUCER

The redesign of the buckle transducer is shown in Fig. 3 (the dimensions are
0.75 x 1.5 x 5 [mm]). The width of the frame is equal to the width of the
tissue strip, so that possible interaction with the surrounding tissue is
minimized. Guides provide proper alignment of the transducer on the
strip. When installing the transducer the ends of the (disposable) U-formed
crossbar are bent over the frame (see Fig. 3) with the aid of a specially
developed installation tool (Fig. 4).
The minimum dimensions of the frame are limited by the applied straingage
(Micro Measurements SA-09-030CG-120)) that is cut to its minimum dimensions
(1.4 x 2.1 [mm]). Mounting of the straingage is done with a special tool
because of the curved plane on which it is bonded. A triple layered coating-
system is used (CETA BEVER Two-part Epoxy bond and Micro Measurement M-Coat
D and M-Coat G) for electrical isolation of the straingage. The duration of

life of the isolation is limited in situations where physiological solutions are used, but it is rather easy to apply a new coating.

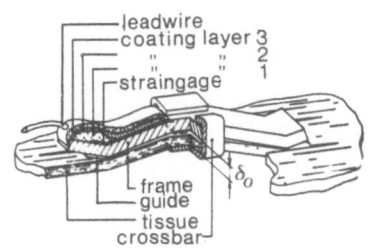

Fig. 3. The installed buckle transducer.

Frame and guides are made of stainless steel while the crossbar is made of Molybdenum. Because of the relatively high Young's modulus of Molybdenum the crossbar combines a high stiffness with minimum dimensions. The transducer is designed for a maximum load of 10[N].

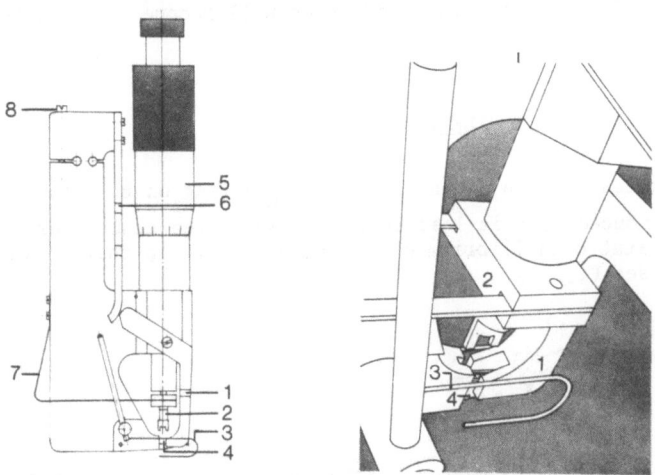

1. Clamp.
2. Crossbar bending stamp.
3. Auxiliary clamp springs.
4. Crossbar bearer.
5. Micrometer spindle.
6. Spring for clamping.
7. Spring for bending stamp positioning.
8. Spring-force adjuster.

Fig. 4. Installation tool (left) and detail (right) of the part where the transducer is clamped and the crossbar is bent over the frame

144

In Fig. 5 the situation is illustrated in which frame, crossbar and the tissue strip are brought together in the installation tool. With the aid of the micrometer spindle the bending stamp is moved down so that the ends of the

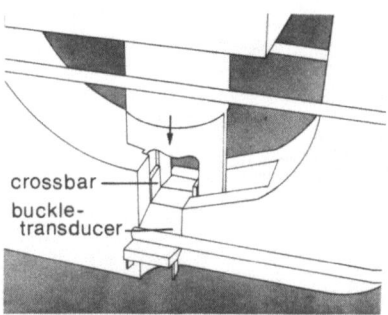

crossbar —
buckle-
transducer —

Fig. 5. Installation tool with buckle transducer just before the crossbar ends are bent over the frame

crossbar are bent over the frame.The micrometer is also used to adjust the distance δ_0 (Fig. 3), the importance of which will become clear from the next sections.

A MECHANICAL MODEL OF THE BUCKLE TRANSDUCER

To derive a relation between a tensile force $\vec{F}_0$ acting on the tissue strip and the strain induced by this load in the frame of the transducer, we consider the quasi-static equilibrium of a part of the strip that is in contact with the frame (see Fig. 6).

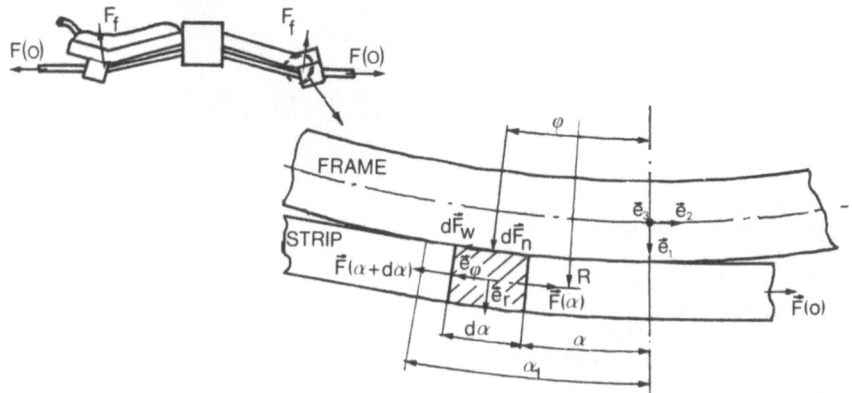

Fig. 6. Forces acting on an infinitesimal part of the strip

It should be noticed that the tissue strip slips over the frame when it is loaded (and slips back in case of unloading). Friction between strip and frame will be taken into account supposing the Coulomb friction model is valid here (which is a simplification). From equilibrium of the forces acting on an infinitesimal part of the strip of tissue it follows that:

$$\vec{F}(\alpha) + \vec{F}(\alpha+d\alpha) + d\vec{F}_w + d\vec{F}_n = \vec{0} \tag{1}$$

On the assumption that full slip occurs the following expression may be derived for the force $\vec{F}_f$ resulting from integration of $d\vec{F}_w + d\vec{F}_n$ and acting on the frame:

$$\vec{F}_f = -F(0)[\exp(\mp\mu\alpha_1) \sin(\alpha_1) \vec{e}_1 - (\exp(\mp\mu\alpha) \cos(\alpha_1) + 1) \vec{e}_2] \tag{2}$$

where μ is the coefficient of friction and the plus and minus signs in front of the powers of the exponential functions apply for the cases with increasing and decreasing load $\vec{F}(0)$, respectively. The unit vectors $\vec{e}_1$, $\vec{e}_2$, and $\vec{e}_3$ constitute an orthonormal vector base. Because the buckle transducer is symmetric it suffices to model one half of it as a cantilever beam (Fig. 7).

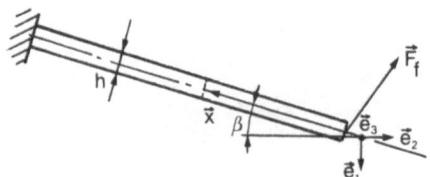

Fig. 7. Reduced model of the frame

Neglecting shear forces, it is easily derived that the maximum principal strain in a cross-section at a position $\vec{x}$ is given by:

$$\epsilon_x = - \frac{F(0)}{AE}[(\frac{6x}{h} \sin(\alpha_1-\beta) + \cos(\alpha_1-\beta)) \exp(\mp\mu\alpha) +$$

$$\frac{6x}{h}\sin(\beta) - \cos(\beta)] \tag{3}$$

with:
        E: Young's modulus and
        A: cross-sectional area.

In equation (3) the angle α_1 has not a constant value but is a function of the load $\vec{F}(0)$ and the momentary thickness d of the tissue strip on which $\vec{F}(0)$ is acting. This means that the load-strain relationship is non-linear.

Writing the deflection due to bending of the frame as $\vec{u}_1 = u_1\vec{n}_1$ with

$\vec{n}_1 = -\cos(\beta)\vec{e}_1 + \sin(\beta)\vec{e}_2$ and the deflection due to bending of the crossbar as

$\vec{u}_2 = u_2\vec{n}_2$ with $\vec{n}_2 = \vec{e}_2$, it follows from Fig. 8 that:

$$\tan(\alpha_1) = \frac{\delta_0 + d - u_1\cos(\beta) - u_2}{L_2 + u_1\sin(\beta)}.$$ (4)

The deflection u_1 is given by:

$$u_1 = \frac{(\vec{F}_f \cdot \vec{n}_1)L_1}{3EI_1}$$ (5)

with I_1 for the moment of inertia of the cross-sectional area of the frame. It can be proven that the deflection u_2 in (4) is negligibly small compared with u_1.

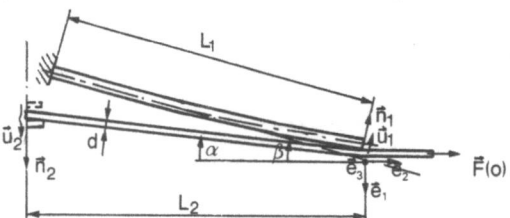

Fig. 8. Deflections of frame and crossbar

The thickness of the loaded tissue strip is in general a function of the load $\vec{F}(0)$ and must normally be specified for each type of material. From equations (4), (5) and (2) it is not possible to derive an explicit expression for the angle α_1 as a function of the load $\vec{F}(0)$. Therefore, for a known load $\vec{F}(0)$, the load $\vec{F}_f$ on the frame and the angle α_1 have to be determined by means of a numerical method. Subsequently, the strain ε_x can be calculated from equation (5).

HYSTERESIS AND OPTIMIZATION

As can be seen from relation (3) a part of the strain in the frame, which is a measure for the load $\vec{F}(0)$ on the tissue strip, depends on whether $\vec{F}(0)$ on the tissue strip is increasing or decreasing. Th'⁻ means that not only the

value of the measured strain should be known for the determination of the
load but also whether the strain is increasing or decreasing. Besides,
relation (3) is based on the assumption that full slip occurs in the
contact-area of frame and tissue strip. Reversal of the load introduces a
certain trajectory in the load-strain relationship (3) during which there is
no or incomplete slip (see Fig. 10). In that case equation (3) no longer
holds and the load-strain relation is unknown.

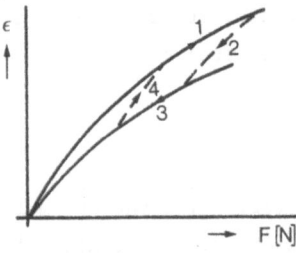

1: increasing load $\overset{\star}{F} > 0$, full slip

2: decreasing load $\overset{\star}{F} < 0$, incomplete slip

3: decreasing load $\overset{\star}{F} < 0$, full slip

4: increasing load $\overset{\star}{F} > 0$, incomplete slip

with $\overset{\star}{F} = \dfrac{dF}{dt}$

Fig. 9. Load-strain relation with hysteresis.

It is desirable that the term in equation (3), accounting for load increase
or decrease, equals zero or, at least, is very small compared with the other
term (we suppose $\mu = 0$ is not possible; in that case the problem would not
occur). In the optimal situation we have:

$$[\frac{6x}{h} \sin(\alpha_1 - \beta) + \cos(\alpha_1 - \beta)] = 0 \tag{6}$$

It should be noticed that in this case equation (3) represents a linear
relation between load and strain. From equation (6) it follows that, for
given x, h and β, the angle α_1 has to satisfy

$$\alpha_1 = \beta - \arctan(\frac{h}{6x}) \tag{7}$$

As β, h and x in equation (7) have fixed values for a given buckle
transducer and the angle α_1 is a function of the load $\overset{\star}{F}(0)$, there exists
only one specific load value for which equation (7) is satisfied. Thus
always some hysteresis will be present. To minimize the hysteresis a
quantity H is defined according to:

$$H = A_h / A_0 \tag{8}$$

with

$$A_h = \int_0^{F(0)_{max}} [\varepsilon_x(\mu \neq 0, \overset{\star}{F} > 0) - \varepsilon_x(\mu \neq 0, \overset{\star}{F} < 0)] dF(0) \tag{9}$$

$$A_0 = \int_0^{F(0)_{max}} \varepsilon_x(\mu=0)dF(0) \tag{10}$$

In words, H is the quotient of the area of the hysteresis loop and the area under the load-strain curve in case there is no friction. A buckle transducer is considered to be optimal when H is minimal. For a given thickness of the tissue strip and a given buckle transducer one of the dimensions δ_0, L_2 or β (see Fig. 8) can be varied. For reasons of simplicity δ_0 is chosen as the adjustable quantity. H is minimal for

$$\frac{dH}{d\delta_0} = 0 \tag{11}$$

It is not possible to derive an explicit expression for the optimal δ_0 but again it is easy to find an approximation for δ_0 where H is minimal by means of a numerical method.
A adventitious important advantage of, what is called an optimal buckle transducer, is the linear relation between load and strain as can be seen from equation (3) if equation (6) is satisfied

TEST RESULTS AND DISCUSSION

The new designed buckle transducer was subjected to three types of tests. All the measurements were made on a linen rope (d = 0.25 [mm]), so test conditions could easily be controlled and specimens were unlimited available.
First, investigation was made into hysteresis phenomena as observed in the pilot experiments and the, with the aid of the mechanical model predicted, possibility to minimize these phenomena, by adjusting the distance δ_0 (see Fig. 3). Fig. 10 shows the measured load-strain curves for two different values of δ_0. The hysteresis is demonstrated (curve 1) and, indeed, it is possible to minimize it (curve 2). It should be noticed that for $\delta_0 < (\delta_0)_{optimal}$ the hysteresis loop is passed through in the clockwise direction and for $\delta_0 > (\delta_0)_{optimal}$ in the reverse direction.

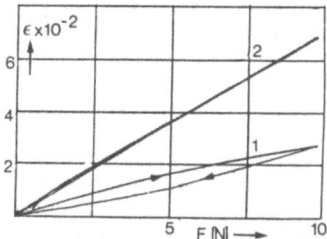

Fig. 10. Measured load-strain curves.

As stated before there exists only one specific load value for which the hysteresis term in equation (3) vanishes. So a little amount of hysteresis will always be left as can be seen in Fig. 10 (curve 2). The almost linear behaviour of the transducer in the optimal situation is also reflected by curve 2 of Fig. 10. Theoretical results have been left out here as the mechanical model is used for a qualitative analysis of the transducer's behaviour. Besides, a quantitative comparison of experimental and theoretical results would require an accurate knowledge of all model parameter values. Each transducer has to be calibrated separately.

Secondly, the interfering input caused by temperature changes, one of the most occurring error sources, was tested. Compensation of temperature effects is, as usual, done with a so-called "dummy transducer" placed so as to assume the same temperature as the active transducer. Long duration measurements (2000 samples over 8 hours) demonstrated an average zero drift of less than 0.3% ($\hat{=}$ 0.03 [N]) of the full scale with a standard deviation better than 0.4% ($\hat{=}$ 0.04 [N]) of the full scale.

Finally accuracy and sensitivity were determined. Five-hundred random load values, which are assumed to be errorless, were chosen within the range of operation (0-10 [N]). The maximum deviation of a straight line (least squares fit) of the corresponding output samples was less than 2,5% of full scale.

As can be seen from equations (2) and (3) the sensitivity of the transducer depends among other things on the distance δ_0 between frame and crossbar.

Thus, the accuracy of crossbar installation determines not only how well the optimal situation, in terms of hysteresis, is achieved but also the accuracy of the sensitivity belonging to that optimal situation. A deviation of the desired sensitivity is a modifying input (see [6]). In the test situation installation-accuracy was better than 0,03 [mm]. From the mechanical model, the deviation of the sensitivity corresponding to the installation accuracy of 0,03 [mm] is calculated to be less than 4%($\hat{=}$ 2.6[N/μstrain]) of the nominal sensitivity. Measured sensitivities of repeated transducer installations for two different transducers are given in Table 1. The maximum difference between the measured sensitivities agrees well with the predicted maximum range.

	Transducer 1.	Transducer 2.
Installation 1	64.5	60
'' 2	63.5	63.5
'' 3	67	58
'' 4	-	58

Table 1: Measured sensitivities and the calculated sensitivity for a given transducer.

CONCLUDING REMARKS

The buckle transducer is a suitable and valuable tool for direct static tensile force measurements in rope- or membrane-like structures, specially when the mechanical properties of the material under consideration are

150

unknown. With the reduced dimensions, the improvements concerning the proper mutual alignment of frame, crossbar and tissue strip and the controlled way of installation, the new buckle transducer has shown to be fairly accurate taking into account the rather complex situations of its application: local force measurement in soft (biological) tissue.
It may be clear that for rational and optimal design a mechanical model is indispensable. Even though it is based on some coarse simplifications the model gives a good insight into the behaviour of the transducer and leads to solutions for established problems.

REFERENCES

[1] Salmons, S., "The 8th International Conference on Medical and Biological Engineering-Meeting Report," Bio-medical Engineering, Vol. 4, 1969, pp.467-474.

[2] Barnes, G.R.G., and Pinder, D.N., "In Vivo Tendon Tensions and Bone Strain Measurements and Correlation," Journal of Biomechanics, Vol. 7, 1974, pp. 35-42.

[3] Walmsley, B., Hodgson, J.A., and Burke, R.E., "Force Produced by Medial Gastrocnemius and Soleus Muscles During Locomotion in Freely Moving Cats," Journal of Neurophysiology, Vol. 41, 1978, pp. 1203-1216.

[4] Lewis, J.L., and Frasier,G., "On the Use of Buckle Transducers to Measure Knee Ligament Forces," Proceedings of the 1979 ASME Biomechanics Symposium, 1979, pp. 71-74.

[5] Lewis, J.L., Lew, W.D., and Schmidt, J., "A Note on the Application and Evaluation of the Buckle Transducer for Knee Ligament Force Measurement," Journal of Biomechanical Engineering, Vol. 104, 1982, pp. 125-128.

[6] Doeblin, E.O., "Measurement Systems, Application and Design," Third Edition, 1983, Mc Graw-Hill International Book Company.

DETERMINATION ON FORCES AND MOMENTS IN PIPE CROSS-SECTIONS : FITTING OF EXPERIMENTAL STRAIN MEASUREMENTS ON A MECHANICAL ANALYTICAL MODEL

A. CAUSSE - J.L. TROLLE

ELECTRICITE DE FRANCE - Direction des Etudes et Recherches - Département R.E.M.E. - 25, Allée Privée - Carrefour Pleyel - 93206 SAINT DENIS

1. INTRODUCTION

The higher unit power of nuclear plants compared with conventional thermal power plants, leads to a considerable increase in steam flow rate. Pipe diameters must be bigger, thus exerting not-well-known forces and moments on the main components of the feedwater heating system particulary on pumps. It is useful to verify that these efforts are in agreement with the designing requirements of the secondary system.

The method described in this paper deals with the determination of forces and moments in pipe cross-section. Two major difficulties occur in performing this method :

- since the values of forces and moments cannot be directly measured, they are inferred from strain measurements taken on the outer wall of the piping ; foil strain gauges are used.

- strains are small, and thermal conditions vary according to pump operating conditions. Temperature-induced apparent strain should be accurately estimated, since the measured strains are of the same order of magnitude.

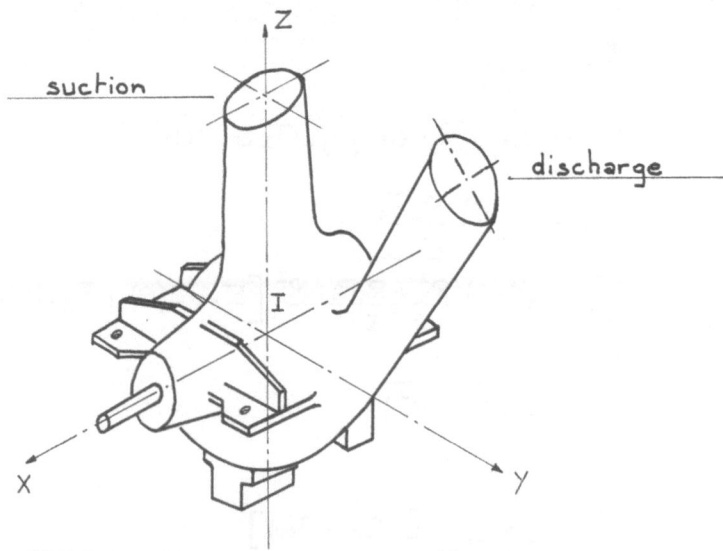

FIGURE 1 - DIAGRAM OF A FEEDWATER PUMP

As an example, some results are shown regarding the efforts on the suction and discharge pipes of a feedwater pump of a 900 MW PWR nuclear power plant.

2. MECHANICAL ANALYSIS

The efforts are given by a theoretical model of strain distribution in a pipe section. The relationship between the efforts and the strains is based upon :

- the Euler-Bernouilli beam theory for the components of forces and moments (tension-bending-torsion),
- the thin shell theory for hydrostatic and thermal loading (lineic expansion-through-wall temperature profile).

In a cylindrical coordinate system (r, θ, z), the theoretical strains $\varepsilon \varepsilon i$ are sinusoidal functions of angle θ on which the measured strains ε mi of a rosette Rj (figure 2) are fitted.

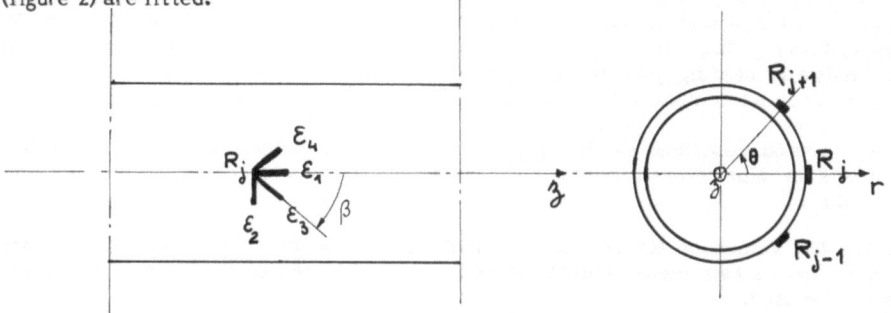

FIGURE 2 - DIRECTION OF THE STRAINS

Seven coefficients C_i are defined. Each coefficient characterizes a single loading. One obtains the following equations :

$$C_1 = \frac{E_1}{ES}$$

$$C_2 = \frac{R_e}{EI} \left[T_2 \sin(\theta) - T_3 \cos(\theta) \right]$$

$$C_3 = (1+\vartheta) \frac{R_e T_1}{E I_o}$$

$$C_4 = -(1+\vartheta) \frac{R_e^2 + R_e R_i + R_i^2}{3EI} \left[E_3 \cos(\theta) + E_2 \sin(\theta) \right]$$

$$C_5 = \frac{2 P_i R_i^2}{E(R_e^2 - R_i^2)}$$

$$C_6 = \varepsilon_{ap}$$

$$C_7 = \frac{\alpha}{2(1-\vartheta)} \left[T_{int.} - T_{ext.} \right]$$

Thus, the theoretical strains ε_i are expressed as a function of these coefficients in table 1. When the pipe is heat-insulated, the effect of a through-wall temperature profile can be neglected.

	Direct. β	Tension C_1	Bending $C_2(\theta)$	Twisting C_3	Shear $C_4(\theta)$	Pressure C_5	Appar. strain C_6	Temp. profile C_7
$\varepsilon 1$	0	1	1	0	0	$-\nu$	1	$-\nu$
$\varepsilon 2$	$\pi/2$	$-\nu$	$-\nu$	0	0	1	1	1
$\varepsilon 3$	$\pi/4$	$\dfrac{1-\nu}{2}$	$\dfrac{1-\nu}{2}$	1	1	$\dfrac{1-\nu}{2}$	1	$\dfrac{1-\nu}{2}$
$\varepsilon 4$	$-\pi/4$	$\dfrac{1-\nu}{2}$	$\dfrac{1-\nu}{2}$	-1	-1	$\dfrac{1-\nu}{2}$	1	$\dfrac{1-\nu}{2}$

TABLE 1 - THEORETICAL STRAIN AS A FUNCTION OF COEFFICIENTS Ci

3. MEASUREMENT PROCEDURE

Analysis of the theoretical strains leads to the following conclusions regarding the measurements and statistical procedure to be performed :

- It is better to measure the axial strain $\varepsilon m1$ than the hoop strain $\varepsilon m2$. The axial strain is more sensitive to the axial and bending loading.
- the tension and bending components are calculated using the axial strain
- the shear and twisting components are calculated using the difference ($\varepsilon m3 - \varepsilon m4$)
- the theoretical strains ε_i and ($\varepsilon 3 - \varepsilon 4$) inferred from mechanical analysis are sinusoidal functions of angle θ.
- a least square sine curve fit is performed on the measured strains $\varepsilon m1$ and ($\varepsilon m3 - \varepsilon m4$) obtained for each rosette Rj. Then the coefficients Ci and their uncertainties are directly computed (figures 4 and 5). The values of efforts can therefore be predicted with a known confidence level.

To improve the accuracy of the estimate of efforts a minimal number of measurements must be available.This choice is based upon the Student distribution for a 90 % confidence level. (table 2)

Number of rosettes	Student Variable 90%	Uncertainty
4	2.92	1 (reference)
6	2.13	0.73
8	1.94	0.67
10	1.86	0.64

TABLE 2 - STUDENT VARIABLE

Eight measurements are a reasonable choice taking into account possibly damaged gauge.

This method requires the following measurements :
- a static pressure reading
- a temperature measurement on the outer wall of the pipe
- strains measurements obtained by eight rosettes of three gauges (figure3).

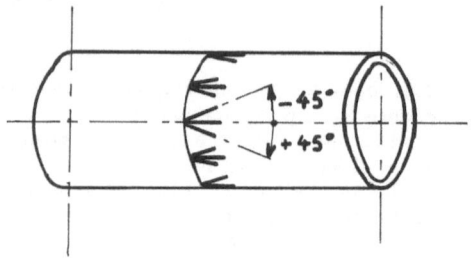

FIGURE 3 - DIRECTION AND LOCATION OF MEASUREMENTS POINTS

4. STATISTICAL ANALYSIS PROCEDURE

Strain measurement analysis procedure consists in :
- a sine curve fit on the eight measurements couples (m_i, O_i) using a least square method.
- the determination of possibly deviating points using the Dixon test.
- the estimate of the confidence intervals.
- the estimate of the uncertainty on the coefficients of the regression.

The strain sine curve is estimated using the following formula :

$$\varepsilon(\theta) = A \pm t \; \sigma(A) + [\; B \pm t \; \sigma(B) \;] \cos(\theta + \phi) \quad (eq.1)$$

The variances $\sigma^2(A)$ and $\sigma^2(B)$ are calculated from the residual variance of the regression $\sigma^2 r$, the variance on the strain measurement $\sigma\varepsilon^2mi$ and the variance on the apparent strain $\sigma\varepsilon^2ap$.

This method has been applied to the determination of the efforts on the suction and discharge pipes of a feedwater pump. The results of a test (figures 4 and 5) are presented as an illustration. Figures 4 and 5 show the regression obtained on the axial strain $\varepsilon m1$.

AXIAL STRAIN FIT AND CONFIDENCE INTERVAL

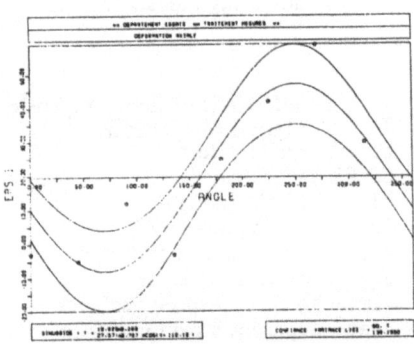 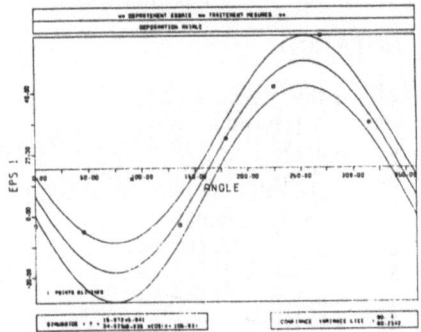

FIGURE 4 : BEFORE REJECTION   FIGURE 5 : AFTER REJECTION

Regression on the complete set of experimental points leads to figure 4. The standard deviation is high (14 µ m/m, ie, twice the standard deviation of the apparent strain). The Dixon test is performed to detect possibly deviating measurements, leading to elimination of the 90° strain gauge (it was later verified that this gauge had been damaged). Once this 90° point is rejected, the standard deviation from the regression is 8 µm/m (figure 5).

Finally the values of the bending moments and the tensile effort can be computed from table 1 and eq 1 and are shown in table 3. Same analysis is performed on strain ($\varepsilon m3$ - $\varepsilon m4$) to determine the shear efforts and the twisting couple.

Conf. level	A µm/m	tσ(A) µm/m	B µm/m	tσ(B) µm/m	φ degree	E1 N	T2 mN	T3 mN
90 %	16	6	.35	7	106	17.10^4 $\pm7.10^4$	-5.10^4 $\pm1.10^4$	15.10^3 $\pm3.10^3$

TABLE 3 - AXIAL STRAIN ANALYSIS RESULTS

5. CONCLUSION

The determination of efforts and moments in pipe cross-section for the different cases of stationary and static loads is based upon a theoretical analysis of the mechanical problem.

In order to compute each one on the components of forces and moments, with their associated uncertainty within a desired confidence level, the described method specifies :
- the measurements to perform
- their location and number
- the relevant statistical analysis procedure.

On site control of the quality of a just carried out test was made possible by implementing this method on a mini-computer. This also allowed for early localization of fautly sensors by automatic detection of statistically deviating measurements.

NOTATION

εi	: theoritical strains	α	: coefficient of lineic expansion
εmi	: measured strains	Te	: external temperature
ε	: curve fit on strain measurements	$E1$	: tensile effort
Rj	: rosette of j-th location	$E2\ E3$	: shear efforts
β	: direction of theoritical strain	$T1$	: twisting couple
E	: young modulus	$T2\ T3$	: bending moments
ν	: poisson coefficient	A, B	: coefficient of the regression
S	: cross-section area	σp^2	: residual variance
I	: moment of inertia	$\sigma \varepsilon mi^2$	: strain measured variance
Io	: twisting moment	$\sigma \varepsilon ap^2$	: apparent strain variance
Ri, Re	: inner and outer radius	$\sigma \varepsilon ap^2$	: apparent strain variance
pi	: internal fluid pressure	t	: student variance

REFERENCES

1. Timoshenko S.P. : Résistance des matériaux - tomes 1 et 2 - Edition Dunod
2. Kerguignas - Caignaert : Résistance des matériaux - Edition Dunod
3. C.E.A. : Statistique appliquée à l'exploitation des mesures - tomes 1 et 2 Edition Masson
4. Flaman M.T. - Shah N.N. : Measurements of stresses, Forces and Moments on nuclear steam generators - Experimental Mechanics Mars 83.

PRESSURE MEASUREMENTS ON SPATIALLY CURVED ELASTIC SURFACES DURING CRASH TESTS

Dr. H. Marwitz, J. Stecher

Daimler-Benz AG, Abt. V1PS, Mercedesstraße,
7000 Stuttgart 60, West-Germany

1. INTRODUCTION

One of the aspects in designing interior equipment in passenger vehicles is to minimize the risk of injuries. Among others this HEAD INJURY CRITERION (HIC) was developed for evaluation and comparison of variants as well as for establishment of a basis for legislation. The HEAD INJURY CRITERION is a combined evaluation factor which includes the acceleration on the inside of a dummy head as a primary characteristic. As described in (1) this factor does not satisfactorily describe the actual risk of injury. This becomes obvious when local rigidity variations and maxima indicate a considerable risk of skull fracture, while the measurement of the acceleration with its integrating effect shows comparatively low values simultaneously. To date, a better measuring procedure is not available, so that this method still represents the state of the art. The purpose of this paper is to present a method, using the well known pressure foil procedure for measuring the pressure to which the dummies are subjected in crash tests, in order to obtain information on the injury risk presented by interior equipment in passengers cars as well as information on the effect of protective systems.

2. PROCEDURE

2.1 The Sensor

The well-known, two layer pressure foil from Fuji (2,3,4) is positioned between the pressure surfaces (Fig. 1). When subjected to pressure the microcapsules in foil layer A burst, the liquid contained therein flows out and is absorbed by foil C, resulting in chemical coloration reactions. The foil is presently available on the market with four different sensitives.

2.2 Dependency of the Coloration Particularly on Impact Loads

2.2.1 Quasi-Static Loads

In addition to the already known

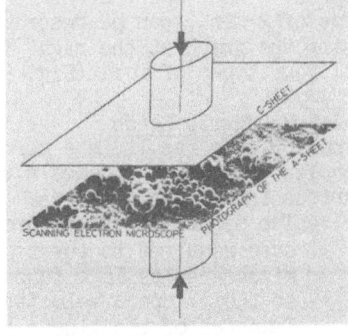

FIGURE 1

dependencies on the temperature and humidity (2), the studies showed that, contrary to the data from the manufacturer, it was necessary to limit the effective measurement range for the individual types of foil considerably for quasi-static application. The following measurement ranges resulted when the limit for the maximum permissible pressure is established as the value at which a pressure increase of 10 % results in a change in the grey level* of only 1 % in the corresponding measuring ranges:

Type of Foil	Lower Application Threshold N/mm^2	Upper Threshold N/mm^2
very low	0,5	1,8
low	1,5	6,0
medium	2,5	12,0
high	10,0	36,0

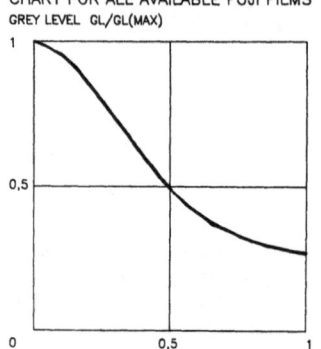

A GENERALIZED GREY LEVEL—PRESSURE CHART FOR ALL AVAILABLE FUJI FILMS
GREY LEVEL GL/GL(MAX)

PRESSURE P/P(MAX)

FIGURE 2

In the selection of these measurement limits it became apparent that the calibration curves also coincide with a corresponding, generalized representation (Fig. 2) within the scope of the measurement accuracy. The dependence of the grey level on the duration of the load in the quasi-static range, i.e. between approx. 1 sec. and the reference time of 2 min., can be described using the graph at the right for all four types of foil (Fig. 3).

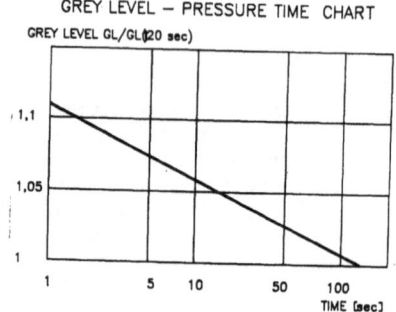

GREY LEVEL — PRESSURE TIME CHART
GREY LEVEL GL/GL(20 sec)

TIME [sec]

FIGURE 3

2.2.2 Dynamic Loads
Dynamic loads result in a more complex coloration process with various interrelationsships:
- The magnitude of the pressure required to burst the microcapsules is a function of their impact velocity.
- Transfer of the liquid from one foil to the other is dependent on the duration as well as the magnitude of the pressure.
- Exact repositioning is a prerequisite for corresponding reproducibility.

For all of these reasons generalization is not possible as in the case of static pressure. Quantitative valuation of foils subjected to dynamic loads

*) The grey level is defined as: unpressed foil: 250 GL
                                 black (covered TV camera: 5 GL
                                 graduation curve, gamma: 1

using digital image processing, requires densitometric data on the foil coloration. This data can be obtained only with dynamic calibration tests, performed under constant climatic conditions (temperature & humidity) with the same production issue, impact velocity, sensitivity of foil and duration of contact in each case.

2.2.3 Calibration

A dynamic calibration device was installed in a pendulum ram impact testing machine, whose impact rate can be increased as required by a pneumatic system (Fig. 4). Essentially, this device consists of an index die with a defined area and a adjustable anvil in a common guide cylinder. In a manner similar to the static calibration (3), the foil pair is passed through the machine between the impact die and the anvil. The way and the impact velocity of the die, whose mass and degree of damping were varied for each experiment, and the anvil acceleration were measured electronically.

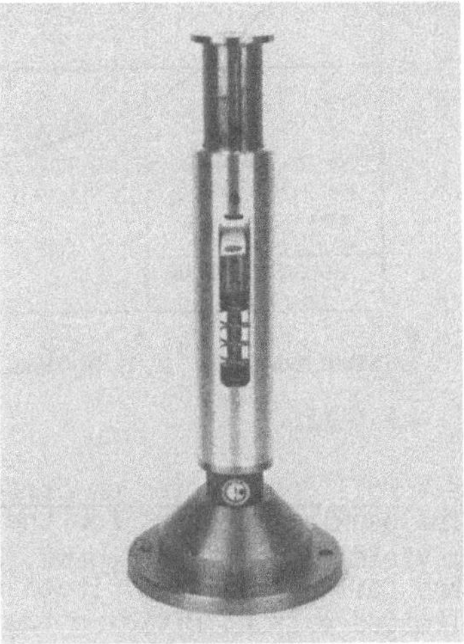

Results:
The results do not correspond exactly to a simple correlation between the impact velocity and a dynamic desensitivity factor.
Instead it is seen that
- the magnitude of acceleration
  (causing the pressure which is a
  function of both the constant anvil mass and the defined stamp area)
- and the contact time during impact
is for all types of foils very significant.

FIGURE 4

It is possible to unite the results of the dynamic calibration in one generalized grey level-pressure/duration chart (Fig. 5).
For a practical application it is important to know the time of duration of an impact. A typical acceleration (or force, or pressure) time chart is presented in Fig. 6.
According to (7), the duration of an impact is a function of:

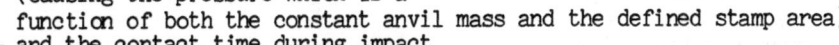

$$\left[\frac{m_1 \times m_2}{m_1 + m_2}\right]^{\frac{2}{5}} \qquad \left[\frac{c_1 \times c_2}{c_1 + c_2}\right]^{-\frac{2}{5}} \quad \text{and} \quad \left[v\right]^{\frac{1}{5}}$$

It is obviously, that the impacting masses and their elasticities are dominating factors for the duration.
On the basis of this working assumption, the following procedure was used: To achieve the ruling impact duration a pendulum dummy face (3.1.) and an idealised anvil (made out of similar elements like the effective parts of the steering wheel) was used.

160

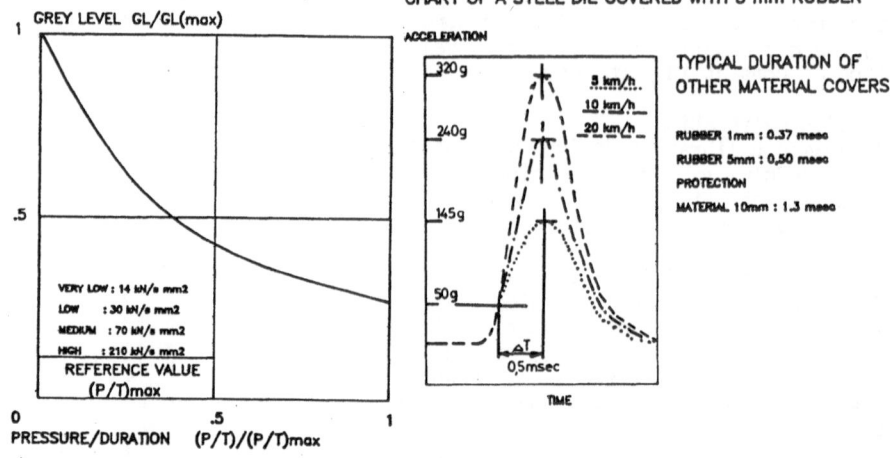

A GENERALIZED GREY LEVEL — PRESSURE/DURATION DEFINITION AND A TYPICAL ACCELERATION — TIME CHART (VALID FOR ALL FOILS) CHART OF A STEEL DIE COVERED WITH 5 mm RUBBER

FIGURE 5   FIGURE 6

2.4 Foil Positioning on Surfaces with Double Curvature

The backing material for the foils is plasic with a thickness of 90 μm and a yielding point of approx. 60 N/mm^2. For this reason the deformability of the foil only allows it to cover double curvatures without wrinkels to a limited extent. Dividing the foil into segments proved to be a solution. These curvatures can be covered when the ratio of the radius of curvature to the segment length of both is greater than 6. Differences in the geometry of these plane segments on the curved surface are compensated with gaps. Creation of optimum arrangements for faces with high- ly varying curvatures is imagi- nable. On the other hand, the amount of time required for this is extremely high. The two foils must be cut separately to elimi- nate marks on the C foil result- ing from the shearing force during cutting. This reduces the possi- bilities for fabrication to simple forms, which can be cut or stamped out on machines, for congruency reasons (Fig. 9).

FIGURE 7

2.5 Method for Protecting the Foils

Before such cut foils can be attached to a face, the A and C foils must be put together to form working pairs. Spot welding with a soldering iron proved to be effective for this purpose. The soldering iron creates spot welds, which cannot be used later for evaluation; however, since the spot diameters are smaller than 0,5 mm, less than 0,4 % of the information is lost. Other methods, such as gluing the foils, resulted in considerably poorer effective surfaces.

These working pairs were then attached to the surface of the head coated with an adhesive in such a manner that the open area between the segments is minimized and the geometrical differences between the flat foil surfaces and the curved face are compensated without overlapping.

A PVC foil with a thickness of 0,5 mm is positioned over the entire system with a deep drawing machine to protect these foils, which are very finely attached to on another, from flying away or damage from wrinkling during preparation for the test and the actual crash test. The protective foil deforms plastically and bubble-free around the face and measuring foil segments, under the effect of hot air followed by a vacuum between the face and protective foil. Following this operation the foil remains highly elastic and capable of transfering pressure to the surface of the measuring foils normally without deformation. On the other hand, due to the close contact and molding effect, the foil is supported against the parts of the face not covered by the measuring foils to such an extent that the foils are not displaced even under the effect of high shearing forces between the impact surfaces. This is a primary prerequisite for use of these measuring foils in crash tests, because relative movement of the foils would destroy the microcapsules and therefore lead to incorrect data. The vacuum subjects the measuring surface itself to a corresponding normal pressure; however, since this load cannot exceed a value of 0,1 N/mm^2, it remains below the sensitivity limit even for the "very low" foil. Great care is required during normal handling, because heavy finger pressure can show an effect on the most sensitive foils even through the protective foil. The short-term effect of the heat when the heated foil is pulled over the dummy face has no negative effect.

3. APPLICATION
3.1 Pendulum Ram Impact Tests

Optimization of interior components in passenger cars in terms of safety engineering is accomplished using pendulum ram impact testing machines. For this purpose dummy heads, prepared in the above manner and mounted on rubber necks, were installed in an impact tester (Fig. 10). The reduced mass of this system is determined experimentally. In conjunction with this, acceleration measurements are performed on the inside of the head and the impact velocities directly in front of the specimen are measured via a light barrier. The impact velocity is determined by the pendulum height. Comparative tests with different steering wheels were performed with this setup. For these tests it was sufficient to attach measurement foils to the forehead only (pressure class "low"). The

FIGURE 8

162

study included various impact positions and angles as well as confirmations of the individual measurements to satisfy the statistical requirements. Following the test the individual foil segments were attached to a flat, white surface in the original positions and further processing was accomplished via digital image processing according to (6): recording of the average and maximum pressure, integration of the pressure over selected surface areas, interpolation for uncovered area, etc. as well as representation of the pressure classes in pseudo colors (Fig. 11 and 12). The results of the above tests document the reduction of the maximum surface pressure relevant for skull fractures by 68 % with the new design in comparison with the old.

However, the value of this procedure is not solely the quantitative evaluation, but also its confirmation of the tendency of the acceleration tests. To a much greater extent the surface representation of the pressure and therefore the rigidity distribution of the steering wheel provide primary information on the local rigidity and the deformation characteristics for the designer. This knowledge is a necessary prerequisite for further improvement.

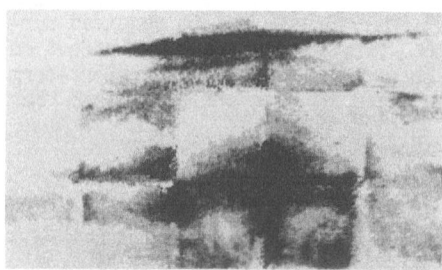 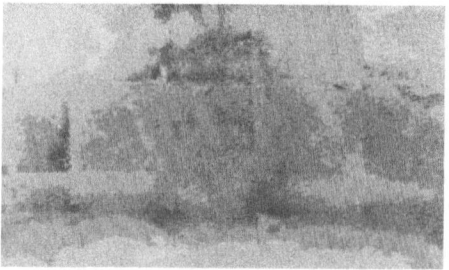

FIGURE 9   FIGURE 10

3.2 Slide Tests
Rubber belt acceleration slides, containing all of the interior equipment with their rigidity distribution relative for passenger car crash tests, were equipped with complete dummies to study the effect of modern restraint systems. Restraint systems such as an airbag (a bag positioned in the centre of the steering wheel under the padded hub which is inflated within milliseconds by a pyrotechnical gas generator in the event of an accident in order to protect the driver from impact with the steeering wheel as well as the windshield) considerably reduce the surface

FIGURE 11

pressures relevant for skull fractures. However, since a measuring system has not existed until now, which is capable of converting these characteristics into objective data,
a comparative evaluation of the direct steering wheel impact and the airbag has not been possible. The criterion (HIC) already mentioned previously is not suitable for this purpose, because characteristic values result, which do not sufficiently describe the true protective characteriatics of the airbag due to the combination of acceleration values with their associated effective durations. The present test procedure can provide data within this context, at least for the maximum surface pressures relevant for skull fractures. As described above, the completely covered head is included in the test (Fig. 13), subjected to the forces during the crash and then evaluated as described previously (Fig. 14). The significantly lower pressures on impact with the airbag in comparison with the steering wheel, made it necessary to use the most sensitive type of foil in the case illustrated.

FIGURE 12

The tests showed that under comparable impact conditions the maximum surface pressure with an airbag is approx. 11 times less than the pressure upon impact with the steering wheel. The type of surface pressure - uniform over a large surface in contrast to highly concentrated pressures - could also be suited for conversion into a safety criterion.

However, this method is not limited to the applications described above. It offers a wide field in the area of safety research:
* Comparative airbag tests
* Differences between airbags for driver and passenger
Other tests and adaptations of this system are planned within the scope of safety research:
* Pressure measurements between seatbelt and dummy chest
* Seatbelt tensioning systems
* Studies on the effect of passive safety elements
In addition this method is useful in the field of seat design, in regard to safety and orthopedic criteria.

REFERENCES

1. Grösch, Dr. Lothar: Injury Criteria for Combined Restraint Systems, Report of the 10th ESV Conference, Oxfort, 1 - 5 July 1985

2. Fuji Photo Film Co., Ltd.: FUJIFILM PRESCALE, INSTRUCTIONS FOR USE, Paper Products Sales Dept. 2-36-30, Nishi-Azabu, Minato-ku, Tokyo, Japan

3. Schöpf, Dr. H.-J., Stecher, J., Karg, E.: Ermittlung von Pressungsverteilungen an Kontakt- und Dichtflächen (Determination of Pressure Distributions on Contact and Sealing Surfaces); Messen und Prüfen V. 16 (1980), I.6, P. 88 ff.

4. Ficker, Dr. E., Hehne, Dr. H.-J., Hultzsch, Dr. W., Jantz, Dr.W: Anwendung der Druckmeßfolie in der Biomechanik: Messung der Druckverteilung in Gelenken des Menschen (Application of Pressure Foil in Biomechanics: Measurement of the Pressure Distribution in Human Joints); Proceedings of the VIIth International Conference on Experimental Stress Analysis, Haifa, Israel, 23 - 27 August 1982

5. Rappoldt, R.: Untersuchung zur Erweiterung des Einsatzbereiches der Druckmeßfolie unter besonderer Berücksichtigung dynamischer Effekte (Study on the expansion of the Area of Application of Pressure Foils with Special Consideration to the Dynamik Effects), Thesis, Technical College, Esslingen, 1986

6. Schöpf, Dr. H.-J.: Messung von Pressungsabdrücken mit rechnergesteuerter Bildverarbeitung (Measurement of Pressure Imprints with Computer-Controlled Image Processing); Preceedings of the VIIth International Conference on Experimental Stress Analysis, Haifa, Israel, 23 - 27 August 1982

7. Hertz, K.: Über die Berührung fester, elastischer Körper J. f. reine und angewandte Mathematik 92 (1881)

INVESTIGATION INTO THE LOOSENING OF RISERS IN CARBON-DIOXIDE CYLINDERS FOR
FIXED FIRE-EXTINGUISHING SYSTEMS ABOARD SHIPS

ing. P. Tegelaar
TNO-IWECO, P.O. Box 29, 2600 AA Delft, the Netherlands

1.   INTRODUCTION

For the sake of safety to man, cargo and ship we find aboard cargo
ships fire-extinguishing systems consisting of a number of integral connected carbon-dioxide cylinders.
Inspectors of the Dutch Director-General of Shipping and Marine Affairs
found in the last few years during the legal inspections aboard ships an
outstanding great number of cylinders with loose risers. To become familiar
with the problem to deal with, figure 1 shows the cylinder and a useful
detail of the quick-release valve. The cylinder is 270 mm wide and 1450 mm
long and made of wrought steel, designed for a test pressure of 150 bar.
The cylinder is shut with the quick-release valve in which a riser (dip
tube) is screwed. The liquid carbon-dioxide (about 45 kg) is contained in
the cylinder with a pressure of about 60 bar at room temperature.

The riser is of utmost importance. After releasing the valve, the
carbon-dioxide has to be transported to the protected room as a liquid. In
the regulations is laid down "when a carbon-dioxide fire extinguishing
installation is put into operation it must, within 2 minutes, be possible
to admit into the room involved 85% of the amount of carbon-dioxide
prescribed" [3]. This rule can only be met with the riser in correct
position. Actually each loose riser means a non-functioning cylinder and a
dangerous sham safety.
In spite of the alarming findings of the inspectors, the people concerned
with the assembling of the cylinders and installation of the complete
systems aboard ships reacted incredulous. Risers that can not be touched by
something nor by someone "have nothing to hold, haven't they?".
In the past not too much attention was given to the fitting of the riser in
the valve. Only a few manufacturers have written instructions giving
mounting torque and use of a sealant. The rest uses transmitted customs
based on craftman's experience.

The question remains if a riser, fitted according to the current and
accepted methods, can work loose after a period of time, thus creating a
perilous situation. Eliminating all possible environmental attacks to the
riser, at the end shock and vibration seem to be the only possible causes
that affect the riser connection with the unwanted results mentioned.

Wieringa, H (ed), Experimental Stress Analysis.
© *1986. Martinus Nijhoff Publishers, Dordrecht.*

2.1 Vibration - Natural Frequencies

In analyzing a shock and vibration problem the first step is to study natural frequencies and damping factors.
The resonance frequency of a one-sided clamped beam can be calculated by [4]:

$$f_n = \frac{a_n}{2\pi} \sqrt{\frac{EI}{\mu l^4}}$$ (1)

Where: E = Young's module (N·cm⁻²)
    I = moment of inertia (cm⁴)
    μ = mass per unit of length (kg·cm⁻¹)
    l = length of riser (cm)
    aₙ = coefficient for the various forms of vibration

a_1 =   3,52
a_2 =   22,0
a_3 =   61,7
a_4 = 121,0

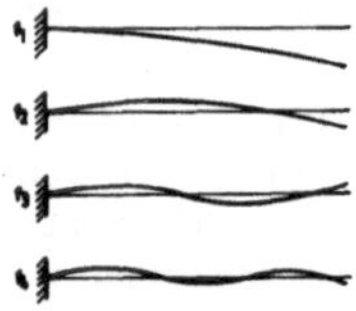

For the experimental verification, a steel riser was firmly clamped to a shaker table and vibrated in the frequency range 5-150 Hz. The response characteristic is given in figure 3. Natural frequencies found: 7,6 Hz, 48,2 Hz and 134,0 Hz with damping factors for the three modes: 0,6%, 1,6% and 0,3%. The damping is so small that large amplification factors can be expected at the resonance frequencies; in the first mode Q = 87. In figure 2 the first two modes are shown.
It may be remarked here that the frequencies corresponding to the first mode are certainly to be expected aboard ships, and passed through anyway during manoeuvring activities.

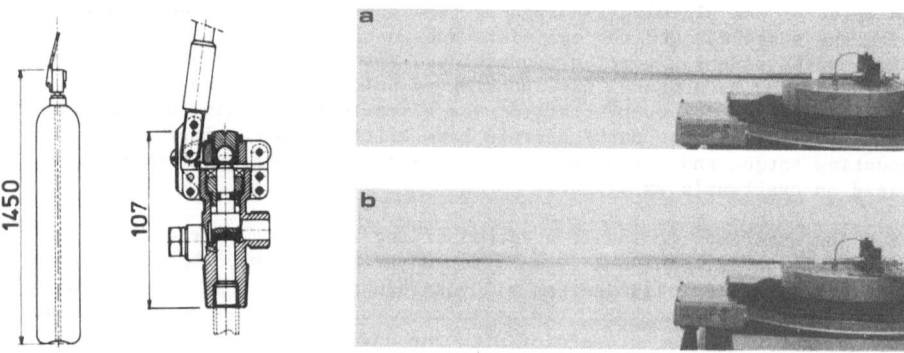

FIGURE 1: fire-extinguisher cylinder with detailed quick-release valve

FIGURE 2: a) first bending mode,   f = 7 Hz
          b) second bending mode, f = 40 Hz

During the first resonance search tests, carried out in a frequency range and at a vibration level, given by the Lloyd's Register of Shipping Requirements for type approval [1] a direct link could be established with the problems which the inspectors came across in practice.
The first (copper) risers were screwed in a well-tightened valve with a small torque. In the frequency sweep none of the risers passed through its natural frequency without becoming loose.

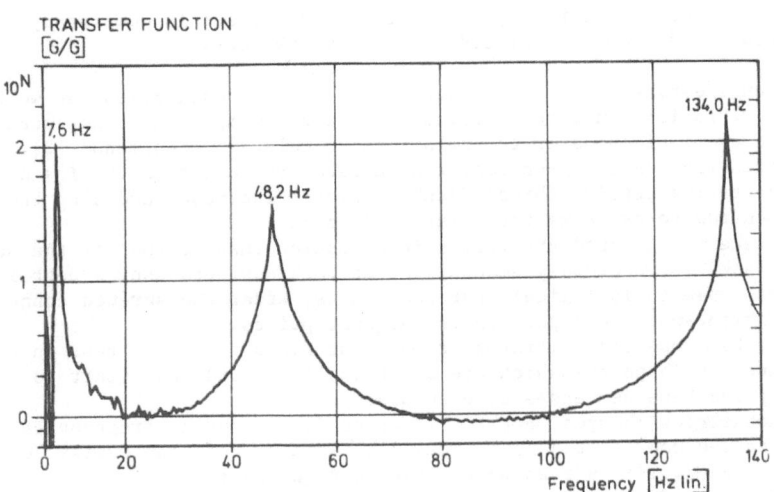

FIGURE 3: transfer function steel riser tube

2.2 Vibration – Damping

The riser, resonating free in the air, shows a very low damping coefficient (0,6% for the first mode). In the fire-extinguisher cylinder, the riser is contained in liquid carbon-dioxide.
Liquid damping is proportional to the kinematic viscosity of the liquid and inversely proportional to the gap in which the liquid is transported in vibration. In a large cylinder, about 250 mm in diameter with a small riser, 16 mm in diameter, there is hardly any question of damping in a suitable gap. Moreover the viscosity of liquid carbon-dioxide is very low and lies in between the viscosities of water and air at room temperature [5].

$$
\begin{aligned}
\text{water} &: 1,0 \quad \text{cP} \\
CO_2 \text{ liq} &: 0,071 \quad \text{cP} \\
\text{air (20)} &: 0,018 \quad \text{cP}
\end{aligned}
$$

Tests with a scale model (scale 1:3) filled with water confirmed the above statements.

3. EFFECTS OF VIBRATION ON SCREW-THREAD CONNECTIONS

With respect to the riser-valve connection, where the male-threaded riser pipe is screwed into the inside thread of the valve body, we distinguish three different situations:

a. the riser for the greater part is screwed in by hand and can be easily rotated, clockwise or anti-clockwise;

b. the riser pipe is tightly screwed up by means of tools to realize a sufficiently high pre-stress in the connection – the connection can only be unscrewed using tools and a particular torque;

c. due to prolonged unscrewing of the riser, the riser is completely unscrewed and become separated from the valve body.

In engineering it is well-known that a screwed connection can work loose by vibration. In Dutch "lostrillen" is one simple verb. Not much theory is available however to explain these relevant phenomenon [2].
By prestressing the screw-thread, the surface pressure from the inside-thread on to the outside-thread flank is highly increased and also the force required to overcome the frictional forces.
When we take for granted the realistic situation that neither of the screw-thread flanks is absolutely smooth, but at both surfaces show a certain roughness, then it is logical that the flanks, after the screwed connection has been tightened, only bear on the highest points.
Now when the connection vibrates or when the riser is highly bent in resonance, those high points which are additionally loaded, may start to flow and the pre-stress decreases or even disappears.
So the connection changes, due to vibration (at resonance frequencies) from a well-secured one (ad b.) to a loose one (ad a.), that can easily be rotated by hand. This situation has two consequences:

1. the damping of the resonating system increases greatly and the effect of resonance is strongly decreased;

2. prolonged vibrations can lead to a complete unscrewing of the riser (ad c.).

The phenomenon, mentioned under 2. cannot be explained theoretically. If in vibration, a threshold level of 1 "g" (9,81 $m.s^{-2}$) is exceeded, irrespective of the frequency of the exciting signal, two parts that are no longer connected under pre-stress, bounce together producing high-frequency vibrations, named "contact-resonances". These contact-resonances introduce a sort of unpredictable spastic motion.
All this theory can be confirmed experimentally very easily. So, at a vibraton level that is 1,5 times the test level indicated by the various classification societies, each unlocked riser connection worked loose in a period of time varying from some minutes to a few hours.
The threshold theory can easily be shown on the shaker table. On a slightly inclined steel surface, see figure 4a, vibrated in a direction corresponding to gravity, a small steel block is placed. At increasing vibration level the small block tends to move when the level crosses the threshold value of 1 "g".
The block does not always move downwards, but moves at random in all directions, possibly due to small imperfections in the touching surfaces.

A comparable experiment can be carried out with a screwed rod with one nut, positioned halfway, see figure 4b. At the crossing of the threshold level of 1 "g" the nut starts rotating in an unpredictable manner, sometimes clockwise, sometimes anti-clockwise.

Summarizing it may be said that two mechanisms, coming into operation one after the other, account for the loosening of a riser under the working of vibrations:

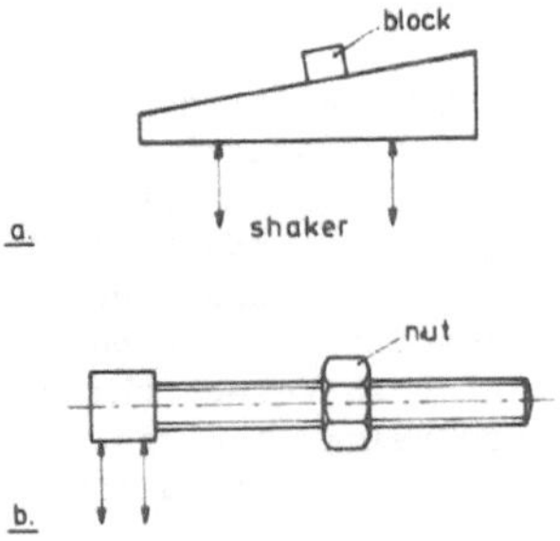

FIGURE 4: experimental verification of threshold theory

a. The pre-stress in the connection disappears through vibration, especially in resonance. When the pre-stress has disappeared we call the riser a loose one, which can easily be turned round manually;
b. if the vibrations at the loose riser exceed the 1 "g" threshold level either for a short or for a longer period, there is the risk of it working loose even until complete disconnection.

To overcome the loosening, two measures have to be taken. The screw-thread connection must be locked, using a chemical sealant (Devcon, Loctite, Permabond, Powerloc, Stalok or Trulok), and the mounting torque must be sufficiently high (20/40 Nm for a 16 mm riser pipe). The chemical sealant must be able to withstand a temperature shock of 80°C per second, as occurs during the first second of the filing of the cylinder with liquid carbon dioxide.

The various variables - riser material, mounting torque and type of chemical sealant - have been tested, using the electrodynamic shaker as a fatigue-testing machine, working at the first resonance frequency of the riser.

A well connected riser (using a suitable torque and a chemical sealant) shows a very low damping factor, which means that in resonance the riser material is highly stressed which can lead to low cycle fatigue and even fatigue fracture of the riser. Especially with copper risers this phenomenon was consistently found in the experiments. So it is advisable to use steel risers and preferably seamless steel precision pipe, chemically treated for corrosion durability.

From the experimental results conclusions can be drawn that agree very well with the findings of the inspectors "in the field":

1. a riser, not adequately connected, works loose by loosing its pre-stress;
2. at a vibration level above the threshold value of 1 "g", the possibility occurs for anti-clockwise rotation of the riser, even until complete separation of riser and valve;
3. this threshold phenomenon does not appear at lower vibration levels, the riser remains then in a good condition as possible resonances are highly damped;

4. a well connected riser will not work loose by vibration but can, due to
the low damping coefficient, be overloaded - sensitive materials with
respect to fatigue, like copper, should be avoided.

4.   MECHANICAL MOBILITY ANALYSIS

From the foregoing it is clear that it would be possible to develop a
method for inspection, with which the proper functioning of the riser could
be established in situ without any dismantling of the cylinder.
This train of thought is based on the philosophy that, if the shock and
vibraton environment could affect the riser inside a hermetically closed
cylinder, it must be possible to reverse the process, using shocks and/or
vibration to detect the riser and the quality of its mounting in the quick
release valve. The mechanical mobility (or admittance) of a complex mechanical structure is defined by the ratio of the response velocity to the
exciting force:

$$M = \frac{\dot{x}_p}{F_p} \qquad \qquad (2)$$

The mobility analysis is carried out with shock excitation. The shock
hammer is instrumented with a piezo-electric force transducer, the response
velocity is measured with an electrodynamic velocity pick-up.
The complex ratio of velocity to force is computed by means of a two channel Fourier analyzer. In figure 5 mechanical mobility characteristics have
been obtained by data processing in a H.P. Fourier analyzer.

In an empty cylinder (not filled with liquid carbon-dioxide) the effect
of tightening is demonstrated. Figure 6a for a well tightened riser, 6b for
a manually tightened riser, 6c an untightened riser (only screwed in over
five threads), 6d a completely loose riser. At the natural frequencies of
the first two modes at 8 and at 49 Hz small discontinuities can be seen in
the diagram, smaller with decreasing damping and not available at all in
the case of the disconnected riser. In the last picture 6e a characteristic
is given for a well-tightened riser in a cylinder filled with 45 kg of
liquid carbon-dioxide. The ratio of the mass of the riser (0,75 kg) and the
mass of the filled cylinder (125 kg) determines in substance the amount in
which the discontinuities are demonstrable.
The horizontal scale of the mobility graph is a calibrated frequency scale
(log or lin), the vertical scale is a logarithmic reference scale in dB.

Using this mobility method, it is clear that the quality of the
riser/valve connection can be established from the outside of the cylinder
without any dismantling of the cylinder. From an economical point of view
this is a great advantage. The usual inspection procedure (5% of the total
number of cylinders for each 5 year's period) is time consuming and costly.
The cylinders have to be transported from the hold of the ship to the deck
and after inspection transported to a carbon-dioxide station to be filled
again and to be replaced.

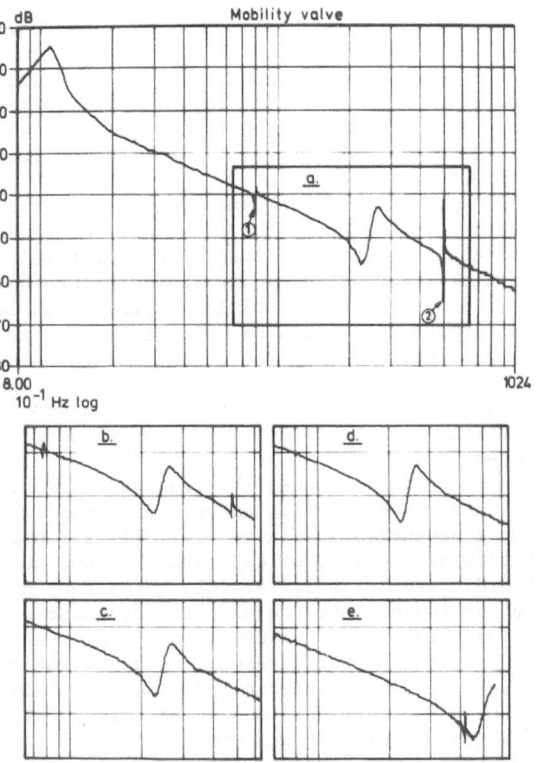

FIGURE 6: inspection method
in situ

FIGURE 5: mobility analysis of a fire-
extinguisher cylinder

5. INSPECTION METHOD IN SITU

Several analyzing methods have been investigated. The most simple one is
now recommended for inspection in situ on board ships.
In fact, the method is a simplified mobility method in which only the velocity response is measured and analyzed in a one channel FFT (fast Fourier
transform) analyzer. To carry out the analysis, the pipe joints to the
manifold are disconnected and a nipple is screwed on, to which the velocity
pick-up can be mounted. The transducer is connected to a one channel FFT
analyzer and the cylinder is excited by a hammer blow, using a rubber
hammer, see figure 6.
The cylinder is excited in the horizontal direction by the rubber hammer,
at the upper side of the cylinder where the cylindric part changes into the

spherical part.

The analyzer is set to record transient events with pre-triggering and with the facility to average the spectrum of successive blows. Figure 8 shows the averaging from 16 successive blows. This averaging serves the noise reduction. There is one trap however. In between two blows the cylinder and the riser must have time to come at rest completely, at least 10 seconds. If the blows are given in a too fast a frequency an unwanted interference will distort the picture.

Figure 7 shows the analyzed response signal for the first four modes of bending vibration of the riser. If the analyzer has the feature of band selectable Fourier analysis (zoom feature), in more detail the first two natural frequencies can be shown (figure 8).

The "zigzag" discontinuities in the graph look like a turned over Z. In conformity with the word "pip" in the audio signal given with the Greenwich time signal, we suggest to name the small discontinuities "zips".

When an unknown cylinder, equipped with a riser made from an unknown material, has to be analyzed the frequency to be expected cannot easily be detected.

One thing is for sure, the frequencies corresponding with the successive vibration modes follow exactly the coefficients in formula (1).

The frequencies in figure 7: 7, 41, 119 and 232 Hz should have been following the theoretical ratio for the successive modes: 6,8 − 42,4 − 118,8 − and 232,9 Hz. Searching the tiny zips is made easier by using a cursor to indicate the expected frequency.

Using this simplified FFT response analysing method, simply establishing zips at the correct frequencies or at the correct frequency intervals, does mean a correct connection of riser and valve body. Hardly demonstrable zips do mean that a riser connection exist, but the quality leaves much to be desired. No zips at all, mean no correct connection and reassembling will be necessary.

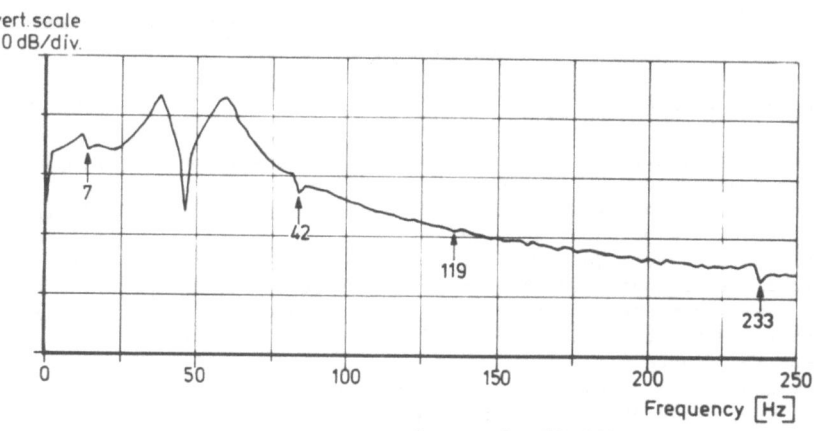

FIGURE 7: FFT analysis of a well-connected riser in a carbon-dioxide cylinder

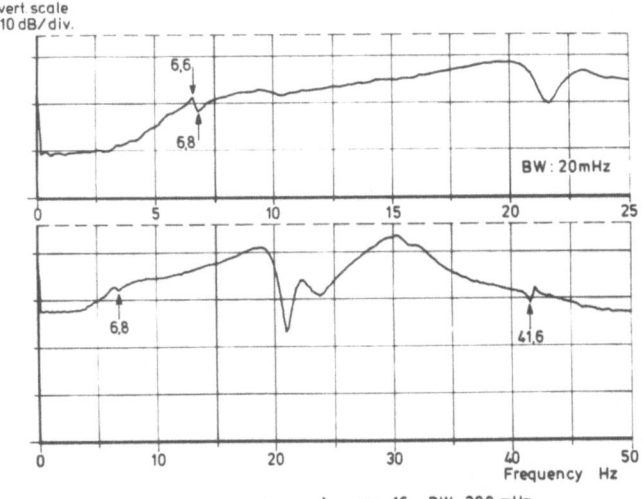

FIGURE 8: zoom recording of figure 7

ACKNOWLEDGEMENTS

The author wishes to thank his TNO-colleagues for their assistance and kind criticism and the staff members of DG Shipping and Marine Affairs for permission to publish this paper and their support during the investigations.

REFERENCES

1. Lloyd's Register of Shipping – Test requirements for the type approval of control and electrical equipment, London 1975.
2. R.E. Batson and J.G. Tokarski – The technology of threaded fasteners, Loctite Bulletin 719, 1974.
3. Voorschriften voor de aanleg, afname en periodieke inspectie van kooldioxyde – brandblusinstallaties – Bekendmaking van de scheepvaart, no. 164/1981 (stcrt 209).
4. J.P. den Hartog – Mechnical vibrations, McGraw–Hill 1956 (appendix V).
5. Handbook of Chemistry and Physics – The chemical rubber Co. 1970/71, 51st ed (p. F38).

INTERACTION OF STRESS CONCENTRATIONS AT PORES IN A MATERIAL PERFORMED BY
PHOTOELASTIC MODELLING

JAN JAVORNICKÝ

Institute of Theoretical and Applied Mechanics, ČSAV, Prague

INTRODUCTION

Material systems, as for instance those of composite materials, cement
concrete, ceramics, etc., are characterized by a greater or smaller number
of pores, void or fluid-filled cavities of various shapes. These shapes are
mostly reminiscent of a deformed sphere, so that in theoretical analysis
the cavities are currently considered to be spherical.

The influence of these pores on the deformation and the failure behaviour
of the material systems has not been interpreted uniquely. Since in the
vicinities of the pores arise concentrations of stresses, it is believed
that the pores act as accelerators of the propagation of cracks, supplying
the cracks with additional mechanical energy. On the other hand, if a crack
passes across the pores, its propagation is decelerated due to the need of
a certain period of time for a sufficient volume of energy to accumulate;
or the crack can be diverted from its course, it can even bifurcate and by-
pass the obstructing pore. Besides, the pores affect the rigidity of the
system, decreasing its modulus of elasticity.

In order that the decision may be reached as to which of these mechanisms
is predominant, an experimental solution has been undertaken. And the present paper intends to show that photoelasticity can supply a rapid answer
to the given question, and that even a simple experiment – in our case a
mere bidimensional one – can well define the conditions for the general
solution of the mechanical problem considered.

PHILOSOPHY OF THE MODEL AND ITS MATERIALIZATION

The process of investigation was governed by the fact that the behaviour
of a material system is given by the amount and the size of its pores.
Owing to the type of processing, the size of the pores in a system does
usually not vary very much, and the amount of the fluid absorbed, mostly
gases, remains constant after compounding /may surface agent be applied or
not/. The basic condition was, therefore, that the volume of all pores in
the area under investigation should be the same, regardless of their size.

Another condition was for the model to exhibit interaction of the pores
with different curvatures, but distances between them constant and given by
their amount. The comparison of these cases had naturally to be undertaken
under equal marginal mechanical conditions.

Two tension strips were thus produced to function as plane models. Figure 1 shows the middle parts of the models. Their total length was

Fig. 1

490 millimeters. They were made from epoxide plates of 6.6 millimetre thickness. These strips represented three regions each and had different holes drilled in them. These holes simulated pores. The surface of each region was the same. A larger zone formed the boundary between these regions, so that approximately uniform loading could be achieved. In every case investigated the area of the holes was the same, viz. an area of constant dimensions had been divided into a number of holes depending on their diameter. The following series of diameters were selected: 1.0; 3.1; 5.3; 10.0; 20.0 and 28.0 millimeters. The number of the rows of the holes in each region had been calculated from the number of the holes to be drilled and the width of the model. Only two cases, those with the largest holes, failed to satisfy the requirement of uniform loading and were therefore subjected to measurement only for the sake of a comparison. Uniform loading was effected with the help of a balanced beam system and four pivots in each end of the model. Loading was equal to 1725 N, for checking purposes to 2415 N. The uniformity of stress distribution was checked upon by measurement of birefringence at each end; perturbation was less than 2 per cent.

MEASUREMENT AND THE RESULTS

Measurement was affected via determination of birefringence along the hole boundaries, particularly of maximum values of birefringence /both in tension and in compression/. Birefringence along the orthogonal axes, viz. the one connecting the hole centres placed above each other, and the other connecting two neighbouring holes, was measured, too.

The other measurements traced the isoclinics in the vicinity of the characteristic hole array. From these the respective isostatics were then constructed.

The figures present various examples of isochromatics. Figure 2 shows the pattern in the case of holes of 1 millimetre diameter, Figure 3 the pattern for 10 millimetre diameter holes. Figure 4 presents plots of nominal stress $\bar{\sigma}$ in the individual cases. It equals 41.7 /40.12/; 44.23 /40.97/; 45.16 /39.66/; 43.12 /34.50/; 57.50, 41.07 MPa, with thickness considered as equal to unity. /The values for the rows with a lower number of holes are quoted in parentheses/. Full consideration is given to the rows with a higher number of holes.

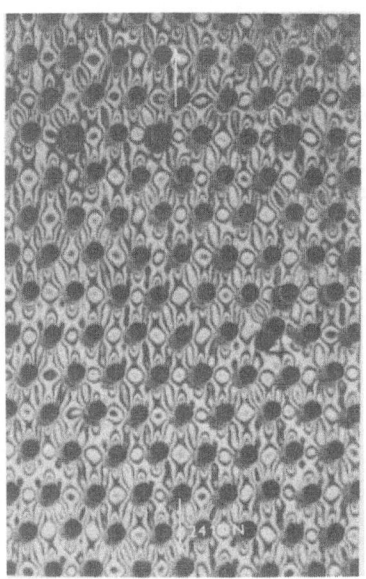

Fig. 2

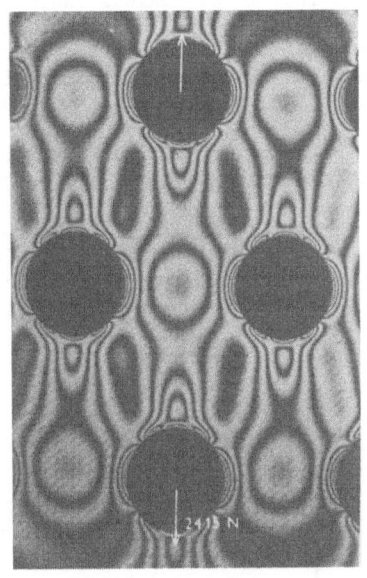

Fig. 3

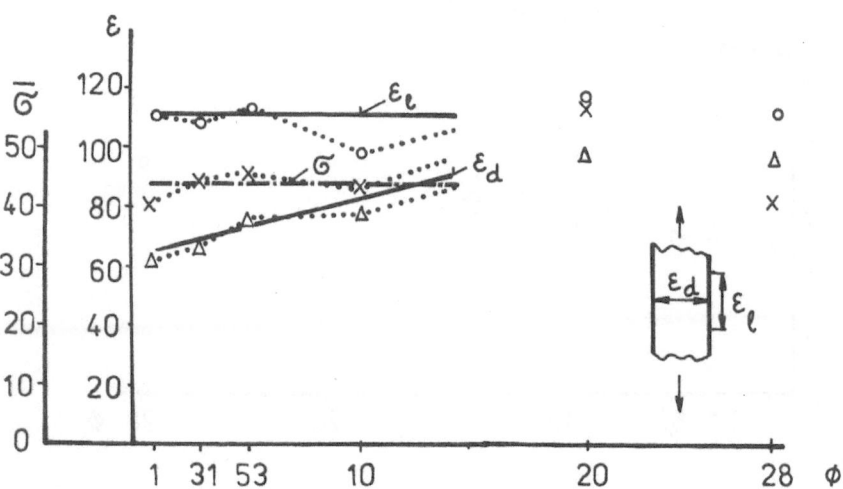

Fig. 4

178

Strain gauges were used to measure changes in the width of the model, ε_d , in individual regions under load, always at several points, and elongation ε_l of the height of each region, always at 5 points. From the diagrams of these strains the average strain was calculated and plotted in the same figure. /The number of rows is too small for to measure more precisely the change in the modulus of elasticity/.

The values of maximum birefringence in tensile /+/ and compressive /-/ stress concentrations are plotted in figure 5.

The difference between the curves of the isostatic families /for small holes/ is small enough to range within the limits of the measuring errors. Only one diagram has therefore been drawn /comp. figure 6/ for practically all the cases of diameters ranging between 1 and 10 millimetres.

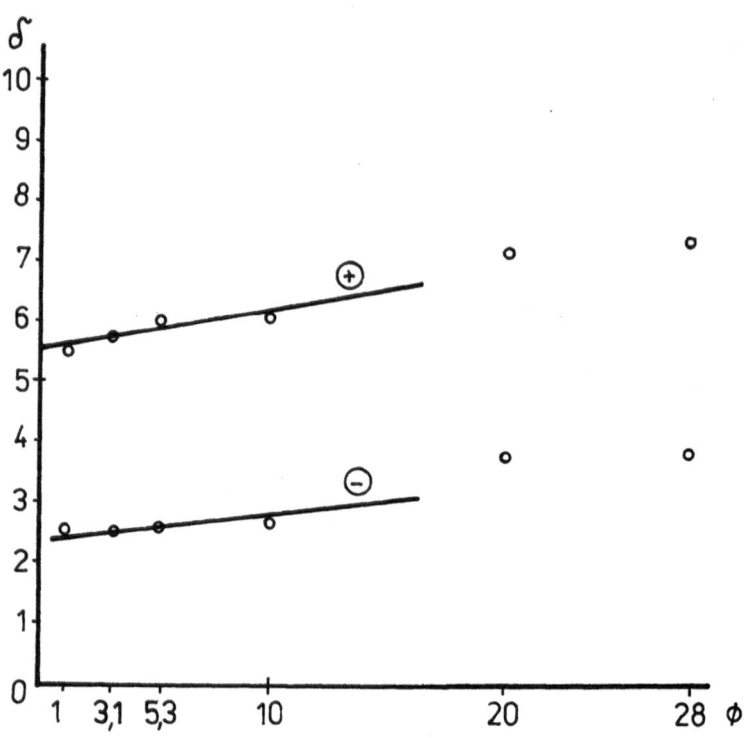

Fig. 5

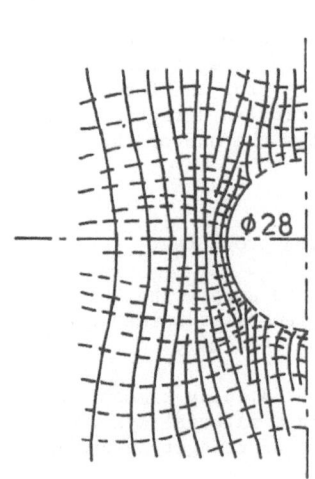
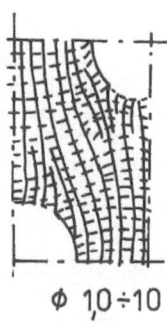

Fig. 6

DISCUSSION OF THE RESULTS

The comparison of tensile stress concentrations shows a slight increase in the maximum values parallelling the growth of the hole diameters.
The compressive stress concentrations differ even much less. There is however a certain displacement of singular points /with zero stress/. With
compressive stresses the circumferential segment shortens with the hole
diameters decreasing. It is obvious that the interaction of the neighbouring holes is relatively less intensive if the holes are smaller. With
nominal stress equal, this explains why the elongation of each region
remains approximately constant. On the contrary, the presence of holes
and the frequency of compressive stress concentrations cause a certain
rise of dilatational deformation /i.e. a change in the width/ with the
diameters of the holes increasing, even though the variation of maximum
compressive stress may be less pronounced.

The isostatics yield the same evaluation. In one family they follow
the direction of the load stress. The smaller holes do not particularly
affect the direction of the load passage. Therefore, also the second
isostatic family does not manifest any significant changes in curvature.

With the material system it means that the influence of smaller pores
is approximately equal and independent of the diameters of these pores.
It is the direction of the propagating crack that is decisive, viz. at
which angle the crack will cross the region of accumulated pores. The
crack propagating in the direction perpendicular to principal tension
will hardly be intercepted: it will propagate straight across the pore.
Its possible bifurcation will depend on the brittle or ductile properties
of the matrix. In such a case the area of compressive stresses may de-

celerate, and in extreme cases even stop, the crack. The crack propagating in the direction of principal tension may be stopped as soon as it reaches the area of compression. Since it will in most cases be a crack caused by shear loading or by unsatisfactory adhesion between the matrix and the re-inforcing fibres, it will traverse the area of tension even gaining an additional share of deformation energy for the formation of new surfaces. The behaviour of the cracks propagating in directions lying between these principal directions will thus follow the one or the other mechanism.

CONCLUSIONS

The results of the investigation prove that

a/   a simple and cheap experiment can supply the quantitative and qualit-ative information required by general considerations concerning mat-erial systems /i.e. statistical systems/;

b/   small pores, single or accumulated, play a much less important role in the propagation of cracks in a material system than it is usually believed. Their ability to stop a propagating crack is practically insignificant.

CEPSTRUM ANALYSIS AS A USEFUL SUPPLEMENT TO SPECTRUM ANALYSIS FOR GEAR-BOX MONITORING

A. DE KRAKER AND M.J.L. STAKENBORG

EINDHOVEN UNIVERSITY OF TECHNOLOGY, THE NETHERLANDS

SUMMARY

Cepstrum analysis is used frequently in data processing and signal analysis. The use of this new technique however in monitoring the mechanical condition of gearboxes is still rather unknown. This paper presents the results of an evaluation of the cepstrum technique as a tool for the interpretation of gearbox spectra. The origin of vibrations in gearboxes and a useful definition of the cepstrum will be presented. The method will be applied for the condition evaluation of a transmission train of 4 gears and it will be shown that the method can be used effectively if some practical considerations and some limitations are kept in mind.

1. INTRODUCTION

In general, most (rotating) machinery such as gearboxes generate vibrations and the corresponding vibration spectra show a characteristic form when the machine is in good mechanical condition. Changes in these vibration spectra are often an indication that the condition of the machine is changing for example due to wear or a toothdamage. A well known and very often used method to evaluate the mechanical condition is normal spectral analysis by means of measuring the vibration of a machine with the help of accelerometers on the housing. This condition monitoring is based on the assumption that, providing the operating conditions have not been changed, any change in vibration behaviour could be an indication of an impending failure. Although this assumption will not always be correct, the condition monitoring has proven to be of great use for the monitoring of e.g. turbine blades, bearings and gearboxes. In 1963 Bogert [1] introduced a new technique in signal analysis called the Cepstrum Analysis as a numerical operation working on measured spectra. This technique is now used frequently in different areas of data processing and signal analysis but the use in gearbox monitoring has left many questions. Some of these questions are dealing with the typical characteristics of gearbox vibration spectra and the origin of vibrations, practical considerations of using the cepstrum technique and the advantage of using the cepstrum in evaluating the changes in vibration spectra of a typical gearbox and the relation of these changes with a possible decreasing condition.

2. VIBRATIONS IN GEARBOXES

The study of the vibrational behaviour of gearboxes has received moderate attention in literature but due to the great complexity of gear dynamics, it remains a poorly understood area. In practice not only the usual out of balance and bearing forces are active in gear dynamics but also the precise geometry of the gear profiles has a crucial effect on the vibrational behaviour. In general flexural vibrations will be more important than torsional

182

vibrations because flexural vibrations are transferred directly to the housing via the bearings. Also coupling between flexural and torsional vibrations is usual and should be taken into account under certain circumstances.
With increasing frequency the nature of the lubricating oil film and the
teeth flexibility become increasingly important. Looking at gearbox vibration mechanisms three of the most important vibration sources are:
- Time variations in the mess stiffness, caused by variation of the number
  of teeth in contact and variation in the stiffness of the individual
  teeth.
- Dynamical effects due to deviations from the ideal tooth profile. In practice all gears contain teeth manufacturing errors, such as errors due to
  the gear cutting process, deviations in the mesh angle, deviations from
  the involute profile, surface roughness of the gears etc.
- Oscillations in the sliding velocity. During the transmission of power
  there will be rolling and slipping in the point of contact and also oscillations may occur because of stick-slip effects.
Due to these mechanisms, an amplitude or frequency modulation of the vibration signal may be caused, resulting in sideband-structures in the spectrum
around the toothmeshing frequency and its harmonics. This is illustrated in
figure 1.

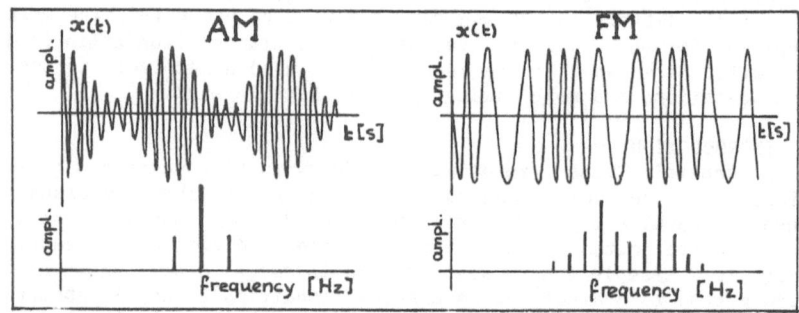

Figure 1. Amplitude- and Frequency modulated signals and their spectra.

Since in gears, not separable combinations of amplitude- and frequency modulation will lead to complex sidebandstructures in measured spectra a cepstrum analysis can offer a tool to improve the interpretation of such a
spectrum.

3. CEPSTRUM FUNDAMENTALS

The cepstrum was first described by Bogert et al. [1] in 1963 as a technique for finding echo arrival times in a composite seismic signal. These
authors defined the (power)cepstrum $C(\tau)$ as "the powerspectrum of the logarithm of the powerspectrum of a random proces", so:

$$C(\tau) = \mid \int_{-\infty}^{\infty} \log S_{xx}(f) e^{-2\pi j f \tau} \, df \mid^2 \qquad (1)$$

where $S_{xx}(f)$ is the two-sided auto powerspectrum of the random proces $x(t)$:

$$S_{xx}(f) = \lim_{T \to \infty} \frac{1}{T} E[X^*(T,f)X(T,f)] \qquad (2)$$

and $X(T,f)$ represents the finite Fourier Transform of $x(t)$:

$$X(T,f) = \int_0^T x(t)\, e^{-2\pi jft}\, dt \qquad (3)$$

For some reasons a number of slightly different definitions for the cepstrum are known in literature, such as

$$\text{Amplitude Cepstrum } C(\tau) = |\int_{-\infty}^{\infty} \log S_{xx}(f)\, e^{-2\pi jf\tau}\, df| \qquad (4)$$

or

$$\text{Amplitude Cepstrum } C(\tau) = |\int_{-\infty}^{\infty} \log S_{xx}(f)\, e^{2\pi jf\tau}\, df| \qquad (5)$$

or

$$\text{Complex Cepstrum } C(\tau) = \int_{-\infty}^{\infty} \log X(T,f)\, e^{2\pi jf\tau}\, df \qquad (6)$$

An important reason for using definition (5) instead of (1) is the relationship between the Cepstrum definition and the autocorrelation function $R_{xx}(\tau)$, defined as:

$$R_{xx}(\tau) = \int_{-\infty}^{\infty} S_{xx}(f)\, e^{2\pi jf\tau}\, df \qquad (7)$$

Comparing this expression with definition (5) we can note that the only difference is that in case of the cepstrum the logarithm of the spectrum values is taken before performing the inverse Fourier Transform. ($\mathcal{F}^{-1}$).
The powerspectrum $S_{xx}(f)$ of a vibration signal on a gearbox-housing is the product of the powerspectrum $S_{yy}(f)$ of the source function and the squared amplitude of the transfer function $H_{yx}(f)$ of the transmission path, i.e.: gears, shafts, housing, etc. So:

$$S_{xx}(f) = S_{yy}(f)|H_{yx}(f)|^2 \qquad (8)$$

Taking the complex cepstrum of S_{xx} we get:

$$C_x(\tau) = C_y(\tau) + C_H(\tau) \qquad (9)$$

where $C_x(\tau)$ respectively $C_y(\tau)$ are the complex cepstra of $x(t)$ respectively $y(t)$ and $C_H(\tau)$ is defined as: $C_H(\tau) = 2\mathcal{F}^{-1}\{\log|H_{yx}|\}$. Eqn (9) shows that the source and transmission path effects are additive in the cepstrum. In general, the amplitude of a transferfunction will have a different cepstrum content (different periodic behaviour) than the cepstrum image of the source function. It will therefore be possible to separate these two influences in the final cepstrum $C_x(\tau)$.

To avoid confusion between spectrum- and cepstrum methodology Bogert et al. introduced the following paraphrased terms,

frequency	...	quefrency
spectrum	...	cepstrum
phase	...	saphe
magnitude	...	gamnitude
filtering	...	liftering
harmonic	...	rahmonic
period	...	repiod

Today, the most utilized terms are cepstrum, quefrency and rahmonic; e.g. magnitude in the cepstrum domain is often called magnitude and not gamnitude. The independent variable of the cepstrum (τ) has the dimension of time and it is called quefrency. This quefrency tells us something about frequency <u>spacings</u> in the spectrum.

4. A NEW CEPSTRUM DEFINITION

Although the cepstrum definitions according eqn (1), (5), and (6) are very useful in analysing speech or seismic signals, the analysis of gearboxes demands a slightly different definition of the cepstrum. To illustrate this need we consider the harmonic function $x(t) = A\cos(2\pi f_0 t + \varphi)$. The autopowerspectrum $S_{xx}(f)$ of this function will be:

$$S_{xx}(f) = \frac{1}{4} A^2 [\delta(f-f_0) + \delta(f+f_0)] \tag{10}$$

According to the cepstrum definition of eqn (5) we get

$$C_x(\tau) = \mathcal{F}^{-1}\{\log S_{xx}(f)\} = \log(\frac{A^2}{2}) \cos(2\pi f_0 t) \tag{11}$$

The functions $x(t)$, $S_{xx}(f)$ and $C_x(\tau)$ are shown in fig 2.

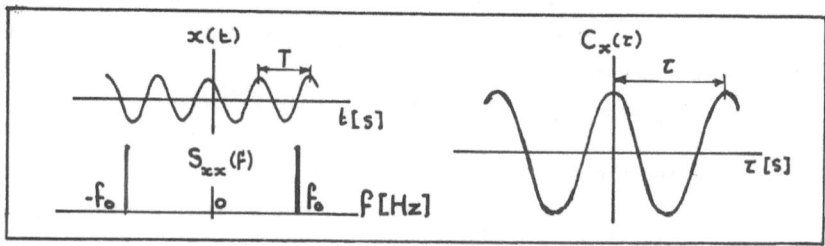

Figure 2. Time signal, spectrum and cepstrum of $x(t) = A \cos(2\pi f_0 t + \varphi)$.

The nonlinear amplitude filtering by taking the log-values of the powerspectrum is done to de-accentuate the larger amplitude frequency components (e.g. carrier frequencies) in order to give some more accent to lower amplitude components (such as modulated sidebands). So one could say that the powerspectrum is flattened, or whitened. Another advantage of the log operation is the separation of source function and transmission path effect as shown earlier.

In analysing speech or seismic signals the amplitude peaks of interest in a powerspectrum will often occur at multiples of some frequency interval, so at frequencies f=kΔ[Hz], k=...,-2,-1,0,1,2,.... In the corresponding cepstrum a distinct peak is then found at τ=1/Δ [s], as shown in fig 3.

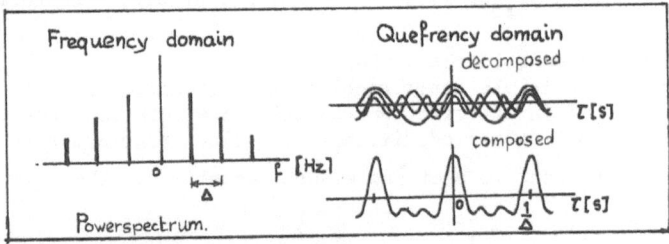

Figure 3. Hypothetical spectrum and its cepstrum.

In the powerspectra of gearbox vibrations a family of sidebands not necessarily go through zero frequency, in other words the first peak of a family of sidebands in the powerspectrum does not need to occur at f=kΔ[Hz], as shown in fig. 4.

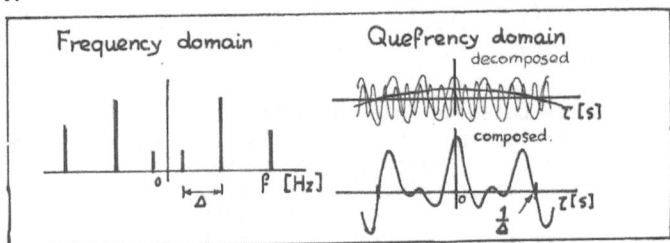

Figure 4. Hypothetical spectrum and its cepstrum.

In the cepstrum of fig. 4 a zero passing occurs at τ=1/Δ instead of a peak as in fig. 3. In relatively complex cepstra these zero passings indicating periodicities in the powerspectrum might be easily overlooked, especially when using absolute values as described in the cepstrum definitions (1) and (5). Another problem occurs when there are also other zero-passings present in the cepstrum which contain no information on sideband spacings as given in fig. 4. This indicates the need for an adaptation of the cepstrum definition when used for analysing gearbox vibration signals. In the formulation of this definition we need the Hilbert transform of which a short explanation will be given first.

A causal function $g(t)$ ($g(t)=0$ for $t<0$) can always be decomposed into an even function $e(t)$ and an odd function $o(t)$. We then can write:

$$e(t) = sgn(t)\, o(t)$$

$$sgn(t) = \begin{cases} 1 & t \geq 0 \\ -1 & t < 0 \end{cases} \qquad (12)$$

$$o(t) = sgn(t)\, e(t)$$

Thus in the time domain there exists a definite relationship between the even and odd part of a causal function. In the frequency domain there is also a relation between the real and imaginary part of the Fouriertransform $G(f)$ of the causal function $g(t)$. This relation is called the Hilbert transform. Let the Fouriertransforms of $g(t)$, $e(t)$ and $o(t)$ be $G(f)$, $R(f)$ and $jX(f)$, (R and X real), then:

$$G(f) = R(f) + jX(f)$$

$$X(f) = -\frac{1}{\pi f} \circledast R(f) \qquad \text{Forward Hilbert Transform} \qquad (13)$$

$$R(f) = \frac{1}{\pi f} \circledast X(f) \qquad \text{Backward Hilbert Transform}$$

where $\circledast$ is the convolution symbol.

We now return to the basic problem: the definition of the cepstrum and the formulation of an efficient way to calculate this cepstrum numerically. The aim was to detect periodicities in the (log-)powerspectrum $S_{xx}(f)$, being a real even function. To meet our wishes, we will use the cepstrum definition:

$$C(\tau) = |\mathscr{F}\{\log S_{xx,1}(f)|^2 \qquad (14)$$

with $S_{xx,1}$ being the 1-sided power spectrum:

$$S_{xx,1}(f) = 2\ S_{xx}(f) \qquad \text{for } f>0$$
$$S_{xx,1}(f) = \quad S_{xx}(f) \qquad \text{for } f=0 \qquad (15)$$
$$S_{xx,1}(f) = \quad 0 \qquad \text{for } f<0$$

This definition is based on three hypotheses (for a prove see [3]):
1) Calculating the cepstrum, the same result is obtained by a forward or inverse transform of the (log-)powerspectrum.
2) It is sufficient to transform the 1-sided spectrum only i.e. the positive frequency values of the (log-)powerspectrum, multiplied by a factor of 2.
3) In some situations, the imaginary part of the cepstrum of a 1-sided powerspectrum, being the Hilbert transform of the real part, can be of great use.

The first two hypotheses are dealing with the use of standard signal analysis equipment for cepstrum analysis, reduction of computertime or the increase of the resolution in the calculated cepstrum by adding zero's to the 1-sided spectrum. The third hypothesis makes it possible to avoid that zeropassings as characteristic points in the cepstrum might be overlooked because zero passings in the real-part of the cepstrum will correspond with peaks in the imaginary-part of the cepstrum.

5. PRACTICAL CONSIDERATIONS USING THE CEPSTRUM ANALYSIS

The cepstrum technique often demands some additional effort and consideration before it can be used succesfully in practice. In the following some practical considerations when using the cepstrum technique for the analysis of the vibration spectrum of gearboxes will be presented. When we start with a spectrum with N spectral lines then a standard FFT algorithm will result in a cepstrum with N/2 significant (complex) values. The resolution in the cepstrum can be increased by appending zeroes to the log spectrum data record before performing the cepstrum computation. A DC-component in the spectrum (resulting for the cepstrum in a value $C(\tau=0)$) has no practical relevance which means that the mean value of the log spectrum can be set to zero before the calculation of the cepstrum. This is likely to reduce the jump (discontinuity) between the last spectrum line and the first appended zero and consequently also will reduce signal leakage. In order to reduce this signal leakage further one might be motivated to use a <u>window-function</u> to the log-spectrum before applying the FFT. Easily can be seen that Hanning-type windows will strongly affect the calculated cepstrum, especially when sideband structures of interest are in the beginning and/or end region of the spectrum. The window should therefore be relatively flat over that por-tion of the log-spectrum containing periodicities. A cosine taper data win-dow after Bingham [3] showed a satisfying behaviour.

In practice there will always be a base noise level in each spectrum which will have a direct effect on the amplitudes of the cepstrum components as indicated in fig. 5.

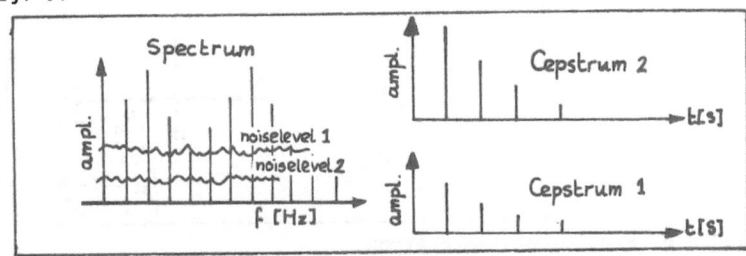

Figure 5. Influence of noise level on cepstrum results.

The consequence of this is that it is only valid to compare cepstra quanti-tatively if the base noise conditions are similar. <u>Bridging</u> occurs when two or more close frequency components in the spectrum are so close that their amplitude peaks cannot be separated and form one broad peak, in other words they are melted together. The possibility of the cepstrum to separate side-band structures with only slightly different sideband spacings consequently will be affected. To be able to detect the smallest modulation frequency f_m in a baseband spectrum the number of spectral lines N_{lines} for cepstrum cal-culation has to be

$$N_{lines} \geq \frac{f_0 Z N_{harm} 8}{f_m} \qquad (16)$$

where: f_0 = shaft rotational speed (often $f_0=f_m$)

      Z = number of teeth

      N_{harm} = number of toothmesh harmonics in the spectrum

If for example, $Z=40$, $N_{harm}=2.5$, $f_0=f_m$ we get $N_{lines} \geq 800$. It will therefore often be necessary to perform a zoom-spectrum analysis in order to obtain sufficient resolution in the original spectrum, before performing the cep-strum analysis. Some analysers have therefore the capability to perform a cepstrumanalysis of a spectrum of up to 4000 lines using a special method of zoom [5].

6. PRACTICAL APPLICATION

A special software program was developed, based on a Nicolet 100A FFT spectrum analyser in combination with a HP9826 desktop computer to perform a cepstrum evaluation. The program uses 4 neighbouring zoomspectra of 400 lines each to get a baseband spectrum with enough resolution. Another 448

188

zero's were added after which a FFT transform on the 2048 spectral data was
carried out, leading to a 1024 amplitude cepstrum according to eqn. (14).
To evaluate the use of the cepstrum technique a gearbox with a transmission
train of 4 gears (see fig. 6) and a load of respectively 1.12, 3.32 and 5.48
KW was analysed.

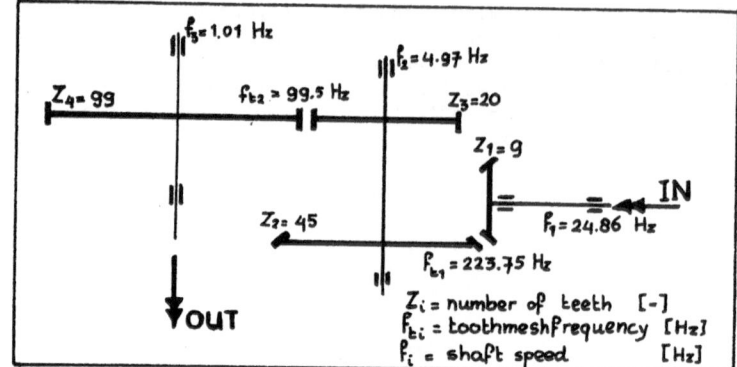

Figure 6. Picture of the gearbox.

Fig. 7a shows a vibrationspectrum at 1.12 KW, measured on the housing of one
of the bearings. High vibration amplitudes can be found at the toothmesh
frequency f_{t_1} = 223.75 [Hz] and a smaller value at the toothmesh frequency
f_{t_2} = 99.50 [Hz]. The corresponding cepstrum is shown in Fig. 7b. The abcis-
sa gives the quefrency [sec] and also the reciproke value of it in [Hz].

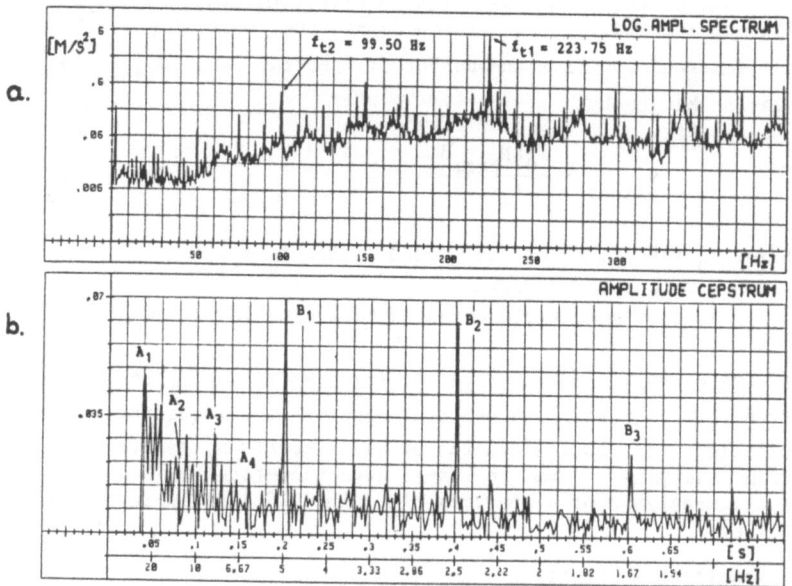

Figure 7. Measured spectrum and calculated cepstrum.

In the cepstrum a high amplitude B_1 at τ = 0.201 [sec] can be detected which

indicates a strong modulation with 4.97 [Hz], which is the shaft rotational speed of shaft 2. The peaks B_2 and B_3 are rahmonics of B_1. The peak A_1 at τ = 0.04 [sec] indicates also a modulation with the shaft rotational speed of shaft 1 (24.86 Hz).
The peaks A_2, A_3 and A_4 are again the rahmonics of peak A_1. The relatively large amplitude of cepstrumcomponent B_1 argues the most serious modulation, namely with frequency f_2 of shaft 2. The real fysical cause for this modulation cannot be located with this cepstrum technique. The observed phenomena however are an indication for:
- An unfavourable toothmeshaction at toothmeshfrequency f_{t1}.
- Mounting errors or excentricity of the shaft with speed f_2.

After opening and inspection of the gearbox it turned out that the pignon wheel 2 was not mounted correctly on its shaft (schrink-fit).
The mounting error resulted in a periodic interference of the pitch circles with a frequency of 4.97 [Hz] leading to this clear amplitude and frequency modulation of the vibration signal. The practical consequence of this condition was a strong wear of the teethprofile which could be observed at the inspection of the gearbox after finishing the measurements.

7. EVALUATION AND CONCLUSIONS
   Apart from the fact that the use of spectrum-cepstrum analysis techniques demand a considerable skill of the user to prevent misinterpretation some complications ofter occur when using this technique on gearboxes in practice:
- The spectra are very sensitive to small load changes and to changes in the rotational frequency,
- There will not always be a definite relationship between changes in the spectrum and the size of a possible mechanical defect,
- It is not always possible to obtain stationary signals for example in case of torque fluctuations, time varying transmission paths as found in epicyclic gearboxes.
- A bad signal to noise ratio complicates the interpretation of the spectra,
- Generalization and standardization is almost impossible. Each gearbox has to be analysed individually.
- It is only possible to compare cepstra qualitatively and not quantitatively.
- The presence of more than one series of rahmonics in the cepstrum might be confusing especially when they are very close.

Due to these complications the enthusiasm on the perspectives of the cepstrum analysis in gearbox monitoring should be somewhat tempered by a healthy scepticism. Nevertheless in many cases the cepstrum will be a useful supplement to the normal spectrum analysis for the detection and evaluation of sidebands in gearbox vibration spectra due to modulation. In order to make these technique more valid, research should be done to get more insight in the effects of mechanical defects or faults on the vibrational behaviour. This points out the need for adequate dynamical models for gearbox systems.

Acknowledgement

The autors wish to thank Ir. Bouwman for the support and the facilities provided by KEMA, Arnhem and Ir. Toersen for his advise and ideas concerning gearbox monitoring and maintenance.

References

[1] Bogert, B.P., Healy, M.J.R., Tukey, J.W., "The Quefrency Analysis of Time Series for Echoes: Cepstrum, Pseudo auto covariance, Cross Cepstrum and Saphe Cracking", Proceedings of a Symposium on Time Series Analysis by M. Rosenblatt, (Ed.), Wiley, N.Y., 1963, pp. 209-243.

[2] Childers, D.G., Skinner, D.P., Kemerait, R.C., "The Cepstrum: A Guide to Processing", Proceedings of the IEEE, Vol. 65, no. 10, oct. 1977.

[3] Stakenborg, M.J.L., "On the Use of Cepstrum Analysis in Gearbox Monitoring", Report WFW 84.025, University Eindhoven, aug. 1984.

[4] Randall, R.B., "A new Method of Modelling Gear Faults", Inl. of Mech. Design, april 1982, Vol. 104/259.

[5] Thrane, N., "Zoom FFT", B&K, Technical Review, no. 2, 1980.

NEW METHOD FOR MEASUREMENT OF TIME CHARACTERISTICS OF MECHANICAL SYSTEMS

D. RUTKOWSKA

TECHNICAL UNIVERSITY OF CZĘSTOCHOWA, POLAND

1. INTRODUCTION

It is well known that mechanical systems can be described by such characteristics as impulse response, step response, frequency response and transfer function. For instance, the suspension system in a car is described by the following transfer function

$$K(s) = \frac{1 + as}{1 + as + bs^2} , \qquad (1)$$

where the positive constants a and b depend on the elasticity, mass and damping coefficient. From (1) we can easily derive the time characteristics - impulse response and step response. Unfortunately, the equality (1) is only a model of the suspension system. This model was established under some assumptions and simplifications (sometimes very rigorous). Consequently, we cannot expect that the step response of the suspension system will be given by

$$L^{-1}(s^{-1} K(s)),$$

where L is the Laplace operator. To avoid such disadvantages we shall propose a new method for direct measurement of time charakteristics of mechanical systems.

2. FORMULATION OF THE PROBLEM

Let us consider a mechanical system described by

$$y(t) = \int_0^t k(l) x(t-l) dl + z(t), \qquad (2)$$

where: $x(t)$ - input signal (e.g. force or turning moment)
$y(t)$ - output signal (e.g. velocity or angular velocity)
$k(t)$ - weighting function (impulse response)
$z(t)$ - measurement noise, such that

$$E[z(t)] = 0, \qquad E[z^2(t)] \leqslant \sigma^2,$$

$$E[z(t') z(t'')] = 0 \quad \text{if} \quad t' \neq t''.$$

It is well known that for the unit-step input the relation (2) becomes

$$y(t) = h(t) + z(t), \qquad (3)$$

where $h(t)$ is the step response.

Suppose that $t \in [0,T]$ and let us partition this interval into n regions $T_1, T_2, \ldots, T_n$, where $T_i = [d_{i-1}, d_i]$, $d_0 = 0$, $d_n = T$ and $\cup T_i = [0,T]$.

Choosing the time points t_i such that $t_i \in T_i$ we may consider the following model

$$Y_i = h(t_i) + Z_i, \qquad i=1,2,\ldots,n, \qquad (4)$$

where: Y_i - measured value of the output signal y at the time t_i

$h(t_i)$ - step response at the time t_i

$Z_i = z(t_i)$ - independent random variables with zero mean and finite variance, i.e.

$$E[Z_i] = 0, \qquad E[Z_i^2] = \sigma_i^2 \leqslant \sigma^2.$$

Our aim is to construct the algorithms for measuring the step response $h(t)$ and the impulse response $k(t)$, given the output observations $Y_i = y(t_i)$, $i=1,2,\ldots,n$.

3. METHOD FOR MEASUREMENT OF THE STEP RESPONSE

Let us expand $h(t)$ in the Fourier series:

$$h(t) \sim \frac{1}{T} + \sum_{j=1}^{\infty} a_j \, \varphi_j(t), \qquad (5)$$

where

$$a_j = \frac{2}{T} \int_0^T h(t) \, \varphi_j(t) \, dt = \frac{2}{T} \sum_{i=1}^{n} \int_{T_i} h(t) \, \varphi_j(t) \, dt \qquad (6)$$

and

$$\varphi_j(t) = \cos \frac{j\pi t}{T}, \qquad j=0,1,2,\ldots. \qquad (7)$$

The following estimator for the step response $h(t)$ will be proposed [1], [3]:

$$\hat{h}_n(t) = \frac{1}{T} + \sum_{j=1}^{N(n)} \hat{a}_{jn} \, \varphi_j(t), \qquad (8)$$

where

$$\hat{a}_{jn} = \frac{2}{T} \sum_{i=1}^{n} Y_i \int_{T_i} \varphi_j(t) \, dt \qquad (9)$$

and $N(n)$ is a sequence of integers, such that $N(n) \xrightarrow{n} \infty$.

Now the convergence properties of this algorithm will be shown. Therefore we consider the mean-square error

$$E(\hat{h}_n(t) - h(t))^2 = \mathrm{var}(\hat{h}_n(t)) + (E(\hat{h}_n(t)) - h(t))^2. \quad (10)$$

Applying the same reasoning as in [3] for the first and the second term of (10) we obtain

$$\mathrm{var}(\hat{h}_n(t)) = \sum_{i=1}^{n} \left(\frac{2}{T} \int_{T_i} \sum_{j=1}^{N(n)} \varphi_j(t) \, \varphi_j(u) \, du \right)^2 \mathrm{var}(h(t_i) + Z_i) \quad (11)$$

and

$$\left| E(\hat{h}_n(t)) - h(t) \right| \leqslant \left| \frac{2}{T} \sum_{j=1}^{N(n)} \varphi_j(t) \sum_{i=1}^{n} \int_{T_i} (h(t_i) - h(u)) \varphi_j(u) \, du \right|$$

$$+ \left| \frac{1}{T} + \sum_{j=1}^{N(n)} \varphi_j(t) \frac{2}{T} \sum_{i=1}^{n} \int_{T_i} \varphi_j(u) h(u) \, du - h(t) \right|. \quad (12)$$

Similarly as in [3] we assume that $h(t)$ satisfies the Lipschitz condition and T_i ($i=1,2,\ldots,n$) fulfil

$$\max_{1 \leqslant i \leqslant n} \left| d_i - d_{i-1} \right| = O(n^{-1}). \quad (13)$$

Using this assumptions for (11) and (12) we have

$$E(\hat{h}_n(t) - h(t))^2 \leqslant c_1 \, n^{-1} \, N^2(n) + (c_2 \, n^{-1} \, N(n))^2$$

$$+ 2 \left| \frac{1}{T} + \sum_{j=1}^{N(n)} a_j \, \varphi_j(t) - h(t) \right|^2, \quad (14)$$

where c_1, c_2 are constants.
From (14) it follows that if

$$n^{-1} \, N^2(n) \xrightarrow{n} 0, \quad (15)$$

then

$$E(\hat{h}_n(t) - h(t))^2 \xrightarrow{n} 0. \quad (16)$$

Thus, the condition (15) imposed on the sequence $N(n)$ assures the convergence of the algorithm (8).

4. METHOD FOR MEASUREMENT OF THE IMPULSE RESPONSE

It is known that the relation between the step response and the impulse response is following:

$$k(t) = \frac{d}{dt} h(t). \tag{17}$$

Let us expand $k(t)$ in the following Fourier series:

$$k(t) \sim \sum_{j=1}^{\infty} b_j \, \psi_j(t), \tag{18}$$

where

$$b_j = \frac{2}{T} \int_0^T k(t) \, \psi_j(t) \, dt \tag{19}$$

and

$$\psi_j(t) = \sin \frac{j\pi t}{T}, \qquad j=0,1,2,\dots \, . \tag{20}$$

It is easily to show that

$$b_j = - \frac{j\pi}{T} a_j. \tag{21}$$

We propose the following estimator for the impulse response $k(t)$:

$$\hat{k}_n(t) = \sum_{j=1}^{M(n)} \hat{b}_{jn} \, \psi_j(t), \tag{22}$$

where

$$\hat{b}_{jn} = - \frac{j\pi}{T} \hat{a}_{jn} \tag{23}$$

and $M(n)$ is a sequence of integers, such that $M(n) \xrightarrow{n} \infty$. This estimator may be easily obtained by taking the derivative of $\hat{h}_n(t)$ [2].

The condition, imposed on the sequence $M(n)$, guaranteeing the convergence of the procedure (22) will be derived by the same way as in section 3. Analogously, for this purpose we consider the mean-square error for $\hat{k}_n(t)$. Thus we have

$$\mathrm{var}(\hat{k}_n(t)) = \sum_{i=1}^{n} \left(- \frac{2\pi}{T^2} \int_{T_i} \sum_{j=1}^{M(n)} j \, \psi_j(t) \, \varphi_j(u) \, du \right)^2 \mathrm{var}(h(t_i)+z_i)$$

$$\leqslant c_3 \, n^{-1} \, N^4(n)$$

and

$$\left| E(\hat{k}_n(t)) - k(t) \right| \leq \left| -\frac{2\pi}{T^2} \sum_{j=1}^{M(n)} j \, \psi_j(t) \sum_{i=1}^{n} \int_{T_i} (h(t_i) - h(u)) \, \varphi_j(u) \, du \right|$$

$$+ \left| -\frac{\pi}{T} \sum_{j=1}^{M(n)} j \, a_j \, \psi_j(t) - k(t) \right|$$

$$\leq c_4 \, n^{-1} \, M^2(n) + \left| \sum_{j=1}^{M(n)} b_j \, \psi_j(t) - k(t) \right|,$$

where c_3, c_4 are constants.

The conclusion concerned with the convergence of the algorithm (22) may be formulated as follows: If $h(t)$ satisfies the Lipschitz condition, (13) is fulfiled, and

$$n^{-1} \, M^4(n) \xrightarrow{n} 0, \tag{24}$$

then

$$E(\hat{k}_n(t) - k(t))^2 \xrightarrow{n} 0. \tag{25}$$

5. FINAL REMARKS

The presented algorithms of measurement of time characteristics may be realized by means of a computer. Based on the measured values of the output signal $y(t_1)$, $y(t_2)$,...,$y(t_n)$, the computer evaluates the step response $h(t)$ or/and the impulse response $k(t)$ according to the procedure (8) or (22), respectively. The sequence $N(n)$ must be choosen by formula (15) and $M(n)$ by (24). As is easily seen, for measuring both characteristics we may choose $N(n) = M(n)$ in accordance with (24). Obviously, the accuracy of the proposed method increases when $n \to \infty$.

REFERENCES

1. Rutkowska D: New methods for estimation of EEG signals, submitted for publication.
2. Rutkowska D: A nonparametric method for measurement of size distribution of martensite plates, submitted for publication.
3. Rutkowski L: On system identification by nonparametric function fitting. IEEE Trans. Autom. Contr., vol. AC-27, No.1, pp.225-227, 1982.

AUTOMATIC DISPLAY OF CAMPBELL DIAGRAMS OF START-UP AND RUN-DOWN PROCESSES

A. LINGENER, Prof.Dr.sc.techn.
G. SCHMIDT, Dr.-Ing.
University of Technology Otto von Guericke, Magdeburg, GDR
Department of Mechanical Engineering

SUMMARY: Evaluations of vibration measurements in frequency domain are often carried out by Campbell-diagrams and order analysis. A method working without speed sensors is presented. Dominating frequency lines are followed automatically by shifting a small frequency window over the speed-depending spectra and finding out the frequency line with maximum power in the window. Applications are shown for a steam-turbine and a turbocompressor with gearbox and asynchronuous motor.

1. INTRODUCTION

A complete experimental investigation of dynamic behavior of machines with rotating and oscillating parts, such as all kinds of prime movers demands large-scale measuring information and its comprehensive evaluation and presentation. The excitations acting on this kind of machines depend mostly on rotational speeds of the rotating parts. Predominantly, there is the question of multiples of a basic speed or in the case of machines with gearboxes of two or some more speeds, connected by the transmission ratio of the gear. External exitation sources and self-excited vibrations additionally may impair smooth operation.

The objektives of experimental investigations are: to discover excitations, their causes and their removal in the case of inadmissible vibrations.

2. CAMPBELL-DIAGRAM AND ORDER ANALYSIS

Campbell-diagrams allow fundamental statements on vibrational behavior of machines and their elements (turbine blades), respectively. The crossings of excitation orders with eigenfrequencies define critical or dangerous speeds respectively. Such kinds of diagrams may be calculated in the frame of dynamic calculations. They point out possible risks. As the amount of excitations is mostly unknown, statements about real risks in the first line by higher orders remain unanswered. Experimental determination of Campbell-diagrams is carried out by vibration measurements at the points of interest during the start-up or run-down process of the machine concerned. Futhermore, it is neccessary that in sufficiently small speed intervals a spectral analysis of the measuring signal is undertaken and the resonance peaks appearing are to be related to the different orders. At the same time, the amplitudes of each excitation order are being obtained.

In fig 1 is presented a measured Campbell-diagram of a steam

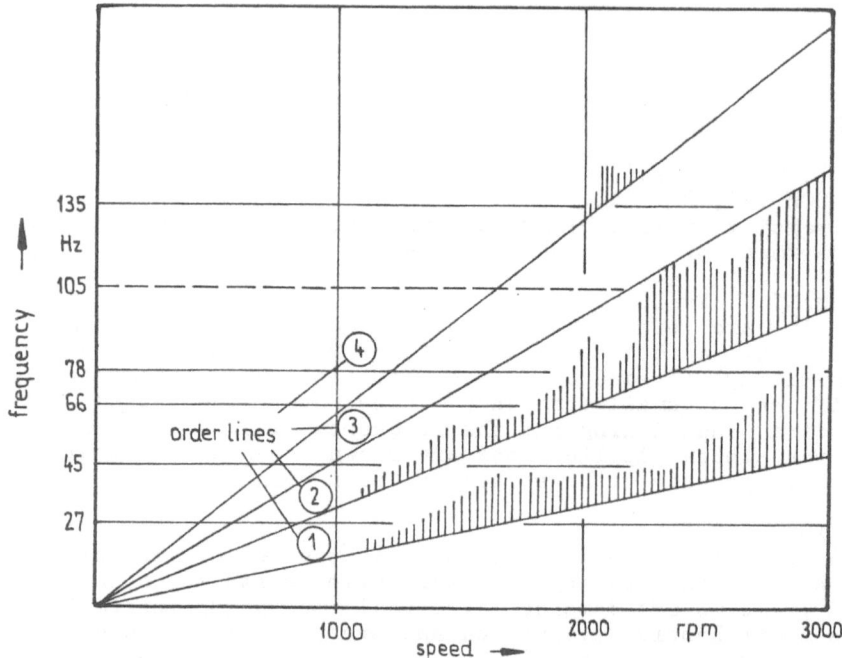

fig. 1   Measured   Campbell - diagramm of a steam turbine

turbine. Additionally the dots along the order-lines represent the amplitudes of the corresponding harmonics.

The spectra required, from which the lines belonging to the speed multiples are taken, are obtained by analysing short cut-outs of the signal. These cut-outs are considered as stationary and transformed into frequency domain by the methods of digital signal processing. The signal is sampled with a fixed frequency, corresponding to the frequency range of interest.

A disadvantage of this method is the "smearing" of spectral lines with high frequencies, as the rotational speed is not exactly constant. For instance, toothmesh frequencies of gearboxes can be detected only with difficulties. This disadvantage is avoided by the so-called order analysis [2]. Here, the sampling frequency is controlled by the variable speed (tracking). For instance, when evaluating the run-down of a machine, the sampling frequency is decreased accordingly. Consequently, a speed-proportional line in the spectrum now appears always at the same point, e.g. at a fixed line-number of the spectrum.

By displaying such kind of order analysis in a diagram, exciting orders become straights parallel to the abscissa axis and constant frequencies appear as hyperbolae (fig 2). The disadvantage of this method is, the upper limiting frequency of anti-alaising filters has to be changed with decreasing sampling frequency. A speed impulse is needed, too.

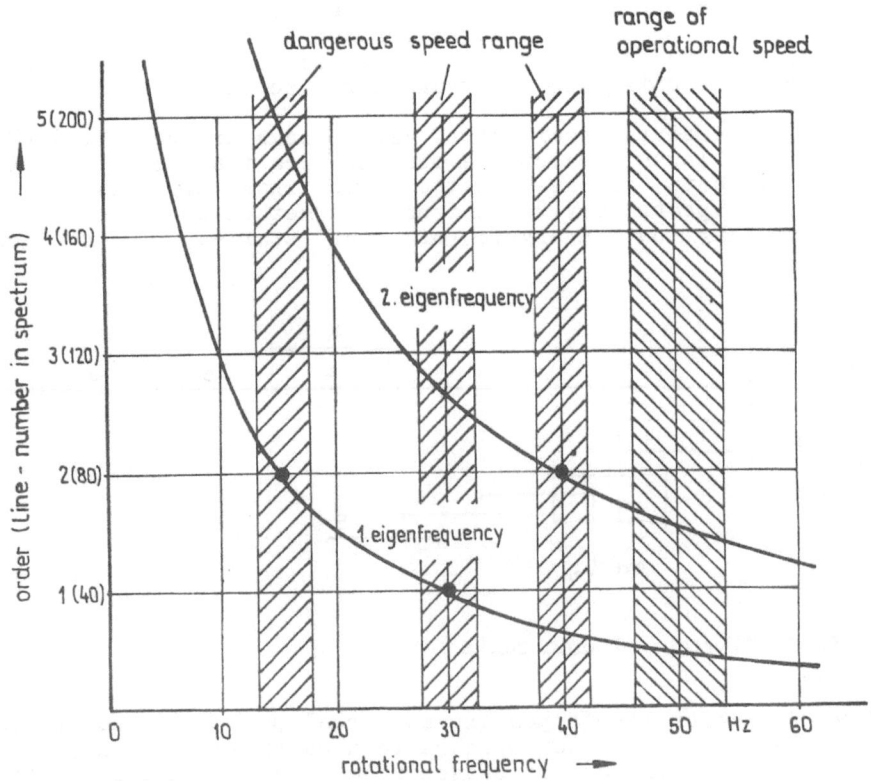

fig. 2 Principle of order analysis with speed-proportional
sampling frequency

3. EVALUTION WITHOUT USING A SEPARATE SPEED IMPULSE

In the following, a method working without an external speed impulse is presented. The neccessity arose from series of measurements in the field of demage diagnostics where the machines were available only for a short time or a speed impulse could be accomplished only with high technological efforts. The method is based on stored time signals measured at the points of interest of the machine investigated. The method is integrated in the program package ASAN [3,4] for analysing systems and analogous measuring signals.

3.1. <u>Implementation on computer</u>. The analogous measured data are read at first via input amplifier and low-pass filter. As the start-up or run-down of the machine may take a longer time interval, because of the limited core storage capacity the measuring signals are divided into blocks of 1024 data and stored on a data disc. An acquisition maximum of 8 blocks in series is possible. The time length T_B of one data block depends on the investigated frequency range defined by sampling frequency. Here applies $T_B = 1024/f_s$. For example using a sampling frequency of 400 Hz processes of a length up to 22 minutes may be analysed. Faster rundows demand higher sampling frequencies. For calculating power or rms-spectra the measured data from the data disc are transferred to the central memory again.

In this procedure, overlaps of data blocks are permitted (fig 3). This allows to calculate more spectra than the original number of data blocks and the time interval between two successive spectra may be dimished, a method used especially for carrying out analysis of fast start-up or run-down processes.

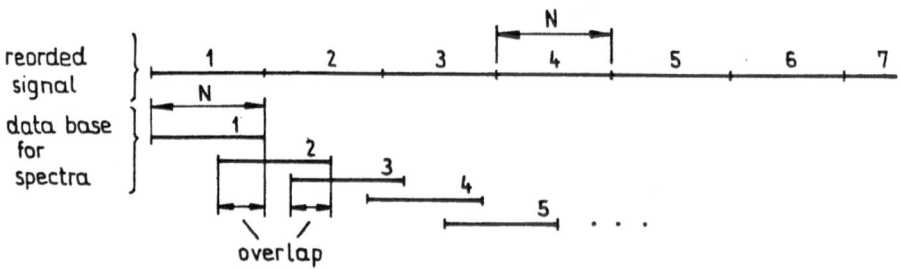

fig. 3   Overlap of data blocks

One frequency line appearing as a real peak (maximum power) and included in all other spectra (caused by an arbitrary speed – proportional share) is chosen from each separate partial spectrum. The numbers of these lines (0,...,511) in each spectrum are marked and stored internally. These numbers are used later as reference frequencies for all simultaneously recorded measurements. If available a speed impulse may be used of course instead of a reference frequency.

3.2. <u>Presentation of results</u>. For the evaluation of all partial spectra (power or rms) it is possible to display them successively on an CRT-display so that start-up or run-down can be watched directly in frequency domain. The shifting of frequencies or change of amplitudes in this way are clearly seen. At the same time it is possible to plot the time-dependent partial spectra in a three-dimensional presentation as the waterfall diagram in figure 4.

3.3. <u>Tracking of spectral lines</u>. An important part of the analysing program consists of automatically tracking defined frequency lines without speed impulse. We have to distinguish between tracking of speed-proportional frequencies (unbalance, reactive forces) and fixed frequencies (eigenfrequencies).

In the case of speed-proportional frequencies from the first partial spectrum the number of the frequency line of interest is extracted. Starting with this number, beginning with the first spectrum a frequency window of defined bandwith is shifted over all partial spectra. At the second spectrum the number of the frequency line with maximum power is seeked within the window. The maximum value is stored with the values of two sidebands being added previously at any time for correcting leakage effects. The position of the window depends on the reference frequency, taken from the reference frequency line.

That means, the window is shifted over the partial spectra independent of the appearance of a real maximum value within the window. By this procedure an aberration in consequence of ossasional disappearence of frequency lines is avoided. The values of single lines in the partial spectra may finally be plotted in dependence on frequency. At the right margin of fig 4 we see three diagrams with the frequency lines f_n, $2f_n$, $4f_n$ with f_n-rotating frequency.

When tracking fixed frequencies the position of the window is not changed. As the result we obtain the power of the fixed frequency line in dependence on the number of the partial spectrum, e.g. time. From such a diagram (fig 7 last diagram) can be seen, at which time or in connection with the reference frequency, at which speed an eigenfrequency is excited. A concentrated computer-controlled representation of all variable and fixed frequency shares in a Campbell-diagram as shown in fig 1 or order analysis diagram (fig 2) is possible, too.

The evidence of the diagrams increases, if the start-ups or run-downs, respectively are carried out continuously and slowly, especially to obtain a good quality of tracked lines at higher frequency ranges.

The method described here permits analyses of machine vibrations during start-up and run-down processes without using speed impulses. It has to be supposed, of course, that during the whole time of measurement at least one frequency line in each spectrum can be distinguished as a separate peak. This supposition is still fulfilled for relatively small values of the peaks, as will be shown in the following applications.

4. APPLICATIONS

4.1. Steam-turbine. During the run-down of a steam-turbine in a power plant vibrations were measured at several bearings and recorded on a tape. The recorded time interval of run-down was 12 minutes. For evalution the recorded signals were digitised with a sampling frequency of 400 Hz and 200 blocks with 1024 data each were stored for each channel.

As a whole, 500 seconds of the run-down process were acquired. As the time interval of one block (2,5 sec) was small relative to the total run-down time, the signals within one block could be treated as stationary, consequently an overlap according to fig 3 was not neccessary. Fig 4 shows the result of the measurement at the bearing of the turbine at the side of the generator. We can see clearly two excitations: The rotating frequency and its double, the latter caused by the two different geometrical moments of inductor shaft. The difference with the Campbell-diagram (fig 1) consists of the fact, that here the order lines appear as curves since braking at higher speeds is more powerful but the blocks were acquired in

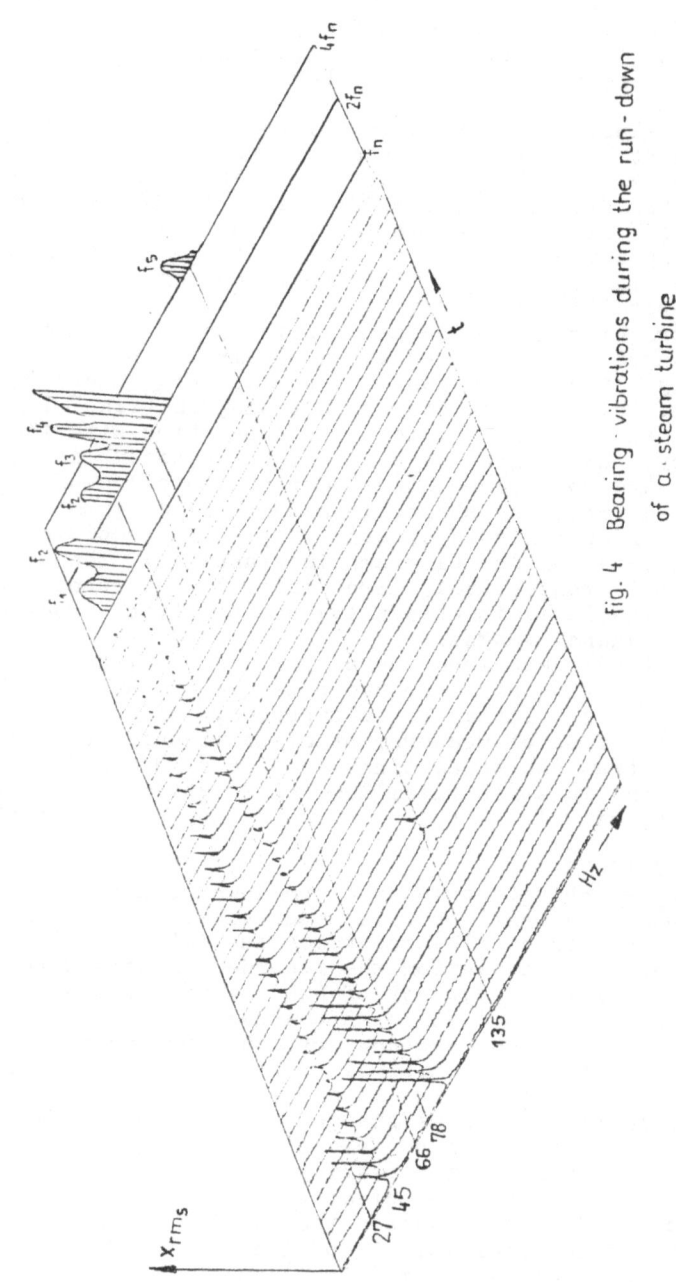

fig. 4 Bearing · vibrations during the run - down
of a · steam turbine

constant time intervals.

At the right margin the resonance diagrams are shown from which we can recognise four eigenfrequencies in the range of operational speeds. The small peak on the 4 th order-lines is of special interest, as it indicates a very high eigenfrequency. The excitation intensity by the various multiples of rotational speed is different, so that the amplitudes of the corresponding resonance peaks are not the same.

The third order of the rotational speed does not appear at all. The operational speed of 50 Hz is situated closely above the second eigenfrequency.

4.2. <u>Axial compressor with electric motor drive</u>. The problem under question concernes an axial turbocompressor driven by an asynchronous motor by means of a toothed gearing. The reason of the investigations were heavy vibrations of the bearing of the compressor at the side of the gear with a 196 Hz frequency. At first a spectral analysis at operational speed (fig 5) was carried out. It shows numerous frequencies from which 196 Hz, corresponding to the fourfold of the motor speed, is dominating. Further we recognise the rotational speed $f_c = 229$ Hz of the compressor.

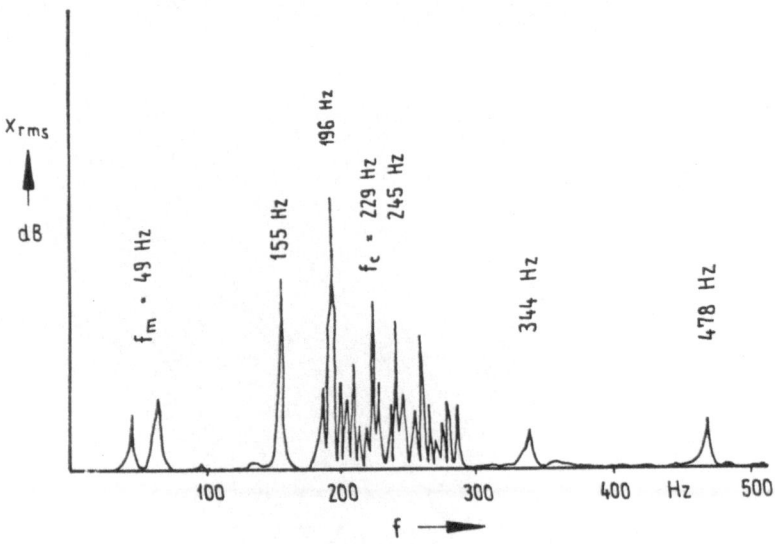

fig. 5 Spectrum at operational speed before run - down

By recording and evaluating a run-down, representing time-dependent spectra (fig 6), statements about speed dependence on single harmonics and excitations could be made. In this figure the frequencies f_m (motor rotational frequency 49 Hz), $3f_m$, $4f_m$, $5f_m$ and f_c (compressor rotational frequency 229 Hz) are visible indeed, but altogether the representation is relatively

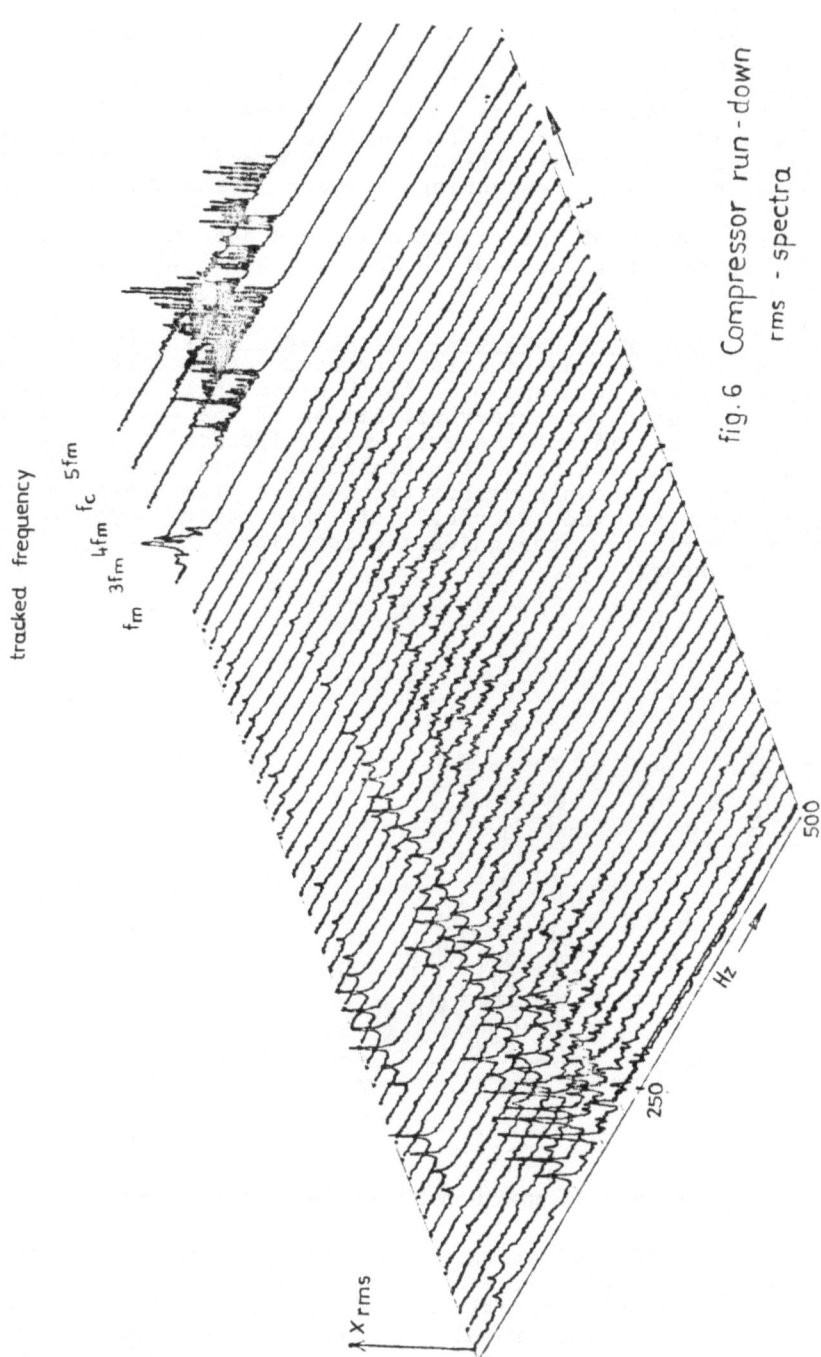

tracked frequency

f_m $3f_m$ $4f_m$ f_c $5f_m$

X_{rms}

Hz

250

500

fig. 6 Compressor run-down

rms - spectra

obscure. On the other hand the tracking of these different lines during run-down provides the desired information. As the resonance diagrams at the margin of fig 6 have a low resolution, these curves were plotted separately (fig 7). The frequency ranges of interest were zoomed in these diagrams.

The frequencies 33, 39, 129, 165, 191 and 233 Hz are marked distinctly as unambious resonances. All resonance diagrams were obtained by tracking the frequency lines observed at operational speed. It is remarkable, that the line 98 Hz = $2f_m$, which is nearly not recognizable in the spectrum, fig 5, could be tracked. After a 10-fold magnification we recognise the lowest eigenfrequencies 33 and 39 Hz and their double.

Fixed frequencies can be tracked in the same way like the manyfolds of the speed-proportional frequencies. This is done in the lower row on the right side in fig 7 with the eigenfrequency of 129 Hz. The power of the frequency band containing this line is represented. The abscissa axis represents time, i.e. a decreasing speed. The peaks of this diagram may be assigned to defined exciting frequencies. By frequencies lower than the 3 rd harmonic of the motor speed this line cannot be excited. The excitation by the compressor speed f_c itself is especially marked.

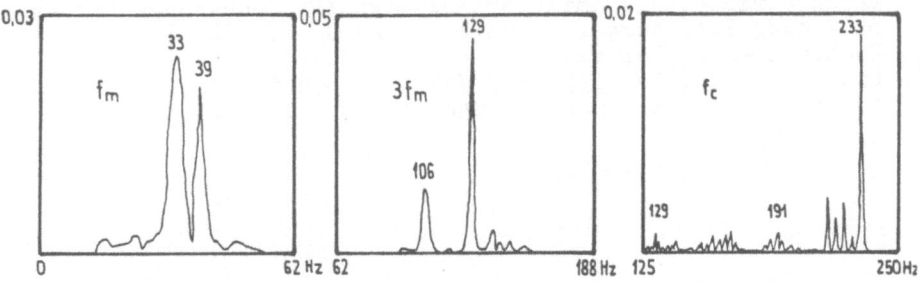

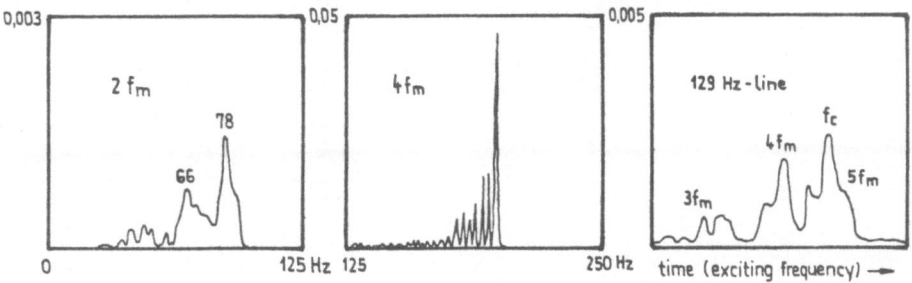

fig. 7 Automatically tracked frequency lines (zoomed)

The essential result of this evaluation was the awareness of the fact that the strong vibrations were caused by an eigenfrequency of the gear close to the fourfold of the motor speed. Its effect was amplified by a fault during the assembly of the coupling. After the correction of this fault a smooth operation could be guaranteed. The position of the resonance at 233 Hz which is situated between the excitations f_c and $5 f_m$ is not favourable, too.

This application demonstrates the efficiency of this method. All essential statements about the machine can practically be made from a single measurement without separate speed impulse. Evaluations of several measuring points give a comprehensive image of the dynamical behavior of the machine under investigation.

REFERENCES
1. Holzweissig, F./Dresig, H. Lehrbuch der Maschinendynamik S.205, Fachbuchverlag Leipzig
2. Herlufsen, H. Order Analysis using Zoom-FFT. Bruel & Kjaer Applikation note. Naerum Denmark 1981
3. Schmidt, G. Kurzbeschreibung eines Programmsystems zur digitalen Verarbeitung analoger Messsignale. Forschungsberichte 6. Tagung Festkoerpermechanik Band C S. XLIII/1-9 Fachbuchverlag Leipzig
4. Duckstein, H., Schmidt, G., Wahl, F. Schwingungsanalyse der Auslaufvorgaenge von Maschinenanlagen. loc.cit. XLI/1-10

DETERMINATION OF MATERIAL CONSTANTS USING EXPERIMENTAL FREE VIBRATION ANALYSIS ON ANISOTROPIC PLATES

W.P. DE WILDE & H. SOL

Free University of Brussels (V.U.B.) - Dept. of Structural Analysis
Pleinlaan 2 ; B-1050 Brussels (Belgium)

ABSTRACT
In this paper, a method is presented which determines anisotropic plate rigidities D_{ij} using vibration data (eigenfrequencies, modeshapes).
The vibration data are experimentally obtained from rectangular testplates with 4 free boundaries. The measured data are compared with the results of an analytical model of the same test plates.
The parameters of this analytical model are the unknown plate rigidities D_{ij}. An eigenvalue problem $[K_{ij}]-\lambda[M_{ij}]$, in which $[K_{ij}]$ is the stiffness matrix and $[M_{ij}]$ the mass matrix (both based on the mathematical model) can be formulated. The solution of this eigenvalue problem yields eigenfrequencies and modeshapes.
The basic idea of the method is to tune the parameters D_{ij} until the eigendata, produced by the mathematical model, fit the experimentally obtained eigendata. Although the method works iteratively, no preliminary knowledge of the plate rigidities is required. For orthotropic or isotropic materials, only the resonance frequencies must be measured. The measurement can be carried out fast and accurately with an accelerometer and a spectrum analyser. For anisotropic materials, also the modeshapes must be measured.
The obtained plate rigidities can be used directly as input data for finite element models, or can be used to calculate engineering elasticity parameters. For layered composite materials with a known stacking sequence, it is possible to calculate the dynamic elasticity moduli E_1, E_2, ν_{12} and G_{12} of a single lamina from the measured D_{ij}.

1. INTRODUCTION
    The use of the finite element method for the analysis of complicated structures becomes more and more successful. The appearance of powerful digital computers allows static and dynamic analysis of finite element models with several thousands degrees of freedom.
A finite element model of an arbitrary structure requires mainly 2 kinds of information blocks : geometrical data (nodal points coordinates, element connectivities and boundary conditions) and material behaviour data. The last one is the most difficult to establish. In most practical calculations, linear material behaviour is assumed.
The analysis of laminated fiber reinforced polymer structures ("composites") requires the knowledge of anisotropic elastic properties for each element. Because the material properties of composites are production system dependent, it is generally impossible to find these properties in tables or databases. The only way to establish the properties is to measure them experimentally.
In this paper we discuss a method which is based on experimentally determined vibration data, taken from completely free rectangular plates.

The vibration data consists of resonance frequencies and their accompanying modeshapes ("modal data").
The test plates were hung up on 2 thin elastic threads (Fig. 1).

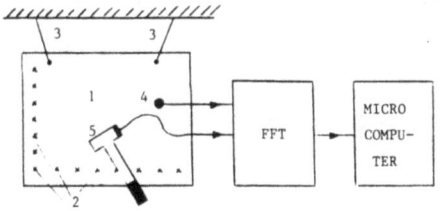

1. Test plate
2. Measuring points
3. Thin threads
4. Accelerometer
5. Impact hammer

Fig. 1. Experimental set-up.

This configuration simulates the completely free boundary conditions fairly well. The modal data can be measured using an experimental modal analysis test set-up. The natural eigenfrequencies and modeshapes are extracted from frequency respons functions in a number of points of the test plate. The frequency respons function in a point can be registrated by taking the fast fourier transform of the displacement respons in the time domain, caused by an impact in another (or the same) point of the plate with a hammer.

For orthotropic or isotropic plates, it is sufficient to measure only the resonance frequencies. These resonance frequencies together with mathematically generated modeshapes [1],[2], are sufficient to start a stable iteration procedure.

In this iteration procedure, the eigenfrequencies calculated with a Rayleigh-Ritz parametrical model of the test plates are compared with the measured eigenfrequencies. The parameters in the analytical model are the so-called plate rigidities D_{ij}.

The iteration cycle stops if the calculated eigenfrequencies of the mathematical model are sufficiently close to the measured eigenfrequencies. After every iteration step the parameters are updated using a sensitivity technique.

For anisotropic plates, the modeshapes used for the start of the iteration, must be measured together with the resonance frequencies.

2. THE ANALYTICAL MODEL FOR AN ANISOTROPIC PLATE WITH 4 FREE EDGES

For the analytical model, only thin plates, subjected to small lateral deflections are considered. If w presents the lateral deflection in a point (x,y) of the plate, the equilibrium equation is given by (1) :

$$D_{11} \frac{\partial^4 w}{\partial x^4} + 4D_{16} \frac{\partial^4 w}{\partial x^3 \partial y} + 2(D_{12}+2D_{66}) \frac{\partial^4 w}{\partial x^2 \partial y^2} + 4D_{26} \frac{\partial^4 w}{\partial x \partial y^3} + D_{22} \frac{\partial^4 w}{\partial y^4} = -\rho \frac{\partial^2 w}{\partial t^2} \quad (1)$$

In this equation of motion the friction with the surrounding medium and the internal damping are disregarded. The solution of (1) can be presented in the form of a product, with separation of time- and displacement variables:

$$w(x,y,t) = W(x,y) \, e^{i\omega t} \quad (2)$$

If we use the Rayleigh-Ritz method to find a solution of (1), we are looking for functions W which satisfy the boundary conditions and represent

an approximate shape of the deflected surface of the vibrating plate.
The global formulation of (1) yield an eigenvalue problem (3) :

$$(K_{ij} - \lambda M_{ij}) \{W_j\} = \{0\} \tag{3}$$

with

$$K_{ij} = \frac{4b}{a^3} D_{11}A_{ij} + \frac{4a}{b^3} D_{22}B_{ij} + \frac{4}{ab} D_{12}C_{ij} + \frac{16}{ab} D_{66}E_{ij} + \frac{8}{a^2} D_{16}F_{ij} + \frac{8}{b^2} D_{26}G_{ij} \tag{4}$$

$$M_{ij} = \frac{ab}{4} H_{ij} \tag{5}$$

The matrices A, B, C, D, E, F, G and H are constant matrices containing
partial derivatives of so-called shape functions, belonging to discrete
points of the plate.
In [3] it was proven that complete lagrangian polynomials are very suitable
as such shape functions.
Fig. 2 shows a 3-dimensional plot of a (7x7)-lagrange polynomial.

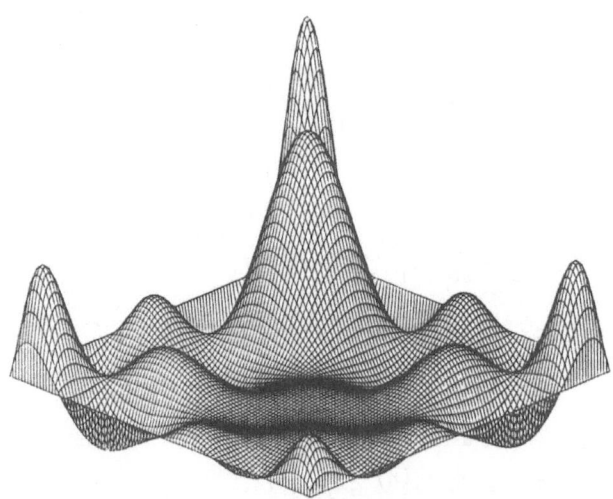

Fig. 2. Lagrangian polynomial of order (7x7).

Once the shape functions are choosen, the matrices A, B, C, E, F and G
can be calculated and it is seen (4), that the stiffness matrix K is a
linear expression of the plate rigidities D_{mn}. The numerical solution of
eigenproblem (3) yields eigenvalues $\lambda^{(k)}$ and their modeshapes $\phi_i^{(k)}$.

3. DETERMINATION OF PLATE RIGIDITIES
    If $\lambda^{(k)}$ and $\phi_i^{(k)}$ are an eigenpair-solution of (6) and if the eigenvec-
tor $\lambda_i^{(k)}$ is normalised as in (7)

$$^\tau\phi_i^{(k)} M_{ij} \phi_j^{(k)} = 1 \tag{7}$$

than

$$\lambda^{(k)} = {}^{\tau}\phi_i \, k_{ij} \, \phi_j^{(k)} \tag{8}$$

Writing (8) for n eigenvalues, and using (4) :

$$\left\{\begin{array}{c} \lambda^{(1)} \\ \lambda^{(2)} \\ \lambda^{(3)} \\ \cdots \\ \cdots \\ \cdots \\ \lambda^{(n)} \end{array}\right\} = \left|\begin{array}{cccccc} S_{11} & S_{12} & S_{13} & S_{14} & S_{15} & S_{16} \\ S_{21} & S_{22} & S_{23} & S_{24} & S_{25} & S_{26} \\ S_{31} & S_{32} & S_{33} & S_{34} & S_{35} & S_{36} \\ \cdots & \cdots & \cdots & \cdots & \cdots & \cdots \\ \cdots & \cdots & \cdots & \cdots & \cdots & \cdots \\ \cdots & \cdots & \cdots & \cdots & \cdots & \cdots \\ S_{n1} & S_{n2} & S_{n3} & S_{n4} & S_{n5} & S_{n6} \end{array}\right| \left\{\begin{array}{c} D_{11} \\ D_{22} \\ D_{12} \\ D_{66} \\ D_{16} \\ D_{26} \end{array}\right\} \tag{9}$$

with

$$S_{k1} = \frac{4b}{a^3} \, {}^{\tau}\phi_i^{(k)} \, A_{ij} \, \phi_j^{(k)}$$

$$S_{k2} = \frac{4a}{b^3} \, {}^{\tau}\phi_i^{(k)} \, B_{ij} \, \phi_j^{(k)}$$

$$S_{k3} = \frac{4}{ab} \, {}^{\tau}\phi_i^{(k)} \, C_{ij} \, \phi_j^{(k)} \qquad (k : 1\ldots n) \tag{10}$$

$$S_{k4} = \frac{16}{ab} \, {}^{\tau}\phi_i^{(k)} \, E_{ij} \, \phi_j^{(k)}$$

$$S_{k5} = \frac{8}{a^2} \, {}^{\tau}\phi_i^{(k)} \, F_{ij} \, \phi_j^{(k)}$$

$$S_{k6} = \frac{8}{b^2} \, {}^{\tau}\phi_i^{(k)} \, G_{ij} \, \phi_j^{(k)}$$

Inspection of (9) shows that S_{kj} physically presents the sensitivities for $\lambda(k)$ for changes of D_{mn}.
In (10) A, B, ..., G are known constant matrices and a and b be easily measured.
So if $\lambda(k)$ and $\phi_i^{(k)}$ can be measured experimentally, the set of equations (9) can be solved for D_{mn}.
Due to the inevitable measurement errors on $\phi_i^{(k)}$, the set of equations (9) must be overdetermined in order to obtain useful results for D_{mn}.
This can be achieved in two ways :

1) measurement of sufficent eigenpairs on 1 test plate
2) measurement of eigenpairs on a number of test plates with the same D_{mn}-value, but with different sizes.

In spite of this overdetermination of (9), the results obtained by immediate injection of experimental values of $\phi_i^{(k)}$ in (10) were found to be
unuseful.
Therefore it was necessary to look for a method in which the immediate
interference of the modeshapes in the result was avoided. The answer was
found in the use of an iteration technique.
In this iteration procedure a first estimate of D_{mn} is obtained by solving
(9) with starting values of $\phi_i^{(k)}$ and the measured $\lambda(k)$, (Fig. 3).

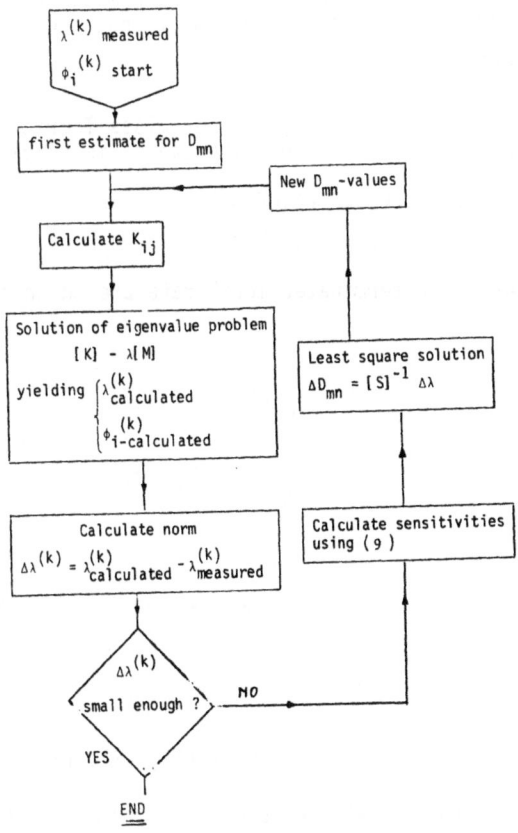

Fig. 3 . Flowchart of material identification program.

With these values of D_{mn}, the stiffness matrix K_{ij} can be evaluated using
(4). The mass matrix M_{ij} can be calculated using (5). The solution of
eigenvalue problem (3) yields values $\lambda_{calculated}^{(k)}$.
The $\lambda_{calculated}^{(k)}$ are compared with $\lambda_{measured}^{(k)}$.
If the norm $\Delta\lambda_{calculated}^{(k)} - \lambda_{measured}^{(k)}$ is sufficient small, the procedure
ends.

If it is not, the sensitivities S of $\lambda^{(k)}$ for changes D_{mn} are calculated using $\phi^{(k)}_{i-calculated}$ and (10).

With these sensitivities and with $\Delta\lambda^{(k)}$, corrections ΔD_{mn} are calculated as the least square solution of (9).

This procedure can be repeated until the calculated values for $\lambda^{(k)}$ approximate the measured values for $\lambda^{(k)}$ sufficiently close.

For orthotropic or isotropic material behaviour, the modeshapes $\phi^{(k)}_j$ used for the start procedure, can be calculated mathematically (see [1] and [2]). In this case, only the resonance frequencies must be measured.

For anisotropic material behaviour also the modeshapes must be measured. After smoothing with bicubic spline functions, these modeshapes can be used for $\phi^{(k)}_{start}$.

3. EXAMPLE

Six layers of a (0°-90°) woven fiber glass fabric impregnated in polyester resin were hot-pressed into transverse isotropic rectangular plates :

plate 1 : (.236 x .209 x .00165 m^3)

plate 2 : (.210 x .209 x .00165 m^3)

The experimentally determined modal data are shown in Table 1 and Fig. 4.

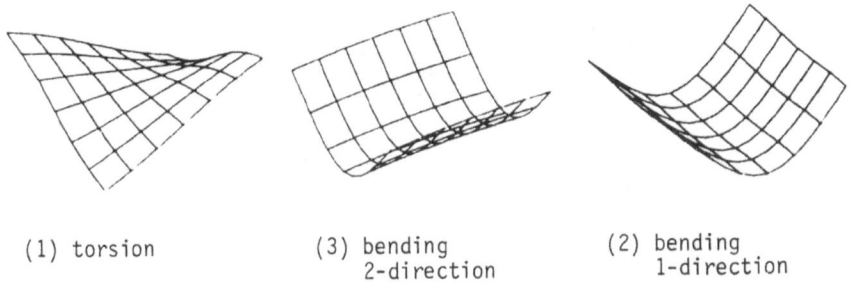

(1) torsion   (3) bending 2-direction   (2) bending 1-direction

Fig. 4. Modetypes (1), (2), (3).

TABLE 1. Experimental values for glassfiber/polyester plates

mode number	frequency plate 1	frequency plate 2	modetype plate 1	modetype plate 2
1	51.33	57.16	(1)	(1)
2	92.53	115.43	(2)	(2)
3	132.87	132.06	(3)	(3)

The plate rigidities were found to be :

$$D_{11} = 6.81 \text{ Nm} \qquad D_{12} = 0.76 \text{ Nm}$$
$$D_{22} = 7.95 \text{ Nm} \qquad D_{66} = 1.46 \text{ Nm}$$

The relations between the plate rigidities and the elastic constants for a transverse isotropic material are :

$$D_{11} = \frac{E_1}{1-\nu_{12}\nu_{21}} \frac{h^3}{12}$$

$$D_{22} = \frac{E_2}{1-\nu_{12}\nu_{21}} \frac{h^3}{12}$$

$$\frac{E_1}{E_2} = \frac{\nu_{12}}{\nu_{21}}$$

$$D_{12} = \frac{12E_2}{1-\nu_{12}\nu_{21}} \frac{h^3}{12}$$

$$D_{66} = \frac{G_{12}}{12} h^3$$

In table 2, the results of the vibration tests are compared with results of tensile tests on an INSTRON-machine :

TABLE 2. Experimental results compared with vibration test results

VIBRATION	INSTRON
$E_1 = 1.80E + 10 \text{ N/m}^2$	$E_1 = 1.82E + 10 \text{ N/m}^2$
$E_2 = 2.10E + 10 \text{ N/m}^2$	$E_2 = 2.11E + 10 \text{ N/m}^2$
$\nu_{12} = 0.096$	$\nu_{12} = 0.095$
$\nu_{21} = 0.112$	$\nu_{21} = 0.128$
$G_{12} = 3.90E + 09 \text{ N/m}^2$	$G_{12} = 3.86E + 09 \text{ N/m}^2$

The comparison between results of the vibration-test and the INSTRON-test shows an excellent agreement.

4. CONCLUSION

Material identification using experimental vibration data from rectangular plates with 4 free edges, promise to become a fast and accurate method to obtain the linear elastic anisotropic data needed for numerical analysis of composite structures.
At this moment, efforts are made to check the accuracy of the results of the method by comparison with classical methods (tensile tests, bending tests, rail shear tests,...) and also to make the experimental extraction of modal data userfriendly and fast.

REFERENCES

1. Sol H: Influence of Poisson ratio on bending modeshapes and resonance frequencies. Internal Report 86/01, Free University of Brussels (VUB), January 1986.
2. De Wilde WP & Sol H: Identification of anisotropic plate rigidities using experimental free vibration data. Internal Report 85/03, Free University of Brussels (VUB), December 1985.
3. Sol H: Calculation of the eigenfrequencies and modeshapes of anisotropic rectangular plates with 4 free edges. Internal Report 85/01, Free University of Brussels (VUB), March 1985.

BIAXIAL FAILURE TESTING OF A TRANSVERSE ISOTROPIC BRITTLE MATERIAL

G. W. Eggeman, Member SEM, Assistant Professor of Mechanical Engineering

Kansas State University
Manhattan, Kansas 66506 U.S.A.

1. ABSTRACT

The paper is concerned with the finite element design of test specimens which were failed in biaxial stress fields. The test procedure used, and the resulting data are included. A fine grained AGOT-EGCR extruded graphite, which is brittle and transversely isotropic, was used in the tests. Data was generated for all quadrants with the emphasis placed on the compression-compression area. Uniaxial compression and tensile tests were also conducted to aid with a failure envelope curve fit.

2. INTRODUCTION

Finite element techniques have been used to analyze and to aid in the design of many complicated parts. In this work, finite element methods were used to enhance the experimental techniques rather than replace them. This was accomplished by using a finite element program to aid in the design of the shape and to investigate the loading pattern in the test specimen. In material tests, a gauge section must be created where the load distribution is known, and the stresses can be accurately calculated. The geometry of the specimen and how the method for holding it affects the load distribution are important for obtaining accurate results.

There are several methods for inducing a biaxial stress field with the pressurized hollow cylinder subject to an axial load being the most popular. The problems encountered using this method are reviewed using AGOT-EGCR extruded graphite as the test material. This material was selected since it is not well behaved, and many different problems had to be addressed. The material is used in high temperature applications such as nuclear reactor shields, electric furnaces and rocket nozzles.

The data developed represents a large experimental base which can be used to verify analytically derived failure theories. A simple strength criterion for multiaxial stress states which can be easily used by the design engineers is needed. It is possible to over-design as well as under-design a part in a biaxial stress field based on uniaxial data. Uniaxial tensile and compression tests were also conducted using the graphite to aid in the development of a curve fit to the biaxial data.

3. TEST SPECIMEN AND MATERIAL

The material used in the tests was AGOT graphite, type EGCR (Experimental Gas-Cooled Reactor) nuclear grade. Specimens were taken from extruded 110 mm x 110 mm x 1300 mm billets. The graphite had a purity of 99.9% with a maximum grain size of 0.8 mm. Its density was 1.72 g/cc versus 2.12 g/cc for solid graphite. This indicated a porous material with 19% air voids present. The extrusion process creates a transversely isotropic condition. This is where the material properties in planes perpendicular to the extrusion axis are different from those parallel to this axis. The material is non-linear and non-conservative in its stress-strain relationships. It is also sensitive to stress concentrations.

In review of the literature, most work [1,2,3,4] with graphite used very small cylinders. The initial decision was to use a quarter of the cross section of the billets as the cylinder diameter. However, this was changed to the use of a larger diameter equal to the billet's cross section so that the wall thickness could be increased. This reduced buckling possibilities, but more importantly it decreased the chance of grain size effecting the test results. The dimensions of the cylinders used are shown in Figure 1.

The cylinders were carefully machined using a single cut pass so that no stress concentrations would be present at a discontinuity. Care was taken in measuring each cylinder for its exact size for use in the calculations. The extrusion axis was used as the center line of the cylinder.

The uniaxial tests were conducted first to determine the material properties in the two directions for use in the finite element model. The compression tests used 0.50 inch diameter solid cylinders with 2.0 and 2.5 length to diameter ratios and were failed on swivel end platens. Strain gages were bonded to the specimens to develop the stress-strain curves as shown in Figure 2. From these graphs, the Modulus of Elasticity was developed, and with readings from a lateral strain gage, Poisson's Ratio was found for each material direction. The compression failure data, along with that from tensile tests [5] using a bonded joint and a cable pull arrangement, were retained for the final results.

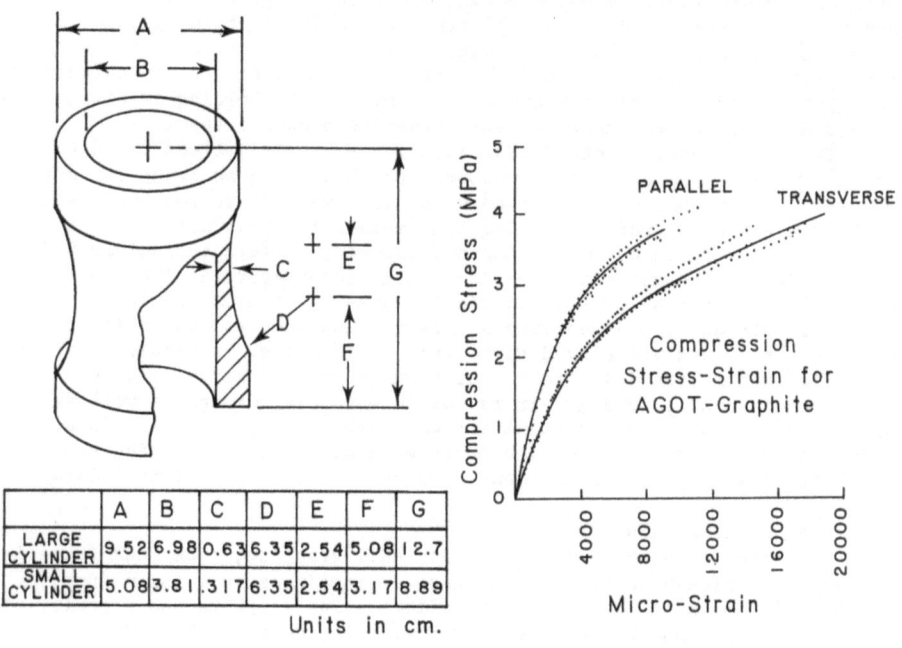

	A	B	C	D	E	F	G
LARGE CYLINDER	9.52	6.98	0.63	6.35	2.54	5.08	12.7
SMALL CYLINDER	5.08	3.81	.317	6.35	2.54	3.17	8.89

Units in cm.

Figure 1. Specimen Dimensions

Figure 2. Compression Stress-
Strain Curve

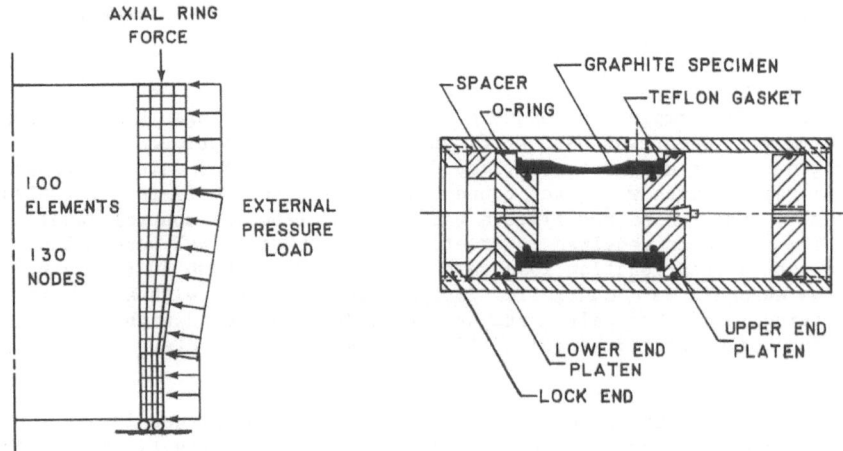

Figure 3. Finite Element Model   Figure 4. Test Fixture

The material data found in the compression tests was used in a finite element model using the ANSYS program. Figure 3 shows the model used which contains 130 nodes and 100 ring elements. Only the upper half of the cylinder was modeled due to symmetry. There are two factors affecting the axial stress at the guage section. A positive pressure on the outer surface causes a tensile stress due to the offset in the surface, and the offset position of the axial load on the flange has a slight effect. The radial stress also varies slightly across the cross section, but the thin wall cylinder equation was deemed sufficient for data reduction. As expected, a third normal stress equal to the pressure was found on the pressurized wall which reduced to zero across the wall thickness.

Figure 4 shows the test fixture used in the tests. To evaluate sealing arrangements, friction and to develop test procedures, strain gages were bonded to six test specimens and a seventh made from aluminum. Failure data were not recorded for these cylinders since they underwent many different load histories.

Several different seating arrangements were evaluated for the axial compression loads. A flat teflon seal with a light coating of silicone vacuum grease was selected. This seal was hard and would not penetrate the graphite, and it spread the load over a large area reducing the possibility of end breakage. This seal was rotated into several different positions, and no hard spots or high areas were found which resulted in unequal loading appearing at the strain gages. An O-ring on the test cylinder's inner surface was used as a back-up system, and it applied a negligible hoop load to the test specimen.

The fixture was a stainless steel cylinder with a highly polished interior surface. The friction between the platens and the hydraulic cylinder was evaluated by cycling the load. The graphite cylinder loading and unloading stress-strain curves followed different paths, but they were repeatable and returned to zero each time indicating negligible friction.

For the axial tensile tests and for the compression tests with low axial loads, the test cylinders were bonded to the platens with approximately 3 mm thick adhesive coating. The adhesive used was Sikadur Hi-Mod Gel, Sikastrix 390 which has a 20 MPa tensile strength after cure.

Figure 5 shows the loading circuit. It consisted of nitrogen gas channeled through two Hoke pressure regulators. The regulator valves retained their last setting after failure so exact readings could be recorded. Since the graphite is porous, the specimen was coated with paraffin. The sample was loaded directly with the nitrogen gas although nitrogen over water was used at first. This was discontinued since nitrogen bubbles were being entrapped in the water. For safety purposes, the nitrogen source was never connected directly to the fixture but bled through a sump. Also the fixture was isolated from the control panel.

The pressures required to create a given stress level were calculated, and a step function load path was created as shown in Figure 6. The two channels were controlled independently to 7 KPa with an always increasing load. The valve rate was at 36 KPa per second with the function being created at about 145 KPa pressure per minute.

4. TEST DATA

During the compression tests, a steel spacer was placed inside the test specimen to prevent post fracture damage. There were two types of failures in the compression-compression quadrant. In the area where the axial stress was predominate, scatter in the data was not pronounced. The failure was a series of longitudinal breaks. In the area where hoop stress was predominate, the gauge section area disintegrated into a series of small pieces. The data also demonstrates more scatter in this area.

The data [6] for failure is shown in Figure 7. The radial stress was calculated using the mean diameter in the thin wall equation. The axial stress was found by distributing the load evenly over the cross section of the gage section. As is the practice in cylinder tests, the pressure normal to the surface was neglected. The curve fit shown is based on tensor polynominal theory by Tsai and Wu [7].

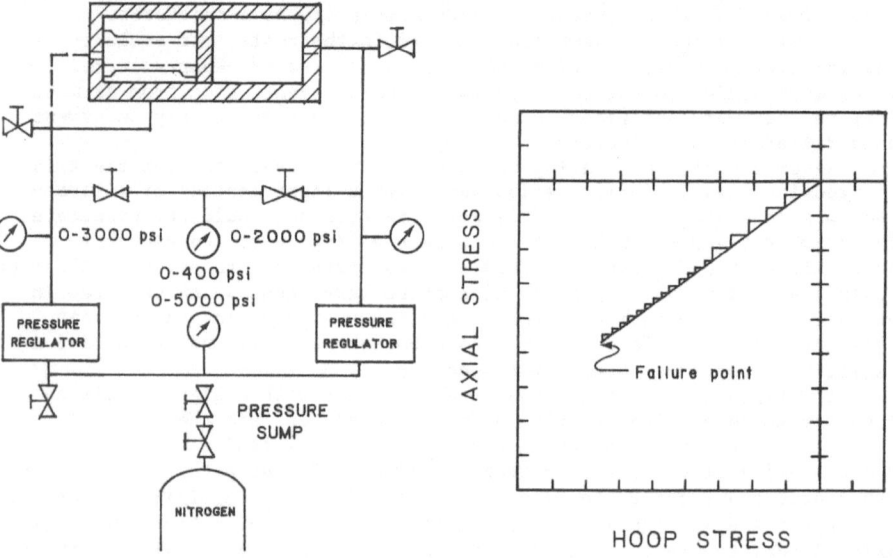

Figure 5. Loading Circuit   Figure 6. Loading Path

Figure 8 is included to show some further work with the same graphite material. The tests were conducted by another investigator [9]. The specimens were the large size described in this work, but some modifications were made on the gauge section.

5. CONCLUSION

The test data for brittle material always shows a degree of scatter. The experimental data, while not enough for statistical averaging, does follow the failure surface in the compression-compression quadrant. The development of enough data for curve fitting was the objective of the test along with developing a reliable test procedure. The load distribution within the specimen gauge section is not constant, but the use of the finite element work is an added advantage in both the design of the specimen and in the data reduction.

REFERENCES

1. Broutman, L. J., S. M. Krishnakumar, and P. K. Mallick, "Effects of Combined Stresses on Fracture of Alumina and Graphite," J. American Ceramic Society, 53, 1970, p. 649.
2. Greenstreet, W. L., J. E. Smith, G. T. Yahr, "Mechanical Properties of EGCR-Type AGOT Graphite," Carbon, Vol. 7, 1969, pp. 15-45.

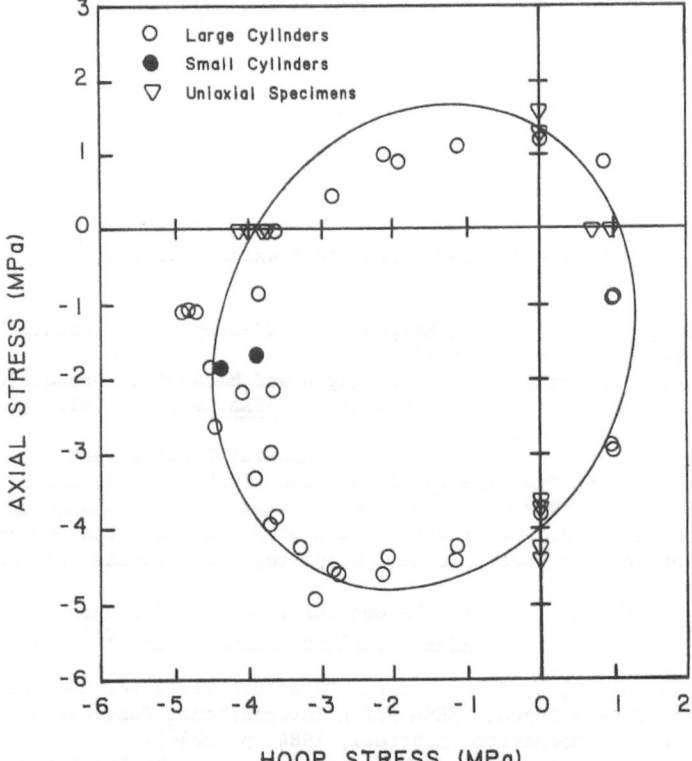

Figure 7. Hackerott-Eggeman Biaxial Failure Data for AGOT Graphite.

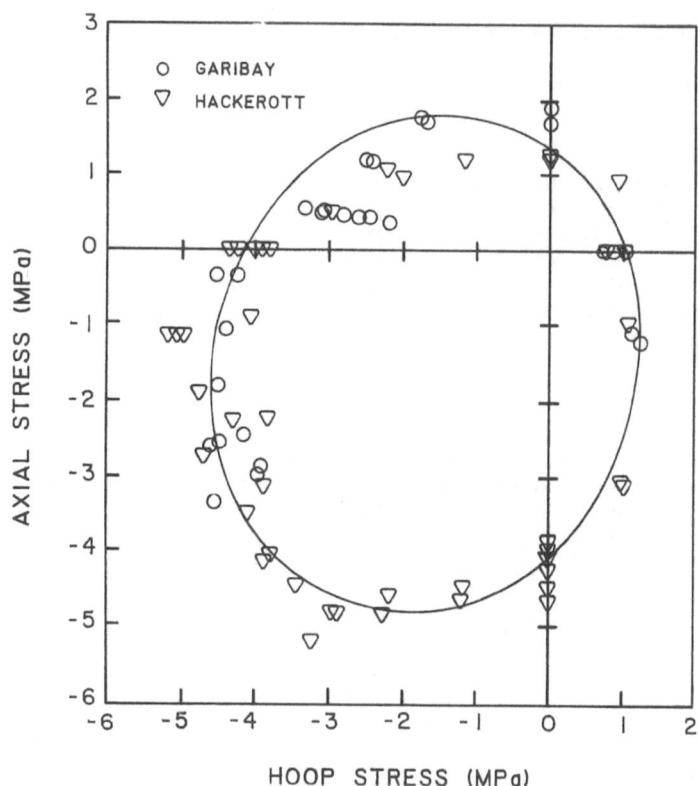

Figure 8. AGOT Graphite Biaxial Failures.

3. Jortner, J., "Multiaxial Response of ATJ-Graphite," Technical Report Afml-TR-71-253. December 1971.

4. Weng, T., "Biaxial Fracture Strength and Mechanical Properties of Graphite-Base Refractory Composites," _ATAA Journal_, Vol. 7, No. 5, May 1969, pp. 851-858.

5. Eggeman, G. W., H. A. Hackerott, "Uniaxial Tensile Test of a Brittle Material," SEM Experimental Techniques, 1985, Vol. 9, No. 7, pp. 18-9.

6. Hackerott, H. A., "Characterization of Multiaxial Fracture Strength of Transversely Isotropic AGOT Graphite," M.S. Thesis, Kansas State University, 1982.

7. Tsai, S. W. and P. M. Wu, "A General Theory of Strength for Aniso-tropic Materials," _Journal of Composite Materials_, Vol. 5, January 1971, pp. 58-80.

8. Eggeman, G. W., H. A. Hackerott, "Fracture Tests for AGOT Graphite in Biaxial Stress Space," SESA Fifth International Congress on Experimental Mechanics, Montreal, 1984, p. 303-7.

9. Garibay, E., "An Improved Characterization of the Multiaxial Fracture Strength of EGCR-AGOT Graphite," M.S. Thesis Kansas State University, 1983.

A SYSTEM FOR CRACK-OPENING-DISPLACEMENT MEASUREMENT AND PHOTOMICROGRAPHY OF CRACKS AT HIGH TEMPERATURES

W. N. Sharpe, Jr., Department of Mechanical Engineering, The Johns Hopkins University, Baltimore, MD 21218, USA

1. INTRODUCTION

It has been observed that very short cracks can propagate at different rates than would be predicted by procedures used for long cracks (1). Often the short cracks propagate faster, which means that designs and analyses based on long-crack behavior would be nonconservative. In addition, developments in nondestructive inspection techniques have lowered the threshold of crack size that can be reliably determined. The definition of a "short crack" is not standardized, but for the purposes of this work an upper limit of 0.5 mm (500 micrometers) can be taken. However, much of the research into short-crack behavior deals with cracks as short as a few micrometers.

It is well-known that cracks grown by fatigue loading exhibit a closure effect; that is, they do not open all the way to the tip until some minimum load is applied. The stress intensity factor (SIF) range applied to the specimen is therefore less than expected. If this closure load can be measured or predicted, then one can compute an effective SIF. It has been hypothesized that closure effects may be less for short cracks, which means that the effective SIF range is actually larger than predicted from long- crack data. This would account for a more rapid growth rate.

The most direct way to measure the closure load is to record the crack opening displacement (COD) versus load. Little displacement will be observed at low loads, but after the crack becomes fully open the curve will be linear and correspond to the compliance of the linear elastic behavior of the specimen. Identification of the load at which the upper part of the plot becomes linear establishes the closure load. COD can be measured a number of ways on larger specimens - microphotography, replicas, interferometry, potential drop, foil strain gages, etc. But use of these techniques becomes difficult, if not impossible, for the very short cracks.

The growth of short cracks is of major concern in those highly sophisticated components or systems for which the increased cost of a more expensive inspection process can be justified. An example of such a system is an air-breathing jet engine, since it operates near its design limits because of weight considerations and subjects its components to high temperatures. The study of short cracks at high temperatures is in its infancy, and it will be appreciated that measurement of COD for short cracks at high temperatures is difficult. This paper describes a system for measuring COD and crack length under these conditions.

A system very similar to the one described here has been developed by Larsen (2). He uses a computer-controlled, laser-based, Interferometric Strain/Displacement Gage (ISDG) originally developed by this author and an automatic photomicrography system to record COD and crack length. The main difference between the system described here and Larsen's system is that this one can be used at high temperatures. High-temperature ISDG use itself is not new; in Sharpe (3), for example, it is applied to nickel-based superalloys at 650C to measure COD under creep and fatigue loading.

Wieringa, H (ed), Experimental Stress Analysis.
© *1986. Martinus Nijhoff Publishers, Dordrecht.*

The new developments are a different optical arrangement enabling high temperature photomicrography and the adaptation of the ISDG for use with an IBM PC minicomputer.

2. BASICS OF THE ISDG

The ISDG has a very small gage length and can measure relative displacement with a resolution of approximately 0.02 micrometer. Only a brief discussion will be included here; for more details, see (4).

Two small indentations are pressed into the specimen surface across the crack using the pyramid-shaped diamond tip of a Vicker's micro-hardness tester. When these two indentations are illuminated with a laser, the reflections from the four sides form interference patterns in space. As the indentations move away from each other, the fringe patterns also move, and this motion is easily associated with the relative displacement, δd:

$$\delta = \frac{\delta m_u + m_l}{2} \cdot \frac{\lambda}{\sin \alpha_o} \tag{1}$$

where λ is the wavelength of the laser and α_o is the angle between the incident laser beam and the reflected fringe pattern. δm_u and δm_u are the relative fringe motions of the two patterns in the plane containing the axis of measurement. λ is equal to 632.8 nm for the He-Ne laser used in this experiment, while αo is approximately 42°. Thus, the calibration factor

$$\frac{\lambda}{\sin \alpha_o} \tag{2}$$

is about 1 μm.

It is therefore necessary to measure fractional fringe motions in order to achieve suitable resolution. This is done with the minicomputer-controlled system described below.

3. OVERVIEW OF THE SYSTEM

Figure 1 is a schematic of the measurement system. All components are fastened to aluminum mounts which are bolted directly to the electro-hydraulic test machine. The Spectra-Physics model 124B laser has a 15-milliwatt output which is more than adequate for the ISDG. The laser is mounted vertically with the beam directed onto the specimen. An adjustable mount holds a mirror designed for 45° reflection. The laser beam passes through a hole in a metal shield which protects the operator from stray reflections. The beam then passes through a corner-cube beamsplitter oriented so that its axis of polarization is horizontal coinciding with the axis of polarization of the laser beam. Nearly 100% of the laser beam is passed through the beamsplitter, which also acts as a mirror so that the optical axis of the camera is at 90° to the laser beam.

The fringe patterns emanating from the indentations on the specimen are captured by two servocontrolled scanning mirrors. These mirrors operate on the galvanometer principle and are each powered by a feedback controller containing a triangular waveform generator. The mirrors are from General Scanning, Inc of Watertown, MA, model G120DT with model CX-660 controllers. The mirrors are mounted on translation and rotation stages which enable one to position them in the fringe pattern.

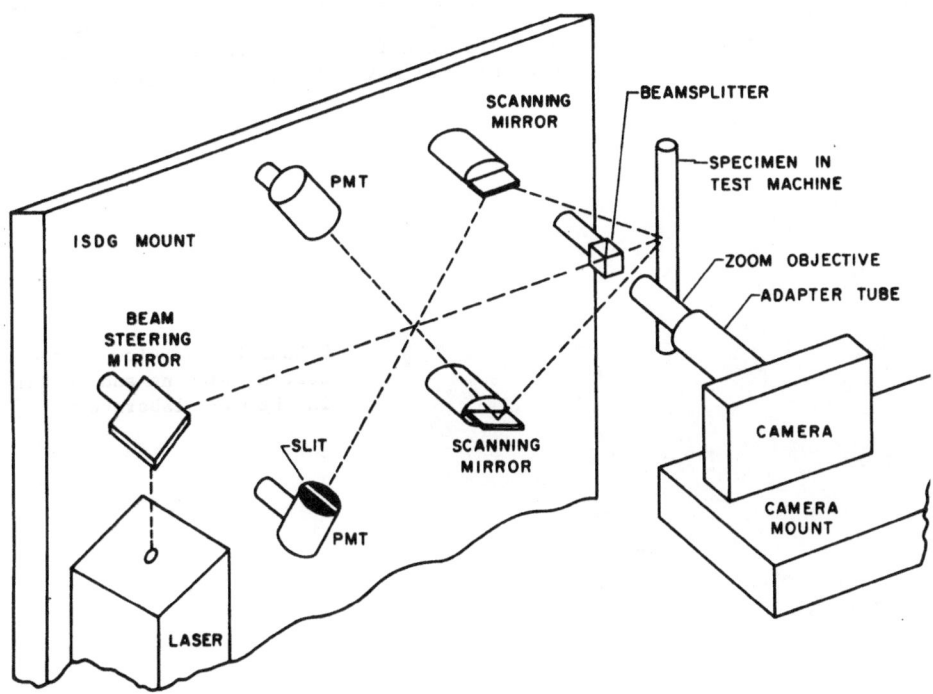

FIGURE 1. A schematic overview of the measurement system.

The mirrors reflect the fringe patterns onto RCA 4840, photomultiplier tubes with RCA PF1042 power supplies requiring 13.5 DC volt input. The aperture of the box covering the tube is approximately 20 mm in diameter and is covered except for a slit roughly 1 mm wide; this slit is narrower than the spacing of the fringes when they impinge upon the tube. The aperture of each tube is covered with an interference filter to block out room lights, and each tube is mounted on a rotation stage so that the filter may be aligned perpendicular to the incident fringe rays. Signals from the photomultiplier tubes pass through amplifiers (Tektronix model AM502), where they are filtered above 300 Hz, inverted, and amplified approximately 50 times.

Figure 2 is a view of the system as it is set up in an industrial testing laboratory. A moderate amount of optical shielding is required around the test machine so that one can see the fringes for alignment purposes; it is not necessary that the rest of the room be dim. The controllers for the mirrors are also shown in Figure 2, mounted in the rack along with the instrumentation associated with the test machine. The two amplifiers and the 13.5 volt power supply are also shown in the rack, as is a combination 2-channel/X-Y oscilloscope used to set up and monitor the ISDG. Figure 3 gives a closer view of the system and includes the camera system, which will be described in more detail below.

4. ISDG SOFTWARE

Since one complete fringe shift corresponds to a relative displacement of approximately 1 micrometer, it is necessary to use some means of measuring fractional fringe motion, preferably on a realtime basis. This has been accomplished in an earlier version of the ISDG using a minicomputer-controlled system to locate a fringe minimum and follow it as it moves in space (3). When that minimum moves out of the range of the

FIGURE 2. Overview of the testing machine and system in the G.E. laboratory.

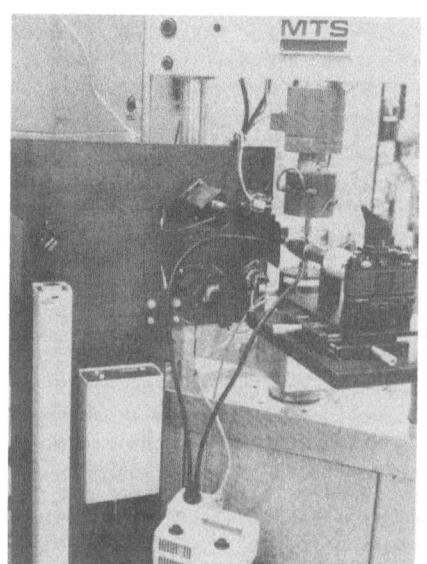

FIGURE 3. Closeup of the ISDG components and camera.

servocontrolled mirrors, the mirrors are automatically shifted to an adjacent minimum moving in the same direction. This particular version of the ISDG controls the scanning mirrors by sending a voltage signal to them and receiving a voltage signal from the PMT. It scans 30 points on either side of the original minimum and processes them to locate the new position of the minimum. The load is then increased by one increment on command from the minicomputer and the displacement measurement is repeated. This ISDG can take one displacement measurement every 0.1 second. Cyclic testing using 60 load increments (and displacement measurements) per cycle may then be conducted at a rate of 10 cycles per minute.

The earlier version uses a DEC MINC laboratory minicomputer which is no longer manufactured. The system described in this paper uses an IBM PC/XT, which has the advantage of being readily available and moderately priced. The data acquisiton board used with the PC/XT is from Data Translation, Inc., model 2801-A; it has a D/A throughput of 25K samples per second. The A/D and D/A routines in a compiled language are much slower than the MINC. This necessitated a change in software strategy. The new approach uses the built-in wave generator of the mirror controllers to control testing and measurement. The General Scanning CX-660 puts out a triangular waveform' with a synchronizing signal at the start of each positive-going portion. That signal is used to trigger the minicomputer to increment the load, scan the fringes, process the acquired data to locate minimums, compute the displacement, and store load and displacement to the hard disk.

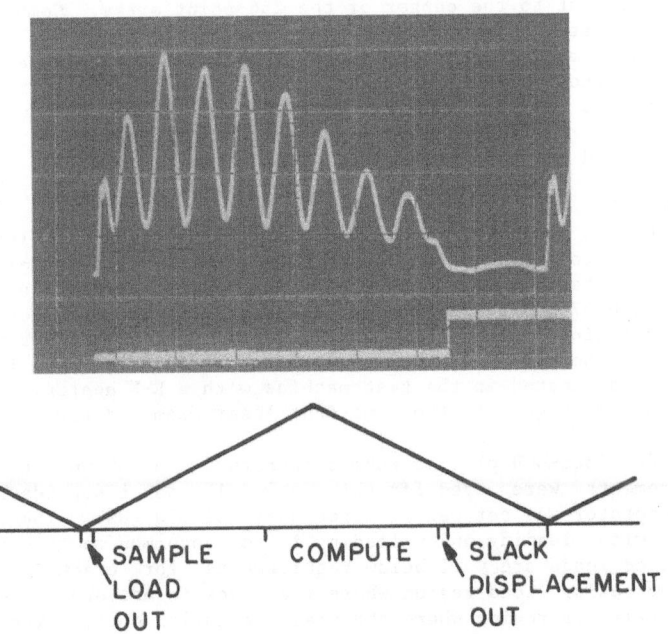

FIGURE 4. Schematic, controlling waveform and oscilloscope photograph of typical fringe pattern signal. Upper trace shows 50mv/div; lower trace shows displacement output, 100mv/div. Sweep speed is 10 ms/div.

The new software allows acquisition of fringe intensity data via one continuous sampling using the direct memory access (DMA) feature of the minicomputer. The fringes are swept across the PMT slits by the triangular waveform as the minicomputer simultaneously samples both channels. Figure 4 helps to explain this strategy. The top schematic represents the triangular waveform, and the oscilloscope photograph shows the intensity signal from one fringe pattern. When the waveform starts to rise, a trigger signal is sent to the minicomputer. It then sends a signal to the electro-hydraulic test machine, changing the load by a prescribed increment; this takes approximately 4 ms. It samples both fringe patterns alternately and continously for about 30 ms, taking 512 data points directly into memory. The odd-numbered data points are from fringe pattern number 1, the even data points from fringe pattern number 0. The stored information is then processed to locate the positions of intensity minimums. One sees from Figure 4 that the second, third, and fourth minimums are used. The oscilloscope display in Figure 4 is directly from one of the PMTs, so it shows the entire fringe pattern, not just the part that is used. The algorithm follows the central (third) minimum and divides its motion by the spacing between minimums to compute a relative fringe motion which is then used in Equation 1. For example, if the original positions of the minimums are memory locations 150, 250, and 350, the next scan might show them at 154, 254, and 354. This would be equivalent to a relative fringe motion of 4/100. If the targeted minimum moves too far from the center position, the algorithm shifts attention to the minimum closest to the center of the 256-point array. Computations and writing of the data to memory takes approximately 30 ms.

The bottom trace in Figure 4 shows displacement output and is used simply for monitoring the experiment. The 0.06 volt step corresponds to a displacement of 0.0063 micron. There is a slack period of roughly 17 ms which allows more time if needed for locating the minimums in the stored data; this period could be reduced to around 5 ms.

5. ISDG RESULTS

The ISDG was evaluated on an edge-cracked specimen of Inconel 718. The specimen cross-section was 10 mm by 4.5 mm with a 1.3-mm-long fatigue precrack. Indentations 100 micrometers apart were placed across the crack at a position 0.67 mm behind the tip of the crack. The specimen was tested to provide a typical material and geometry for the ISDG evaluation. A photograph of the crack and the indentations at 650C is shown in Figure 8. The specimen was heated in the test machine with a R-F heater; the coil was adjusted for passage of the incident laser beam and the exiting fringe patterns.

A typical load-COD plot at room temperature is shown in Figure 5. Sixty increments were used for this cycle; it took 6 seconds because the waveform generator was set at a frequency of 10 Hz. The hysteresis associated with loading and unloading is not uncommon on this scale, even though applied loads are well below requirements for crack growth. The initial increasing load region where the crack is not open is evident, as is the upper linear region where the crack is fully open. The criterion for establishing the opening load (and whether it should be on loading or unloading) is still not established. There is some drift of the measured displacement over 50 cycles at zero load. However, the largest displacement measured is 0.9 micron.

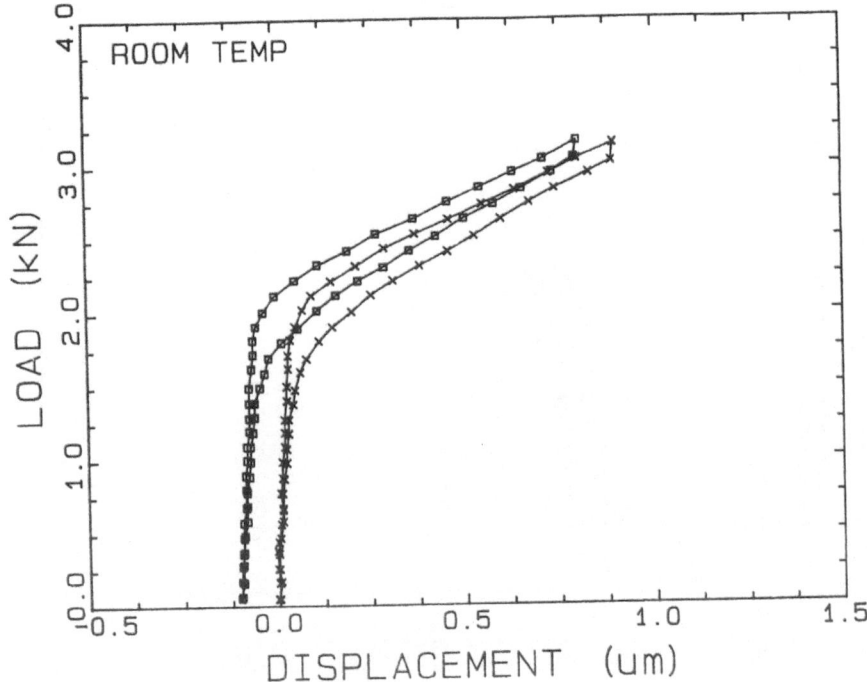

FIGURE 5. Load-COD record at room temperature. Crosses represent the first cycle, squares mark the fiftieth cycle.

Figure 6 is a similar record at 650C. Since the modulus of the material has decreased approximately 20% at this temperature, the slope of the fully open region is very flat for a measurement location this far behind the tip of the crack. It was necessary to limit the maximum load to avoid a displacement increment too large for the ISDG to follow for that plot. It is easy to get around this problem by taking finer load increments (3); however, this lengthens the time required for a measurement. Note that there is less hysteresis for the fiftieth cycle at this temperature; perhaps the oxide film on the crack face has been abraded away by the cycling.

To evaluate the drift of the ISDG at 650C, another 50 cycles of measurement were run with a zero load increment. Figure 7 shows the results plotted as displacement versus time for 3000 data points (50 cycles of 60 points each). There is a total drift of approximately 0.05 micron in this time; this corresponds to 8 bits of output (8 bits X 0.0063 micrometer/bit = 0.05 micrometer). It is not clear therefore whether or not the increase in displacement at zero load in Figure 6 is an effect of the measuring system. Again, note the scale of the measurements.

228

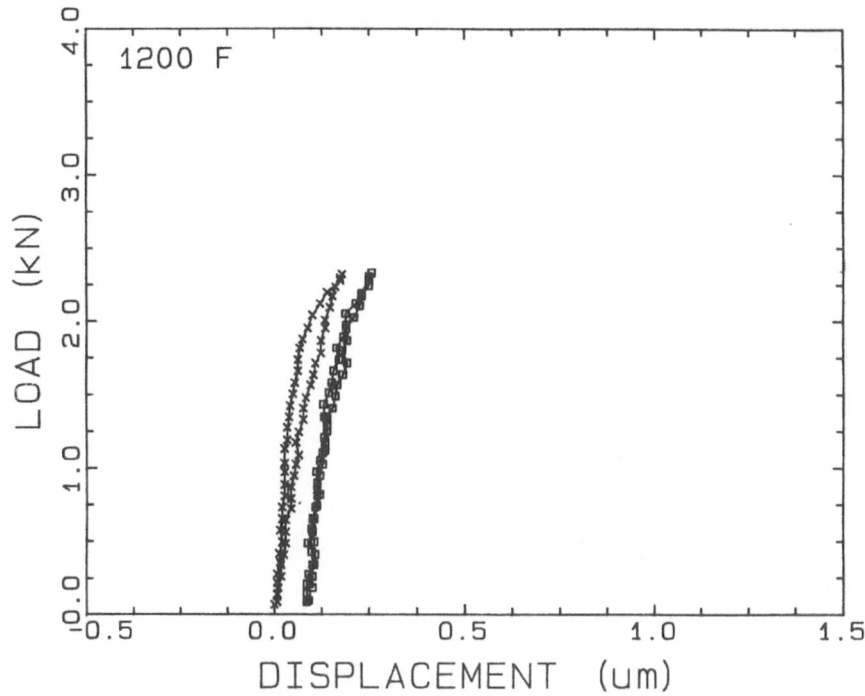

FIGURE 6. Load-COD plot at 650C (1200F). Crosses represent the first cycle, squares mark the fiftieth cycle.

6. PHOTOGRAPHIC SYSTEM AND RESULTS

The two main design problems for the photomicrography system were the temperature of the specimen and the presence of the ISDG laser beam. The general approach was similar to that of Larsen (2) and Papirno and Parker (5), where a motor-driven camera coupled to a microscope was used.

A polarizing beamsplitter cube permits the laser beam to pass through with little attenuation and serves as a 45 mirror for the camera. Illumination for photography comes from a fiberoptics lamp, as shown in Figure 3. The final version of the system will have relay-operated shutters to open and close for the laser beam and the fiberoptics illuminator.

High magnification at remote distances is a vexing optical problem, but a new component on the market solved the problem here. A telescoping objective, Unitron's model 15846 with an extra 1.5X lens, provides a magnification of 67.5X at 50 mm when used with a 10X eyepiece. The "zoom" feature permits continuous reduction to 10.5X, which is convenient as a crack grows. This objective and eyepiece are mounted on a standard 170 mm microscope tube and coupled to a Nikon F3HP camera with an adapter tube. The camera has a motor drive and the capability of accepting a 250 exposure film back. It also has a high magnification view finder with a type M focusing screen; this arrangement is recommended for high magnification photography.

A photograph of the specimen at 650C is presented in Figure 8. The magnification on the film plane of the camera is approximately 24X, similar to that used by Larsen (2). He estimated a crack-length resolution on the order of 1 micrometer with fine-grain film, careful flash lighting, and subsequent magnification for viewing. His results were at room temperature, however, and probably cannot be achieved at 650C with this camera system.

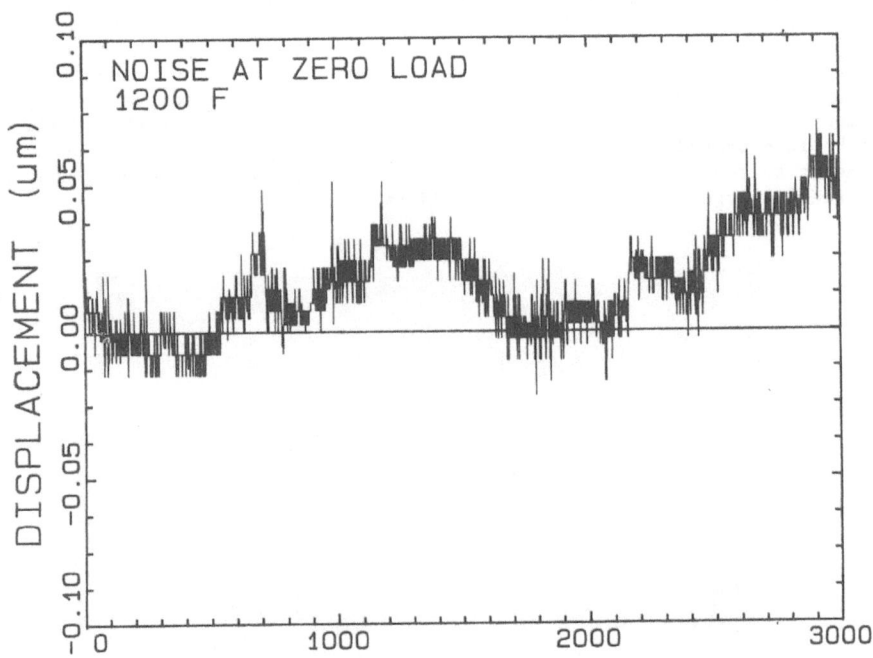

FIGURE 7. ISDG output for zero load, showing the drift of the system over 3000 data points (50 cycles).

7. DISCUSSION

Two elements of a measurement system - the ISDG and the camera - have been described here. They can each be operated under total minicomputer control to provide automated testing of short or long crack behavior a room and elevated temperatures. One would simply write a controlling program for the minicomputer in compiled BASIC, the language used here, and call the ISDG measurement as a subroutine. Calls to stop the fatigue loading for COD measurement and photography would be inserted where needed.

This new version of the ISDG has the same sensitivity as the earlier version (3) with a little more speed. It can be run at 13 samples per second, while the other system operated at 10 samples per second. Limiting factors on the speed here are the minicomputer and the data acquisition board. Programming in assembly language on a faster minicomputer would increase the computational speed, and faster data acquisition boards are available. Since the data acquisition and the computations take about the same amount of time (see Figure 4), both the minicomputer and the board would need up-grading. This system shows a slightly higher drift than the previous one which typically drifts only +/- one bit over 3000 data points. It is believed that this additional drift arises from small variations in the timing of the trigger signal and the minicomputer's response to it.

The photomicrography system appears to have the same capabilities as earlier systems, even though the objective lens is 50 mm from the specimen. The temperature at the beamsplitting cube and at the objective lens was less than 70C; the R-F coil heats only the specimen in the water-cooled grips. Resistance furnaces with quartz ports have been used in other experiments with the ISDG (3), but they would not accomodate the photography system. Photographs can be taken automatically, but they still need to be interpreted. It is difficult for other imaging approaches such as TV cameras to achieve the needed resolution directly, but they could be used to read enlarged projections of the film.

FIGURE 8. Photograph of the specimen at 650C. The indentations are 100 micrometers apart.

8. ACKNOWLEDGEMENTS
The development of this system was sponsored by the General Electric Company of Evendale, Ohio. The assistance of Mr. Terry Richardson and Dr. Robert VanStone of that company is greatly appreciated. The mechanical design and construction of the various components was the work of Mr. F. VanHooijdonk of Hopkins.

REFERENCES

1. Suresh S and Ritchie RO: Propagation of Short Fatigue Cracks. International Metals Reviews, Vol.29, No.6, pp445-476, 1984.
2. Larsen JM: Advanced Experimental Methods for Monitoring the Behavior of Small Cracks. AGARD Conference Proceedings No.393, AGARD, 1985.
3. Sharpe WN: A New Optical Techinque for Rapid Determination of Creep and Fatigue Thresholds at High Temperature. AFWAL-TR-84-4028, Wright Patterson Air Force Base, Ohio, 1984.
4. Sharpe WN: Applications of the Interferometric Strain Displacement Gage. Optical Engineering, Vol.21, No. 3, pp 483-488, 1982.
5. Papirno R and Parker BS: An Automatic Flash Photomicrographic System for Fatigue Crack Initiation Studies. ASTM STP 519, American Society for Testing and Materials, pp98-108, 1973.

EVALUATION OF RECENTLY PROPOSED RECOMMENDATIONS FOR THE DETERMINATION OF
FRACTURE PARAMETERS FOR CONCRETE IN BENDING

S. E. SWARTZ AND S. T. YAP

DEPARTMENT OF CIVIL ENGINEERING, KANSAS STATE UNIVERSITY, MANHATTAN, KS,
USA

1. INTRODUCTION

The development of experimental procedures for determination of
fracture parameters for concrete utilizing bending specimens has reached a
stage where proposals for standardized testing methods are forthcoming.

Herein, these various methods are applied to 102 mm deep beams tested
by Rood (7) and 203 and 305 mm deep beams tested by the writers (11).

2. PROPOSED METHODS

2.1 Test Specimens

All test specimens were beams conforming to the geometry given in
Fig. 1 and all had $\frac{S}{W} = 3.75$ with a beam width B = 76 mm. The measurements
were taken as load-displacement traces using a closed-loop testing system
(MTS) with high stiffness characteristics operating in load control. The
time to peak load generally exceeded 30 sec. As described in reference
10, the resulting traces were generally the same as those obtained using
strain control.

These plots showed the crack-mouth-opening displacements CMOD (Figs.
1 and 2) and the load-point-displacements LPD (Figs. 1 and 3). All beam
test results reported here had these plots which were obtained for each
test simultaneously. The complete set of data may be found in Reference
11.

2.1.1 Material properties. The Rood (7) beams (102 mm) were tested
in July, August 1984 while the 203 and 305 mm beams were tested in January
1986. All beams had the same mix design with the properties summarized at
the bottom of Table 1. It is seen that the ultimate compressive strengths
f' of both sets are in good agreement (average f' = 56.5 MPa $\pm$ 1.2%) as
are the measured values of the modulus of elasticity (average E_c = 41.2
GPa $\pm$ 10.6%). The aggregate was well-graded with a maximum size of 19
mm.

2.2 RILEM Method (6)

This method uses the full P-LPD curve to determine the fracture
energy per unit area of crack extension as

$$G_F = \frac{W_0 \pm mg \; \delta_0}{B(W - a_0)} \tag{1}$$

in which W_0 is the area under the curve, δ_0 is the maximum value of the
vertical displacement and mg is (approximately) the weight of the beam

Wieringa, H (ed), Experimental Stress Analysis.
© *1986. Martinus Nijhoff Publishers, Dordrecht.*

between supports. The 102 mm beams were loaded downward and mg is positive, while the 203 and 305 mm beams were loaded upward and mg is negative.

2.3 Modified RILEM Method

This is proposed by the writers and uses the P-LPD curve up to the point of instability at which the deflection is denoted $\bar{\delta}_0$. The fracture energy is

$$\bar{G}_F = \frac{U \pm mg\ \bar{\delta}_0}{B(W - a_0)}.$$ (2)

In this, U is the energy absorbed by the beam up to the point of instability.

2.4 Direct Energy Method

This is a modification of a method proposed by Go (3) and gives the fracture energy as

$$\bar{G}_{IC} = \frac{U \pm mg\ \bar{\delta}_0}{1.15\ B(W - a)}.$$ (3)

The coefficient "1.15" represents the effect of average surface roughness in the cracked region (3) and the extended crack length a is estimated at the point of instability from a modified compliance calibration technique (3).

2.5 K_{IC} Methods

2.5.1 Jenq/Shah Method (5). With reference to Fig. 2, an equivalent elastic value of the crack-mouth-opening displacement associated with instability is given as $CMOD_e = CMOD_{e1} + CMOD_{e2}$. Herein the effect of $CMOD_{e2}$ is ignored due to lack of data. Using $CMOD_e$ and LEFM, the corresponding value of crack length a_e may be obtained (5).

With this and the value of the maximum load P_m, $K_{IC} = K_{IC}^S$ is determined by LEFM. The formula used here is that given by Go (3) for $\frac{S}{W} = 3.75$. Then, the fracture energy is

$$G_{IC} = \frac{K_{IC}^2}{E_c}$$ (4)

in which the effect of Poisson's Ratio is taken to be zero.

2.5.2 Go Method (3). In this method, the extended crack length a at the point of instability is estimated by the modified compliance technique and $K_{IC} = K_{IC}^G$ determined by LEFM (3). Then, the fracture energy is again given by Eq. 4.

2.6 J_{IC} Method

An approach using the J-integral concept was proposed by Go (3). In this the values of U obtained for each beam are plotted against either the initial crack length or the extended crack length. According to this concept the slopes of the straight lines through each set of data points should be equal and give the fracture energy as

$$J_{IC} = \frac{-(slope)}{1.15 \ BW} \tag{5}$$

2.7 Bazant Three Beam Method (1,2)

In this method, it is necessary to test three beams, or groups of beams, with different spans and depths but with $\frac{S}{W}$ remaining constant. The width B is also kept constant. Then a plot of $(\frac{BW}{P_0})^2$ versus W is made and the best straight line fitted through the three points. In this P_0 is the maximum beam load plus or minus the effect of beam weight, i.e.

$$P_0 = P \pm \frac{1}{2} \ mg \tag{6}$$

The slope of the plot, denoted by A, is used to determine the fracture energy G_f by

$$G_f = \frac{g(\alpha_0)}{E_c \ A} \tag{7}$$

and

$$g(\alpha_0) = (\frac{S}{W})^2 \ \pi \ \alpha_0 \ [1.5 \ F(\alpha_0)]^2. \tag{8}$$

In this,

$$\alpha_0 = \frac{a_0}{W} \quad and$$

$$F(\alpha_0) = 1.089 - 1.746 \ \alpha_0 + 8.231 \ \alpha_0^2$$
$$- 14.22 \ \alpha_0^3 + 14.59 \ \alpha_0^4 \tag{9}$$

for beams with $\frac{S}{W} = 3.75$. For other expressions corresponding to other $\frac{S}{W}$ see Reference 1.

3. EVALUATION OF METHODS

The methods were evaluated using notched beams and precracked beams.

3.1 Notched Beams

Results of tests on six beams with W = 102 mm, two beams with W = 203 mm and three teams with W = 305 mm are presented in Tables 1-3.

The RILEM results in Table 1 show variation with a_0 and also with beam size. This behavior is similar to that reported by Hillerborg (4). The writers have another fundamental disagreement with this method and that is with respect to the use of the full load-displacement curve to determine W_0. Since the crack length changes rapidly after the point of instability, it is not clear why the associated energy (A_2 in Fig. 3) should be correlated with the initial notch length a_0. In addition, the δ_0 value is very uncertain.

An alternative to the RILEM method is suggested as described by Eq. 2. The results obtained from this method, which uses only A_1, or U, are given in Table 2 along with results obtained from the direct energy method (Eq. 3). It is seen that the modified RILEM results are more consistent and also they agree reasonably well--although lower--with the direct energy method. Note that the values of δ_0 are determined reliably.

Also in Table 2 are J_{IC} results which agree well with the other two methods. These were obtained from the plots in Fig. 4. In Table 3 are presented the results based on the K_{IC} approach. The Jenq/Shah results show considerable scatter but it should be remembered that $CMOD_e$ is calculated in an approximate manner. The G_{IC} values are all much smaller than the corresponding values given in Table 1. The results obtained from Go's method are reasonably consistent but are also much smaller than the values in Table 1.

Finally, the Bazant three size method is plotted in Fig. 5 from which it is seen that the points fall on a straight line very well. The value of G_f is also lower than the results in Table 1 but agrees fairly well with the Jenq/Shah results in Table 3.

3.2 Pre-cracked Beams

The only pre-cracked beams tested to date with $\frac{S}{W}$ = 3.75 are those with W = 102 mm. Those tested by Rood (7) are used for the data presented in Tables 4 and 5--twenty six beams. Other beam tests conducted at Kansas State University are given in Reference 11.

The average values from the RILEM method are given in Table 4, in which it is seen that the variation with $\frac{a_0}{W}$ is still present. The same is true for the modified RILEM method but the direct energy method shows good consistency in results. The average values are displayed in Fig. 6.

Also in Table 4 are given values of J_{IC}. These were obtained from plots similar to those in Fig. 4 of U versus $\frac{a_i}{W}$ or $\frac{a}{W}$ in which a_i is the measured, initial crack length and a is the extended crack length (3,7,9). The average J_{IC} value agrees well with the direct energy results.

Finally the K_{IC} results are given in Table 5. The Jenq/Shah results are now much more consistent as are the Go results. These are plotted in Fig. 6.

All of these results are somewhat different from those obtained from the notched beams although the other material properties are eseentially the same. This difference has been remarked upon in other work (3,7,8,9).

4. CONCLUSIONS AND RECOMMENDATIONS

Based upon the results presented here the following conclusions are reached.
1. Pre-cracked beams should be used for fracture testing (possible exception-notched beams seem to work well with Bazant's method).
2. The modified RILEM, direct energy and J_{IC} methods all seem to give equivalent results. The writers prefer the latter method because it may be used with the initial crack length which is easily measured.
3. The K_{IC} methods may work but then should be conducted using strain control-particularly in the case of the Jenq/Shah method in order to measure $CMOD_e$ accurately.
4. The writers recommend, tentatively, the use of beams at least 102 mm deep.

5. ACKNOWLEDGEMENTS

The work reported here has been supported by the National Science Foundation on Grants CEE-8317136 and MSM-8317136. This support is gratefully acknowledged.

REFERENCES

1. Bazant, Zdenek P., "Fracture Energy of Concrete from Maximum Loads of Specimens of Various Sizes," Proposal for RILEM Recommendation, Northwestern Univ., Evanston, IL, 1985.

2. Bazant, Zdenek P., Kim, Jin-Keun and Pfeiffer, Phillip, "Nonlinear Fracture Properties from Size Effect Tests," Journal of Structural Engineering, ASCE, Vol. 112, No. 2, Feb. 1986.

3. Go, Cheer-Germ and Swartz, Stuart E., "Fracture Toughness Techniques to Predict Crack Growth and Tensile Failure in Concrete," Report 154, Engineering Experiment Station, Kansas State Univ., Manhattan, KS, July, 1983.

4. Hillerborg, Arne, "Influence of Beam Size on Concrete Fracture Energy Determined According to a Draft RILEM Recommendation," Report TVBM-3021 Lund Institute of Technology, Division of Building Materials, Lund, Sweden, 1985.

5. Jenq, Y. S. and Shah, S. P., "Two Parameter Fracture Model for Concrete," Journal of Engineering Mechanics, ASCE, Vol. 111, No. 10, Oct. 1985.

6. RILEM Technical Committee 50-FMC, "Determination of the Fracture Energy of Mortar and Concrete by Means of Three-Point Bend Tests on Notched Beams," proposed RILEM recommendation, January 1982, revised June 1982. Lund Institute of Technology of Building Materials, Lund, Sweden.

7. Rood, S., "Fracture Toughness Testing of Small Concrete Beams," M.S. Thesis, Kansas State University, Manhattan, KS, 1984.

8. Swartz, S. E., Hu, K. K. and Fartash, M., "Stress Intensity Factors for Plain Concrete in Bending - Prenotched Versus Precracked Beams," Experiment Mechanics, Vol. 22, No. 11, Nov. 1982.

9. Swartz, S. E. and Rood, S. M., "Fracture Toughness Testing of Concrete Beams in Three-Point Bending: Phase I, Small Beams," Proceedings, 1985 Spring Conference on Experimental Mechanics, Las Vegas, NV, June 9-14, 1985.

10. Swartz, S. E. and Siew, H. C., "Time Effects in the Static Testing of Concrete to Determine Fracture Energy," Report 182, Engineering Experiment Station, Kansas State University, Manhattan, KS, June 1986.

11. Swartz, S. E. and Yap, S. T., "Evaluation of Proposed Methods to Determine Fracture Parameters for Concrete in Bending," Report 181, Engineering Experiment Station, Kansas State University, Manhattan, KS, June 1986.

TABLE 1. Notched Beams, RILEM Method (6)

W	a_0	δ_0	W_0	G_F	Avg G_F
mm	mm	mm	N-m	N-m/m^2	N-m/m^2
102	29.7	0.34	0.45	86	
102	33.0	0.46	0.60	121	
102	46.2	0.34	0.29	74 ⎫	
102	52.3	0.33	0.21	62 ⎬	68
102	63.5	0.33	0.13	51 ⎭	
102	68.1	0.29	0.07	36	
203	102.0	0.80	0.90	87	
203	102.0	0.69	0.89	90	88
305	152.0	0.66	1.17	65	
305	152.0	0.90	1.47	78	72
305	152.0	0.73	1.30	73	

Notes: 1. W/C = 0.50,   C:S:A = 0.41 : 0.69 : 1,   Max, Agg = 19 mm

2. For W = 102 mm,   S = 381 mm,   L = 406 mm,   mg = 7.08 Kg
f'_c = 55.8 MPa, E_c = 36.8 GPa

3. For W = 203 mm,   S = 762 mm,   L = 813 mm,   mg = 28.4 Kg
For W = 305 mm,   S = 1143 mm,   L = 1219 mm,   mg = 63.8 Kg
f'_c = 57.2·MPa, E_c = 45.5 GPa

TABLE 2. Notched Beams, Modified RILEM Method and Direct Energy Method

W	a_0	$\overline{\delta}_0$	U	$\overline{G}_F$	Avg. $\overline{G}_F$	Extended $\dfrac{a}{w}$	$\overline{G}_{IC}$
mm	mm	mm	N-m	N-m/m^2	N-m/m^2		N-m/m^2
102	29.7	0.20	0.27	52.0		0.51	65.3
102	33.0	0.35	0.37	76.0		0.51	91.1
102	46.2	0.21	0.20	50.3 ⎫		0.62	62.7
102	52.3	0.21	0.13	39.2 ⎬	44.8	0.67	50.3
102	63.5	0.20	0.10	38.2 ⎭		0.82	69.2
102	68.1	0.14	0.04	21.2		0.91	67.6
						Avg.	67.7
203	102.0	0.33	0.50	53.3			
203	102.0	0.30	0.52	56.6	55.0		
305	152.0	0.25	0.58	37.0			
305	152.0	0.33	0.75	46.8	43.9		
305	152.0	0.34	0.77	48.0			

Notes: 1. For dimensions and material properties see Table 1.

2. J_{IC} = 82.7 N-m/m^2 based on a_0 (3).

J_{IC} = 76.4 N-m/m^2 based on extended a (3).

Avg J_{IC} = 79.5 N-m/m^2.

TABLE 3. Notched Beams, Jenq/Shah (5) and Go (3) Methods

W mm	P_m N	$CMOD_e \times 10^{-2}$ mm	$\dfrac{a_e}{W}$	K^S_{IC} $\times 10^{-3/2}$ kN-m	G_{IC2} N-m/m	Avg. G_{IC2} N-m/m	extended $\dfrac{a}{w}$	K^G_{IC} $\times 10^{-3/2}$ kN-m	G_{IC2} N-m/m
102	2540	1.2	0.308	590	9.5		0.51	1029	27.5
102	2540	1.5	0.360	681	12.6		0.51	1037	28.0
102	1650	3.1	0.568	791	17.0	13.0	0.62	964	24.2
102	1290	2.1	0.542	575	8.9		0.67	875	20.0
102	730	0.9	0.698	287	2.3		0.82	1025	27.5
102	420	2.3	0.720	350	3.3		0.90	1363	48.4
							Avg.	1049	29.3
203	2850	2.5	0.475	739	11.9	18.8			
203	2940	5.2	0.595	1088	25.8				
305	3600	6.2	0.589	1070	25.0	25.5			
305	4000	5.3	0.547	1042	23.6				
305	3780	6.6	0.592	1132	27.9				

Note: For other dimensions and material properties, see Table 1.

TABLE 4.   Precracked Beams (7), RILEM (6), Modified RILEM and Direct Energy Methods, W = 102 mm

Initial $\frac{a_i}{W}$	$\overline{\delta}_0$ mm	U N-m	$\overline{G}_F$ N-m/m²	Avg. $\overline{G}_F$ N-m/m²	Extended $\frac{a}{W}$	Avg. Extended $\frac{a}{W}$	$\overline{G}_{IC}$ N-m/m²	Avg. $\overline{G}_{IC}$ N-m/m²
0.30	0.11	0.172	33.1		-----		-----	
0.36	0.11	0.244	51.0		-----		-----	
0.45	0.14	0.236	57.5		-----		-----	
0.28	0.14	0.246	45.6		0.45		52.4	
0.33	0.13	0.221	44.7	51.9	0.40		43.1	
0.31	0.18	0.353	68.2	(124.9)	0.43	0.44	72.0	56.2
0.32	0.20	0.313	61.7		0.48		70.6	
0.33	0.18	0.199	40.5		0.45		43.1	
0.30	0.16	0.323	61.3		-----		-----	
0.40	0.18	0.247	55.7		-----		-----	
0.60	0.10	0.210	69.7		-----		-----	
0.50	0.15	0.171	47.1		0.58		48.4	
0.52	0.21	0.163	74.3		0.59	0.59	48.7	47.8
0.52	0.16	0.171	48.9	51.9	0.62		53.6	
0.52	0.16	0.141	41.3	(87.1)	0.58		40.6	
0.59	0.20	0.195	66.4		-----		-----	
0.59	0.15	0.125	42.6		-----		-----	
0.68	0.19	0.101	45.7		0.77		55.4	
0.69	0.15	0.069	33.1	38.0	0.79		42.6	
0.67	0.14	0.082	36.3	(72.7)	0.82	0.79	57.8	50.9
0.67	0.18	0.081	37.0		0.78		47.8	
0.82	0.15	0.047	40.6		-----		-----	
0.81	0.09	0.024	20.1	32.3	-----		-----	
0.79	0.12	0.044	32.2	(67.6)	-----		-----	
0.75	0.15	0.059	35.6		-----		-----	
0.81	0.17	0.036	33.1		-----		-----	

Notes: 1.   For dimensions and material properties see Table 1.

2.   Initial $\frac{a_i}{w}$ is based on dye measurements (3).

3.   ( ) values are obtained from RILEM method (6).

4. J_{IC} = 47.3 N-m/m² based on initial a (3),

J_{IC} = 41.9 N-m/m² based on extended a (3),

Avg J_{IC} = 44.6 N-m/m².

TABLE 5. Precracked Beams (7), Jenq/Shah (5) and Go (3) Methods

		Jenq/Shah							GO			
P_m N	$CMOD_e$ mm × 10^{-2}	$\dfrac{a_e}{W}$	K_{IC}^S kN-m$^{-3/2}$	G_{IC} N-m/m^2	Avg. G_{IC} N-m/m^2	Avg. $\dfrac{a_e}{W}$	P_m N	extended $\dfrac{a}{w}$	K_{IC}^G kN-m$^{-3/2}$	G_{IC} N-m/m^2	Avg. G_{IC} N-m/m^2	Avg. $\dfrac{a}{W}$
3870	1.59	0.29	858	20.0			4072	0.45	1395	52.7		
3470	1.92	0.34	882	21.0			4030	0.40	1220	40.3		
4030	1.66	0.29	893	21.5			4250	0.43	1387	51.5	47.9	0.44
4050	2.29	0.35	1057	30.3	23.9	0.32	3960	0.48	1458	57.6		
4270	2.08	0.32	1029	28.7			3427	0.45	1174	37.3		
3960	2.24	0.35	1034	28.9								
3470	1.91	0.34	882	21.0								
3850	1.59	0.29	854	19.8								
2180	4.52	0.58	1089	32.2			2136	0.58	1084	31.9		
2140	3.94	0.56	1003	27.3			2336	0.59	1197	30.9		
2310	4.11	0.56	1087	32.1	25.4	0.56	1736	0.62	955	24.7	27.9	0.59
1690	3.51	0.58	845	19.3			1980	0.58	974	25.8		
2050	3.20	0.54	905	22.3			2092	0.56	982	26.1		
1890	3.10	0.54	837	18.9								
980	5.99	0.73	845	19.3			1001	0.77	1062	30.5		
850	5.56	0.74	763	15.8			850	0.79	982	26.1		
980	6.05	0.74	883	21.2	18.6	0.76	980	0.82	1360	50.5	32.6	0.79
730	6.12	0.77	767	15.9			850	0.78	930	23.5		
850	6.91	0.77	882	21.0								
710	7.21	0.79	827	18.6								

Notes: 1. For all beams, W = 102 mm

2. For other dimensions and material properties see Table 1.

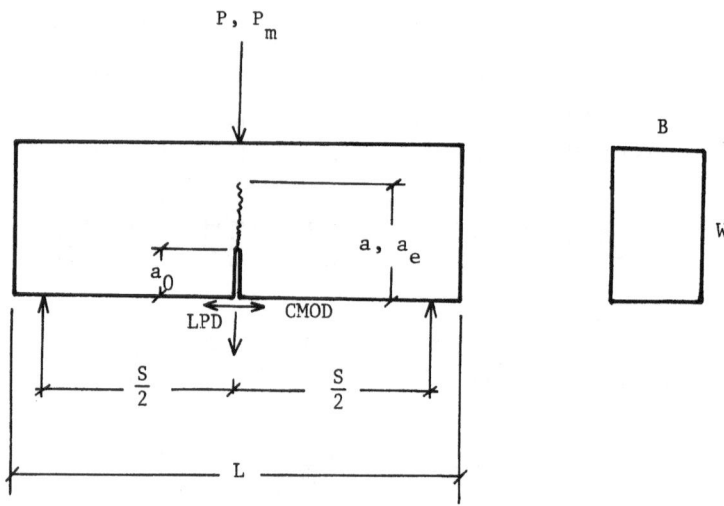

Fig. 1   Test Beam Geometry

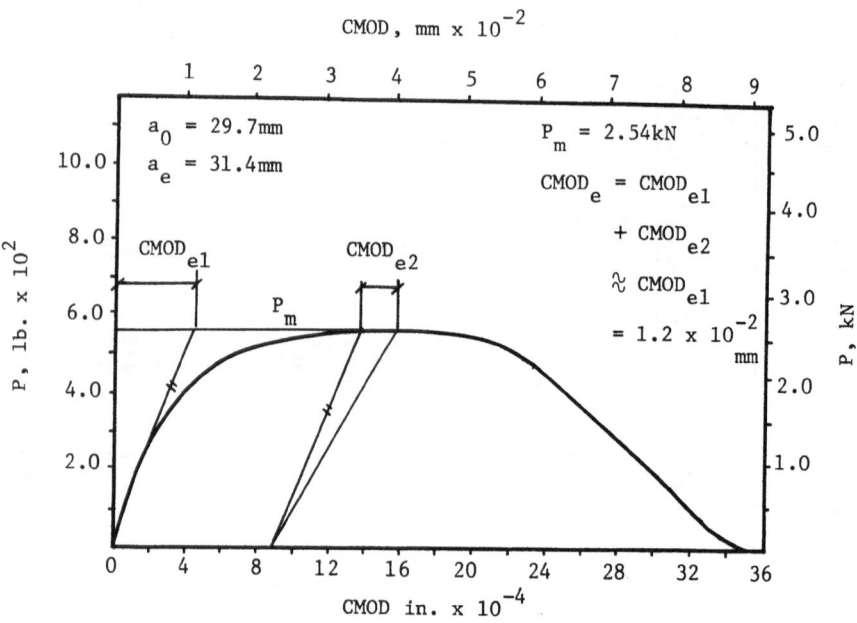

Fig. 2   P vs CMOD, 102mm Deep Beam,
Load Control, C-15 (7,11)

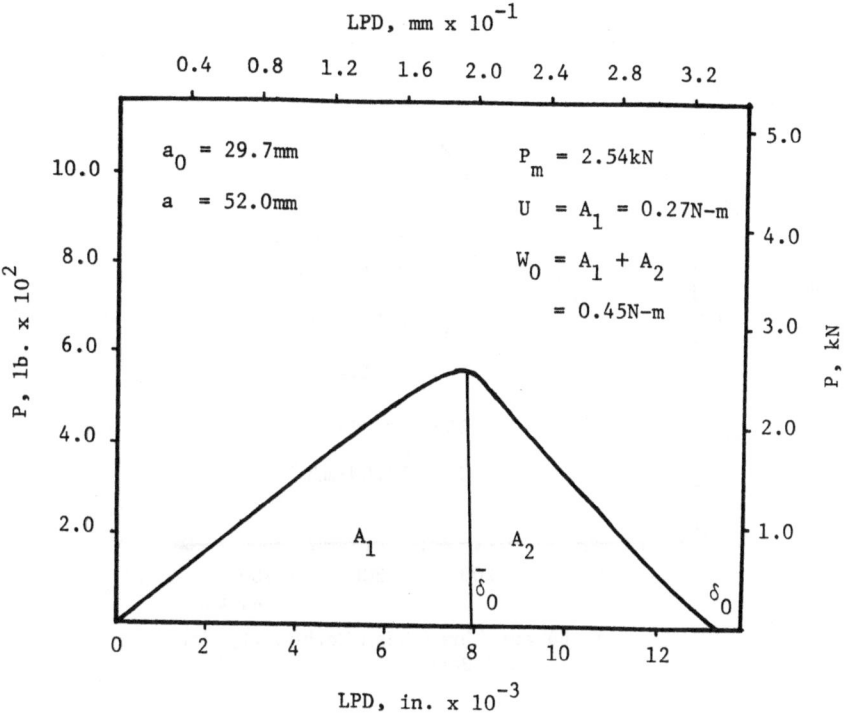

Fig. 3 P vs LPD, 102mm Deep Beam,
Load Control, C-15 (7,11)

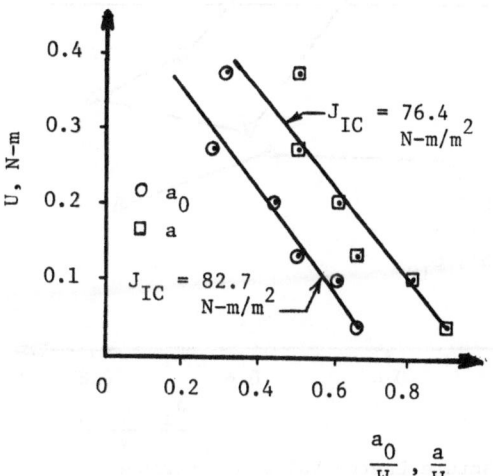

Fig. 4 J-Integral Method (3),
Notched Beams, W = 102mm

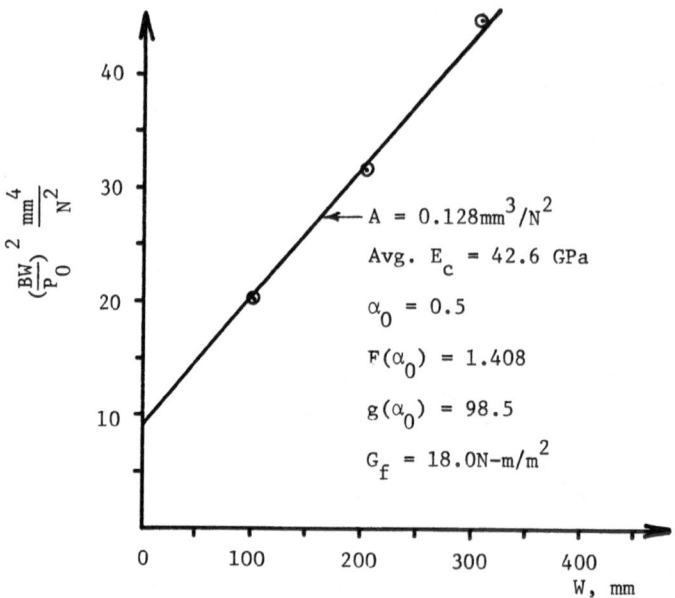

Fig. 5   Bazant Three Beam Method (1) for
         Notched Beams

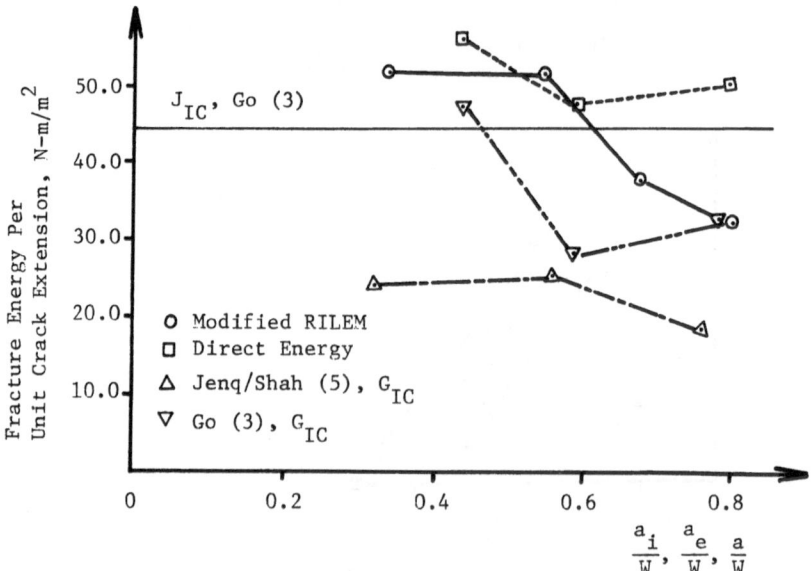

Fig. 6   Fracture Energy Values by Various
         Methods for Precracked Beams
         Using Averaged Results

DETERMINATION OF DYNAMIC MODE I AND MODE II FRACTURE MECHANICS PARAMETERS
FROM PHOTOELASTIC DATA

A. Shukla* and R. Chona**

*
Department of Mechanical Engineering and Applied Mechanics
University of Rhode Island
Kingston, RI   02881

**
Department of Mechanical Engineering
University of Maryland
College Park, MD   20742

1. INTRODUCTION

The stress field representation needed to completely describe the
stress state around stationary cracks in fracture specimens of finite
dimensions has been studied extensively (1,2,3) with some attention having
been paid to the dynamic counterpart of this problem (4,5). However, in
most cases, extraction of dynamic K values from experimental mechanics
data still relies on a one or two parameter representation of the stress
field (6,7). In this paper it is shown that stress field representation
incorporating terms additional to the singularity are necessary if the
stress state surrounding the tip of a running crack in a finite plate
specimen is to be modeled adequately. A failure to include these terms
in the analysis model can seriously affect the accuracy of K determination.
    This paper first presents the development of the generalized elasto-
dynamic stress field equations for both mode I and mixed mode crack propagation. These equations are then applied to dynamic photoelastic experiments and solved in a least square sense (8) to obtain fracture mechanics
parameters.

2. DYNAMIC STRESS FIELD REPRESENTATION

Irwin (9) has shown, that for a crack tip stress field translating
in the positive x-direction at a fixed speed, c, the dilatation, Δ, and
rotation, ω, can be expressed without loss of generality as

$$\Delta = \frac{\partial u}{\partial x} + \frac{\partial v}{\partial y} = \alpha \ (1 - \lambda_1^{\ 2}) \ \text{Re} \ \Gamma_1(z_1)$$

$$\omega = \frac{\partial v}{\partial x} - \frac{\partial u}{\partial y} = \beta \ (1 - \lambda_2^{\ 2}) \ \text{Im} \ \Gamma_2(z_2)$$

(1)

where λ_1, λ_2, z_1, and z_2 are defined in Figure 1. Using Hooke's Law

$$\sigma_{xx} = \mu[\alpha \ (1 + 2\lambda_1^2 - \lambda_2^2) \ \text{Re} \ \Gamma_1 - 2\beta \ \lambda_2 \ \text{Re} \ \Gamma_2]$$

$$\sigma_{yy} = \mu[-\alpha \ (1 + \lambda_2^2) \ \text{Re} \ \Gamma_1 + 2\beta \ \lambda_2 \ \text{Re} \ \Gamma_2] \tag{2}$$

and $\quad \tau_{xy} = \mu[-2\alpha \ \lambda_1 \ \text{Im} \ \Gamma_1 + \beta \ (1 + \lambda_2^2) \ \text{Im} \ \Gamma_2]$

where the constants α and β have to be evaluated so as to satisfy the boundary conditions for the crack problem of interest and depend upon the choice of the stress function, Γ_1 and Γ_2.

For the opening mode case, one logical choice for Γ_1 and Γ_2 is

$$\Gamma_1 = Z_1 = \sum_{n=0}^{n=N} A_n \ z_1^{n-1/2} \ \text{with} \ \Gamma_2 = Z_2 = \sum_{n=0}^{n=N} A_n \ z_2^{n-1/2} \tag{3}$$

The leading coefficient, A_0, is related to the opening-mode stress intensity factor, $K_I = A_0\sqrt{2\pi}$, and the leading term is the familiar inverse-square-root stress singularity. The symmetry condition requires that $\beta = 2\lambda_1\alpha/(1+\lambda_2^2)$.

For the static problem, it has been demonstrated (1,10) that a second stress function is required to completely describe the stress field in specimens with finite boundaries. The analogous choice for the dynamic problem is

$$\Gamma_1 = Y_1 = \sum_{m=0}^{m=M} B_m \ z_1^m \ \text{with} \ \Gamma_2 = Y_2 = \sum_{m=0}^{m=M} B_m \ z_2^m \tag{4}$$

In this case, the boundary condition on the crack faces requires that $\beta = \alpha(1+\lambda_2^2)/2\lambda_2$. The leading term, B_0, gives rise to a superposed constant stress in the direction of crack propagation which is similar to the σ_{ox}-term in Irwin's static near-field equations (11).

The constant $\alpha\mu$ can be determined from the definition of the opening mode stress intensity factor

$$K_I = \lim_{\substack{r \to 0 \\ \theta = 0}} \sigma_{yy} \ \sqrt{2\pi r} \tag{5}$$

as $\alpha\mu = (1+\lambda_2^2)/[4\lambda_1\lambda_2 - (1+\lambda_2^2)^2]$. The general solution to the constant speed opening mode elastodynamic crack problem is thus obtained by superposition of solutions obtained by both equations 3 and 4.

The shear mode solution can be obtained from the following considerations. First of all, equations 1 and 2 are general solutions that are not restricted to the opening mode crack problem. Secondly, the major difference between mode I and mode II stress fields is the antisymmetric nature of the stress field resulting from shear mode loading. A logical first step then is to select the stress functions, Γ_1 and Γ_2, as

$$\Gamma_1 = Z_1^* = \sum_{n=0}^{n=N} -i \ C_n \ z_1^{n-1/2} \ \text{with} \ \Gamma_2 = Z_2^* = \sum_{n=0}^{n=N} -i \ C_n \ z_2^{n-1/2} \tag{6}$$

where multiplying by $i = \sqrt{-1}$ provides the desired antisymmetry and the minus sign has been introduced for computational convenience. In this case, the leading coefficient, C_0, is related to the shear-mode stress intensity factor, $K_{II} = C_0\sqrt{2\pi}$, and the leading term is once again an inverse-square-root stress singularity. The boundary conditions for the mode II problem require that, for this choice of Γ_1 and Γ_2, $\beta = \alpha(1+\lambda_2^2)/2\lambda_2$.

A second choice for Γ_1 and Γ_2 that follows from finite specimen boundaries in a manner analogous to equation 4 is

$$\Gamma_1 = Y_1^* = \sum_{m=0}^{m=M} -i\, D_m\, z_1^m \quad \text{with} \quad \Gamma_2 = Y_2^* = \sum_{m=0}^{m=M} -i\, D_m\, z_2^r \tag{7}$$

with the boundary conditions being satisfied when $\beta = 2\lambda_1\alpha/(1+\lambda_2^2)$. The definition of the mode II stress intensity factor

$$K_{II} = \lim_{\substack{r\to 0 \\ \theta=0}} \tau_{xy}\, \sqrt{2\pi r} \tag{8}$$

yields $\alpha\mu = 2\lambda_2/[4\lambda_1\lambda_2 - (1+\lambda_2^2)^2]$. The solution to the shear mode problem is thus obtained by superposition of the solutions obtained by equations 6 and 7.

Finally, keeping in mind that the problem being considered is linear elastic, solutions for both mode I and mode II can be combined to provide the complete solution to the mixed-mode elastodynamic problem for a crack propagating at constant speed in a finite geometry. Similar results have been obtained by other investigators also (12,13).

It should be noted that for the case of a curving crack, the x and y directions must be taken as the instantaneous tangent and normal to the crack path, respectively.

3. PARAMETER DETERMINATION USING THE LEAST SQUARES METHOD

Equations 2 - 8 when combined with the stress optic law can be used to relate the fringe order, N, at any point in an isochromatic field with the unknown real coefficients, A_n, B_m, C_n and D_m

$$\left(\frac{Nf_\sigma}{2t}\right)^2 = \tau_{max}^2 = 1/4\,(\sigma_{yy} - \sigma_{xx})^2 + \tau_{xy}^2 \tag{9}$$

where f_σ is the fringe sensitivity of the model material and t is the model thickness. The first step in the analysis of an isochromatic pattern is to take a region around the crack tip from the experimental pattern being analyzed, extract a large number of individual data points and determine the coordinates and fringe order at each point. These data points are then used as inputs to an over-determined system of non-linear equations of the form of equation 9 and solved in a least-squares sense for the unknown coefficients. As a final check, the best fit set of coefficients is used to reconstruct the fringe pattern over the region of data acquisition to ensure that the computed solution set does, in fact, predict the same stress distribution as that observed experimentally.

When analyzing dynamic stress patterns, the data acquisition region is usually restricted to that portion of the stress pattern that can be seen to translate with only moderate changes in order to approximate the constant crack-speed assumption. In cases where the crack tip is approaching a specimen boundary, the data region should be restricted to no more than 1/4 to 1/3 the distance to the boundary. The number of coefficients necessary for an adequate representation of the stress field over the data acquisition region can be estimated by examining, as a function of the number of parameters, the average fringe order error, the values of the leading coefficients, and the reconstructed (computer-generated) fringe pattern corresponding to a given set of coefficients. Stability of the leading coefficients, as well as good visual match between experimental and reconstructed patterns indicates convergence to a satisfactory solution.

4. ILLUSTRATIVE EXAMPLES

The methodology outlined above is illustrated through the analyses of two different crack propagation experiments. The first experiment was performed using a Homalite 100 Ring segment whose geometry and loading are shown in Fig. 2. A total of 16 isochromatic photographs associated with the running crack were recorded using a Cranz-Schardin type high speed camera system. Fig. 3 shows the isochromatic pattern 145μs after crack initiation. A set of 60 data points was taken for analysis purposes. This data set was input to the least squares algorithm and analysed using successively higher order models. Figure 4 shows the changes in the error term and in the coefficients, A_O, B_O, and A_1, that occurred as the order of the analysis model was increased from two parameters (A_O, B_O) to six parameters (A_O, B_O, A_1, B_1, A_2, B_2). Figure 5 compares the experimental fringe pattern over the data acquisition region with the computer-generated patterns corresponding to the coefficient set from each model. It can be seen that six parameters are required before the reconstructed pattern matches the salient features of the input experimental pattern. The same conclusion could be reached from an examination of the error term in Figure 4, where the stabilized error of 2.5% associated with a six-parameter model has been found to be typical of the error at which a good match is achieved.

A similar analysis procedure was followed for all sixteen frames, and Figure 6 shows the instantaneous crack-tip stress intensity factor as a function of crack tip position for this experiment. The decreasing K-field is typical of this particular geometry and loading, and the results shown were obtained from analyses with parameter models ranging from four to seven, depending upon the crack tip position and the size of the data region relative to the specimen boundaries.

The second experiment was performed with a cross shaped specimen (14) also fabricated from Homalite 100. The geometry of the model is shown in Fig. 7. The specimen was eccentrically loaded such that the crack propagated in a curved path. A sequence of frames obtained during the experiment is shown in Fig. 8. The fringes are unsymmetric about the crack tip showing the mixed mode nature of the stress field. A set of sixty data points was again taken and the pictures analyzed with different parameter models. The changes in the fringe order error with an increase in the order of the analysis model were then examined and typically showed the trends illustrated in Fig. 9, with the error stabilizing to about 3% of the input average fringe order once the stress state has been modeled well over the region of data acquisition. The leading coefficients of each series were also found to stabilize once a good fit had been achieved and this is also shown in Fig. 9. Each of the coefficient sets obtained from the analysis was used to reconstruct the fringe pattern over the region of data acquisition so as to visually confirm that the stress state around the crack tip had in fact been properly modeled by the analysis model selected as being "correct". The sequence of reconstructed fringe patterns corresponding to 2 through 5-parameter models from frame 12 is compared to the experimental pattern in Fig. 10 and the 5-parameter solution can be seen to closely match the stress state over the region of data acquisition. Similar analyses were performed for each frame of the experiment. The variation of the stress intensity factor as a function of time is shown in Fig. 11.

5. SUMMARY

Dynamic stress field equations are developed and a parameter extraction methodology is presented for both mode I and mixed mode crack propagation problems. The method is applied to examples of crack propagation in a ring and a cross shaped specimen. The results show that more than two parameters in the stress field representation are needed to adequately model the state of stress around the crack tip. Both the singular and the non singular terms are seen to vary systematically with the crack tip position.

ACKNOWLEDGEMENTS

This work has been supported by Grant No. MEA-8415494 from the National Science Foundation to the University of Rhode Island and Subcontract No. ORNL/sub-74778 from the Oak Ridge National Laboratory Heavy-Section Steel Technology Program to the University of Maryland.

REFERENCES

1. Liebowitz, H., Lee, J.D., and Eftis, J., Engineering Fracture Mechanics, 10(3), pp. 315-335 (1978).
2. Sanford, R.J., Mechanics Research Communications, 6, pp. 289-294 (1979).
3. Chona, R., Irwin, G.R. and Sanford, R.J., Proceedings of the 14th National Symposium on Fracture Mechanics, ASTM STP 791 (1983).
4. Rossmanith, H.P. and Irwin, G.R., Department of Mechanical Engineering Report, University of Maryland (1979).
5. Kobayashi, A.S. and Mall, S., Experimental Mechanics, 18(1), pp. 11-18 (1978).
6. Beinert, J. and Kalthoff, J.F., Mechanics of Fracture VII, G.C. Sih, ed. (1982).
7. Ramulu, M. and Kobayashi, A.S., Experimental Mechanics, 23(1), pp. 1-9, (1983).
8. Sanford, R.J. and Dally, J.W., Engr. Frac. Mech., 11, pp. 621-633 (1979).
9. Irwin, G.R., Lehigh University Lecture Notes on Fracture Mechanics (1968).
10. Sanford, R.J., Mechanics Research Communications, 6, pp. 289-294 (1979).
11. Irwin, G.R., Proceedings SESA, XVI, pp. 93-96, (1958).
12. King, W.W., Malluck, J.F., Aberson, J.A. and Anderson, J.M., Mechanics Research Communications, 3, pp. 197-202 (1976).
13. Nishioka, T. and Atluri, S.N., Engineering Fracture Mechanics, 18, pp. 1-22 (1983).
14. Shukla, A. and Anand, S., ASTM STP, 17th National Symposium on Fracture, Albany, NY 1984.

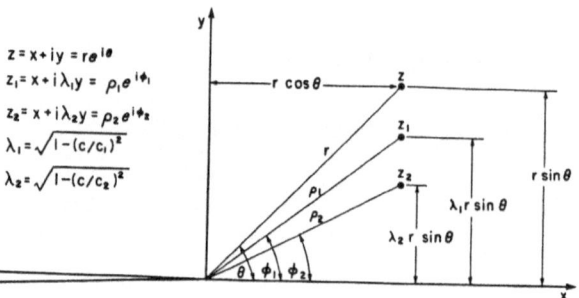

Figure 1 – The coordinate systems and transformation relations used in the constant-crack-speed analysis

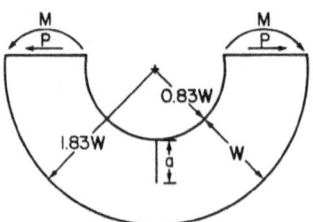

Figure 2 – The geometry and loading of the ring segment used

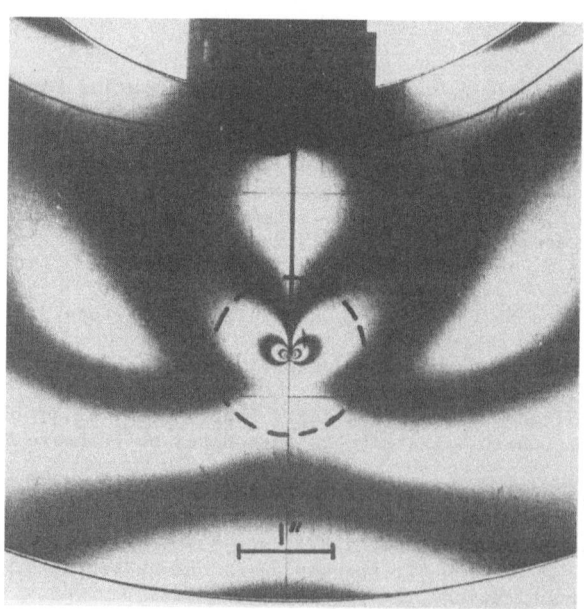

Figure 3 – The isochromatic fringe pattern 145 μs after crack initiation in the ring segment; $a/W = 0.52$; $c/c_2 = 0.31$

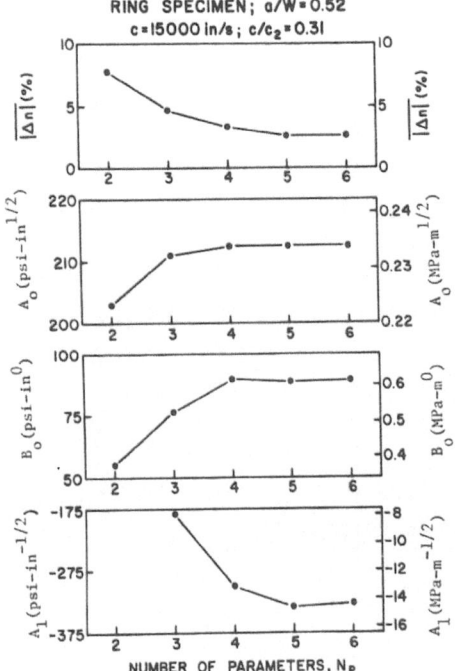

RING SPECIMEN; a/W = 0.52
c = 15000 in/s; c/c_2 = 0.31

Figure 4 - The error term and
leading coefficients versus the
number of parameters in the
analysis model

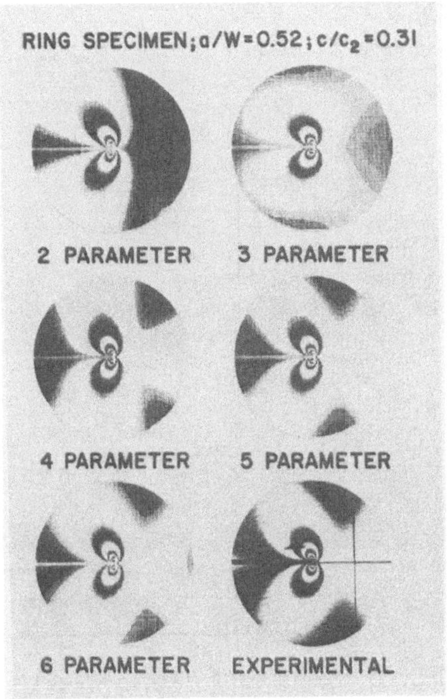

RING SPECIMEN; a/W = 0.52; c/c_2 = 0.31

2 PARAMETER   3 PARAMETER

4 PARAMETER   5 PARAMETER

6 PARAMETER   EXPERIMENTAL

Figure 5 - Experimental and
reconstructed fringe patterns from
different order models for the
fringe pattern of Figure 3

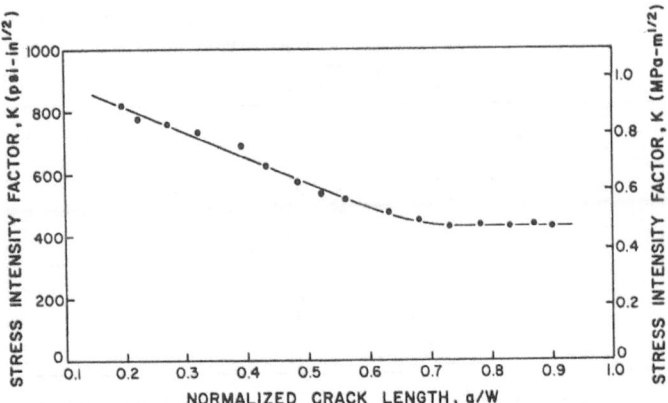

Figure 6 - Instantaneous stress intensity factor
as a function of crack length for a crack propagating
across a ring segment

252

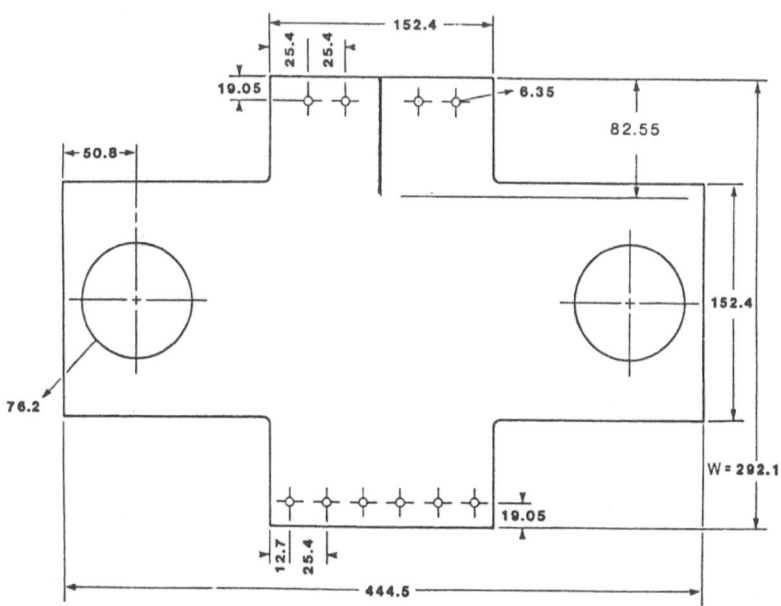

ALL DIMENSIONS ARE GIVEN IN MILLIMETERS.

Figure 7 - Specimen geometry and loading of the cross-shaped biaxial fracture specimen used in this study

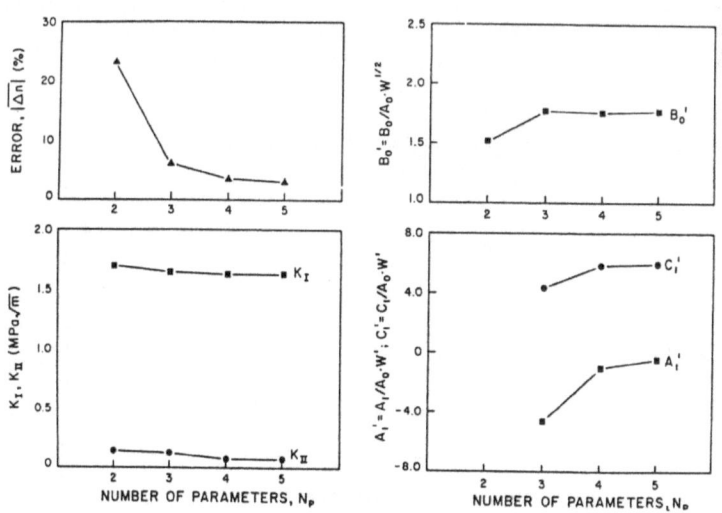

Figure 9 - The changes in the fringe order error and leading coefficients of the series stress field representation as a function of the number of parameters

* Figure 8 appears on following page

FRAME 9, t=87μs   FRAME 10, t=96μs   FRAME 11, t=144μs

FRAME 12, t=154μs   FRAME 14, t=172μs   FRAME 15, t=180μs

FRAME 16, t=190μs   FRAME 17, t=201μs   FRAME 18, t=212μs

Figure 8 - A sequence of frames showing the stress field
surrounding the rapidly propagating curving crack

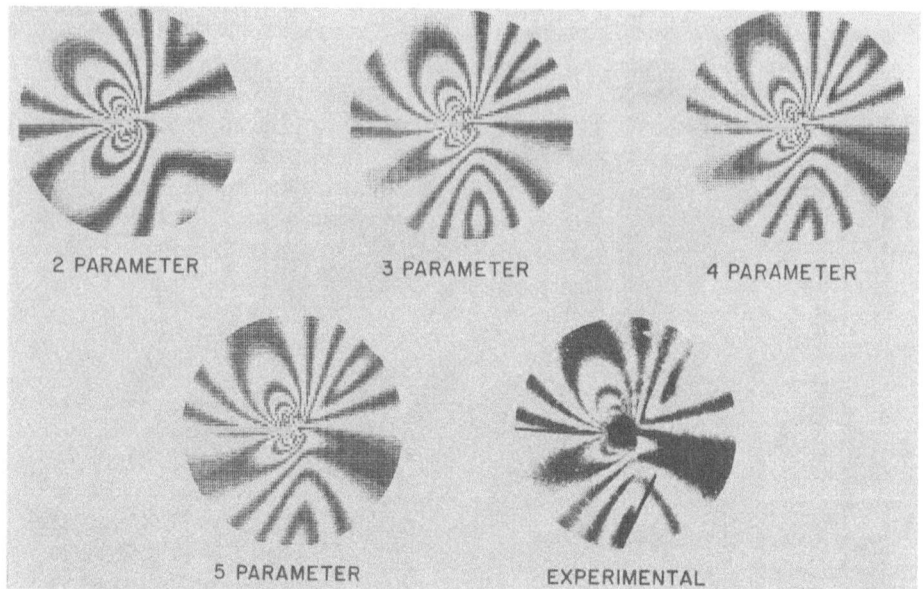

Figure 10 - The experimental fringe pattern over the data acquisition region of radius equal to 13 mm and the reconstructed (computer generated) isochromatic patterns corresponding to 2, 3, 4, and 5-parameter models

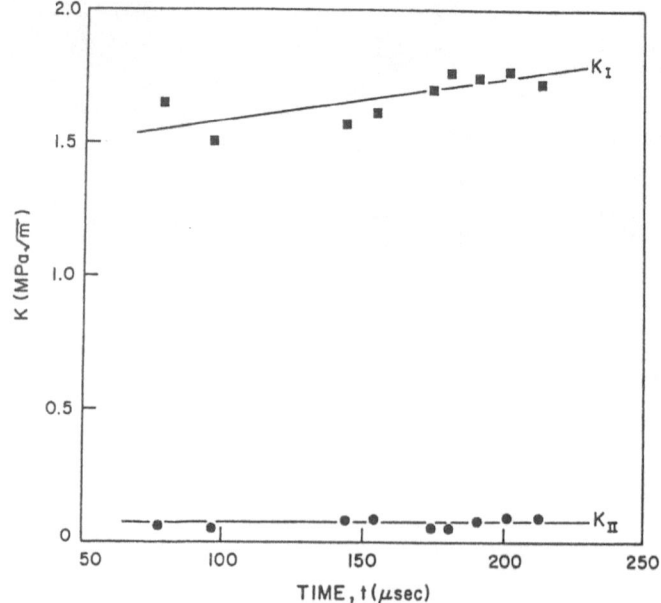

Figure 11 - Opening and shear mode stress intensity factors as a function of time after crack initiation

MODE II AND MODE III FRACTURE TESTING OF ADHESIVE JOINTS USING A STIFF ADHEREND SPECIMEN

V. WEISSBERG AND M. ARCAN

FACULTY OF ENGINEERING, TEL-AVIV UNIVERSITY, TEL-AVIV 69978, ISRAEL

1. INTRODUCTION

Bonded structures are economic in nature, the fabrication cost being reduced by design simplicity and by replacement of the process of assembly by adhesive bonding. However, there are many areas, especially those concerned with the joint failure criteria, where the knowledge is inadequate. Until such aspects are better understood by experiments and analysis, the full potential of adhesives as a means of joining materials will not be realized. Two testing approaches to characterize adhesive-bonded joint strength will be considered: "flexible adherends" and "stiff adherends" test specimens.

Flexible adherends testing is achieved with relatively thin adherends, which undergo important deformations. Examples of flexible adherend specimens are: the popular tension lap shear ASTM D1002-72, the double lap shear ASTM D3528-76, and the cracked long lap shear [1, 2]. These test specimens originate from real structures, but since they yield complicated stress states under load, it is difficult to derive some basic adhesive properties, including fracture properties, from this type of test. For example, a lap shear specimen (Fig. 1), fails, due to a non-uniform, complicated combination of tension and shear [3], [4].

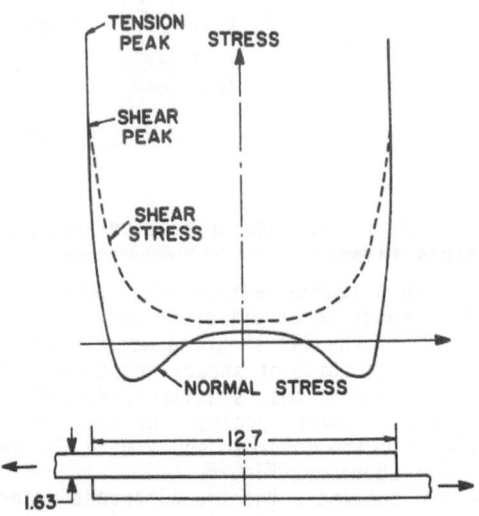

FIGURE 1. Distribution of shear and normal stress in lap shear joint, ASTM D1002-7 [4].

Stiff adherend testing consists of thick plates joined by an adhesive layer. The problem is reduced to the fracture of a weak layer in a homogeneous body. As a result, certain fracture parameters can be isolated and studied. For instance, the thick adherend test specimen (lap shear), which is used to measure the stress-strain curve of adhesives [5], has a nearly uniform state of shear, but with important secondary stresses. Other stiff adherend test specimens are the double section compact shear [6], scarf joint [7], cone test [8], butt joint and napkin ring[9]. In all these tests, no special attention was paid to the uniformity and purity of the stress state in the adhesively bonded joints.

In this paper, we shall describe a family of stiff adherend test specimens as proposed by Arcan et al.[10,11], designed for a pure and uniform state of shear stress, free of tension and compression, where Mode II and ModeIII can be accurately obtained. An opening mode could also be superimposed [12]. Furthermore, several series of tests were performed in order to obtain accurate characteristic stress-strain curves of adhesives, as well as to study some fracture parameters (e.g.influence of adhesive thickness on strength).

2. PURE SHEAR TEST SPECIMENS

Two types of test specimens were used in this program. A first specimen with a rectangular section was used to obtain stress-strain curves of adhesives, (Fig. 2).

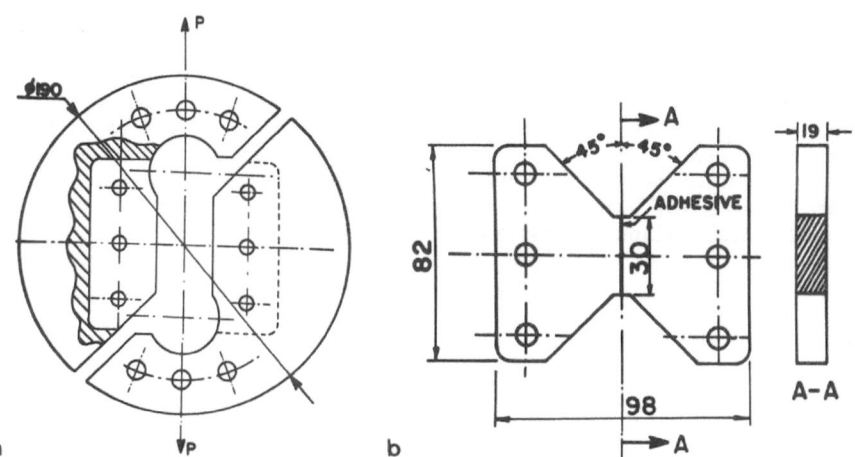

FIGURE 2. Testing system for rectangular section specimen. a) Loading frame. b) Specimen, dimensions in mm.

A second specimen with a square section (Fig. 3) was used to study the fracture strength as a function of different parameters. In this second specimen, the behaviour of the two shearing modes could be compared.

It appears that a uniform state of shear, practically free from tension and compression, exists in both test specimens, Fig. 4.

The advantage of the pure shear specimen for adhesive testing is emphasized by the comparison of its deformed shape with the deformed shape of the thick adherend test specimen, Fig. 5.

In the thick adherend specimen, the two adherends undergo opposite bending, which gives rise to secondary tension and large stress concentrations. By contrast, in the proposed pure shear specimen, because isostatics transfer stresses through the adhesive layer at ± 45°, the two adherends slip almost parallel, inducing a pure and uniform shear in the significant section.

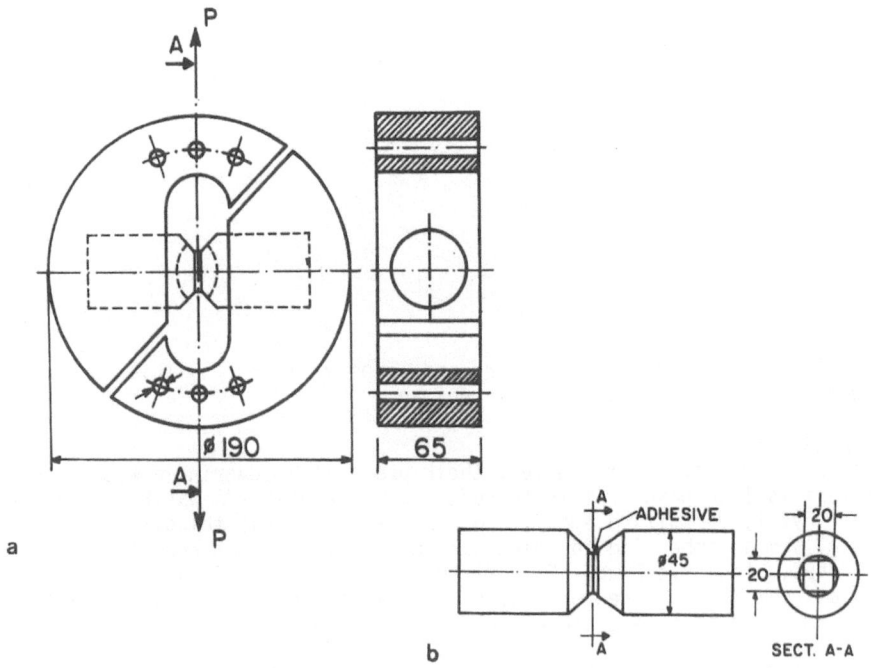

a

b

FIGURE 3. Testing system for square section specimen. a) Loading frame. b) Specimen,dimensions in mm.

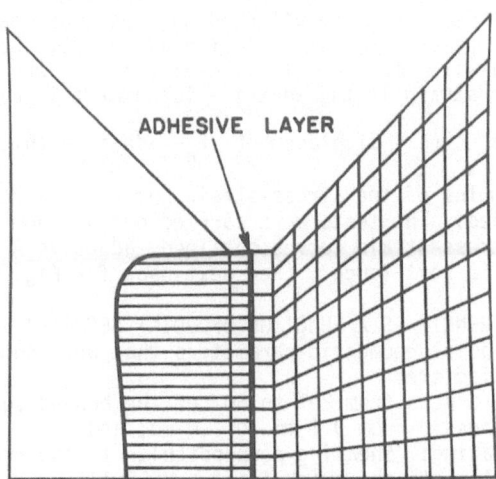

FIGURE 4. Distribution of shear stress in the rectangular section specimen - finite element analysis.

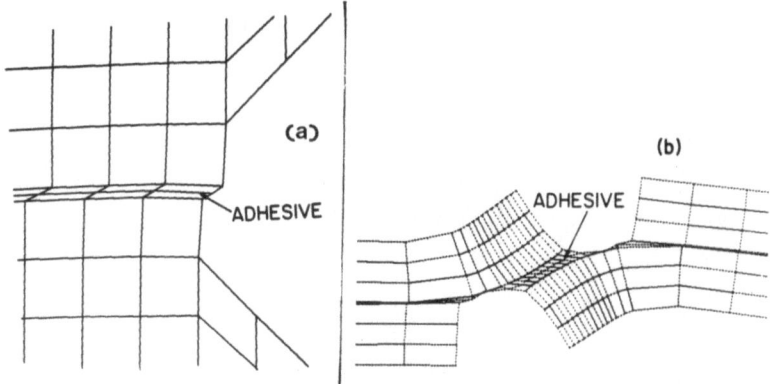

FIGURE 5. Deformed shape of adherend and adhesive layer. a) Thick adherend. b) Rectangular section pure shear specimen.

3. MATERIALS

Two epoxy adhesives were tested, their properites covering a wide range of materials from hard-brittle to soft-tough. Adhesive A was EA 934NA (Dexter Hysol) unmodified epoxy, and consists of a gray thixotropic paste (part A) and an amber liquid amine curing agent, cured at room temperature for 7 days. Adhesive B was EA 9321 (Dexter Hysol) modified 2-part epoxy cured at room temperature for 7 days. The adherend material was 2024-T3 Aluminium. The surface preparation for bonding was by chromic acid etching.

4. EXPERIMENTAL EVALUATION OF THE ADHESIVE STRESS-STRAIN CURVE

The stress-strain curve may very well characterize mechanically the adhesive systems. In order to obtain the stress-strain curves, tests were carried out, using the rectangular section specimen, equipped with a 2670-114 C.O.D. Gauge (5 mm base), which measures movements with an accuracy of 0.0015 mm in the Instron Machine. The C.O.D. Gauge was mounted between 2 small blades, each one attached to 1 part of the adherend. The C.O.D. measures the relative movement of the adherends. Because the adherends move almost parallel to each other, giving a pure and uniform state of shear, the shear strain in the adhesive is accurately calculated by:

$$\gamma = \delta/t_a \quad (1) \quad (\delta - \text{displacement}; t_a - \text{adhesive thickness})$$

Fig. 6 shows examples of the stress-strain curves obtained for the two adhesives studied. The tests were carried out at room temperature. According to the stress-strain curves obtained, adhesive A is hard-brittle, whereas adhesive B is soft-touch. (See next page for Fig. 6).

5. THE FRACTURE STRENGTH AS A FUNCTION OF JOINT GEOMETRY

5.1. Basics. The joint geometric parameters that were numerically and experimentally studied are:
  a. Crack Length: Cracks from 20% to 60% of the bonded surface;
  b. Adhesive thickness: From 0.1 mm to 1.0 mm; and
  c. The mode of loading: Shearing perpendicular to the crack front - MODE II , Parallel to the crack front - MODE III.
According to finite element analysis of the stiff adherend pure shear test specimen [13], the state of uniform pure shear is practically maintained in the whole range of the parameters, above mentioned for an adhes-

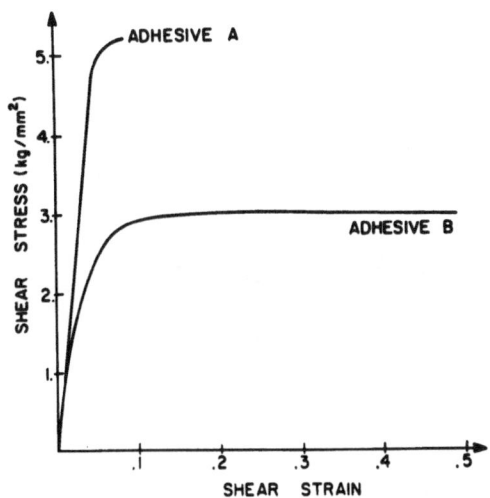

FIGURE 6. Shear stress-strain curve for two adhesives.

ive modulus of elasticity up to 300 Kg/mm^2. This particular property of the pure shear test specimen makes it very useful for the research of adhesively bonded joints. Varying the geometrical parameters, and keeping the state of stress practically unchanged, the variation in fracture strength can be well-related to the adhesive or adhesion characteristics.

5.2. Fracture tests were carried out, using the square section specimen as follows:

5.2.1. The cracks were introduced by inserting a very thin teflon tape. Three cracked configurations were tested, viz. 20%, 40%, and 60% of the total bonded surface. The cracks were placed symmetrically on the edges of the test specimen. The results obtained show that the fracture stress is practically constant, which indicates a similar behaviour with an ideal infinitely rigid parallel strip [14, 15].

5.2.2. The adhesive thicknesses were produced, varying between 0.1 mm and 1.0 mm. The strength was found to be a decreasing function of the adhesive thickness and was plotted against $t_a^{-1/2}$. The fracture strength of adhesive A is practically a straight line. The adhesive B is a more complex function of $t_a^{-1/2}$, as shown in Fig. 7 (see next page).

5.2.3. Some spot point tests of Mode III were performed in order to check the testing versatility of the proposed system. The fracture strength was 10 - 15% higher than the one obtained in Mode II testing.

6. DISCUSSION AND ANALYSIS

The thickness effect was selected as a significant parameter of adhesive joints to be analyzed under pure shearing modes. The finite element analysis [13] shows that the stress distribution in the adhesive layer is not affected by the variation in adhesive thickness. On the other hand, the test results show the fracture strength to be a strong function of thickness. An energy failure criterion was used to find the relationship between thickness and fracture strength. It is assumed, as in [16], that the fracture initiation occurs when a characteristic strain energy per unit interface area W_i is stored in the deformable adhesive. The energy per unit

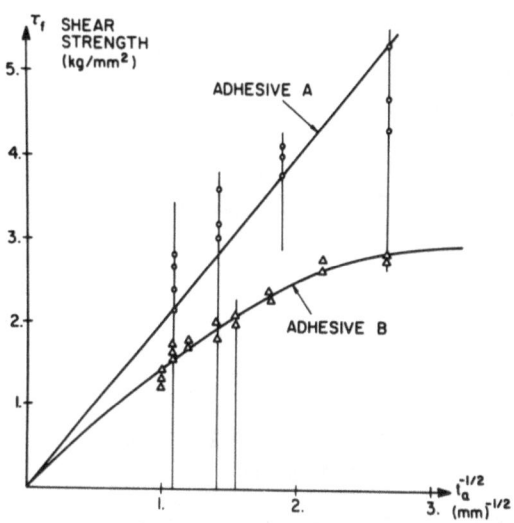

FIGURE 7. Fracture strength against $t_a^{-1/2}$ (t_a - adhesive thickness).

volume of the adhesive to fracture initiation is denoted by W_f. In the pure shear test specimen, stress being uniform, the relationship between W_i and W_f is:

$$W_i = t_a \cdot W_f \quad , \tag{2}$$

or

$$W_i/t_a = \int_0^{\gamma_f} \tau d\gamma \quad , \tag{3}$$

by integrating the stress-strain curve. As adhesive A is hard-brittle, the relationship is linear; then,

$$W_i/t_a = \tau_f^2 /2G_a \quad , \tag{4}$$

where G_a is the shear modulus and τ_f is the shear stress at fracture initiation; according to our experiments, τ_f is always on the part of τ-γ characteristic curve, where $d\tau_F/d\gamma_F > 0$. Hence,

$$\tau_f = (2G_a W_i)^{1/2} t_a^{-1/2} \quad . \tag{5}$$

As one can see, τ_f is a linear function of $t_a^{-1/2}$, which agrees well with the test results, Fig. 7.

Adhesive B is soft-tough; the stress-strain curve is well-approximated by a sine function:

$$\tau^* = \sin \pi\gamma^*/2 \tag{6}$$

where τ^* and γ^* are normalized values as defined in Fig. 8.

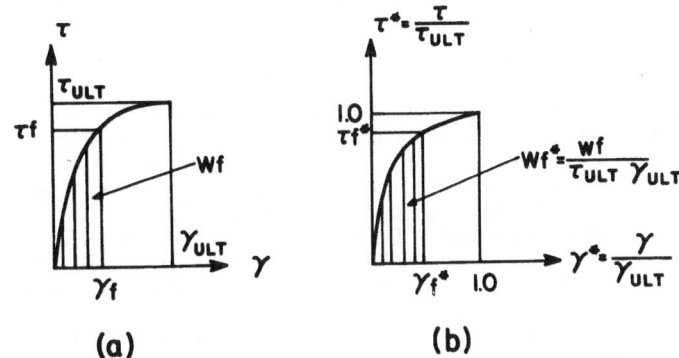

(a)   **(b)**

FIGURE 8. τ - γ curves of soft-tough adhesives. a) Typical. b) Normalized diagram.

Hence,

$$W_f^* = \int_0^{\gamma_f^*} \sin \pi\gamma^*/2 \cdot d\gamma^* \quad , \tag{7}$$

$$W_f^* = 2(1 - \cos \pi\gamma_f^*/2) \quad , \tag{8}$$

$$W_f^* = 2[1 - (1 - \sin^2 \pi\gamma_f^*/2)^{1/2}]/\pi \quad . \tag{9}$$

Substituting (6) in (9), we get

$$W_f^* = 2[1 - (1 - \tau_f^{*2})^{1/2}]/\pi \quad . \tag{10}$$

Hence,

$$\tau_f^* = (\pi W_f^* - \pi^2 W_f^{*2}/4)^{1/2} \quad , \tag{11}$$

$$W_i^* = W_i/\tau_{ult} \gamma_{ult} = W_f^*/t_a \quad , \tag{12}$$

and finally,

$$\tau_f^* = (\pi W_i^*/t_a - \pi^2 W_i^{*2}/4t_a^2)^{1/2} \quad . \tag{13}$$

A characteristic value of $W_i^* = 0.07$ mm^{-1} was found to fit well with the experimental fracture strength of adhesive B as presented in Fig. 7. For both hard-brittle and soft-tough adhesives, good correlation was found between the energy fracture initiation criterion, as represented, respectively, by expressions (5) and (13), with the test results (Fig. 7).

7. CONCLUSIONS

A new methodology of Mode II and III fracture testing of adhesive joints was developed, using a stiff adherend specimen, inducing uniform pure shear stresses in the gauge area. Finite element analysis, as well as experimental determinations, allowed us to evaluate the stresses in the adhesive layer as being uniform over the tested volume, and to obtain stress-strain diagrams, using specific instruments.

A series of tests was performed with untoughened EA 934 NA and EA 9321 Epoxy adhesives for different crack lengths and adhesive thicknesses. For adhesives, good correlation between the energy fracture initiation criterion and the test results was found.

REFERENCES

[1] Brussat TR, Chiu ST and Mostovoy S: Fracture Mechanics for Structural Adhesive Bonds - Final Report. AFML - TR - 77 - 163, 1977.
[2] Dattaguru B, Everett Jr. AR, Whitcomb DJ and Johnson SW: Geometrically Nonlinear Analysis of Adhesively Bonded Joints. J. of Engineering Materials and Technology, Vol. 16, January 1984.
[3] Cooper PA and Sawyer JW: A Critical Examination of Stresses in an Elastic Single Lap Joint. NASA TP-1507, 1979.
[4] Adams DR and Wake CW: Structural Adhesive Joints in Engineering. Elsevier Applied Science Publishers, Ltd., 1984.
[5] Krieger RB: Stiffness Characteristics of Structural Adhesives for Stress Analysis in Hostile Environment. American Cyanamid Company, Havre de Grace, Maryland, 1975.
[6] Chisholm DB and Jones DL: An Experimental Stress Analysis of a Practical Mode II Fracture Test Specimen. Experimental Mechanics, 17, 7-13, 1977.
[7] Trantina GG: Fracture Mechanics Approach to Adhesive Joints. J. of Composite Materials, 6, 192-207, 1972.
[8] Anderson PG, Bennett SJ and DeVries LK: Analysis and Testing of Adhesive Bonds. Academic Press, Inc., 1977.
[9] Ishai U and Dolev G: Mechanical Characterization of Bonded and Bulk Adhesive Specimens as Affected by Temperature and Moisture. 26th SAMPE Symp. April 1981.
[10] Arcan M, Hashin Z and Voloshin A: A Method to Produce Uniform Plane-Stress States with Application to Fiber-Reinforced Materials. Experimental Mechanics, 18, 141-146, 1978.
[11] Arcan M. and Banks-Sills L: Mode II Fracture Specimen-Photoelastic Analysis and Results. Proc. 7th Int. Conf. on Experimental Stress Analysis, Israel, 187-201, 1982.
[12] Banks-Sills L, Arcan M and Bortman Y: A Mixed Mode Fracture Specimen for Mode II Dominant Deformation. Engineering Fracture Mechanics, 20, 1, 145-157, 1984.
[13] Weissberg V and Arcan M: A Uniform Pure Shear Testing Specimen for Adhesive Characterization. A.S.T.M. International Symposium on Adhesively Bonded Joints, September 1986 (to be presented).
[14] Williams JG: Fracture Mechanics of Polymers. Ellis Horwood Series in Engineering Science, 1984.
[15] Andrews EH: Fracture in Polymers. Oliver and Boyd, Ltd. 1968.
[16] Gent A: Fracture Mechanics of Adhesive Bonds. Rubber Chem. Technol. 47, 202, 1974.

AN INTEGRATED SOFTWARE/HARDWARE APPROACH TO EXPERIMENTAL STRESS ANALYSIS

Howard J. Howland

Probably most "seasoned" engineers associated with design, testing, or failure analysis of mechanical components or structures can recount instances where a design or measurement problem could not be solved, at least in a cost-effective way, due to lack of access to a computer or computerized test instrumentation.

For example, new product designs that should have been subjected to a rigorous testing program may instead have been designed with excessive safety factors, increasing material costs and weight with no real guarantee of increased reliability. Or, if an existing design failed, the time required for failure analysis to determine the root cause of the failure might have been prohibitive, and the new component simply made "larger and stronger" increasing cost and weight with no guarantee of increased performance or service life.

Experimental Stress Analysis is often called the quality control of design. The most widely used method of experimental stress analysis involves the use of electrical resistance strain gages adhesively bonded or spot welded at discrete points on the test part surface. The measured strain values are used to determine the magnitudes and directions of principal stresses on the surface when the test part is subjected to various test loads. Analysis of a large or complex part or structure may necessitate measurement at numerous test points, under a large number of actual or simulated in-service loading conditions, resulting in enormous amounts of data being acquired.

In smaller organizations without access to a computer, the time required for data acquisition and analysis may simply be prohibitive. In larger organizations, where a computer is available, data may be manually collected and presented to the computer department for data reduction using software written by the test engineer. But this may mean lengthy time delays between the acquisition of data and analysis of test results. Further, the test engineer might have no indication of faults in the test instrumentation system that might produce useless data, or worse, the applied test loads might produce dangerously high stresses in the test part that would not be realized until subsequent data reduction. In either case, costly re-testing may be required.

Wieringa, H (ed), Experimental Stress Analysis.
© *1986. Martinus Nijhoff Publishers, Dordrecht.*

A partial solution then, for both large and small organizations, is to have a low-cost personal computer available for data reduction at the test site. But even this approach is significantly limited because the measurement instrumentation generally is not directly compatible with the available (computer) hardware, and test data must first be acquired by the test instrumentation and then entered, by hand, into the computer for reduction and analysis. Furthermore, appropriate software for data analysis and presentation is not generally commercially available, and must be specially designed by the test engineer.

While some test engineers have become, or have access to, expert computer programmers, accurate strain gage data reduction requires an extensive understanding of the behavior of the strain gage and all possible error sources that must be accounted for to produce accurate test data. And in addition, these connection factors must be applied at precisely the appropriate point in the data reduction process.

For example, measured strain data may contain significant errors due to thermally induced "apparent strain," and a change in strain gage sensitivity, or gage factor, resulting from a change in temperature of the test part. Apparent strain will be a constant value at a given temperature regardless of the magnitude of any load-induced stress or strain in the test part. And therefore, if not accounted for, the percentage of error introduced into the measurement varies inversely with the magnitude of the test load. The change in gage factor of the strain gage will be manifested as a change in calibration with a change in temperature of the test part, referred to as temperature coefficient of gage factor. The error magnitude will be a fixed percentage of the measured strain value at a given test temperature, but will vary in direct proportion to test temperature.

These temperature-induced errors are repeatable and predictable (See Figure 1), and must be accounted for to obtain accurate stress and strain data when a change in temperature will occur during the test sequence.

Another common error source is the transverse sensitivity of the strain gage grid. The strain gage is designed for maximum sensitivity to surface strain in the test part parallel to the primary sensing axis of the strain gage grid, and minimum sensitivity to strains transverse to the primary sensing axis. While this transverse sensitivity is generally a small percentage of the strain gage gage factor, it can introduce significant errors in strain measurements obtained in a biaxial stress field where the transverse strain may be several times greater than that along the gage primary sensing axis. (See Figure 2).

The strain gage is most commonly connected as a single, variable-resistance arm of a Wheatstone bridge circuit. This "quarter-bridge" configuration is a nonlinear measuring circuit, and the percentage of error increases with the measured strain magnitude. This non-linearity is a fourth example of an error source that must be accounted for during data reduction for accurate stress and strain data. (See Figure 3).

It is important to re-emphasize that these and other corrections must be applied at the appropriate points in the data reduction process. The corrected strains are then reduced using the appropriate values for Young's Modulus and Poisson's ratio to provide desired stress information, including stress magnitudes and the principal stress directions on the test part.

In addition to the knowledge and care exercised during data reduction, accurate strain measurement requires care in the selection of the measurement instrumentation. For example, the measured data is directly affected by the stability and magnitude of the Wheatstone bridge excitation voltage supply. The instrumentation should incorporate a highly stable, adjustable excitation voltage to maximize signal-to-noise ratio, without applying unnecessarily high voltages to the strain gage grid that would produce gage self-heating and instability.

The measuring system should provide for "auto zero" of the instrument amplifier, and automatic shunt calibration to verify accuracy of the instrumentation. It should include bridge completion and "dummy" resistors for quarter, half and full bridge circuits, and provide automatic bridge balance capability.

In addition, because accurate strain gage data reduction often requires measurement of test temperature, the instrumentation system should also acquire and reduce temperature data from a wide range of temperature measurement devices such as thermocouples, RTD's and thermistors. And, as measurement of displacement, load or some other engineering variable might also be required as part of the test sequence, it should accept inputs from a wide range of sensor types, and reduce the measured data and present it directly in engineering units as data is acquired.

It is apparent from the preceding discussion that measurement of strain data is best accomplished by use of a dedicated stress analysis measurement system, with the ability to address the specific characteristics of the strain gage itself. And that design of such an instrumentation system requires extensive knowledge of the specific behavior of the strain gage and associated error sources. Attempts to interface general-purpose instrumentation with general-purpose computer hardware, and then designing specialized software to measure, reduce and present data, generally results in unsatisfactory performance at very high cost in time and capital investment.

Designed specifically for test engineers and technicians, the Measurements Group System 4000 is a dedicated stress analysis data system incorporating measurement instrumentation, computer hardware and a comprehensive software package for data reduction and presentation. (See Figure 4). A major design objective was to develop an integrated system of hardware and software for maximum performance at minimum cost, in an easy-to-use instrumentation system with direct application to a very broad range of stress analysis test measurement programs.

System 4000 acquires, reduces and presents test data from up to 750 measurement channels of sensor inputs such as strain gages, transducers, thermocouples, RTD's potentiometers and voltage sources. All set-up information and constants for data reduction are readily entered in plain English language -- no computer programming is required. All measurement channels are automatically zeroed and calibrated, and strain gage measurements are corrected for temperature effects, Wheatstone Bridge non-linearity and transverse sensitivity; strain data is further reduced to provide principal strain and stress magnitudes and directions. If data exceeds user-defined test or control limits, out-of-limit channels are immediately printed or recorded on floppy disk, and a limit relay is provided to initiate some alerting or corrective action, all AS THE DATA IS BEING ACQUIRED. (See Figures 5 and 6).

Fully reduced data is available, as it is being acquired, in tabular form on the system monochrome monitor or printer, or in graphical form on the color monitor. Data may also be stored on floppy disk for later analysis or manipulation. The software package includes specially designed software to generate test reports, and optional software is available to plot data on a flatbed plotter, or for residual stress analysis using the strain gage hole drilling method. For use with customer-designed data analysis or presentation software, reduced data stored on floppy disk is IBM-PC compatible. (See Figures 7 and 8).

In summary, System 4000 is a comprehensive, dedicated stress analysis data system designed for test engineers and technicians. The measurement instrumentation, computer hardware and a comprehensive software package are integrated to provide accurate, fully reduced data on-site for immediate review or analysis, in an extremely cost-effective computer-controlled test measurement system.

Figure 1

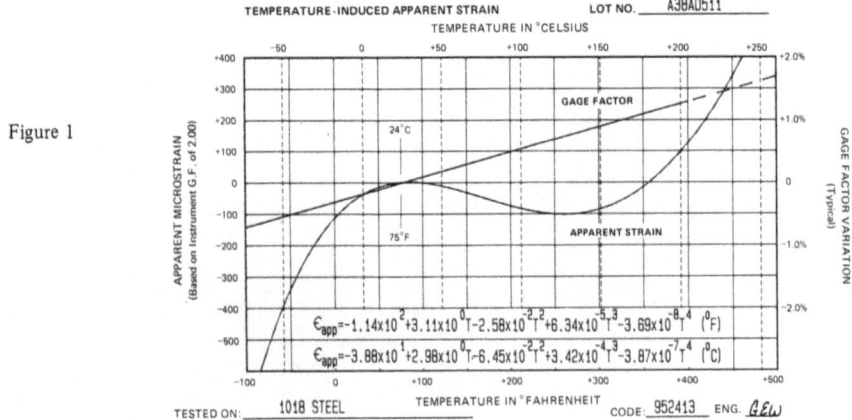

Figure 2

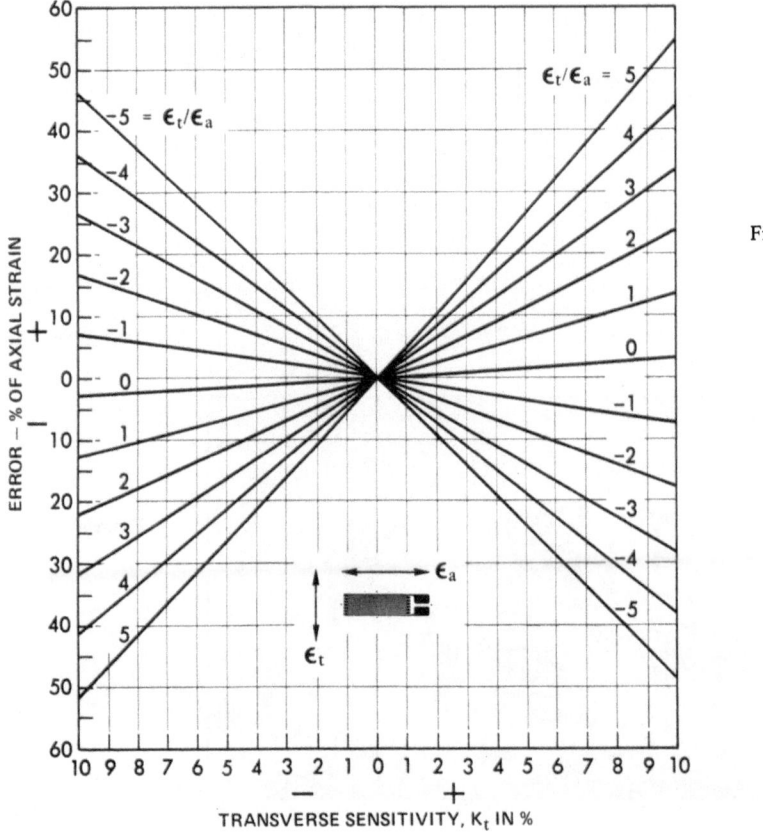

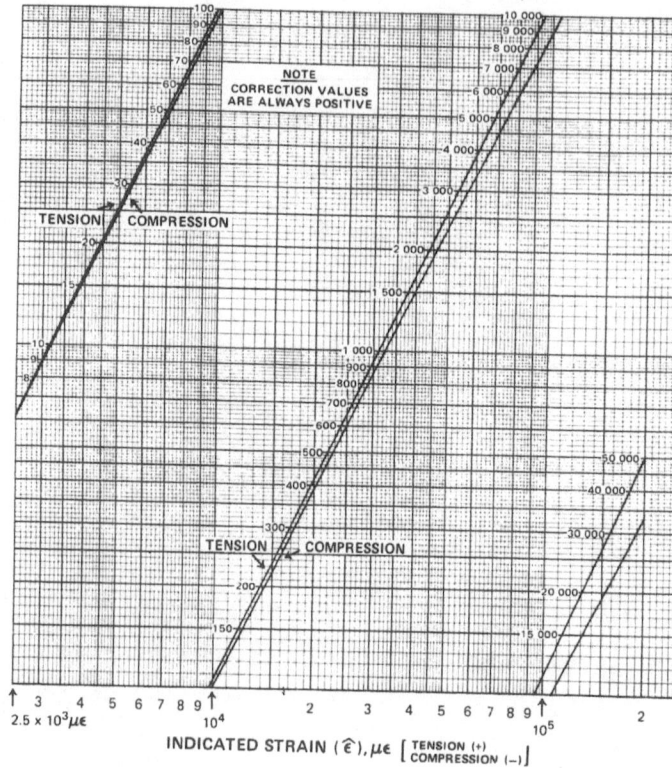

Figure 3

Figure 4

Figure 5

```
                        PROGRAM MENU

        <CC>  Channel Assignment Prog    AUXILIARY PROGRAMS
        <II>  Info Setup Programs
        <LS>  Limit Setup               <DT>  Date/Time Read/Reset
        <IS>  Info Summary Menu         <CT>  Change Temp Scale
        <RI>  Record Info               <PR>  Print Data Record Times
                                        <SS>  Scan Sequence
        <SZ>  System Zero/Cal           <PZ>  Print Zeros & Cal Values
        <DS>  Display Channel(s)        <AR>  App Strain Coef Regression
        <GP>  Graphics Plot of Data     <RR>  RTD Coef Regression
                                        <CI>  Clear Info
                                        <CD>  Change Data Disk
        <SR>  Scan & Record Data        <ED>  Erase Data Records
        <SP>  Scan & Print Results      <PD>  Prepare Data Disk
        <RD>  Reduce Data               <TY>  Type Remarks


                        ENTER LETTERS
```

```
                    INFO SETUP PROGRAMS
        CHANNEL INFO                    GENERAL INFO
    - - - - - - - - - - STRAIN GAGE SCANNER - - - - - - - - - -
                                ROSETTES
<GF>  Gage Factor                <RO>  Type & Elements
<TC>  Temperature Corrections    <KT>  Kt (Transverse Sensitivity)
<SC>  Shunt Calibration (ue)     <MA>  Material/Channel Assignment
<AA>  Active Arms                <MP>  Material Properties

<MM>  =MM= Temperature Sensor       GAGE LOTS
<ST>  S.G. Transducer            <GC>  Temp Coef of GF by Lot
                                 <AC>  Apparent Strain Coef's by Lot


    - - - - - - - - - - UNIVERSAL SCANNER - - - - - - - - - -
<HT>  High-Level Transducer
<LV>  LVDT  (AC)                    CURVE
<DC>  DCDT  (DC LVDT)            <RC>  Special RTD Coef's
<RT>  RTD or Thermistor
<PO>  Potentiometer
<VO>  Volts or mV
                    ENTER LETTERS
```

Figure 6

270

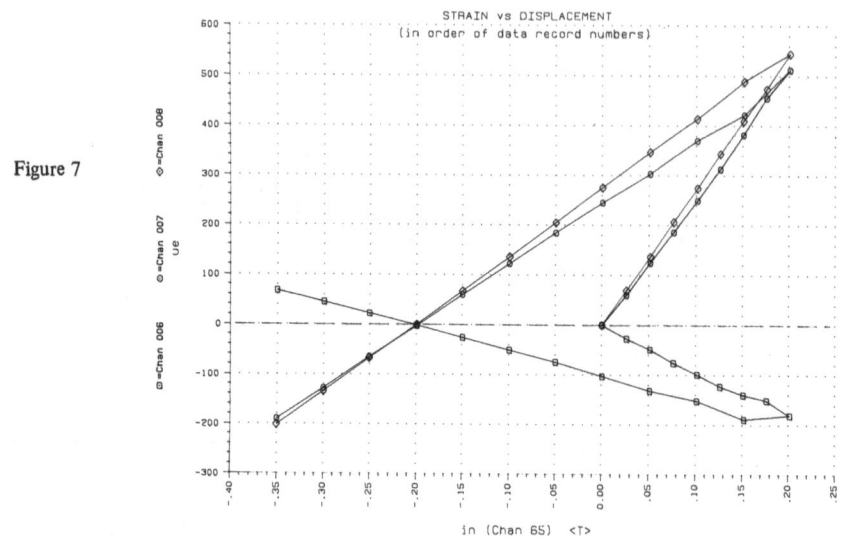

Figure 7

STRUCTURAL EVALUATION
Upper and Lower Control Arms

Model No: I-5 (PRO)   Serial No: 0056-34
Test Number: 23-987
Loading: Static
Environment: Light rain, calm
Test Engineers: W.C. Michaels, D.P. Carroll

Figure 8

Scan Record Number	Elapsed Time (minutes)	Scan Temp. (°F)	σ_{xx} (psi)	σ_{yy} (psi)	Angle ∅ from Grid 1	Percent of Yield Stress
The following Records are dated 23 Jul 85						
Record Source: Data Disk # 2025.371						
A1	0.00	+70.	+17.	+16.	U/C	+0.
A2	0.60	+87.	+602.	-84.	+69.	+2.
A3	1.27	+87.	+1184.	-199.	+69.	+4.
A4	1.85	+87.	+1775.	-305.	+68.	+6.
A5	2.53	+72.	+2368.	-397.	+68.	+7.
A6	3.17	+71.	+2961.	-489.	+68.	+9.
A7	3.60	+86.	+3532.	-576.	+68.	+11.
A8	4.20	+72.	+4123.	-682.	+68.	+13.
A9	4.72	+84.	+4703.	-761.	+68.	+15.
A10	5.23	+71.	+5275.	-849.	+68.	+16.
A11	5.78	+82.	+5856.	-912.	+68.	+18.
A12	6.30	+71.	+6428.	-1000.	+68.	+20.
A13	6.70	+71.	+5767.	-1223.	+71.	+18.
A14	7.03	+71.	+5206.	-1097.	+71.	+16.
A15	7.53	+71.	+4649.	-958.	+71.	+15.

Average: +77.

Notes/Comments:

1. Negative values indicate compressive strains

2. Material: 1018 steel; E = 30×10⁶ psi; ν = 0.285; a = 6.7 ppm/°F;
      Yield Stress = 32,000 psi

3. All strain gages are =MM= CEA-Series

4. Refer to section VI for gage locations

EXPERIMENTAL-NUMERICAL HYBRID TECHNIQUE FOR STRESS ANALYSIS OF PLATES WITH HOLES IN POST BUCKLING STATE

H. KOPECKI and J. SMYKLA
TECHNICAL UNIVERSITY OF RZESZÓW
POLAND

1. INTRODUCTION

The paper presents method of determination of combined, membrane and bending state of stress in thin-walled structures, on the basis of the model investigation. The problem is illustrated on the example of the rectangular plate weakened with two holes. The plate subject to shearing is present in post-buckling state is made of photoelastic material. Membrane stress state is obtained using photostress analysis. Distribution of deflections is determinated with accuracy of 0.1 mm. Discrete results of tests are interpolated by bicubic polynomial splines. Distribution of deflections and combined effective stress is presented in the form of the computer maps.

2. DESCRIPTION OF EXPERIMENT

2.1. Stress optic tests

Model of the plate made of photoelastic material possesses the following parameters: length $2a=350$ mm, width $2b=200$ mm, thicknes $t=1$ mm, Young modulus of the material in temperature $291K - E=2800$ MNm^{-2}, Poisson ratio $\mu=0.38$.

Membrane stress state is identificated by photostress method using normal and oblique incidence. For this purpose rectangular grid of the lines is drown on the model with 10 mm spaces. Measurement of the isoclinic parameter Θ, fringe order of isochromatic N and N_Θ in normal and oblique incidence respectively, is done in the joints of the grid. The model is fastened in the instrument reproducing the fixation of the plate. Loading of the model constituting shearing force $F=300$ N, is getting the plate into post-buckling state.

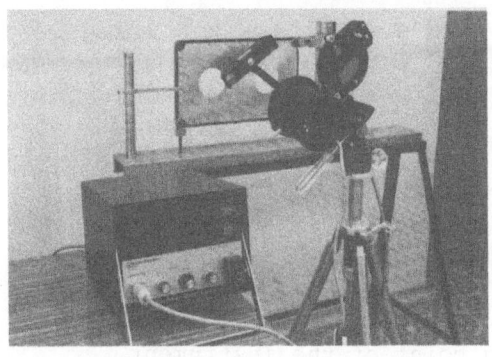

FIGURE 1. Model fastened in the instrument.

Membrane state stress components can be received using following formulas

$$(1) \quad \delta_{2m} = \frac{E}{1-\mu^2} \; f \; \left[C(1+\mu)N_\Theta - (D+\mu F)N \right],$$

$$(2) \quad \delta_{2m} = \frac{E}{1-\mu^2} \; f \; \left[C(1+\mu)N_\Theta - (F+\mu D)N \right],$$

where f is elastooptical constant (fringe value). Constants
C,D,F are dependend on Poisson ratio and angle Θ between the
normal of the surface of the test piece and the light ray [4].
In our case C=1.45, D=0.95, F=1.95, f=0.00096.

2.2. Deflection test

Deflections of the plate are measured by means special instrument equipped with gauging point. It is possible to receive results in digital form and to register them on the recorder. Measuring is done along the lines of the rectangular grid.

FIGURE 2. Instrument for deflections testing.

FIGURE 3. Gauging point of the instrument.

3. INTERPOLATION OF TEST RESULTS
3.1. Deflection interpolation

In order to obtain bending stress components of the plate the derivatives of deflection: $w_{,xx}$, $w_{,yy}$, $w_{,xy}$, should be known in arbitrary point of analysed area. Analytical form of presentation of deflection function, which enables the calculation of the above mentioned derivatives can be received by interpolation methods. Over the post years the theory of interpolation has been enriched with new part of methods, which are splin functions [1], [2], [3]. These methods will be applied to consider the problem.

Deflection interpolation of rectangular plate constituting on the plane x,y rectangle $D = \{(x,y) : -a \leqslant x \leqslant a,\ -b \leqslant y \leqslant b\}$ in which we discriminate grid $\mathcal{JT} = \{(x_i, y_j) : -a = x_0 < x_1 < \ldots < x_n = a,\ -b = y_0 < y_1 < \ldots \ldots < y_m = b\}$, tends to denominate function w(x,y) in considered area D. Let $w \in C^2(D)$. We assume, that function w(x,y) is polinomial of degree 3 in every mesh of rectangular grid

$$(3) \quad w(x,y) = w_{i,j}(x,y) = \sum_{k=0}^{3} \sum_{l=0}^{3} a_{kl}^{ij} (x_i - x)^k (y_j - y)^l .$$

Inside the area the plate can enclose arbitrary quantity of geometrically smooth holes. Effective contour of holes is described by means of generalized rational splines given in the parametric form. As a parameter we assume the length of the broken line contour S (Fig. 4).

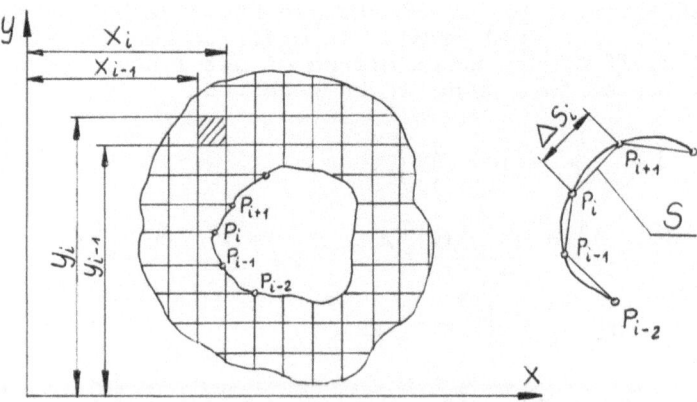

FIGURE 4. Fragment of the plate including the hole; broken line contour S.

Parameter equations of the contour we assume in the form

$$(4) \quad x(S) = A1_i (1-t) + B1_i t + C1_i (1-t)^3 / (1+p_i t) + D1_i t^3 / (1+q_i(1-t)),$$

(5) $y(S) = A2_i(1-t) + B2_i t + C2_i (1-t)^3 / (1+p_i t) + D2_i t^3 / (1+q_i (1-t))$,

where $S_i < S < S_{i+1}$, $t = (S - S_i)/\Delta S_i$, $\Delta S_i = \sqrt{(x_{i+1}-x_i)^2 + (y_{i+1}-y_i)^2}$,
and $p_i > -1$, $q_i < \infty$ are so selected in order to damp the members
of eqs. (4) and (5) presenting nonlinear parts on the   linear
segment of the contour.

3.2. Boundary conditions
We assume the following boundary conditions for the plate:
On the free bound   (holes contour) the following equation   is
fulfilled:

(6) $\tilde{w},_{nn} + \mu \, \tilde{w},_{tt} \Big|_b = 0$

where n - normal and t - tangent to the hole contour. According
to assumption concerning the fixation of the plate, for the external bound of the plate we have

(7) $\tilde{w},_x \Big|_{\substack{y=-b \\ y=b}} = 0$, $\tilde{w},_y \Big|_{\substack{x=-a \\ x=a}} = 0$.

The above conditions comply with determination   of   the
splin coefficients. We assume that we know the values   of the
function w(x,y) in grid joints. In order to receive the above
mentioned coefficients three groups of set linear   equations,
of the following form ought to be resolved:

(8) $\left[A\right]_1 \left\{\tilde{w},_{xx}\right\} = \left[H\right]_1 \left\{w\right\}_1$,

(9) $\left[B\right]_k (\tilde{w},_{yy})^T = \left[G\right]_k (w)^T_k$,

(10) $\left[B\right]_k (\tilde{w},_{xxyy})^T_k = \left[G\right]_k (\tilde{w},_{xx})^T_k$,

where $\left[A\right]_1$, $\left[H\right]_1$, $\left[B\right]_k$, $\left[G\right]_k$ are three - diagonal   matrixes,
where elements are dependent upon joint coordinates of tge grid,
on the horizontal - l and vertical - k lines respectively:

$$[A]_1 = \begin{bmatrix} (h_{BN(1)+1}+h_{BN(1)+2})/3 & h_{BN(1)+2}/6 & 0\dots\dots\dots\dots\dots 0 \\ h_{BN(1)+2}/6 & (h_{BN(1)+2}+h_{BN(1)+3})/3 & h_{BN(1)+3}/6\dots\dots 0 \\ 0 & h_{BN(1)+3}/6 & (h_{BN(1)+3}+h_{BN(1)+4})/3.0 \\ \dots\dots\dots\dots\dots\dots\dots\dots\dots\dots\dots\dots\dots\dots\dots\dots \\ \dots\dots\dots\dots\dots\dots\dots\dots\dots\dots\dots\dots\dots\dots\dots\dots \\ 0\dots\dots\dots\dots\dots\dots h_{EN(1)-1}/6 & (h_{EN(1)-1}+h_{EN(1)})/3 \end{bmatrix}$$

(11)

$$[H]_1 = \begin{bmatrix} 1/h_{BN(1)+1} & (-1/h_{BN(1)+1}-1/h_{BN(1)+2}) & 1/h_{BN(1)+2}\dots\dots 0 \\ 0 & 1/h_{BN(1)+2} & (-1/h_{BN(1)+2}-1/h_{BN(1)+3})\dots 0 \\ \dots\dots\dots\dots\dots\dots\dots\dots\dots\dots\dots\dots\dots\dots\dots\dots \\ \dots\dots\dots\dots\dots\dots\dots\dots\dots\dots\dots\dots\dots\dots\dots\dots \\ 0\dots\dots\dots\dots\dots (-1/h_{EN(1)-1}-1/h_{EN(1)}) & 1/h_{EN(1)} \end{bmatrix}$$

(12)

$$\{\tilde{w},_{xx}\}_1 = \begin{Bmatrix} \tilde{w},_{xx}(x_{BN(1)+1},y_j) \\ \tilde{w},_{xx}(x_{BN(1)+2},y_j) \\ \cdot \\ \cdot \\ \cdot \\ \tilde{w},_{xx}(x_{EN(1)-1},y_j) \end{Bmatrix}, \quad \{w\}_1 = \begin{Bmatrix} w(x_{BN(1)},y_j \\ w(x_{BN(1)+1},y_j) \\ \cdot \\ \cdot \\ \cdot \\ w(x_{EN(1)},y_j) \end{Bmatrix},$$

(13)

$$
\begin{bmatrix} B \end{bmatrix}_k =
\begin{bmatrix}
(v_{BM(k)+1}+v_{BM(k)+2})/3 & v_{BM(k)+2}/6 & 0\ldots\ldots\ldots\ldots\ldots 0 \\
v_{BM(k)+2}/6 & (v_{BM(k)+2}+v_{BM(k)+3})/3 & v_{BM(k)+3}/6\ldots\ldots\ldots 0 \\
0 & v_{BM(k)+3}/6 & (v_{BM(k)+3}+v_{BM(k)+4})/3\ldots 0 \\
\ldots\ldots\ldots\ldots\ldots\ldots\ldots\ldots\ldots\ldots\ldots\ldots\ldots\ldots\ldots\ldots\ldots\ldots\ldots \\
\ldots\ldots\ldots\ldots\ldots\ldots\ldots\ldots\ldots\ldots\ldots\ldots\ldots\ldots\ldots\ldots\ldots\ldots\ldots \\
0\ldots\ldots\ldots\ldots\ldots\ldots\ldots\ldots v_{EM(k)-1}/6 & (v_{EM(k)-1}+v_{EM(k)})/3
\end{bmatrix}
\tag{14}
$$

$$
\begin{bmatrix} G \end{bmatrix}_k =
\begin{bmatrix}
1/v_{BM(k)+1} & (-1/v_{BM(k)+1}-1/v_{BM(k)+2}) & 1/v_{BM(k)+2}\cdots\cdots 0 \\
0 & 1/v_{BM(k)+2} & (-1/v_{BM(k)+2}-1/v_{BM(k)+3})\cdots 0 \\
\ldots\ldots\ldots\ldots\ldots\ldots\ldots\ldots\ldots\ldots\ldots\ldots\ldots\ldots\ldots\ldots\ldots\ldots\ldots \\
\ldots\ldots\ldots\ldots\ldots\ldots\ldots\ldots\ldots\ldots\ldots\ldots\ldots\ldots\ldots\ldots\ldots\ldots\ldots \\
0\ldots\ldots\ldots\ldots\ldots\ldots (-1/v_{EM(k)-1}-1/v_{EM(k)}) & 1/v_{EM(k)}
\end{bmatrix}
\tag{15}
$$

$$
(\widetilde{w}'_{yy})^T_k =
\begin{Bmatrix}
\widetilde{w}'_{yy}(x_{BM(k)+1}, Y_j) \\
\widetilde{w}'_{yy}(x_{BM(k)+2}, Y_j) \\
. \\
. \\
. \\
. \\
\widetilde{w}'_{yy}(x_{EM(k)-1}, Y_j)
\end{Bmatrix}^T
,\quad
(w)^T_k =
\begin{Bmatrix}
w(x_{BM(k)}, Y_j) \\
w(x_{BM(k)+1}, Y_j) \\
. \\
. \\
. \\
. \\
w(x_{EM(k)}, Y_j)
\end{Bmatrix}^T
,
\tag{16}
$$

Here $h_i = x_i - x_{i-1}$ — on the horizontal line — 1, $v_j = y_j - y_{j-1}$ — — one the vertical line — k, $h_{BN(1)}$ $h_{EN(1)}$ are the lengths of the first and the last step on the line 1. Whereas $v_{BM(k)}$ and $v_{EM(k)}$ are the lengths of the first and the last step on the line k respectively. Matrixes $\{w\}_1$ and $\{w\}_k^T$ include measured value of deflections in grid joints, lengthwise 1 and k lines. Matrixes $\{\tilde{w},xx\}_1$, $(\tilde{w},yy)^T$, $(\tilde{w},xxyy)^T$ include discrete aproximate values of derivatives in grid joints, which are exploved coefficients of the bicubic polynomial.

Resolving the set of equations (8), (9), (10) with respect to coefficients $[\tilde{w},xx]$, $[\tilde{w},yy]$, $[\tilde{w},xxyy]$ we can present function $w(x,y)$ in the following form

$$(17) \quad w(x,y) = \tilde{w},xx(x_{i-1},y)\left\{[(x_i-x)^3 - h_i^2(x_i-x)]/6h_i + \right.$$

$$+ \tilde{w},xx(x_i,y)\left\{[(x-x_{i-1})^3 - h_i^2(x-x_{i-1})]/6h_i + \right.$$

$$+ \tilde{w}(x_{i-1},y)(x_i-x)/h_i + \tilde{w}(x_i,y)(x-x_{i-1})/h_i,$$

where

$$(18) \quad \tilde{w},xx(x_{i-1},y) = \tilde{w},xxyy_{i-1,j-1}\left\{\frac{1}{6v_j}[(y_j-y)^3 - v_j^2(y_j-y)]\right\} +$$

$$+ \tilde{w},xxyy_{i-1,j}\left\{\frac{1}{6v_j}[(y-y_{j-1})^3 - v_j^2(y-y_{j-1})]\right\} +$$

$$+ \tilde{w},xxyy_{i-1,j-1}\frac{(y_j-y)}{v_j} + \tilde{w},xx_{i-1,j}\frac{(y-y_{j-1})}{v_j},$$

$$(19) \quad \tilde{w},xx(x_i,y) = \tilde{w},xxyy_{i,j-1}\left\{\frac{1}{6v_j}[(y_j-y)^3 - v_j^2(y_j-y)]\right\} +$$

$$+ \tilde{w},xxyy_{i,j}\left\{\frac{1}{6v_j}[(y-y_{j-1})^3 - v_j^2(y-y_{j-1})]\right\} +$$

$$+ \tilde{w},xx_{i,j-1}\frac{(y_j-y)}{v_j} + \tilde{w},xx_{i,j}\frac{(y-y_{j-1})}{v_j},$$

(20) $\tilde{w}(x_{i-1},y) = \tilde{w}'_{yy_{i-1,j-1}}\left\{\frac{1}{6v_j}\left[(y_j-y)^3 - v_j^2(y_j-y)\right]\right\} +$

$+ \tilde{w}'_{yy_{i-1,j}}\left\{\frac{1}{6v_j}\left[(y-y_{j-1})^3 - v_j^2(y-y_{j-1})\right]\right\} +$

$+ w_{i-1,j-1}\frac{(y_j-y)}{v_j} + w_{i-1,j}\frac{(y-y_{j-1})}{v_j}$

(21) $\tilde{w}(x_i,y) = \tilde{w}'_{yy_{i,j-1}}\left\{\frac{1}{6v_j}\left[(y_j-y)^3 - v_j^2(y_j-y)\right]\right\} +$

$+ \tilde{w}'_{yy_{i,j}}\left\{\frac{1}{6v_j}\left[(y-y_{j-1})^3 - v_j^2(y-y_{j-1})\right]\right\} +$

$+ w_{i,j-1}\cdot\frac{(y_j-y)}{v_j} + w_{i,j}\cdot\frac{(y-y_{j-1})}{v_j}$.

Equations (17) constitute the base to determine bending stress components, from formulas

(22) $\sigma_{xb} = \frac{12M_x}{t^3}z, \quad \sigma_{yb} = \frac{12M_y}{t^3}z, \quad \tau_b = \frac{12M_{xy}}{t^3}z,$

where

$M_x = -D(w'_{xx}+\mu w'_{yy}), \quad M_y = -D(w'_{yy}+\mu w'_{xx})$

(23)

$M_{xy} = -D(1-\mu)w'_{xy}, \quad D = \frac{Et^3}{12(1-\mu^2)}.$

4. RESULTS

Results of tests and calculations are presented in the form of computer maps, which illustrate contour lines of constant deflections (Fig. 5) and constant combined (membrane and bending) effective stress (Fig. 6), for external surface of the plate (coordinate z = t/2).

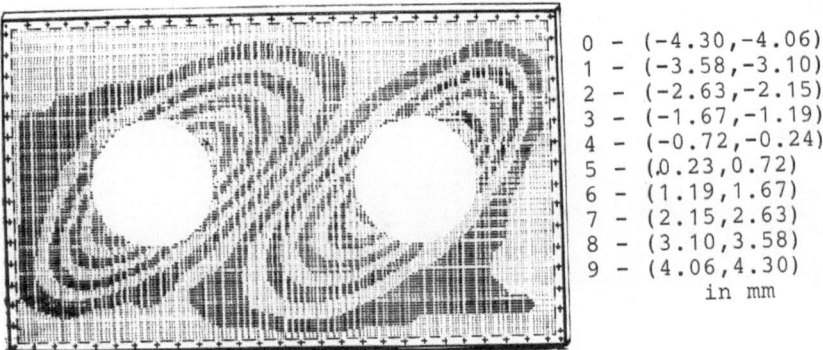

```
0 - (-4.30,-4.06)
1 - (-3.58,-3.10)
2 - (-2.63,-2.15)
3 - (-1.67,-1.19)
4 - (-0.72,-0.24)
5 - (0.23,0.72)
6 - (1.19,1.67)
7 - (2.15,2.63)
8 - (3.10,3.58)
9 - (4.06,4.30)
         in mm
```

FIGURE 5. Constant deflection zones.

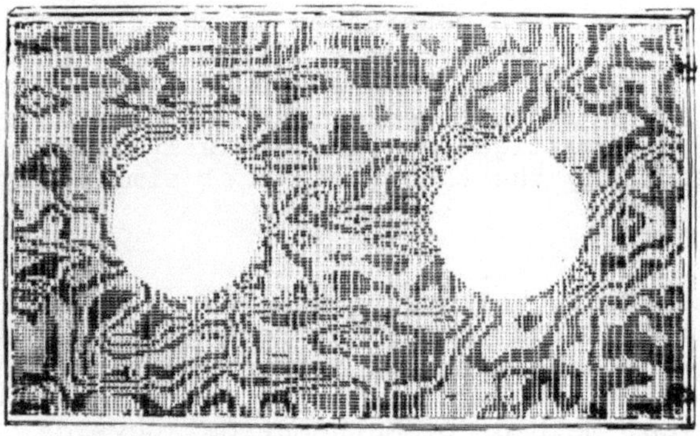

```
0 - (0.32,0.66)
1 - (1.35,2.04)
2 - (2.70,3.40)
3 - (4.11,4.80)
4 - (5.48,6.17)
5 - (6.86,7.55)
6 - (8.24,8.93)
7 - (9.62,10.3)
8 - (10.9,11.7)
9 - (12.4,12.7)
      in MNm$^{-2}$
```

FIGURE 6. Distribution of effective stress value, in combined (bending and membrane state) according Huber-Mises hypothesis.

For the purpose of comparison ve present distribution of effective stress value, only in membrane state (Fig. 7).

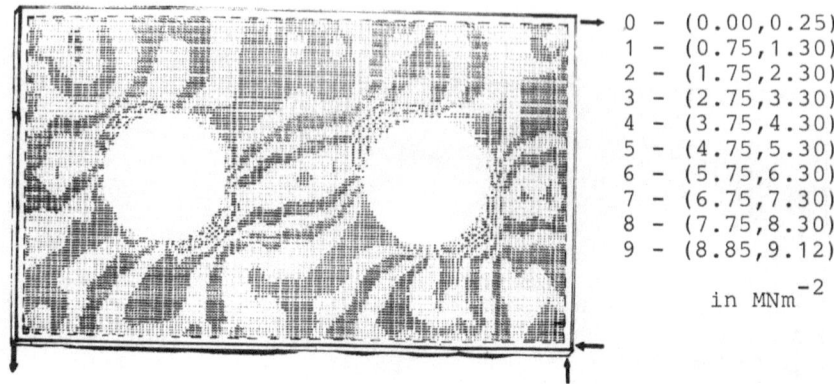

0	-	(0.00,0.25)
1	-	(0.75,1.30)
2	-	(1.75,2.30)
3	-	(2.75,3.30)
4	-	(3.75,4.30)
5	-	(4.75,5.30)
6	-	(5.75,6.30)
7	-	(6.75,7.30)
8	-	(7.75,8.30)
9	-	(8.85,9.12)

in MNm^{-2}

FIGURE 7. Distribution of effective stress value in membrane state of the plate

5. CONCLUSIONS

The methodology of determination of membrane and bending stress state of the thin-walled structures in post-buckling state has been presented in the paper.

Although the problem has been considered on the example of rectaugular plate with holes, however the presented method seems to be possible to be applied in the cases of more compliated structures, having local attenuations and stiffeners.

The results of model investigation can be casily transposed on the real structures using the theory of model similarity.

REFERENCES

1) Ahlberg J.H., Nilson E.N., Walsh J.L. The Theory of Splines and Their Applications, Acad. Press New York 1967.
2) Prenter P.M. Splines and Variational Methods, John Wiley & Sons, New York 1975.
3) Marczuk G.I. Numerical Methods of Mathematical physics, Moscow PWN Warsaw 1983.
4) Instruction Manual for 030 Series Reflection Polariscope, Photolastic Inc.

DIGITAL IMAGE PROCESSING IN EXPERIMENTAL MECHANICS

A. PIETRZYK

RESEARCH ASSISTANT, INSTITUTE FOR STRUCTURAL MECHANICS, WARSAW UNIVERSITY OF TECHNOLOGY, WARSAW, POLAND

1. GENERAL REMARKS

In several methods used in experimental mechanics an image is obtained as a result of an experiment. There are displacement measurement methods based on the phenomenon of light interference, like various moire techniques, holographic interferometry or speckle. In these methods a pattern of fringes appears on the specimen surface or in the camera. The fringes are contour lines of some specific deformation characteristics like in-plane displacements (e.g. intrinsic moire) or off-plane displacements (e.g. holographic interferometry). In the reflexion moire technique the contour map of the derivatives of the off-plane displacements is obtained. In all these methods usually specimen made of real structural materials are tested. Thus the fringe pattern frequently appears on the inhomogenously bright background and image filtration would be much-desired. There are also stress measurement methods based on the phenomenon of birefringence - various versions of photoelasticity. In these methods a pattern of fringes is also obtained. It is usually a contour map of the differences of principal stresses. Models made of special transparent materials are frequently tested instead of real structures or structural members. Thus, it is possible to assure homogenous background illumination and in many cases image filtration is not necessary. Further analysis is usually more complicated than in the previously mentioned methods, however, for fringe patterns in general the accurate determination of fringe axes is very important. The brittle lacquer technique can also be regarded as one giving visual information, although of a different type. The analysis of these images consists in the determination of densely cracked areas, thus being similar to texture analysis. At the end of this short review we should also mention the methods used in non-destructive material testing such as ultrasonography, X-raying, gammascopy, thermovision and microstructure analysis. The traces of faults or imperfections of some kind (e.g. discontinuities) or of intrusive objects are searched in the image. The analysis of these images resembles the analysis of similar images in medicine.

Digital computers are presently commonly used in experimental data analysis. In several applications they are also used for data logging and the control of testing machines (e.g. strain gauges data aquisition systems). The papers on the application of digital image processing to the analysis of images obtained in experimental mechanics have appeared in the last few years. It is connected with the fact, that till late seventies the minicomputers memory capacity was not big enough to handle the amount of data common in image analysis and the sufficiently fast analog-to-digital (A/D) converters were not easily available. The wide dissemination of personal computers with memories ranging at 1MB and the introduction of integrated, single-chip, "flash" A/D converters with sampling frequency exceeding 20MHz enables the automatization of visual information processing.

The foundations of digital image processing may be found in (1), which has
already become classical. Digital coding and processing of images can eliminate tedious manual work and fine mechanics devices. But more important,
it can make the process of data analysis more objective. In some cases it
can also, even by means of serial computers speed-up the analysis. The
following stages can be distinguished in the digital processing of experimental patterns: image coding, noise filtration, pattern recognition, classification and final analysis of the experimental data or postprocessing.
These stages will shortly be described in the following chapters specially
regarding the digital noise filtration. Some published examples of the
application of digital image processing to the analysis of experimental results in mechanics are given below. In (2) the fringes patterns obtained in
photoelastic models testing were considered. The image was coded in the
form of 512x290 pixels matrix with the resolution of 256 grey levels.
Noise filtration was omitted. Fringes were recognized by constant threshold technique. The finite element method was used for the separation of
principal stresses. In (3) the similar problems were considered, but the
preliminary image filtration was employed. The image was coded in the form
of 1024x512 pixels matrix with 64 grey levels resolution. Finite element
method was also used for postprocessing. In (4) the digital spatial frequencies filtering was used for smoothing the light intensity distribution
in speckle images. However, an image was analyzed along lines, not as the
full field. In (5) a very interesting method for the determination of phase
ditribution in moire fringe patterns was proposed. It is based on the spectrum analysis of the pattern brightness along the lines approximately perpendicular to the fringes. A CCD line sensor with 1024 elements providing
256 grey levels resolution was used. The Fast Fourier Transform (FFT) algorithm was used for spectrum samples calculation. In (6) an interesting
method of the determination of plastic zones extent without actually
measuring strains in the tested structural members was described. The
roughness of the initially polished specimen surface was an indicator of
plastic deformation development. This was revealed by random interference
(the appearance of speckles) in some regions of the coherently illuminated
specimen surface. The analysis proceeded, however, again along lines. In
(7) the digital processing of intrinsic moire fringe pattern was presented.
The image, coded in the form of 128x128 pixels matrix with 16 grey levels
resolution, was treated as a full field.

2. DIGITAL IMAGE FILTRATION
2.1 Spatial frequencies filtration

This method consists in band-pass filtering of the image's spatial spectrum. It can be used for filtering out both the low frequency components
caused by the slow variations of background brightness due to inhomogenous
illumination and the high frequency ones introduced by the coding system
or due to fine structure of the information bearing objects in the image.
It can, however, also be used to enhence in the image object's edges and/or
other elements with relatively fast varying grey levels distribution if a
high-pass filter is used. On the other hand, by letting only low frequency
components pass through the filter, a reference image of the background
brightness can be obtained. This can be substracted from the original image
to enhence the objects of interest. The two-dimensional (2D) Discrete
Fourier Transform (DFT) of a coded image is calculated as follows. First the
rows of the matrix representing the image are replaced by their one-dimensional (1D) Fourier transforms. Thus a matrix of 1D transforms is formed.
Then the columns of this matrix are replaced by their 1D DFT's. In this
manner a 2D transform is calculated by performing the "number of rows + num-

ber of columns" 1D DFT s. The preprocessing includes multiplication of the image representing matrix by a window function (e.g. Hamming window function). Postprocessing includes the rearrangement and normalization of the results for showing them on the display in the form similar to that obtained by performing this process optically. For 1D DFT calculations an FFT algorithm (see (8)), which can be hard-wired to form a co-processor for a serial computer if the time of computation plays important role. Having the filtered spectrum an inverse 2D DFT is calculated in the very similar manner. Thus the filtered image is obtained. The filter form, however, should be well fitted to the spectrum. For the 2D case it is not easy to detrmine a band-pass filter form automatically. Therefore an interactive mode of image processing system s operation should rather be suggested. The method described in (5) makes Fourier transform based processing still more attractive. The multiprocessor parallel computers, which will become popular in the future, promise substantial reduction in calculation time. Using such acomputer the 2D DFT of a typical image can be done in the real time. As an example of this type of image filtration let us consider the digitally coded fragment of an intrinsic moire fringe pattern shown in Fig. 1a. It consists of 128x128 pixels coded with 16 grey levels resolution. In this picture several imperfections can be seen including slow background brightness variations from upper right towards bottom left corner and dark spots between fringes, as well as, bright spots disconnecting the fringes. The 2D spatial frequencies spectrum of this pattern is shown in Fig. 1b. The indicated parts of the s-like object situated near the center of this picture represent the fringe pattern. We can now let only the indicated fragment of this spectrum pass through the filter. In Fig. 1c the positive elements of the real part of the recovered complex image (inverse transform of the indicated spectrum fragments) are shown. Absolute values of the recovered image complex samples optical analogy are presented in Fig. 1d. This way fringes multiplication is obtained.

2.2 Non-isotropic image filtration

Another efficient tool for image filtration and features extraction is the relaxation method proposed in (9) and explicated in (10) and in (11). In this method local features of grey levels ditribution in small windows are examined with respect to some global rule. The initial probabilities of local features are iteratively corrected in order to obtain the consistent, in global terms, set of locally most probable features. The method is best known from its application to fringe patterns filtration. In this case the local features are the directions of ranges or slopes in the grey levels surface. The initial directions probabilities are obtained by matching the model range or slope fragments with the image fragments within each window. The model template is applied in several directions and its correlation with the actual image is examined. The higher the correlation, the more probable it is that the fringe runs through the window in the given direction. The set of initially most probable directions is usually inconsistent due to noise. However, from the global point of view, some arrangements of directions in adjacent windows are improbable, while the others are possible. Thus, by defining the compatibility function between the directions of fringes in adjacent windows (favouring, for example, fluent lines to sharp angles) the correction process is defined. After performing a few steps of iteration one arrives at the consistent set of most probable directions. This set can then be used for controlling image filtration. This process takes place in small windows. Each pixel can be, for example, replaced by the mean or median value of its specially designed neighbour-

284

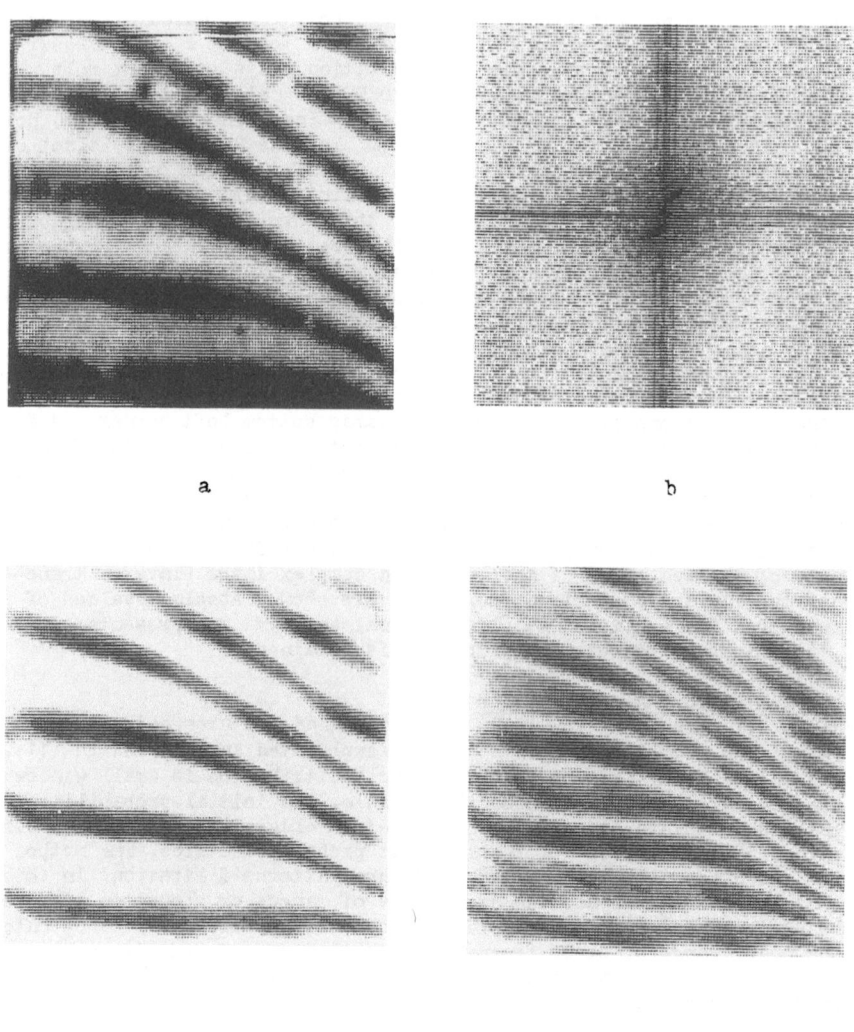

a

b

c

d

FIGURE 1. a) Fragment of the digitally coded moire fringes pattern.
b) Spacial frequencies spectrum of the image shown in a.
c) Image after filtration I, d) Image after filtration II.

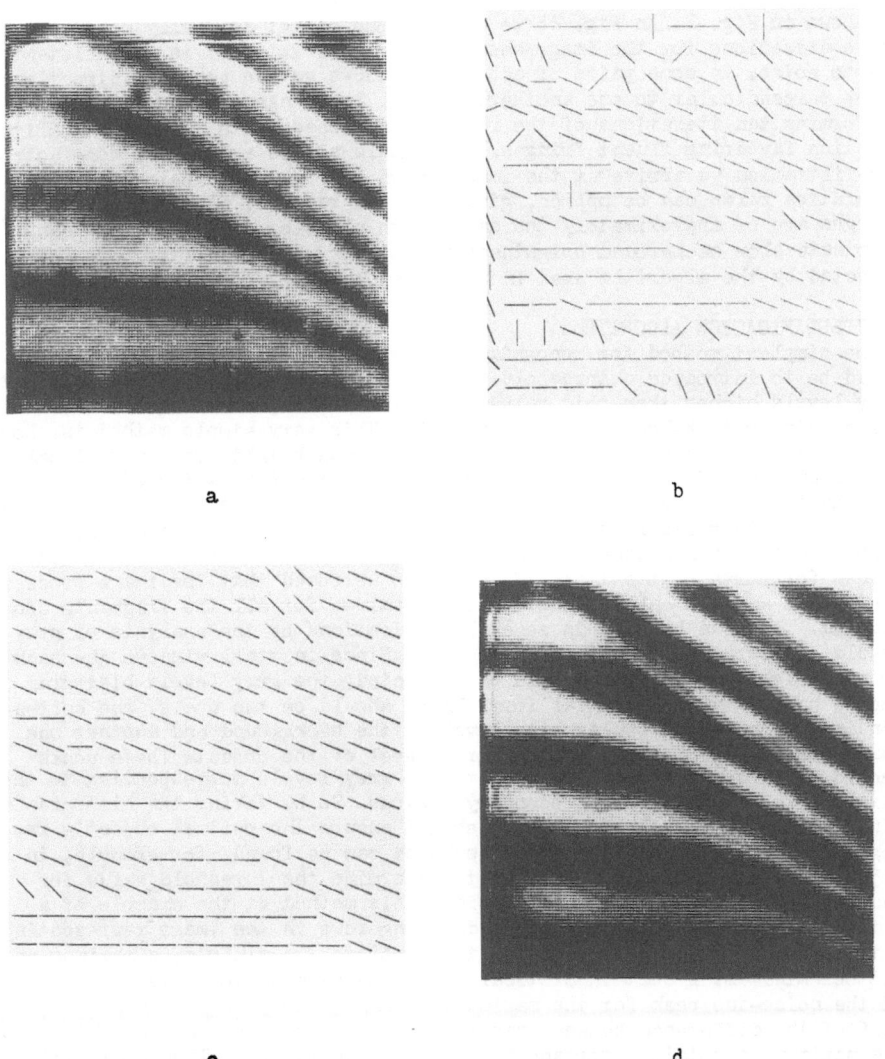

a

b

c

d

FIGURE 2. a) Fragment of the digitally coded moire fringes pattern.
b) Initially most probable directions for the image shown in a.
c) Most probable directions after the relaxation process.
d) The filtered image.

hood. The neighbourhood should be elongated in the most probable direction for the given window. Thus, gaps in the fringes would be filled and dark spots connecting separate fringes would be washed away. An example of the application of the described method to the intrinsic moiré fringes pattern is shown in Fig. 2. In Fig. 2b we can see the set of initially most probable directions for the image shown in Fig. 2a. This set is inconsistent due to noise. In some pairs of adjacent windows we can see this directions to be perpendicular to one another. In Fig. 2c we can see the result of performing ten iterations of the relaxation process with the compatibility function favouring fluent fringes. Using this set as a mask for non-isotropic filtration we arrive at the image shown in Fig. 2d. Other types of filtration rules can be applied, giving better results, especially if the fringes are of approximately the same width all over an image. They can eliminate slow background brightness variations which, in fact, are not affected by the algorithm used in the given example.

3. IDENTIFICATION ALGORITHMS

The simplest method for recognizing objects in digitally coded image would be to introduce a threshold value and to consider all the pixels with grey levels higher than this value belong to the objects in question, while the remaining to belong to the background. This very simple method is, however, sensitive to slow variations of background brightness and to local faults in the objects. Thus, it can be used for the processing of good quality images, like the ones obtained in photoelastic testing of thin plane models. Otherwise it should be preceeded by noise filtration. The more sophisticated method consists in the determination of various threshold values for parts of the image. It would be welcomed that the image processing system automatically finds threshold values for all the fragments considered. This can be done in several applications by the analysis of grey levels histogram of the given image part. Since in small windows the background brightness variations can be neglected, the grey levels histogram should have a typical bimodal form. There should be two peaks, one corresponding to the most popular grey level in the background and another one corresponding to the most popular grey level of the object. These peaks should be separated by a trough of medial grey levels corresponding to the pixels lying in the objects boundary regions. So by taking the histogram's minimum for the threshold value we should ensure the optimal object's recognition. Further details on this subject can be found, for example, in (13). The most complicated idea is to determine the threshold value for each object separately. Let us consider this method at the example of a fringe pattern. The analysis proceeds along rows in the image representing matrix. We determine a sequence of three points – a minimum, a maximum and another minimum. A certain critical value can be used to avoid the taking of the noise-due peak for the maximum. Having these points we can calculate half of the difference between maximum and minimum. Substracting this from the maximum we obtain a threshold value cutting the so-called half-width of the fringe. We repeat this process along columns to detect the fringes runnung along rows. Having the binary image of recognized fringes it is possible to determine the fringes axes by thinning. Several thinning algorithms have been described, but the problem is still actual especially regarding parallel processing.(see (14) or (15)). Two examples of fringe recognition and thinning of binary pictures are presented. In Fig. 3a the result of applying a constant threshold to the image shown in Fig. 1c is plotted. The thinning algorithm described in (15) was used to obtain the pattern shown in Fig. 3b. In Fig. 4a the result of applying the technique of determining a threshold for each fringe separately to the image shown

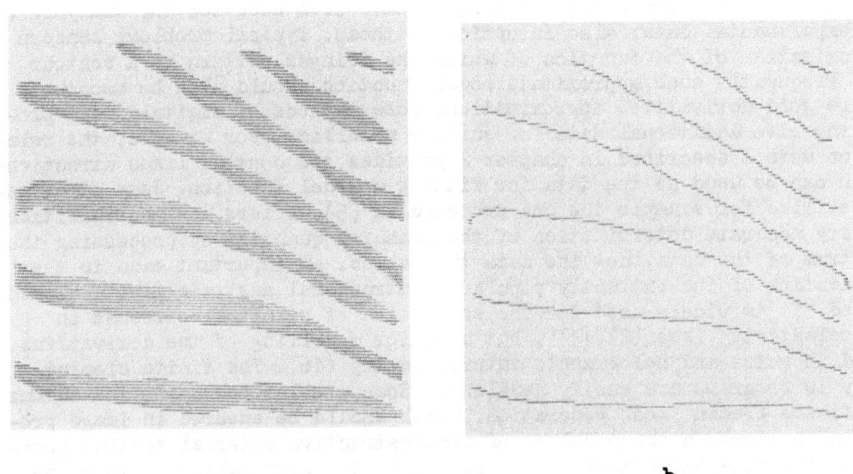

FIGURE 3. a) Recognized fringes. b) Thinned fringes.

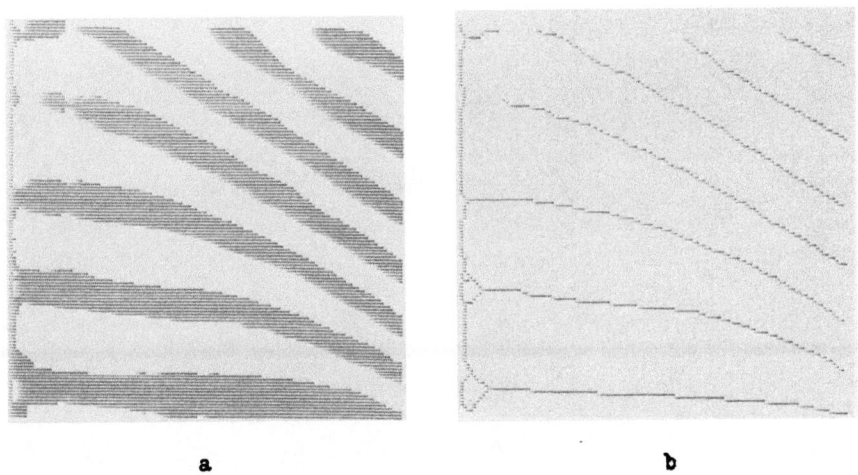

FIGURE 4. a) Recognized fringes. b) thinned fringes.

in Fig. 2d can be seen. The same thinning algorithm transformed this picture into the pattern shown in Fig. 4b.

4. POSTPROCESSING

Since several years digital computers have been used for the analysis of the experimental data, also in optical methods. Typical problems concern approximation of the function of which the fringes pattern is a contour map. Frequently such approximate model function should also be smooth to ensure good derivatives approximation. Some methods of digital image processing give additional data for surface modelling. For example, the relaxation method described in chapter 2 provides the contour lines directions which can be used as the data for fitting a model function. Some other methods, like for example the one proposed in (5), offers new possibilities of very accurate determination of the measured quantity by processing the spectrum of the data, not the data themselves. An important case is the application of the extremely popular in structural analysis finite element method. It is widely used for the separation of principal stresses in photoelasticity (see (2),(3)), but also for modelling of the derivatives field in moire and holographic interferometry (16). The finite element analysis programs are easily available. Some problems concerns the automatic finite element mesh generation, which should be ensured in image processing systems. In the problems of non-destructive material testing postprocessing consists mainly in the automatic decision making whether the recognized faults do disqualify the object being tested or not. However, in some problems the computer-aided stereometric reconstruction of an object from several images is required. As an example the least squares fit of the polynomial of two variables to the data shown in Fig. 3b is presented. The contour lines of the model function are given in Fig. 5a. Its partial derivative is mapped in Fig. 5b.

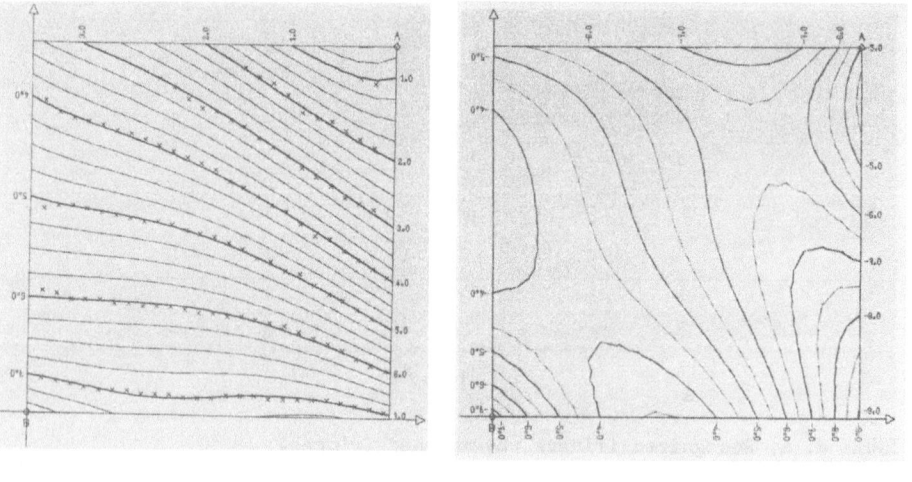

a   b

FIGURE 5. a) Contour lines of the model function of displacements.
b) Contour lines of the partial derivative of the model function.

5. LABORATORY COMPUTER SYSTEMS

In most of the references minicomputer based systems were used for image coding and processing. Efficient, real time digital image processing requires dedicated, multiprocessor, parallel computers. Such machines are still at the research stage and they are very expensive. One of the most impressive prototypes – MPP:1 (see (17)) – consists of over 16000 processors. An interesting discusion on the optimal architecture for image processing systems can be found in (18) and (19). The so-called "pyramid" architecture is recommended, but no prototype exists. These machines seem to be, for some time at least, not yet available. The IBM-PC/AT class microcomputers seem to be the low cost solution for laboratory applications. With the directly addressed memory of over 1MB and high speed these machines fullfil the minimum requirements for image processing systems. Easily available peripheral devices like Winchester-type disc drives and floppy disc drives provide useful extension of the directly addressed memory. For hard copies generation dot printers can be used. The most important reason for their laboratory application is the rich and easily available software library. They can work in the local network mode, be connected to the more powerful host computer, but they can also accept several specialized co-processors. Several companies offer the "flash" A/D converters cards for IBM-PC. This computer can also be used to control other data aquisition systems e.g. strain gauges . The FFT routines used for image processing can also be employed to vibrations analysis. With the new i80386 32-bit processor, software compatible with the previous i80286, the new perspectives are set for the IBM-PC class. For all these reasons it seems justified to recommend this computer to those, who would like to start the computer-aided image processing investigations in theirs laboratories.

ACKNOWLEDGEMENTS

The author is deeply indebted to Dr Z. Kulpa from the Institute for Biocybernatics and Biomedical Engineering of Polish Academy of Sciences for the permision to perform image coding on the CPO-2/K-202 image processing system. Encouragement and valuable help of T. Gajewski from the Institute of Roads and Bridges of Warsaw University of Technology is sincerely appreciated.

REFERENCES

1. Rosenfeld A, Kak AC: Digital Picture Processing. Academic Press, New York, 1976.
2. Muller RK, Sackel LR: Complete Automatic Analysis of Photoelastic Fringes. Experimental Mechanics, vol.19(1979)/7.
3. Seguchi Y, Tomita Y, Watanabe M: Computer-aided Fringe-pattern Analyzer A Case of Photoelastic Fringe. Experimental Mechanics, vol.19(1979)/10.
4. Chambless DA, Broadway JA: Digital Filtering of Speckle-photography Data. Experimental Mechanics, vol.19(1979)/8.
5. Takeda M, Ina H, Kobayashi S: Fourier-transform Method of Fringe-pattern Analysis for Computer-based Topography and Interferometry. Journal of the Optical Society of America, vol.72(1982)/1.
6. Lee C, Peters WH, Sutton MA, Ranson WF: A Study of Plastic Zones Formation by Digital Image Processing. Proceedings of the Vth International Congress on Experimental Mechanics, Montreal, Canada, 1984.
7. Pietrzyk A: Digital Processing of the Moire Fringes Pattern. Kurzreferate 1. Internationale Fachtagung Automatische Bildverarbeitung, Berlin, DDR, 1985.

8. Oppenheim AV, Schafer RW: Digital Signal Processing. Prentice-Hall Inc., New Jersey, 1975.
9. Rosenfeld A, Hummel RA, Zucker SW: Scene Labeling by Relaxation Operations. IEEE Transactions on Systems, Man and Cybernatics, vol.SMC-6 (1976)/6.
10. Zucker SW, Hummel RA, Rosenfeld A: An Application of Relaxation Labeling to line and Curve Enhencement. IEEE Transactions on Computers, vol.C-26(1977)/4.
11. Hummel RA, Zucker SW: Foundations of Relaxation Labeling Processes. IEEE Transactions on Pattern Analysis and Machine Intelligence, vol.PAMI-5(1983)/3.
12. Granlund GH: In Search of a General Picture Processing Operator. Computer Graphics and Image Processing, vol.8(1978) pp. 155-173.
13. Feiveson H: Classification by Thresholding. IEEE Transactions on Pattern Analysis and Machine Intelligence, vol.PAMI-5(1983)/1.
14. Arcelli C, Cordella LP, Levialdi S: From Local Maxima to Connected Skeletons. IEEE Transactions on Pattern Analysis and Machine Intelligence, vol.PAMI-3(1981)/2.
15. Arcelli C, Sanniti di Baja G: A Thinning Algorithm Based on Prominence Detection. Pattern Recognition, vol.13(1982)/3.
16. Segalman DJ, Woyak DB, Rowlands RE: Smooth Spline-like Finite-element Differentiation of Full Field Experimental Data over Arbitrary Geometry. Experimental Mechanics, vol.19(1979)/12.
17. Kushner T, Wu AY, Rosenfeld A: Image Processing on MPP:1. Pattern Recognition, vol.15(1982)/3.
18. Rebel B, Wilhelmi W: Trends in Image Processing Architecture and Implementation. Kurzreferate 1. Internationale Fachtagung Automatische Bildverarbeitung, Berlin, DDR, 1985.
19. Rosenfeld A: Parallel Algorithms for Image Analysis. Kurzreferate 1. Internationale Fachtagung Automatische Bildverarbeitung, Berlin, DDR, 1985.

COMPARATIVE STUDY OF THE SENSITIVITY OF VARIOUS MEASUREMENT
TECHNIQUES ON "GLASSES"-SHAPED ELASTIC ELEMENT MODELS ANA
LYSED BY THE FINITE ELEMENT METHOD

D.M. ŞTEFĂNESCU

Strength of Materials Chair (Chairman: Prof.Dr.Ing. M. BUGA),
POLYTECHNIC INSTITUTE OF BUCHAREST, ROMANIA

1. INTRODUCTION

The idea to study the transducer elastic elements by the
finite element method (FEM) is a fairly novel one. Mitchell
(6) has obtained good agreement between strain gauge and FEM
results for the elastic column type, the difference being only $2 \cdot 10^{-6}$. Bray (5) combined the theoretical (analytical and
numerical calculation) and experimental (photoelasticity and
strain gauges) techniques carrying out a complete research of
a circular ring. Barbato (3) has studied by different methods
the "square ring", plotting the strain diagrams for various
sets of parameters.

Elastic elements from transducers for the measurement of
mechanical quantities can be provided with various parametric
sensitive elements: resistive, inductive, capacitive or with
variable frequency. There is a mechanical sensitivity, given
by the deflection under the applied load, and an electrical
one, specific to the chosen measurement technique, aiming to
obtain a maximum global sensitivity.

Stimulated by what Abel (1) calls "interdisciplinary FEM"
our aim is to compare the sensitivities of some measurement
techniques using different elastic elements of rather complicated shapes.

This study was performed for the rib-stiffened membrane
(9) the main performances of the measurement techniques used
being summarized in Table 1.

In this paper an analogue study is done for various models
of binocular or "glasses"-shaped elastic elements. They have
the following advantages (2):
- ease of manufacturing (parallelipipedic block with two
transverse holes connected by a slot);
- high sensitivity due to the bending of the elastic element;
- loading capability in both directions using adequate
mountings;
- relatively small height;
- good cost/performance ratio.

This element is encountered as cantilever beam in the
technical literature about load cells published by several
companies (10 c,d,f). It is less utilized in the loading mode
along the small symmetry axis (10 b). Even not manufactured
as a "black box" (7), it has sufficient "secrets" in order to

Wieringa, H (ed), Experimental Stress Analysis.
© *1986. Martinus Nijhoff Publishers, Dordrecht.*

TABLE 1. Comparative performances of the measurement methods.

Performance	Vibrating wire	Variable inductance	Variable capacitance	Strain gauges
Rangeability	6:1	8:1	6:1	10:1
Sensitivity factor	300	100	250	2
Accuracy (% of span)	0,2	0,35	0,25	0,5
Non-linearity error (maximum per cent)	0,02	0,5	0,5	0,02
Temperature effects (% of span / 50 °C)	1	1,2	1,2	0,5
Stability (% per year)	0,01	0,5	0,05	0,5
Frequency response- f_{max} (kHz)	0,1	0,5	0,1	1
Price ratio	1	0,4	0,9	0,3

ensure a high quality measurement. We shall try to point out some of these in the following.

Four "glasses" models (other also exist), to each of which four measurement techniques are applied, are schematically presented in Figure 1. Hence 16 possible combinations result. The comparative study of the sensitivity of various measurement techniques, as well as within the same measurement principle, can be done using the finite element method.

2. CALCULATION MODEL

Several fixed dimensions have been chosen for the glasses models: the holes have 40 mm diameter and 70 mm distance between centres, being connected by a 2 mm width slot, while the minimum frame width is 5 mm. The overall dimensions are 120 mm x 50 mm x n mm.

Due to the symmetry of the elastic structure, it is sufficient to model only a quarter of it. A plane stress analysis is performed on n "slices" of unit thickness. Discretization (Fig. 2) contains 63 two-dimensional elements and 87 nodes. (For a discretization 3 times finer, the calculation differences are less than 5 %.) A 50 N concentrated load is applied at the central node No. 1. Due to the symmetry of loading, nodes No. 1, 17, 31, 39, 56, 64, 72 and 80 are constrained on the vertical and nodes No. 85, 86 and 87 are guided on the horizontal.

The transition from one glasses model to another is done with minimum changes. Transition from rectangular glasses to those (half)octagonal is accomplished by "elimination" of some elements, actually by replacing the steel by another material, with more elastic characteristics. Transition from half-octagonal to circular glasses is done by the slight "displacement" of some nodes.

The calculation using the computer program SAP IV (4)

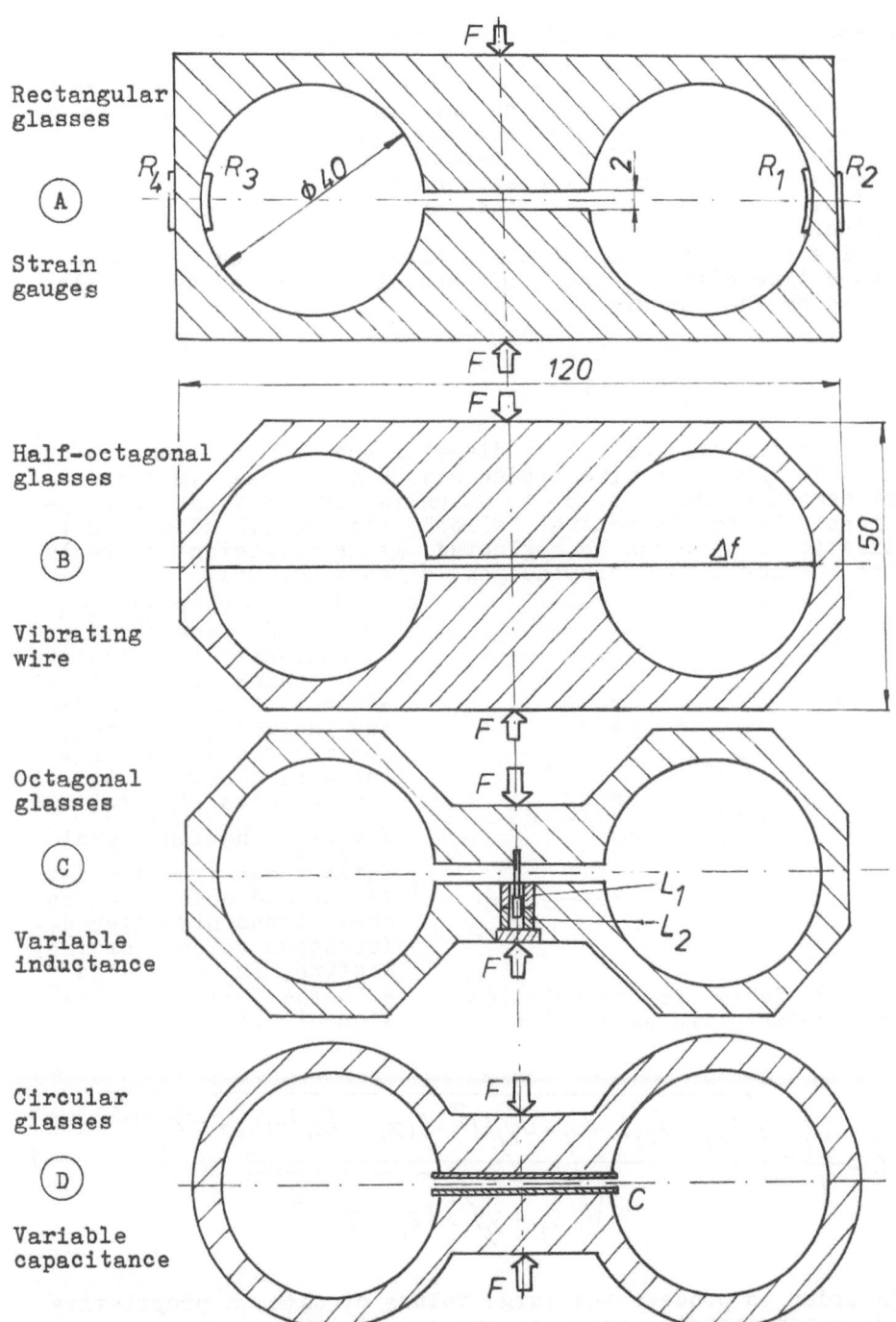

FIGURE 1. "Glasses" models — different measurement principles.

gives as results the node displacements and stresses from finite elements. Taking into account that, in order to convert
mechanical strains into electrical signals, different transducers have to be mounted on the glasses, it is necessary to
further process the data from the listing.

Thus, for inductive and capacitive transducers and analogously for mechanical measurement devices with dial gauges
(10 a,e) the vertical displacement of node No. 80 is directly
noted.

The effect of the vibrating wire is modelled by a truss or
bound-type element, the stiffness of the latter being calculated by the formula

$$k = \frac{F}{\Delta \ell} = \frac{E \cdot A}{\ell} \qquad (1)$$

For a steel wire of 0.2 mm diameter and $\ell = 55$ mm one obtains
$k = 120$ N/mm. The wire produces a 25 % decrease in the displacement of node No. 85 to which is fixed but, as this applies only to the central "slice", its overall effect on the
elastic element can be neglected. It is sufficient to register the horizontal displacement of node No. 85.

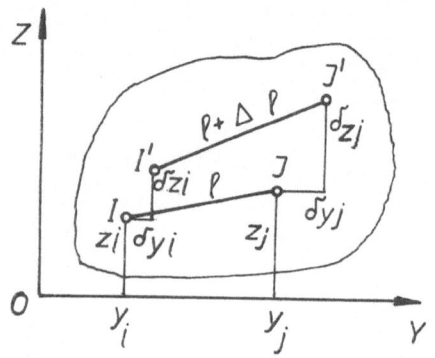

FIG. 3. Scheme for calculation
of ε between two adjacent nodes.

For strain gauge ins-
trumented transducers it
is necessary to calculate
the strains of the elastic
element. For a homogeneous
and isotropic material,
obeying the Hooke's law,
the strain between two adjacent nodes $I(y_i, z_i)$ and
$J(y_j, z_j)$, having coordinates known from discretization and node displace-
ment components after de-
formation taken from the
listing, is calculated
with the following rela-
tionship (Fig. 3)

$$\varepsilon = \frac{\Delta l}{l} = \frac{\sqrt{[(y_i + \delta y_i) - (y_j + \delta y_j)]^2 + [(z_i + \delta z_i) - (z_j + \delta z_j)]^2}}{\sqrt{(y_i - y_j)^2 + (z_i - z_j)^2}} - 1$$

In order to process the large volume of data, a proprietary
(8) postprocessing computer program is used.

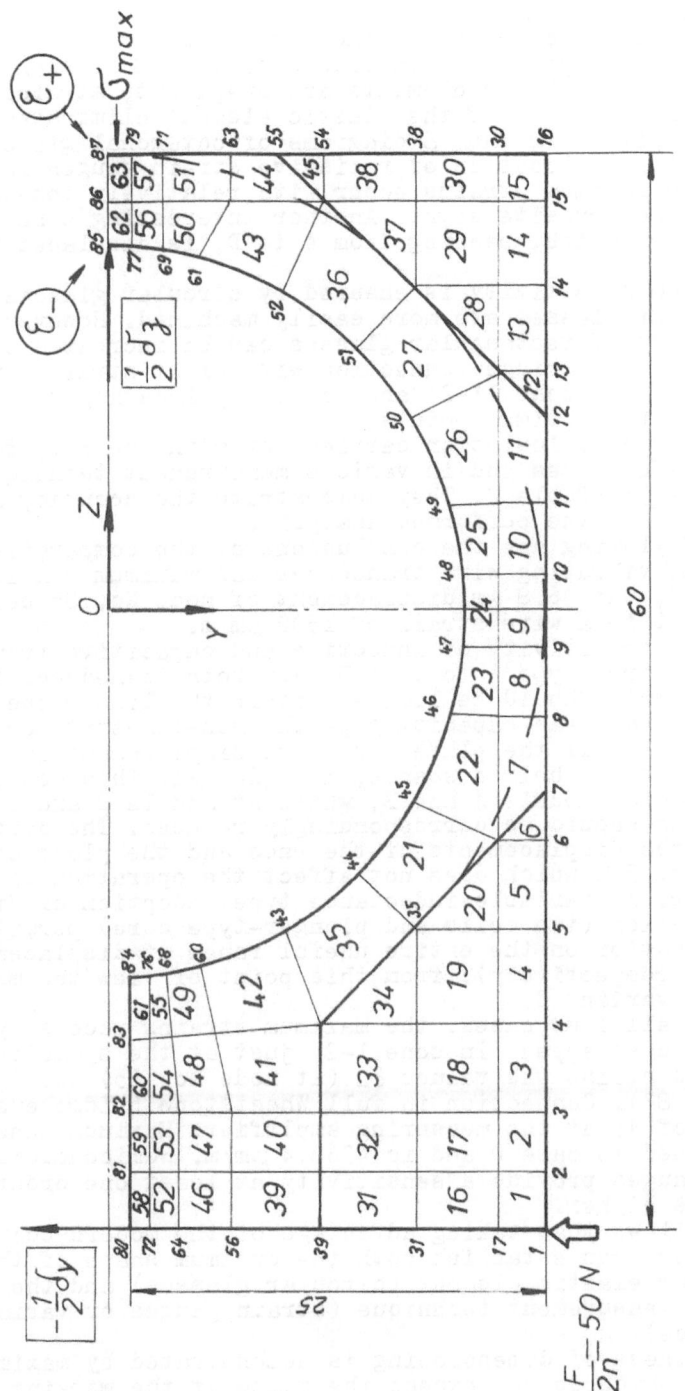

FIGURE 2. Finite element computation model.

3. NUMERICAL RESULTS

Strain diagrams on inner and outer sides of the four glasses models are presented in figure 4. It can be seen that, going from A to D, as elements are swept out, strains increase and rounding of the elastic element eliminates the discontinuities in the $\mathcal{E}$ diagrams of octagonal glasses.

Traditional location of resistive strain gauges is in zone 1-2 where maximum strains occur with relatively constant magnitudes and opposite signs. Another interesting zone (1'-2') also occurs which, passing from A to D, is displaced towards the slot.

Maximum sensitivity is ensured by circular glasses while rectangular glasses are more easily machined. However, the sensitivity of rectangular glasses can be increased choosing the Wheatstone bridge connection with eight strain gauges, using also the zone 1'-2' and carefully locating the strain gauges in the optimal points.

Results of calculation carried out with the FEM, for the four glasses types and in various measurement techniques, are summarized in Table 2. They demonstrate the accuracy and high resolution of the performed analysis.

The following are the conclusions of the comparative study:

1) The vibrating wire transducer has maximum sensitivity in case B, the 96.8 μm displacement of node No. 85 being equivalent to a wire strain of 1500 μm/m.

2) The sensitivity of inductive and capacitive transducers increases from type A to type D. For both transducer types displacements should be limited within the linear operating range. At variable capacity type the non-linearity error is less than 1 % if the plates relative displacement is not more than 1/10 from their distance, i.e. 200 μm. This condition is fulfilled for models A and B, while at models C and D the nominal load should be correspondingly reduced. The differences between the displacements of the ends and the plate centre do not exceed 3 % which does not affect the operation of plane condenser. At variable inductance type, adoption of differential solution (two coils and plunger-type core) permits linear operation on the entire useful range of displacements (forces, respectively); from this point of view the most sensible is variant D.

3) In all four cases, the maximum strains necessary for strain gauges appear in zone 1-2, just at the symmetry axis. Table 2 contains the values $\mathcal{E}_-$ (at node No. 85) and $\mathcal{E}_+$ (at node No. 87). Connection in full Wheatstone bridge ensures a reading of 4$\mathcal{E}$ at the measuring amplifier. Maximum sensitivity is obtained in case D and is 4368.4 μm/m. Semiconductor strain gauges provide a sensitivity at least one order of magnitude higher.

It follows that taking advantage of the modern computer science one can establish both the optimum shape of the transducer elastic element (circular glasses) and the most adequate measurement technique (strain gauges or variable inductance).

Corectness of dimensioning is demonstrated by maximum stresses which do not exceed the value of the working stress:

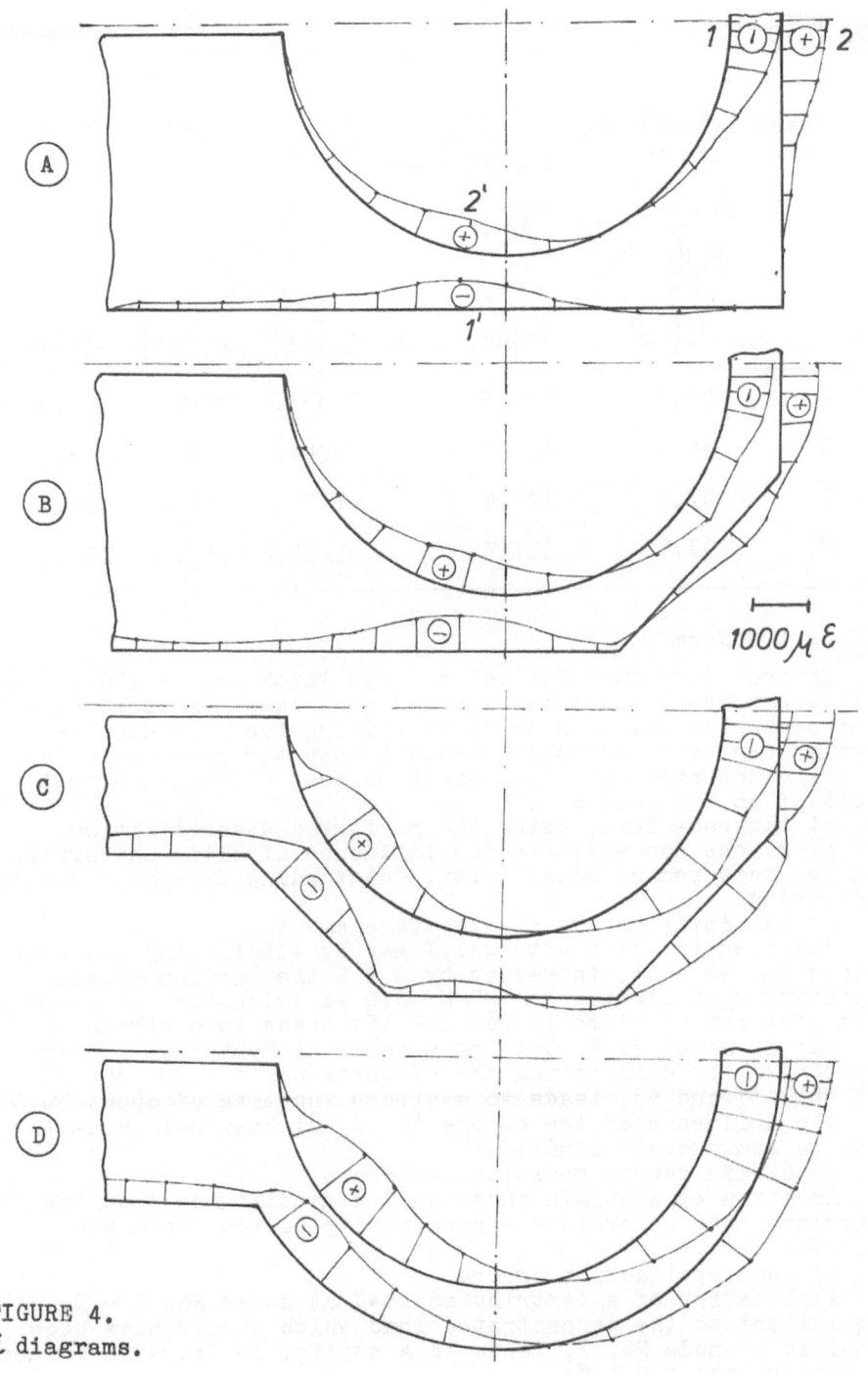

FIGURE 4.
ε diagrams.

TABLE 2. Comparative sensitivities for different measurement techniques applied to various models of "glasses".

"Glasses" models	Vibrating wire	Variable inductance/ capacitance	Strain gauges – Bonded foils		Check of σ_{max}
	$\delta_z [\mu m]$	$\delta_y [\mu m]$	$\varepsilon_- [\mu\varepsilon]$	$\varepsilon_+ [\mu\varepsilon]$	$\sigma [N/mm^2]$
A	90,7	170,5	-923,9	780,2	103,3
B	96,8	197,2	-784,0	645,7	88,3
C	67,1	486,9	-1152,5	992,8	127,4
D	93,5	557,7	-1199,4	984,8	125,1

$\sigma_a = 200 \ N/mm^2$.

Nominal load on a "slice" of unit thickness is 100 N. Load cells can be obtained for a broad load range by increasing the number of slices n or by decreasing the hole diameter. Design can be based on an adequate computer program, establishing the necessary slot width so that it should play the role of an overload stop.

At the same time, using the performed discretization (Fig. 2) one can estimate the influence of different errors on the measurement sensitivity. Outstanding drawbacks are the following:

a) Machining errors of elastic element

Widening the slot with 2x1.4 mm, by eliminating the elements No. 58 – 61, increases by 0.5 % the maximum strain. Removing a horizontal "chip" of 2.6 mm thickness, by eliminating the row of elements No. 1 – 15, leads to a strain increase of about 30 %. Cutting a vertical "chip" of 2.6 mm thickness, by eliminating the elements No. 15, 30, 38, 44, 45, 51, 57 and 63, leads to a strain increase of about 50 %.

The influence of the change in any dimensional parameter can be analogously studied.

b) Strain gauges positioning errors

Location of a strain gauge at 2.4 mm distance from the symmetry axis determines a sensitivity decrease of about 0.5 %.

c) Load application errors

Application of a distributed load at nodes No. 1 – 3, equivalent to the concentrated load which should have been applied at node No. 1, leads to a negligible decrease of sensitivity (about 0.1 %).

Effects of eccentric or inclined application of load can be assessed only on the complete model of the elastic element cross-section.

4. CONCLUSIONS

The finite element method, the most versatile technique of structural analysis, permits both the global approach of the problem and the detailed investigation in any point of the structure; it is a rapid computation technique allowing a graphical and/or tabular presentation of results.

The method permits the parametric determination of design characteristics and the establishment of the optimal shape of the transducer elastic body.

Moreover, since the design stage, simple or combined influences can be studied on sets of configurations, resulting from machining, transducer location, load application etc.

In this paper different "glasses" models have been analysed in order to make a comparative study regarding the sensitivity of various measurement techniques: variable resistance, inductance or capacitance, vibrating wire.

This multidisciplinary method can help to build new types of elastic elements for transducers, to build high sensitivity sensors for the measurement of different mechanical quantities.

ACKNOWLEDGEMENTS

Thanks are due to all authors, publishing houses and firms who kindly furnished documentation, as well as to my colleagues M. Radeş and M. Găvan for their useful suggestions formulated during the preparation of this paper.

REFERENCES

1. Abel JF et al (ed): Interdisciplinary Finite Element Analysis. Proceedings of the U.S.-Japan Seminar. Ithaca: Cornell University, NY, 1978.
2. Bahra CS, Evans JW: Strain Gauge Loadcell Design and Use. Report of the Transducer Tempcon Conference, London, 1983.
3. Barbato G et al: Load-cell-design Developments by Numerical and Experimental Methods. Experimental Mechanics, Vol. 21, No. 9, 341-348, 1981.
4. Bathe KJ et al: SAP IV - A Structural Analysis Program for Static and Dynamic Response of Linear Systems. Berkeley: University of California, 1973.
5. Bray A: The Role of Stress Analysis in the Design of Force-standard Transducers. Experimental Mechanics, Vol. 21, No. 1, 1-20, 1981.
6. Mitchell RA et al: Formulation and Experimental Verification of an Axisymmetric Finite-Element Structural Analysis. Journal of Research of NBS - C. Engineering and Instrumentation, Vol. 75C, No. 3+4, 1971.
7. Pontius PE, Mitchell RA: Inherent Problems in Force Measurement. Experimental Mechanics, Vol. 22, No. 3, 81-88, 1982.

8. Ştefănescu DM: Untersuchung des Verformungszustands eines elastischen Rohrelementes mit Spalten, anwendbar beim Bau der Kraftaufnehmer mit Dehnungsmeßstreifen. Revue Roumaine des Sciences Techniques - Série de Mécanique Appliquée, Tome 29, No. 5, 519-533, 1984.
9. Ştefănescu DM: Comparative Study of the Sensitivity of Various Measuring Techniques on the Model of a Stiffened Membrane Analysed by the Finite Element Method. Report of the Conference "Sensoren - Technologie und Anwendung", Bad Nauheim, 1986.
10. xxx Technical documentation from: a) Dillon; b) Enertec-Schlumberger; c) Hottinger Baldwin Messtechnik; d) Instron; e) MFL Prüf- und Meßsysteme; f) Vishay Micro-Measurements.

VIBRATION ANALYSIS USING MOIRE INTERFEROMETRY

A. ASUNDI AND M.T. CHEUNG
DEPARTMENT OF MECHANICAL ENGINEERING
UNIVERSITY OF HONG KONG, HONG KONG

1. INTRODUCTION

Moire interferometry extends the applications of conventional moire into a realm of higher sensitivity. It has been quite extensively used for in-plane displacement measurement and an excellent review of some of this work is given in ref. 1. In these applications, a high frequency (600 ℓ/mm or 1200 ℓ/mm) phase grating is replicated on the specimen via a thin layer of epoxy. This is then interrogated with an optically generated reference grating of twice the frequency to reveal the in-plane isothetics with a sensitivity of 0.83 μ/fringe or 0.417 μ/fringe. The use of an optically generated reference grating facilitates the creation of mismatch fringes, whch enables optical differentiation of two patterns to remove any initial pattern or for strain determination.

An extension of the technique for out-of-plane displacement measurement utilizing these high frequency gratings was proposed recently (2). A specularly reflecting objbct was placed perpendicular to a high frequency grating. On illumination by a collimated beam of laser light, incident at a predetermined angle, contours of out-of-plane displacement are obtained. In this paper, we have further extended this technique for out-of-plane vibration studies.

2. EXPERIMENTAL METHODOLOGY

The experimental set-up, schematically shown in fig. 1, is similar to

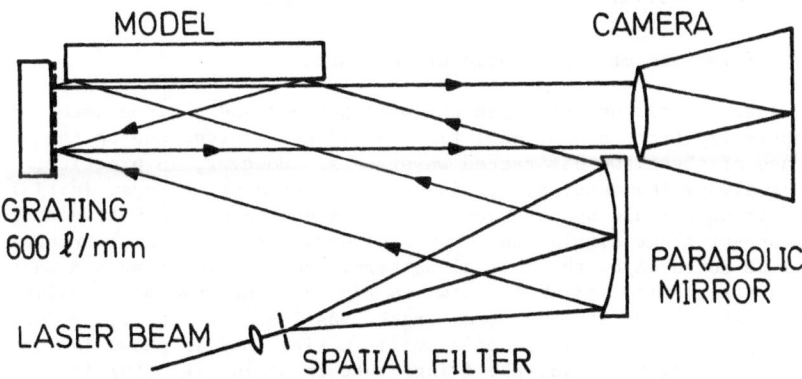

FIGURE 1. Schematic of set-up for out-of-plane displacement measurement using moire interferometry.

the one used in the in-plane moire interferometric method. In the present
method, however, the diffraction grating is placed perpendicular to the
model and mounted on a stage which can be rotated about three perpendicular
axes. The model (or specimen) has a specularly reflecting surface. The
surface of commercial grade perspex is sufficiently specular to be used as
is, but for other materials, it may be necessary to deposit a thin layer of
aluminium to achieve this end. A collimated beam of laser light illuminates
both the specimen and grating, at an angle of incidence as determined by
the diffraction equation

$$\sin \alpha = \frac{\lambda f}{2} \tag{1}$$

where λ is the wavelength of light and f is twice the frequency of the
grating.

The light reflected of the specimen strikes the grating at an angle '$-\alpha$'.
The grating diffracts the two beams into the different diffraction orders
(fig. 2). One order from each beam propogates approximately normal to the

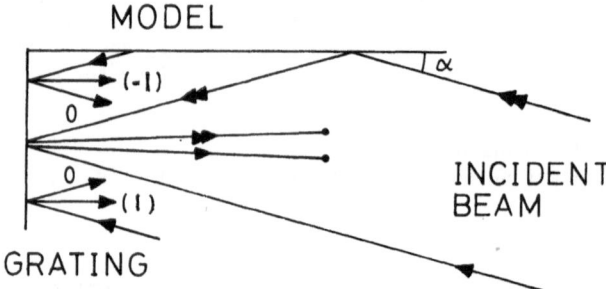

FIGURE 2. Diffraction of light incident on grating.

grating plane. If all optical elements were perfect and the specimen
surface optically flat, a near null field would be observed due to the
interference of these two diffracted wavefronts. However, in practice,
this is not always the case, nor is it deemed necessary; hence an initial
pattern of fringes would be observed. If the specimen were now loaded, the
resulting out-of-plane displacement would give rise to additional fringes
which will be added on to the initial pattern. An example of such a pair
of patterns for static loading is shown in fig. 3. The specimen, a simply
supported perspex beam with a central crack, was loaded at its mid-point.
The initial pattern, which included a small initial load is shown in fig.
3(a). On increasing the load, the fringe pattern as in fig. 3(b) is
observed. The difference between these two patterns would give the out-of-
plane displacement due to load increment alone. The difference can also be
obtained optically using the conventional moire method. In this method,

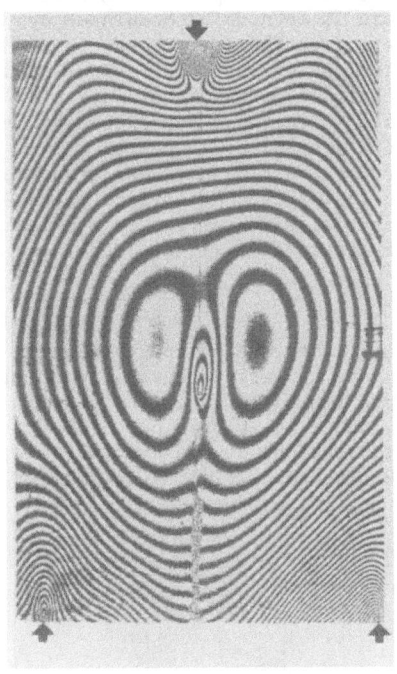
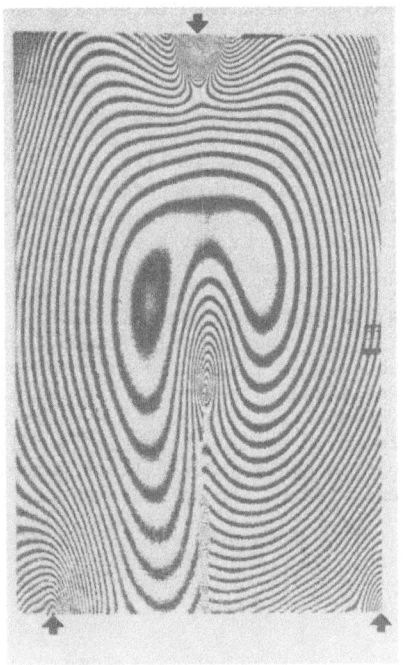

a)   b)

FIGURE 3. a) Initial pattern of simply supported cracked beam with small
initial load. b) Pattern obtained by increasing the load.

when two almost similar geometric patterns are superimposed, the resulting
moire fringes give the difference between the two patterns. In order to
obtain good contrast fringes the two initial patterns should have a high
fringe density (greater than 5 ℓ/mm). To increase the fringe density of the
patterns, as in fig. 3, a bias or carrier pattern is added to both the
patterns. This carrier can be generated either by changing the angle α or
by rotating the grating in the plane. The exact value of the carrier
fringes is not important so long as it is the same in both patterns. Superposition of these two patterns will give rise to the moire fringes which is
difference between the two patterns and corresponds to the out-of-plane
displacement due to load alone. The pattern can further be enhanced by
optical filtering (1).

3. VIBRATION ANALYSIS

The method can be further extended to obtain displacement contours for
a vibrating plate. The set-up is similar to the static case, except that
the model is attached to a shaker causing the plate to vibrate sinusoidally
resulting in a time varying out-of-plane displacement. An initial pattern
is necessary in this application and since a finer pattern is more amenable
to optical filtering, a carrier is also added. The plate is then excited
to one of its resonant frequencies and an exposure made on film. The

exposure time is usually much greater than the vibration period. A time average image results and the original pattern is seen to be modulated by the characteristic Bessel function J_0. Inherent to the J_0 fringes, the lowest order has the best contrast, which also define our nodal points, and higher orders have correspondingly reduced contrast. Thus, the initial pattern is sharpest in regions where the plate is rest and corresponding to the higher zeros of the Bessel function the visibility is poorer. In other areas the initial pattern is washed out by the time averaged exposure. The contrast of the higher fringe orders can be improved by optical filtering, such that in regions where we have an initial pattern, we see a bright fringe and the parts with no initial pattern provides no light to the filtered image.

4. EXPERIMENTAL DEMONSTRATION

To experimentally demonstrate this result, a square perspex plate was mounted at its centre onto the arm of mechanical shaker. A carrier pattern was added to the initial pattern to facilitate the filtering. The plate was excited at various resonant frequencies, and time average pictures recorded on high sensitivity film. Due to limitations of the available optics, only about half of the plate could be recorded. Fig. 4(a) is an enlarged portion of part of the plate vibrating plate, showing zones where

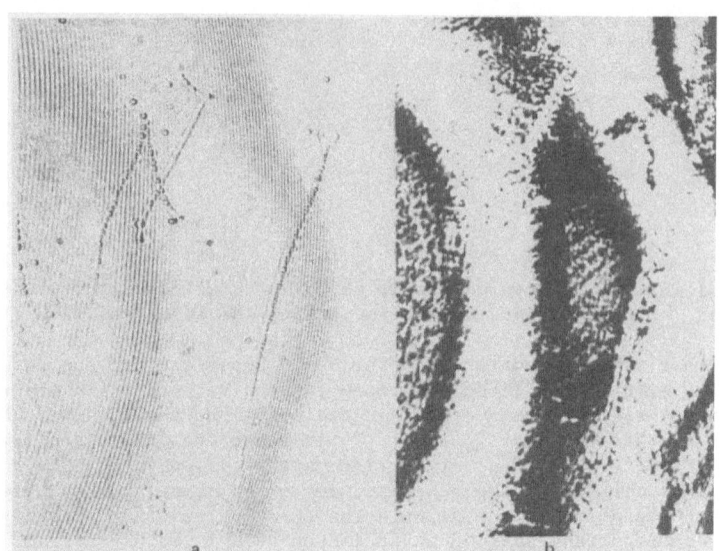

FIGURE 4. Enlarged region of a vibrating plate. (a) Unfiltered image, (b) Filtered image.

the initial pattern is visible and other areas where they have been washed out. The filtered image (fig. 4b) shows the improved contrast as obtained by optical filtering. Fig. 5 are the filtered patterns of the plate vibrating at various resonant frequencies. The brightest fringe corresponds to the zero order Bessel functions, and the fringe contrast diminishes for higher orders as expected. Finally fig. 6 is the out-of-plane displacement patterns of the plate vibrating at 4 kHz with different amplitudes. Remarkable improvement was observed by filtering as in the unfiltered image the fringes were barely visible.

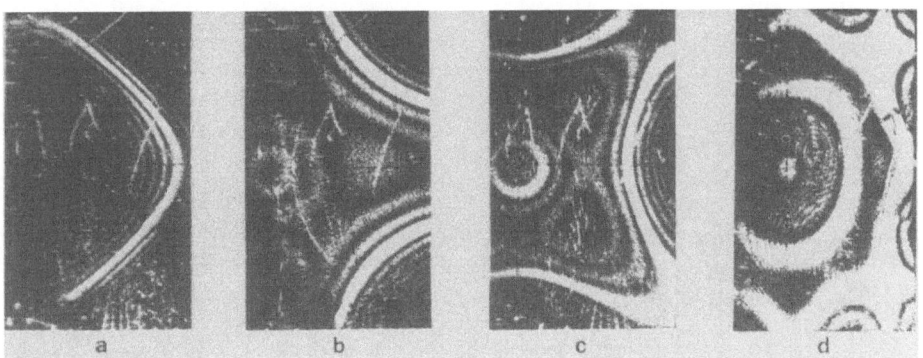

FIGURE 5. Filtered patterns of plate vibrating at different resonant frequencies. (a) 1.1 kHz, (b) 1.6 kHz, (c) 4 kHz, (d) 8.4 kHz.

5. CONCLUSION

We have demonstrated the use of moire interferometry for low amplitude vibration analysis. Contours of out-of-plane displacement with sensitivity comparable to that in holography can be achieved. However, unlike time-average holography no prior knowledge of the resonant frequency is necessary. If the plate were not vibrating at its resonance, the initial pattern would not be visible anywhere.

REFERENCES

1. Post D: Developments in Moire Interferometry. Optical Engineering, 21(3), p. 458 (1982).

2. Asundi A and MT Cheung: Moire Interferometry for Out-of-Plane Displacement Measurement. Journal of Strain Analysis (to be published).

306

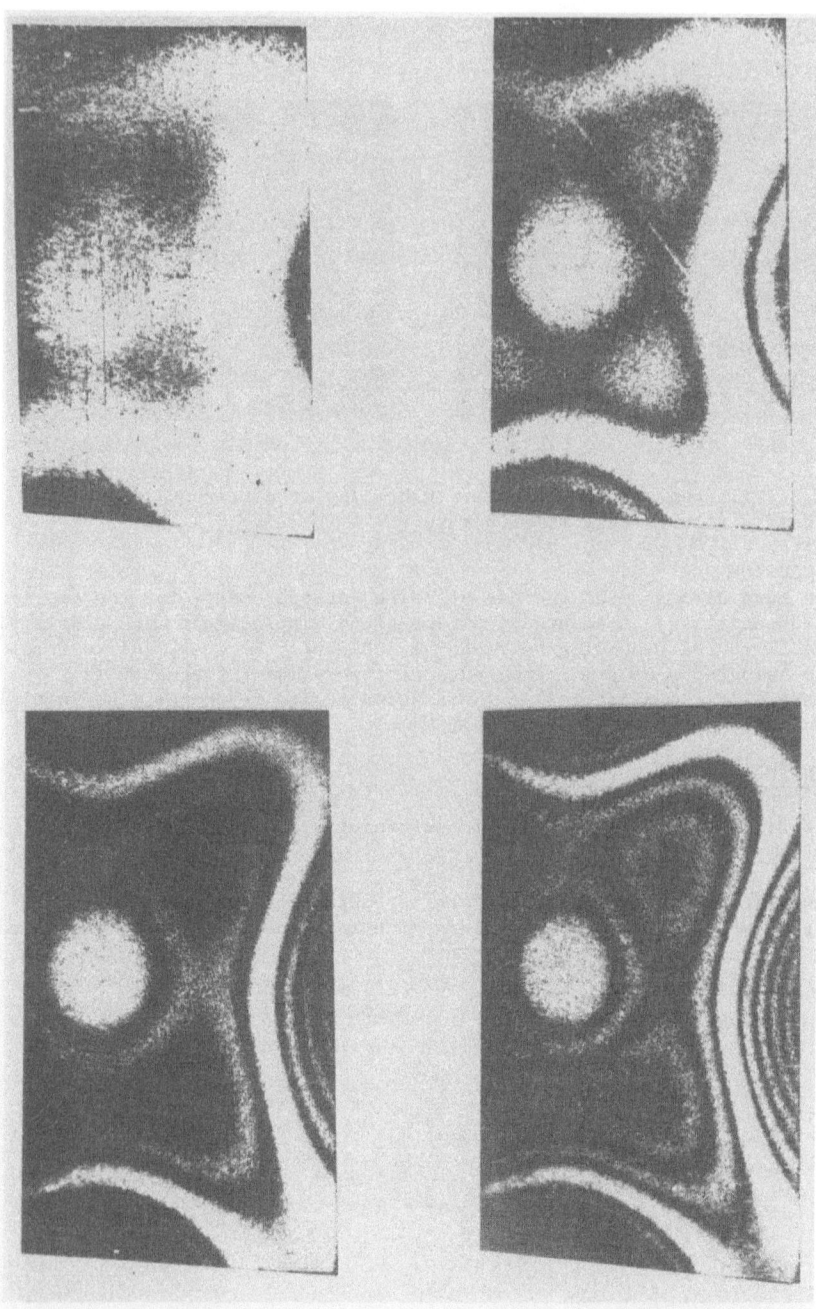

FIGURE 6. Filtered patterns of plate vibrating at 9 kHz at different amplitudes.

HETERODYNE MOIRE INTERFEROMETRY - SOME EXPERIENCE OF THE DEVELOPMENT OF A
RAPID MEASUREMENT SYSTEM

C A WALKER, P MacKENZIE, J McKELVIE

UNIVERSITY OF STRATHCLYDE

1. INTRODUCTION
    Developments in the grid application and optical systems used in moire
interferometry have led to a powerful technique for whole-field stress
analysis, with practical applications in elastic/plastic strain
distributions, fracture mechanics and finite-element code verification[1].
To a large extent the problems associated with the use of the moire
technique now relate to the analysis of the interferograms; this may be a
task of some magnitude, since the productivity of the current systems is
such that many hundreds of interferograms may be generated in a short space
of time.
    An alternative approach to whole-field photography plus subsequent
analysis is the use of a point-measurement of strain directly by electro-
optic techniques. The heterodyne principle applied to moire gratings was
first described some years ago[2]; this paper discusses operational
difficulties which arose with the original system and describes a programme
of improvements which have overcome many of these deficiencies.

2. PRINCIPLES OF HETERODYNE MOIRE INTERFEROMETRY
    It is a basic principle of all moire interferometers that a grating is
illuminated by beams of collimated, coherent light. The diffracted orders
interfere, and the resulting interference patterns form contours of
deformation of the underlying specimen. These interference patterns are
recorded photographically; the strain at any point is measured by
identifying fringe spacing at that point, and inserting it in the
relationship:-

$$\text{strain} = \frac{\text{grating spacing}}{\text{fringe spacing}}$$

    In a heterodyne system, the fringes are made to sweep across the field
of view by modulating the frequency of one beam relative to the other. At
any one point in the field, a sinusoidal variation of intensity with time
will be observed. The strain at that point can be assessed by measuring
the phase shift between the outputs of two photodiodes sampling the moving
fringes, the gauge-length of measurement being defined by the spacing of
the photodiodes.
    The aim of the original system[2] was to measure all three components
of strain simultaneously, and to keep the optics in registration with the
grating with an automatic alignment system which used spare diffraction
orders for sensing.

308

Experience, however, showed up the following problems with the original concept:-

1) The complex optical path, (Figure 1), though compact, gave rise to excessive light loss; in consequence, one was faced with the choice either of increasing the laser power by a factor of ~ 10, thereby jeopardising the portability of the complete instrument, or of accepting a disadvantageous signal : noise ratio, with severe implications for the precision and repeatability of the strain measurements.

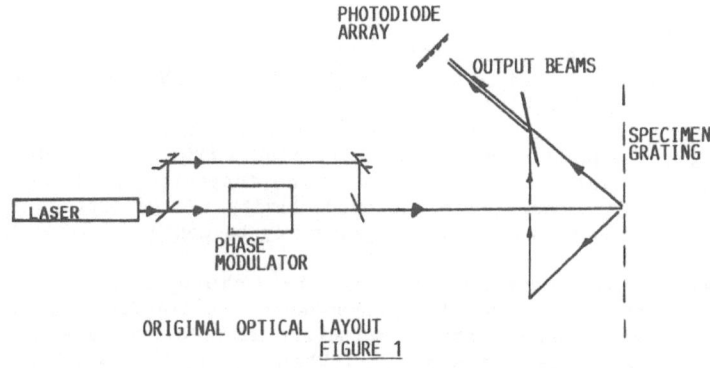

ORIGINAL OPTICAL LAYOUT
FIGURE 1

2) The reverse-optics layout, (Figure 1), in which one beam impinged normally on the surface, and the diffracted orders were recombined, was sensitive to minor misalignments. More seriously, overlap between the interfering beams could not be maintained except over a restricted range of strain measurement, since the strain caused the beams to rotate in opposite directions.

3) As a further consequence of the low light levels, the alignment system was sensitive to ambient light levels. This further exacerbated the misalignment sensitivity discussed in the previous paragraph.

3. REVISED DESIGN CONSIDERATIONS
As a result of the above limitations, the original concept was modified along the following lines:-

3.1 Optical Layout
In line with the previous discussion, the original optical principle, employing a single beam impinging normal to the surface, showed a disappointing sensitivity to misalignment. In the interval, since the previous design study, a large body of experience had been amassed using a portable moire interferometer[3], in which the problems of alignment have been completely analysed and are now well understood. Accordingly, a similar system was adopted for the redesigned heterodyne interferometer,

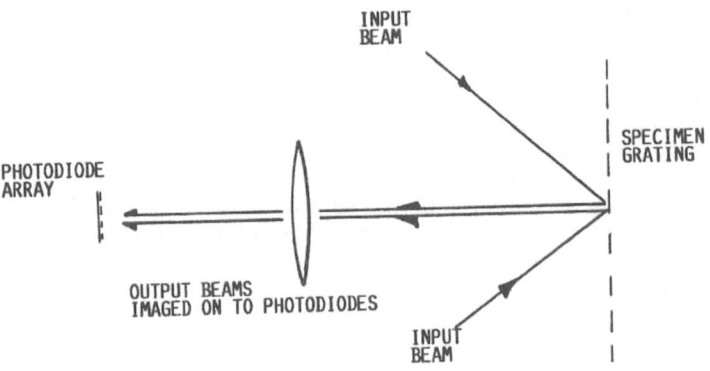

REDESIGNED OPTICAL LAYOUT

FIGURE 2A

(Fig 2a). While this layout does have the drawback that all three combined output beams appear in the same direction, the three components can be separated in a straightforward manner, for example, by sequential shuttering, or by using different heterodyne frequencies. It is important to note that, while the diffracted beams show the same sensitivity to strain as in the original layout, the use of a lens to image the illuminated area avoids all tendency for the diffracted beams to shear away from each other (at least up to about 10% strain).

3.2 Beam Diameters
The diameter of the beam used was increased from 0.6 mm to 6 mm by passage through a 10X beam expander. This increase improves the collimation, and, essentially, provides a flatter wavefront at the plane of interference. The raw output beam from the laser has a divergence of 0.6 mrad, which gives rise to apparent strain fringes if the beams are sheared over each other.

The consequence of this expansion is that the specification on control may be eased; the penalty, however, is that the light intensity level is reduced. Should this turn out to be a serious limitation, the output beams may be demagnified, or larger photodiodes used as detectors, and, while either of these options does increase the gauge length, this does remain within acceptable limits (i.e. < 1 mm).

3.3 Frequency Modulation
A dual-beam 3-pass electro-optic modulator was used to modulate both beams in antiphase. This proved to be an option preferable to the original idea of using a miniaturised modulator on only one beam. The problems of aligning the beam through the narrow plate spacing more than offset the apparent gain from having only one beam to align rather than two (Fig 2).

For the same reason, the design opted, in the name of convenience, for two separate collimators rather than recombining the beams and passing them through a single system. Since the modulator could handle only narrow beams, the light enters the beam expanders after being modulated.

3.4 Assembly and alignment

In view of the experience gained from the use of a 3-beam live-fringe moire interferometer, the heterodyne system was assembled as a unit and mounted on the same baseplate as a standard interferometer Fig 2b. In this way, the adjustment and alignment were carried out using well-understood and familiar procedures – both for major rotations and translations of the complete unit and for fine adjustment of individual beams. This proved to be one of the most important improvements over the original system, since, once the modified optics were assembled and aligned with the standard baseplate, it was the work of a few moments to align a specimen grating and make measurements.

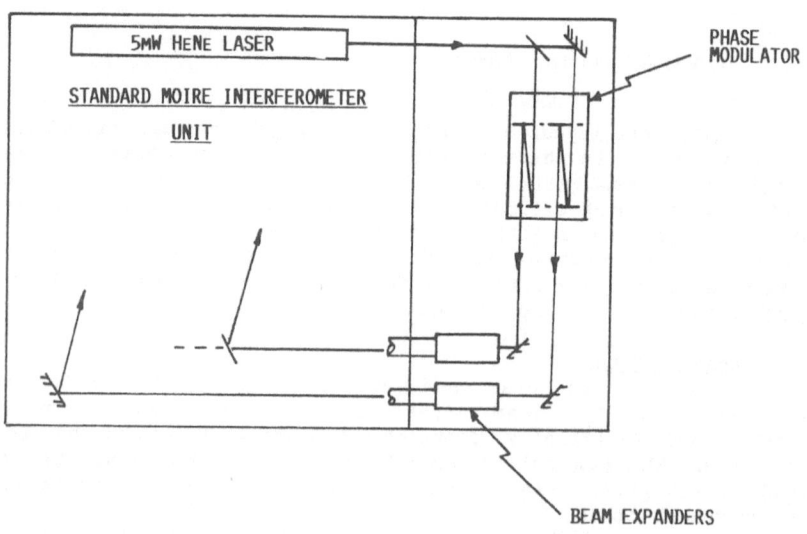

REDESIGNED OPTICAL LAYOUT-RELATIONSHIP WITH MOIRE INTERFEROMETER UNIT

FIGURE 2B

3.5 Electronics and data processing

The measurement process is initiated by the triggering of the modulator drive ramp voltage (Figure 3). The photodiode outputs are monitored, and the period between zero-crossings is counted. At the start of the ramp, however, there is a transient effect on the fringe position, and so to avoid false transients, the start of the detection process is delayed by 30 microseconds after the start of the ramp, to avoid the counting of any false zero crossings.

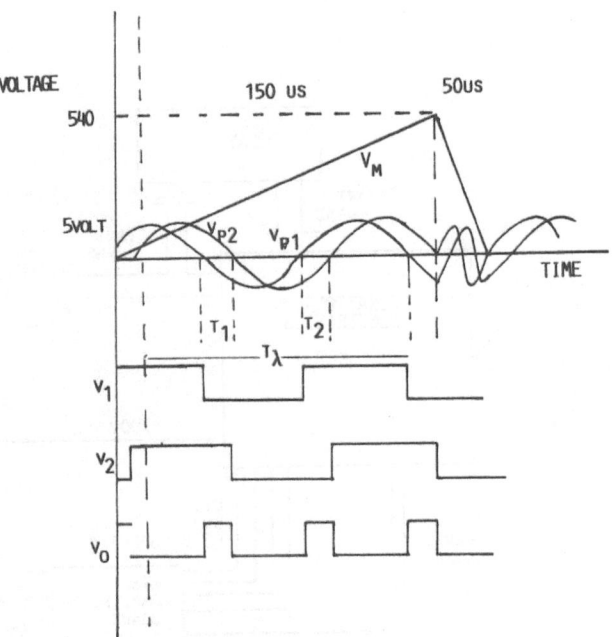

MODULATOR,PHOTODIODE,LIMITER,&COUNTER VOLTAGES

V_M=MODULATOR DRIVE VOLTAGE

V_{P1},V_{P2}= PHOTODIODE VOLTAGES

V_1,V_2= LIMITER OUTPUT VOLTAGES

V_0 = PHASE COUNTER VOLTAGE

FIGURE 3

It was found, too, that at the operating frequency of 5 kHz, the capacitance of the photodiode leads caused an attenuation of the signal. The use of low-capacitance leads restored the signal : noise ratio to its low-frequency level of 10:1.

The basic fringe period, t_λ (Figure 3), and the phase lag ($t_1 + t_2$) were held in registers, and then stored on the discs of a microcomputer. From these values, a corresponding strain is calculated, (Figure 4).

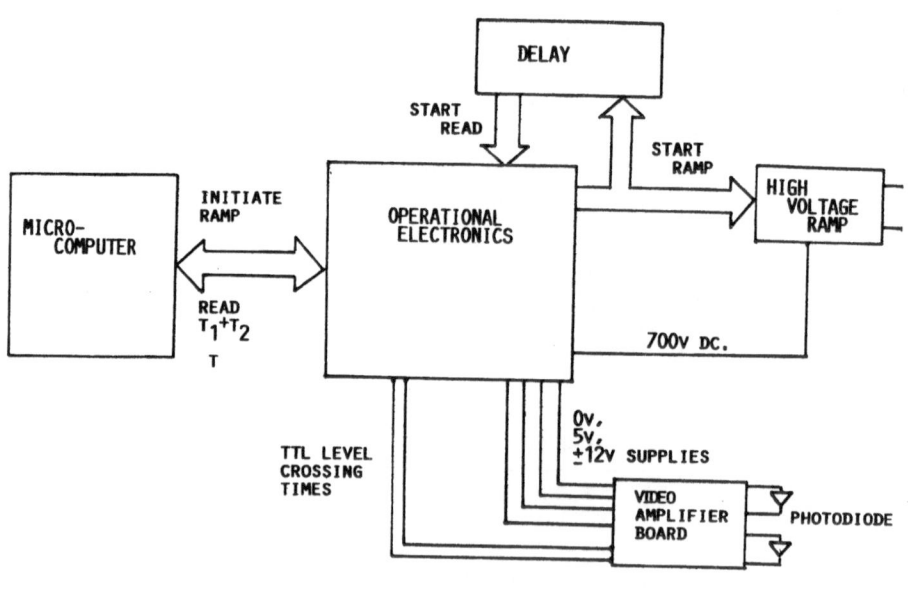

FIGURE   4

This 'one-shot' measurement process gives a spectrum of values for a nominally constant strain. It was found that a process of averaging several readings, together with a rejection of rogue values remote from the mean, gave a repeatability within 10 microstrain.

4.   EXPERIMENTAL TRIALS

The system was evaluated by comparison with a strain gauge attached to a plain specimen loaded in tension in the range 0 - 1000 microstrain. The diode separation (i.e. the gauge length) was 0.5 mm. For simplicity, one axis of strain measurement was used, parallel to the axis of loading.

It was found that, even with the beam expansion, residual effects due to lack of collimation could be measured. Careful control of the specimen-optics distance eliminates this effect, by ensuring that the wavefront is not sheared over itself before interference.

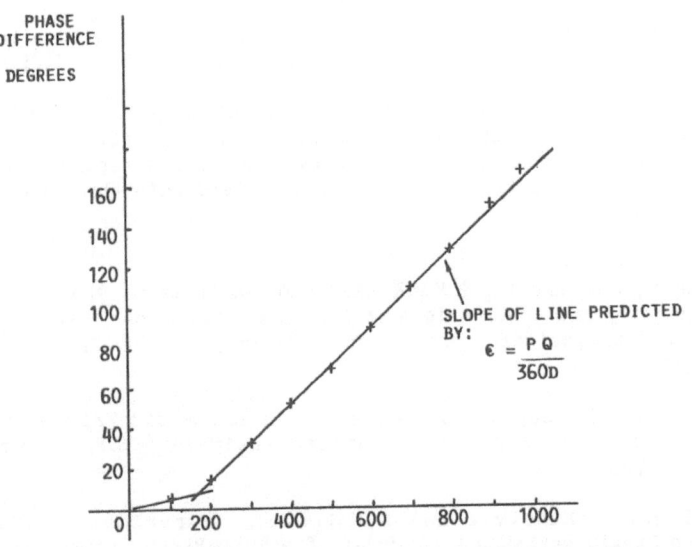

STRAIN GAUGE READING--MICROSTRAIN

CORRELATION BETWEEN HETERODYNE INTERFEROMETER & STRAIN GAUGE
FIGURE 5   MEASUREMENTS.

The correlation between heterodyne and strain gauge is shown in Figure 5. It will be seen that, in accordance with the relationship between strain and phase difference:-

$$\epsilon = \frac{p\,Q}{360D}$$ where p = grating pitch
Q = phase difference
D = gauge length

the plot is a straight line.

Below 150 microstrain, it will be seen that this linear law breaks down. This is ascribed to the algorithm used for rogue rejection; since the system at present deals only with tensile or compressive strain, all negative values are rejected as 'rogues' and so, at present, low strain levels are not properly averaged. In subsequent developments, in which both tension and compression are handled this non-linearity will not occur.

The slope of the linear portion agrees with the value indicated by equation 1, using the prescribed values for grating pitch and gauge length.

The zero shift of ~ 120 microstrain has not yet been satisfactorily explained. In a real system one would be able to calibrate out this shift.

5.   DISCUSSIONS
The principle of heterodyne moire interferometry has been shown to be stable and accurate by comparison with a strain gauge measurement. The

alterations to the optical layout, have greatly eased the alignment of the instrument with the specimen, while the revision of the electronic data processing has resulted in increased range and greater accuracy.

One may now envisage building a practical instrument with a range of properties - i.e. gauge length, speed, portability, which is difficult to match.

When it is recalled that one of the most attractive features of the moire technology is its ability to assess strain distributions, it will be realised that such an instrument, to achieve its full potential, would be mounted on a scanning mount to cover the whole field point-by-point.

6.   REFERENCES

1.   A McDonach, J McKelvie, P MacKenzie and C A Walker.   Improved Moire Interferometry and Applications in Fracture Mechanics, Residual Stress and Damaged Composites.   Experimental Techniques, Vol 7, No 6, 20-24, (June 1983)

2.   J McKelvie, D Pritty and C A Walker.   An Automatic Fringe Analysis Inteferometer for Rapid Moire Stress Analysis.   S.P.I.E. Vol 164, Utrecht, 1978.

3.   C A Walker, J McKelvie and A McDonach.   Experimental Study of Inelastic Strain Patterns in a Model of a Tube-Plate Ligament using an Inteferometric Moire Technique.   Experimental Mechanics, Vol 23, No 1, 21-29, March 1983.

ON THE LIMITS TO THE INFORMATION OBTAINABLE FROM A MOIRE FRINGE PATTERN

J McKELVIE

UNIVERSITY OF STRATHCLYDE

1. INTRODUCTION

The moire method of strain measurement, together with modern derivatives, is now well established (see, for example, [1]). As more and more detailed information is sought about strain-fields, for example in relation to crack-tip behaviour, so do the limitations of the classical method of fringe interpretation (the 'Discrete Moire Law'), appear irksome. In consequence, a variety of fringe interpolation methods have been proposed and applied.

This work addresses the question of how far such techniques may be pushed in the quest for detail. The analysis will be particularly concerned with strongly heterogeneous fields. (The strain fields described in [2] would be extreme examples).

Although written from the point of view of strain measurement, the conclusions will have relevance for moire's employed for other purposes.

2. FORMATION AND CLASSICAL INTERPRETATION.

2.1 The Formation of a Moire in Incoherent Light.

We consider the case of a moire formed by superimposition of a specimen grating and a master grating. Any regular grating parallel to the y-axis has an intensity profile whose transmission function is described by

$$T_u(x) = a_o + \sum_{n=1}^{\infty} a_n \cos(\frac{2\pi nx}{p} + \alpha_n) \qquad \dots\dots\dots\dots\dots(1)$$

where p is the grating pitch and the a_n's and α_n's are determined by the profile.

With a displacement function u(x) we have the transmission function T_d of the deformed grating, in the original coordinates,

$$T_d(x) = T_u(x - u(x))$$

$$= a_o + \sum_{n=1}^{\infty} a_n \cos[\frac{2\pi n}{p}(x-u(x)) + \alpha_n] \qquad \dots\dots\dots\dots(2)$$

The illumination of these two gratings in superposition gives

$$I_e = I_i\Big[a_o^2 + a_o \sum_{n=1}^{\infty} a_n \cos[\frac{2\pi nx}{p} + \alpha_n] + \sum_{m=1}^{\infty} a_m \cos[\frac{2\pi m}{p}(x-u(x)) + \alpha_m]$$

$$+ \sum_{n=1}^{\infty} \sum_{m=1}^{\infty} \frac{a_n a_m}{2} \cos[\frac{2\pi}{p}\{(n+m)x - mu(x)\} + \alpha_n + \alpha_m] +$$

$$+ \cos[\frac{2\pi}{p}\{(n-m)x + mu(x)\} + \alpha_n - \alpha_m] \Big] \dots(3)$$

where I_i and I_e are the intensities of the illuminating and emergent beams respectively.

This is of the general form

$I_e = I_1 + I_2 + I_3$ where I_1 = a constant term

I_2 = a series of cosine terms whose argument contains $2\pi kx/p$, $k \neq 0$

I_3 = a series of cosine terms whose argument consists solely of $2\pi m u(x)/p$, $m=1,2,3..$ arising when $n = m$.

We now 'filter out', or 'average out', or 'smooth out' the high frequencies leaving us with the moire intensity,

$$I_m = I_0 + b_1 \cos\{2\pi u(x)/p + \beta_1\} + b_2 \cos\{2\pi.2u(x)/p + \beta_2\} + b_3 \cos... \quad ..(4)$$

the b's and β's being constants which depend on the grating profile. Eqn.(4) is the basis for classical moire interpretation. It has been developed at length here because the derivation process is of subsequent importance. It can be shown that the same form of equation, achieved by 'filtering out' the high frequencies, represents the intensity corresponding to the superimposition of two gratings in the following cases:-

i) superimposition in coherent light,
ii) master grating projected onto specimen grating,
iii) specimen grating imaged onto master grating,
iv) observation in reflection.

(The formation of a moire by double exposure, requires an intermediate optical process to render the moire visible. However, the end result is essentially the same as Eqn.(4), after a similar filtering process).

2.1 The Discrete Moire Law

This arises from the fact that every term in I_m is periodic in $u(x)/p$ so that I_m itself has period $u(x)/p$. Therefore two adjacent maxima represent a change of p in $u(x)$.

The problem with the use of the Discrete Moire Law is that it provides no information about the regions between the maxima.

A variety of techniques have been devised and utilised to overcome this unfortunate fact and provide more detail.

3. THE MEANING OF 'DETAIL'.

It is appropriate at this point to define what we mean by 'the quest for detail'.

We consider the strain distribution $\varepsilon(x)$ to be described in the usual manner by a series (or continuum) of sinusoids.

$$\varepsilon(x) = F^{-1}E(v) \quad ....(5)$$

where F^{-1} is the inverse Fourier transform operator and $E(v)$ is the Fourier transform of $\varepsilon(x)$.

Since $u(x) = \int \varepsilon(x)dx$, $\quad ....(6)$

we have, by the properties of Fourier transforms,

$$u(x) = F^{-1}\left[\frac{1}{j2\pi v}.E(v) + \frac{E(0)}{2}.\delta(v)\right] \quad ....(7)$$

the usual nomenclature obtaining.

The 'quest for detail' is the attempt to elucidate the structure of $\varepsilon(x)$, especially at the higher frequencies.

We will consider as disadvantageous any method or process that increases the uncertainty in our estimate of $\varepsilon(x)$, or deforms $E(v)$ in some ill-defined manner, or reduces the detectability of spatial frequencies (or obliterates them altogether).

In relation to conventional strain measurement, the 'quest for detail' can be considered to be equivalent to the attempt to achieve the smallest possible gauge length.

4.METHODS UTILISED TO YIELD MORE DETAIL.

The various methods used to elicit more detail, and their respective restrictions, are now summarised.

4.1 Interpolation by curve fitting

The device of fitting some mathematical function to the experimental points in a displacement-vs-position plot has been widely reported. The use of particular mathematical forms or splines, whether 'smoothed' or not, is in reality a totally deceptive stratagem, - it appears to yield a mass of additional information whereas in fact it provides none, since the information so provided is different if different functional forms are fitted.

4.2 The Displacement/Light-Intensity Law

This was derived as follows:-

Considering equation (4):- 'it is possible to filter out the harmonics of order higher than the first' (ref.[4]), leaving, neglecting β_1, which is a function of an arbitrary choice of origin,

$$I_m = I_o + b_1\cos\{\tfrac{2\pi u(x)}{p}\} \qquad \dots\dots\dots\dots\dots\dots(8)$$

we therefore have
$$u(x) = 1/2\pi.\text{arc }\cos\frac{I_m(x) - I_o}{b_1} \qquad \dots\dots\dots\dots\dots(9)$$

This is the continuous Displacement/Light-Intensity Law, allowing us, having ascertained I_o, - the local mean intensity, - and b_1, - the local amplitude, -to measure $u(x)$ at any point between maxima simply by measuring I_m.

In fact, as will now be shown, the conditions under which the necessary filtering can be carried out so as to lend validity to (8) are very restricted, and will not in fact prevail in normal Moire as so far described:-

The process used to arrive at equation (3) can be represented in the Fourier domain:- Equn.(3) was obtained from

$$I_e = \frac{I_o}{2}\left[a_o + a_n\cos\{\tfrac{2\pi nx}{p} +\alpha_n\}\right]\left[a_o + a_m\cos\{\tfrac{2\pi m}{p}(x - u(x) + \alpha_m\right]\dots(10)$$

Each pure cosine is represented by a pair of δ-functions standing at frequencies $v = \pm n/p$ (Fig 1a). Each modulated cosine is represented by a pair of broadened spectra positioned around the same frequencies (Fig 1b).

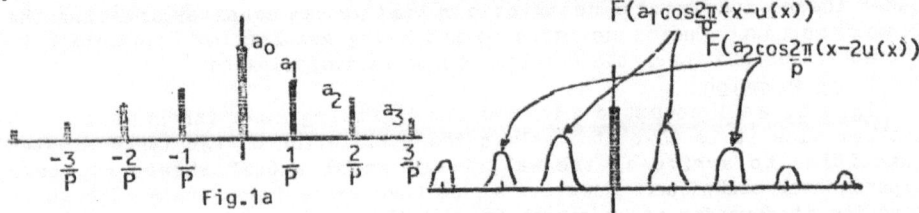

Fig.1a

Fig.1b

Multiplication in ordinary space becomes convolution in the Fourier domain, so that in the representation of equation (10), every broadened spectrum is convolved with every δ-function. Convolution of a function $f(v)$ by a δ-function standing at v_o results simply in $f(v)$ being translated by v_o, together with multiplication by the δ-function amplitude (Fig 2a), i.e. $\qquad f(v) * \{A.\delta(v - v_o)\} = A.f(v - v_o)$

318

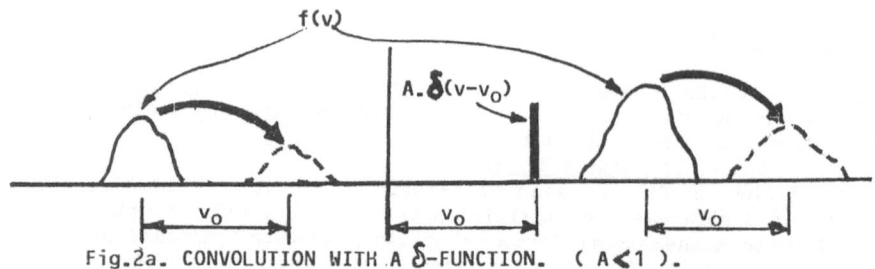

Fig.2a. CONVOLUTION WITH A δ-FUNCTION. (A<1).

The result of a few of the whole set of convolutions representing the set of multiplications in equation (10) is illustrated in Fig 2b.

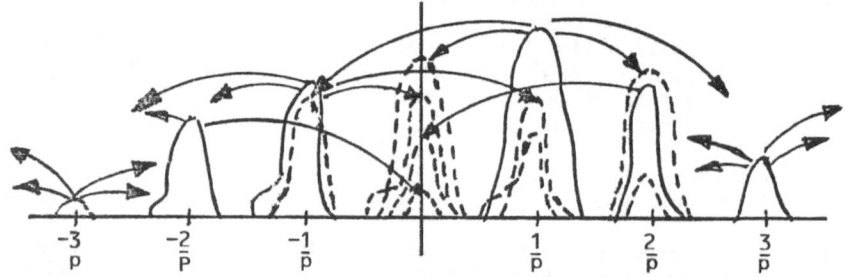

Fig.2b. CONVOLUTION OF SERIES OF SPECTRA WITH SERIES OF δ-FUNCTIONS

The sum of the broadened spectra which are translated to the zero frequency region is the transform of the moire. It will now be clear that whereas it is practicable, at least for spectra of the breadth shown, to remove the high frequency spectra by filtering so as to leave the moire, there is, because of the superimposition, no way in which the terms in 2u(x), 3u(x), etc, can be eliminated from the moire so as to leave only the u(x) term, - as is required to obtain the continuous law.

However, it will also now be obvious that the condition that there can be no overlapping (so that the term in u(x) can therefore be separated) is....."if, and only if, either the specimen grating or the master grating (or both) are reduced to single sinusoids (plus a d.c. term)". It is only under these experimental restrictions that we can consider applying the Continuous Law. Once a moire is formed using gratings not conforming to the above restrictions, the Continuous Law is inapplicable.

4.3 Pitch Mismatch

This is well accepted as a means of getting more information. The problem with it is that it involves the subtraction of two large similar quantities to arrive at the relatively small actual strain - a poor experimental technique. However, we do have extra information, though in practice it degrades as we search for detail.

4.4 Fringe Shifting

Fringe shifting is achieved by introducing a known relative translation between master and specimen grating, this causes the moire fringes to move (or to 'shift'), and the shifts of the maxima are measured. This technique undoubtedly provides extra information, provided we can relate one picture to the next, this may make considerable demands on the shift mechanism. The use of optical shifting [4] is an alternative to mechanical translation.

The technique appears to permit, in principle, of infinite interpolation.

4.5 Phase Measurement

This is a development of fringe shifting using the continuous displacement law so that the analysis is not confined to the maxima:- from the complete form of equation (8), if we shift the master an amount x_s we get

$$I_m(x) = I_0 + b_1 \cos\{2\pi u(x)/p - 2\pi x_s/p + \beta_1\} \quad \ldots\ldots\ldots\ldots(12)$$

If at a point x_1 we make measurements of I_m at two different positions of the master, we can solve for $u(x_1)/p + \beta_1$. If we then do the same at a point x_2, we get $u(x_2)/p + \beta_1$. This yields $u(x_1) - u(x_2)$. The method is superior to other methods in that it allows of better discrimination between signal and noise, because much of the noise will not move systematically with the fringes.

An additional feature is the fact that it is possible to distinguish the sign of the strain using phase measurement. The considerations applicable to the continuous law and to fringe shifting apply in combination.

4.6 Heterodyning

In this technique the phase is varied continuously and linearly in time. Reference [6] describes such an arrangement for strain measurement, where the x_1 and x_2 measurements are made continuously in time, with phase comparison of the two resultant sinusoids. Reference [7] describes a system for surface topography.

4.7 Fringe Multiplication

By various techniques involving filtering in the focal plane of a lens, the higher harmonics of the specimen grating can be utilised to form moire's of effective grating frequency equal to an integer multiple of the fundamental, - see for example [8]. This provides many more maxima than the simple moire and thus provides more information. In practice it is superior to mismatch also to simple fringe-shifting.

4.8 Moire Interferometry

In moire interferometry [9] the master grating is replaced by a 'virtual' grating created by two overlapping beams of coherent light. With efficient gratings and optics, operation in everyday surroundings is practical, - a fast exposure freezing the fringes [10].

It is to be noted that moire interferometry meets the continuous-law restriction, the master grating consisting in essence of a pure sinusoid.

4.9 Combinations

Various methods are capable of combination. For example we can fringe shift, or mismatch, with moire interferometry or fringe multiplication.

5. ASPECTS OF INFORMATION AND COMMUNICATION THEORY.

Certain aspects of Information Theory and Communication Theories are relevant to moire formation and analysis.

5.1 Gabor's Uncertainty Principle

This says that if a function is subjected to a filtering process so that it is band-limited to frequencies below, say, v_f, then there is no information within a dimension $1/v_f$.

This can be expressed as

$\Delta x . v_f \geq 1$ where Δx is the limit of resolution.

The relationship is derivable from the Heisenberg Uncertainty principle - see Ref.[11].

It also holds for a band-pass system where v_f is then the band-width.

5.2 Phase and Frequency Modulation

Ref [12] and the Appendix of [13] contain material of relevance to this section. A phase modulated wave is one in which a sinusoidal carrier wave $V\cos2\pi f_c t$ is modulated by a signal $s(t)$ to produce a wave $p(t)$ according to

$$p(t) = V\cos\{2\pi f_c t + k.s(t)\} \quad \dots\dots\dots\dots\dots\dots\dots(13)$$

where k is a constant.

In frequency modulation the signal is integrated first and the wave transmitted is

$$\xi(t) = V\cos2\pi\{f_c t + k'.\int_0^t s(\lambda)d\lambda\} \quad \dots\dots\dots\dots\dots(14)$$

If we consider $s(t)$ itself to be a pure sinusoid,

$$s(t) = s_1\cos2\pi f_{s1} t , \quad \dots\dots\dots\dots\dots(15)$$

and putting $k's_1/f_{s1} = \mu$ (the 'modulation index'),

then

$$\xi(t) = V\cos(2\pi f_c t + \mu\sin2\pi f_{s1} t) \quad \dots\dots\dots\dots\dots(16)$$

which can be expressed as the series

$$\xi(t) = V\sum_{n=1}^{\infty} J_n(\mu)\cos2\pi(f_c + nf_{s1})t \quad \dots\dots\dots\dots\dots(17)$$

where $J_n(\mu)$ is the n-th order Bessel function of argument μ. The spectrum of $\xi(t)$ therefore extends to infinity quasi-symmetrically about f_c. However, for $\mu > n+1$, $J_n(\mu)$ approaches zero very rapidly and therefore a limited number of sidebands can describe $\xi(t)$ in practice. Fig [3] depicts this.

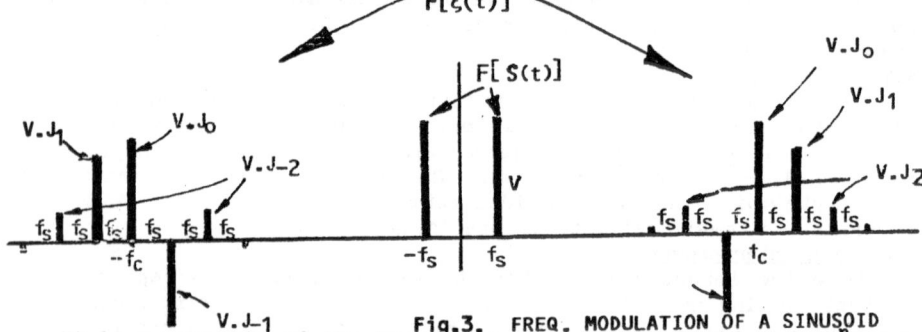

Fig.3. FREQ. MODULATION OF A SINUSOID

It is a property of the Bessel function that $J_{-n}(\mu) = (-1)^n J_n(\mu)$. Using this, and letting $f_c=0$, we obtain

$$\cos(\mu\sin2\pi f_s t) = J_0 + 2\sum_{n=1}^{\infty} J_{2n}(\mu)\cos4\pi nf_s t \quad \dots\dots\dots\dots(18)$$

i.e., the fundamental and all odd harmonics are eliminated.

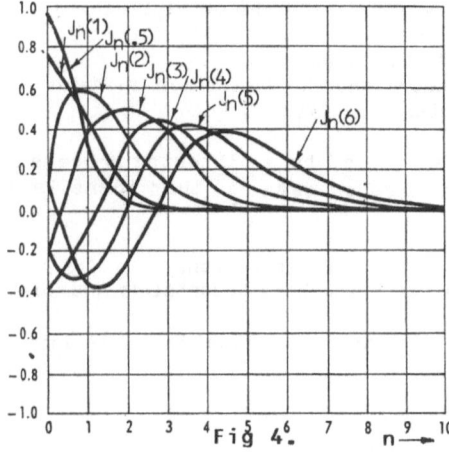

Fig 4.

Fig[4] illustrates the behaviour of $J_n(\mu)$, and Ref [14] gives tables of values over wide ranges of n and μ.

The spectrum of $\xi(t)$ when $s(t)$ contains many sinusoids is very complicated. For example, with three sinusoids, say

$$s(t) = s_1\cos2\pi f_{s1}t + s_2\cos2\pi f_{s2}t + s_3\cos2\pi f_{s3}t \quad\dots\dots\dots\dots(19)$$

putting $\mu_r = k's_r/f_{sr}$ we obtain

$$\xi(t) = \sum_l \sum_m \sum_n J_l(\mu_1)J_m(\mu_2)J_n(\mu_3)\cos2\pi(f_c+lf_{s1}+nf_{s2}+mf_{s3})t \dots\dots(20)$$

It is clear that the bandwidth required could expand considerably beyond that of $s(t)$ if the coefficients are not small.

5.3 Sideband Foldover and Spectrum Overlap

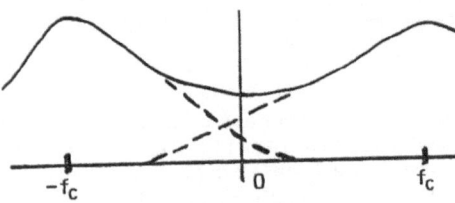

When the side-band-width exceeds the carrier frequency then what is known as "side-band foldover" occurs. This is illustrated in Fig [5a], and is clearly a condition of ambiguity (and therefore of uncertainty), of a spectrum.

Foldover is a particular example of spectrum overlap, illustrated in Fig [5b]. Obviously if we wish to know the separate spectra we will, again, be faced with uncertainty. We shall exclusively use the term 'foldover' whenever overlap occurs at zero frequency.

Fig.5a   SIDEBAND FOLDOVER

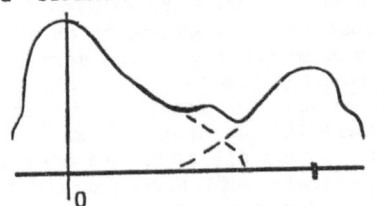

Fig.5b. SPECTRUM OVERLAP

6. APPLICATION OF INFORMATION AND COMMUNICATION THEORIES

6.1 Specimen Grating as a F.M. Signal

Considering equations (2) and (13) it is now obvious that the specimen grating is a veritable analogue of a multiplicity of phase modulated carriers of frequencies n/p.

Furthermore, since $u(x) = \int \varepsilon(x)dx$, we can equally consider the terms to be frequency-modulated waves in which the signal is $\varepsilon(x)$ and k' is $-n/p$

If now we consider the effect of $\varepsilon(x)$ containing a d.c. term, we have the Fourier representation of $\varepsilon(x)$,

$$\varepsilon(x) = \varepsilon_0 + \sum_r \varepsilon_r\cos(2\pi f_r x + \gamma_r) \quad\dots\dots\dots\dots\dots\dots(21)$$

Therefore $u(x) = \varepsilon_0 x + \sum_r \dfrac{\varepsilon_r}{2\pi f_r}\sin(2\pi f_r x + \gamma_r) + u_0 \quad\dots\dots\dots\dots(22)$
where the summations will in general be integrals.

For the case where only the fundamental of the specimen grid is significant, and neglecting α, and u_0, we now have,

$$T_d = a_0 + a_1\cos\left[\frac{2\pi}{p}\{x - \varepsilon_0 x - \sum_r \frac{\varepsilon_r}{2\pi f_r}\cdot\sin(2\pi f_r x + \gamma_r)\}\right] \quad\dots\dots\dots(23)$$

and it is clear that the effect of the d.c. term is effectively to change the carrier frequency from 1/p to $1/p(1 - \varepsilon_0)$, leaving us with another wave of the F.M. form of equation (14), where the signal is expressed as a sum of sinusoids.

This is illustrated in Figure (6), where $\varepsilon(x)$ is taken to have a continuous transform of maximum frequency v_m .

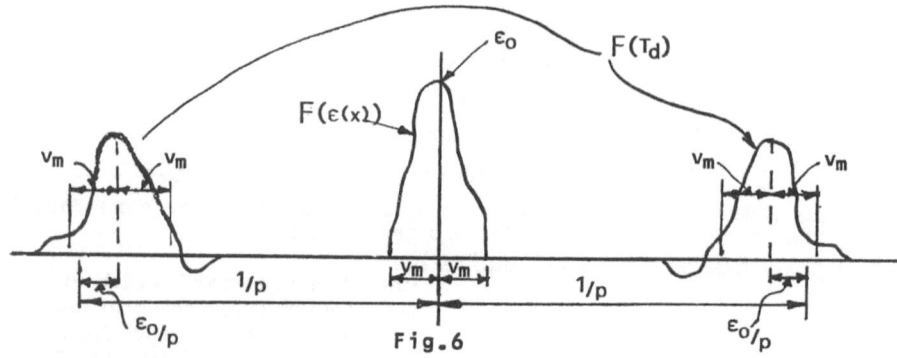

Fig.6

6.2 Effect of Filtering

As has been shown above, it axiomatic in the formation of a Moire that a low-pass filtering process be applied. Conventionally this is done by having a lens of suitably restricted aperture (for example, the eye), or possibly defocussed [15]. Alternatively the light-intensity may be converted to electrical signals which may then be filtered by analog or digital means. The filtering reduces equation (3) to the desired form (4), and it is clear that the filter unit must reject frequencies at and above $1/p$ to get rid of the terms due to the master grating, – and very probably lower, – as shown in Fig (6). This means that in the moire there is, by Gabor's principle, complete uncertainty within, (at least), a distance p.

If now we consider the condition where the spectrum of $\varepsilon(x)$ contains a compressive d.c. component sufficient to keep all specimen frequencies higher than $1/p$, so that $1/p$ is the appropriate filter band-width for a proper moire formation, then if two fringe maxima, (no mismatch), are P apart, we have the uncertainty on the measured fringe spacing is

$$\Delta P = \pm p \qquad \dots\dots\dots\dots\dots(24)$$

The uncertainty is readily appreciated by considering the location of a maximum:– a missing sinusoid could, if it had not been filtered out, change the position of a maximum to anywhere within half its (the sinusoid's) own period either side of the apparent location, depending upon the magnitude and phase of the sinusoid.

The strain is therefore measured with an uncertainty expressed by

$$\varepsilon + \Delta\varepsilon = \frac{p}{P\pm p} = \frac{p}{P}(1 \pm \frac{p}{P})^{-1} = \varepsilon(1 \pm \frac{p}{P}) \qquad \text{for } P\gg p \quad \dots\dots\dots(25)$$

i.e.

$$\frac{\Delta\varepsilon}{\varepsilon} = \pm \frac{p}{P} \qquad \dots\dots\dots\dots(26)$$

That is, the relative uncertainty is equal to the mean strain, (e.g., there is a $\pm 5\%$ uncertainty on a strain of 5%).

To examine now the behaviour for interpolated values, let us consider a "perfect" shift mechanism, which produces a shift of P' in a fringe position for a master grating translation of u'. The distance between the fringe in the shifted pattern and in the unshifted is clearly subject to the uncertainty $\pm p$.

We would calculate the average local strain as $\varepsilon' = \frac{u'}{P'}$, and

$$\varepsilon' + \Delta\varepsilon' = \frac{u'}{P'\pm p} \qquad \dots\dots\dots\dots(27)$$

It follows that as we reduce the 'gauge-length' P', so the relative uncertainty increases. In fact as P' tends to p, so an apparent value of infinite strain becomes possible, (i.e. the fringe might not have moved at all as far as we can tell), - and measurements become meaningless.

It is stressed at this point that complete ideality has been presumed throughout.

We therefore conclude that a Moire is not infinitely interpolable. There is a relationship of a truly Heisenbergian type:- the more we try to improve positional exactitude, (viz, "reduce gauge-length", "elicit higher frequency content", "look for detail"), the greater is the uncertainty in the measured value of strain.

For example, if we try to interpolate to 1/10th of a fringe in the case of a 5% average strain level, we would have, with $P' = P/10$, from (24),

$$\varepsilon' + \Delta\varepsilon' = \frac{u}{P'(1 \pm p/P')} = \frac{\varepsilon}{(1 \pm 10p/P)} = \frac{\varepsilon}{1 \pm .5} = \varepsilon' \{ ^{+100\%}_{-33.3\%} \} \quad \ldots\ldots (28)$$

In fact, the above considerations are academic if, in a particular case, we can be _certain_ (from, say a knowledge of the physics) that the moire spectrum does not extend to 1/p. The filtering action cannot then exclude any moire components, and thus the maxima can, under the above idealised conditions, be located with certainty.

The condition where the specimen grating spectrum does not extend below 1/p will be encountered in only a minority of cases, and the question arises in most cases as to what is the optimum filter cut-off frequency:- too high and extraneous terms will be included in and, distort, the moire,- too low, and we will exacerbate the uncertainty problem.

6.3 Consistency with Microscopy and the Sampling Theorem

If we now consider an idealised bar and space line spectrum grating, of pitch p, to be examined with an idealised travelling microscope, we can see immediately that, traversing across the lines, we could measure the position of every edge and thus specify the displacement twice every wavelength of the original, undistorted, grating. (Equally obviously, we could tell absolutely nothing about what had happened between two adjacent edges.) By the Sampling Theorem [3], we could reconstruct u(x), - and from it $\varepsilon(x)$, - provided it had no frequencies higher than 1/p. If u(x) had higher frequencies, we could not know about them.

We therefore conclude that even with idealised measurement conditions there would be a limit to the detail available in the specimen grating itself. The earlier postulation that by making a moire we thereby produce a continuous signal which can, in principle, be then subdivided ad infinitum is seen to be untenable.

6.4 A Moire as a Convenience

Of course, the suggested microscopy would be tedious in the extreme and we may consider a Moire to be a mere convenience which very easily supplies us with an overall picture of the deformations, but only at the cost of a loss of information. If we can resolve the grating lines, and record them in the original and distorted conditions, and, by suitable means, such as computer processing of video images, readily compare the two conditions on a grating-line position basis, then there is no point in forming a Moire.

6.5 Further Restrictions in the Incoherent Illumination Case

In the case where the illumination is incoherent, there is bound to be in the master grating a d.c. component of intensity whose magnitude a_0 is at least equal to a_1. The effect of this is to 'leave' the spectrum of the specimen grating situated around $\pm 1/p$, (as well as translating it to around zero frequency). We have so far imagined this spectrum to be moved up in frequency (by the convenient presence of a compressive mean strain), leaving only the master grating δ-functions standing at $\pm 1/p$. In general this will not be the case, and the untranslated spectrum in question will have significant frequencies below $1/p$. If we now filter at $1/p$ we will not form a proper moire. It wi'l be distorted by the inclusion of components of the specimen grating and the maxima will be shifted from their 'correct' positions. If the strain distribution has frequencies above $1/2p$, then we __must__ have overlap of the spectra. We therefore have no ground to stand on, unless we can say something, with some certainty, about the absence of significant components of the strain distribution at periods smaller than two grating pitches. The point then arises as to what extent this begs the whole question in the search for detail.

6.6 Improvements possible with coherent illumination

With a suitable master grating, (for example in Moire Interferometry), it is possible, with coherent illumination, to have no d.c. term (in complex amplitude) and pure sinusoidal form. There is therefore no specimen spectrum left standing around $1/p$. We can thus narrow the overlap question from periods of two pitches to periods of only one, – a considerable gain.

6.7 The Moire itself as a F.M. signal

It follows from the interpretation of Moire formation as a frequency translation of the specimen grating spectra, that the Moire itself has a Frequency-Modulated structure, the carrier frequency f_c (equn(14)) now being (nominally) zero. Just as it was shown (equation 23) that a mean strain ε_0 results in a change ε_0/p in grating carrier frequency, so in the moire the new carrier will not be zero but ε_0/p.

If a mismatched master grating of frequency $1/p_m$ is used, the result obtained from equns (23) and (8), is

$$I_m = I_0 + b_1\cos\left[2\pi x(1/p - 1/p_m) - \frac{2\pi}{p}\{\varepsilon_0 - \sum_r \frac{\varepsilon_r}{2\pi f_r}\sin(2\pi f_r x + \gamma_r)\}\right] \quad\ldots\ldots(29)$$

The carrier frequency is now

$$f_{co} = (1/p - 1/p_m - \varepsilon_0/p) \qquad \ldots\ldots\ldots\ldots\ldots(30)$$

This is illustrated in Fig.7.

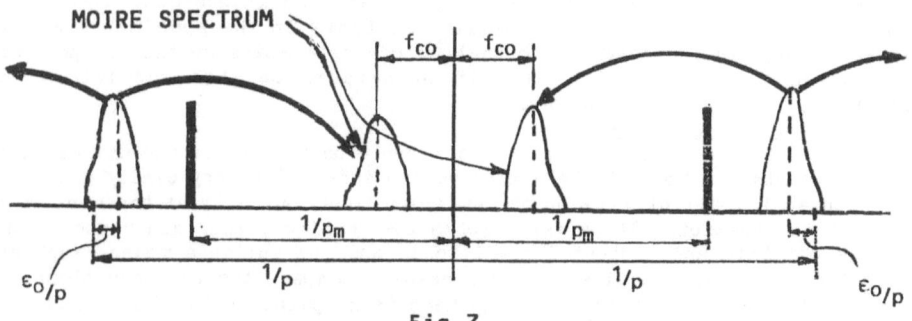

Fig.7

6.8 Sideband Foldover in the Moire

This is quite the most serious restriction yet encountered in the search for detail. From Fig.7 it will be appreciated that if the spectrum of $\varepsilon(x)$ contains significant components at frequencies higher than the eventual effective carrier f_{co}, then foldover will occur. This means that we cannot, from the moire, establish the individual specimen spectra at $1/p$ and $-1/p$, and therefore we cannot establish $\varepsilon(x)$.

This is not to say that the Discrete Moire Law no longer applies. Despite foldover it will still be true that, from equation (4), the maxima will occur at intervals of $u(x) = p$. The problem arises when trying to interpolate between maxima, and may be related to the ambiguity in the arcosine in equation (9). We are now required, if we wish to interpolate, to be quite certain that $\varepsilon(x)$ contains no significant frequencies higher than f_{co}, – a most serious limitation indeed.

An obvious palliative is to make f_{co} as high as possible by introducing the maximum possible mismatch.

6.9 Possible Solutions to Foldover and Overlap

i)   Foldover: a large mismatch will, as noted above, help to avoid the condition: furthermore, the finer the grating pitch, the higher the mismatch frequency can be.

ii) overlap: as already pointed out, the use of a master grating with no d.c. alleviates the overlap problem and the reduction of grating pitch to the practicable minimum manifestly assists in this aspect also.

7. PRACTICAL CONSIDERATIONS: DETECTABILITY AND NOISE

So far we have considered only ideal systems. We have presumed that the gratings are perfectly formed and that we have a noise-free intensity-measuring instrument of infinite dynamic range.

Considering firstly the restrictions imposed by practical instruments:-

7.1 Relative levels of intensity fluctuation

As pointed out above, the moire has the structure of a frequency-modulated wave. It is important to be able to establish typical values of the Bessel functions, which we can do, since

$$\mu_r = \frac{ks_r}{f_{sr}} \qquad \text{and} \qquad k = \frac{1}{p} \qquad \dots\dots\dots\dots(31)$$

(again we restrict consideration to gratings consisting only of the d.c. plus a fundamental).

We have, in the terminology of equation (21)

$$\mu_r = \frac{\varepsilon_r}{pf_{sr}} \qquad \dots\dots\dots\dots(32)$$

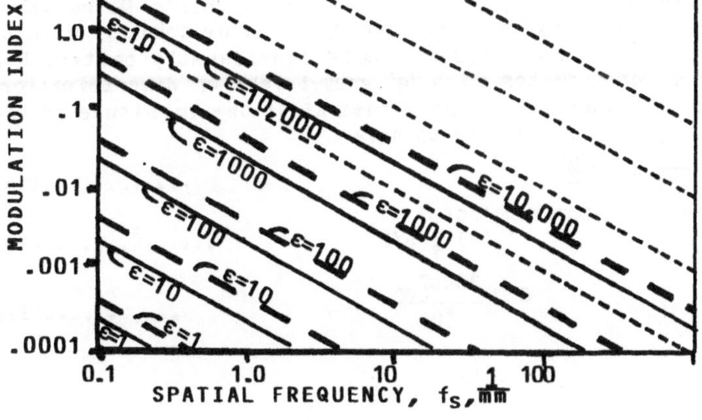

Fig.8

Equation (32) is of considerable significance in assessing the detail detectable in practice. It allows us to determine the relative magnitudes of the variation in I_m due to a component of $\varepsilon(x)$ of any amplitude and frequency, for any grating pitch:-

Values of μ for various values of these parameters are shown in Fig 8

Table I shows J_o, J_1, and J_2 for various values of μ.

TABLE I

μ	.001		.01		.05		.1		.5			1.2			1.5		
	J_o	J_1	J_o	J_1	J_o	J_1	J_o	J_1	J_o	J_1	J_2	J_o	J_1	J_2	J_o	J_1	J_2
	1	10^{-3}	1	.06	1	.025	1	.05	.94	.24	.03	.67	.05	.06	.5	.56	.23

We can see from the table and from Fig (4) that, for $\mu < 1$, as the value of μ decreases, so does the amplitude of the sidebands (J_1) in comparison to the carrier (J_o).

From equation (29) and the principle behind equn.(20), it would be possible to deduce the exact form of the harmonic structure of I_m. We shall presume that the values of μ are small enough that double and higher products (e.g., terms in $J_n J_m$ etc) are significant only for $n=0$, $m=0$ or ± 1, $m=0$, $n=0$ or ± 1, etc and $J_o(\mu_r) \doteqdot 1$ for all r. This is the 'low-index case' [12], and givs the simplified result, from equation (28),

$$I_m = I_o + b_1 \left\{ \cos 2\pi f_{co} x + \sum_r [J_1(\mu_r)\cos\{2\pi(f_{co} + f_{sr})x + \gamma_r\} + J_{-1}(\mu_r)\cos\{2\pi(f_{co} - f_{sr})x - \gamma_r\}] \right\} ..(33)$$

Because $J_1(\mu) = -J_{-1}(\mu)$, the fluctuation due to a single frequency f_{sr}, is given by

$$A_r = 2J_1(\mu_r)\sin 2\pi f_{co} x . \sin(2\pi f_{sr} x + \gamma_r) \qquad \ldots\ldots\ldots(34)$$

Recalling the origin of I_o and b_1, from equation (3), viz,

$$I_o = I_1 a_o^2$$

$$b_1 = I_1 a_1^2/2$$

and that, for an amplitude grating $a_o \geq a_1$, we have,

$$I_o \geq 2b_1$$

To detect in the moire any effect whatsoever of a component of $\varepsilon(x)$ of frequency f_{sr} we must therefore have a system capable of discriminating relative intensity levels of $2J_1(\mu_r)$, (equation 34), within a Moire whose maximum intensity level will exceed, relatively, 3. If we actually wish to know anything about it we must discriminate considerably better. For example if we use a densitometer with 'g' grey-levels of discrimination, and we wish not to make an error of more than E% in the amplitude of the component of frequency f_{sr}, we require to have

$$g = \frac{3}{2J_1(\mu_r)} \times \frac{100}{E} \qquad \ldots\ldots\ldots\ldots(35)$$

For $\mu < 0.5$, we have a good approximation,

$$J_1(\mu) \doteqdot \mu/2 \qquad \ldots\ldots\ldots\ldots(36)$$

This gives us

$$g = \frac{300 p f_{sr}}{\varepsilon_r} \qquad \ldots\ldots\ldots\ldots(37)$$

Thus we now have a specification for the performance requirement of the interrogation system in terms of the detail we wish to elicit. (Or alternatively we have set limits on the detail available from a particular system). Again we can note, from equn.(37), the relative attenuation as the spatial frequency increases, and the benefit of a finer grating pitch).

Equation(37) will, in general, yield a <u>minimum</u> necessary value for g. A knowledge of the a_o's and a_1's would be required in order to be more precise. Other components too will tend to increase the maximum intensity, so helping to submerge further our component at f_{sr}

However, in the particular case when the master grating has no d.c. component, $I_o = b_1$. Also, in this case, we effectively half the specimen grating pitch. This gives

$$g' = \frac{100pf_{sr}}{\varepsilon_r \cdot E} \qquad \ldots\ldots\ldots\ldots\ldots(38)$$

an improvement of 3 for the same specimen grating pitch p.

Whether the low-index case holds in a particular instance can only be determined by a spectrum analysis. If there is general non-compliance then the analysis is virtually unmanageable. It transpires that strong low-frequency components (but not mean strain or mismatch) will conspire to 'hide' high frequencies.

7.2 The effects of noise

"Noise" in a Moire can be a considerable problem, especially when coherent illumination is used or if the grating has degraded, – for example after exposure to high temperature environment. It is tempting to filter out the raggedness, but we then lose information concerning frequencies of $\varepsilon(x)$ which may exist at and above the cut-off frequency, and we introduce more uncertainty. On the other hand the noise,if not removed, distorts the moire, – and so a dilemma is posed.

8. THE DISTORTING EFFECTS OF INCOHERENT LENS FILTERING

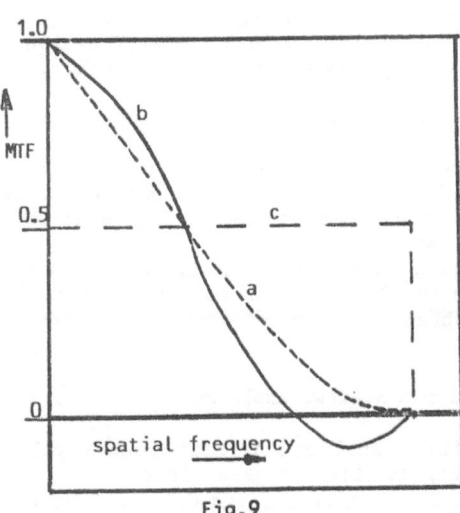

Fig.9

As earlier indicated, a lens may be used to perform the filtering necessary for the extraction of a moire. This may be done either by using a small aperture, or by defocussing a lens with a large aperture. These actions are represented in Figs 9a and 9b, (after [16]), which are typical MTF's for these conditions. It is clear that if the spectrum of the Moire contains anything but very low frequencies, then the higher frequencies are going to be relatively attenuated. This will result in a distortion of the moire so that even the Discrete Law will not hold.

It is clear that incoherent lens filtering is an unsatisfactory method of producing a moire. Fig.9c shows the MTF of a lens when coherent filtering is used in the focal plane: obviously this solves the above problem of relative attenuation.

9. CONCLUSIONS

The moire phenomenon has been examined with regard, in particular, to its suitability as a tool for eliciting detail of a strain field. It is found that

i) A moire involves, intrinsically a filtering process, and as a result, it is not in general an infinitely interpolable continuum in any meaningful sense.

ii) A moire is a mere convenience: the deformed grating contains much more information than the moire, and if there is instrumentation which allows convenient, direct, interrogation of the grating, then this approach is much to be preferred.

iii) The setting of the filter cut-off frequency may be a compromise between uncertainty on the one hand and distortion due to spectrum overlap on the other. Some prior knowledge of the upper frequency limit of $\varepsilon(x)$ is helpful in this regard.

iv) A moire is particularly susceptible to the phenomenon of sideband foldover and the concomitant ambiguity. Foldover can cause serious restriction on the detail that is available. The extent of foldover is minimised by high mean strain and/or the use of maximum mismatch. Finer grating pitch allows higher mismatch frequencies.

v) The problem of spectrum overlap is minimised by the use of the smallest possible specimen grating pitch, and by using a master with no d.c. term.

vi) It is the nature of a moire that the variations in intensity resulting from the higher frequencies of $\varepsilon(x)$ are attenuated relative to those due to the lower frequencies, thus militating against the elucidation of detail.

vii) It is possible at least in the 'low index' case, to specify the detectivity of an interrogating instrument against a specification of the amplitudes and frequencies of the components of $\varepsilon(x)$ required to be resolved.

vii) The use of coherent light and a master with no d.c. term improves the detectability of all components.

ix) The presence of noise in the moire introduces either uncertainty or distortion. There is no complete solution to the noise problem, other than to minimise it in the first place, but the use of fringe-shifting – or related methods – can be of assistance.

xi) Filtering by a lens with incoherent light is an unsatisfactory method of moire formation, causing even the Discrete Law to fail. The use of coherent light overcomes the problem.

10. REFERENCES

1. Sciamarella CA: The Moire Method – A Review. Expl. Mech, 22,(11), 418–433 (1982).
2. Sciamarella CA, Rao, MPK: Failure Analysis of Stainless Steel at Elevated Temperatures. Expl. Mech, 19,(11), 389–398 (1979).
3. Gaskill JD: Linear Systems, Fourier Transforms, and Optics. J Wiley and Sons (1978).
4. Sciamarella CA: Basic Optical Law in the Interpretation of Moire Patterns Applied to the Analysis of Strains – Part I. Expl. Mech,

5,(5), 154–160 (1965).

5. Sciamarella CA, Lurowist N: Multiplication and Interpolation of Moire Fringe Orders by Purely Optical Techniques. Jnl. Appl. Mech. 425–430, (June 1967).

6. MacKenzie PM, Walker CA, McKelvie J: Experience with a Heterodyne Moire Interferometer. These Proceedings.

7. Reid GT, Dixon RC, Nesser HI: Absolute and Comparative Measurements of Three-dimensional Shape by Phase-Measuring Moire Topography. Optics and Laser Technology 315–319, (Dec 1984).

8. Post D: Moire Fringe Multiplication with a Non-symmetrical Doubly Blazed Grating. App. Opt. 10 (4), 901–907 (1971).

9. Post D: Moire Interferometry at VPI and SU. Exp-Mech, 23 (2), 203–210, (1983).

10. McDonach A, McKelvie J, MacKenzie PM, Walker CA: Developments in Moire Interferometry and Applications.

11. Yu FTS: Optics and Information Theory. J Wiley and Sons, (1976).

12. Rowe HE: Signals and Noise in Communication Systems. D Von Nostrand Co. Inc. (1965).

13. Bennett W R: Introduction to Signal Transmission. McGraw-Hill Book Co. (1970).

14. British Association Mathematical Tables. Volume X. Bessel Functions, (Parts I and II). Cambridge Univ. Press 1952.

15. Sturgeon DLG: Analysis and Synthesis of a Moire Photo-Optical System. Exp. Mech. 7 (8), 346–352, (1967).

16. Born M, Wolf E: Principles of Optics. Pergamon Press, (1980).

INTERPRETATION OF MOIRÉ EFFECT FOR CURVATURE MEASUREMENT OF SHELLS

R. RITTER and M. HAHNE

Technical University of Braunschweig, Federal Republic of Germany

1. INTRODUCTION

For experimental dimensioning of objects by optical whole field methods, procedures are preferred, which directly lead to strain and curvature values in order to get stress or bending moment distribution.

One well-known principle |1|-|4| consists of superposing two identical grating images, related to the deformed state of the considered object, which are shifted relatively to each other by a small amount.

In the case of curvature measurement the reflection moiré effect will be applied for example. Thereby the records of two grating images are superposed, which come into existence observing a reference grating via the reflective surface of the object.

The published arrangements and relationships, which are based on this procedure, are suitable for analysis of beam and plate problems, whereby the slope w_x of the object surface is assumed to be so small, that for example the curvature

$$\kappa_x = \frac{w_{xx}}{(1 + w_x^2)^{2/3}} \tag{1}$$

can be approximated by the second derivatives w_{xx} of the contour function $w(x,y)$ of this surface. Then it is sufficient to produce only one field of moiré fringes with grating lines, which are orientated perpendicular to the shifting direction for determining w_{xx} or w_{yy} and parallel to this, if w_{xy} is desired.

In the following an interpretation of this moiré effect is presented, which is applicable for analysis of shell problems |5|.

2. EXPERIMENT

Fig. 1 shows a cross-section of the optical arrangement based on the reflection moiré principle. It consists of the shell S, the grating G and a recording camera C. For describing the geometrical relationship between these three elements, a Cartesian x-y-z-reference co-ordinate system (R.P.) is introduced. The points and their co-ordinates in the grating plane (G.P.) are signed by bars, in the image plane (I.P.) by double bars. The z-axis presents the common normal of the grating and image plane. The deflection $w(x,y)$ of the shell vertical to the reference plane is measured by the z-component.

Furthermore three distances are noted: d (reference plane - grating plane), a (reference plane - objective plane (O.P.) of the recording camera) and a' (objective plane - image plane).

The grating consists of alternating black and transparent (white) straight lines of the same width, which are orientated parallel to the

Wieringa, H (ed), Experimental Stress Analysis.
© *1986. Martinus Nijhoff Publishers, Dordrecht.*

$\bar{x}$- or $\bar{y}$-axis (grating pitch $\bar{p}_x$ respectively $\bar{p}_y$), Fig. 2.

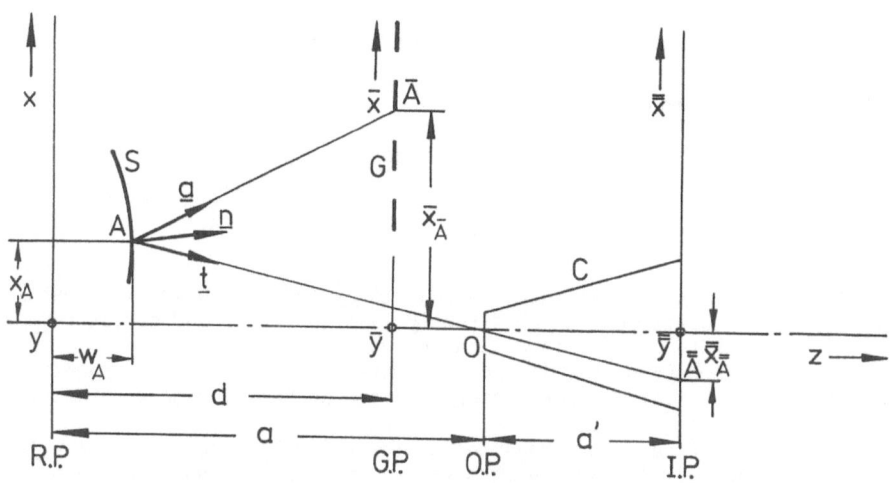

FIGURE 1. Optical system.
R.P. reference plane, G.P. grating plane, O.P. objective plane of the recording camera, I.P. image plane

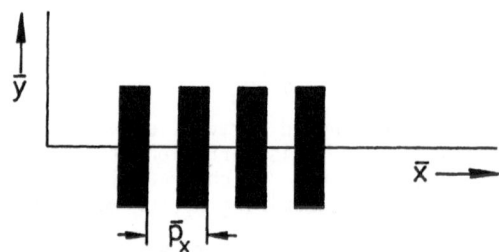

FIGURE 2. Line grating.

A should be the considered point of the object. Then from the point $\bar{\bar{A}}$ of the image plane the point $\bar{A}$ of the grating is observed via the reflecting surface of the object point A.

Now the curvature moiré effect comes into existence, if two identical grating images are superposed and shifted relatively to each other in the direction of the $\bar{x}$- respectively $\bar{y}$-axis by a small amount $\Delta\bar{x}$ or $\Delta\bar{y}$.

Dark moiré fringes can be observed at the places, where dark grating lines from one of both images and transparent lines of the other are superposed. On the other hand bright moiré fringes result from the super-position of dark and transparent lines of both grating images. In the following the centers of the bright fringes are considered.

3. MOIRÉ EFFECT

If one of both grating images is shifted for example in the direction of the $\bar{\bar{x}}$-axis of the image plane by $-\Delta\bar{\bar{x}}$, the observed moiré fringe will be related to the point $\bar{A}_0(\bar{\bar{x}},\bar{\bar{y}})$ of the unshifted image and the point $\bar{A}(\bar{\bar{x}} + \Delta\bar{\bar{x}},\bar{\bar{y}})$ of the shifted one, Fig. 3. Therefore the fringe cannot be attached clearly to one image point but only to the intervall of the shifting direction, which in this case means the value $\Delta\bar{x}$.

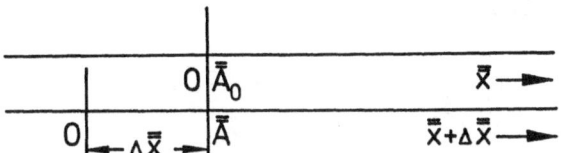

FIGURE 3. Co-ordinates of the unshifted and the shifted grating images.

Now the grating lines are defined by the shelf numbers

$$N_x = \frac{\bar{x}}{\bar{p}_x} \tag{2}$$

and

$$N_y = \frac{\bar{y}}{\bar{p}_y} \tag{3}.$$

From the difference of the shelf numbers of the superposed grating lines follows the order of the resulting moiré fringe.

If for example the grating lines are orientated perpendicular to the $\bar{x}$-axis and one grating image will be shifted in the direction of the x-axis, the order M_{xx} of the araising moiré fringe is given by the difference of the shelf number N_x of the shifted and N_{xo} of the unshifted image

$$M_{xx} = N_x - N_{xo} \tag{4}.$$

The first index of M marks the direction perpendicular to the grating orientation and the second index the direction of the shifting.

Eq.(2) substituted into eq.(4) leads to

$$M_{xx} = \frac{1}{\bar{p}_x} \left[\bar{x}(\bar{\bar{x}} + \Delta\bar{\bar{x}},\bar{\bar{y}}) - \bar{x}(\bar{\bar{x}},\bar{\bar{y}}) \right] \tag{5}.$$

For the cases of the other combinations between grating line orientation and shifting direction one gets

$$M_{xy} = \frac{1}{\bar{p}_x} \left[\bar{x}(\bar{\bar{x}},\bar{\bar{y}} + \Delta\bar{\bar{y}}) - \bar{x}(\bar{\bar{x}},\bar{\bar{y}}) \right] \tag{6}$$

$$M_{yx} = \frac{1}{\bar{p}_y} \left[\bar{y}(\bar{\bar{x}} + \Delta\bar{\bar{x}},\bar{\bar{y}}) - \bar{y}(\bar{\bar{x}},\bar{\bar{y}}) \right] \tag{7}$$

and

$$M_{yy} = \frac{1}{\bar{p}_y} \left[\bar{y}(\bar{\bar{x}},\bar{\bar{y}} + \Delta\bar{\bar{y}}) - \bar{y}(\bar{\bar{x}},\bar{\bar{y}}) \right] \tag{8}.$$

The moiré fringes describe differences of the grating co-ordinates. Perpendicular to the shifting direction of the grating images the fringes can be localized and in the direction of them they are observable at different co-ordinates of the unshifted and shifted state. Then only an intervall can be specified, to which the moiré fringe is related. The limits of this intervall are equal to the co-ordinates, where a fringe appears.

Dividing the eqs.(5)to(8) by the shifting values, then for example follows from eq.(5) with $\Delta \bar{x}$

$$\frac{M_{xx}}{\Delta \bar{\bar{x}}} \bar{P}_x = \frac{\bar{x}(\bar{\bar{x}} + \Delta \bar{\bar{x}}, \bar{\bar{y}}) - \bar{x}(\bar{\bar{x}}, \bar{\bar{y}})}{\bar{\bar{x}} + \Delta \bar{\bar{x}} - \bar{\bar{x}}} \qquad (9).$$

Furthermore eq.(9) can be transformed by the mean value theorem of the differential calculus:

$$\frac{M_{xx}}{\Delta \bar{\bar{x}}} \bar{P}_x = \frac{\partial \bar{x}}{\partial \bar{x}(\bar{\bar{x}}_o, \bar{\bar{y}})} \quad \text{with } \bar{\bar{x}}_o \in [\bar{\bar{x}}, \bar{\bar{x}} + \Delta \bar{\bar{x}}] \qquad (10).$$

The same procedure applied to the eqs.(6)to(8) leads to

$$\frac{M_{xy}}{\Delta \bar{\bar{y}}} \bar{P}_x = \frac{\partial \bar{x}}{\partial \bar{y}(\bar{\bar{x}}, \bar{\bar{y}}_o)} \quad \text{with } \bar{\bar{y}}_o \in [\bar{\bar{y}}, \bar{\bar{y}} + \Delta \bar{\bar{y}}] \qquad (11)$$

$$\frac{M_{yx}}{\Delta \bar{\bar{x}}} \bar{P}_y = \frac{\partial \bar{y}}{\partial \bar{x}(\bar{\bar{x}}_o, \bar{\bar{y}})} \qquad (12)$$

$$\frac{M_{yy}}{\Delta \bar{\bar{y}}} \bar{P}_y = \frac{\partial \bar{y}}{\partial \bar{y}(\bar{\bar{x}}, \bar{\bar{y}}_o)} \qquad (13).$$

The related orders of the moiré fringes and the differential quotients of the eqs.(10)to(13) can be combined to the functial matrix

$$\underline{\underline{F}} = \begin{pmatrix} \partial \bar{x}/\partial \bar{\bar{x}} & \partial \bar{x}/\partial \bar{\bar{y}} \\ \partial \bar{y}/\partial \bar{\bar{x}} & \partial \bar{y}/\partial \bar{\bar{y}} \end{pmatrix} = \begin{pmatrix} M_{xx}\bar{P}_x/\Delta \bar{\bar{x}} & M_{xy}\bar{P}_x/\Delta \bar{\bar{y}} \\ M_{yx}\bar{P}_y/\Delta \bar{\bar{x}} & M_{yy}\bar{P}_y/\Delta \bar{\bar{y}} \end{pmatrix} \qquad (14).$$

4. GEOMETRICAL RELATIONSHIPS BETWEEN GRATING, OBJECT AND IMAGE

The relationship between the incident and the reflected light, Fig. 1, depends on the law of deflection and can be described by the direction vectors $\underline{a}$ from A to $\bar{A}$ and $\underline{t}$ from A to $\bar{\bar{A}}$ via O including the normal vector

$$\underline{n} = (-w_x, -w_y, 1) \qquad (15)$$

on the surface of S in A by

$$\underline{a} = 2\underline{n} \frac{\underline{n} \cdot \underline{t}}{\underline{n} \cdot \underline{n}} - \underline{t} \qquad (16)$$

or $$\underline{a} = h \underline{n} - \underline{t} \qquad (17)$$

if $$h = 2 \frac{\underline{n} \cdot \underline{t}}{\underline{n} \cdot \underline{n}} \qquad (18).$$

From the local vectors

$$\bar{\underline{A}} = (\bar{x}, \bar{y}, a+a') \qquad (19)$$

and $$\underline{0} = (0,0,a) \qquad (20)$$

follows the direction vector

$$\underline{t} = \bar{\underline{A}} - \underline{0} \qquad (21).$$

Eqs.(19) and (20) substituted into eq.(21) yields with the unit length for the local distance a'

$$\underline{t} = (\bar{\bar{x}}, \bar{\bar{y}}, 1) \qquad (22).$$

Now the local vector

$$\underline{A} = (x, y, w(x,y)) \qquad (23)$$

can be expressed by

$$\underline{A} = \underline{0} - a * \underline{t} \qquad (24).$$

From the z-component of eq.(24) follows with the eqs.(20) and (22)

$$a* = a - w \qquad (25).$$

The local vector

$$\bar{\underline{A}} = (\bar{x}, \bar{y}, d) \qquad (26)$$

of the considered grating point depends on $\underline{a}$ and $\underline{A}$ by

$$\bar{\underline{A}} = \underline{A} + \mu a \qquad (27).$$

Now from the z-component of eq.(27) follows

$$\mu = \frac{d - w}{h - 1} \qquad (28)$$

or $$\mu = \frac{d*}{h - 1} \qquad (29)$$

if $$d* = d - w \qquad (30).$$

Eqs.(16),(24),(25) and (28) substituted into eq.(27) leads to

$$\bar{\underline{A}} = \underline{0} - a * \underline{t} + \frac{d*}{h - 1}(h \underline{n} - \underline{t}) \qquad (31).$$

The x- and y-components of eq.(31) describe the wanted relationships between grating, object and image co-ordinates:

$$\begin{bmatrix} \bar{x} \\ \bar{y} \end{bmatrix} = -a* \begin{bmatrix} \bar{\bar{x}} \\ \bar{\bar{y}} \end{bmatrix} + \frac{d*}{h-1} \left[h \begin{bmatrix} -w_x \\ -w_y \end{bmatrix} - \begin{bmatrix} \bar{\bar{x}} \\ \bar{\bar{y}} \end{bmatrix} \right]$$ (32)

or

$$\begin{bmatrix} \bar{x} \\ \bar{y} \end{bmatrix} = -(d-w)\frac{h}{h-1} \begin{bmatrix} w_x \\ w_y \end{bmatrix} - \left((a-w)+\frac{d-w}{h-1}\right) \begin{bmatrix} \bar{\bar{x}} \\ \bar{\bar{y}} \end{bmatrix}$$ (33).

The co-ordinates of the grating point $\bar{A}$, the image of it is the point $\bar{\bar{A}}$, are nonlinear functions of the slopes and his z-component respectively the deflection w of the object S.

5. SECOND DERIVATIVES OF THE CONTOUR FUNCTION

Now the relationship between the second derivatives of the contour function and the orders of the moiré fringes is deducted.

The derivatives of the grating co-ordinates $\bar{x}$ and $\bar{y}$ in eq.(33) by the image co-ordinates $\bar{\bar{x}}$ and $\bar{\bar{y}}$ lead to the following functional matrix

$$\underline{\underline{F}} = \underline{\underline{A}} + \underline{\underline{B}}\underline{\underline{W}}$$ (34)

with the wanted matrix

$$\underline{\underline{W}} = \begin{bmatrix} w_{xx} & w_{xy} \\ w_{yx} & w_{yy} \end{bmatrix}$$ (35)

which includes the second derivatives of the contour function. $\underline{\underline{A}}$ and $\underline{\underline{B}}$ are matrices, which depend on $w,w_x,w_y,\bar{\bar{x}}$ and $\bar{\bar{y}}$. Eq.(34) yields

$$\underline{\underline{W}} = \underline{\underline{B}}^{-1}(\underline{\underline{F}} - \underline{\underline{A}})$$ (36).

If $\underline{\underline{A}}$ and $\underline{\underline{B}}$ are known and $\underline{\underline{F}}$ is determined from the orders of the moiré fringes, the values of $\underline{\underline{W}}$ follow from eq.(36).

The most complicated procedure consists in computing $\underline{\underline{A}}$ and $\underline{\underline{B}}$. For this purpose several other matrices are needed: first

$$\underline{\underline{U}} = (U_{ij}) = ((w_i + \bar{\bar{x}}_i)w_j) = \begin{bmatrix} (w_x + \bar{\bar{x}})w_x & (w_x + \bar{\bar{x}})w_y \\ (w_y + \bar{\bar{y}})w_x & (w_y + \bar{\bar{y}})w_y \end{bmatrix}$$ (37)

and the unit matrix

$$\underline{\underline{1}} = \begin{bmatrix} 1 & 0 \\ 0 & 1 \end{bmatrix}$$ (38)

which lead to

$$\underline{\underline{A}} = -(a^* + \frac{d^*}{h-1})\underline{\underline{1}} - (a^* \frac{h}{h-1} + \frac{2d^*}{n^2(h-1)^2})\underline{\underline{U}} \tag{39}$$

whereby $n^2 = \underline{n} \cdot \underline{n}$. Furthermore

$$\underline{\underline{R}} = (R_{ij}) = ((w_i + \bar{\bar{x}}_i)\bar{\bar{x}}_j) = \begin{bmatrix} (w_x + \bar{\bar{x}})\bar{\bar{x}} & (w_x + \bar{\bar{x}})\bar{\bar{y}} \\ (w_y + \bar{\bar{y}})\bar{\bar{x}} & (w_y + \bar{\bar{y}})\bar{\bar{y}} \end{bmatrix} \tag{40}$$

from which together with $\underline{\underline{U}}$ of eq.(37) and $\underline{\underline{1}}$ of eq.(38) the matrix

$$\underline{\underline{B}} = a^* d^* \frac{h}{h-1}\underline{\underline{1}} + \frac{2d^*}{n^2(h-1)^2} a^*(\underline{\underline{R}} + h\underline{\underline{U}}) \tag{41}$$

can be determined.

In the special case of small deformations and deflections of the object surface which often is given by plate problems, and small angles between the optical axis and the projection rays follow with

$$\left. \begin{array}{ll} w \ll d & w \ll a \\ w_x \ll 1 & w_y \ll 1 \\ \bar{\bar{x}} \ll a' & \bar{\bar{y}} \ll a' \end{array} \right\} \tag{42}$$

the simplifications

$$\left. \begin{array}{ll} a^* = a & d^* = d \\ \underline{n} \cdot \underline{n} = 1 & \underline{n} \cdot \underline{t} = 1 \\ & h = 2 \\ \underline{\underline{U}} = \underline{\underline{0}} & \underline{\underline{R}} = \underline{\underline{0}} \\ \tilde{\underline{\underline{A}}} = -(a+d)\underline{\underline{1}} & \tilde{\underline{\underline{B}}} = 2ad\,\underline{\underline{1}} \end{array} \right\} \tag{43}.$$

Then the matrix $\underline{\underline{W}}$ of eq.(36) changes into

$$\tilde{\underline{\underline{W}}} = \frac{1}{2ad}(\underline{\underline{F}} + (a+d)\,\underline{\underline{1}}) \tag{44}.$$

This relationship is very practicable, as it consists only of values, which are independent of the considered point.

6. NUMERICAL SIMULATION

The developed theory was tested by applying the eqs.(36)-(41) to a numerical example. This procedure is advantageous compared to an experiment, as it needs no optical arrangement and avoids measurement or analysis mistakes. Furthermore the transfer from the experimental data for the analysis in a computer is not necessary.

In this case the example consisted in the surface function of a sphere, which is known by

$$w(x,y) = z_0 + \sqrt{R^2 - x^2 - y^2}$$ (45).

R means the radius of the object and z_0 describes the position of the sphere's center. From eq.(45) follow the first and second derivatives w_x, w_y, w_{xx}, w_{xy} and w_{yy}.

For determining the elements of the functional matrix $\underline{\underline{F}}$ the eq.(33) is applied to a triple of the points $\bar{\bar{A}}_0, \bar{\bar{A}}_1$ and $\bar{\bar{A}}_2$ of the image plane, Fig. 4, the position of them is given by the co-ordinates $(\bar{\bar{x}}, \bar{\bar{y}})$, $(\bar{\bar{x}} + \Delta\bar{\bar{x}}, \bar{\bar{y}})$ and $(\bar{\bar{x}}, \bar{\bar{y}} + \Delta\bar{\bar{y}})$.

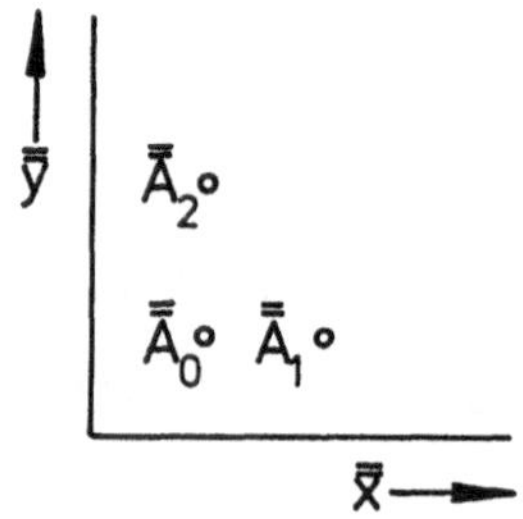

FIGURE 4. Considered points for determination of the elements of $\underline{\underline{F}}$.

The values $\Delta\bar{\bar{x}}$ and $\Delta\bar{\bar{y}}$ mean the simulated shifting. For example follows F_{xx} in the same manner by $(x(\bar{\bar{x}} + \Delta\bar{\bar{x}}, \bar{\bar{y}}) - x(\bar{\bar{x}}, \bar{\bar{y}}))/\Delta\bar{\bar{x}}$ as in the case of the moiré fringe order.

The calculation of the values in eq.(33) is not problematic, as they are determined from eq.(45) and their derivatives.

Fig. 5 shows the results of the simulation. They are related to the moving of the point triple along the line $\bar{\bar{y}} = 0$. The radius was chosen to be 0.05 a and the centre of the sphere at the point -0.05 a.

In the figure the top curve means the function w of the object point, observed from the image point $\bar{\bar{A}}_0(\bar{\bar{x}}, 0)$. Below this the function w_{xx} is plotted, which was derived from eq.(45).

The results of the determination of the functional matrices in eq.(36) respectively eq.(44) are shown by two systems of symbols. The horizontal line describes always the amount of the simulated image shifting which is reduced to the tenth part from figure to figure. The type of the evaluation is signed by a vertical line for the results from eq.(36) and a lying cross for the results from eq.(44). Although the curvature κ is constant in this case, the value w_{xx} becomes greater with increasing distance from the sphere's center, as it depends by eq.(1) on the slope of the object surface. For $\bar{\bar{x}} = 0$ follows $w_x = 0$ and $w_{xx} = \kappa = 20/a$. The plot of a symbol is always given at that point, where the simulated moiré fringe order gets an integer.

The figures show, that the results from the simplificated relationship of eq.(44) much earlier differ from the theoretical function than the values determined by the eq.(36) of the presented new theory.

Furthermore a decrease of the shifting amount causes an improvement of the results from eq.(36) but it does not improve the second derivatives of w, approximated by eq.(44).

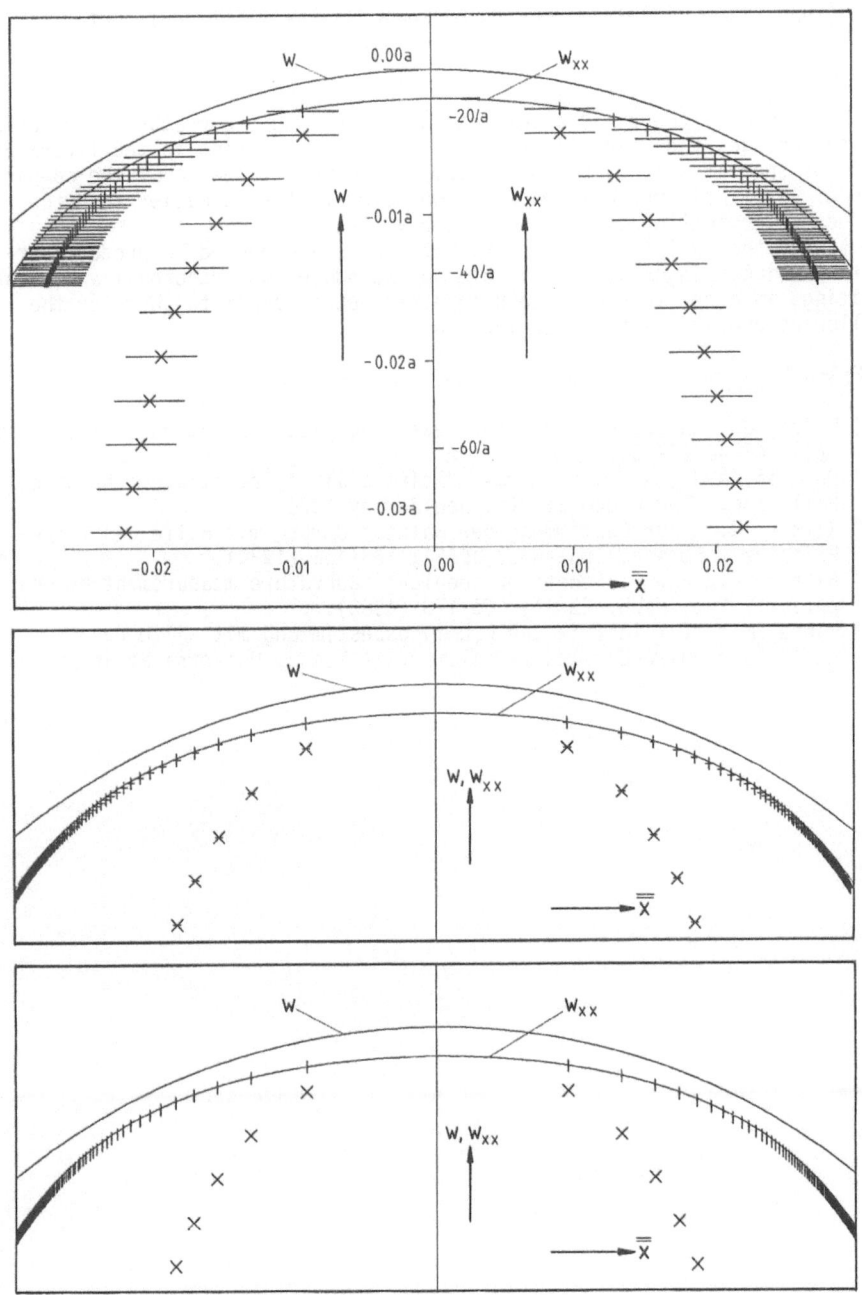

FIGURE 5. Numerical results of the simulated experiment.

To get the same fringe density by reducing the shifting amount, the density of the grating lines must increase.

7. CONCLUSION

The proposed theory is a complete analysis of the curvature moiré effect, which includes the simplificated solutions, one can find in the literature for the special case of plate problems. It demonstrates, that the comparison of a nominal and real image of two moiré fields is easier than the direct determination of the second derivatives.

As for the analysis now two grating fields are needed to produce curvature moiré fringes it is suitable to introduce two orthogonal and colored gratings in order to record both together but to separate them for the following operations by filtering process.

REFERENCES

|1| Heise, U.: "A moiré method for measuring plate curvature", Exp. Mech. 7(1), 47-48 (1966).

|2| Durelli, A.J., and V.J. Parks: "Moiré analysis of strain", Prentice-Hall, Inc., Englewood Cliffs, New Jersey 1970.

|3| Ritter, R.: "Zur Bestimmung der Balkenkrümmung mit Hilfe des Moiré-Prinzips", Forsch. Ing.-Wes. 46(5), 164-166 (1980).

|4| Ritter, R., and R. Schettler-Koehler: "Curvature measurement by moiré effect", Exp. Mech. 23(2), 165-170 (1983).

|5| Hahne, M.: "Zur Theorie der Krümmungsbestimmung mit Hilfe des Reflexions-Moiré-Effekts", Diplomarbeit Techn. Universität Braunschweig (1984).

OPTICAL METHOD OF STRAIN MEASUREMENT. APPLICATION TO STUDY OF CIRCULAR BENDING OF A BEAM IN THE LARGE STRAIN RANGE.

F. BREMAND and A. LAGARDE

LABORATOIRE DE MECANIQUE DES SOLIDES - Unité Associée au C.N.R.S. 40, Avenue du Recteur Pineau   86022 POITIERS CEDEX FRANCE

1. INTRODUCTION

For a long time researchers have shown interest in the measurements of large deformations on the surface of an object.

The technique of wire resistance extensometers was extended to the measurements of relatively large deformations of annealed constantan. Three-gauge rosettes enabled the access to the three parameters of Mohr Circle of deformation with a good linearity and a good sensitivity for up to 20 %. However they present the inconvenience of not being able to resist successive alternating deformations and in addition the measurement base is quite large (in the regions of 4×10^2 mm^2).

The techniques of moiré have also been used with the aim of measuring large deformations. These deformations are obtained by different ways [1] : graphic derivation, rotation and variation of the pitch of the reference grating, moiré of moiré. These methods which are easy to use for small deformations became extremely difficult to apply in the case of large deformations.

In 1964 Durelli and Parks [2] studied the relation displacement-deformation with the aim of extending the possibilities offered by the method of moiré in the domaine of large deformations. Three years later Duffey and Mesmer [3] obtained the components of the deformation tensor and the rotation of the rigid solid by calculating the partial derivatives of the displacements.

In 1970, Ebbeni proposed three different methods based on the use of figures of moiré, in the determination of large deformations [4]. He assumes a domain called "Linearisation of displacements" whereby the deformations of the model are taken to be homogeneous and surrounding at least four mesh of the initial crossed meshing.

The first method is based on the observation of two crossed gratings (reference grating and deformed grating) by means of two optical systems, sufficiently withdrawn and carefully oriented enabling the projection of the deformed grating in two directions chosen in advance and therefore simplifying the calculation. In the image plane of the optical system we can observe fringes of moiré between the gratings. Hence, the two figures of moiré together, enable us to determine easily the deformation by calculating the mean of the results obtained from each moiré. The inconvenience of this method is that two families of fringes appear simultaneously ; on the other hand the shape of the deformed gratings can vary largely from one region to another therefore becoming difficult to obtain a visible image of the fringes on the surface of the model. In addition the effects of perspective have to be taken into consideration since the plane of the deformed gratings is not perpendicular to the axis of the optical system.

The second method described is based on the observation of the model following the three perpendicular directions and uses the same relations determined in the first approach. This method enables us to use each

Wieringa, H (ed), Experimental Stress Analysis.
© *1986. Martinus Nijhoff Publishers, Dordrecht.*

mesh separately (this is the method of grid) in which we study the figures of moiré by superposing the reference grating over each projection of the deformed grating. This technique is all the easier to apply as the choice of the reference grating can be adapted to the form of the deformed grating.

In the same period Martin and Ju developed the theory of moiré applied to fields of large plane deformations [5], using the generalisation of the indicial representation of figures of moiré described by Oster, Wasserman and Zwerling [6]. They extended the equations of moiré to the case of gratings, reference grating and deformed grating, of arbitrary pitches and arbitrary orientation. This allows the choice of the pitch and the orientation of the initial grating during measurement thus allowing a better adjustement with the geometry of the deformed gratings. The authors applied this method, in large homogeneous deformations, on a shearing test. The experimental results were in good agreement with the calculated values (with an error of 1 %) and this for principal strains of the order of 20 %.

One year later, the same authors studied [7] a non-homogeneous deformation field by modifying the equations obtained from first method. They tested their new approach on a circular bending of a parallelepipedic beam and obtained excellent results compared to the kinematic.

The method that is widely used for the measurements of large deformations is that of grid which consists of engraving a series of orthogonal lines on a group of circles on the surface of the specimen. The use of circles give directly the orientation of the values of the principal strains. The measuring base may be much smaller than in the previous case and the dispersion of results may go up to 15 %.

The solution we are proposing is based on the use of two gratings of parallel orthogonal lines (10 lines per mm) marked on the surface of the specimen of which the photographic film is being analysed by the diffraction procedure. This method has already been applied [8] in the same way, with 10 lines per mm, and the analysis of the points by photodetectors made possible the measurement of deformations of the orders of 10^{-3}.

2. THE PRINCIPLE OF THE METHOD

We hereby describe the physical aspect of the deformation. Let us consider two families of parallel lines and of the same pitch p engraved or marked on the surface of the specimen supposed to be locally plane. The gratings therefore obtained follow perfectly the deformations of the specimen each initial square therefore becoming a parallelogram. Let p_1 and p_2 represent the new corresponding pitches of each family and let α_1 and α_2 represent the angles giving the orientation of these families with respect to a reference direction (fig. 1). We can therefore consider

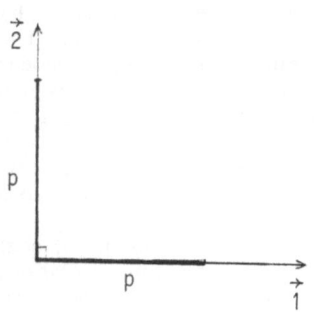
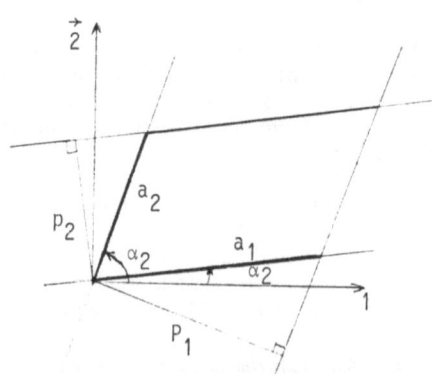

Figure 1

locally the dilatation and the variation of angles of an initial right angle of which the deformed sides have widths of a_1 and a_2.

$$a_1 = \frac{p_1}{\sin (\alpha_2 - \alpha_1)} \qquad a_2 = \frac{p_2}{\sin (\alpha_2 - \alpha_1)}$$

The physical interpretation of the right strain tensor of Cauchy Green $\bar{\bar{C}}$ leads us to write in lagrangian description, in the initial base 1 and 2 associated to the non deformed gratings, the representative matrix of $\bar{\bar{C}}$

$$C = \begin{bmatrix} (\frac{a_1}{p})^2 & \frac{a_1 a_2}{p^2} \cos (\alpha_2 - \alpha_1) \\ \frac{a_1 a_2}{p^2} \cos (\alpha_2 - \alpha_1) & (\frac{a_2}{p})^2 \end{bmatrix}$$

We hereby recall some results :

- we note $\bar{\bar{c}}$ the left tensor of Cauchy Green (in eulerian description). It is well known that the tensors $\bar{\bar{C}}$ and $\bar{\bar{c}}$ have similar proper values

- the expressions of $\bar{\bar{F}}$ tensor (the tensor of Green-Lagrange) and $\bar{\bar{e}}$ (the tensor of Euler Almansi) are

$$\bar{\bar{F}} = \frac{1}{2} (\bar{\bar{C}} - \bar{\bar{I}}) \qquad \text{and} \qquad \bar{\bar{e}} = \frac{1}{2} (\bar{\bar{I}} - \bar{\bar{C}}^{-1}) .$$

It is therefore evident that a diagonalisation made on tensor $\bar{\bar{C}}$ will give us proper values C_1 and C_2 allowing therefore the determination of the principal strains of $\bar{\bar{E}}$ (E_1 and E_2) in lagrangian description and the principal strains of $\bar{\bar{e}}$ (e_1 and e_2) in eulerian description.

The polar decomposition $\bar{\bar{F}} = \bar{\bar{R}} \bar{\bar{U}}$ of tensor $\bar{\bar{F}}$ gradient of the transformation such as $\bar{\bar{C}} = {}^t\bar{\bar{F}} \bar{\bar{F}}$ and where $\bar{\bar{R}}$ represent the rotation tensor of a rigid solid and $\bar{\bar{U}}$ the right pure deformation tensor has led to the choice of $\bar{\bar{U}}^2 = \bar{\bar{C}}$. It is therefore easy to show that tensors $\bar{\bar{U}}$ and $\bar{\bar{C}}$ have similar proper vectors.

Let λ' and U' be the proper value and the proper vector of $\bar{\bar{U}}$. We have $\bar{\bar{U}} \mu' = \lambda' \mu'$ and multiplying by $\bar{\bar{U}}$

$$\bar{\bar{U}}^2 \mu' = \lambda' \bar{\bar{U}} \mu' = \lambda'^2 \mu'$$

from which $\bar{\bar{C}} \mu' = \lambda'^2 \mu'$.

The diagonalisation previously done therefore leads to the orientation γ' of the principal directions of pure deformation. This angle γ' is generally different from the angle γ (visualising the principal direction of the deformation tensor) and the difference represents the rotation of the rigid solid R. We therefore have the relation $R = \gamma - \gamma'$.

The angle γ represents the orientation of the ellipsoid axis of Cauchy. The use of Cauchy's theorem which indicates that as an initial right angle transforms itself in conjugated diameters of an ellipsöid in the deformed state, leads us to determine analytically the direction of the axis, in other words, the value of the angle γ . We are now in a position to deduce easily the value of the rotation of the rigid body.

We also show the possibility of obtaining the orientation and the value of the principal extensions and the rotation of the rigid solid from the knowledge of four parameters (two pitches p_1 and p_2 and two angles α_1 and α_2). These values are obtained using the procedure of diffraction on photographic negatives representing the deformed state of the gratings studied.

The diffraction phenomena of a parallel beam of coherent light in a plane grating is well known [9, 10]. The hypothesis made in the case of a phenomena of diffraction of Fraunhofer (infinite diffraction giving regularly spaced points), allow the determination of the pitch of the grating knowing the wavelength λ of the radiation, the distance L between the screen (E) and the photographic negative, the distance d between two consecutive points of diffraction

$$p = \frac{\lambda L}{d}$$

This relation assumes small angles of diffraction, in other words a large value of distance L with respect to d. When this hypothesis is not verified we use the relation

$$p = \frac{\lambda\, m}{\text{Arctg } \dfrac{d^m}{L}}$$

We have represented on figure 2 the diffraction image of a grating of parallel crossing lines. We notice that the direction formed by the diffraction points are perpendicular to the orientation of the family of corresponding lines

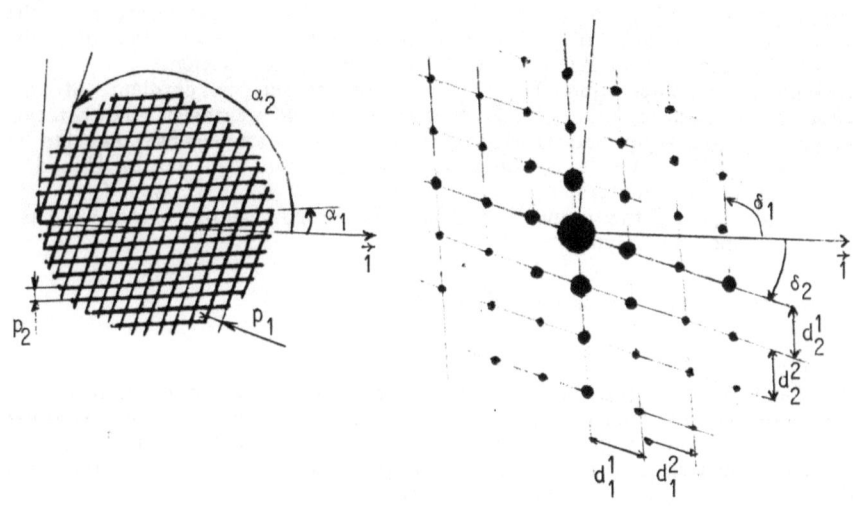

Fig. 2 : Crossed grating and their diffraction image

It is now easy to describe P_1, P_2, α_1, α_2 as functions of $d_1^{\,m}$, $d_2^{\,m}$, δ_1, δ_2.

$$P_1 = \frac{\lambda\,m}{Arctg\,\dfrac{d_1^{\,m}}{L}} \qquad\qquad P_2 = \frac{\lambda\,m}{Arctg\,\dfrac{d_2^{\,m}}{L}}$$

$$\alpha_1 = \delta_1 - \frac{\pi}{2} \qquad\qquad\qquad \alpha_2 = \delta_2 + \frac{\pi}{2}$$

The polar coordinates of these point are read numerically by digitalising using a digital table HP 9111 A. In order to minimise the uncertainties over the four parameters the centre of each point in taken three times. A numerical analysis done on a micro computer gives the statistical analysis of the data and calculates the deformation values.

3. SIMULATION

In order to test the validity of this measuring method we have made a simulation of a field of homogeneous deformations. In this case we consider a grating, in a deformed state, composed of two same families of parallel lines equidistant of pitch p inclined one with respect to another with an angle $\pi/2 - \delta$ (fig. 3). We suppose that the two families are initially perpendicular. The transformation which changes from an initial state (X_1, X_2) to the deformed state (x_1, x_2) has the following expression :

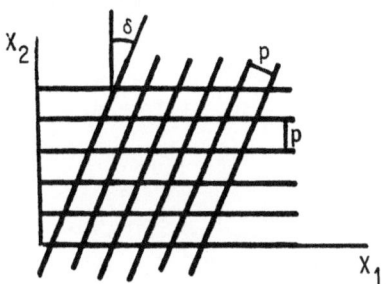

$$x_1 = \frac{1}{\cos\,\delta}\,X_1 + tg\,\delta\,X_2$$

$$x_2 = X_2$$

Fig. 3 : Deformed state

We deduce the expression of the gradient of transformation tensor $\overline{\overline{F}}$:

$$F = \begin{bmatrix} \dfrac{1}{\cos\,\delta} & tg\,\delta \\[2mm] 0 & 1 \end{bmatrix}$$

and as a result the tensor $\overline{\overline{C}}$ (right Cauchy Green) and $\overline{\overline{c}}$ (left Cauchy Green) are written as :

$$C = \frac{1}{\cos^2\delta}\begin{bmatrix} 1 + \sin\,\delta & 0 \\[2mm] 0 & 1 - \sin\,\delta \end{bmatrix} \qquad c = \begin{bmatrix} 1 + \sin\,\delta & 0 \\[2mm] 0 & 1 - \sin\,\delta \end{bmatrix}$$

from which

$$E_1 = \frac{1}{2}\,\sin\,\delta\,\left(\frac{1 + \sin\,\delta}{\cos^2\delta}\right) \qquad\qquad e_1 = -\frac{1}{2}\,\sin\,\delta$$

$$E_2 = \frac{1}{2}\,\sin\,\delta\,\left(\frac{1 - \sin\,\delta}{\cos^2\delta}\right) \qquad\qquad e_2 = \frac{1}{2}\,\sin\,\delta$$

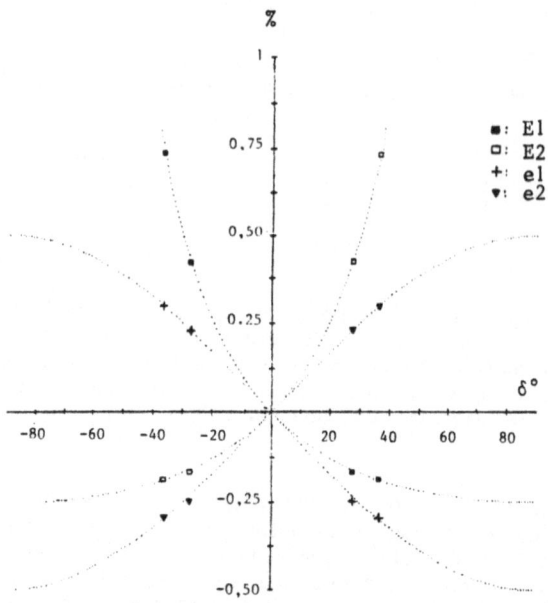

Fig. 4 : Comparison Between Measured and Exact Principal Strains

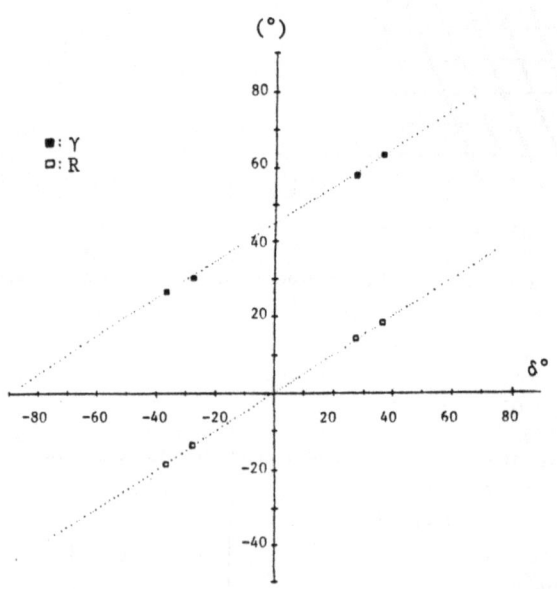

Fig. 5 : Comparison Between Measured and Exact Rotation and
Principal Strain Direction

Considering the physical representation of the transformation, we know that the principal directions in the spatial representation are diagonals of the lozenges therefore

$$\gamma = \frac{\delta}{2} + \pi/4$$

The diagonalisation of $\bar{\bar{C}}$ leads to a value of δ'

$$\gamma' = \pi/4$$

from which $R = \delta/2$.

We note that the rotation is equal to the variation of the orientation of a diagonal.

The gratings used are identical, with the pitch being equal to 0.042 mm. Four tests were done with values of δ equal to -36.5°, -27.5°, 27.5° and 36.5°, the distance L being equal to 1095 mm while the wavelength of the laser beam being equal to 632.8×10^{-6} mm. The experimental results (fig. 4 and fig. 5) compare very well with the theoretical results. It is worth noting that the gratings were of excellent quality and the diffraction points were circular and with a good contrast.

4. CIRCULAR BENDING

The aim of this study is to specify the kinematic description in large deformations of a beam supposed to be elastic, isotropic and incompressible, from well known written works.

A cylindrical coordinate system (r, θ, z) (fig. 6) is used to describe the deformed state and the point $M(X, Y, Z)$ transforms itself in $m(r, \theta, z)$ following the transformation obtained by the following hypothesis :

- each plan initially orthogonal to the X axis is transformed into part of a cylinder with Z as the axis

- plans that are normal to the Y axis remain plan and they intersect on the same straight line parallel to the Z axis

- the rectangular section X Z transforms itself into a trapezium.

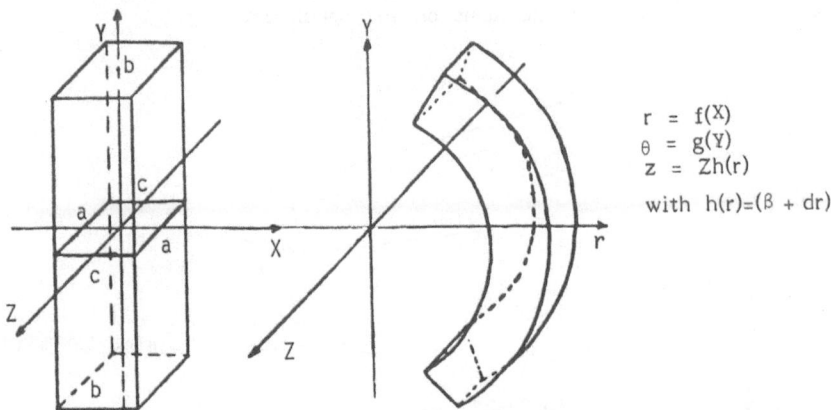

$$r = f(X)$$
$$\theta = g(Y)$$
$$z = Zh(r)$$

with $h(r) = (\beta + dr)$

Fig. 6 : Geometry of the bending specimen in the initial state and deformed state.

The function $h(r)$ represents the equation of the sides of the trapezoïdal section X Z. If $\alpha = 0$ then $h(r) = \beta$ and we obtain the calculation obtained by Green and Zerna [11] and Eringen [12].

We can now determine the tensor gradient of the transformation then the tensor $\bar{\bar{C}}$

$$
C = \begin{bmatrix} f'^2 \, (1+z^2\alpha^2) & 0 & Z\alpha f' \, (\beta+\alpha f) \\ 0 & g'^2 & 0 \\ Z\alpha f' \, (\beta+\alpha f) & 0 & (\beta+\alpha f)^2 \end{bmatrix}
$$

The incompressibility equation det $\bar{\bar{C}} = 1/r^2$ leads to the calculation of f and g functions

$$
g(Y) = \frac{AY}{\beta} + c'
$$

$$
\frac{\beta f^2(X)}{2} + \frac{\alpha f^3(X)}{3} = \frac{\beta}{A} X + B
$$

whereby A, B, C', α and β are determined from the limit values of the functions f and h knowing the geometry of the beam in the deformed state.

The previous relationship does not allow us to know the analytical expression of the function f. A numerical calculation is therefore necessary to determine the values taken by f for different values of r. In these points we can easily know the values of f' and finally the components of $\bar{\bar{C}}$.

We note that $\bar{\bar{C}}$ is not diagonal in the reference X Y Z while it is in the reference X Y. The measurements taken by our method are easily expressed in the later reference and an immediate comparison of results can therefore be done in the Lagrangian description.

The specimen used was made of polyurethane PL3 and had the following dimensions 2a = 10.1 mm, 2b = 60 mm and 2c = 10.6 mm. The specimen was prepared by moulding. The gratings of pitch 0.1 mm engraved at the base of the mould are reproduced on the surface of the specimen during moulding.

The specimen is then embedded at the two extremities over two sliding chariots which can move over two articulated arms. The apparatus allows us to obtain a circular bending in which the centre is the axis of circulation of the arms, and the radius, the distance between this centre and the specimen. The displacements of the chariots is done by a bolt and nut link. The apparatus is used horizontally so as to eliminate the effects of the force of gravity of the arms on the specimen.

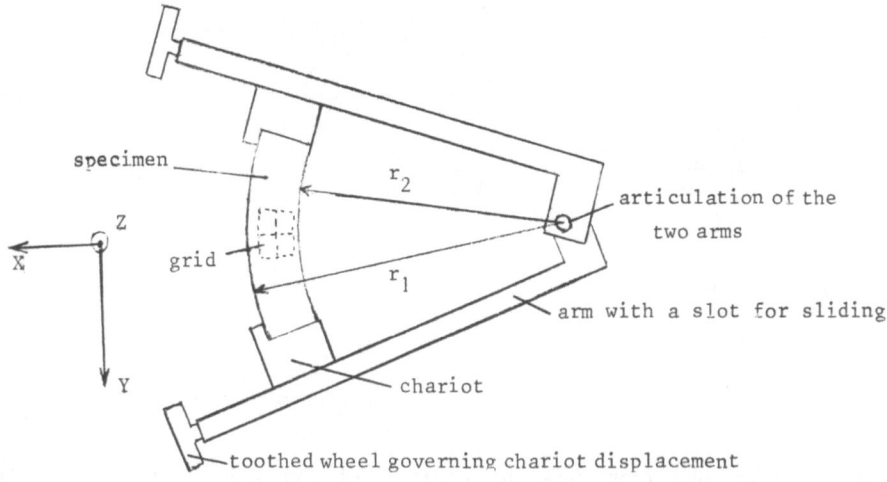

Fig. 7 : Schematic representation of the apparatus

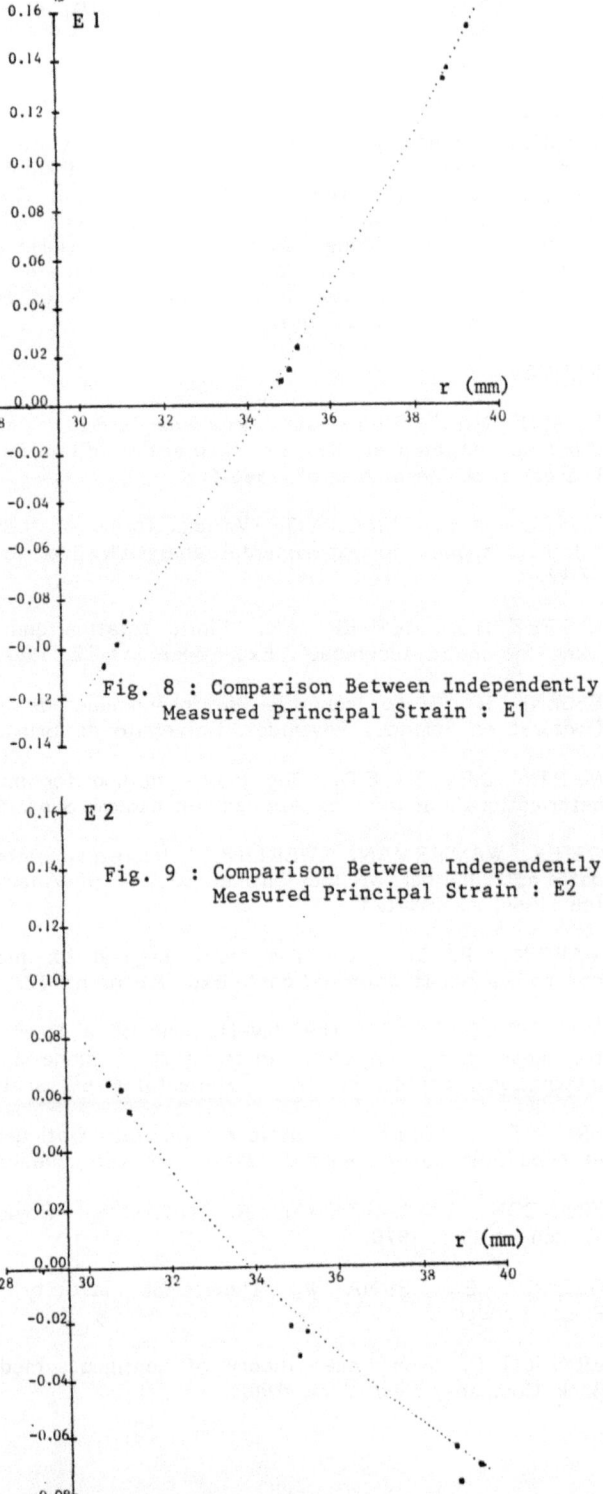

Fig. 8 : Comparison Between Independently
Measured Principal Strain : E1

Fig. 9 : Comparison Between Independently
Measured Principal Strain : E2

In figs 8 and 9, the lagrangian strains E_1 and E_2 are represented. The experimental points compare very well to the curves obtained from the kinematic.

CONCLUSION

This optical method leads to the measurements of the four parameters which are of interest to a mechanical engineer (orientation and values of the principal strains ., rotation of a rigid solid). The method enables us to measure at the same time small deformations as well as large deformations, without being limited to a measurement base which is relatively small, of the order of a square millimeter in the case of our study. We note that the use of a simple system of analysing the images allows measurements in real time.

REFERENCES

[1] DANTU M.P. "Moiré du deuxième ordre : méthode permettant d'obtenir directement les lignes d'égales dilatations linéiques". Revue Française de Mécanique n° 1966-6.

[2] DURELLI A.J., PARKS V.J. "Various forms of the strain displacement relations applied to experimental analysis". Exp. Mech. n° 2, 1964, p. 37-47.

[3] DUFFEY H.J., MESMER G.K. "Finite rotation and strain measurement using the moiré technique". Exp. Mech. n° 12, 1967, p. 537-540.

[4] EBBENI J. "Etude générale du phénomène de moirure". Thèse de Doctorat en Sciences Physiques, Université de Bruxelles, 1969.

[5] MARTIN L.P., JU F.D. "The moire method for measuring large plane deformations". Journal of Applied Mechanics, Sept. 69, p. 385-391.

[6] OSTER, WASSERMAN, ZWERLING "Theoretical interpretation of moiré patterns". Journal of the Optical Society of America, Vol. 54, n° 2, feb. 1964, p. 169-175.

[7] MARTIN L.P., JU F.D. "The moiré method for measuring large plane non homogeneous deformations". Exp. Mech. n° 12, 1970, p. 521-528.

[8] SEVENHUIJSEN P.J. "The development of a laser grating method for the measurement on strain distribution in plane opaque surfaces". 6th international conference on experimental stress analysis, 1978, München.

[9] BRUHAT J. "Cours de Physique Générale - Optique". 6e édition revue et complétée par A. Kastler, Masson et Cie, 1965.

[10] FRANCON M., MARECHAL A. "Diffraction - Structures des images". Masson et Cie, 1970.

[11] GREEN A.E., ZERNA W. "Theoritical elasticity". Oxford University Press, London.

[12] ERINGEN C. "Non linear theory of continuous media". Mac Graw Hill Book Company, New York, 1962.

THE PHASE SHIFT METHOD APPLIED TO REFLECTION MOIRÉ PATTERN

K. ANDRESEN and R. RITTER

Technical University of Braunschweig, Federal Republic of Germany

1. INTRODUCTION

The moire methods are well-known for analysing the contour and the deformation of an object |1|. The moiré fringes do not lead only to the wanted values but represent also a qualitative overview to the nature of the deformation. However their automatic evaluation is relatively diffi- cult by digital picture processing: First the fringes are generally of no regular structure and moreover their order cannot be determined without additional information of the related experiment. Usually the input of the order must be done interactively.

The mentioned problems can be avoided by applying a new method developed from the phase shift principle of the holographic interferometry |2|. Instead of analysing the geometric contours of the moiré fringes their grey intensity values related to three phase shifted moiré patterns in one point are used for calculation of the displacement of a deformed pattern compared with a known reference pattern. By this procedure described in detail in |3| one gets the grating displacement not only on the moiré fringes but also between them in distances of the grating pitch. Moreover generalized orders of the fringes follow by this process automatically.

The phase shift method needs only a simple picture processing algorithm. Furthermore the proportion of postprocessing is smaller than in the case of direct moiré evaluation.

In the following a modified phase shift method is applied to reflection moiré patterns as an example of the different moiré techniques.

2. PHASE SHIFT METHOD

Regarding holographic interferometry the phase shift of an actual beam against a reference one, contains the information needed for further calculation. A constant phase shift belongs to interferometric patterns similar to those of the moiré technique with corresponding difficulties concerning automatic digital picture processing. The idea of this phase shifting consists in considering the basic and two shifted positions of a reference beam. Usually the shifting amount is chosen to be the third part of the wave length λ. In every case the reference beam is superposed on the deformed beam with the result of three phase shifted patterns. Then the grey values of the related intensity distributions are taken at the same local point, yielding the desired phase shift |4|.

This procedure can be transferred to the moiré principle. In that case the wave length λ is replaced by the pitch p of the reference grating. The correspondence of λ and p is based on the model, that the intensity distribution of the grating can be approximated by a trigonometric funct- ion. Instead of shifting the reference beam now the reference grating is moved by an amount of $\pm$ p/3.

Then the local intensity can be described by

Wieringa, H (ed), Experimental Stress Analysis.
© *1986. Martinus Nijhoff Publishers, Dordrecht.*

$$I = I_G + I_D \cos \varphi \qquad (1),$$

whereby I_G means the local mean value of the measured intensity and I_D the amplitude of the cosine function with the phase angle

$$\varphi = \frac{v_1}{p/2} \qquad (2).$$

In eq.(2) v_1 represents the local displacement of the deformed grating against the reference grating.

The local intensity I in eq.(1) is assumed to be measured in a moiré image without observable generating grating lines. If the grating lines are visible, a mean value I integrated over an intervall of the pitch p

$$\bar{I} = \frac{1}{p} \int_p I \, dx \qquad (3)$$

must be calculated.

Regarding the local intensity I at the same local point with the co-ordinates x and y of the mentioned three patterns shifted by $\pm p/3$ respectively $\pm 120°$ leads to the equations

$$I_o = I_G + I_D \cos \varphi \qquad (4)$$
$$I_m = I_G + I_D \cos(\varphi - 120°) \qquad (5)$$
$$I_p = I_G + I_D \cos(\varphi + 120°) \qquad (6).$$

I_o, I_m and I_p are the measured local intensity values, while I_G, I_D and φ respectively related to eq.(2) the displacement v_1 are to be determined by eqs.(4)-(6). Applying the addition theorems of harmonic functions it follows:

$$I_G = (I_o + I_m + I_p)/3 \qquad (7)$$

$$I_D = \sqrt{(2I_G - I_p - I_m)^2 + (I_m - I_p)^2/3} \qquad (8)$$

$$v_1 = \frac{p}{2} \arctan \left(\frac{(I_m - I_p)\sqrt{3}}{2I_o - I_m - I_p} \right) \qquad (9).$$

Eq.(9) leads to the local displacement v_1 of the deformed grating against the reference grating related to the main value.

The absolute displacement has the form

$$v = v_1 + k \, p \qquad (10),$$

where k is an integer, which must be determined from the sequence of v_1 values, Fig. 1a. Every jump of v_1 from p to 0 means an increase of k by 1 and vice versa. Applying the procedure to v_1, the absolute displacement v in Fig. 1b is derived which now represents a continuous function.

The phase shift equation (9) is exact only if the generating functions are harmonic of the same period. On the other hand the moiré effect results from the difference in the period of the deformed and the reference grating. In |3| it is shown, that a difference of the period of about 10 % yields a phase error of only 2 %.

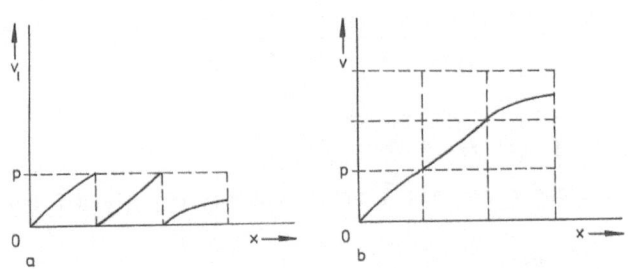

FIGURE 1. Phase shift calculation, (a) local values v_1, (b) absolute values v.

3. REFLECTION GRATING METHOD

In the following the phase shift method is applied to the reflection grating principle, Fig. 2.

The point $\bar{A}$ of the reference grating G is reflected from the object surface O in A and recorded in $\bar{A}$ of the image plane IP.

Refering to the optical and geometrical constants of Fig. 2 the slopes w_{xA} and w_{yA} of the object surface can be determined from the co-ordinates $\bar{x}_A$ and $\bar{y}_A$ of the image plane and the orders n_x and n_y of the grating lines. They are related to $\bar{x}_A$ and $\bar{y}_A$ by

$$\bar{x}_A = n_x \, \bar{p}_x \tag{11}$$

and

$$\bar{y}_A = n_y \, \bar{p}_y \tag{12}.$$

with the pitches $\bar{p}_x$ and $\bar{p}_y$ of the reference grating.

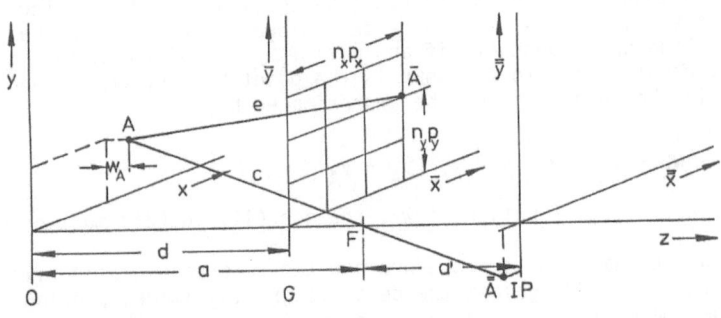

FIGURE 2. Reflection grating principle.
G reference grating, O object, IP image plane.

The coordinates are given by

$$x_A = - \frac{\bar{\bar{x}}_A}{a'} (a - w_A) \tag{13}$$

and

$$y_A = - \frac{\bar{\bar{y}}_A}{a'} (a - w_A) \tag{14}.$$

With

$$e = |\underline{e}| = \sqrt{(\bar{x}_A - x_A)^2 + (\bar{y}_A - y_A)^2 + (d - w_A)^2} \qquad (15)$$

and

$$c = |\underline{c}| = \sqrt{x_A^2 + y_A^2 + (a - w_A)^2} \qquad (16)$$

from the vectors $\underline{e}$ and $\underline{c}$ between $\bar{A}$ and A respectively A and F one gets nominating

$$\frac{d - w_A}{e} + \frac{a - w_A}{c} = \kappa \qquad (17)$$

the desired slopes

$$w_{xA} = (\frac{x_A - \bar{x}_A}{e} + \frac{x_A}{c})/\kappa \qquad (18)$$

and

$$w_{yA} = (\frac{y_A - \bar{y}_A}{e} + \frac{y_A}{c})/\kappa \qquad (19).$$

The last two equations are related to a cross grating. On the other hand the phase shift method is based on a moiré effect generated by grating lines only in one direction, for example the x-axis. Consequently the slope w_{xA} is not available because $\bar{x}_A$ along a line is unknown. Therefore the slope w_{xA} is assumed to be equal to zero yielding the grating co-ordinate $\bar{x}_A$ from eq. (18) to

$$\bar{x}_A = x_A(1 + \frac{e}{c}) \qquad (20)$$

The error with respect to w_{yA} is relatively small if the distances a and d are large compared with the size of the object.

The second unknown value in the eqs. (14) to (16) and (19) is the deflection w_A of the object in A. It can be determined by integrating the slope w_{yA} along a line x_A = constant, if an initial $\mathring{w}_A$ is given. Assuming the slope and the deflection to be known in the point P_i, the corresponding deflection in the point P_{i+1} can be approximated by

$$w_A^{i+1} = w_A^i + 0.5 (w_{yA}^{i+1} + w_{yA}^i) (y_A^{i+1} - y_A^i) \qquad (21)$$

As w_{yA} for itself is a function of w_A, the eqs. (14) to (21) must be executed iteratively.

Regarding the phase shift procedure now the reference grating image must be shifted by $\pm p/3$ against the deformed grating image, p being the pitch in the image plane. v_1 and v are determined by eqs. (9) and (10). If $\bar{y}_A = n_y \bar{p}_y$ is observed in the undeformed state of the object, then after deforming it at the same place of the image plane the coordinate

$$\bar{y}_{Ad} = \bar{y}_A + \frac{v}{p} \bar{p}_y = (n_y + \frac{v}{p}) \bar{p}_y \qquad (22)$$

must be considered.

Replacing the eq. (12) by eq. (22), the described iterative integration process yields the slope and the deflections of the object.

4. EXPERIMENTAL RESULTS

A quadratic plate (200 mm x 200 mm) was considered, Fig. 3. It was clamped at the edges of a quadratic cutout in the centre and loaded verti- cally to the undeformed plane of the object in the middle of its upper edge. The surface was mirrored. Object, grating and image plane were ad- justed parallel to each other for a simple measurement of the geometrical and optical data.

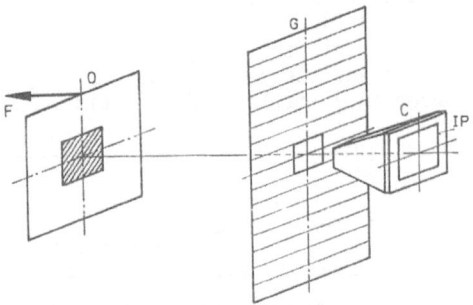

FIGURE 3. Experimentel setup.
O mirrored plate, G line grating, C recording camera

Through a hole in the grating plane pictures of the reflected grating lines were taken at different loads. The reference grating was generated for the unloaded state.

The negatives of a deformed grating and the reference grating were fixed on two frames, which could be moved against each other by turning a micrometer screw, yielding the desired phase shift. This procedure proved to be very sensitive, because the pitch in the image was very small. In future a piezo electric transducer controlled by a computer should be used.

Appropriate shifting yielded 3 moiré images which were scanned by a CCD-camera, digitized and put into an image store of 512 x 512 pixels à 8 bit. In this case the linear transformation between pixel and image coordinates was not taken into account, because it didn't influence the basic relationships.

Fig. 4 shows one of these moiré images, the left side of the plate being nearly undeformed. The fringes on the right side illustrate the growing deflection in the direction of the load.

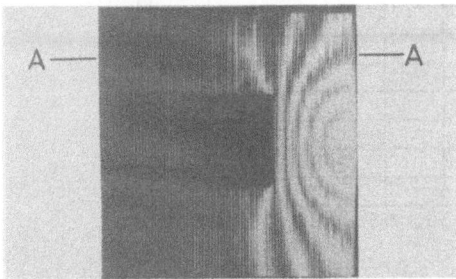

FIGURE 4. Moiré fringes in the deformed state of the plate.

356

The evaluation of the phase shift values was started from the three mentioned images.
   It proceeded in the following steps:
a) Read out one row of the pixel store perpendicular to the grating lines
   for each of the three images.
b) Integrate the intensity functions over distances of the pitch p in
   order to get I_0, I_m, and I_p.
c) Calculate the local displacement v_1 by eq (9).
d) Calculate the absolute displacement v by eq.(10) according to Fig. 1.
e) Insert the displacements into the grating reflection by eqs.(12) to
   (22) yielding the slope and the deflection.
The steps a) to d) needed only little computing time. They were performed

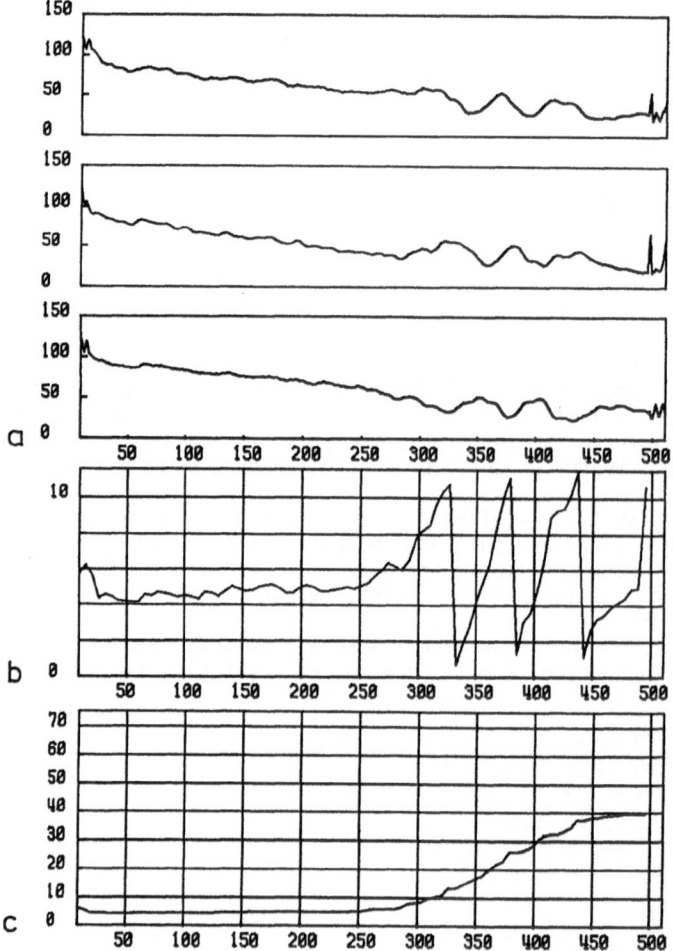

FIGURE 5. Grating displacement of the section A-A of Fig. 4. a) Intensity
function of three phase shifted patterns, b) local displacement v_1 and
c) absolute displacement v.

automatically, which shows the preference of the phase shift method compared with the geometric evaluation of moiré fringes.

In Fig. 5a there are plotted three phase shifted intensity functions of
the section A-A of Fig. 4. The generating grating lines were already removed by a nonrecursive symmetric low pass filter. The significant oscillations on the right hand side are related to the moiré fringes. The local
displacement v_1 is shown in Fig. 5b and the absolute displacement v in
Fig. 5c.

Altogether 30 sections were considered each of them containing 85 discrete phase shift values v_1. Figs. 6a/b show the contour lines of the
slope and deflection determined by the mentioned procedure.

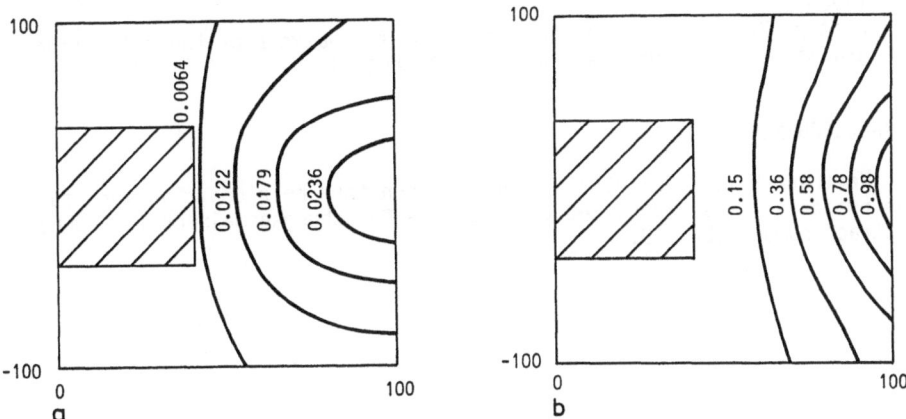

FIGURE 6. Calculated contour lines of the slope and deflection of the
deformed plate, a) slope, b) deflection.

5. CONCLUSION

The phase shift principle provides a method, particularly suited for
automatic digital processing of moiré-images. Instead of looking for the
coordinates of the moiré fringes only the grey intensity values of three
phase shifted patterns are taken to calculate the absolute displacements
of a deformed grating against the reference grating. These displacements
determined in distances of the pitch in an arbitrary number of sections
perpendicular to the generating grating lines contain the information
needed for further processing. Applying the method to reflection moiré
patterns the slope and the deflection of an object is derived in a
corresponding number of local points.

The phase shift method can be regarded as an extension of the general
grating methods for those cases, where the grating lines cannot be
resolved directly by the image digitizing system. By use of the moiré
effect which implies an optical preprocessing, very small deformations
of high density gratings can be evaluated.

REFERENCES

|1| Ritter, R.: "Reflection moiré methods for plate bending studies",
    Opt. Eng. 21, p. 663-671 (1982).
|2| Daendlicker, R., Thalmann, R., and J.-F. Willemin: "Fringe interpola-
    tion by two-reference-beam holografic interferometry: Reducing
    sensitivity to hologram misalignment", Optics Communications 42,
    p. 301-306 (1982).
|3| Andresen, K.: "The phase shift method applied to moiré image proces-
    sing", Optik 72, p. 115-119 (1986).
|4| Breuckmann, B., and W. Thieme: "Ein rechnergestütztes Holografie-
    system für den industriellen Einsatz", VDI-Berichte 552, p. 27-36
    (1985).
|5| Ritter, R., and R. Hahn: "Contribution to analysis of the reflection
    grating method", Optics and Lasers in Eng. 4, p. 13-24 (1983).

ACKNOWLEDGEMENTS

We thank Mrs. K. Spieler and Mr. V. Hohn for their contributions to
the project. It was supported by "Deutsche Forschungsgemeinschaft" (DFG),
contract number SFB 319.

DETECTION OF FRINGE PATTERN INFORMATION USING A COMPUTER BASED METHOD

CESAR A. SCIAMMARELLA and MANSOUR A. AHMADSHAHI
Mechanical and Aerospace Engineering
Illinois Institute of Technology, U.S.A.

1. INTRODUCTION

Information encoded in fringe patterns requires a process of extraction to put it in a form accessible to the human observer. In the middle sixties, to obtain an economical and accurate method of fringe pattern analysis, research work in the use of analog and digital computers for this purpose, was started. To this date, numerous papers have been published with contributions in this field. Important developments in the computer technology in the 1970's and 1980's have made possible very important advances in this area, leading to fully automated methods.

Early attempts to utilize digital computer to process fringe patterns were made by one of the authors in 1966 [1]. Successive improvements were introduced in [2], [3], [4], [5], and [6]. In this paper, new developments both in methodology to process fringe patterns and in the hardware required to implement it, will be described.

2. FRINGE PATTERNS AS FREQUENCY MODULATED SIGNALS

Fringe patterns analyzed as signals along a reference axis, for example the x-axis can be represented by an equation of the form,

$$I(x) = I_o(x) + I_1(x) \cos \phi(x) + I_n(x) \tag{1}$$

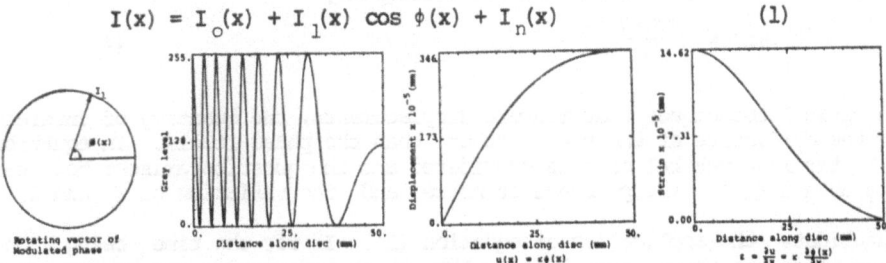

FIG.1. Diagram showing the steps performed in the digital processing of data.

where the desired information is contained in the phase $\phi(x)$. $I_o(x)$ is the background illumination which shows variations due to uneven illumination or nonuniform light reflection. $I_1(x)$ is the fringe amplitude, which also has variations of amplitude due to the same causes pointed out in the case of $I_o(x)$. $I_n(x)$ is a term that represent changes of illumination unrelated to the quantity of interest and will be called noise. The arguement $\phi(x)$ can be represented by

$$\phi(x) = \frac{2\pi\delta(x)}{m} + \alpha \tag{2}$$

Wieringa, H (ed), Experimental Stress Analysis.
© 1986. Martinus Nijhoff Publishers, Dordrecht.

where $\delta(x)$ is an optical path difference, m is a quantity that depends on the particular optical technique under consideration. In the moire method, m is the grating pitch, in speckle interferometry the spatial frenquency of the sampled signal, in holography, the wave length of the light, in photoelasticity the fringe optical constant. The α term, is an arbitrary phase term. Equation (1) can be considered the equation of a frequency modulated signal. Assuming I_o and I_1 constant, Fig. 1, the

signal can be considered, as being generated by a rotating vector. The phase

$$\phi(x) = \text{arc cos } \frac{I(x) - I_o}{I_1} \qquad (3)$$

is the phase of the vector I_1, and is the measure of a certain optical

path change, which through the application of (2) can be transformed in the quantity of interest. Equation (3) can be considered as the generalization of the concept of fringe order, which through this definition becomes a continuous quantity. Through the application of Fourier optics concepts equation (3) is a rigorous generalization, limited only like any other signal by the perturbation of the noise term. In equation (1) the signal is assumed to have only one harmonic. This may not be true as it happens with moire patterns or patterns that have been recorded on film. The presence of other harmonics is not a limitation of the theory, since as it will be shown, higher harmonics can always be eliminated. Since the signal is a frequency modulated signal, one must introduce the notion of instantaneous frequency

$$\omega(x) = \frac{d\phi(x)}{dx} \qquad (4)$$

In optical techniques that measure displacements, the quantity of interest is the derivative of the phase rather than the phase itself. In equation (1), the most general case is considered and the question arises: how can one obtain $\phi(x)$? The presence of noise and the variation of I_o and I_1,

invalidate the application of equation (3). If for the time being, one puts aside the noise term $I_n(x)$, the phase information can be separated

from the amplitude information through the use of the notion of signals in-quadrature [2]. If the background term I_o and the noise term are

removed, equation (1) reduces to the form

$$I_f(x) = I_1(x) \cos \phi(x) \qquad (5)$$

From the above signal, called inphase signal, one can generate another signal, called inquadrature signal,

$$I_q(x) = I_1(x) \sin \phi(x) \qquad (6)$$

From (5) and (6) one gets,

$$\phi(x) = \text{arc tg} \ \frac{I_q(x)}{I_f(x)} \tag{7}$$

Equation (7) provides the phase of the vector with independence of the changes of its amplitude. The same process can be applied to the derivatives, dividing (5) and (6) by I , one gets,

$$s_1(x) = \cos \ \phi(x) \tag{8}$$

and

$$s_2(x) = \sin \ \phi(x) \tag{9}$$

taking derivatives with respect to x

$$s_1'(x) = -\phi'(x) \ \sin \ \phi(x) \tag{10}$$

$$s_2'(x) = \phi'(x) \ \cos \ \phi(x) \tag{11}$$

By squaring and adding

$$\phi'(x) = \omega(x) = \sqrt{[s_1'(x)]^2 + [s_2'(x)]^2} \tag{12}$$

which again is independent of the amplitude changes.

In the preceding derivations the variations along one coordinate axis have been considered. The same procedure can be applied to sections perpendicular to that axis, and one can obtain $\phi(y)$ and $\omega(y)$.

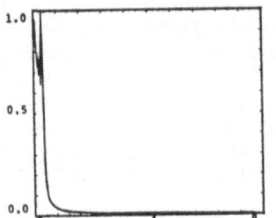

FIG.2. Fourier spectrum of the signal.

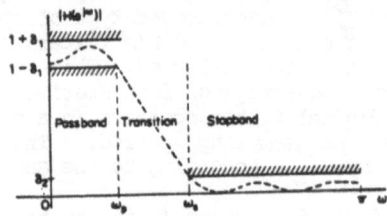

FIG.3. Parameters defining a digital filter.

3. DIGITAL PROCESSING OF THE SIGNALS

The basic concepts required to separate phase and phase derivative information from amplitude changes have been outlined. The output signal contains, besides the phase and frequency information, other changes of irradiance, which distort the signal and have been called noise. One important source of signal distortion is the transfer function of the digitizing system, or of the film if the fringes are recorded in film. The effect of the nonlinear transformation has been addressed in [3]. The effect of the nonlinearity of the transfer function is to generate for each component of the signal an infinite number of high order harmonics of the argument.

The basic problem that one confronts in the process of fringe data reduction is to remove noise from the signal. This is a classical problem in the theory of signal processing. The methodology to separate a signal from noise is called filtering. There are many classes of filters. A particular choice for a given signal depends on the information available concerning the signal and the noise. If nothing is known concerning the signal and the noise, adaptive filtering techniques can be applied.

The approach presented in this paper is based on the spectrum analysis of the signal. Fig. 2 shows the spectrum of the signal of Fig. 1. From the Fourier spectrum it is possible to estimate the frequency range where the signal is contained.

By means of a filtering operation, one can separate the background component and the noise not contained within the frequency range of the signal. The filtering can be performed in two alternative ways. One can apply a digital band pass filter. The digital filter operates in the sampled version of I(x) [2]. An alternative approach [7] is to filter in the Fourier space by setting all the amplitudes outside the band pass equal to zero and then performing an inverse Fourier transform.

Although this two alternatives look equivalent from the theoretical point of view they are not in the actual application. Experience gained through application shows that the Fourier inversion technique yields much poorer results than digital filtering technique. The reasons for this behavior will become clearer later.

A digital filter can be represented in the frequency space as shown in Fig. 3. The dashed curve represents the frequency response of a filter. There is a pass band region where the magnitude response must approximate 1 with an error $\pm\delta_1$. There is a stop band in which the magnitude response must approximate to zero with an error $\pm\delta_2$. Between these two regions there is a transition region of non-zero width where the magnitude response drops smoothly from 1 to zero. Given a set of specifications, the desired frequency response can be approximated in two ways, either by a rational function, infinite impulse response function (IIR), or by a polynomial, finite impulse response (FIR).

Since in the envisaged type of application, the phase response of the filter is essential and FIR filters have an excellent phase linearity, FIR filters have been selected. This type of filters also, allow the user to take full advantage of fast Fourier transform computational speed.

The digital filter operates in a sampled version of $I(x) = I(n\Delta s)$, where s is the sampling period. The reciprocal of Δs, is the sampling frequency f_s. According to the sampling theorem [8] to uniquely describe

$I(x)$, $f_s > 2f_p$, where f_p is the highest frequency contained in $I(x)$, then

$I(x)$ must be band limited. The filtering operation is expressed by

$$\tilde{I}(k\Delta s) = \sum_{p=-m}^{p=+m} h_p \, I[(k+p)\Delta s] \tag{13}$$

p is the running index, the tilde over the quantity represents the filtered version of $I(x)$ and the h_p are the filter weights, m is the number of filter weights. The filter operation is a running weighted average that replaces the value of the function at a given point with the weighted average of its value with the values of the (m-1) points preceding and following it. Consequently the values of $I(k\Delta s)$ are only influenced by these points. If the filtering is performed in the Fourier plane, the influence is extended over all the interval. For this reason, filtering in the Fourier plane by direct inversion can be the source of large errors.

The filtered version of $I(x)$ will be of the form of (8), to obtain the phase $\phi(x)$, the inquadrature signal is needed. Given the real function $I(x)$, the Hilbert transform

$$I_q(x) = \frac{1}{\pi} \int_{-\infty}^{+\infty} \frac{I(\varepsilon_1)}{\varepsilon_1 - x} \, d\varepsilon_1 \tag{14}$$

yields the inquadrature signal. A finite response approximation of the Hilbert transform or discrete transform is applied, which has a form identical to equation (13), with the appropriate filter weights. The Hilbert transform filter can be generated by the same program that generates the passband filter.

In moire fringes, holography and speckle, the derivatives of the displacements are required. The derivatives can be computed by means of differentiating filters [6]. The differentiating filters can be also represented by an equation of the form identical to (13), except that the filter weights are different. As is the case in the Hilbert transform the same computer program that generates the weights of the passband filter can generate be differentiating filters.

Equation (7), assumes that there is a way to obtain the phase angle from the fringe order information. A simple algorithm can be based on the computation of

$$-\pi < \text{arc tg} \, \frac{\tilde{I}_q(n\Delta s)}{\tilde{I}_f(n\Delta s)} < \pi \tag{15}$$

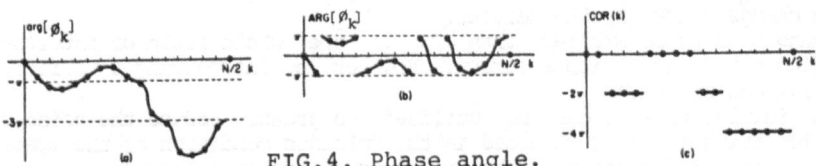

FIG.4. Phase angle.
a) continuous phase, b) values computed by sub-routine, c) correction.

which can be achieved using standard tangent routines available in most computers.

Fig. 4(a) show the continuous phase, Fig. 4(b) shows the discontinuous phases obtained by applying equation (7) and Fig. 4(c) shows the required corrections to obtain a continuous curve. To compute the phase $\phi(k\Delta s)$, one must add an appropriate multiple of 2π. The appropriate value of 2π can be obtained if the sample interval Δs is small enough so that the discontinuities can be detected. Since $\phi(k\Delta s)$ is the phase of a rotating vector one can follow the quadrant position of the vector through the difference between two consecutive points

$$\Delta\phi(k\Delta s) = \phi[k\Delta s] - \phi[(k-1)\Delta s] \tag{16}$$

If the sampling rate is such that aliasing is avoided, the absolute difference of the phase should be smaller than 2π, except at the points where jump occurs. Criteria can be established for a phase change, such that only values larger than an adopted maximum will be considered as points of jump. If only relative values of phase are determined, the point of zero phase can be arbitrarily selected. From the point of zero phase, relative phases can be established by sequentially adding 2π values at the points of jump.

If $\Delta\phi(k\Delta s)=-2\pi$ a phase jump of 2π should be added to $\phi(k\Delta s)$, if $\Delta\phi(k\Delta s)=2\pi$ a phase jump of -2π should be added. When the whole plane is analyzed, the relative phases of the different lines must be coordinated. One can select a particular fringe which is labeled zero order fringe. The above outlined procedure applies to techniques that measure relative displacements. For example, in the case of photoelasticity, where absolute orders are required, the order of the reference fringe must be provided by an independent procedure. In both cases for the program to provide the correct phases, a sub-routine for fringe order determination must be introduced. Utilizing differentiating filters, lines of zero derivative can be located, and relative orders can be found.

4. INTRODUCTION OF A CARRIER SYSTEM OF FRINGES

The form of signal given in equation (1) is not the most advantageous for fringe pattern processing. Additional information can be gained if a system of carrier fringes is added

$$I(x) = I_o(x) + I_1(x) \cos [2\pi\omega_c + \phi(x)] + I_n(x) \qquad (17)$$

where ω_c is the frequency of the carrier. The carrier shift the signal spectrum to a position determined by its frequency f . Changes of phase and frequency with respect to the initial phase and frequency can be established, consequently relative signs are automatically obtained. Absolute signs can also be determined by knowing the sign of the optical path change caused by the carrier.

Changes of phase smaller than 2π in the whole field of interest are difficult to detect with accuracy. However, by introducing a carrier, the problem can be solved.

The initial carrier can be utilized to greatly reduce the noise term. Carrier fringes can be recorded in the unloaded condition of the specimen. The specimen is loaded and the frequency modulated carrier is recorded. Both patterns can be multiplied and a beat frequency pattern can be obtained. The beat frequency pattern has a phase which is the difference of the phases of the two patterns. The noise term common to both patterns will be removed.

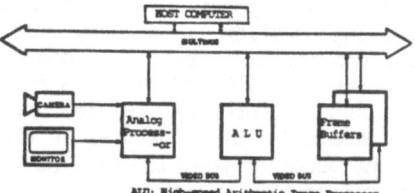

FIG.5. Schematic representation of the computer processing system.

5. CONFIGURATION OF A BASIC SYSTEM FOR FRINGE PATTERN ANALYSIS

The basic system utilized in the research work presented in this paper is schematically shown in Fig. 5. The basic components of the system are the following: 1) an image-scanner, a TV camera. Three pieces of information are provided by the camera, instantaneous coordinate values and the light intensity at the scanning element position; 2) The image processing system where the output of the camera is handled; 3) The host computer, where the basic commands controlling the system are generated, and at the same time the most complex tasks of the data input processing are performed. The imaging processing system supports the following basic functions: 1) image digitization, conversion of the video camera input to digital form; 2) image storage-frame buffers to store the digital information in the memory; 3) internal image processor - a system for high speed image processing; 4) Interfaces - bus systems; 5) image display.

The analog-to-digital, digital-to-analog converter is a flash converter capable of digitizing 10 million samples per second.

The frame buffers hold a full video frame signal in their 25 kbyte RAM as 512 lines of 512 pixels. The frame buffers transfer the images at a rate of 30 frames per second. The high speed image processor can perform 10 million operations per second and a full video frame can be operated in real time. The host processor has 512 Kbyte RAM memory, a 20 megabyte hard disk and 1.2 megabyte floppy disk, the computer system is an advanced form of UNIX.

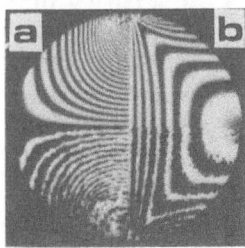

a) v displace-
ment (theory
and experiment)
b) u displace-
ment (theory
and experiment)

FIG.6. Disc under diametral compression.

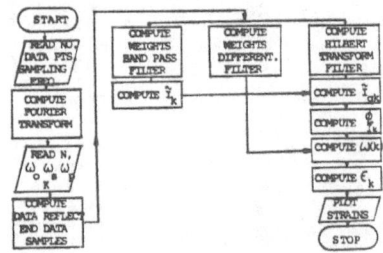

FIG.7. Flow chart of the computer program.

6. EXAMPLE OF APPLICATION

In order to illustrate the pattern processing hardware and software developed to run it, the problem of a disk under diametral compression has been selected. A holographic-moire pattern of the disk under diametral compression has been processed. The principle of holographic moire consists [9], [10], [11], [12], in illuminating the specimen with two coherent collimated beams, symmetrically situated with respect to the normal to the surface where the displacements are projected. The displacements are mapped as moire patterns of the two holographic interferograms produced by the two illumination beams. Fig. 6 shows the u and v displacements (displacements perpendicular and parallel to the loading direction). Each fringe corresponds to a displacement of 0.4 micron. The recorded moire patterns all identical to the patterns that would have been observed by printing 2500 lines/mm.

To perform the band pass filtering, the Hilbert transform, and the differentiation, a filter design program developed by McClellan, Parks and Rabiner was utilized [13]. The band pass filtering operation requires to specify the central frequency ω_o, the pass band cutoff frequency ω_p and the stop band cutoff frequency ω_s, Fig. 3, the ratio $K=\frac{\delta_1}{\delta_2}$ and N, the number of filter weights. The filter design parameters were obtained from the fast Fourier spectra of the corresponding lines.

In the process of filtering the data the following problem arises. The fringe patterns give only a finite length data sample, consequently at both ends of the interval, points required by the filtering process are missing. To solve this problem, the signals have been artificially extended. This has been achieved by data reflection and data reflection following by inversion, depending on the case [4]. Fig. 7 shows the flow chart of the computer program.

In view of the symmetry of the disk, only one quadrant of the disk was processed. A total of 16 lines in each direction were chosen for processing both the u and v patterns. Because all the lines contain zero order frequencies, low pass filter were utilized to remove the high frequency noise. A Hilbert transform was applied to obtain the

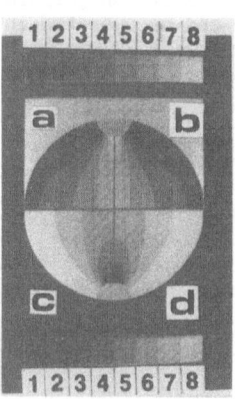

TOP:

1. -212.5×10^{-6}
2. -125.0×10^{-6}
3. -37.57×10^{-6}
4. $+49.89 \times 10^{-6}$
5. $+137.4 \times 10^{-6}$
6. $+224.8 \times 10^{-6}$
7. $+312.3 \times 10^{-6}$
8. $+399.8 \times 10^{-6}$

BOTTOM:

1. -979.0×10^{-6}
2. -834.9×10^{-6}
3. -690.9×10^{-6}
4. -546.8×10^{-6}
5. -402.8×10^{-6}
6. -258.7×10^{-6}
7. -114.7×10^{-6}
8. $+29.40 \times 10^{-6}$

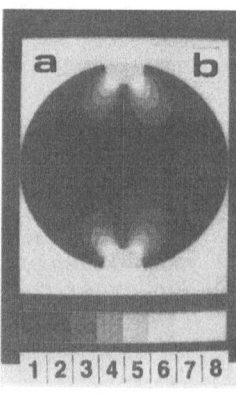

1. $+000.0$
2. $+140.2 \times 10^{-6}$
3. $+280.4 \times 10^{-6}$
4. $+420.6 \times 10^{-6}$
5. $+560.8 \times 10^{-6}$
6. $+701.0 \times 10^{-6}$
7. $+841.2 \times 10^{-6}$
8. $+981.6 \times 10^{-6}$

FIG.8.   Disc under diametral compression   FIG.9.
a) ϵx theory, b) ϵx experiment   a) γxy theory, b) γxy experiment
c) ϵy theory, d) ϵy experiment

inquadrature signals. An all pass differentiating filter was utilized to obtain the derivatives. Four low pass filters designs were utilized with N=31, the Hilbert transform filter had N=99 and the number of terms for the differentiating filter was N=31. The u and v patterns were differentiated in the two orthogonal directions, and values of

$$\epsilon_x = \frac{\partial u}{\partial x} \tag{18}$$

$$\epsilon_y = \frac{\partial v}{\partial x} \tag{19}$$

$$\gamma_{xy} = \frac{\partial u}{\partial y} + \frac{\partial v}{\partial x} \tag{20}$$

were obtained for the points of intersection of the two orthogonal systems of lines. To get a continuous set of values from the discrete values a cubic interpolation function was utilized. From the strains, the stresses were computed by applying Hooke's law. Fig. 8 shows a comparison of the theoretical and experimental normal strains ϵ_x and ϵ_y, Fig. 9 provides the

theoretical and experimental shear strains. Fig. 10 shows the principal stresses σ_1, and σ_2, and Fig. 11 the isoclinic lines. The overall agreement theoretical and experimental results is excellent. The accuracy of the experimental results is such that differences between the two sets of results can be attributed to the differences between the mathematical model and the physical problem.

7. CONCLUSIONS

The fringe processing method described in this paper, the hardware and software designed to implement it provide an accurate and fast method to extract information encoded in fringe patterns. In the particular example presented in this paper, the processed fringes belong to a holographic moire pattern, but the same method can be applied to speckle patterns, photoelastic fringes or any other type of fringe patterns. The combination of the technology for fringe pattern processing and the different optical techniques provides a powerful tool to solve problems in a number of areas of stress analysis, materials and nondestructive testing.

8. ACKNOWLEDGEMENT

This investigation was supported by the National Science Foundation of the U.S.A. through Grants: MSM-8413198 and CEE-8405431.

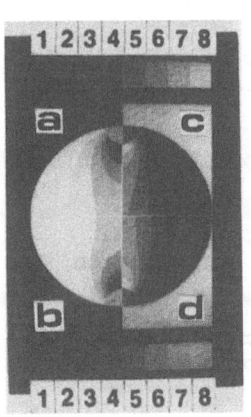
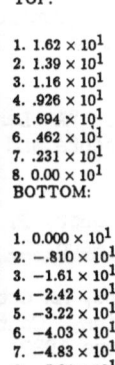
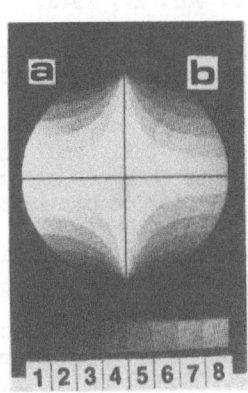

(N/mm^2)
TOP:

1. 1.62×10^1
2. 1.39×10^1
3. 1.16×10^1
4. $.926 \times 10^1$
5. $.694 \times 10^1$
6. $.462 \times 10^1$
7. $.231 \times 10^1$
8. 0.00×10^1

BOTTOM:

1. 0.000×10^1
2. $-.810 \times 10^1$
3. -1.61×10^1
4. -2.42×10^1
5. -3.22×10^1
6. -4.03×10^1
7. -4.83×10^1
8. -5.64×10^1

degrees
1. 0.00
2. 12.9
3. 26.0
4. 38.6
5. 51.4
6. 64.3
7. 77.1
8. 90.0

Disc under diametral compression

FIG.10.
a) σ_2 theory, b) σ_2 experiment
c) σ_1 theory, d) σ_1 experiment

FIG.11.
a) isoclinics theory
b) isoclinics experiment

9. REFERENCES

(1) Sciammarella, C.A., Sturgeon, D.L., "Substantial Improvements in the Process of Moire Data by Optical and Digital Filtering", 3rd Int. Cong. Exp. Stress Anal., Berlin, March 1966.

(2) Sciammarella, C.A., Sturgeon, D.L., "Digital Filtering Techniques Applied to the Interpolations of Moire-Fringe Data, "Experimental Mechanics, 7 (11), Nov. 1967.

(3) Sciammarella, C.A., Doddington, C.W., "Effect of Photographic Film Non Linearities on the Processing of Moire-Fringe Data", Exp. Mech., 7 (9), Sept. 1967.

(4) Sciammarella, C.A., "A Numerical Technique of Data Retrieval from Moire on Photoelastic Patterns", Proceedings of the S.P.I.E. Seminar-In-Depth Pattern Recognition Studies, 18, 1969.

(5) Sciammarella, C.A., Rowlands, E., "Numerical and Analog Techniques to Retrieve and Process Fringe Information", Proceedings of the 5th International Conference on Experimental Stress Analysis, Udine, Italy, 1974.

(6) Sciammarella, C.A., Narayanan, R., "The Determination of the Components of the Strain Tensor in Holographic Interferometry", Exp. Mech., 24 (4), December 1984.

(7) Takeda, M., Ina, H., Kobayashi, S., "Fourier-transform Method of Fringe Pattern Analysis for Computer-based Topography and Interferometry", J. Opt. Soc. Am. 72 (1), January 1982.

(8) Shannon, C.E., "A Mathematic Theory of Communications, Bell System Tech. J., Vol. 27, 1948.

(9) Sciammarella, C.A., Gilbert, J.A., "Holographic Moire Technique to Obtain Separate Patterns for Components of Displacement, Exp. Mech. 16 (6), June 1976.

(10) Sciammarella, C.A. and Chawla, S.K., "A Lens Holographic Moire Technique to Obtain Displacement Components and Derivatives", Exp. Mech. 18 (10), October 1978.

(11) Sciammarella, C.A., Rastogi, P.K., Jacquot, P. and Narayanan, R., "Holographic Moire in Real Time", Exp. Mech. 22 (2), February 1982.

(12) Sciammarella, C.A., "Holographic Moire, An Optical Tool for the Determination of Displacements, Strains, Contours and Slopes of Surfaces, Optical Engin. 21 (3), May/June 1982.

(13) McClellan, J.H., Parks, T.W. and Rabiner, L.R., "FIR Linear Phase Filter Design Program", Programs for Digital Signal Processing, IEE Press, 1979.

FATIGUE LIFE PREDICTION OF SPOT-WELDED LAP JOINTS BY MOIRE INTERFEROMETRY

G. NICOLETTO

Istituto di Meccanica Applicata alle Macchine
University of Bologna, Viale Risorgimento,2
40136 Bologna - Italy

1.INTRODUCTION

Although numerical methods of stress analysis are often preferred over experimental methods, the value of the experimental approach increases with the complexity of loading conditions, structural geometry and material response, [1]. Fatigue in notched components typically involves material non-linear inelastic response and severe deformation gradients.

In this paper, after a brief introduction to a strain-based fatigue life prediction method and an analysis of the fatigue failure mode of spotwelds, an elastic-plastic strain concentration study of a spot-welded lap joint by moire interferometry is described. The experimental results are subsequently used in predicting the fatigue life of spotwelds.

2.STRAIN CONTROL FATIGUE CONCEPTS

Notches are often initiation sites of fatigue failures. Stress and strain concentrations are responsible for localized material degradation. In the strain control fatigue approach (also local strain approach),[2], the fatigue life of notched components is evaluated by correlating the stresses and strains at the critical location of the component with constant amplitude fatigue test results obtained from small unnotched specimens that are tested in completely reversed strain control.

The approach requires an accurate applied load vs. maximum elastic-plastic strain relationship for the notch root, which may be obtained with a nonlinear Finite Element code, [3]. Alternatively, approximate relationships such as Neuber's Rule can be selected for practical calculations. In this case, the theoretical stress concentration factor, Kt, of the component is equated to the geometric mean of the stress and the

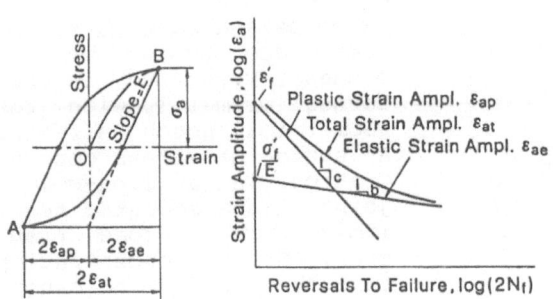

Fig. 1 - Typical cyclic material response and strain-life curve

strain concentration factors in the plastic regime, [2].

Material characterization for the strain control approach includes the determination of a cyclic stress-strain curve accounting for strain hardening and softening response during cyclic strain-control testing. The fatigue resistance of metals is often characterized by strain-life curves. A total strain amplitude, ε_{at}, vs. number of reversal to failure, $2N_f$, curve is schematically shown in Fig. 1. According to the definitions of the same Fig. 1, the total strain amplitude consists of two components, namely, the elastic and the plastic strain amplitudes, ε_{ae} and ε_{ap}. The fatigue-life curve of a material usually has the following form, [2],

$$\varepsilon_{at} = \varepsilon_{ae} + \varepsilon_{ap} = \frac{\sigma_f'}{E}(2N_f)^b + \varepsilon_f'(2N_f)^c \qquad (1)$$

in which σ_f' and b are the fatigue strength coefficient and exponent, and ε_f' and c are the fatigue ductility coefficient and exponent. The first term of Eq. (1), which relates the elastic strain amplitude to endurance, is separately referred to as the Basquin's equation. The second term, relating the plastic strain amplitude to endurance, is the well-known Manson-Coffin equation, [4].

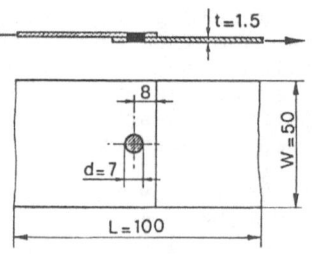

Fig. 2 - Lap joint with single spot weld

3. FATIGUE FAILURE MODE OF A SPOT-WELD

Tension-shear lap joints are often used for characterizing the fatigue resistance of spotwelds, [5]. The joint is obtained by welding two metal sheets by means of a single spot, (see Fig.2). The fatigue resistance of spotwelds is notoriously poor for various reasons, namely, geometrical notch effects, empoverished material response and altered stress state due to the welding process. This study is concerned primarily with the geometrical notch effects. A source of stress concentration in a spot-welded joint is associated to a limited area of load transfer. Stress concentration through the thickness arises for the crack-like notch and, and in the case of a lap joint, on the secondary bending stress

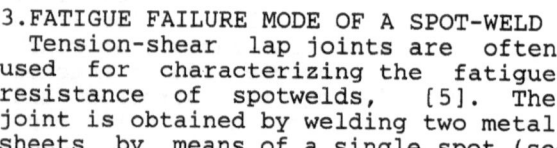

Fig. 3 - High cycle fatigue failure mode of a spotweld

contribution introduced by the off-axis traction.

Various fatigue failure modes are found in spot-welded joints, [5]. The most common in the case of high cycle fatigue is associated to crack initiation at the joint inter- face near the nugget (see Fig. 3). The crack evolution is given by propagation through the heat-affected-zone (HAZ) and final growth to failure as a through crack.

Numerical and analytical studies of a spot-welded lap joint have appeared in the literature, [5,6]. The author is not aware of a previous experimental elastic and plastic strain concentration analysis for this type of joint.

4.EXPERIMENTAL PROGRAM
4.1.Equivalent specimen for moire interferometric study

Since moire interferometry is a surface method and the most highly strained location of a spot-welded lap joint is near the nugget at the joint interface, an alternative specimen geometry, which is shown in Fig. 4, was initially devised. Through-the- thickness strain visualization in the plane of crack initiation was thus made possible.

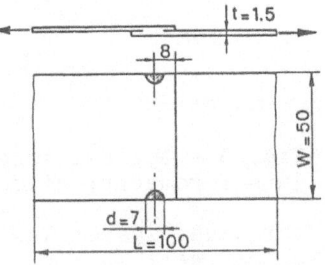

Fig. 4 - Alternative specimen geometry

Support to the equivalence assump- tion between the two specimen geome- tries of Figs. 2 and 4 may intuiti- vely draw from symmetry arguments and fluid flow analogy, [7]. More rigo- rously, the elastic analysis of a strip containing a spot under constant distributed shear, [6], evidences maximum radial and negligible tangential stres- ses at the point of crack initiation on the weld circumference where the free surface is introduced. The modified geometry for the moire interferometric study, however, conceivably alters the residual stress distribution and provides reduced constraint to plastic zone formation with respect to the original geometry.

The specimen material is a cold rolled mild steel for deep drawing operations (Euronorm EU 27 denomination: Fe PO4). The welding parameters for the 1.5-mm-thick sheet metal are : welding current 10 kA, diameter of fused area 7 mm, electrode force 2.7 kN. The basic mechanical properties of the sheet material are : yield stress σy = 221 MPa, ultimate stress σu = 332 MPa and reduction of area R.A. = 67 %.

4.2.Moire interferometry

Moire interferometry has developed in the recent years as a powerful technique in experimental mechanics, [1]. By combi- ning concepts of physical optics such as diffraction and interference, moire interferometry visualizes the relative in- plane displacements on the specimen surface with a sensi- tivity in the subwavelength range, viz. 0.417 μm per fringe order for the present application. It is a full-field techni- que and its geometrical nature renders it sensitive to nonlinear material response. Details of theory and procedures involved in moire interferometry are covered in depth in [10]. In the present study, only the longitudinal displacement component, which has prevalent effect at the point of crack initiation, was considered. A reflective phase grating of frequency f/2 (i.e. f=2400 l/mm in this study) was replicated on the lateral surface of the specimen. After specimen preparation, the equivalent specimen was mounted in a specially designed loading rig. It was monotonically loaded at increasingly higher levels by tightening a

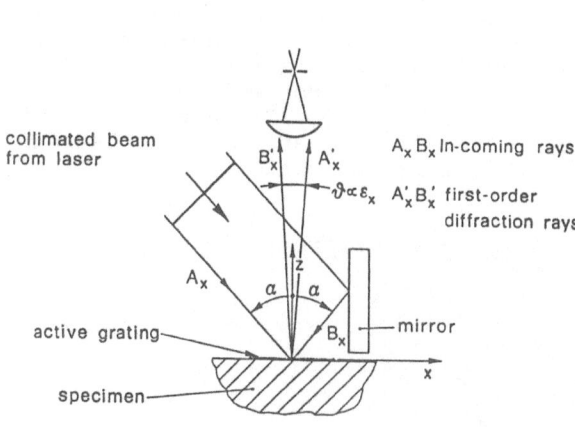

Fig. 5 - Optical scheme for moire interferometry: direct strain case

screw-nut system while simultaneously monitoring the load with a load cell in series with the specimen. Intermediate unloadings were also performed for the visualization of the residual plastic deformation.

In moire interferometry, a simple optical system creates two intersecting collimated beams of coherent light, Ax and Bx in Fig. 5 and 6, in front of the diffraction grating at angles given by

$$\sin \alpha = f \lambda / 2 \tag{2}$$

where λ is the wavelength of the light. Prior to specimen loading, the first diffraction orders, Ax' and Bx', emerge normally to the grating surface, [10]. During loading, however, the active grating distorts thus affecting the emerging beams. Interference of two warped wavefronts generates fringe patterns related to in-plane relative displacements and rotation of the specimen surface. In [11], conveniently simple relationships between in-plane strains or rotations and moire fringe gradients have been derived using the three-dimensional diffraction grating equations and the small angle assumption. When the specimen is subjected to a direct strain ε_x, as in Fig.5, the resultant interference pattern has fringes normal to the x-axis and their frequency F_x is given by

$$\varepsilon_x = F_x / f \tag{3}$$

If the specimen grating rigidly rotates about the z-axis by a small angle Φ as in Fig.6, the resultant interference pattern has fringes normal to the y-axis and their frequency F_y is given by

$$\Phi = F_y / f \qquad (4)$$

Finally, if the active grating is subjected only to a shear strain γ_{xy}, the resulting interference fringes are still perpendicular to the y-axis and their frequency F_y is given by

$$\gamma_{xy} = 2F_y / f \qquad (5)$$

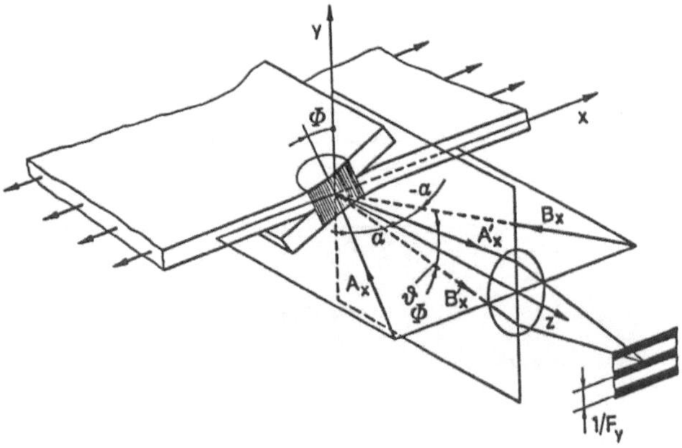

Fig. 6 - Ray analysis for the case of prevalent rigid body rotation

These relationships have been used in the quantitative analysis of the moire interferometric patterns.

5.RESULTS AND DISCUSSION
5.1.Through-the-thickness fringe patterns

Coherent optics techniques respond with high sensitivity to strain-induced displacements as well as rigid body motions. The nugget of a spot-welded joint considerably rotates about the z-axis of Fig. 6 upon loading because of the secondary bending moment. Not surprisingly, therefore, moire interfero- metric patterns were affected by this relatively large rigid rotation. A dominant fringe pattern of rotation (i.e. when $\Phi = 1°$, Eq. (4) gives a moire fringe frequency Fy $\approx$ 42 1/mm) submerged the relative in-plane displacement information. The conventional optical system of moire interferometry, schema- tically shown in Fig.5, allowed control of the fringes of rotation. They were eliminated by analogous tilting of the plane containing the intersecting beams in the undeformed condition (see Fig. 6). The amount of rotation was established to give vertical fringes in the nugget. While this rotation inevitably altered shear strain fringes (i.e. Eqs. (4) and (5) are reletive to similar patterns), direct strain measurements were not affected.

Typical depurated fringe patterns associated to the displa- cement component parallel to x-axis and for various loading conditions are shown in Fig. 7. From inspection of these magnified fringe patterns, the nugget appears to behave much as a rigid inclusion. As expected, the displacement fringes in the loaded sheets show superimposed bending and elongation, [7]. Direct strains are found to prevail at the point of crack initiation (see Fig. 3), although shear strain concentrations are localized at the nugget periphery.

Live-load elastic configurations were used in the determina- tion of the theoretical stress concentration factor Kt = 6.3, based on the following definition of nominal stress $\sigma n = P/dt$. A previous analytical elastic stress analysis of the tensile- shear joint was reported in [6]. The moire interferometric

374

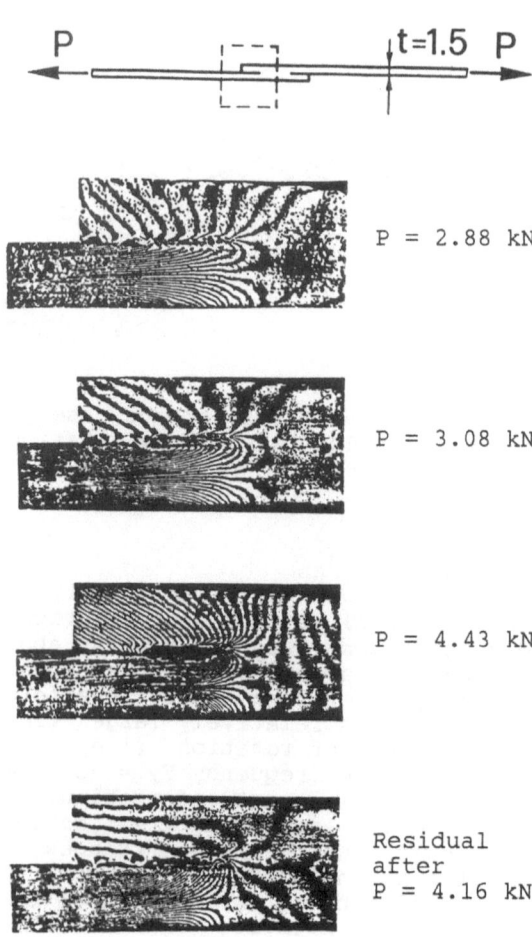

P   ⟶   t=1.5   P ⟶

P = 2.88 kN

P = 3.08 kN

P = 4.43 kN

Residual
after
P = 4.16 kN

Fig. 7 - Typical displacement component in the x-direction

estimate , however, is found to be significantly higher than the analytical counterpart (in Ref. [6], Kt = 4.2) for the present geometry. A cause of discrepancy may derive from secondary bending effects, which are neglected in the mathematical model, ·[6-7].

5.2. Applied load vs. maximum elastic-plastic strain curve

On the basis of the fatigue failure mode, the direct strain ε_x at the location of crack initiation was identified as the representative continuum mechanics parameter to be used in the life predictions. The applied load vs. maximum total strain results obtained by moire interferometry are summarized in Fig. 8. Significantly, the range of deformation, which has been investigated spanned over two orders of magnitude.The plastic strain components, obtained by intermittent specimen unloadings, are also given in Fig. 8 as a function of maximum applied load.

In Fig. 8, the load at first yield, Py, is identified . It is the load at which measurable residual strains begin to develop at the point of crack initiation. This first yield load is obviously dependent on the sensitivity of the experimental technique, which in the present case is remarkable. Since no residual strains were found for an applied load P = 1.3 kN and a residual strain of 270 µε was induced by an applied load P = 1.6 kN, the Py in Fig. 8 is determined as the average of these two values.

5.3. Fatigue life predictions

The goal of the present study was an assessment of the strain control approach to the fatigue life prediction of spot-welded lap joints. The life predictions to be discussed examine only the completely reversed constant amplitude loading case, (i.e. R =σ min/σ max = -1). The fatigue test results of Ref. [8] relative to nominally identical material and welding parameters and equivalent specimen geometry provide the benchmark fatigue curve for the spot-welded joint.

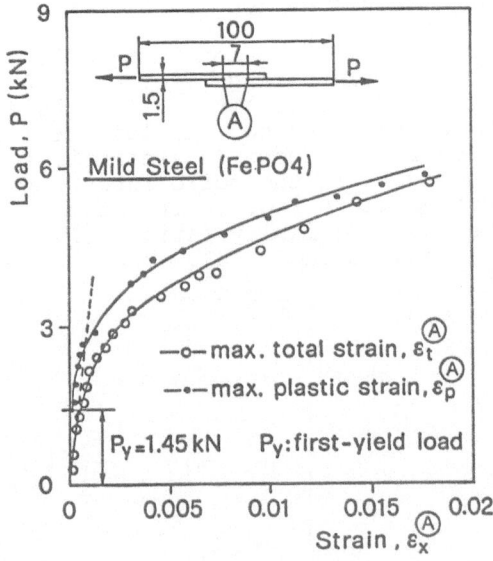

Fig. 8 - Load vs. maximum strain data for a spot-welded lap joint

In the previously described experimental strain analysis, the monotonic relationship between load vs. maximum strain has been determined. In order to estimate the fatigue life of the component from a material strain-life curve such as the one shown in Fig. 1, the monotonic load-maximum strain curve of Fig. 8 was assumed equivalent to an actual load amplitude vs. strain amplitude relationship. The rational behind this assumption is that uniformly repeated load cycles impose strain-controlled cycles to the material at notch root, as long as most of the net section remains elastic, which is the case in the high cycle fatigue range, [2].

The strain control approach requires a fatigue life curve of the component material, (viz. Eq. (1)). Fatigue properties for this class of materials have been reported previously in the literature. Reference fatigue curves for mild steel used in the automotive industry are found in [9], while coefficients and exponents of Eq. (1) for many steels are listed in [2]. Moire-based fatigue life predictions and actual test results are condensed in Fig. 9. The εat-based load amplitude vs. endurance curve refers to the material characterization of [9] for a cold-rolled mild steel. In this case, life predictions appear always conservative.

More interesting, however, is the discussion of the εap-based load amplitude vs. endurance curves. The validity field of the Manson-Coffin Equation is usually considered as confined to the low cycle fatigue range (according to [4], it extends up to about 10^5 cycles). In [12], this relationship is demonstrated to hold up to a number of cycles to failure of the order of 10^7 for a low carbon steel. Such long lives correspond to plastic strain amplitudes εap of the order of 10^{-3}. In the present study, the Manson-Coffin Equation was assumed to hold for both high and low cycle fatigue in accordance with [12]. Selection of the plastic strain range as the

376

fundamental damage parameter in fatigue would explain the
presence of the fatigue limit (also endurance limit, [4], the
stress amplitude conventionally associated to the ultra high
cycle fatigue range i.e. Nf > 10^7) observed in low strength
steels, [2,12]. It could be associated to the load amplitude
immediately below the load giving yield initiation at the
notch root. These considerations are supported by the present
findings. If the moire interferometric first-yield load Py is
introduced in Fig. 9, it indeed approximates remarkably well
the 50% reliabily fatigue limit load obtained according to
staircase method, [8].

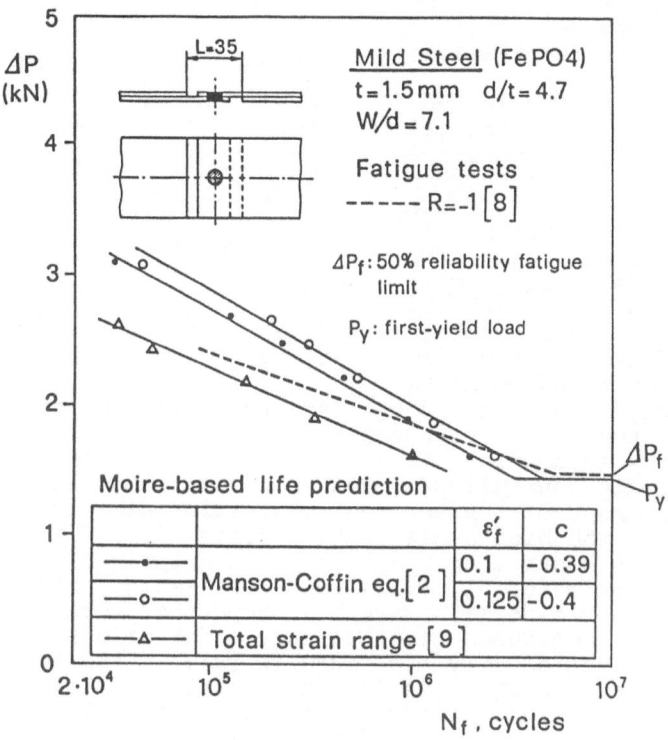

Fig. 9 - Moire-based and experimental fatigue curves for a
spot-welded lap joint under completely reversed load cycling

Inspection of Fig. 9 reveals, also, that the εap-based life
predictions are accurate in the high cycle range but are
unconservative at high load levels. Relatively minor variations in the fatigue ductility coefficient and exponent as
listed in [2] for mild steel show contained effect on the life
predictions. A possible explanation of the different slopes of
the predicted and the experimental fatigue curves for the
spot-welded joint may be attributed to the reduced span between the loading grips, (see Fig. 9). Differently from the
specimen of the present study, (see Fig. 3), the specimens

used in the fatigue tests were provided with supporting
plates to eliminate instability in the compression portion of
the fully reversed fatigue cycle, [8]. The reduced span
conceivably increased the secondary bending effects,which, on
the other hand, increased the effective strain concentration.

6.CONCLUSIONS

Moire interferometry has been used in the application of the
strain control approach to the fatigue life prediction of
spot-welded lap joint of mild steel. Applied load vs. maximum
elastic-plastic strain data for the most critical location
were obtained from moire interferometric fringe patterns. An
estimate of the theoretical stress concentration factor Kt
for this type of joint has been determined. The experimental
results in the elastic-plastic range were used in the life
prediction of spotwelded joints under fully-reversed constant
amplitude loading. The moire-based life predictions correlated
well with fatigue test results, especially at long life and
when using plastic strain range as design parameter. Furthermore, the moire interferometric estimate of load at first
yield closely approximated the fatigue limit for this type of
joint found in fatigue tests.

The present work, also, demonstrates the versatility of
moire interferometry when applied to complex problems such as
severe strain concentrations, inelastic material response,
rigid body motions and complex specimen geometry. Enhanced
understanding of fatigue mechanics and development of reliable
design parameters will hopefully result from future applications of this emerging technique.

ACKNOWLEDGMENTS

Drs. Blarasin and Castagna of the FIAT Research Center,
Orbassano, Italy, are gratefully acknowledged for providing
test specimens. Financial support to the present study was
provided by the C.N.R.-funded Progetto Finalizzato Trasporti.

REFERENCES

1. I.M. Daniel, "Experimental Methods in Applied Mechanics",
    Trans. ASME,J. of Applied Mechanics, Vol.50, 1983,963-976
2. H.O. Fuchs, R.I. Stephens,"Metal Fatigue in Engineering",
    J.Wiley and Sons, New York, 1980
3. D.F. Socie,"Fatigue Life Estimates for Bluntly Notched
    Members", Methods for Predicting Material Life in Fati-
    gue, ASME Conf. Procs., 1979, 25-39
4. S.S. Manson, "Predictive Analysis of Metal Fatigue in the
    High Cycle Life Range", Methods for Predicting Material
    Life in Fatigue, ASME Conf. Procs., 1979, 145-183
5. Y.R. Kan, "Fatigue Resistance of Spotwelds - An Analyti-
    cal Study", Metals Engineering Quarterly,Nov. 1976, 23-36
6. H. Oh, "Fatigue Life Prediction for Spotweld using Neu-
    ber's Rule", Design of Fatigue and Fracture Resistant
    Structures, ASTM STP 761, 1982, 296-309
7. G. Nicoletto "Analisi con l'interferometria moire del
    modo di rottura per fatica di un giunto saldato per
    punti" Procs. Nat. Conf. AIAS (Italian Assoc. for Exp.

Mechanics), 89-103, 1985 (in italian)

8. A. Blarasin, M. Castagna "Proprieta meccaniche di giunti in acciaio altoresistenziale saldati a punti per applicazioni automobilistiche", Centro Ricerche FIAT Report, 1984 (in italian)

9. A.R. Krause, R.W. Landgraf, B.T. Crandall, "Fatigue Properties of Cold Rolled Sheet Steels", SAE Paper 790461, 1979

10. D. Post, "Moire Interferometry", Chap. 7, Handbook on Experimental Mechanics, A.S. Kobayashi ed., Prentice-Hall, 1986,(in press)

11. R. Czarnek, D. Post, "Moire Interferometry with +/- 45-deg Gratings", Experimental Mechanics, March 1984, 68-74

12. M. Klesnil, P. Lukas, "Fatigue of Metallic Materials", Elsevier Scie. Publ. Co., Amsterdam, 1980

EVALUATION OF FINITE ELEMENT CALCULATIONS IN A CURVE-FRONTED CRACK BY
COHERENT OPTICS TECHNIQUES

G.H. KAUFMANN[(+)], A.M. LOPERGOLO[(+)], S. IDELSOHN[(++)] and E. BARBERO[(++)]
(+) Instituto de Física Rosario -IFIR- (CONICET and UNR), Rosario,
     Argentina.
(++) Instituto de Desarrollo Tecnológico para la Industria Química -INTEC-
     (CONICET and UNL), Santa Fe, Argentina.

ABSTRACT
   Two techniques, speckle photography and holographic interferometry, are
used.to test three-dimensional finite elements calculations in an internally pressurized cylinder with an external part-circular crack. Opening displacements along the crack line are measured by speckle photography and
radial displacements are obtained from holographic fringe patterns. Good
agreement between experimental and numerical data is obtained. Stress-inten
sity factor variations along the crack front are calculated from numerical
results.

INTRODUCTION
   Engineering estimates for stress-intensity factors in cracked cylinders
under internal pressure have been transformed in a very important problem
in fracture mechanics. Failure of nuclear piping systems subjected to a
cyclid load condition has been traced to fatigue growth of surface cracks.
The stress-intensity factor is required to predict the rate of fatigue crack
growth for a crack shape which changes continuously because of its growth.
For this reason, an extensive work has been recently done to calculate
stress-intensity factors for various shapes of semi-elliptical and semi-
circular surface cracks in pressurized cylinders. Three-dimensional numerical analysis using the boundary-integral equation method has been reported
by Heliot et al.[1]. Atluri and Kathiresan[2], Niyazaki et al.[3] and Raju and
Newman[4] have used three-dimensional finite element methods. Variations of
the stress-intensity factor along the crack front have been obtained for a
range of crack shapes and sizes for external and internal cracks. In spite
of the considerable importance of the experimental evaluation of numerical
methods, only few works are available in literature.
   In the past few years, coherent optics techniques such as holographic
interferometry and speckle photography have been used to determine the displacement fields adjacent to the crack tip[5-9]. Several authors use these
measurements to calculate stress-intensity factors for different types of
cracks. The use of both techniques in fracture mechanics offers several
advantages. As other optical methods, they are noncontact techniques and
the model surface needs no preparation. They do not require materials with
special optical characteristics, so they can be directly applied to any
opaque materials such as metals or composites.
   The purpose of this paper is to show the usefulness of both techniques in
fracture mechanics to evaluate calculations obtained by means of numerical
methods. As an example, the case of an external surface part-circular crack
in a thick cylinder is analyzed. Finite elements calculations are performed
and good agreement is found between experimental and numerical results.
Variations of the stress-intensity factor along the crack front calculated

by the finite element idealization are compared with results by Raju and Newman[4].

EXPERIMENTAL TEST

- Specimen

The circular cylinder used in this study was made of PMMA. This material was chosen because it is a brittle one, so no plastic flow will be expected around the crack tip. Hence, experimental results could be compared with those calculated using linear elastic fracture mechanics. The cylinder has a wall thickness t of 0.5 cm and its internal radius R_i is 2.98 cm, so the thickness to radius ratio t/R_i is 0.168. Its length 2b is 20 cm, with a 1 cm clamping on both edges, as it is shown in Fig. 1.

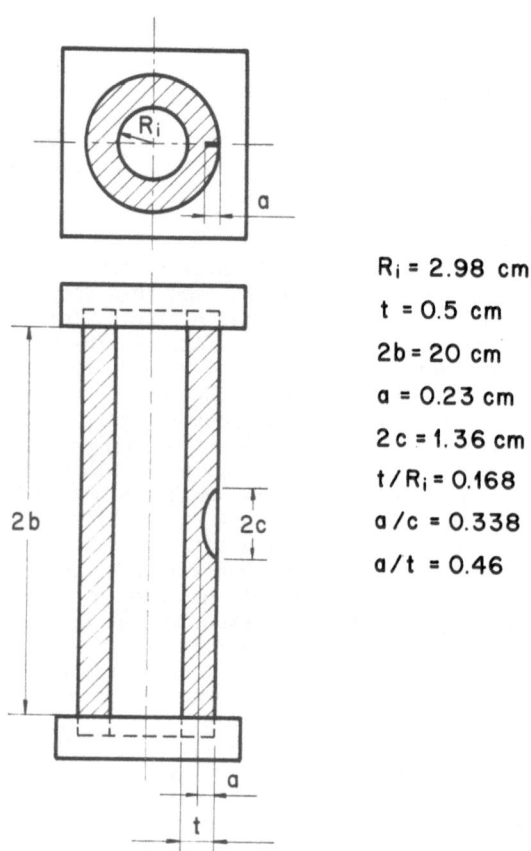

R_i = 2.98 cm

t = 0.5 cm

2b = 20 cm

a = 0.23 cm

2c = 1.36 cm

t/R_i = 0.168

a/c = 0.338

a/t = 0.46

FIGURE 1. Circular cylinder with a part-circular crack.

After the cylinder was cut into size, a central axial part-circular slot was inserted on the external surface of the specimen with a circular saw. The crack length 2c is 1.36 cm and the crack depth a is 0.23 cm, which gives an a/c ratio of 0.338 and a/t ratio of 0.46. The slot is 0.02 cm wide and terminated in a vee notch with a very small radius.

Then the specimen was heat treated to relieve stresses. Afterwards, the area surrounding the cracks was sprayed with a thin coat of matt white paint.

Two square-end steel plates were made to clamp the edges of the cylinder. On each end-plate there is a circular groove 1 cm depth which fitted nicely with the cylinder. Before the two ends of the specimen were sliden into the groove, a cyanoacrylate cement was poured, thus clampling the ends of the cylinder to the end-plates. Four rods parallel to the cylinder and fastened to both end-plates made sure the clamping not to be unglued. Pressure was applied by means of an hydraulic system and it was measured with a mano-meter. A valve was used to maintain the pressure at a constant value during the exposure time.

The values of the elastic constant $E = 2920$ MPa and $\nu = 0.33$ were ob-tained from a calibration of a bar of the same material using electrical strain-gages.

- Speckle photography
The speckle photography technique[10] was used to measure the crack opening (in-plane) displacements u_θ along the slot line. They were obtained by measuring the crack aperture along two lines parallel to the slot direction, one at each side of the crack and at a distance of 0.025 cm from it. The u_θ displacements were calculated as one half of the difference between the cracks apertures corresponding to a pair of points at each side of the slot.

The sensitivity of this technique is limited by the average speckle sizes. It can be shown that displacements to be measured must be greater than[10]

$$\delta_o = 1.22 \ \lambda F \ (1 + m)/m \qquad (1)$$

where λ is the wavelength of the laser, F is the aperture ratio of the recording optical system and m is its magnification. Nevertheless, dis-placements observed along the crack line were very small (about a few microns) which were outside the sensitivity of the technique given by eq. (1). This could be increased, however, by superimposing a small known artificial displacement. For our application, a rigid body displacement of about three times the speckle diameter was produced by mounting the photo-graphic camera on a vernier table. This displacement was cancelled when the difference between the crack apertures corresponding to a pair of opposite points along the slot was done.

When the cylinder was pressurized between both exposures, its surface moved in the radial direction. It is well known that out-of-plane displace-ments produce speckle decorrelation which causes a decrease in fringe con-trast and introduces errors in the measurements[10]. The Rayleigh criterion establishes the movement which can be tolerated without any appreciable de-gradation of the fringes. This movement must be less than

$$\Delta = 4 \ \lambda \ F^2 (1 + m)^2/m^2 \qquad (2)$$

For the experiments presented in this paper $\Delta \sim 500$ µm. Taking into account that the radial expansion for the maximum pressure utilized was about 30 µm, fringes with good contrast were obtained without using any compensation procedure.

To simplify data reduction, the cylinder surface was imaged by means of a well corrected camera lens on to a Kodak SO-253 photographic plate mounted parallel to the specimen. A 21 cm focal length of $F = 8$ aperture was used at m = 2.8 magnification. The specimen was illuminated by an He-N$_e$ laser. Double-exposure specklegrams were analyzed using the pointwise technique. By measuring the spacing and inclination of the fringes, aperture displacements along the crack line were calculated.

- Holographic interferometry

Holographic interferometry was used to measure the radial (out-of-plane) displacement u_r along the crack line. The usual holographic set up to measure out-of-plane displacements was utilized[11]. The cylinder was normally illuminated with a collimated beam of He-N$_e$ laser light and it was mounted parallel to the holographic plate. Several double-exposure holograms were recorded on Agfa-Gevaert 8E75 plates for different pressures. After processing the hologram, the image showing deformation fringes was re constructed and photographed at normal incidence. Under these conditions, any point on the axial crack line has a radial displacement given by[11]

$$u_r = N \lambda/2 \tag{3}$$

where N takes on integer values at the center of each bright fringe. From each pattern, the change in radial displacement along the crack line was ob tained and was compared with numerical predictions.

NUMERICAL ANALYSIS

Three-dimensional Finite Elements Calculations were performed to be compared with the experimental results. Fig. 2 shows the discretization used. It employed 275 elements with 4107 degrees of freedom. Two types of elements were used to model the cylindrical vessel: a layer of colapsed singular elements[12] around the crack front and 20-nodes isoparametric elements with 60 dof elsewhere. The curvature of the model was represented through second grade parabolaes. 3D elements were used because the model was a thick cylinder. The dimensions of the elements in the neighbours of the crack were determined by means of a convergence study on a known strain crack problem. As it is shown by reference 12, colapsed singular elements are the best choice among the singular isoparametric 3D elements to model the singular crack problem.

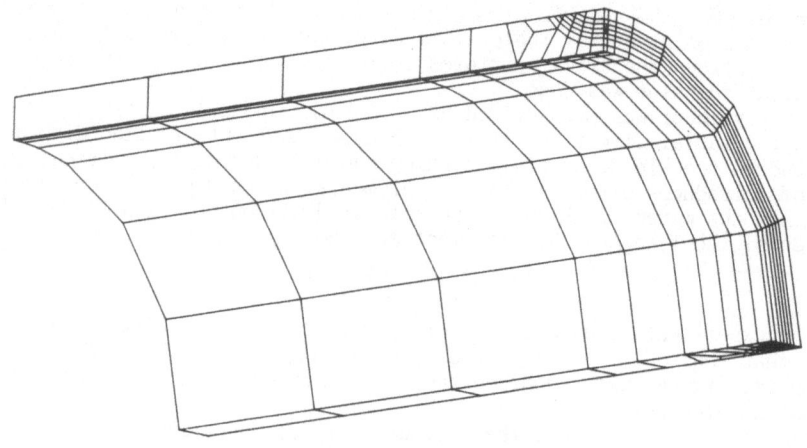

FIGURE 2. Finite element mesh

The model idealizes one-eighth of the vessel, so that it simulates a vessel with two diametrically located surface cracks, 180 degree apart. As it is shown by reference 4 this does not introduce appreciable error.

Symmetry boundary conditions were applied on the $x = 0$, $y = 0$, $z = 0$ planes, and clamping restrictions on the $z = b$ plane. Internal pressure was introduced by means of a system of energetically equivalent radial forces. The computer program SAMCEF[13] was used for the analysis.

Mode I elastic stress intensity factors were calculated by means of the crack-opening displacement method, assuming plane strain condition.

RESULTS

Opening displacement along the crack line for an applied pressure $p = 1.2$ MPa are shown in Fig. 3. Radial displacement along the same line for $p = 0.2$ MPa are plotted in Fig. 4. Satisfactory agreement between experimental and numerical results was obtained.

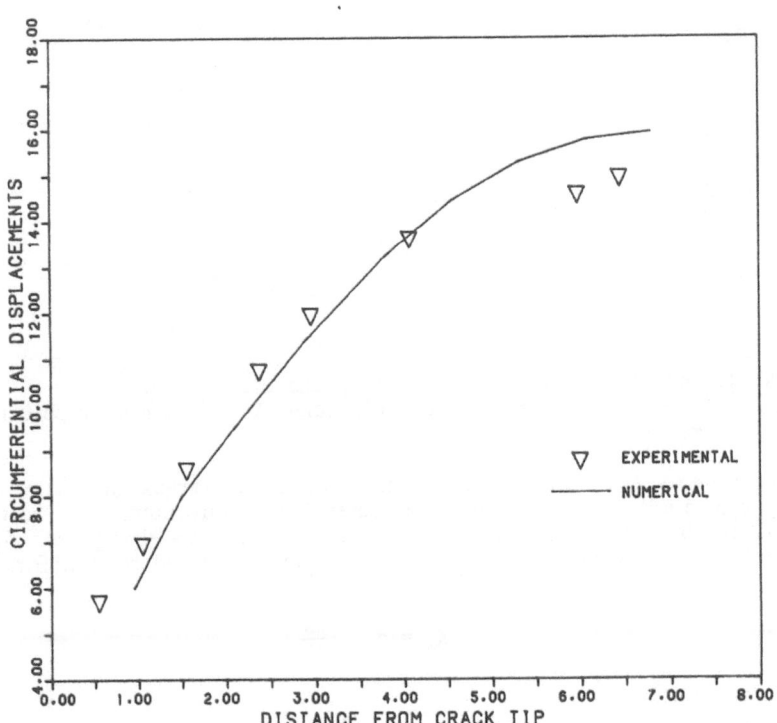

FIGURE 3. Opening displacement u_θ along the crack line.
——— , finite element calculations; ∇ , from speckle measurements.

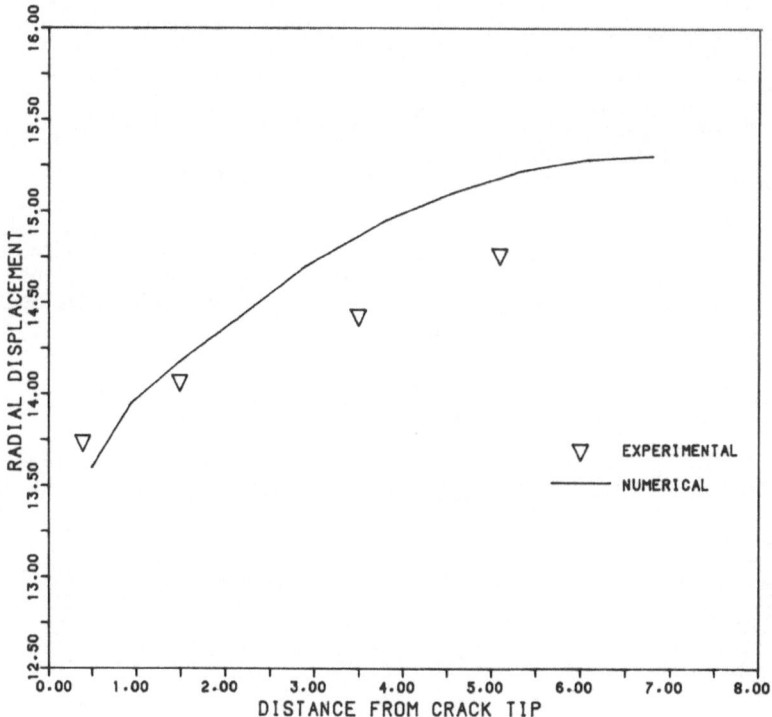

FIGURE 4. Radial displacement u_r along the crack line.
———— , finite element calculations; ∇ , from holographic measurements.

Fig. 5 shows the distribution of nondimensional stress-intensity factors along the crack front obtained from numerical calculations. The results are normalized by the stress-intensity factor K_0 at $\phi = 90$ degrees for an elliptical crack embedded in an infinite body subjected to a uniform tension stress σ_0

$$K_0 = \sigma_0 \sqrt{\frac{\pi a}{Q}}$$

with

$$\sigma_0 = \frac{2 \, R_i^2}{R_o^2 - R_i^2} P$$

where R_0 is the external radius of the cylinder, and Q is the square of the complete elliptic integral of second kind approximated by $Q = 1 + 1.464 \, (a/c)^{1.65}$. The results of Raju and Newman[4] for a semi-elliptical crack of $t/R = 0.1$, $a/c = 0.4$ and $a/t = 0.5$ and the results from ASME

Boiler & Pressure Vessel Code Section XI are also depicted for comparison in Fig. 5.

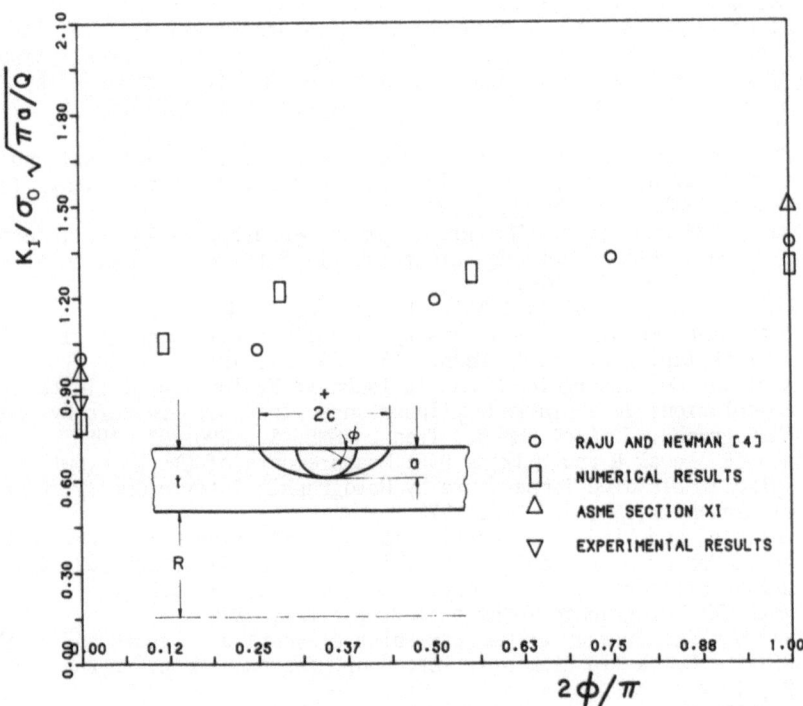

FIGURE 5. Stress-intensity factors distribution.
□ numerical results (part-circular crack, $t/R_i = 0.168$,
$a/c = 0.338$, $a/t = 0.46$); o Raju and Newman results[4] (semi-elliptical crack, $t/R_i = 0.1$, $a/c = 0.4$, $a/t = 0.5$); △ from ASME Boiler & Pressure Vessel Code Section XI.

CONCLUSION

Concluding, the use of the speckle photography technique for the measurements of displacements surrounding a crack tip in an internally pressurized cylinder was successfully demonstrated. Once displacements are determined, fracture parameters such as stress-intensity factor can be deduced. This work also demonstrates the practicability of using speckle photography techniques to evaluate calculations obtained by means of numerical methods, such as finite elements. Using this technique, future investigations will be directed to the measurement of stress-intensity factors for curve-fronted cracks, problem in which much work remains to be done in order to test different numerical methods.

REFERENCES

1. Heliot J, Lahbens RC and Pellissier-Tanon A: Semi-Elliptical Cracks in

the Meridional Plane of a Cylinder Subjected to Stress Gradients. Fracture Mechanics, ASTM STP 677, American Society for Testing and Materials, 341-364, 1979.

2. Atluri SN and Kathiresan K: 3 D Analysis of Surface Flaws in Thick-Walled Reactor Pressure Vessels Using a Displacement-Hybrid Finite Element Method. Nucl. Eng. Des., 51, 163-176, 1979.

3. Miyazaki N, Shibata K, Watanabe T and Tagata K: Stress-Intensity Factor Analysis of Surface Cracks in Three-Dimensional Structures. Comparison of the Finite Element Solutions with the Results obtained by the Simplified Estimation Methods. Int. J. Pres. Ves & Piping 15, 37-59, 1984.

4. Raju IS and Newman JC: Stress-Intensity Factors for Internal and External Surface Cracks in Cylindrical Vessels. J. Pres. Ves. Techn. Trans. ASME, 104, 293-298, 1982.

5. Barker DB and Fourney ME: Displacement Measurements in the Interior of 3-D Bodies Using Scattered-Light Speckle Patterns. Experimental Mechanics, 16, 209-214, 1976.

6. Chian FP and Asundi A: A White Light Speckle Method Applied to the Determination of Stress-Intensity Factor and Displacement Field Around a Crack Tip. Eng. Fract. Mech., 15, 115-121, 1981.

7. Kaufmann GH, Lopérgolo AM and Idelsohn S: Evaluation of Finite Element Calculations in a Cracked Cylinder under Internal Pressure by Speckle Photography. J. Appl. Mech., Trans. ASME, 50, 896-897, 1983.

8. Hsu TR, Lewak R and Wilkins BJS: Measurements of Crack Growth in a Solid at Elevated Temperature by Holographic Interferometry. Experimental Mechanics, 18, 297-302, 1978.

9. Ennos AE and Virdee MS: Application of Reflection Holography to Deformation Measurement Problems. Experimental Mechanics, 22, 202-209, 1982.

10. Erf RK(ed): Speckle Metrology. Academic Press, 1978.

11. Vest CM: Holographic Interferometry. Wiley, 1979.

12. Barsoum RS: Triangular Quarter-Point Elements as Elastic and Perfectly-Plastic Crack Tip Elements. Int. Jour. for Num. Meth. in Eng., 11, 85-98, 1977.

13. SAMCEF: Systeme D'Analyse des Milieux Continus par Elements Finis. Manuel D'Utilisation. Universite de Liege.

REAL-TIME MOIRE-HOLOGRAPHIC ANALYSIS OF PLATES WITH SHARP V-NOTCHES IN TENSION

F. GINESU[(*)], C. PAPPALETTERE[(§)]

(*) Dipartimento di Meccanica, University of Cagliari, Italy
(§) Dipartimento di Progettazione e Produzione Industriale, University of Bari, Italy

1. INTRODUCTION

Fatigue design methods of structures are based on either maximum values or the stress field near the shape change. The former is usually applied when the change of shape has a fillet-radius and relies on a stress concentration and a notch sensitivity factor (1). The latter is used for cracks with a zero bottom radius and a small opening angle (2).

Recently a new method based on stress field criteria (similar to that of Linear Elastic Fracture Mechanics) has been proposed for structures with high stress concentrations, i.e. with a very small or zero fillet radius (discontinuities). The technique has been validated by numerical and experimental analysis of cruciform welded joints with a small fillet radius (use of normal electrodes) at the weld toe (3),(4) and for lap joints with a sharp change of shape (fillet radius of zero) (5).

Using numerical techniques (FEM), Atzori (6) has recently confirmed the possibility of employing stress field criteria for cracks with a bottom radius of zero, or very small, and an opening angle different from zero, i.e. V-notches.

In the present paper the strain distribution around the discontinuity point of V-notches is studied by means of an optical technique. In particular the Holographic-Moiré method (7),(8) is used as it allows a whole field analysis of the specimens and sensitivity can be easily varied. The latter technique has been used by the authors (9),(10) in studying mechanical structures.

A numerical analysis is then developed to check experimental data. Two different finite element programs are used for this purpose (STRUDL-IUG version - ANSYS).

2. EXPERIMENTAL

Seven specimens, whose geometrical dimensions are shown in Fig.1 were cut from 1.52 mm thick Kodak Photoplast plates. As can be seen from the figure (S), three specimens had a notch on just one edge and a β angle of 45°, 90° and 135°, whereas the other specimens (D) had a notch on both edges with a β angle of 45°, 30° and 15°.

All specimens had a 10 mm deep notch apart from the one with a β angle

Wieringa, H (ed), Experimental Stress Analysis.
© *1986. Martinus Nijhoff Publishers, Dordrecht.*

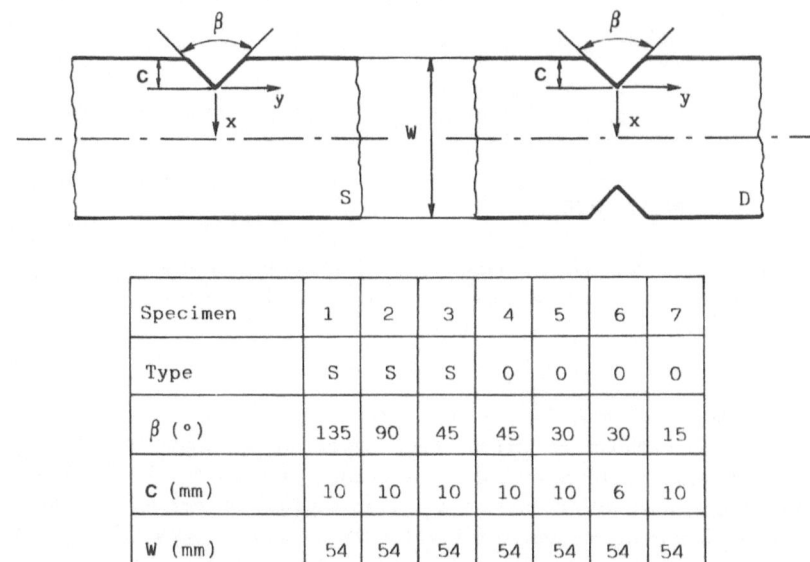

FIGURE 1. Specimen geometries.

Specimen	1	2	3	4	5	6	7
Type	S	S	S	O	O	O	O
β (°)	135	90	45	45	30	30	15
C (mm)	10	10	10	10	10	6	10
W (mm)	54	54	54	54	54	54	54

of 30° which had a depth of 6 mm. V-notches were cut using a milling mach-ine with the exception of those specimens with opening angles of 15° and 30° where a 100 W CO_2 laser was employed.

A crossed grating was photoengraved on the models by means of a 5 W Argon laser. The resulting graticules had a frequency of 80–133 lines/mm.

The experimental set-up is shown in Fig.2. Obviously a pitch equal to $\lambda/(2 \cdot \sin \theta)$ (λ = 488 mm is the wavelength of the light) was obtained.

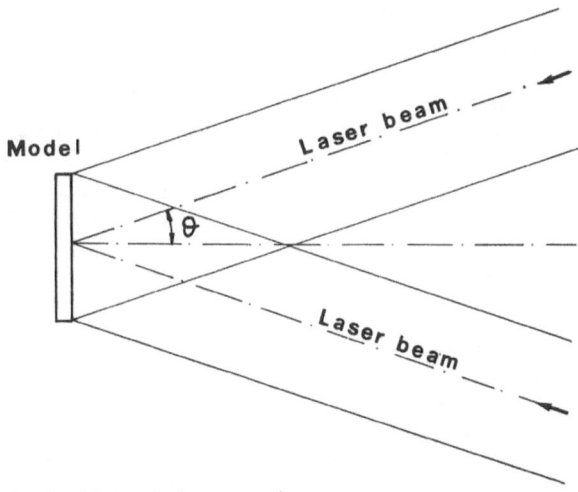

FIGURE 2. Grating photoengraving.

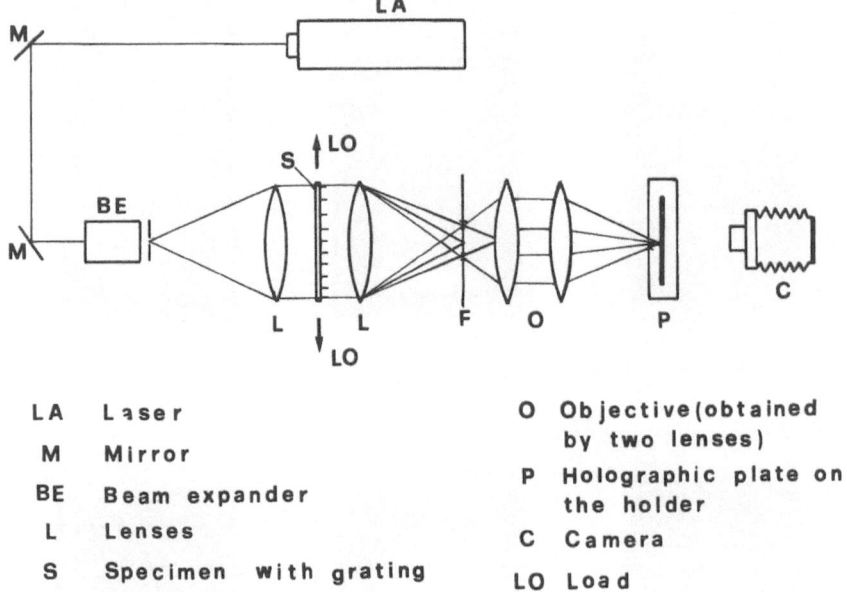

LA	Laser	O	Objective(obtained by two lenses)
M	Mirror		
BE	Beam expander	P	Holographic plate on the holder
L	Lenses		
S	Specimen with grating	C	Camera
F	Filter	LO	Load

FIGURE 3. Moiré-holographic set-up.

As Fig.3 shows, the models were placed in a Fourier transform optical system. Moiré fringes were obtained using either an Argon or a Helio-Neon laser. Only two symmetrical diffraction orders, +1 and -1, were taken into account. In the case at hand the sensitivity of the method was about $5 \cdot 10^{-3}$ mm/fringe which proved sufficient for obtaining significant results.

All tests were performed by real time analysis. To avoid any rigid displacements a film holder-processer, which provided in-place development, was employed.

With the first exposure a hologram was taken on an Agfa holotest film while a common FP4 plate by Ilford was utilized to record moiré fringes.

The moiré-holographic technique is well known (7),(8); it consists here in placing the model between the two lenses L (see Fig.3) and taking a first exposure. The film plate (P) is then developed and another exposure taken. Next the model is loaded and this second exposure, superimposed on the first, produces the moiré fringes.

Figure 4 shows some pictures of moiré-holographic fringes. All these pictures were obtained using a He-Ne laser. Only load direction displacements are given in the figure as they were of more importance in the present analysis.

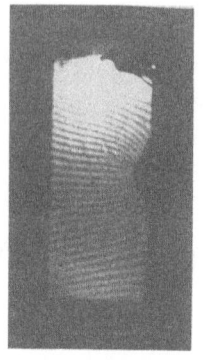 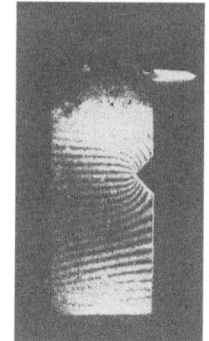 

single-notched specimens

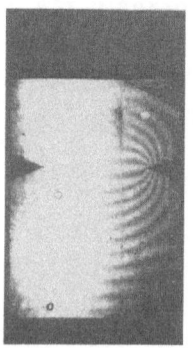 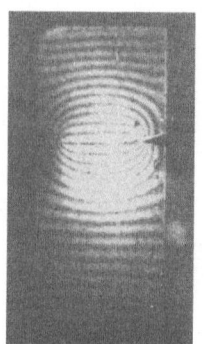 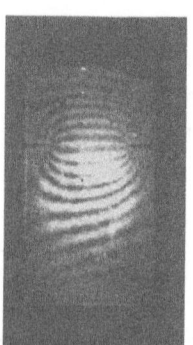

double-notched specimens

FIGURE 4. Moiré-holographic fringes

3. NUMERICAL ANALYSIS AND RESULTS

To check the experimental data two different finite element programs were used: the STRUDL IUG version and the ANSYS codes, implemented on an IBM 4341 computer and a 300 Apollo graphic station respectively.

Figure 5 shows one of the finite element mesh , which was obtained by quadratic-isoparametric elements, and the displacement plot for the double notched specimen with opening angle of 45°.

Figures 6 and 7 give the numerical and experimental results respectively for the single V-notched specimens while Figures 8 and 9 show similar results for the double V-notches.

From these figures it emerges that numerical and experimental data appear to be in good agreement.

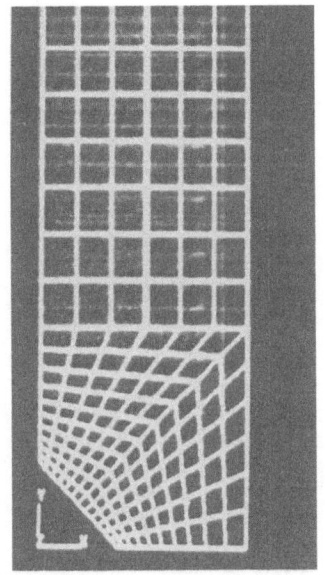 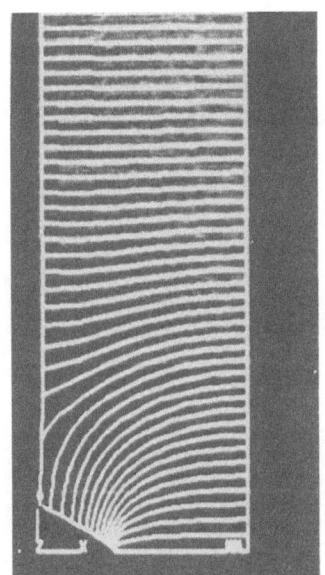

FIGURE 5. Example of FEM mesh and displacement plot.

4. CONCLUSIONS

In the graphs of Fig.6-9 logarithmic scales were used to verify field criteria in fatigue design of V-notched components. According to these criteria strain intensity distribution near the sharp fillets is linear and its slope varies with opening angle (2). Both numerical and experimental results, which are plotted in the same figures against the y coordinate, have linear slope and they seem to confirm the above mentioned criteria.

The graph of Fig.10 summarizes the relationship between the opening angle of V-notches and the $\varepsilon_y/\varepsilon_n$ ratio. This graph can of course be very useful in fatigue design of V-notched mechanical parts.

All results are reported in this diagram and they are very close to those found by Atzori (6).

According to (11),(12),(13) stress ratio can be expressed, using the notation of Fig.1, for small opening angles of the crack, as

$$\frac{\sigma_i}{\sigma_n} = \frac{F}{\sqrt{2}} \sqrt{\frac{C}{r}} \, f_i(\theta)$$

where σ_i is $\sigma_x, \sigma_y, \tau_{xy}$
σ_n is the nominal stress
F is the shape factor

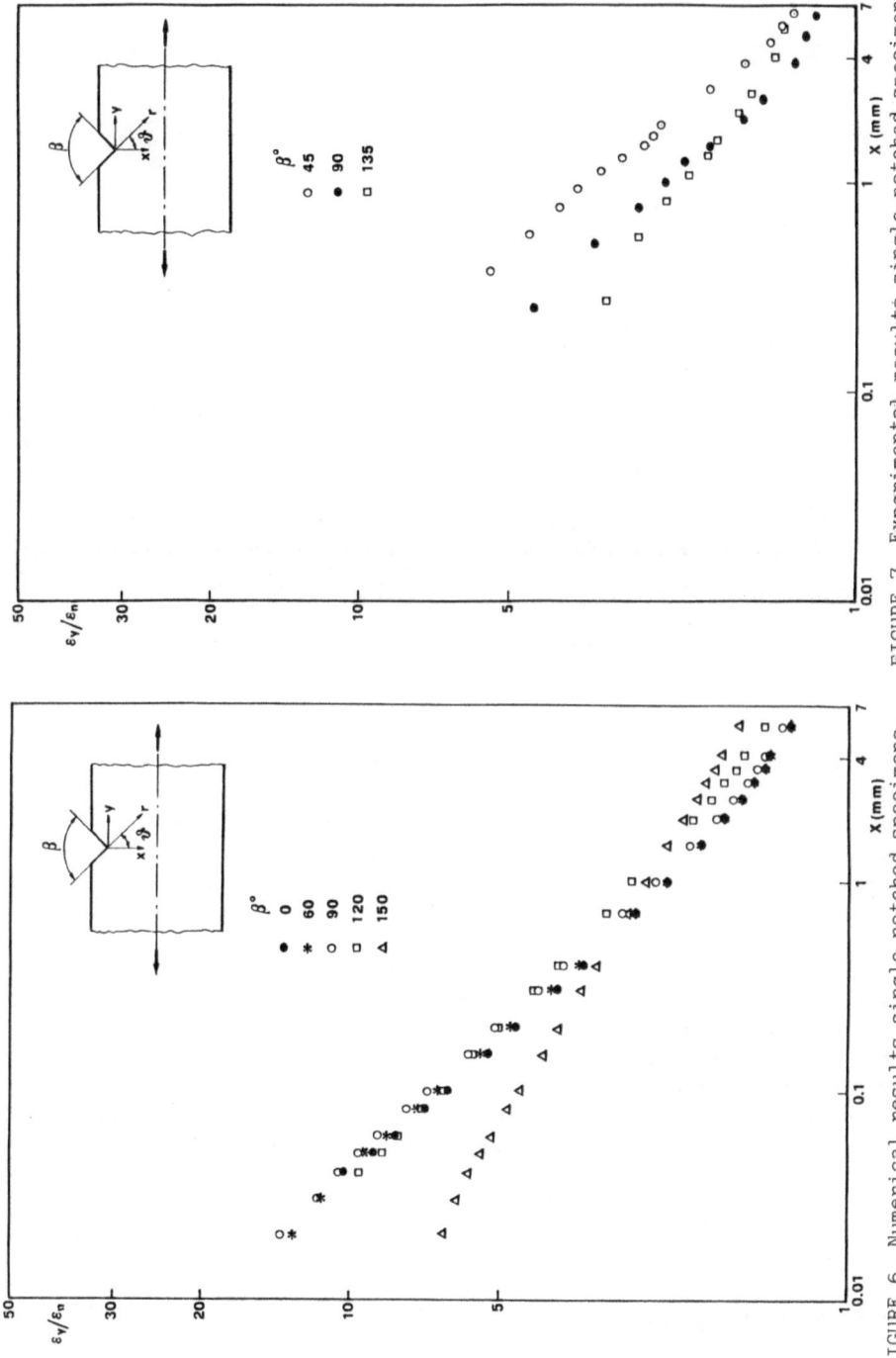

FIGURE 6. Numerical results single-notched specimens.

FIGURE 7. Experimental results single-notched specimens.

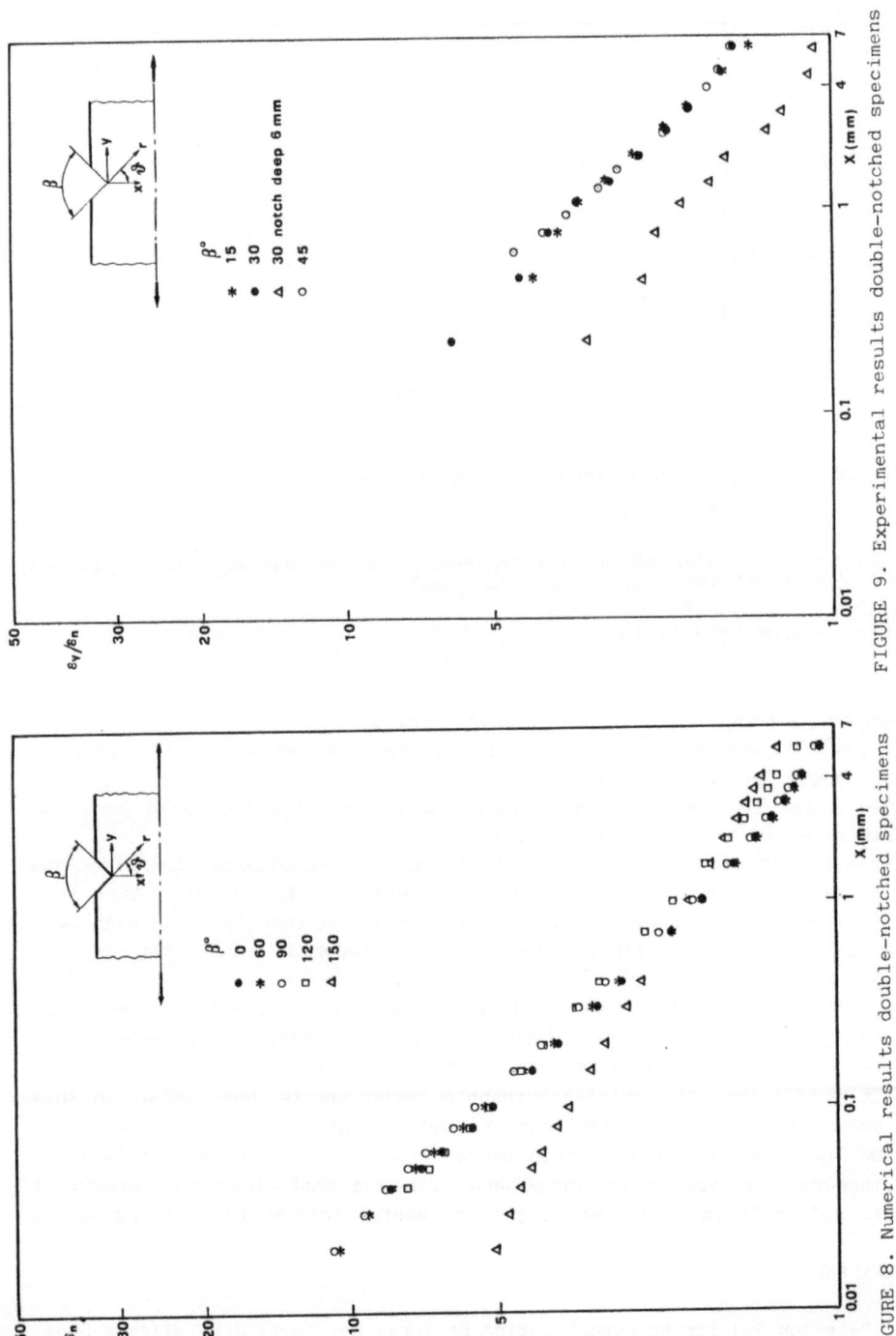

FIGURE 9. Experimental results double-notched specimens

FIGURE 8. Numerical results double-notched specimens

394

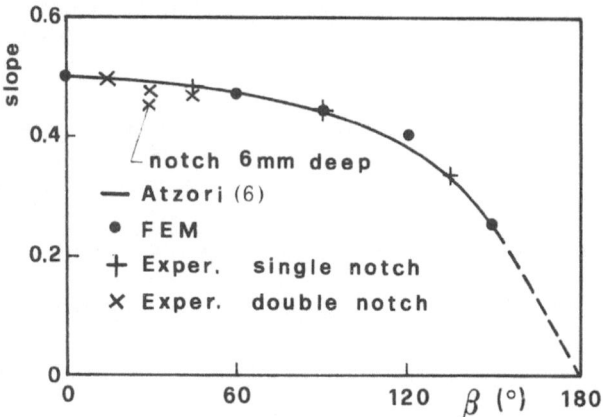

FIGURE 10. Comparison between numerical and experimental results.

In the case under study this expression can be appropriately rewritten

$$\frac{\sigma_i}{\sigma_n} = \text{const.} \left(\frac{c}{r} \right)^\alpha \cdot f_i(\theta)$$

where α is extracted from the graph of Fig.10.

With this modification Linear Elastic Fracture Mechanics Analysis can be adopted for this kind of problem.

In regard to the experimental and numerical techniques used here the authors would point out the following:

- the FEM is certainly easier to use than the experimental technique and different geometries can be analysed very quickly. On the other hand it simplifies and idealizes the structures and sometimes, through such idealization, the real behaviour of the component under study is mis- understood;

- optical experimental techniques generally allow whole field measurement of specimen deformations. Often the only idealization introduced is the use of a model but measurements are made directly.

In particular, the moiré-holographic technique is very useful in this case because it gives displacements and strains can be easily read.

On the other hand with strain gauge techniques for instance only local measurements can be performed and with the photoelasticity method it is difficult in this case to extract useful information for engineers.

REFERENCES

1. Peterson RE: Stress Concentration Factors. New York: John Wiley & Sons, 1974.

2. Frost NE, Marsh KJ, Pook LP: Metal Fatigue. Oxford: Clarendon Press, 1974

3. Atzori B, Haibach E: Deformazioni Locali al Piede dei Cordoni di Saldatura d'Angolo e loro Correlazioni con la Vita a Fatica del Giunto. Il Progettista Industriale, n.1, March 1981, pp.70-81.

4. Atzori B, Blasi G, Pappalettere C: Evaluation of Fatigue Strength of Welded Structures by Local Strain Measurements. Experimental Mechanics June 1985.

5. Atzori B, Pappalettere C: Applicazione di Tecniche Numeriche e Sperimentali alla Progettazione a Fatica dei Giunti a Sovrapposizione. XIII National Congress of Italian Society for Stress Analysis (AIAS), Bergamo, September 1985.

6. Atzori B: Meccanica della Frattura o Effetto d'Intaglio nella Progettazione a Fatica. XIII National Congress of Italian Society for Stress Analysis (AIAS), Bergamo, September 1985.

7. Sciammarella CA: The Moiré Method - a Review. Experimental Mechanics, November 1982.

8. Di Chirico G: Advances in Moiré Strain Measurements of Diffracted Wavefronts. IUTAM Symposium, Poitiers, September 1979.

9. Di Chirico G, Ginesu F: Analisi Moiré-Olografica su Modelli di Lastre Rinforzate. IX National Congress of Italian Society for Stress Analysis (AIAS), Trieste 1981.

10. Di Chirico G, Ginesu F, Pirodda L: Optical Differentiation with White Light Diffracted Wavefronts. 13th Congress of the International Commission for Optics, Sapporo 1984.

11. Westergaard HM: Bearing Pressures and Cracks. J. of Applied Mechanics, 61, June 1939.

12. Irwin GR, Kies JA, Smith HL: Proceedings ASTM, 58, 640, 1958.

13. Liebowitz (ed): Fracture. New York: Academic Press, 1968.

ACKNOWLEDGEMENT

Thanks are due to Ing. F. Cordero of Valfivre, Florence, for laser cutting of some of the specimens.

INVESTIGATION ON CRACK CLOSURE BY REAL-TIME HOLO-INTERFEROMETRY

F.M. FURGIUELE, M.L. LUCHI, A. POGGIALINI

DIPARTIMENTO DI MECCANICA, UNIVERSITA' DELLA CALABRIA - ITALY

1. INTRODUCTION

The plastic deformation left in the wake of a crack causes fracture surfaces to be pressed together by compressive residual stresses thus producing the well-known crack closure phenomenon which can affect the crack propagation rates whenever the minimum load in the cycle is insufficient to open the crack.

Since first observed by Elber, the phenomenon has been investigated both numerically and experimentally /1,2,3,4,5,6,7,8,9,10/. Although most of the investigations confirm crack closure and its effect on fatigue crack growth, a completely clear picture has not as yet emerged.

This paper describes experiments conducted on a high strength steel standard CT specimen.

The holographic-interferometry technique employed /11/ allows the transverse displacements due to plastic and/or elastic deformation to be detected on the surface of the specimen with a resolution of about 1/4 μm. The real time version of the holographic technique enables the residual deformation ahead of and behind the crack front to be estimated during propagation. In fact, by using one or more reference holograms and by repositioning both the hologram and specimen on the holographic bench during fatigue tests, it is possible to observe interference fringe patterns related both to the plastic deformation occurring during propagation and to the elastic deformation due to the compressive load caused by crack closure.

Furthermore, if the cracked specimen is loaded at different levels on the holographic bench after a hologram has been recorded, fringe patterns arise which contain information on the elastic deformation due both to the applied load and to the release of the compressive residual stresses thus enabling the crack opening load to be estimated /9/ .

2. EXPERIMENTAL PROCEDURES

2.1. Holographic System

The optical layout of the holographic system employed in the real-time interferometry experiments is shown in Fig.1. The light beam from an Argon-Ion laser is split by a semireflecting mirror; the reference beam is obliquely (22°30') directed to the holographic plate after being expanded to cover its surface; the object beam, suitably expanded and collimated, is directed to the specimen. The front surface of the specimen is illuminated at a constant angle of 22°30'. A mirror, placed behind the specimen and parallel to it, enables the back surface to be illuminated at the same constant angle.

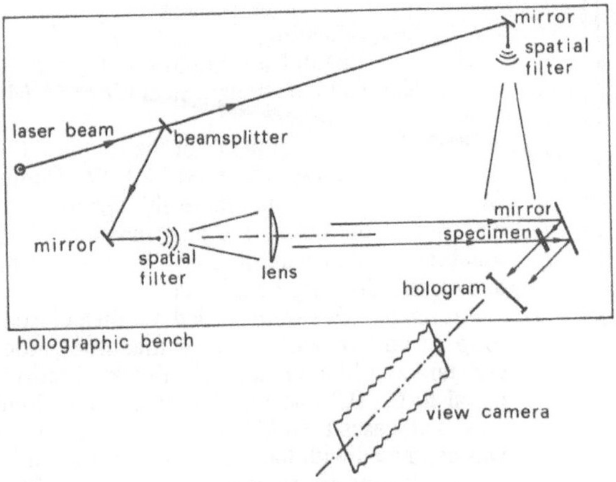

When the line of sight and the direction of illumination are symmetrically orientated about the normal to the surface, the fringe pat-

FIGURE 1. Layout of the holographic system.

Wieringa, H (ed), Experimental Stress Analysis.
© *1986. Martinus Nijhoff Publishers, Dordrecht.*

FIGURE 2. Positioning and loading of the specimen on the holographic bench.

terns observed are generated by out of plane displacements only. The contour interval between the fringes corresponds to a displacement of λ/cos 22°30', where λ is the wavelength of the laser light (488nm). The same line of sight can be used for both front and back surfaces, the latter being viewed through the mirror employed for the illumination.

The fringe patterns are recorded without perspective distortion using a view camera. The focal length of the lens (300 mm) and the magnification (~2/1) make the direction of observation practically constant over the whole area investigated. In order to reduce speckle noise the photographs are taken at the lowest f number of the lens (f/9) whenever the in-plane displacements are sufficiently low.

A holder consisting of a clamping device and a repositioning base designed on kinematic principles is used to reposition accurately (within λ/4) the specimen once the fracture has been grown on the testing machine (see Fig. 2). Three V-grooves are milled on the bottom side of the base which rests on three balls fixed on a post mounted on the optical bench.

A similar repositioning holder allows the holographic plate to be relocated with same accuracy once it has been processed. The plate is glued by cyanoacrylate adhesive to an L-shaped aluminum support screwed to the kinematic base.

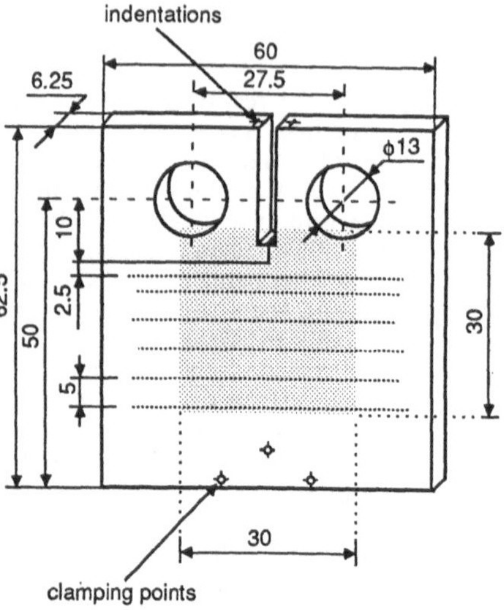

Since several holograms can be recorded during fatigue tests to compare two or more different loading conditions or crack lengths, the full potential of the real-time technique is exploited.

2.2. Method of testing.

The experimental investigations have been conducted on a high strength steel (28Ni Cr Mo V 12).

Fig.3 shows the geometry of the standard CT specimen tested (ASTM E 647-83, W=50mm, B/W=1/8). The points where the specimen is clamped to the holder are also indicated. The machined notch was obtained by sawcutting and electro-discharge machining.

All fatigue tests were carried out on a closed-loop servo-hydraulic testing machine under constant amplitude loading, with a load ratio R equal to 0, 0.2 and 0.5. The maximum load was in all cases P=10kN. The crack tip position was estimated with the precision of 0.05mm by using a low power travelling microscope (25x) mounting a reticle.

FIGURE 3. Geometry of the CT specimen (the area taken in the photos is evidenced).

No particular device is required to detect residual deformation in the zero load specimen. Using as reference a hologram of the uncracked specimen the total lateral contraction which develops ahead of and behind the crack tip after the crack has advanced can be observed by simply replacing the specimen on the optical bench at different stages of propagation. A hologram of the cracked specimen can also be employed to detect more accurately the deformation in the wake. In order to release contact loads, the crack is advanced for a length such that contact takes only place ahead of the position of the crack tip in the holographed specimen. Residual stresses can alternatively be released by cutting the specimen (e.g. by wire electro-discharge machining) into two halves through the crack front along a plane normal to the bottom face. When half of the specimen is relocated on the optical bench, the fringe patterns arising relate to the elastic deformation both behind and ahead of the crack tip.

In order to estimate the opening load, the cracked specimen is loaded after a hologram has been recorded, processed and replaced on the optical bench. The load is manually applied with a purpose designed device supported by the specimen through two pins fastened to the holes (see Fig.2). The extremities of the two pins can be independently moved apart by means of two lever arms, mounted on roller bearings and hinged on a ball placed below the plane of the pin axes. A spherical tipped micrometer acts on each couple of lever arms. It is thus possible to

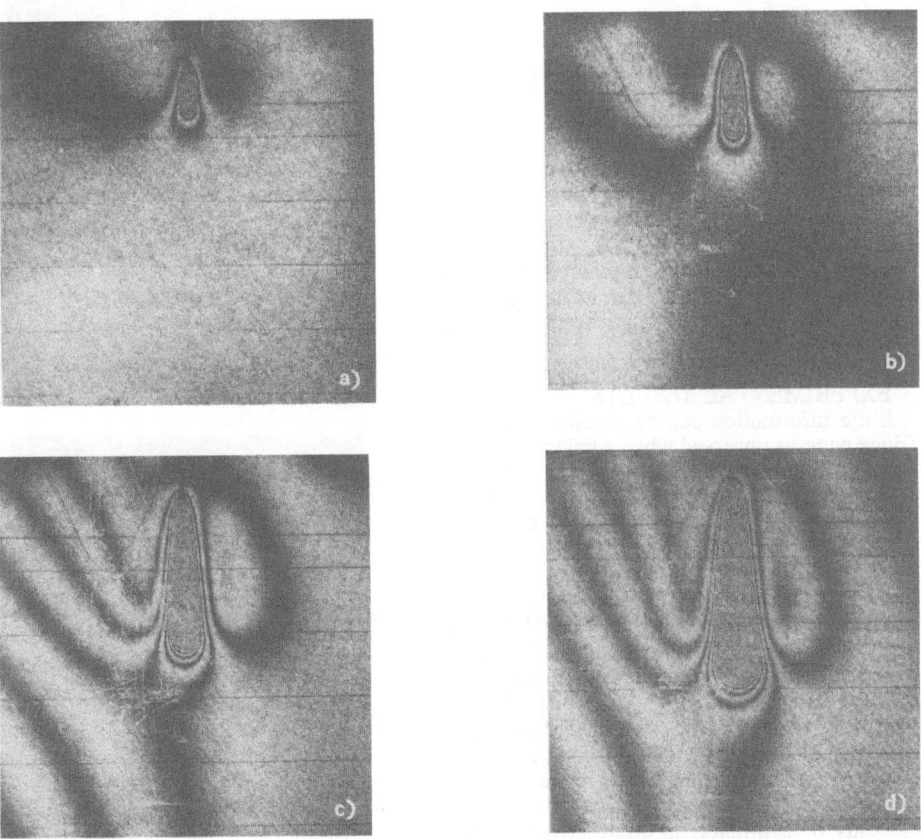

FIGURE 4. Fringe patterns, observed using a hologram of the uncracked specimen, at different crack lengths: a)15mm, b)17.5mm, c)20mm, d)22.5mm (R=0.5).

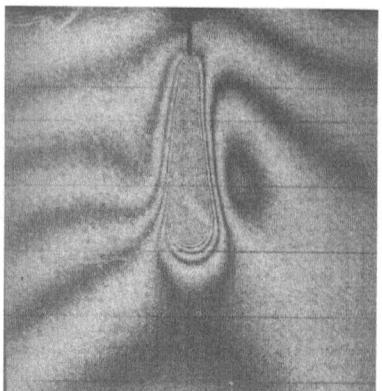

FIGURE 5. Fringe pattern observed using a hologram of the uncracked specimen (R=0.2, a=22.5mm).

apply heavy loads manually, without introducing undesired tearing components. The absence of bending can be monitored by comparing the fringe patterns on the front face with those arising on the back face. A single cantilever gauge, held to the specimen by tension springs and located by needle points which fit in the indentations shown in Fig.2, allows crack opening displacement to be measured with good accuracy. The applied load can be derived from COD if a calibration curve has been previously plotted on the testing machine.

3. EXPERIMENTAL RESULTS

Little information can be obtained from the fringe patterns observed when a hologram of the uncracked specimen is used as reference. The four photographs in Fig.4 refer to a specimen fatigued at R=0.5 to the crack lengths a=15mm, 17.5mm, 20mm and 22.5mm. Apart from the widely spaced fringes caused by slight tilting and bending of the specimen, closed contour fringes can be observed around the crack tip. However, a quantitative interpretation is practically impossible since fringes are very closely spaced due to the high gradient of the transverse displacement field in the vicinity of the crack line. Moreover, it cannot be excluded that a loss in contrast occurs as a result of permanent changes in the structure of the surface layer. The size of the plastic zone can only be estimated approximatively, since elastic deformations due to residual stresses are superimposed. The effect of residual compressive stresses in the wake of the crack can be noticed from the widening of the contours as crack

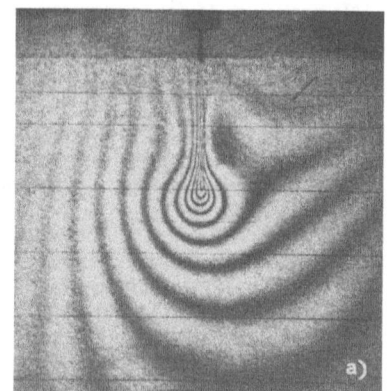

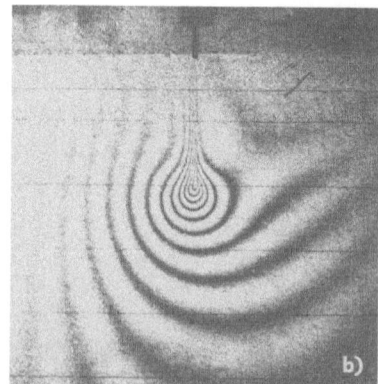

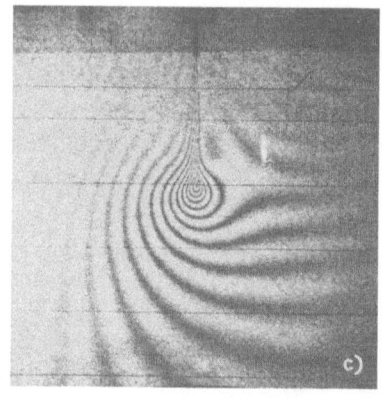

FIGURE 6. Fringe patterns at different loads: a)3.8kN, b)4.4kN, c)4.9kN using a hologram at zero load (R=0, a=20mm).

advances. Since the fatigue tests were conducted at a constant Pmax, the size of the plastic zone, as expected, varies linearly with the crack length, i.e.with Kmax². The dependence of the plastic zone on Kmax only is confirmed by comparing the photo in Fig.5, which refers to a crack grown at R=0.2 up to a length of 22.5mm, with that in Fig.4 taken at the same crack length.

The opening load Pop was estimated for the three different R-ratios R=0, 0.2 and 0.5. In all three cases a hologram was recorded once the fatigue crack had advanced to a length of 20mm.

Fig.6 shows photographs referring to the crack produced at R=0; the fringe patterns shown arise when the specimen is loaded at 3.8kN, 4.4kN and 4.9kN, respectively. Besides the widely spaced fringes which appear over the entire surface of the specimen as a result of the bending caused by the applied load, it is also possible to observe closely spaced fringes along the crack line which are due to the release of contact stresses (cf photograph a in Fig.6). The asymmetry in the disposition of the fringes about the crack line is due to the slanting of the fracture surfaces in the surface region. As the load increases and the point of last contact gets closer to the crack tip, more fringes move forward to enclose it. When all the fringes surround the tip the crack can be considered fully open (photo b in Fig.6). By increasing the load still further, crack tip singularity develops and new fringes arise from the tip (photo c in

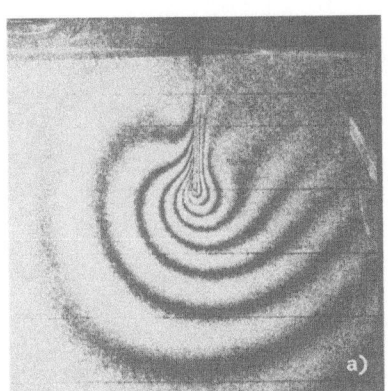

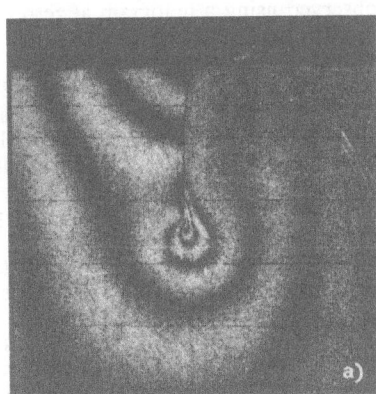

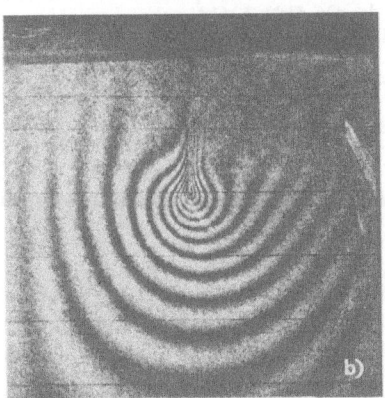

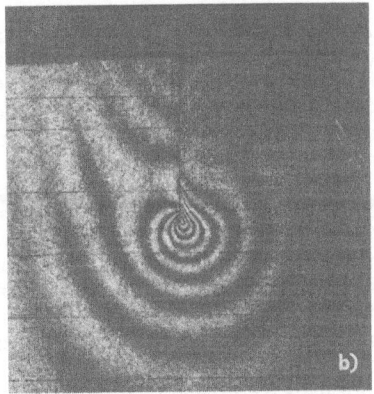

FIGURE 7. Fringe patterns at different loads: a)4kN, b)5kN, observed using a hologram at zero load (R=0.2, a=20mm).

FIGURE 8. Fringe patterns at different loads: a)3.5kN, b)4.8kN, observed using a hologram at 2kN (R=0.2, a=22.5mm).

402

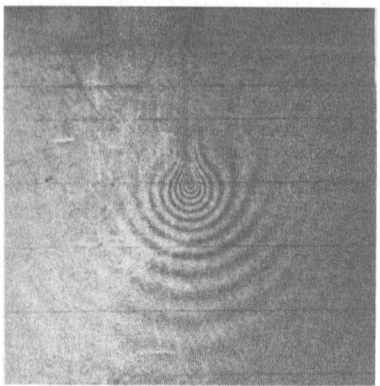

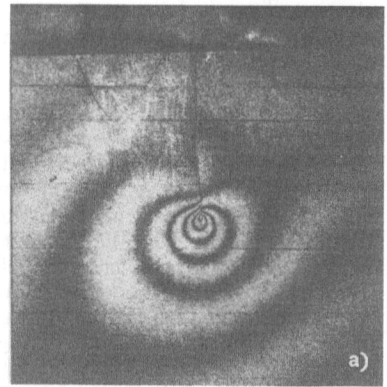

FIGURE 9. Fringe pattern at 5kN
observed using a hologram at zero load
(R=0.5, a=20mm).

Fig.6). In the present case (R=0, a=20mm) the
crack opening load Pop was estimated to be
about 4.4kN. This value must actually be considered a lower limit for Pop; in fact, although no
singularity fringes are clearly present at 4.4kN,
the slow transition from one fringe pattern to
another makes evaluation uncertain.

The photographs in Fig.7 show the fringe
patterns arising when loading the second
specimen (R=0.2, a=20mm) at 4kN and 5kN,
respectively. The crack, which was partially
closed at 4kN, appears to be fully opened at
5kN. In order to obtain greater accuracy in
evaluating the crack opening load, which could
only be estimated in the range 4.5kN to 5kN, a
reference hologram of the specimen loaded at
2kN was employed in a second loading test
(a=22.5mm). In this way, fewer and more wide-

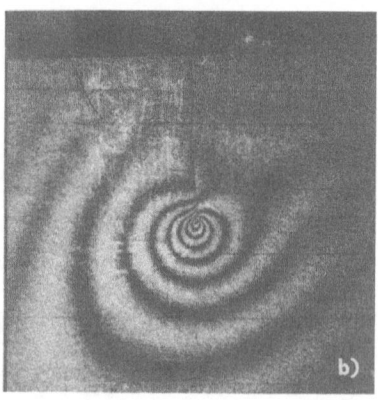

FIGURE 10. Fringe patterns at different
loads: a)5.5kN, b)6kN, observed using a
hologram at 3kN (R=0.5, a=22.5mm).

ly spaced fringes are obtained under the same loads. The fringes generated by the release of
contact stresses no longer run along the crack line (cf Fig.8a) since the contact stresses were already partially released when the hologram was taken. They do, however, still completely surround the crack tip at the opening load (cf Fig.8b) which was estimated at around 4.8kN.

The loading test conducted on the specimen fatigued at R=0.5 using as reference the
hologram taken at zero load only enabled a lower limit of 5kN to be established for Pop
(photograph in Fig.9). Interpretation was in fact particularly awkward due not only to the large
number of fringes but also to the fact that the negligible slanting which developed in the fracture
surface brought about a nearly symmetrical fringe pattern. A second loading test was carried out
on the specimen containing a 22.5mm long crack; the reference hologram used was recorded
under a load of 3kN. This second test enabled Pop to be evaluated in the range 5.5 kN- 6kN; the
photographs in Fig.10 show the last (inmost) fringe due to the release of contact stresses which
surrounds the tip and the first singularity fringe, respectively. A third test was carried out by
unloading the specimen at the same crack length, after taking a hologram under a higher applied
load (7kN) than the opening load previously estimated. On unloading the specimen, fringes
arise from the tip which are due to crack tip singularity (cf photo a in Fig.11, taken at 5.7kN).
As soon as crack closure starts new fringes can be observed which close on the crack line

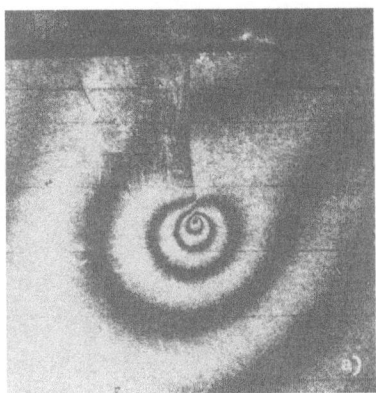

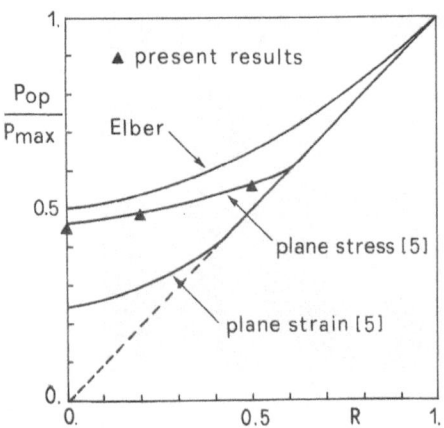

FIGURE 12. Variation of the crack opening load with R-ratio.

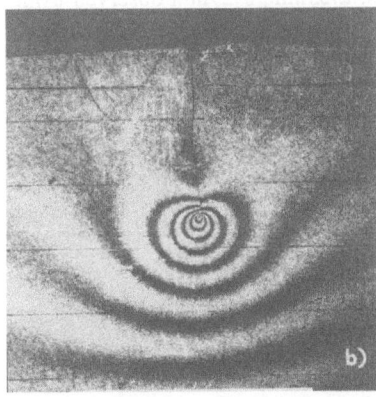

FIGURE 11. Fringe patterns at different loads: a)5.7kN, b)4.7kN observed using a hologram at 7kN (R=0.5, a=22.5mm).

rather than on the tip (photo b in Fig.11). The unloading test did not prove any more accurate than the previous loading test and Pop was estimated to be about 5.5kN for R=0.5 and a=22.5mm.

Fig.12 compares the values presently estimated for Pop with both those given by Elber's relation and those obtained from Newman's numerical model /5/ using the mechanical properties of the steel under testing. The values are in good agreement with those calculated in plane stress by Newman's algorithm.

Contact stresses in the wake of a crack of any length at zero load can be detected by using a reference hologram of the cracked specimen. The fringe patterns observed when the crack grows can in fact be related to the lateral contraction due to contact stresses along that portion of the crack line over which they have been released. The photographs in Fig.13 were taken once the crack had advanced from 20mm (at which length the hologram was recorded) to 25mm and 29mm respectively. The crack appears to be completely closed. The asymmetry in the patterns is imputable to the slanting of the crack surface in the surface region. From the difference between the two patterns in the region behind the position of the crack tip at the time when the hologram was recorded it can be assumed that contact stresses were not completely released at 25mm. Intermediate observations suggested that release could be considered completed at 29mm; unfortunately the influence of clamping can no longer be neglected at this length.

The photograph in Fig.14 shows the fringe patterns obtained by splitting the same specimen (a=29mm) into two halves. Besides the lateral contraction due to the compressive stresses in the wake, it is also possible to observe the transition from compressive to tensile stresses ahead of the crack tip. The effects of clamping are also noticeable.

The pronounced difference between the fringe patterns in Figs.13b and 14 can be partly ascribed to the different specimen region examined (the crack length is 20mm in the first case and 29mm in the second). It should be also pointed out that any permanent alterations suffered by the surface layer of the specimen in contact while the crack was grown from 20mm to

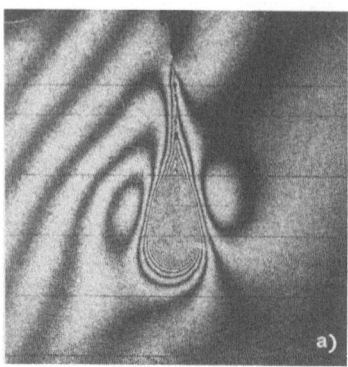
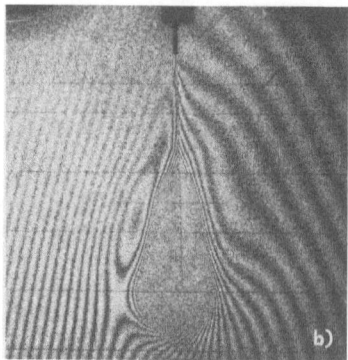

FIGURE 13. Fringe patterns at different crack lengths: a)25mm, b)29mm observed using a hologram of the cracked specimen (R=0, a=20mm)

FIGURE 14. Fringe pattern due to the release of residual stresses observed on half of the specimen using a hologram taken before the specimen was split (R=0, a=29mm).

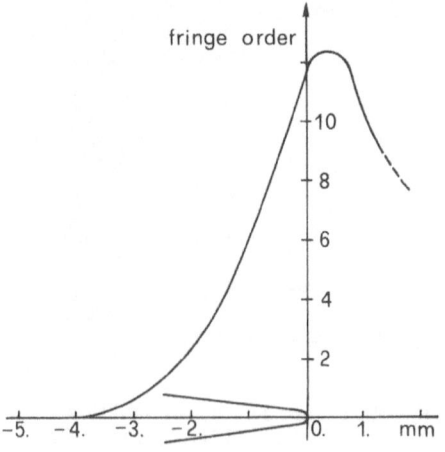

FIGURE 15. Fringe order versus distance from the crack tip.

29mm, would be revealed by the technique employed in the first case. It should be mentioned here that no shear lip was present in the side of the specimen examined.

If contact stresses are evaluated from the transverse displacements measured on one or other of the patterns assuming that contact develops through the whole thickness of the specimen, the resulting crack opening load (calculated as the load giving the same stress intensity factor as contact stresses) represents such a high fraction of the maximum load that it must be assumed that the contact thickness decreases along the crack line until only the surface layers are involved.

Fig.15 plots the variation of transverse displacement ahead of and behind the crack tip which can be derived from the fringe pattern in Fig.14. On observing that the fractional fringe order at the crack tip corresponds to the yield stress (650 MPa), the size of the reversed plastic zone can be estimated at about 0.9mm from the width of the contour fringe. Estimating at 4.5mm the size of the monotonic plastic zone from the fringe pattern in Fig.13b, the ratio (size of reversed plastic zone / size of monotonic plastic zone) is in good agreement with the 0.25 value generally assumed in the literature.

4. CONCLUSIONS

Although the tecnique employed enables the transverse displacement in the crack region to be detected with very high accuracy, the evaluation of the crack opening load has proved not only cumbersome but also somewhat uncertain.

It is our opinion that only the surface layers are closed even when the applied load is considerably lower than the opening load. The variation of the contact length across the thickness could be observed by gradually removing the surface layers of the specimen. Whether and how the crack closure phenomenon, when limited to the surface region, can tangibly affect crack growth rate should be verified. A possible variation of Pop/Pmax with Pmax, i.e. with the crack length, should also be investigated.

The technique might prove convenient in studying the effects of overloads on crack growth. However, before it is employed in further experiments, the technique should be made more reliable and easier to employ.

REFERENCES

1. Newman JC Jr: A Finite-Element Analysis of Fatigue Crack Closure. Mechanics of Crack Growth, ASTM STP 590, Philadelphia, 1976.
2. Dill HD, Saff CR: Spectrum Crack Growth Prediction Method Based on Crack Surface Displacement and Contact Analyses. Fatigue Crack Growth under Spectrum Loads, ASTM STP 595, Philadelphia, 1976.
3. Budianski B, Hutchinson JW: Analysis of Closure in Fatigue Crack Growth. Journal of Applied Mechanics, Trans.ASME, Vol.45, 1978.
4. Führing H, Seeger T: Dugdale Crack Closure Analysis of Fatigue Crack under Constant Amplitude Loading. Engineering Fracture Mechanics, Vol.11, 1979.
5. Newman JC Jr: A Crack Closure Model for Predicting Fatigue Crack Growth under Aircraft Spectrum Loading. Methods and Models for Predicting Fatigue Crack Growth under Random Loading,ASTM STP 748, Philadelphia, 1981.
6. Sehitoglu H: Crack Opening and Closure in Fatigue. Engineering Fracture Mechanics, Vol.21, No.2, 1985.
7. Vazquez JA, Morrone A, Gasco JC: A Comparative Experimental Study on the Fatigue Crack Closure Behaviour under Cyclic Loading for Steels and Aluminum Alloys. Fracture Mechanics, ASTM STP 677, Philadelphia, 1979.
8. Sharpe WN Jr, Grandt AF Jr: A Preliminary Study of Fatigue Crack Retardation Using Laser Interferometry to Measure Crack Surface Displacements. Mechanics of Crack Growth, ASTM STP 590, Philadelphia, 1976.
9. Marci G, Packman PF: The Effects of the Plastic Wake Zone on the Conditions for Fatigue Crack Propagation. International Journal of Fracture, Vol.16, No.2, 1980.
10. Macha DE, Corbly DM, Jones JW: On the Variation of Fatigue-crack-opening Load with Measurement Location. Experimental Mechanics, Vol.19, No.6, 1976.
11. Poggialini A: Rilevazione degli Stati di Deformazione all'Apice di una Frattura mediante Interferometria Olografica., Proceedings of 13th AIAS Conference, Bergamo, Italy, 1985.

EXPERIMENTAL INVESTIGATION OF DYNAMIC CONTACT PROBLEMS BY MEANS OF THE METHOD OF CAUSTICS

H.P.Rossmanith*, R.E.Knasmillner* and A.Shukla**

* *Institute of Mechanics, Technical University Vienna, Karlsplatz 13, A-1040 Vienna, Austria*

** *Department of Mechanical Engineering and Applied Science, University of Rhode Island, Kingston, RI 02881, USA.*

1. INTRODUCTION

General dynamic contact of two elastic bodies during collision or impact represents an extremely complex dynamical physical process due to the creation of time-dependent contact area the extension of which is not known apriori and therefore it becomes part of the solution of the problem. In general, numerical modelling techniques have to be employed for the study of transmission, reflection, refraction and diffraction of elastic waves across the contact zones and about the time-dependent moving contact edge during increasing and receding contact /1/.

Photomechanics as a means for visualization of stress distribution and transfer of load across the contact zone of elastic bodies in contact has been utilized successfully for wave propagation studies in connection with granular materials /2/. The method of caustics or shadow patterns is particularly suited for determination of contact pressure distribution /3,4/.

In this contribution the method of caustics in conjunction with high-speed photography is employed to study explosive impact wave propagation across dynamic contacts and the time variation of the contact pressure distribution. The generation of a local transient contact caustic at the contact zone produced during impact forms part of the information for the solution of the problem. Pseudo-caustics appear that are associated with singularities and discontinuities in the load distribution and with the projection of the deformed boundary of the elastic bodies in the range of interest. In addition to the transient contact pressure caustic a moving Rayleigh-wave caustic can be observed in the recordings from which the space-time distribution of the contact pressure can be determined. Series of sequentially recorded shadow patterns provide pertinent information about the dynamic event.

2. THE METHOD OF CAUSTICS

In recent years the experimental method of caustics or shadow spot technique has been extensively used for determination of quantities of interest in elastic and elasto-plastic crack-and contact problems, such as stress intensities and load distributions. The technique originally introduced by Manogg /5/ for transparent materials has been adapted by Theocaris and his coworkers /6,7/ for nontransparent materials. The general equations of caustic for plane static and dynamic elasticity theory may be found in Refs./8-10/.

408

The physical principle of the method of caustics is the inhomogeneous deflection of parallel light rays during their passage through a plate specimen due to two effects: the reduction of the thickness of the specimen and the change of the refractive index of the material as a consequence of stress intensification. The transmitted and/or reflected light rays form a shadow space and the intersection of this shadow space with a screen produces a shadow area (shadow spot) surrounded by a bright curve (caustic) /11/. The formation of this shadow space and and the shadow region on the image-plane (screen) at a distance behind the specimen for a half-plane subjected to a concentrated edge-load is schematically shown in Figure 1 for the transmitted light method.

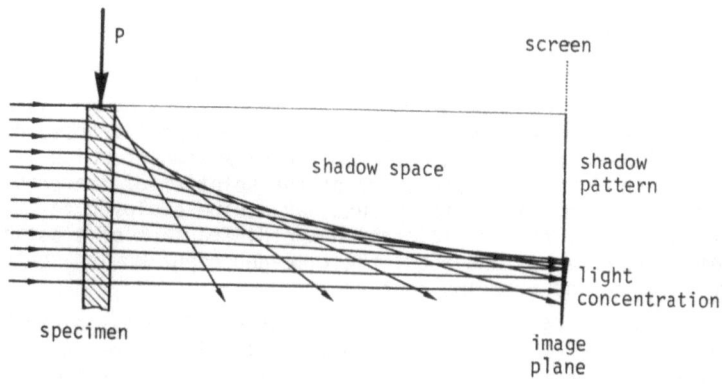

Figure 1 : The transmission method of caustics applied to contact problems

For nontransparent materials with a mirrored surface the reflection-light method is utilized where qualitatively similar shadow patterns can be observed. It has been shown /9/ that for caustic analysis the cases of plane stress and plane strain, transmitted light and reflected light are basically similar and differ only by the values of the elasto-optical parameters. In a third method, where part of the light is reflected at the front surface and part of the light is reflected at the back surface the resulting caustic the caustic is a combination of the caustics obtained by transmission and simple reflection.

3. ANALYSIS

When a normally impinging light beam traverses the unloaded specimen at the point $P(r,\theta)$ in the object plane, its image on the shadow image plane is defined by the vector $\vec{r}$ (Fig.2). Load application induces deflection of the beam to the point $P'(x',y')$. With the deflection $\vec{w}=\vec{w}(r,\theta)$ the vector of the image point P' is given by

$$\vec{W} = \vec{r} + \vec{w} . \tag{1}$$

This deflection gives rise to the formation of a shadow space formed by the deflected light rays upon passage of the object. The caustic is a singular curve of the image equ.(1) and is generated by the intersection of the image plane with the shadow space and ray field. A necessary condition for the existence of the singular caustic curve is that the Jacobian functional determinant, J, of equ.(1) vanishes:

$$J = \frac{\partial x'}{\partial r}\frac{\partial y'}{\partial \theta} - \frac{\partial x'}{\partial \theta}\frac{\partial y'}{\partial r} = 0 \tag{2}$$

for all points $P_c'(r,\theta)$ of the caustic. Hence, the fundamental problem of the method of caustic is to solve equ.(2) under certain conditions imposed by the real physical problem.

An elasto-optical analysis yields a relation between light deflection, the stress field parameters, the elasto-optical material parameters and the geometry of the experimental set-up:

$$\vec{w} = -dhc\ grad\{(\sigma_1+\sigma_2) \pm \varepsilon(\sigma_1-\sigma_2)\} \tag{3}$$

where h is the model thickness, $\sigma_{1,2}$ denote the principal normal stresses, d is the distance between model and screen, c is an elasto-optical parameter and ε accounts for the optical anisotropy of the material. For optically isotropic or inert material (e.g. Plexiglass) $\varepsilon=0$ and one obtains one single caustic. Optically anisotropic or birefringent materials (e.g.

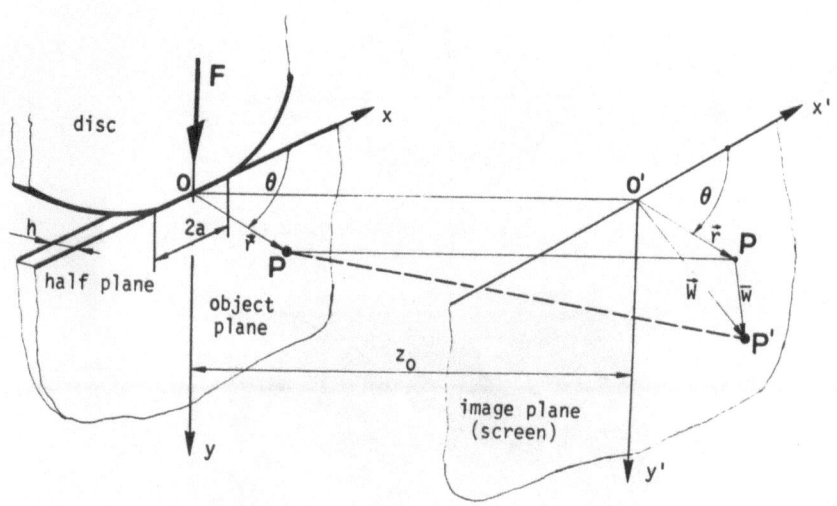

Figure 2 :   Light ray deflection associated with the contact
             of two elastic bodies

plate glass) give rise to a double-caustic. In addition, the mapping of the deformed boundaries of the elastic bodies in contact produces socalled 'pseudo-caustics'.

4. CAUSTICS APPLIED TO CONTACT PROBLEMS

Flamant's problem of the normal action of a point load P on a half-plane is governed by the stress function /12/ $\Phi(z)=iP/2\pi z$ and yields the following equations for the singularity caustic

$$W = x'+iy' \; ; \; x' = x - CP \frac{xy}{(x^2+y^2)^2} \quad , \quad y' = y + CP \frac{x^2-y^2}{2(x^2+y^2)^2} \quad (4)$$

with $C=dhc4/\pi$ and $x^2+y^2 = (CP)^{2/3}$ where C is proportional to the geometrical and optical parameters. The deformed boundary is mapped on the pseudo-caustic given by

$$x' = x - CP \frac{xv(x)}{\{x^2+v(x)^2\}^2} \quad , \quad y' = v(x) + CP \frac{x^2-v^2(x)}{2\{x^2+v^2(x)\}^2} \quad (5)$$

where $v(x)$ is the equation of the deformed boundary. Figure 3, taken from Ref./12/ shows the caustics formed by the reflected light rays from the

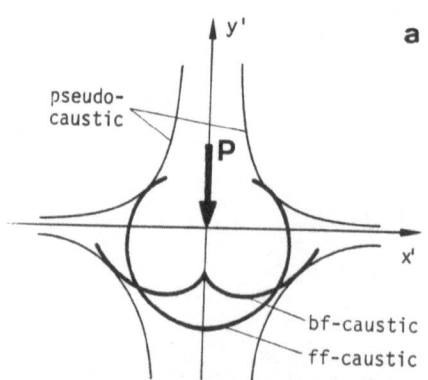

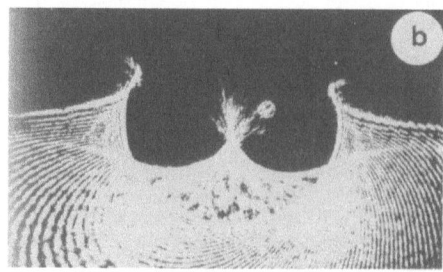

Figure 3: Caustics formed by reflected light rays from the front face and back face as well as pseudo-caustics: a) analytically generated and (b) experimentally recorded /12/.

front (ff) and from the back (bf) faces of a semi-infinite plate loaded by a concentrated normal load, P.

The light ray deflection due to an extended surface load p(x) can be written in the form

$$\mathsf{W} = z + \frac{iC}{2}\,\overline{\frac{d\Phi(z)}{dz}} \quad , \quad \Phi(z) = \int_{-a}^{a} -\frac{p(\zeta)}{\zeta-z}\,d\zeta \qquad (6)$$

where the variable ζ pertains to points on the boundary.

The initial curve of the caustic is given by eqs.(2) and (4), whereas the initial curve of the pseudo-caustic is expressed by the equation of the deformed boundary. The initial curve of the caustic is always located in the interior of the field.

5. DYNAMIC CONTACT CAUSTIC

General dynamic contact of two elastic bodies during collision or impact represents an extremely complex dynamical process. Wave motion across contacts is associated with wave transmission, reflection, refraction and diffraction about the time-dependent moving contact edge during increasing and receding contact. Complete numerical and/or analytical treatment of this complex wave propagation problem is not feasable at present. In order to simplify the dynamic situation attention is focussed on the particular but representative special problem, where a half-plane is in contact with a disc which is dynamically loaded by a small charge of explosive in its center. Upon detonation of the explosive the expanding compressive incident P-wave is reflected along the non-contacting section of the boundary of the disc, and transmitted across and diffracted about the contact. Lateral deformation (and change of refractive index of the material) gives rise to the formation of a time-dependent contact caustic. This can clearly be observed in the sequence of photographs shown in Fig.6. Part of the energy and momentum of the wave has been transferred across the contact. Wave diffraction about the contact yields a system of bulk waves (P-and S-waves) in the body of the two contacting solids. In addition, two pairs of Rayleigh-surface waves that originate at the contact propagate along the free surfaces. These waves carry most of their energy concentrated within a thin layer just underneath the free surface and propagate with no dispersion along the surface. This energy concnetration within the boundary layer causes appreciable thickness variations at points close to the boundary of the specimen during the passage of the surface wave. Formation of a caustic is expected.

Experiments suggest that after an initial phase of contact establishment the dynamic contact problem may be decomposed into a local dynamic contact problem and a global far-field contact wave propagation problem. Lamb's solution for the line disturbance suddenly applied to a half-space /13/ is the starting point for the analytical analysis. The structure of the resulting stress equations given in Ref./13/ on page 224, at a time t after application of a line force exhibit moving and local contributions to the solution which pertain to the global impact wave propagation problem and the local dynamic contact process, respectively. A more general solution suitable for time-dependent extended contact pressure may be obtained by integrating Lamb's solution with respect to time and space. Work on local dynamic contact phenomena is in progress at the Technical University Vienna.

6. THE CAUSTIC OF A RAYLEIGH-WAVE

During earlier experimental work in dynamic photoelasticity the existence of a Rayleigh-wave has been observed by the first author. In the experiments an elastic half-plane fabricated from a sheet of clear polyester
has explosively been loaded at a point located on the free edge of the plate.

If the loading function $P(\zeta)$ is known, then the stress field associated with the Rayleigh-wave may be represented in terms of the stress function

$$\Phi'(\zeta_j) = \delta \int_{-\infty}^{\infty} \frac{P(\zeta)}{\zeta_j - \tau} \, d\tau \quad . \tag{7}$$

The stress field is then given by

$$
\begin{aligned}
\sigma_x &= \alpha \, \mathrm{Re}\Phi''(\zeta_1) + \beta \, \mathrm{Re}\Phi''(\zeta_2) \\
\sigma_y &= \beta \, \mathrm{Re}\Phi''(\zeta_1) - \beta \, \mathrm{Re}\Phi''(\zeta_2) \\
\sigma_{xy} &= 2\gamma\{\mathrm{Im}\Phi''(\zeta_1) - \mathrm{Im}\Phi''(\zeta_2)\}
\end{aligned}
\tag{8}
$$

where the coefficients α, β, γ and δ depend on the wave propagation speeds
and the complex variables ζ_j with respect to a constant speed moving coordinate system (see Figure 4)

$$\zeta_j = \xi + i\eta_j = (\tau - \frac{x}{c_R}) + iy\sqrt{c_R^{-2} - c_j^{-2}} \quad (j=1,2) \tag{9}$$

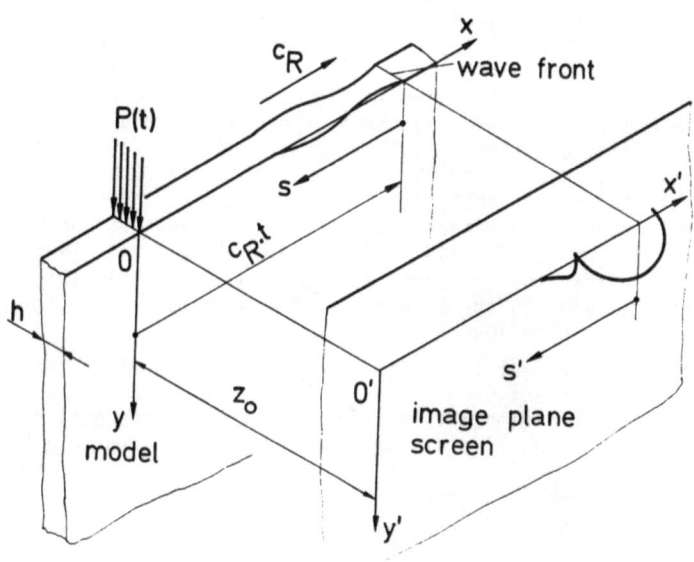

Figure 4 : Optical arrangement and coordinate systems for the
Rayleigh-wave caustic

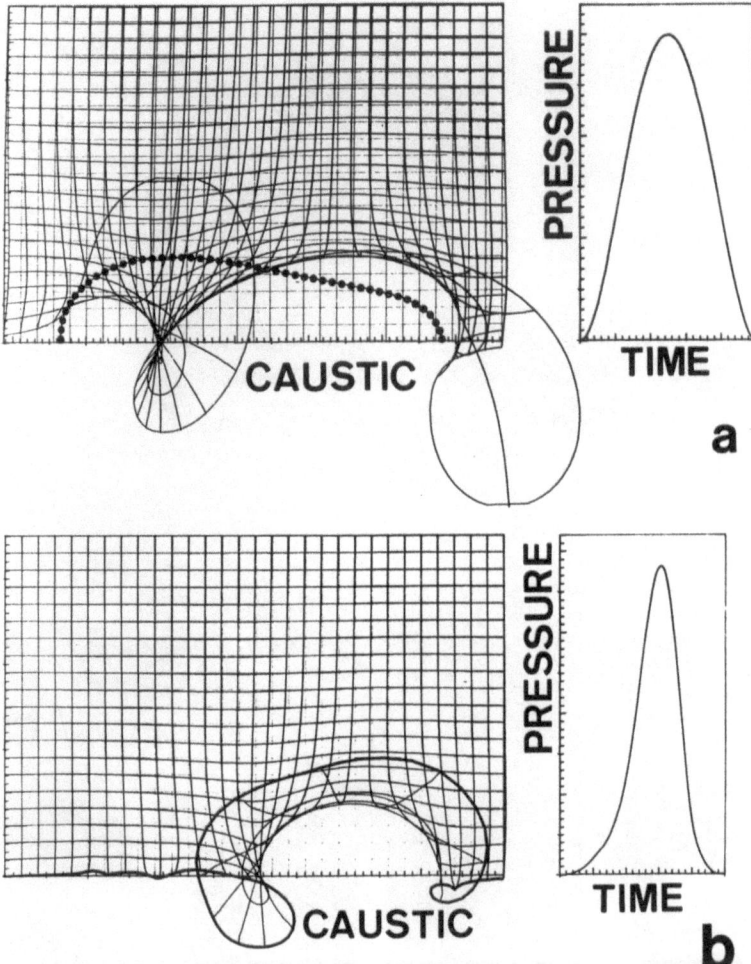

CAUSTIC

CAUSTIC

Figure 5 :   Caustics, pseudo-caustics and initial curve for a
a) symmetrical and a (b) asymmetrical loading pulse
P(t) obtained with front-face reflection technique

414

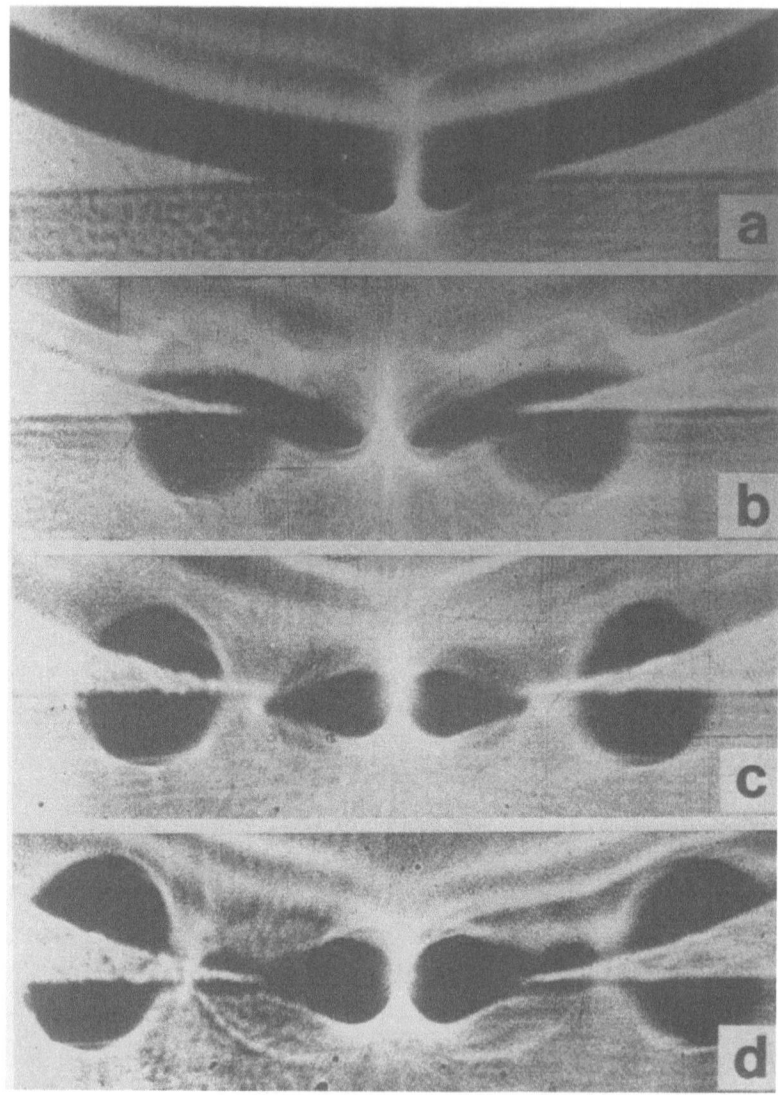

Figure 6 : Sequence of high-speed photographs of progressive
dynamic contact development showing contact caustic
and Rayleigh-wave caustic taken at times t=46 μsec
(a), 58 μsec (b), 63,5 μsec (c) and 69 μsec (d).

depend on the Rayleigh-wave speed c_R, plate wave speed c_1 and shear wave speed c_2.

The caustic image equations are given by $\vec{W} = \vec{z} + \vec{w}$,

$$x' = x + C_1 \mathrm{Re}\Phi''' = x + C_1 \int_{-\infty}^{\infty} P''(\tau) \frac{\eta_1}{(\xi-\tau)^2 + \eta_1^2} \, d\tau$$

$$y' = y + C_2 \mathrm{Im}\Phi''' = y + C_2 \int_{-\infty}^{\infty} P''(\tau) \frac{\xi-\tau}{(\xi-\tau)^2 + \eta_1^2} \, d\tau$$

(10)

where $P''(\tau)$ is the second time-derivative of the loading function.

The pseudo-caustic represents the mapping of the boundary of the half-plane $y=0$, i.e. $\eta_1 = 0$ and is associated with the image equation by

$$x' = x + D_1 P''(\zeta) \quad , \quad y' = D_2 \int_{-\infty}^{\infty} P''(\tau) \frac{d\tau}{\zeta - \tau} \ . \tag{11}$$

Because of $P''(\tau)=0$ for $\tau<0$ and $\tau>T$, the upper and lower limits of the Cauchy-integral in equ.(11) may be replaced by T and 0, respectively. The initial curve for the caustic is given by the solution of the Jacobian,

$$J = \frac{\partial(x',y')}{\partial(x,y)} = 1 - \varepsilon \, \mathrm{Re} \, \Phi^{IV} - \omega \, |\Phi^{IV}|^2 = 0 \ , \tag{12}$$

where the constants ε and ω depend on the optical and elastical parameters.

Even for the most simple form of the loading function $P(t)$ the initial curve can be determined only numerically. The Rayleigh-wave caustics and pseudo-caustics and the associated initial curves for asymmetrical and a nonsymmetrical loading pressure profile $P(t)$ are shown in Figure 5.

The size of the caustic is directly proportional to the load magnitude and inversely proportional to the fourth-power of the pulse width in the central-peak region. The caustics obtained in reflection depend only on the elastic parameters. For a given load level the back-face reflection method not only yields the complete caustic but also the largest caustic.

In the experiments a circular plate of diameter $d=15$cm and thickness $h=10$mm which is in contact with a half-plane is centrally loaded by a contained charge of about 120mg PETN. Upon detonation a rotationally symmetrical wave system radiates from the center of the disc and these waves establish dynamic contact upon reaching the disc-half-plane interface. Contact is readily established by dynamic extension of the contact area. The sequence of photographs shown in Figure 6 illustrates the formation of local dynamic contact and Rayleigh surface wave propagation along the free edges. The individual phases of the dynamic contact process are associated with (a) formation of contact, wave reflection and diffraction about extending contact edge; (b) generation of Rayleigh-wave caustics and decomposition of interaction process; (c) separation of Rayleigh-caustics from classical contact pressure caustic, and (d) maximum contact load transmission.

7. CONCLUSION

The dynamic contact caustic and the Rayleigh-wave caustic may be utilized to determine the time-varying contact pressure distribution during impact. Evaluation of a sequence of high-speed photographic caustic recordings at discrete times allows the generation of contact pressure history and energy and momentum transfer across the contact. From each particular caustic photograph the pressure distribution P(t) may be determined by means of a multipoint-overdeterministic least-squares data reduction technique /14,15/. In order to accomplish this task, the load function P(t) is represented in form of a polynomial with time-dependent coefficients. The data set associated with the caustic curve in each photograph yields the value of these time-dependent coefficients at one particular instant of time. An iterative procedure is recommended, where the initial pulse is estimated and th computer-generated caustics are compared with the experimental recordings.

ACKNOWLEDGEMENT

The first two authors would like to acknowledge the financial support granted by the Austrian Science Foundation under project number # 4532.

REFERENCES

/1/ Feda, J.: *Mechanics of Particulate Materials. The Principles.* Developments in Geotechnical Engineering 30, Elsevier Sci.Publ.(1982)

/2/ Rossmanith, H.P. and A.Shukla: Photoelastic investigation of dynamic load transfer in granular media. *Acta Mechanica* 42,211-225 (1982)

/3/ Theocaris, P.S.: Stress singularities at concentrated loads.*Experimental Mechanics* 13,511-518 (1973)

/4/ Theocaris, P.S. and Stassinakis, C.A.: The elastic contact of two discs by the method of caustics. *Experimental Mechanics* 18,409-415 (1978)

/5/ Manogg, P.: Die Lichtablenkung durch eine elastisch beanspruchte Platte und die Schattenfiguren von Kreis- und Rißkerbe. *Glastechnische Berichte* 39,323-329 (1966)

/6/ Theocaris, P.S. and Gdoutos, E.: An optical method for determining opening-mode and edge-sliding-mode stress intensity factors. *J.Appl. Mechanics* 39,91-97 (1972)

/7/ Theocaris, P.S.: The method of caustics applied to elasticity problems. In: *Development in Stress Analysis* (G.Hollister, Ed.), 27-63 (1979)

/8/ Theocaris, P.S. and Ioakimidis, I.N.: Technical note on the equations of caustics for cracks and other plane elasticity problems. *Engg.Fract. Mechanics* 12,613-615 (1979)

/9/ Beinert, J. and Kalthoff, J.F.: Experimental determination of dynamic stress intensity factors by the method of shadow patterns. In: *Mechanics of Fracture*, Vol VII (Ed.G.C.Sih) (1979)

/10/ Rossmanith, H.P.: The method of caustics for plane elasticity problems. *J.of Elasticity* 12,193-200 (1982)

/11/ Poston, T. and I.Stewart: *Catastrophe Theory and its Application. Chapter 12: Optics and Scattering Theory.* Pitman Publ., London/San Francisco/Melbourne (1978)

/12/ Theocaris, P.S. and C.Razem: Deformed boundaries determined by the method of caustics. *J.Strain Analysis* 12,223-232 (1977)

/13/ Fung,Y.C.: *Foundations of Solid Mechanics.* Prentice Hall, Inc. New Jersey (1965)

/14/ Sanford, R.J.: Application of the least-squares method to photoelastic analysis. *Experimental Mechanics* 20,192-197 (1980)

/15/ Klein, G.: *Bestimmung von Spannungsfaktoren bei gemischter Beanspruchungsart am Beispiel eines Risses in der Umgebung eines Kreisloches.* IWM (IFKM)-Report, Freiburg, BRD (1964)

PHOTOMETRIC METHODS OF CAUSTICS

A.A. Sukere

Mechanical and Aerospace Engineering
University of Missouri-Columbia
Independence, Missouri 64050

1. INTRODUCTION

Failure in structures and machine components are known to be caused by flows or crack like defects which ar considered to be initially present in the components or initiated by repeated stressing and grown finally to critical size.

In order to assess the structural reliability of cracked structures it is necessary to know both the strength of the cracked component and the rate at which the cracks grow under in-service fatigue loads. Both the strength and crack growth depend upon the stress intensity factor K.

In recent years the experimental method of caustics has proved to be a simple yet powerful and most accurate of methods for the determination of stress intensity factors at crack tips in plane isotropic and aniosotropic elastic media. Several investigations have been carried out using the method of caustics, some of which are: Theocaris and Katsamanis [1], Kalthoff et al [2], Ravi Chandar and Knauss [3] and Beinert [4] used the method of caustic in the determination of dynamic stress-intensity factors and to study the crack arrest phenomena. Obata et al [5] and Theocaris [6,7] used the method of caustics to evaluate stress intensity factors in yielding materials. Theocaris [8] used the method to determine the stress intensity factor for a visoelastic material. More recently Lee and Knauss [9] used the method of caustic to study the crack healing phenomena.

This paper presents photometric methods of caustics for rotating and stationary structure applications. Basically the photometric method consists of sensing the spatial light-intensity variation, at the reference plane of a typical optical arrangement for observing caustics, with a single or linear array of optical sensors. These sensors convert variations in light intensity into variations in electrical potential or resistance. With a suitable signal conditioner, these variations can be displayed or recorded in a convenient form in a storage oscilloscope, recorder, or a digital computer.

2. PRINCIPLE OF THE METHOD OF CAUSTICS

The basis of the method is that a light ray passing through a stressed plate is deviated from its straight path partly due to thickness variation and partly due to the change in refractive index caused by the stress optic effect. If the plate contains a crack, due to the strong thickness and refractive index variation at the region close to the crack tip, the rays are scattered and concentrated along a strongly illuminated curve on the reference plane placed some distance from the specimen. This singular curve whose precise geometry with reference to the crack tip can be determined is variously called caustic, stress corona, or shadow spot. The method of caustics was introduced by Manogg [10], then later developed further by Theocaris [11].

418

Considering the optical configuration in Figure 1, the equation of the caustic for the case of a straight edge crack in a tensile field is given

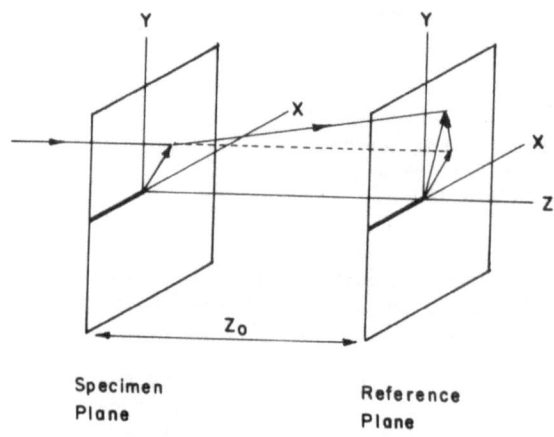

FIGURE 1. Optical configuration for the method of caustics.

by [11]

$$x = \lambda\ r_o(\cos\theta + \frac{2}{3}\cos\frac{3\theta}{2})$$

$$y = \lambda\ r_o(\sin\theta + \frac{2}{3}\sin\frac{3\theta}{2})$$

(1)

where

$$r_o = (\frac{3}{2}\frac{C}{\lambda})^{2/5}$$

(2)

$$C = \frac{1}{\sqrt{2\pi}}\ z_o t \frac{\nu}{E} K_I$$

Here r_o is the radius of the initial curve in the specimen plane, λ is the image magnification factor, z_o is the distance between the object and the reference plane, K_I is the opening mode stress intensity factory and t, ν and E are respectively the thickness, Poisson's ratio and modulus of electricity of the specimen.

From equation 1 it can be shown that the transverse diameter, D_t, and the longitudenal diameter, D_ℓ, of the caustic are related to the radius, r_o, through the following relation

$$r_o = \frac{D_t}{3.17\lambda} = \frac{D_\ell}{3.01\lambda}$$

(3)

Figure 2 shows the theoretical shape of a caustic image which is drawn using equation 1 for $-\pi < \theta \leq \pi$. The ratio of the front and back radii of the caustic along the direction of the crack can be shown to be equal to 5/4. Thus, the position of the crack tip can be readily determined from the position and shape of the caustic.

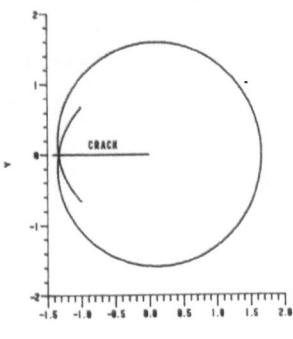

FIGURE 2.   Theoretical shape of shadow spot image.

Substituting this value of r_o in relation (1) one obtains an expression for the stress intensity factor in terms of the caustic diameters

$$K_I = \frac{1.67E}{z_o t \nu} \left(\frac{1}{\lambda}\right)^{3/2} \left(\frac{D_t}{3.17}\right)^{5/2}$$

(4)

$$K_I = \frac{1.67E}{z_o t \nu} \left(\frac{1}{\lambda}\right)^{3/2} \left(\frac{D_\ell}{3.01}\right)^{5/2}$$

Thus, the determination of K_I is a simple matter of measuring either caustic diameter.

3.   PHOTOMETRIC SYSTEM DESCRIPTION
3.1   Rotating Structure Application

An example of the experimental set-up for application to rotating structures is shown in Figure 3. The setup consists of a disk specimen, and driving plate attached to the shaft of a 25,000 rpm grinder motor, a 5 mm HeNe laser, a photomultiplier tube, a trigger circuit and an oscilloscope. A Dayton model 4X797B speed control unit was used to infinitely vary the motor speed. The angular speed of the disk was monitored using an HP 5300 universal counter.

The configurations of the disk specimen and drive plate are shown in Figure 4. The specimen in the present investigation is a 305 mm diameter, 3 mm thick solid disk, machined of PMMA cast acrylic sheet with mass density of $\rho = 1.19 \times 10^3$ kg/m^3. The radial crack was simulated by 0.3 mm wide radial saw cuts. The disk specimen was rotated by the drive plate

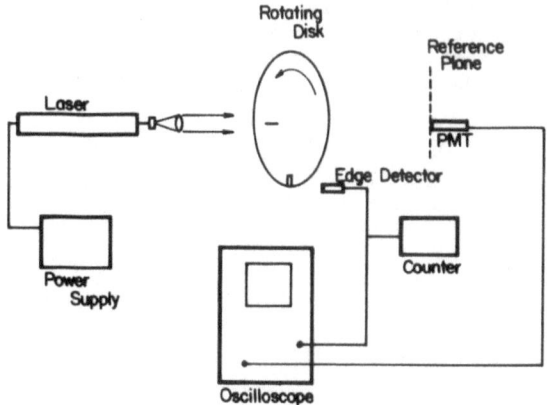

FIGURE 3. Schematic of the experimental setup for rotating structures.

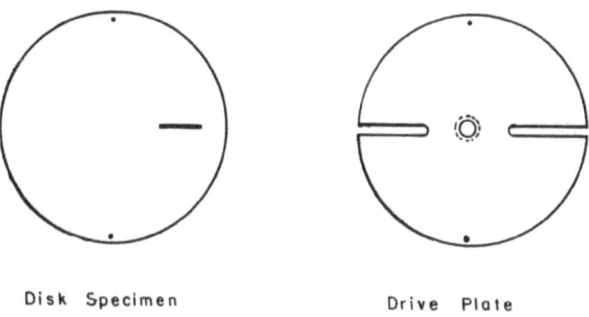

Disk Specimen   Drive Plate

FIGURE 4. Configurations of the disk specimen and drive plate.

via two steel pins which transmitted rotation through the drive holes in the specimen.

As mentioned earlier when a light beam impinges on the loaded specimen in the close vicinity of the crack a shadow spot with a bright halo is formed on the reference plane as shown in Figure 5. In determining the stress intensity factor one need only measure the transverse diameter of the shadow and the distance between the specimen plane and the reference plane. The transverse diameter of the caustic is monitored by a photomultiplier tube whose window is covered except for a thin slit with a width far less than the width of the crack image on the reference plane. As the shadow spot passes over this slit, the intensity seen by the photomultiplier tube varies and the result is an intensity profile of the shadow spot as shown in Figure 6. The photomultiplier tube used in this experiment is an RCA type 4840 tube at an operating voltage of 1200 v.

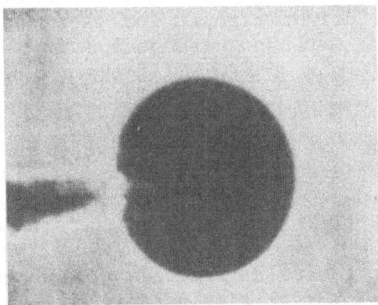

FIGURE 5.   Shadow spot surrounding a mode I loaded crack tip.

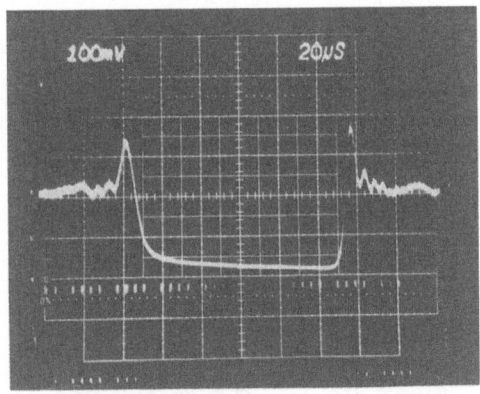

FIGURE 6.   Intensity profile of a shadow spot.

The oscilloscope was used to display and manipulate the intensity profile of the shadow spot. The oscilloscope is triggered using a TRW type OPB703A reflective object sensor which sensed the edge formed by a contrasting aluminum foil and black tape radially glued on the back of the disk on a line perpendicular to the plane of the crack.   Since the measurement of the transverse diameter of the shadow spot is critical one needs to expand the time scale of the trace for better resolution, and the improvement would be maximized if the intensity profile of the shadow spot is at the beginning of the trace.   This function was accomplished by placing the reflective edge sensor on a travelling stage.

3.2 Stationary Structure Application

The instrumentation to implement the measurement strategy consists of the following components: an edge cracked PMMA specimen mounted in a tensile testing machine, a light source for illumination, a photodiode

422

linear array for scanning images formed by the crack and the shadow spot on the reference plane, a driver circuit for the sensors, a locally built video processing circuit and an oscilloscope for displaying the video. Figure 7 shows the test setup and Figure 8 shows a schematic of the sensor system.

Each element of the photodiode array is an independent silicon photocell, operating in the photo conductive mode. Operationally, the image sensor converts incident light to electric charge which is integrated and stored until readout. The integrated charge is directly proportional to the intensity of the light impinging on the sensing elements. Readout is initiated by a periodic start pulse. The charge information is then sequentially readout at a rate determined by clock pulses applied to the image. The output is a discrete time analog representation of the spatial distribution of light intensity across the array.

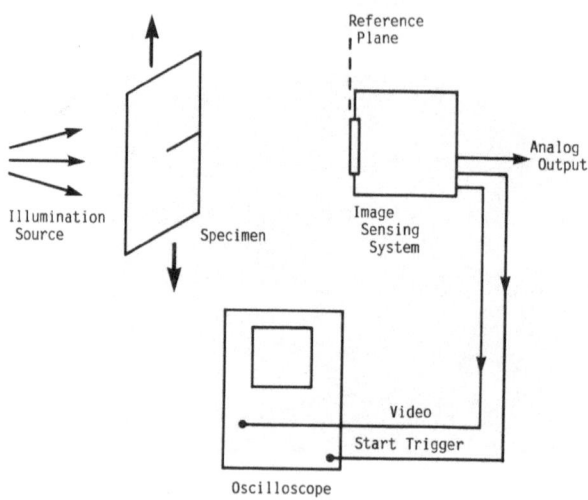

FIGURE 7. Schematics of the experimental setup for stationary structures.

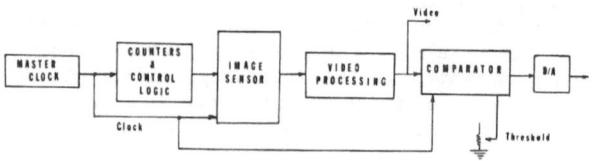

FIGURE 8. A block diagram showing the major components of the image sensing system.

If the photodiode array is positioned transverse to the crack plane in the reference plane, a video output wave-form as shown in Figure 9 results. The darkened pixels could relate to the diameter of the shadow spot in Figure 5. Since the pixels are on precise geometrical centers, the diameter of the shadow spot can be calculated by multiplying the number of elements in the dark times their center-to-center spacing times the optical magnification factor. It should also be noted that the image of the crack at the reference plane is a dark line.

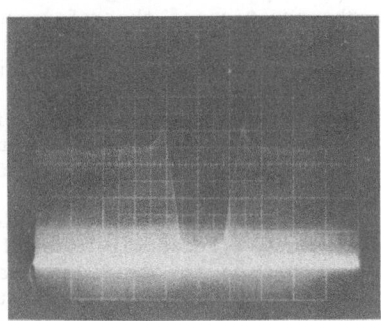

FIGURE 9. Transverse intensity profile of a shadow spot.

The linear photodiode array used is a Reticon Type RL1024G which consists of 1024 photodiode elements spaced at 25 μ-m intervals along a line of 25.6 mm length so that the resolution is one part of 1024. The array is mounted on a 76 mm square driver/amplifier PC board which is attached to a mother board controller.

Figure 10a shows a video waveform resulting when the linear array is positioned such that the scan line is in the plane of the crack for the specimen under no load condition and Figure 10b shows the intensity profile when the specimen is loaded. Note that the oscilloscope display does not show a very sharp drop in power, but rather shows that a few elements

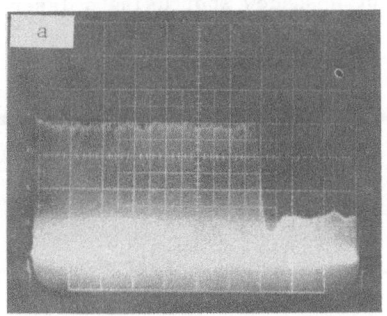
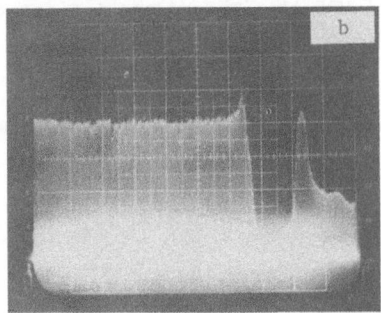

FIGURE 10. Logitudinal intensity profile of a crack image (a) under no load (b) under load (Note that the crack tip is masked by the shadow spot).

partially respond to light even though they should be in darkness. This effect is caused by overlapping sensitivity of the photodiode elements. Using a calibrated threshold level one can precisely determine the size of a shadow spot or the shadow cast by a crack profile by simply counting the number of pixels with an intensity level below the threshold level.

Note that the sharpness of the projected image can be improved by using a laser as a light source as shown in Figure 11. But this introduces irregularities in the waveform due to the unavoidable speckles in laser lighting. The size of the speckles change with the position of the reference plane. Since the position of the reference plane is determined by the desired size of the singular curve this problem is unavoidable.

The data collection procedure which involves monitoring crack growth and caustic diameter is demonstrated schematically in Figure 12. It should be noted here that to avoid errors introduced by rigid body motion in the measurement of crack growth, one needs to monitor the relative distance between the shadow cast by the crack tip and a fiducial mark on the specimen.

The system as is can be used to gather caustic diameter and crack growth data, but this would involve recording 1024 voltage levels per scan. A more desirable approach would be to obtain an analogue voltage output proportional to the size of the shadow projected on the linear array. If the analogue video is compared to a threshold voltage, then digital pulse data can be generated. By counting the number of clock pulses in the

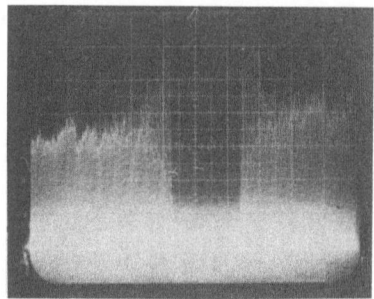

FIGURE 11. Transverse intensity profile of a shadow spot using a laser as a light source.

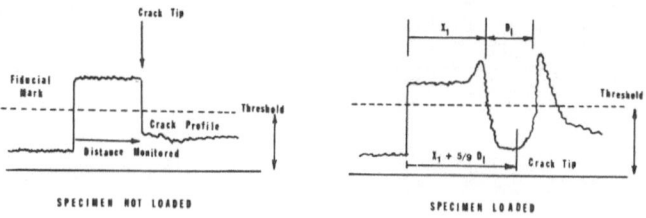

FIGURE 12. Procedure for measuring crack growth and caustic diameter.

digital data, the number of pixels in the shadow spot can be determined. A circuit was locally constructed that provides a voltage output proportional to the number of pixels in the dark. Currently the circuit is calibrated to give a full scale output of 10.24 volts for a field of view of 25.6 mm with a resolution of .025 mm. A typical trace of the circuit output as a function of time is shown in Figure 13. This corresponds to the change in the transverse caustic diameter for a specimen under cyclic loading at the rate of 1.56 Hz. The clock rate of the array in this case was set so that a complete scan would take place at the rate of 40 Hz.

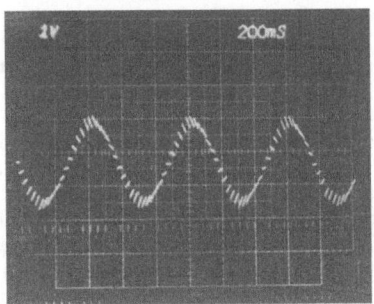

FIGURE 13. Variation of the transverse caustic diameter for a cyclicly loaded specimen measured by the image sensing system.

4. RESULTS AND DISCUSSION

Considering the geometry and coordinates shown in Figure 14, in the absence of a crack the stresses in a rotating solid disk take the form [12]

$$\sigma_\theta(r) = (\frac{3+\nu}{8}) \rho \omega^2 R^2 [1 - \frac{1+3\nu}{3+\nu} \frac{r^2}{R^2}] \tag{5}$$

$$\sigma_r(r) = (\frac{3+\nu}{8}) \rho \omega^2 R^2 [1 - \frac{r^2}{R^2}] \tag{6}$$

Here, R is the radius of the disk, ν is Poisson's ratio, ρ is the mass density of the disk material r the radial distance and ω is the angular velocity. For radially oriented through the thickness cracks the loading by the hoop stress $\sigma_\theta(r)$ is of the simple mode I type but varies along the

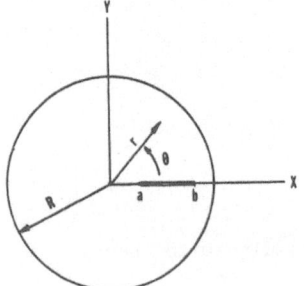

FIGURE 14. Rotating disk containing a radial crack.

crack length (b–a) and is in general, unsymetric about the crack center with the stress intensity at tip a, $K_{I,a}$, being greater than $K_{I,b}$ at tip b.

The procedure used to obtain the transverse caustic diameter is as follows: A circular tab of known diameter is radially affixed at the same radial distance as the crack tip. As the image of the shadow spot and the calibration tab pass the slit of the photomultiplier tube they each create a pulse. Thus, measuring the width of the two pulses is sufficient in determining the caustic diameter.

The experimental data was obtained by sequentially incrementing the angular speed a small amount, waiting for the speed to get stable, recording the pulse width of the shadow spot and the calibrator and the corresponding frequency. Once the transverse diameter, D_t, of the caustic has been determined, the values of the stress intensity factors, K_I, can be computed using relation (4).

Figure 15 shows a comparison of the experimental $K_{I,a}$ and $K_{I,b}$ results with the theoretical results obtained using the procedure outlined in Reference [13]. These results are for a crack of length 50 mm with tip a and b 81 mm and 132 mm from the center of the disk.

The technique for stationary structure application is tested by carrying out simultaneous photometric and travelling microscope measurements of transverse caustic diameters of an edge cracked PMMA acrylic sheet specimen loaded in tension. A comparison between typical results obtained is presented in Figure 16. Agreement between the two data sets is good.

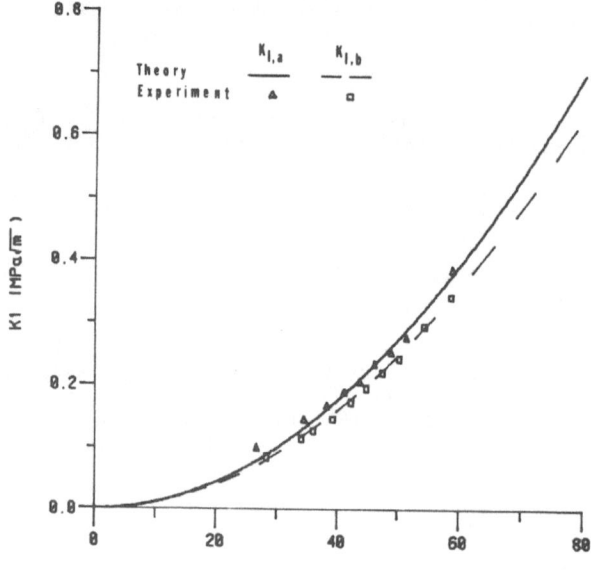

FIGURE 15. Comparison of experimental results with theory.

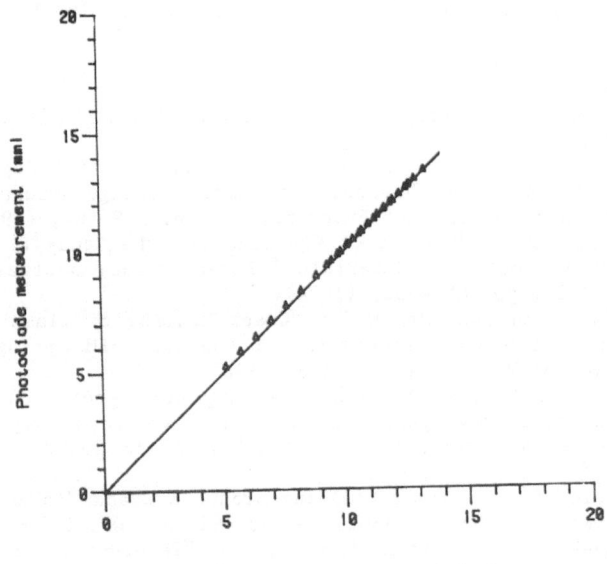

FIGURE 16.   Calibration results.

5.   CONCLUSIONS

Photometric methods of caustics for applications to rotating and stationary structures are presented. Both the accuracy and ease-of-use of these techniques are demonstrated. The present investigation also shows that with the linear photodiode array positioned with its axes in the plane of the crack one can determine not only the stress intensity factor but crack growth as well. Furthermore, the method offers a means of propagating cracks under constant alternating stress intensity, because the feeback signal from caustic diameter, which is proportional to the stress intensity, can be made to control the stress applied by a servo-controlled machine so as to maintain constant K conditions automatically.

REFERENCES
1.   Theocaris, P.S., and Katsamanis, F., "Response of Cracks to Impact by Caustics," Engineering Fracture Mechanics, 10, pp. 197–210, (1978).
2.   Kalthoff, J.F., Beinert, J., Winkler, S., and Klemm, W., "Experimental Analysis of Dynamic Effects in Different Crack Arrest Test Specimens," Crack Arrest Methodology and Applications, ASTM STP 711, pp. 109–127, (1980).
3.   Ravi Chandar, K., and Knauss, W.G., "Dynamic Crack-tip Stresses Under Stress Wave Loading," International Journal of Fracture, 20, pp. 209–222, (1982).
4.   Beinert, J., "Experimental Methods in Dynamic Stress Analysis," Modern Problems in Elastic Wave Prpagation, ed. J. Miklowitz and J.P. Achenback, John Wiley & Sons, pp. 23–43, (1977).

5.   Obata, M., and Shimada, H., "Applicability of a Caustic in Reflection to Detection of Internal Crack Initiation in Steel-Determination of J_{IC}," Proceedings of the V International Congress on Experimental Mechanics, Montreal Canada, June 10-15, pp. 117-121, (1984).

6.   Theocaris, P.S., "Stress Intensity Factors in Yielding Materials by the Method of Caustics," International Journal of Fracture, 9, pp. 195-197, (1973).

7.   Theocaris, P.S., and Gdoutos, E.E., "The Modified Dugdale-Barenblatt Model Adapted to Various Fracture Configurations in Metals," International Journal of Fracture, 10, pp. 549-564, (1974).

8.   Theocaris, P.S., "Method of Caustics for the Study of Cracked Plates Made of Viscoelastic Materials," International Journal of Mechanical Science, 16, pp. 855-863, (1974).

9.   Lee, O.S., and Knauss, W.G., "Crack Healing of Glassy Polymers - An Experimental Study," Proceeding of the 1985 SEM Spring Conference on Experimental Mechanics, pp. 1-8, (1985).

10. Manogg, P., "Schattenoptische Messung der Specifischen Bruchenergie Während des Bruchvorgans bei Plexiglas," in Proceedings, International Conference on the Physics of Non-Crystaline Solids, Delft, p. 481, (1964).

11. Theocaris, P.S., "Local Yielding Around a Crack Tip in Plexiglas," J. Appl. Mech. 37, Trans. ASME, Series E92, p. 409, (1970).

12. Timoshenko, S., and goodier, J.N., "Theorgy of Elasticity," 3rd edition, McGraw-Hill, New York, (1970).

13. Rooke, D.P. and Tweed, J., "The Stress Intensity Factors of a Radial Crack in a Finite Rotating Elastric Disc," Intnl. J. Engrg. Sci., 10, 323-335, (1972).

EVALUATION OF STRAIN GAUGE MEASUREMENTS IN ELASTO-PLASTIC AREA

Václav DOLHOF

Central Research Institute
ŠKODA, Concern Enterprise, Plzeň

1. INTRODUCTION

The evaluation of strain gauge measurements in the elasto-plastic area gains in importance as new demands for performance, efficiency, weight and maximum utilization of material arise.

During hydraulic tests of pressure vessels there occurred the problem of evaluating strain gauge readings up to 0.2 % of plastic strains in the elasto-plastic area. The HBM strain gauges, type 6/120 XY 11 were selected as the most suitable for these tests because of their short length and grid arrangement. In accordance with the manufacturer's recommendations these strain gauges can be installed using the Z 70 one-component adhesive, which sets the upper temperature limit for static measurements at 100 $^\circ$ C.

The calculations of the principal stresses from the strain gauge readings were performed on the PAC-16 Control Computer of the Compulog Two data logger using Hooke's extended equations for biaxial-stress conditions. After tightening the bolted connection the pressure vessel was heated to 80 $^\circ$ C and in some locations of the flange connections the principal stresses reached the limit of proportionality of the vessel material. The inapplicable fictitious stresses which exceeded the values of the yield stress or tensile strength of the vessel material in these locations were calculated during the hydraulic test. However, it is more difficult to obtain the real principal stresses from experimental data and, therefore, their calculation is either carried out by time-consuming hand computation following the strain gauge measurements or, often, not at all.

The possibility of evaluating the real principal stresses in the elasto-plastic area on the available data logger computer using a suitable method during the strain gauge measurements was the main objective in working on these problems.

An overview of some methods of evaluating the strain gauge data in the elasto-plastic area is presented in this paper. The problems related to the evaluation proper of the plastic strains up to 0.2 % are discussed and the numerical results of the fictitious and the real principal stresses evaluated from the strain gauge readings are presented.

2. METHODS OF EVALUATING STRAIN GAUGE MEASUREMENTS IN ELASTO-PLASTIC AREA

Several methods exist for evaluation of data from strain gauge measurements and for stress analysis in the elasto-plastic area. The stresses and the residual stresses arising during loading and unloading, respectively, depend on the material coefficients in the linear and the non-linear areas of the stress-strain curve for the construction material as obtained from the uniaxial tensile test, as well as on the plasticity condition used. The method based on the von Mises hypothesis and on the method of small elasto-plastic strains proved to be the most suitable for computer processing.

2.1. Method based on the von Mises hypothesis

The general solution of stress analysis in a biaxial stress state is based on the von Mises yield criterion

$$\sigma_v^2 = \sigma_1^2 + \sigma_2^2 - \sigma_1 \sigma_2 \tag{1}$$

and on the two Prandtl-Reuss equations

$$d\varepsilon_1 = \frac{3}{2} \frac{\sigma_1'}{\sigma_v} d\varepsilon_p + \frac{1}{E} d\sigma_1 - \frac{v}{E} d\sigma_2 \tag{2}$$

$$d\varepsilon_2 = \frac{3}{2} \frac{\sigma_2'}{\sigma_v} d\varepsilon_p + \frac{1}{E} d\sigma_2 - \frac{v}{E} d\sigma_1 \tag{3}$$

where σ_1 and σ_2 are the principal stresses, σ_v is the equivalent stress, σ_1' and σ_2' are the deviatoric stresses, ε_1 and ε_2 are the principal strains, and ε_p is the plastic strain. A continuous recording of the loading and the strain signals is necessary. In order to calculate the post-yield stresses in a biaxial stress field, it is necessary to solve repeatedly the two Prandtl-Reuss equations (2) and (3) at each load step provided the strain increments are small. Theoretical problems are treated in great detail in Refs. (1) and (2).

Derived from the general solution, a simplified solution is given in Ref. (2). An analysis of the elasto-plastic state is carried out in order to find an equivalent point $(\sigma_v, \varepsilon_v)$ on the stress-strain curve, determined from the uniaxial tensile test (Fig.1a), which corresponds to the principal strains ε_1, ε_2 found at the point of measurement.

By introducing variable characteristics (the secant modulus $S=f(\varepsilon_v)$ instead of E and the combined Poisson's ratio v_g instead of v) the elasto-plastic and the linear-plastic material behaviour can formally be described in the same way.

Then the calculation of the real principal stresses, for elasto-plastic strains, can formally be carried out by using Hooke's equations for biaxial-stress conditions:

$$\sigma_1 = \frac{S}{1 - v_g^2} (\varepsilon_1 + v_g \varepsilon_2) \tag{4}$$

$$\sigma_2 = \frac{S}{1 - V_g^2} \left(\epsilon_2 + V_g \epsilon_1 \right) \qquad (5)$$

while the combined Poisson's ratio V_g is given as

$$V_g = 0.5 - (0.5 - V) \frac{S}{E} \qquad (6)$$

and the equivalent strain as

$$\epsilon_v = \sqrt{N_q(\epsilon_1^2 + \epsilon_2^2) + N_g \ \epsilon_1 \ \epsilon_2} \qquad (7)$$

where

$$N_q = \frac{1 - V_g (1 - V_g)}{(1 - V_g^2)^2}$$

and

$$N_g = \frac{-1 + V_g(4 - V_g)}{(1 - V_g^2)^2}$$

are the characteristics of the material tested.

Equations (6) and (7) and the dependence between S and ϵ_v, which is given by the stress-strain curve obtained from the uniaxial tensile test, provide all the relations necessary for the analysis of the elasto-plastic state. The equations contain three unknowns (S, V_g, ϵ_v) which can be solved analytically.

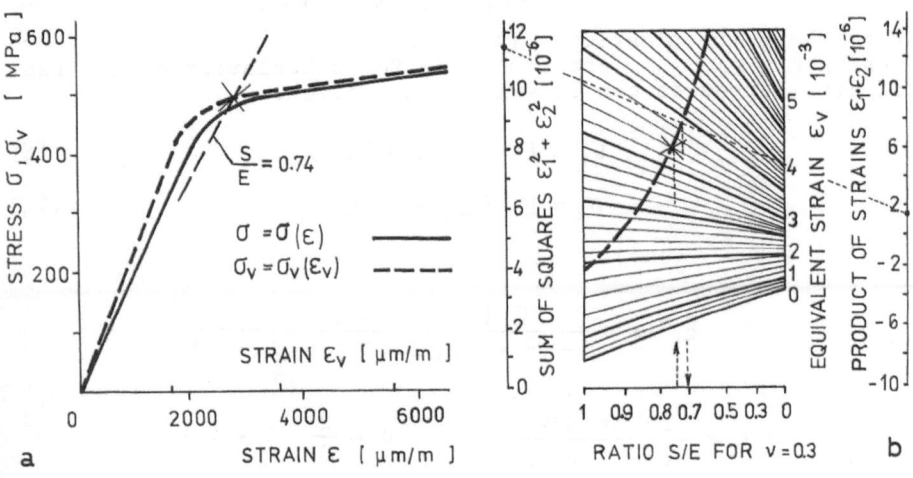

FIGURE 1.a) Stress-Strain curves. b) Nomograph for the deter-mination of the equivalent strain ϵ_v.

Another solution, a graphic one, using the Troost nomograph (Fig. 1b) is very practical. In the first step, the stress-strain curve $\sigma_v = \sigma_v(\varepsilon_v)$ from Fig. 1a must be transferred point by point into the nomograph. This procedure is indicated by the transfer of the point (x). From the measured principal strains we calculate the sum of the squares $\varepsilon_1^2 + \varepsilon_2^2$ and the product of the strains $\varepsilon_1 \cdot \varepsilon_2$. We plot these values on the appropriate vertical scales in the nomograph and connect the plotted points by a straight line (shown in Fig. 1b). The point of intersection of this straight line with the transferred stress-strain curve of the measured material determines the equivalent strain ε_v, and the perpendicular dropped to the horizontal scale for Poisson's ratio $v = 0.3$ determines the value S/E. We obtain the combined Poisson's ratio v_g by solving equation (6), and the equivalent stress σ_v by solving the equation $\sigma_v = S \cdot \varepsilon_v$. Hereby the analysis is complete transforming the biaxial-stress state (triaxial state of deformation) at the point of measurement into an equivalent uniaxial state characterized by ε_v and σ_v.

In the second step, the real principal stresses at the point of measurement can be calculated from equations (4) and (5). This graphic method of solution was used to verify the analytical method described above.

2.2. Method of small elasto-plastic strains

The method of gradual approximation is given in great detail in Ref. (3). The principal strains measured at the point of measurement are ε_1, ε_2 and it is necessary to determine the value of the strain ε_3 as

$$\varepsilon_3 = -(\varepsilon_1 + \varepsilon_2) \frac{3v + \varphi(1+v)}{3(1-v) + \varphi(1+v)} \qquad (8)$$

where φ is the plasticity function.

Using ε_1, ε_2, ε_3, the equations for calculation of the real principal stresses can be written as follows

$$\sigma_1 = E^*(\varepsilon_1 + v^*\varepsilon_2) \qquad (9)$$

$$\sigma_2 = E^*(\varepsilon_2 + v^*\varepsilon_1) \qquad (10)$$

where

$$E^* = \frac{E(1 + c\varphi)}{1 - v^2 + c\varphi(2-v) + \frac{3}{4}c^2\varphi^2}$$

$$v^* = \frac{2v + c\varphi}{2(1 + c\varphi)} \qquad c = \frac{2}{3}(1+v)$$

The equivalent strain ε_v is defined as

$$\varepsilon_v = \frac{\sqrt{2}}{3}\sqrt{(\varepsilon_1 - \varepsilon_2)^2 + (\varepsilon_2 - \varepsilon_3)^2 + (\varepsilon_3 - \varepsilon_1)^2} \qquad (11)$$

By introducing the auxiliary function F defined as

$$F_{(\nu,\varphi)} = \frac{3\nu + \varphi\,(1+\nu)}{3(1-\nu)+\varphi\,(1+\nu)} \qquad (12)$$

and after substituting ε_1, ε_2, ε_3, and re-arranging equation (11), the expression for the equivalent strain can be written

$$\varepsilon_v = \frac{2}{3}\,\sqrt{(\varepsilon_1+\varepsilon_2)^2(F^2+F+1)-3\,\varepsilon_1\varepsilon_2} \qquad (13)$$

The function $F(\nu,\varphi)$ takes values within the limits $0.429 \leqq F \leqq 1$ for φ within the limits $0 \leqq \varphi \leqq \infty$, ($\nu = 0.3$ for steels). Then the real values of the equivalent strain ε_v can be evaluated by the method of gradual approximation, using expressions (12), (13) and the idealized stress-strain curve $\sigma_v = \sigma_v(\varepsilon_v)$ (Fig. 2a). In the first approximation, the equivalent strain ε_v can be calculated from equation (13) for the

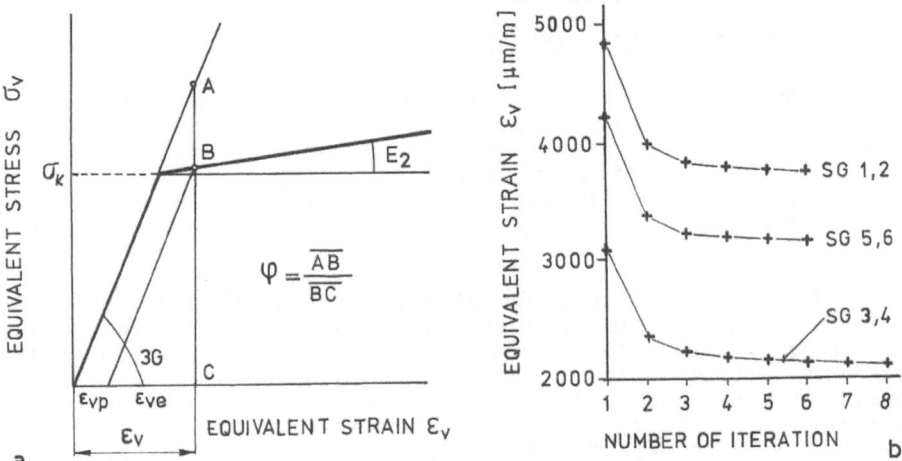

FIGURE 2. a) Idealized stress-strain curve for the determination of the plasticity function. b) Corvengence of gradual results.

limit value $F(\nu,\varphi)=1$ and, since $\varepsilon_v = \varepsilon_{ve}+\varepsilon_{vp}$, the value of the plasticity function φ can be determined from Fig. 2a by solving

$$\varphi = \frac{\overline{AB}}{\overline{BC}} = \frac{(\varepsilon_v - \dfrac{\sigma_k}{3G})(3G - E_2)}{\varepsilon_v E_2 + (3G - E_2)\dfrac{\sigma_k}{3G}} \qquad (14)$$

More exact values for ε and φ are obtained, if equations (12), (13) and (14) are solved and repeatedly corrected after each calculation, until the predetermined accuracy is achieved.

The following accuracy check has been made:

$$\frac{\varepsilon_v^{N+1} - \varepsilon_v^{N}}{\varepsilon_v^{N}} \lesseqgtr \text{predetermined accuracy} \qquad (15)$$

where N is the number of the repeated calculation. In using this technique it has been found that convergence to a high degree of accuracy (about 0.1 %) is obtained within five to ten iterative cycles (Fig. 2b). Finally the real principal stresses are calculated from equations (9) and (10).

2.3. Other methods of evaluation

Two graphic-numerical methods, unsuitable for computer processing, are mentioned for the sake of completeness: the method of small elasto-plastic strains with a simplified calculation (Ref. 4) and the graphic-numerical method based upon Neuber's principle (Ref. 5).

3. PROBLEMS CONNECTED WITH THE EVALUATION

The above mentioned evaluation of strain gauge readings up to 0.2 % of plastic strain is necessary otherwise the application of Hooke's extended equations would lead to inapplicable fictitious stresses exceeding the values of the yield stress or tensile strength of the measured material.

The problems connected with the evaluation are the following:

a) The stress-strain curve of the material obtained from the uniaxial tensile test is usually not known before and during the strain gauge measurement. The main material characteristics (yield stress, tensile strength, etc.) are not sufficient for the evaluation. In addition, in most cases, the stress-strain curve cannot be used for their evaluation because of inadequate resolution of tensile diagrams as obtained on tension testing machines in the material testing laboratories.

b) Bulky components made of the same material may show considerably differing values of the yield stress (20 MPa or more).

c) Residual stresses at the point of measurement are nearly always unknown because of the impossibility to use sectioning techniques or partly nondestructive methods (e.g. the hole drilling method, the ring core method) for their evaluation. The values of residual stresses influence the accuracy of the evaluation directly.

d) The limit of proportionality of the construction material is usually exceeded at places of stress concentration where elasto-plastic strains originate in a small bounded area or in a part of its cross-section, which implies that the plasticity condition is not valid in the whole range.

e) It should be noted that the methods of evaluation are not quite accurate. For example, an idealized stress-strain curve $\bar{\sigma}_v = \bar{\sigma}_v(\epsilon_v)$ is made use of in the method 2.2.

4. EXAMPLE OF APPLICATION

By means of the methods mentioned an evaluation of pricipal stresses based on strain gauge readings (see Fig. 3)

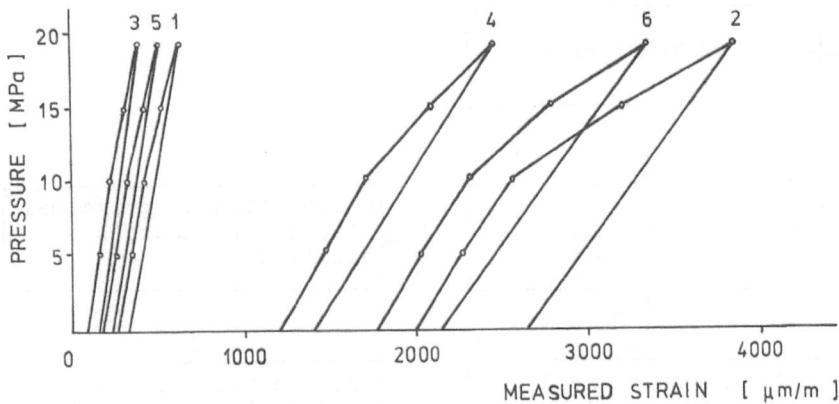

FIGURE 3. Loading pressure vs. measured strain.

in the elasto-plastic area was carried out for three definite points on the flange connection. The results of the evaluation are shown in Table 1.

TABLE 1. Fictitious and real principal stresses.

Strain gauge	Measured strains	Residual strains	Stresses Ficti- tious	Real stresses Method 2.1	Method 2.2	Method 2.3
1	597	75	404	288	292	282
2	3831	1040	918	566	574	553
3	387	60	261	241	249	240
4	2467	215	591	528	558	524
5	502	75	349	275	277	271
6	3349	565	804	564	569	550

Values of strains   — μm/m
Values of stresses — MPa

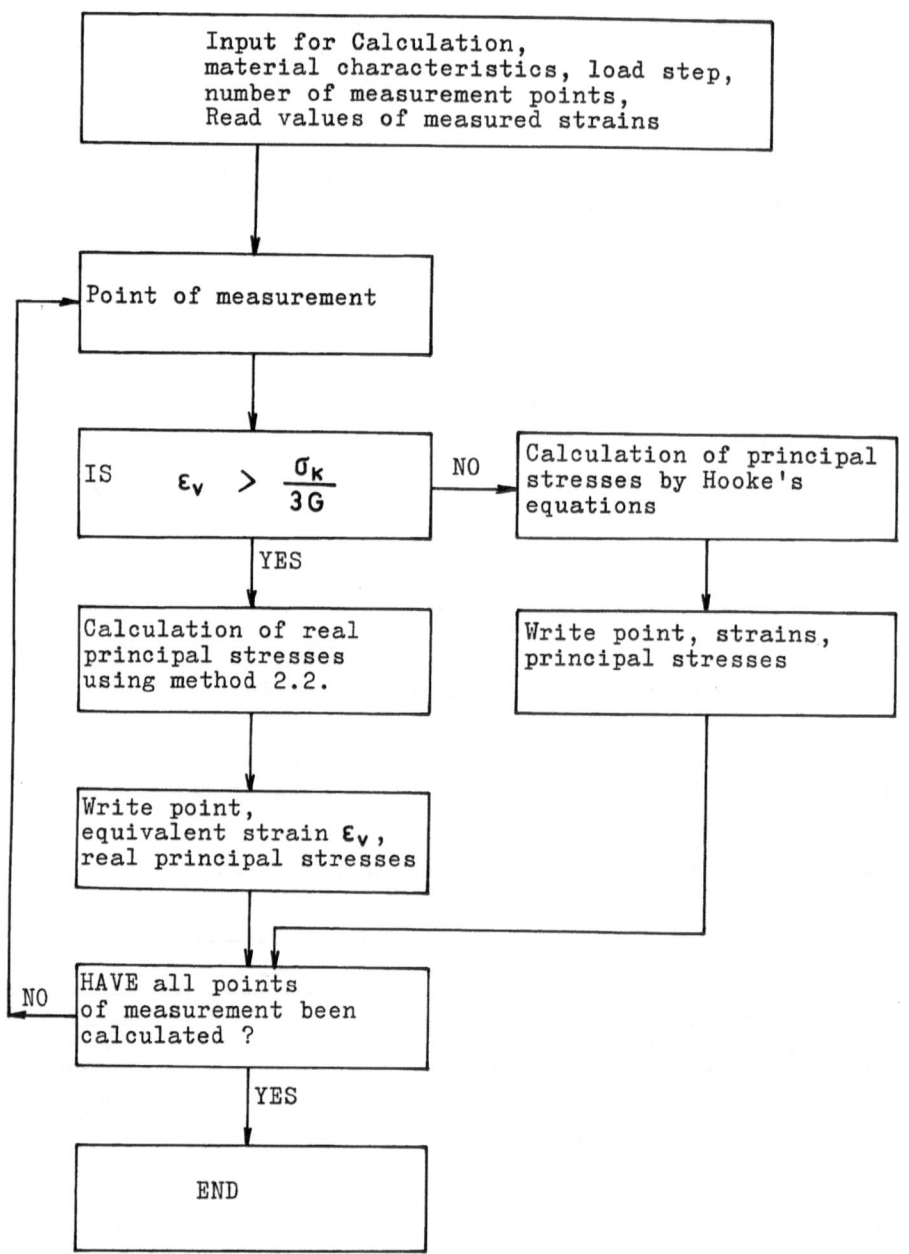

FIGURE 4. Flow diagram for the principal stress and the real principal stress calculation program.

The stress-strain curve of the construction material, which was determined for three test specimens of a circular cross-section, 10 mm in diameter, by means of the HBM 3/120 LY 11 strain gauges, was used for the evaluation of the principal stresses (see Fig. 1a). The loading cycles corresponding to the Czechoslovak State Standard ČSN 42 03 10 were adhered to.

The evaluation of the real principal stresses following methods 2.1. and 2.2. was carried out on a computer, method 2.3. was evaluated manually.

The results obtained for method 2.1. were the most accurate. This method takes into account the transient part of the stress-strain curve, from the linear into the non-linear zone. The differences of the real principal stress values, evaluated for each of the three methods to be compared, are less than 6 %.

5. CONCLUSION

Method 2.2., the method of small elasto-plastic strains with gradual approximation, has been selected and shown to be suitable for evaluating real principal stresses in the elasto-plastic area on a data logger computer. The flow diagram for the principal stress and the real principal stress calculation program is shown in Fig. 4. The program will be used in our laboratory to evaluate all future strain gauge measurements, especially in the elasto-plastic area.

REFERENCES

1. Crips R.J.: A Computer Program to Calculate Post-Yield Stresses and Shakedown Loads from Strain Gauge Readings. CEGB Report No. RD/B/N912,1967.
2. Kiel S., Benning O.:On the Evaluation of Elasto-Plastic Strains Measured with Strain Gauges. Exper.Mech. , August 1979.
3. Lébl O.: An Elasto-Plastic State of Stress in a Symmetrical Pressure Vessel. Research Report SVUSS No.78-03014, 1978. (In Czech).
4. Kuliš Z.: Problems of Evaluation of Stress State in the Range of Plastic Strains. Acta Polytechnica, II, Praha 1976.(In Czech).
5. Vejvoda S. et al.: Strain Gauge Measurements on Bodies of Pressure Vessels for the VVER Type Nuclear Power Stations. Research Report ÚAM Brno, 1978.(In Czech).

ERRORS ASSOCIATED WITH THE USE OF STRAIN GAGES ON COMPOSITE MATERIALS

C.C. HIEL

VRIJE UNIVERSITEIT BRUSSEL - Faculty of Engineering - Composites Group (KB)
Pleinlaan 2, B-1050 Brussels (Belgium)

1. INTRODUCTION

The use of strain gages on composite materials is an essential element in most mechanical property characterization methods. A recent summary, on the state of the art in strain gage technology, was written by C.C. Perry, [1], one of the leading authorities in the field. Dr. Perry concludes that there are fringe areas in strain gage technology which have been only partially explored such as large strains, reinforcement effects, thermally and hygroscopically anisotropic test materials, etc. They require a much more thorough examination if accuracies of ± 10 % or better are to be reliably achieved in strain measurements on plastic based structural materials.

Now that composites are going to be used more and more in strength critical applications the precise measurement of the materials response is even more important because it is essential that the response which precedes failure be understood so that a larger proportion of the potential material strength can be utilized in design. Indeed a typical strain value currently used in design is 0.4 % which may be compared to the fiber breaking strain of about 1.3 %.

Research along these lines has also been initiated by Brinson & Tuttle, [2].

2. BASIC STRAIN GAGE RESPONSE

2.1. The uniaxial strain gage by itself

The gage which is schematically shown in Fig. 1 is basically a conductor with resistivity , cross-section S and total length . The resistance R between the terminals A and B is given as a function of these three independent variables by Eq. 1.

$$R = \rho \cdot \frac{\ell}{S} \tag{1}$$

This linear relationship is well established in the theory of conductive media [3].

The strain gage wiring in Fig. 1 will undergo a stress when it is pulled upon. This of course can change the conductive properties due to piezoelectric effects.

Alternatively when the gage is exposed to temperature, thermoelastic and thermoresistivity effects do contribute to a change in R.

The relative change in resistance is given by Eq. 2

$$\frac{dR}{R} = \frac{1}{E_g} (1 + 2\nu_g + E_g \cdot \gamma_g) \, d\sigma + (\beta_g - \alpha_g) \, dT \tag{2}$$

with E_g : Young modulus, ν_g : Poissons ratio, γ_g : piezoresistivity, β_g : thermoresistivity, α_g : thermal expansion.

Fig.1. Strain-gage -- Schematic

Eq. 2 can also be written as

$$\frac{dR}{R} = K_1 \, d\sigma + k_2 \, dT \tag{2'}$$

with

$$K_1 = 1 + 2\beta_1 + E_1\gamma_1$$
$$K_2 = \beta_1 - \alpha_1$$

The resistance change has two contributions as can be clearly seen in Eqs. 2 and 2'. It is important to estimate their relative importance. For example for Constantan (45 % Ni, 55 % Cu)

$$K_1 = 2.1$$
$$K_2 = 3.3 \, 10^{-6}/°C \tag{3}$$

A mechanical deformation of $5 \, 10^{-6}$ (5 microstrain) causes the same change in gage resistance as a temperature change of 3°C.

It is important to recognize at this stage that Eq. 2 is a statement about the strain-gage wire itself. The situation in engineering practice is one where the wire is bonded onto a deforming structure, which in our case is a composite material. Generally there is a thermal mismatch between the composite and the gage because the thermal expansion of Constantan wire is $6 \, 10^{-6}/°C$ while the thermal expansion for graphite epoxy varies between $.1 \, 10^{-6}/°C$ in the fiber direction and $15 \, 10^{-6}$ transverse to the fiber direction.

2.2. The uniaxial gage bonded onto a unidirectional reinforced material

Suppose a strain gage-bonded onto a fiber reinforced plastic as schematically shown in Fig. 2.

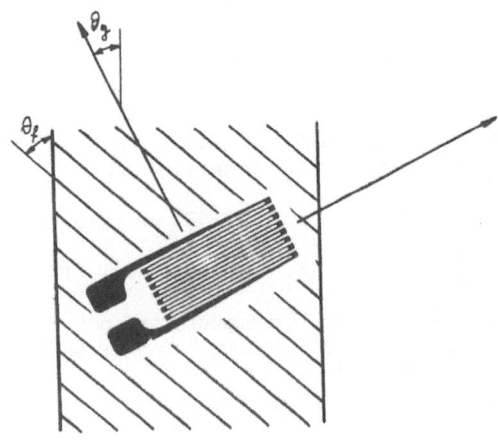

Fig. 2. Strain-gage and fiber orientation.

The coefficient of thermal expansion along the fiber direction is given by α_L is much smaller than α_T, the thermal expansion perpendicular to the fibers. Eqs. 4 then give the thermal expansion coefficients in the x and y directions

$$\alpha_x = m^2\alpha_L + n^2\alpha_T \tag{4a}$$

$$\alpha_y = n^2\alpha_L + m^2\alpha_T \tag{4b}$$

$$\alpha_{xy} = 2(mn\,\alpha_L - mn\,\alpha_T) \tag{4c}$$

with $m = \cos\theta_f$ and $n = \sin\theta_f$.

The coefficient of thermal expansion which is relevant for the strain-gage bonded at an angle θ_g is given by Eq. 5.

$$\alpha_c = \alpha_x \cos^2\theta_g + \alpha_y \sin^2\theta_g + \alpha_{xy} \sin\theta_g \cos\theta_g \tag{5}$$

For the special case of an anisotropic material $\alpha_x = \alpha_y = \alpha$ and $\alpha_{xy} = 0$. Thus

$$\alpha_c = \alpha \tag{6}$$

We subsequently consider the situation were the strain gage is bonded onto the composite material. The characteristics of the composite are it's modulus in the direction in which the strain gage is bonded given by E_c and it's coefficient of thermal expansion α_c as given in Eq. 5.

When the hypothesis is made of a perfect adhesion between the strain gage and the composite, we may write :

$$\begin{bmatrix} \text{mechanical} & + & \text{thermal} \\ \text{deformation} & & \text{deformation} \end{bmatrix}_{\text{strain-gage}} = \begin{bmatrix} \text{mechanical} & + & \text{thermal} \\ \text{deformation} & & \text{deformation} \end{bmatrix}_{\text{composite}}$$

or

$$\frac{\Delta\sigma_g}{E_g} + \alpha_g \, \Delta T = \frac{\Delta\sigma_c}{E_c} + \alpha_c \, \Delta T \tag{7}$$

$$\underbrace{\phantom{\frac{\Delta\sigma_g}{E_g} + \alpha_g \Delta T}}_{\varepsilon_g} \qquad \underbrace{\phantom{\frac{\Delta\sigma_c}{E_c} + \alpha_c \Delta T}}_{\varepsilon_c}$$

The Δ's in Eq. 7 represent the variations with respect to the "stress free" state which existed at the moment of adhesion between gage and composite.

Upon subsitution of Eq. 7 into Eq. 2' we obtain

$$\frac{\Delta R}{R} = K_1 \, \varepsilon_c + (K_1 \, (\alpha_c - \alpha_g) + K_2) \, \Delta T \tag{8}$$

The effects of strain and temperature are not fully separated in Eq. 8 because K_1 turns out to be a function, however weak, of temperature. It is measured in an experimental set up.

Substitution of Eq. 5 into Eq. 8 gives

$$\frac{\Delta R}{R} = K_1 \varepsilon_c + [K_1 (\alpha_x \cos^2\theta_g + \alpha_y \sin^2\theta_g + \alpha_{xy} \sin\theta_g \cos\theta_g - \alpha_g + K_2] \Delta T \tag{9}$$

There are thus three ways to assure thermal compensation :

1) $\Delta T = 0$

This is the trivial case, which is more of an academic value. One area of practical interest though are measurements on shock wave transients or very fast dynamic strains.

2) $K_1 (\alpha_x \cos^2\theta_g + \alpha_y \sin^2\theta_g + \alpha_{xy} \sin\theta_g \cos\theta_g - \alpha_g) + K_2 = 0$ (10)

It is sufficient to solve Eq. 10 for ν_g. A strain gage mounted along this angle will be completely compensated. This does not bring us any further though because the strain itself is a function of θ_g due to the anisotropic nature of the material.

3) The technique of the "dummy gage"

The dummy gage is only subject to thermal variations, thus

$$\left(\frac{\Delta R}{R}\right)_{dummy} = [K_1 (\alpha_c - \alpha_g) + K_2] \, \Delta T \tag{11}$$

A wheatstone circuit wired in half bridge enables us to separate out the mechanical contribution

$$\left(\frac{\Delta R}{R}\right)_{active\ gage} - \left(\frac{\Delta R}{R}\right)_{dummy} = K_1 \, \varepsilon_c \tag{12}$$

The hypothesis which is implicit in the derivation of Eq. 20 is that the active gage and the dummy gage both are bonded on two perfectly identical composite structures. This hypothesis is violated as soon as there is a small gage misalignment of the dummy with respect to the active.

2.3. The influence of transverse sensitivity

The strain gage, which is shown mounted on the composite in Fig. 2, has a certain amount of wire in the perpendicular direction which is indicated by the t-axis in the figure.

Consequently the response of the strain gage is not only influenced by the axial strain ε_a but also by ε_t (note that a more consistent notation would be to use ε_{ac} and ε_{bc}. We will implicitly assume however that from now on the material is a composite unless we specify it otherwise. We thus drop the c index). This statement can be mathematically written as

$$\frac{\Delta R}{R} = F_a \, \varepsilon_a + F_t \, \varepsilon_t \tag{13}$$

Eq. 13 can also be written as

$$\frac{\Delta R}{R} = F_a \, (\varepsilon_a + K_{ta} \, \varepsilon_t) \tag{14}$$

This equation can be written more explictly to model the experiments in the gage calibration phase. We then know precisely what the biaxial strain field is, since ν_0 is known thus

$$\frac{\Delta R}{R} = \underbrace{[F_a \, (1 - K_{at} \, \nu_0)]}_{K_1} \, \varepsilon_a \tag{15}$$

For all other situations in which we actually want to characterize the material we can not measure the true strain immediately but we measure an apparent strain which is given as

$$(\varepsilon_a)_{\text{apparant}} = \frac{\Delta R / R}{K_1} \tag{16}$$

by substituting the results from Eqs. 14 and 15 we obtain

$$(\varepsilon_a)_{\text{apparant}} = \frac{(\varepsilon_a + K_{ta} \, \varepsilon_t)}{(1 - K_{at} \, \nu_0)} \tag{17}$$

Eq. 17 contains two unknowns consequently we need a second equation which means that a second measurement has to be taken in the t-direction as schematically indicated in Fig. 3.

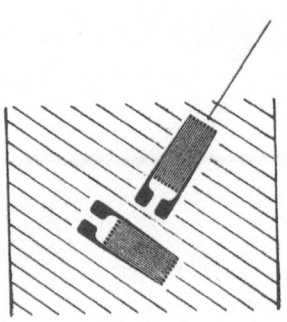

Fig. 3. Gage configuration in fiber- and transverse direction.

We thus obtain

$$(\varepsilon_t)_{apparant} = \frac{\varepsilon_t + K_{at}\,\varepsilon_a}{1 - K_{at}\,\nu_o} \tag{18}$$

Eqs. 17 and 18 can be used to determine the percentage error in the axial strain and in the transverse strain respectively. The majority of materials characterization tests is done on coupons which are loaded in the fiber direction and transverse to the fiber direction respectively. The percentage error E in the axial and transverse strain indications while loading in the fiber direction is expressed in Eq. 19 and 20 :

$$E_L(\varepsilon_a) = \frac{K_{at}(\nu_o - \nu_{12})}{1 - K_{at}\,\nu_o} \times 100\;\% \tag{19}$$

$$E_L(\varepsilon_t) = \frac{K_{at}(\nu_o - 1/\nu_{12})}{1 - K_{at}\,\nu_o} \times 100\;\% \tag{20}$$

The percentage error E in the axial and transverse strain indications while loading transverse to the fiber direction is expressed in eqs. 21 and 22

$$E_T(\varepsilon_a) = \frac{K_{at}(\nu_o - \nu_{21})}{1 - K_{at}\,\nu_o} \times 100\;\% \tag{21}$$

$$E_T(\varepsilon_t) = \frac{K_{at}(\nu_o - 1/\nu_{21})}{1 - K_{at}\,\nu_o} \times 100\;\% \tag{22}$$

Eqs. 19 and 21 are represented graphically in Fig. 4. The percentage error in the axial strain is plotted as a function of Poissons ratio.

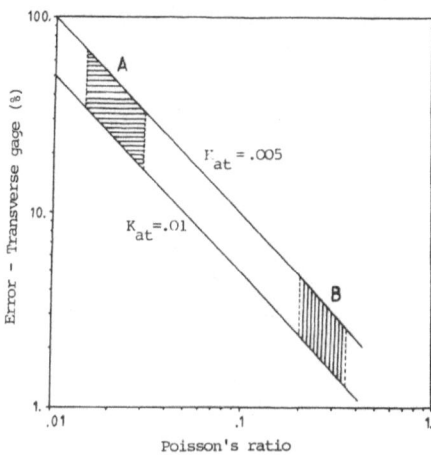

Fig. 4. The percentage error in the transverse strain-gage readings.

The values for K_{at} which we choose to plot Fig. 4 are representative for the best strain gages which are available today. The conclusion which can be drawn is that the errors are only of the order of a few percent. The two shaded areas in the figure are representative for the error on E_{11} and E_{22} respectively.

Fig. 5 graphically represents Eq. 20 and 22 where we choose the same values for K_{at} as in Fig. 4. Again we marked two shaded areas A and B. It should be emphasized that Fig. 5 is plotted on a logarithmic scale.

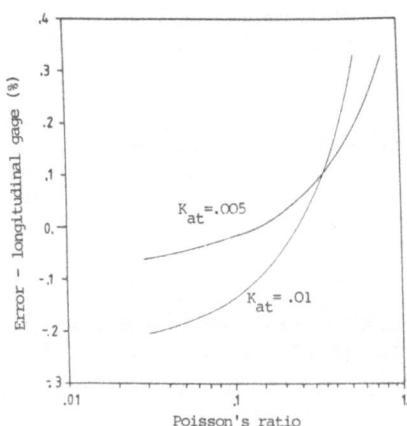

Fig. 5. The percentage error in the transverse strain-gage readings.

For the minor Poissons ratio v_{21}, which lies in the rage .015 till .03 we can define the shaded error region A. The error apparently varies between 16 % and 68 % depending on the transverse sensitivity of the gage that is actually used and the specific composite material on which it is bonded.

The major Poissons ratio v_{12} usually lies in the range .2 - .35. This enables us to define the shaded region B. The transverse error for this type of measurement varies between 1.3 % and 4.7 %.

In a practical materials testing-type situation, we don't actually know what the values of the major and minor Poissons ratio's are. The correction for transverse sensitivity can be made easily however when we solve Eqs. 17 and 18 for ε_a and ε_t. We thus obtain

$$\varepsilon_a = \frac{(1 - v_o K_{at})(\varepsilon_{am} - K_{at}\,\varepsilon_{tm})}{1 - K_{at}^2} \tag{23}$$

$$\varepsilon_t = \frac{(1 - v_o K_{at})(\varepsilon_{tm} - K_{at}\,\varepsilon_{am})}{1 - K_{at}^2} \tag{24}$$

3. CONCLUSIONS

For every material which has directionally dependent coefficients of therman expansion, there exists a direction for which the strain gage is totally compensated.

Corrections for transverse sensitivity should be done routinely. Especially the errors on the transverse strain measurements should be considered. We get the largest error when we attempt to measure the minor Poisson ratio of a composite.

The correction procedures remain important for strain measurements on laminates.

Important errors due to misalignment and the non-coincidence of principal stress and principal strain directions will also be discussed during the presentation.

4. ACKNOWLEDGEMENTS

The author acknowledges support from NATO in the form of a summer-fellowship.

Special thanks are due to Myriam Bourlau at the Free University of Brussels for her excellent typework.

5. REFERENCES

[1] Perry, C.C. "The Resistance Strain Gage Revisited", Journal of the Society for Experimental Mechanics, vol. 24, no. 4, 1984, pp. 286-299.

[2] Tuttle, M.E. and Brinson, H.F. "Resistance Foil Strain-Gage Technology as Applied to Composite Materials", Journal of the Society for Experimental Stress Analysis, vol. 24, no. 1, 1984, pp. 34-65.

[3] Thompson, W. (Lord Kelvin) "On the Electrodynamic Qualities of Metals", Proc. Roy. Soc., 1856.

HOW TO AVOID ERRORS CAUSED BY HEAT EFFECTS IN STRAIN GAGE
MEASUREMENTS WHEN USING SCANNING UNITS

MANFRED KREUZER

HOTTINGER BALDWIN MESSTECHNIK GMBH, DARMSTADT

Excitation voltages applied to strain gages cause heat effects
which may produce signal changes and measuring errors. This
paper shows the most important results of many measurements
which had to be taken to determine the amplitudes and time
constants of these shifts of signals for different strain gage
types, excitation voltages, adhesives and construction materials. In its second part the paper discusses methods of avoi-
ding measuring errors due to heat effects and describes a
scanning unit which allows to take measurements almost insensitive towards heat effects, voltage drops across the measuring leads and switching elements, or induced error voltages.

1. INTRODUCTION
   If many strain gage signals are to be measured, scanning
units usually do this job. These scanning units normally contain only one central measuring amplifier and signal conditioning unit which are connected to individual strain gages one
by one using highly sophisticated switching circuits. This is
qiite difficult because errors due to voltage drops produced
by switch- and wire-resistances must be avoided.
This feat is especially difficult to accomplish when applying
strain gages in half- and quarter-bridge circuits because
changes in measured resistances are extremely small. The paper
shows how to design switching circuits, so that voltage drops
across the measuring leads and switching elements cannot fal-
sify the readings. Since the individual measuring points are
switched one by one, the excitation voltage supplies the strain
gages only during measurement. Therefore, the temperatures of
the strain gages do not increase significantly. This is especially important if strain gages are adhered to plastic, wood
or other materials with low thermal conductivity. Heat effects
cannot be neglected completely, however, because the signals
dependent on temperature change very rapidly immediately after
exciting the strain gages. So if measuring time varies, the
readings may also vary significantly. There are several reasons
why the measuring time may vary. High scanning rates and high
signal resolution cannot be achieved simultaneously. To get
stable strain gage measurements with high resolution, the
strain gage signals have to be integrated for an extended time
period, e.g. 100 ms. But 100 ms is about 4 times the thermal
heat-up time constant of a medium sized strain gage adhered to
a steel test objekt.
Thus, the readings obtained when working with higher scanning
rates (> 40 s^{-1}) will differ from the readings obtained when

working with lower scanning rates and extended integration
times. So, if scanning rates are lowered and integration time
increases, you do not only get higher resolution but also
changing measurments as a negativ result.
Most scanning units provide an automatic zero balancing facility. It may occur that measuring values taken immediately
after zero balancing will differ from zero, though there was
no change in load. This is likely to happen under the conditions, when zero balancing procedures take longer than the
actual measurements and when small strain gages are excited
with high bridge supply voltages. These deviations can become
expecially large if the measuring points are selected manually and excited for a longer time period.

2. SIGNAL SHIFTS OF STRAIN GAGES INDUCED BY HEAT EFFECTS

When a strain gage is to be measured with the help of a
scanning unit, the strain gage has to be energized in a first
step. When the excitation voltage is applied to a strain gage, the temperature of the strain gage starts to rise, until
the flow of electrical excitation power is balanced by the
absorption of the heat power flow into the specimen and to a
much smaller part into the tap wires and the surrounding air.
The amplitude and the time function of the heat-up period
depend on several parameters, as excitation voltage, size and
resistance of the strain gage, the adhesive and the thickness
of the glueline, the material and sometimes also on the mass
of the specimen. This paper covers only test results of strain
gages adhered to massive ferritic steel and aluminium specimen,
where the temperature increase of the specimens could be neglected, because of their mass.
Temperature dependent changes of the specific strain gage grid
resistance occur in the applied gage owing to the linear thermal expansion coefficients of the grid and specimen materials
and the thermal coefficient of the specific electrical resistance. These resistance changes pretend mechanical strain in
the specimen.
The temperature coefficient α_a of an applied strain gage can be
calculated [1] according to equation (1):

$$\alpha_a = \alpha_s - (\alpha_g - \alpha_r/k) \qquad\qquad \text{in } m \ m^{-1} \ K^{-1} \qquad (1)$$

α_s = thermal expansion coefficient of the specimen
α_g = thermal expansion coefficient of the strain gage
    grid matrial (15 ppm with Constantan)
α_r = thermal coefficient of the specific electrical
    resistance of the strain gage grid material
k = gage factor

In order to keep apparent strain due to temperature changes as
small as possible, the strain gages are matched during the
production to a certain linear thermal expansion coefficient,
e.g. matched to steel, aluminium, quartz, or plastics. This is
done by changing the thermal coefficient of the specific electrical resistance. The resulting temperature coefficient of the
applied gage, which is obtained after the application of the
matched strain gage to the appropriate measurement object, is

approximately zero ($\alpha_a = 0$).
In this case, according to equation (1) the strain of the specimen caused by temperature change is compensated for the apparent strain of the applied strain gage.
This leads to equation (2):

$$\alpha_s = \alpha_g - \alpha_r/k \qquad (2)$$

In cases where stain gages heat up due to the energizing voltage while the temperature of the specimens remain approximately constant because of their good thermal conductivity the
lenghth of the specimens remain unchanged while due to heating
the grids of the strain gages would expand if they had not
been fixed to the specimens and that is the reason why the
strain gages show negativ apparent strain. This apparent strain
can be calculated according to equation (3):

$$\varepsilon_a = (\alpha_g - \alpha_r/k) \star \Delta T \qquad (3)$$

ΔT = difference between the temperature of the strain
gage grid and the temperature of the specimen

In cases when strain gages and specimens are thermally matched, equations (2) and (3) lead to equation (4):

$$\varepsilon_a = \alpha_s \star \Delta T \qquad (4)$$

Vice versa, the difference ΔT between the temperature of the
strain gage grid and the temperature of the specimen material
is easily obtained when the apperent strain of the applied
strain gage can be measured and the values of α_g, α_r and k are
known. Rearranging equation (3) leads to:

$$\Delta T = \varepsilon_a / (\alpha_g - \alpha_r/k) \qquad (5)$$

In cases when the strain gages are thermally matched to the
thermal expension coefficient of the specimen the difference
ΔT of the temperatures is even easier to obtain by rearranging
equation (4):

$$\Delta T = \varepsilon_a / \alpha_s \qquad (6)$$

Equations (3) and (4) show an interesting piece of information. If the applied strain gages are matched to materials
with $\alpha_s = 0$, heat effects would not result in any apparent
strain, even, if the thermal expansion coefficient of the specimen is quite different from zero.
Fig. 1 shows the apparent strain ε_a as a function of time
after switching on the energizing voltage. The strain gage is
a foil type 3/120 LY11 (see also table 1; No.4...6) matched to,
and adhered on, a block of ferritic steel. The measured curves
fit pretty well into the expected exponential behaviour.

$$\varepsilon_a = \varepsilon_{ao} \star (1 - \exp(-t/\tau)) \qquad (7)$$

ε_{ao} = apparent strain after heat-up has stabilized

τ = Time constant of the measured curve immediately after switching on the excitation voltage

The theoretical approximation of the measured curves can be improved a bit, if two exponential functions with different time constants and different amplitudes are superimposed. The parameters are shown in equation (8):

$$\varepsilon_a = \varepsilon_{ao} * (0.8*(1-\exp-(6/7*t/\tau))+0.2*(1-\exp-(3*t/\tau))) \quad (8)$$

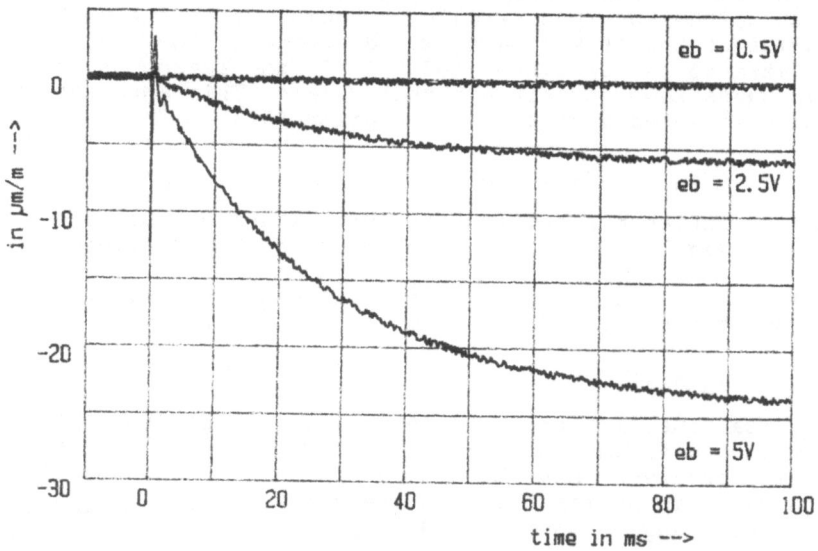

FIGURE 1. Apparent strain α_a of an applied strain gage type 3/120 LY11 when bridge supply voltage is switched on

Table 1 gives a selection of results with several measuring parameters varied. Obviously, the amplitude of the apparent strain is proportional to the square of the bridge supply voltage. The smaller the active grid area, the bigger the apparent strain induced by heat effects, but also the shorter the time constants. The time constant of a heat-up period depends also on the type of adhesive, the thickness of the glueline and the material of the specimen. The hot curing adhesive EP250 provides better thermal conductivity than the cold curing adhesives Z70 and X60. That's why the time constants of the tests of No.7 to No.12 are especially short and the apparent strains are pretty low. With Z70, a one component adhesive, thinner gluelines can be achieved than with X60. That is the reason why the tests with Z70 show shorter time contants and lower apparent strain amplitudes than the tests with X60, a two component adhesive. But it has to be mentioned that the results depend also strongly on how carefully the application was made. The given values in table 1 are average test results of several carefully applied strain gages. The thermal conductivity of the specimen has also some influence on the heat effects. This becomes obvious if test No. 7 is compared with test No.10. The temperature difference ΔT raises to 1,4 K, if strain gages are adhered with EP250 on steel specimens compared to only 1,1 K if adhered under the same conditions on aluminium specimens. A possible explanation could be that the

TABLE 1. HEAT UP EFFECT DEPENDENCES ON VARIOUS EXPERIMENTAL CONDITIONS

No.	strain gage type	grid length [mm]	specimen material	adhesive	e_b [Vrms]	ε_{ao} [μm/m]	ΔT [K]	τ [ms]	settling time [ms] <10μm/m	<1μm/m
1	(3/120 LY 11) matched to steel with ≈11·10⁻⁴/ K [6.1ppm/°F]	3	steel St37	X60	5	-35	3.2	30	35	150
2					2.5	-9	0.8		0	75
3					0.5	< -1	0.03		0	0
4				Z70	5	-25	2.2	20	15	60
5					2.5	-6	0.5		0	35
6					0.5	< -1	0.02		0	0
7				EP250	5	-15	1.4	15	5	40
8					2.5	-4	0.35		0	20
9					0.5	< -1	< 0.02		0	0
10	(3/120 LY 13) matched to aluminium with = 23·10⁻⁶/ K [12.7ppm/F]	3	AL	EP250	5	-26	1.1	12	10	40
11					2.5	-6.5	0.3		0	25
12					0.5	< -1	< 0.02		0	0
13	(1/120 KY 11) strain gage chain matched to steel with ≈11·10⁻⁶/ K [6.1ppm/°F]	0.6	St37	Z70	5 [1]	-160	14.5	8	25	100 [2]
14					2.5 [1]	-40	3.6		12	30
15					0.5	-1.5	0.15		0	3
16	(10/120 LY 11) matched to steel with ≈11·10⁻⁴/ K [6.1ppm/°F]	10	St37	Z70	5	-3	0.25	50	0	50
17					2.5	< -1	0.06		0	0
18					0.5	≈ 0	< 0.01		0	0

1) Only e_b ≤ 1.5V allowed
2) Owing to creep effects because of overheating

area of the steel specimen directly below the strain gage grid may have heated up for about 0.4 K, wheras the area of the aluminium specimen may have heated up for only 0.1 K due to the thermal conductivity of aluminium, four times superior than that of steel.
All test results show surprisingly short time constants. It can be stated, that 100 ms after powering the strain gages, the remaining error is lower than 1 μm/m in almost every case. When supplying the bridge with 0,5 V, the apparent strains generated by heat effects are near or far below 1 μm/m .

3. HEAT EFFECTS AND COUNTERMEASURES

Though the heat effects settle in about 100 ms, they cannot be ignored when measuring rates rise to tens or even hundreds of measurements per second. A measurement taken at high speed immediately after switching on the excitation voltage, may differ from continuously taken measurements up to apparent strains of ε_{ao} in table 1. That is why fast measurements taken immediately after zero balancing may differ considerably from zero if zero balancing of the measuring points is done manually or in a longer automatic balancing procedure.

If a group of strain gages join the same compensation strain gage, the measurements of the different measuring points may differ from one another though they have been zero balanced in a correct short measuring circle just a short time before and no mechanical stress has been applied in the mean time. The source for these deviations can be traced to heat-up effects of the compensation strain gage.

When the first strain gage of the group is being switched to the excitation voltage, the dummy gage or compensation gage will also be switched to the excitation voltage for the first time and therefore will show about the same heat-up effects as the measured strain gage, whereas the heat-up effects of the compensation strain gage settle during the following measurements. That is why the heat-up effects of the strain gage switched on first in a group may show nearly no apparent strains, whereas the others show increasingly apparent strain amplitudes.

Now, what can be done to avoid measuring errors due to these heat effects? Different methods may be adopted. Application of one of the following hints may be sufficient. Here are some suggestions:

* Use strain gages with thermal coefficients $\alpha_a = 0$ and try to achieve temperature compensation with the help of an unstressed compensation gage close to the common group of strain gages
* Use a separate compensation gage for each measured strain gage
* Choose bridge supply voltages of $=< 0,5$ V
* Delay each measurement for about $=>50$ ms until heating has settled
* Perform zero balancing and measurments in equally time periods
* If a measuring point is to be displayed continuously the energizing voltage should be pulsed
* Supply the strain gages with carrier frequency to suppress thermo coupled voltages generated by heat effects

Fig. 2 shows a scanning unit of the type UPM60 which provides all of the above mentioned features. The strain gages can be powered with 225 Hz carrier frequency bridge supply voltages of 5 V or 0,5 V. Measurements can be taken up to speeds of 100 measuring points per second and the zero balancing time is always equal to the measuring time. When selecting the measuring points manually, the UPM60 powers the strain gages only for about 40 ms every 500 ms interval time to reduce heating.

4. UNIVERSALLY APPLICABLE MULTIPOINT MEASURING UNIT

Apart from the heat-up effects, many other error sources exist if measuring strain gages, especially if they have to be measured in quarter bridge circuits. The biggest error sources may be voltage drops across the measuring leads and switching elements, because strain gage signals are very low and voltage

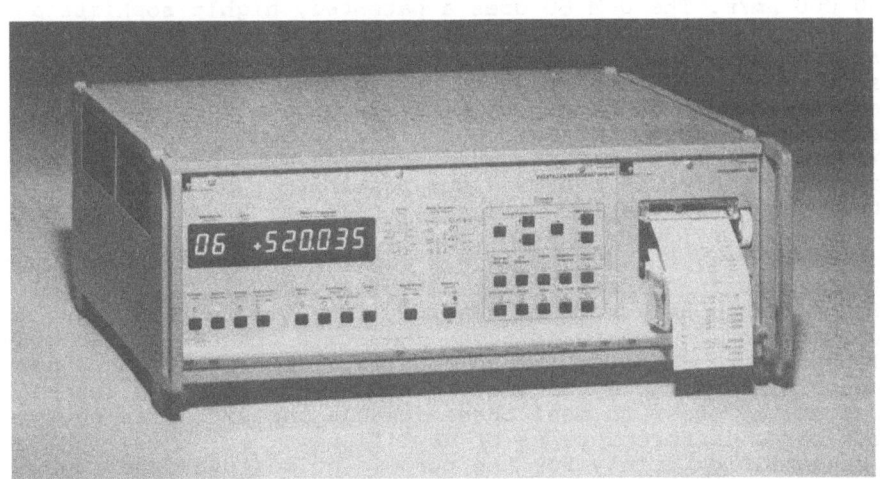

FIGURE 2. Scanning unit equipped with dc and carrier frequency amplifiers to provide highest precision with various transducer types

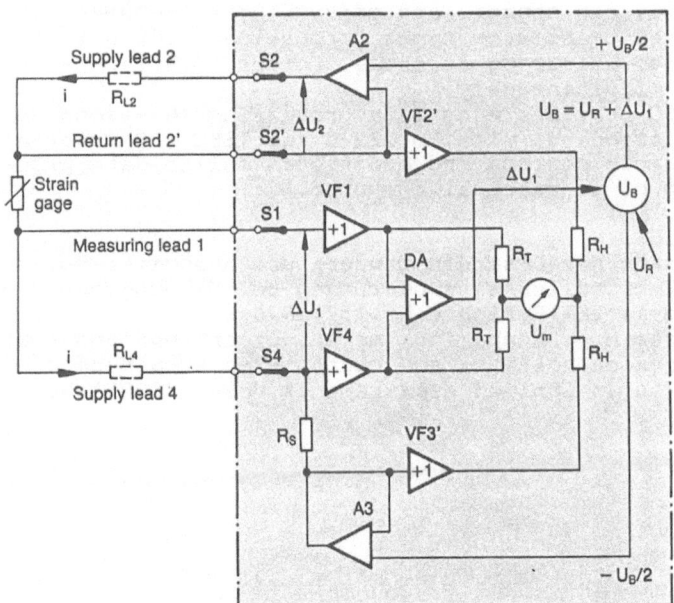

FIGURE 3. New bridge circuit reduces zero and sensitivity errors significantly with internal feedback circuits which correct for voltage drops across the leads and switches

drops may exceed equivalent strain gage measuring signals of
100 000 µm/m. The UPM 60 uses a patented, highly sophisticated intermediate circuitry [2] .Errors due to voltage drops
are avoided almost completety. Fig. 3 shows a simplified schematic diagram of the bridge circuit of the UPM 60.
Feed back technics have been adopted which allow to choose
reliable semiconductor switches (FET), though they have got
higher "switch on" resistances, compared with mechanical contacts. This circuit provides the possibility of connecting
strain gages in quarter bridge circuits which can be located
up to 500 ... 1000 m apart from the UPM60.
But of course, the UPM 60 is not limited to measurements with
strain gage in quarter bridge circuits only. It is equipped
with suitable selector modules for connecting strain gages in
quarter-, half- and fullbridge, inductive transducers as displacement transducers, thermo couples, Pt100 resistance thermometers, resistance transducers and any voltage and current
sources. In order to meet these demands the UPM 60 is equipped
with three different types of amplifiers:
a dc-amplifier mainly for the purpose of measuring thermo
couples and voltage or current sources or for high speed strain
gage measurements,
a 225 Hz carrier frequency amplifier for high precision measurements with strain gages and resistance transducers,
a 5 kHz carrier frequency amplifier to provid the possibility
to measure inductive transducers.
The UPM 60 can connect and measure up to 60 measuring points
with mixed transducer types to meet the modern test requirements of measuring stresses, forces, displacements and temperatures simultaneously.
The UPM60 provides many data processing functions and is equipped, as standard, with RS 232-C and IEEE 488 interfaces to
allow remote control and additional online data processing with
the help of an external computer.

REFERENCES
1. Hoffmann K.: Ursachen temperaturabhaengiger Nullpunkts-
   und Empfindlichkeitsänderungen bei Dehnungsmeßstreifen-Auf-
   nehmern. VDI-Berichte Nr.137,1970
2. Kreuzer M. : Comparing the effect of lead and switch resistances on voltage- and current-fed strain-gage circuits.
   Reports in Applied Measurements Vol.1 (1985) No.1

COMPENSATION OF LEADWIRE EFFECTS WITH RESISTIVE STRAINGAUGES IN MULTI-CHANNEL STRAINGAUGE INSTRUMENTATION.

W.J. VERSNEL.

INTRODUCTION:

Despite the now widespread acceptance and use of theoretical methods for stress-analysis and for statistical work associated with it, experimental stress-analysis and strain-measurement still continue to exist and even expand in being decisive factors in verification and in matters of complex loading.

More than ever experimental results are used in combination with theoretical predictions, to aid in obtaining more complete and more realistic final results.

The equipment for experimental stress-analysis on larger structures, on scaled-models and on representative test-specimen, used now, in this era of micro-computers, usually takes the form of a straingauge-datalogger, coupled to or integrated in a micro-computer. The straingauge-datalogger is the specialised electronic module or system, where the actual scanning and measurement of the connected straingauges takes place.

Straingauge dataloggers are used, where a fairly large number of straingauge inputs have to be measured and where the time-domain requirements of the test will permit the scanning of the signals of the connected straingauge inputs. This is in general mostly the case with quasi-static or slow dynamic tests.

The actual theme covered in this presentation, is to describe the relations in and the specification of a universal straingauge input-scanner and measurement system, used with multi-channel inputs of resistive straingauges and linear variable displacement transducers, employing the AC-carrier frequency method to excite and measure the straingauges.

DEFINITIONS:

Present-day multichannel straingauge measurement-systems use one Wheatstone bridge for each connected straingauge input, from which only the outputs of the Wheatstone bridges are scanned. Although this scheme is straight-forward and simple, it displays several disadvantages, such as differences in sensitivity, very limited flexibility and it requires considerable space and cost.

The new method of scanning, presented here, uses only one and the same Wheatstone bridge for all straingauge input channels. The actual scanning of the external straingauges is done inside the single Wheatstone bridge, which is constructed to accept all known configurations of straingauge inputs. Modular equipment is used, consisting of a number of simple multichannel input-scanners and one signal conditioner, which contains the actual single Wheatstone bridge.

The signal-conditioner completes the straingauge input into a compensated full-bridge, and presents this to the input of a AC-carrier frequency amplifier, which performs the signal measurement in combination with a digital voltmeter.

Through the scanning units the excitation voltage is applied to the actual straingauge(s) connected to the scanned input channel. Excitation is only applied to one channel at a time.

The advantages of this system are:
. one single sensitivity for all connected channels
. complete flexibility to accept straingauge inputs in all known configurations of straingauge connection.
. elimination of apparent strain caused by contact- and cabling resistances.
. elimination of loss of sensitivity from cabling resistances.
. minimised gauge heat dissipation due to scanned excitation.
. reduced space requirements, since scanner units only require relay-switches and input-connectors.
. moderate cost.

SWITCHING in the WHEATSTONE BRIDGE:
The problems associated with switching active and passive straingauges into the arms of a Wheatstone bridge can best be shown by the diagram fig.1, where the scanning contacts connect a 1/4 bridge single straingauge into the Wheatstone bridge. The small, but varying contact resistance of the scanner-switch prohibits a precise measurement

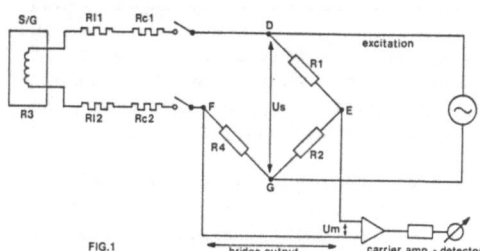

FIG.1

of the resistance and hence of the resistance-change of the active straingauge. Also, since the varying contact-resistance appears in series with the cable resistance, the combined effect of both these parasitic resistances can introduce considerable errors in strain-measurement.

ANALYZING the SOURCES of ERROR:
The contact-resistance appears in series with the cable resistance and the straingauge itself. In order to produce a succesfull accuracy, the variation of the parasitic resistances has to be less than 240 micro-Ohm, given a gauge resistance of 120 Ohms and a resolution if one microstrain. Neither relay-contacts nor semi-conductor switches can offer such a constant switch-resistance. Cable resistances are known to vary with the temperature of their environment.

It is possible to analyse the individual components of the error in strain-measurement, caused by both the parasitic contact- and cable-resistances. There are two aspects:
1. the variations in the parasitic resistances will cause varying apparent strain in the Wheatstone bridge in the form of non-reproducable readings and instable zero-readings.
2. the parasitic resistances, whether variable or constant, will cause an non-predictable loss of sensitivity for both active and passive straingauges in the Wheatstone bridge, resulting in loss of calibration.

The elimination of the effect of the cable- and contact resistances requires an arrangement shown in the diagram of fig.2, where a small, high quality transformer is used. The primary winding of this transformer senses the error voltage-drop over the varying parasitic resistances.

Given a constant amplitude excitation voltage at points D and G, the error signal Ve is experienced in the bridge output leads as a deviation of Ve/2, since the gauge resistance R3 and R4 are in practice equal and the parasitic resistance is small compared to the gauge resistance.

To compensate for the voltage drop at point F of approximately Ve/2, the voltage at point E is dropped by an equal amount, by connecting a secondary winding of transformer TR1 with half the number of turns in series with it to provide point E', so it also drops by Ve/2.

This takes care of the complete elimination of the apparent strain in a continuous dynamic way, by providing an unchanged output from the Wheatstone bridge, despite changes in the parasitic resistances.

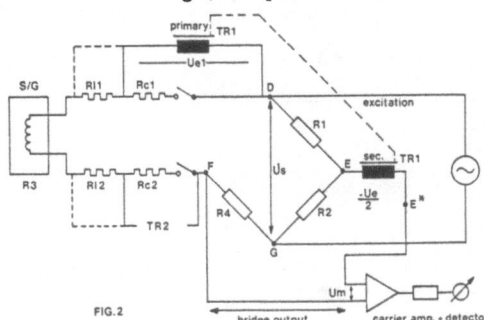

FIG.2

Since the primary winding of the compensation transformer senses the actual voltage drop over the contact- and cable-resistances, it is also possible to enter the total of the voltage drop in a correction scheme to reconstruct the effective excitation voltage at the measurement device to compensate for the loss in sensitivity and thus restore almost perfect calibration.

APPARENT STRAIN:

The TR1-A secondary winding has half the number of turns of the primary winding, so the compensating signal is also half the magnitude of error signal Ve. Thus the portion of the error signal experienced at F, has been compensated by subtraction of the transformer output at point E,to provide an unchanged output at points E' and F.

In theory the influences of the switch-contact and cable-resistances have been eliminated in a dynamic manner. Due to the approximations made and the non-ideal transformer characteristics, the compensation of the error signal is a little less than ideal, but the original effect is attenuated between 400 and 1000 times. So an actual parasitic resistance of 1 Ohm, is at bridge output level reduced to the effect of an uncompensated change of 2.5 to 1 milliOhm. In practice all errors due to changes in the temperature of the connection cable will be eliminated to appear as apparent strain.

When a similar transformer as TR1 is used in connection with parasitic resistance Rc2 (see figs.1 and 2), the total effect of switching the R3-arm of the Wheatstone bridge will be eliminated in the same way as with Rc1, thus compensating the return leadwire and the associated contact as well by transformer TR2.

SENSING the LOSS in SENSITIVITY:

By using the compensation transformers TR1 and TR2 to sense the voltage drop over the connection cable to and from the active straingauge R3, a quantative measure can be obtained of the loss in excitation for the active straingauge, due to the parasitic resistances in the connecting leadwires. The voltage drops over the cable resistances attenuate the effect of the ∆R/R-changes in the actual bridge output-signal.

By using the principle of substitution, the parasitic resistances Rc1 and Rc2 can be effectively shown to appear in series with eachother. From the original excitation voltage Vs appearing between points D and G, the EFFECTIVE excitation voltage Vs`is found by actual subtraction of Ve1 and Ve2 from Vs, by using additional 1:1 secondary windings on TR1 and TR2 (see fig.3).

ANALYSIS of HALF-BRIDGE S/G connection:

A detailed analysis of the compensation for apparent strain and the compensation for the loss in sensitivity can be made, using simple mathematics in the example of a half bridge using two active straingauges, experiencing equal but opposite changes in gauge-resistance (the classical example of bending). See fig.3.

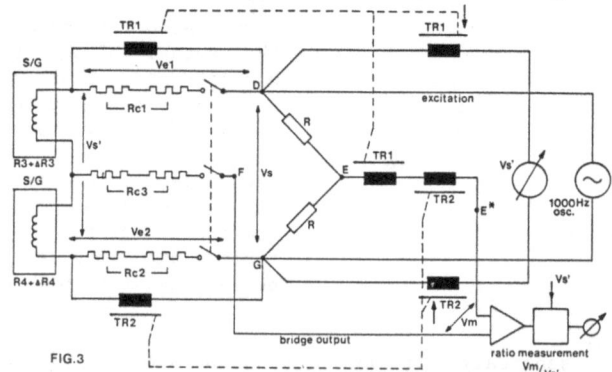

FIG.3

Here the two resistances Rc1 and Rc2 represent the parasitic resistances of scanner-contacts as well as the leadwires to connect both gauges into the bridge.

The factors kT1 and kT2 represent the effective signal transfer-ratios of the compensation transformers TR1 and TR2.
To view the compensation effect, the ratio of bridge output signal to excitation voltage is calculated : Vm/Vs.

with:

$$\frac{Vm}{Vs} = -\frac{R4+ \Delta R4+Rc2}{R3+R4+Rc2+Rc1+\Delta R3+\Delta R4} + \frac{R}{2*R} + \frac{kT2 * Rc2 - kT1 * Rc1}{Rc1+Rc2+R4+R3+\Delta R3+\Delta R4}$$

assuming ∆R3= - ∆R4 and R3=R4=R
the equation condenses to:

$$\frac{Vm}{Vs} = -\frac{\Delta R4}{2 * R + Rc1 + Rc2} + \frac{Rc1 * (1- 2 * kT1) - Rc2 * (1- 2 * kT2)}{4 * R + 2 * Rc1 + 2 * Rc2}$$

The first term contains the actual strain signal, caused by the straingauges. The second term shows the resulting error in apparent strain caused by the contact and cabling resistances.

if now:

$$1 - 2 * kT1 = 0$$
$$1 - 2 * kT2 = 0 \quad ; \quad \text{hence } kT1 = kT2 = 0.5 \text{ (transfer-ratio } 2:1 \text{)}$$

then we find:

$$\frac{Vm}{Vs} = \frac{- \Delta R4}{2 * R + Rc1 + Rc2} \qquad (1)$$

The effect of apparent strain caused by the influence of the varying contact- and cabling resistances has been completely eliminated.

What stays, is a loss in sensitivity caused by the series-effect on the gauge resistanc of the parasitic resistances, which will influence calibration.

Looking at the APPLIED excitation voltage Vs and at the EFFECTIVE excitation voltage Vs' , then we can find their ratio :

$$\frac{Vs}{Vs'} = \frac{Vs}{Vs - Ve1 - Ve2} = \frac{R4 + R3 + Rc1 + Rc2}{R4 + R3}$$

we can simplify, taking R4 = R3 (= R)

$$\frac{Vs}{Vs'} = \frac{2 * R + Rc1 + Rc2}{2 * R}$$

by introducing the value for Vs in equation (1), we get as final result:

$$\frac{Vm}{Vs'} = \frac{- \Delta R4}{2 * R} \qquad (2)$$

Equation (2) shows that all effects of the parasitic resistances Rc1 and Rc2 have disappeared. So there is no apparent strain and no loss of sensitivity, provided that in equation (2) Vs' can be constructed, by utilising the additional secondary windings on TR1 and TR2 to subtract Ve1 and Ve2 from Vs.

Note, however, that in order to realise the algorithm of equation (2) a ratio-measurement of two voltages is required. In practice this is effected by using a digital voltmeter with an external reference-voltage input, besides the signal-voltage input . The external reference voltage for the digital voltmeter is fed with the reconstructed effective excitation voltage, experienced by the active straingauges.

QUARTER BRIDGE connections:

For the connection of a quarter bridge active straingauge, using either a resistor or an external common-dummy straingauge as the completion element, similar circuits as used with the compensated half-

460

bridge have been developed. These circuit are using 3-, 4-, or 5- wire
connections from the single active gauge to the scanning equipment. Each
of these connection-methods exhibits characteristic properties.

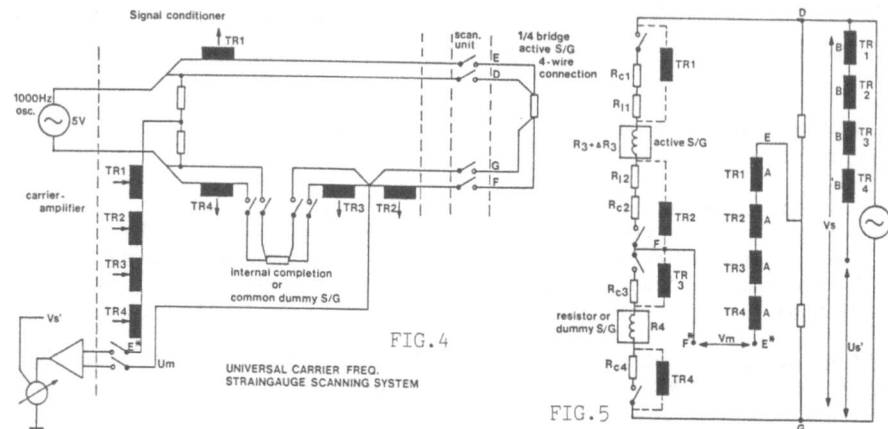

FIG.4
UNIVERSAL CARRIER FREQ.
STRAINGAUGE SCANNING SYSTEM

FIG.5

The arrangement used with 4-wire connection of the active straingauge is
shown in fig.4. The positions of all the parasitic contact- and cable-
resistances and the compensation transformers are shown in fig.5 . Four
compensation transformers are needed in total, to compensate all switches
and cables. Two transformers cater for the connections to the active
gauge, and two further transformers are needed to compensate for the
internal connection of the completion-resistor into the bottom arm of the
bridge.

After assuming $R3 = R4 = R$ and expressing the small imperfectness of
the transformers with B , the ratio Vm/Vs' can be calculated to be:

$$\frac{Vm}{Vs'} = \frac{\Delta R_3}{4R + 2\Delta R_3} - \frac{\sum\limits_1^4 B_n \cdot R_{cn}}{4R + 2\Delta R_3} \qquad (3)$$

here $B = 1 - 2 * kTn = 0.003 -- 0.001$

Again, the first term is responsible for the true strainsignal, due to
compression or elongation experienced by the test-piece, the second term
shows the total effect of all parasitic resistances and their variation
contributing to apparent strain. The cable resistances Rwl and Rw2 have
been included in the total of Rcl and Rc2, since the primary windings of
TR1 and TR2 include both the cable- and the contact-resistance.

$$\text{In the term } \sum_{n=1}^{4} B_n * R_{cn}$$

B1 and B2 are expressed with a negative sign , while B3 and B4 are
positive. Since also Rcl and Rc2 are significantly greater than Rc3 and
Rc4, because the first contain mainly cable resistances, it is true
that with a certain small imperfectness in the transformers, some
apparent strain theoretically can occur.

The practice to connect straingauges with long thin wires, gives large variations in cable-resistance with temperature. Although the nett effects of the variations are attenuated by the factor B, ranging from 0.003 to 0.001, the 4-wire compensated connection is in principle somewhat less in performance than a 5-wire connection.

In equation (3) the effective excitation voltage Vs' is reconstructed in precisely the same way as with the half-bridge circuit, but in this arrangement for 1/4-bridge, requires four additional 1:1 secondary windings (one on each transformer), to cover all of the uneffective sections of the configuration (see fig.5).

QUARTER-BRIDGE connection with 3- or 5- WIRES:

The 5-wire connection used in this new compensation scheme is a further development of the popular 3-wire connection used in traditional straingauge installations. As may be remembered, the cable resistances of the connection wires then appear in opposite arms of the Wheatstone bridge. The nett effect of the variations of the cable resistances with temperature on apparent strain should be zero, provided exactly the same resistance variations occur in both excitation leads. In practice, the resulting apparent strain is not negligible, when long lengths of small diameter cable are used for connection. This effect is caused by unequal resistances in the leadwires, due to unavoidable tolerances in wire diameter and variations in conductancy of the copper material used in the wires.

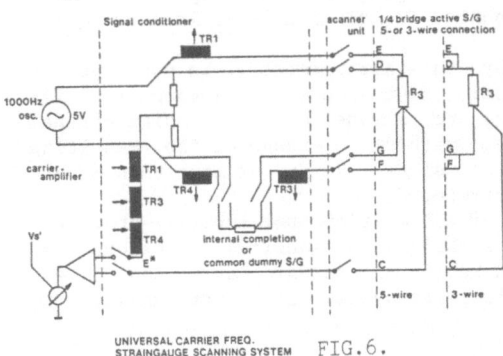

UNIVERSAL CARRIER FREQ.
STRAINGAUGE SCANNING SYSTEM FIG.6.

Deviations of up to 5 % have been experienced in practice in this aspect.

The 5-wire connection is shown in fig.6 and the scheme of compensation is shown in fig.9.

$$\frac{Vm}{Vs'} = \frac{\Delta R_3}{4R + 2\Delta R_3} - \frac{\sum_{1}^{3} B_n \cdot (R_{cn} + R_{ln})}{4R + 2\Delta R_3} \qquad (4)$$

Again the first term in equation (4) represents the true strain signal, while the second term shows the total effect of the all parasitic resistances on apparent strain.

Since in the second term: $\sum_{n=1}^{3} B_n * (R_{cn} + R_{wn})$,

B1 is expressed with a negative sign and B2 and B3 with a positive, the cable resistances Rw1 and Rw2 appear in such a way that they counteract each other, even if this effect is not complete, it is greatly reduced by the compensation factor B.

The total nett effect is approaching the ideal very close of having negligible apparent strain even from long, thin leadwires. As shown

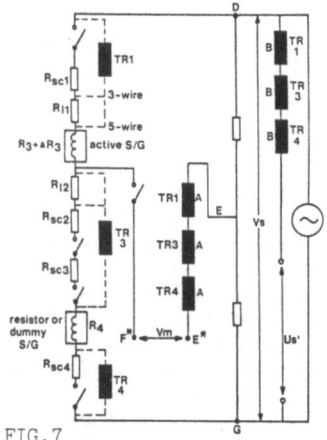

before, the loss of sensitivity in the 5-wire connection is compensated also by the use of the reconstructed effective excitation voltage at the measurement-device to perform the ratio-measurement Vm/Vs'.

An interesting feature can be observed in fig.7, where between excitation points D and G all leadwires, connectors and contacts are compensated. The only items allowed to contribute to the nett strain signal are the active straingauge and the completion resistor mounted in the signal-conditioner equipment.

By this arrangement it is possible to transfer the connected straingauge and its cabling to anyone of the multi-channel inputs, without a change of the resulting absolute signal Vm/Vs' of more than + - 1 microstrain.

FIG.7

This almost absolute equality of all multichannel inputs, shows the large effect of this compensation technique on leadwires, contacts and connectors. It is now possible to make use of standard cables and connectors to form the complete leadwire-system from straingauge to measurement-device.

PRACTICAL REALISATION:
Transferring the principle-drawings

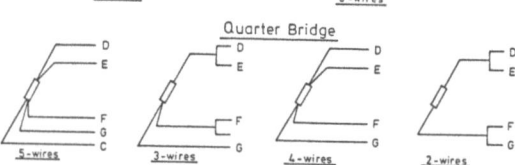

figs.3, 4, and 6 into a layout were all the possibilities of using 1/2-, 1/4- and full bridge can be selected by by the user, results in a scheme where the complete multi-channel straingauge scanner is constructed from a few functional units.

The structure is such, that one rather complicated signal-conditioner is used with a large number of 10-channel straingauge scanner modules. The scanner-modules are extremely simple, since they only contain relay-switches and input-connectors. In the signal-conditioner all the before-mentioned straingauge configurations are provided, to be selected by the user of the straingauge datalogger during

FIG.8

connection of the straingauge inputs. The signal-conditioner contains the 4 compensation transformers, the bridge completion resistors and an arrangement to insert one of 5 possible compensated common-dummy straigauges as the completion element in quarter bridges. Apart from the typical straingauge modes, several other input-connection modes are present to be utilised, such as DC-signal, thermocouple, potentiometers and linear variable displacement transducers, all in combination with the straingauge connection modes.

By separating the scanning-functions from the actual signal conditioning, a relatively simple lay-out is obtained with the possibility to maintain a good compatibility in case of changes and adaptations.

OVERVIEW of the combined MEASUREMENT-EQUIPMENT:

An overview of the total set-up on equipment, necessary for obtaining a digital representation of the logged and scanned straingauge signal is shown in fig.9. In this figure a 1/4-bridge 4-wire connected straingauge input is shown, with all compensations applied, both for apparent strain and for the loss in sensitivity. Compensation for apparent strain utilises the four secondary-A windings, while for the reconstruction of the effective excitation voltage Vs', the secondary-B windings are in use for transformers TR3 and TR4.

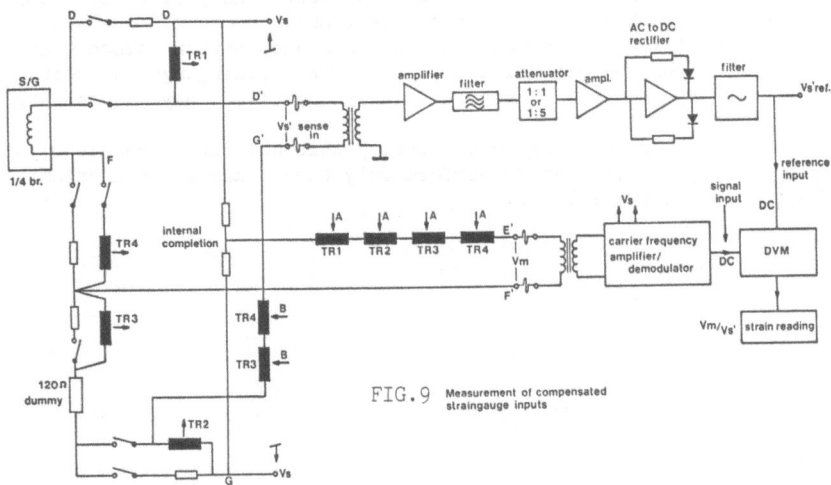

FIG.9   Measurement of compensated straingauge inputs

However, this still produces a voltage Vs', with the waveshape and frequency of the 1000 Hz carrier-excitation. In order to to be converted into a suitable signal for taking part in the ratio-measurement with the digital voltmeter, the effective excitation voltage has to be converted into a filtered and fully conditioned DC-signal.

The effective excitation voltage Vs' is connected to a module to obtain a properly usable external reference voltage for the digital voltmeter, so that the ratio-measurement Vm/Vs' can be obtained from it.

Through an input transformer, a bandpass-filter, an attenuator and a precision rectifier followed by another low-pass filter, finally the reference-voltage Vs'(ref) is obtained. The electronic processing of the Vs'-signal is rather extensive and not simple, because unaccounted sources of error could destroy the measure of accuracy and stability obtained in the compensated Wheatstone bridge.

STRAINGAUGE DATALOGGERS:

Typical applications of the universal straingauge scanning system, together with the carrier amplifier, the digital voltmeter and the module to process the effective excitation voltage for the ratio-measurement, require data-acquisition logic to perform the automatic actions for which they have been designed.

Suitable interfaces exist for direct coupling to a digital processor or micro-computer in the form of using IEEE-488 instrument bus or via a simple terminal-connection.

From the central processor or micro-computer the typical commands are transmitted to the measurement system, to perform the actions of scanning and measurement. The application-program in the micro-computer performs the typical routines essential in straingauge datalogging: software- autobalancing, various scale- and gauge-factor corrections and all other well-known stress-analysis routines. Many of these programs exist, each with their typical specialisations.

The specific requirements of the data-acquisition procedures are embedded in these application programs. The straingauge scanning and measurement system should operate in a standardised way, according to normalised procedures.

This was also the reason to develop this universal scanning system, which is strictly limited to perform only those hardware-functions not replacable by software logic.

RECENT DEVELOPMENT OF THE SPATE TECHNIQUE FOR MEASURING STRESS IN
STRUCTURES LOADED WITH COMPLEX WAVEFORMS

D.E. OLIVER, OMETRON INC.
W.R.S. WEBBER, SIRA LTD.
G. JOSEPH, SIRA LTD.
J. GILBY, SIRA LTD.

1. INTRODUCTION
    In 1982/3 the SPATE 8000 Stress Analyzer was launched by Ometron.
Since then, SPATE 8000 has continued to bring substantial benefits to the
cost and time effectiveness of product development and troubleshooting.
In fact, some of the work would have been impossible by conventional
methods.   Engineers in four continents are now using this technology on
components ranging from turbine blades to large offshore oil rig node
joints.
    The SPATE 8000 system requires uniform amplitude cyclic load at a
single frequency between 0.5Hz and 20KHz for the duration of each scan.
There are, however, many test conditions where these simple load waveforms
do not exist such as: 'in-service' running plant engines and machinery,
playback of recorded real-time automotive road load data, road bridges
under traffic loading, block by block flight programs in aerospace and
modal analysis. Also, in the testing of complex geometry structures at a
single relatively high frequency outside the quasi-static range, a single
phase shift setting of the SPATE 8000 correlator may be correct for only a
fraction of the area scanned.
    A new SPATE development by Ometron, with the technical assistance of
the British National Engineering Laboratory (NEL) enables stress patterns
to be obtained from structures undergoing broad-band random, pseudo-random
and repetitive impulse loading.

2. FUNCTIONS MEASURED AND DISPLAYED BY THE NEW SPATE DEVELOPMENT
    To operate the new SPATE development, load waveforms applied to the
test part may comprise many frequencies and many amplitudes occurring
together and changing from one measurement location to the next during a
scan. These random, pseudo random and repetitive impulse loads are found
applied to structures under laboratory control to simulate service
conditions, or exist in many in-service structures.
    0.5Hz – 500Hz is the maximum frequency range for complex stress
waveforms. Five narrower ranges are available. Up to 400 frequency lines
can be analyzed in any frequency range giving a frequency resolution to
38mHz.
    To aid readers not familiar with random signal processing as employed
in modal analysis, the following is a brief description of several
functions available from the new SPATE development.
    Frequency Response Function:
    The complex Frequency Response Function (FRF) as displayed by the new
SPATE development is a measure of the output (SPATE infrared stress
signal) to input (reference signal) ratio. The reference signal may be

Wieringa, H (ed), Experimental Stress Analysis.
© *1986. Martinus Nijhoff Publishers, Dordrecht.*

derived from, for example, load cells and strain gages. The FRF describes the level of stress obtained for a given reference input level at each of a range of stress frequencies assuming the system is linear and there is sufficient signal to overcome noise.

The block diagram (Figure 1) shows how the FRF is derived. Two different estimates of the FRF may be calculated. In displaying $H_1(f)$, the new SPATE development assumes extraneous noise is more likely to be found at the system output (detector signal).

Coherence Function:

The Coherence is a function which on a scale from 0 to 1 measures the degree of system linearity (input to output) and absence of uncorrelated noise in the measurements.

Autospectrum:

The Autospectrum is sometimes called the Power Spectrum. It is a real quantity (no phase information) and gives the distribution of power (or energy) in the signal as a function of frequency. It is a squared term if the signal is periodic, e.g. Stress^2. Continuous random signals are characterized by the Power Spectral Density (PSD) when units are scaled, in for example Stress^2/Hz. Another quantity presented by the new SPATE development - the Spectral Density (SD) is simply defined as $\sqrt{PSD}$ and is scaled, in for example Stress/$\sqrt{HZ}$.

Cross Spectrum:

The FRF and Coherence are computed through the Cross Spectrum. The Cross Spectrum is a description of the common correlated frequency composition in each signal and has real and imaginary parts. This function is not displayed by the new SPATE development.

Window Functions:

The FFT algorithm of the Discrete Fourier Transform (DFT) is applied to blocks of time history of length T. The inherent restriction in the use of the DFT is that the measured signal R(t) or D(t) should be periodic in each period T. Most signals will not be. A non-periodic signal leads to leakage which modifies the shape of the spectrum. Weighting or window functions are applied to the time histories in order to shape the waveform into becoming more periodic. For example, the Hanning window is normally used when analyzing continuous random signals. The Hanning window is simply one period of a cosine of length T lifted to start and end at zero. Period T is also referred to as one average. The quality of data can usually be improved by increasing the number of averages over which measurements are taken at each location. Other window functions available to the new SPATE development users are Flat Top, Hamming, Rectangular and Exponential. See Figure 2.

Triggering:

Signals triggering the measurement period are needed especially when sporadic bursts of energy appear. The trigger slope can be positive or negative going, or bidirectional. Trigger voltage levels are selectable.

The following functions are displayed:
- o Frequency Response Function (FRF) Magnitude
- o FRF Magnitude and Phase
- o FRF 'in phase' Magnitude
- o FRF 'in phase' Magnitude displayed in bipolar form
- o FRF Phase only
- o Coherence Function

o Autospectrum (for reference or for infrared detector)
  Normalized to Power Spectral Density (PSD_R or PSD_D)
o Spectral Density = $\sqrt{PSD_R}$ or $\sqrt{PSD_D}$
Display Modes available are:
o Graph of function selected from above versus frequency
o Selected function over area as a color map at selected frequency
o Selected function along line as a graph at selected frequency
o Frequency versus distance along scanned line displayed as color
  map. Color level showing magnitude of selected function
o Nyquist plot. Real versus imaginary data for prescribed
  frequency
Spatial, frequency and ensemble averaging are available to minimize
random errors in data.

Apart from these functions the new SPATE development also adds a
number of powerful software features to those available in SPATE 8000.
For example, line plots can be interrogated, data can be edited to save
memory space and X, Y coordinate information from the scanning mirrors is
available. The new SPATE development has very powerful data manipulation
capabilities with access to a 40 MByte Winchester hard disk.

3. APPLICATIONS
This development revolutionizes the range of applications to include
monitoring the integrity of 'in-service' structures. In addition,
structures and components may be tested in the laboratory with broad band
load waveforms representative of service conditions.
  In-Service - Examples include:
o Running of chemical power and production plant
o Structures on board ships
o Wind induced stresses in utility structures; TV, radio and radar
  antennae
o Traffic and railroad loaded bridges
o Other running machinery
Engineers may for example monitor 'in-service' integrity of structures
with Power Spectral Density information that will help obtain an accurate
assessment of remaining lifetime. When necessary, stress patterns will
help engineers design a suitable repair prior to failure; a benefit that
adds to confidence in the installed product.

The ability of the coherence function to detect system non-linearities
may lead to sensitive detection of defects, yielding, etc.

There is a distinct trend in many industries towards more accurate
laboratory testing with waveforms representative of stresses that are
actually seen in-service.
  Lab-Testing - Examples include:
o Playback of recorded real-time road load data in the automotive
  industry
o Running automotive, marine and aero engines. Stresses induced
  by them in other components
o Block by block flight programs in the aerospace industry
o Simulated sea-state loading of offshore oil rig structures
The new SPATE development also has considerable potential for modal
analysis work in terms of stress - a parameter often more valuable than
traditional accelerometer data. From a broad-band load input and a single
scan of an area, the new SPATE development will display stress distributions at each of the modes of resonance without surface contact.

4. RESULTS
4.1. Steel Plate With Central Hole

A steel test piece 40mm wide by 6mm thick with an 11mm diameter central hole shown in Figure 3 was loaded with a broad band random signal between 10 and 30Hz. The reference spectral density is shown in Figure 4. The reference is the signal driving the servo hydraulic loading machine. The stress signal is taken from the infrared detector.

A black and white photograph of the color display of the frequency response function pattern around the hole may be seen in Figure 5. A line of data can be extracted from memory. Figure 6 shows a horizontal stress plot through the center of the hole. Figure 7 is a phase plot along the same horizontal line.

4.2. Bending Beam

A straight beam shown in Figure 8 and fixed at one end to an electrodynamic shaker was tested with broad band random loading between 5 and 120Hz (Figure 9 shows the reference spectral density of the shaker input signal).

Figure 10 shows the FRF versus frequency at a point near the beam root. The loads are setting up modes of resonance at 12Hz, 41Hz, and 82Hz. Figure 11 shows the phase distribution of these modes-measurements taken at the same point. Figure 12 is very similar to Figure 10 except the vertical axis is now position along the beam, and the intensity of the display indicates the magnitude of the FRF. This information could also have been displayed in bipolar form to distinguish between areas of different phase as well.

The next three figures are plotted versus distance along the X axis. Figure 13 demonstrates a high level of coherence where there is a substantial amount of stress. Coherence here was plotted at approximately 82Hz (the fourth bending mode). The fourth mode is also shown in Figure 14 - this time as a plot of in-phase FRF versus distance. Finally, Figure 15 shows a good quality plot of the 3rd resonance mode FRF versus distance.

5. CONCLUSIONS

SPATE 8000 has already established itself as a major new development in experimental stress analysis. An optional module for SPATE 8000, the new SPATE development, offering complex processing of stresses induced by pseudo-random, random and repetitive impulse loads has now been developed and early results reported here are very encouraging. A full field, non-contact technique for monitoring the mechanical stress integrity of in-service structures is now commercially available.

6. ACKNOWLEDGEMENT

Ometron gratefully acknowledge the technical assistance of the National Engineering Laboratory (NEL) of East Kilbride with the new SPATE development.

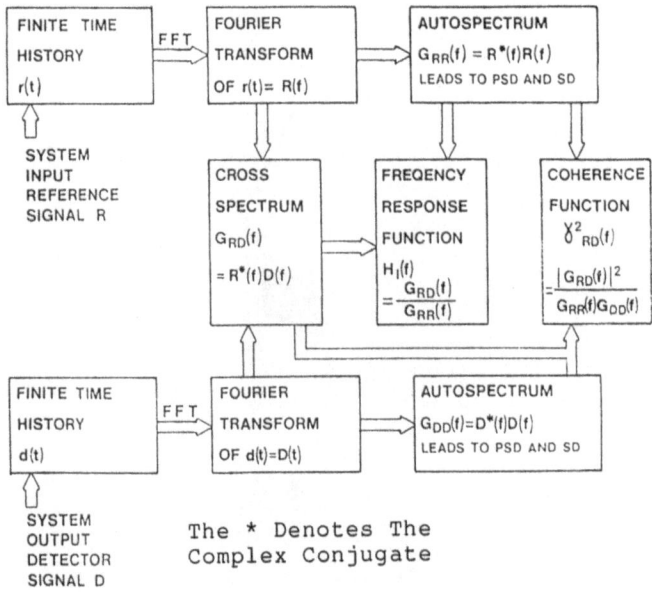

FIGURE 1 – BLOCK DIAGRAM OF THE NEW SPATE DEVELOPMENT RANDOM DATA PROCESSING

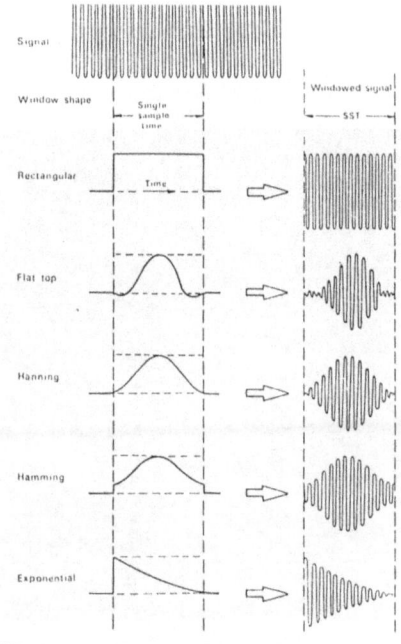

FIGURE 2 – WINDOW TYPES

FIGURE 3 – HOLE IN PLATE SPECIMEN UNDER BROAD BAND RANDOM LOADING

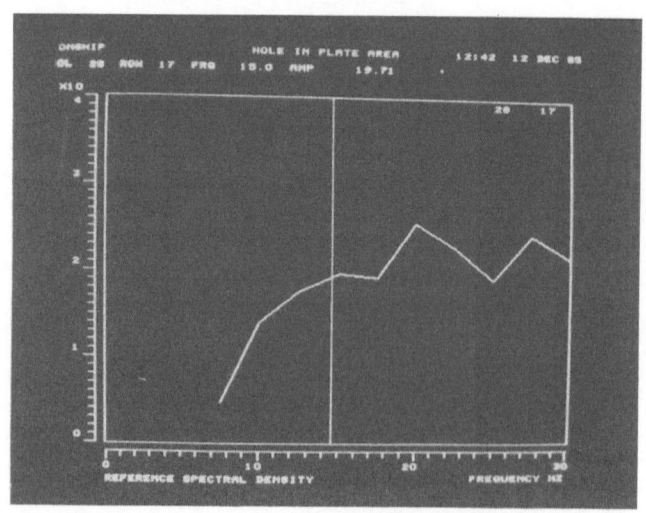

FIGURE 4 – HOLE IN PLATE REFERENCE SPECTRAL DENSITY

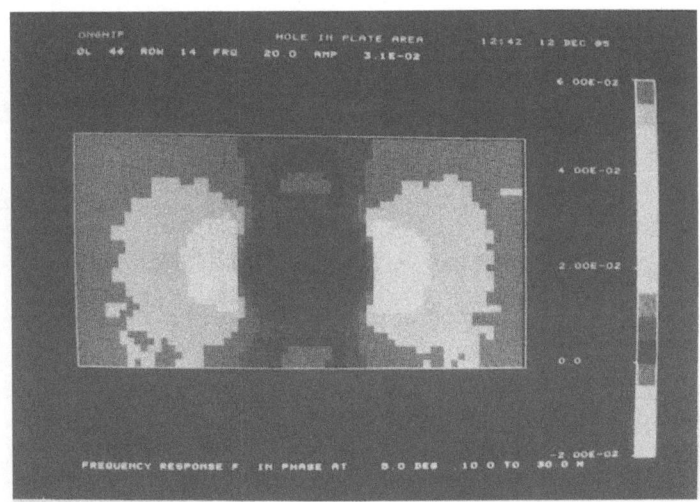

FIGURE 5 — FRF DISTRIBUTION AROUND HOLE IN PLATE SPECIMEN

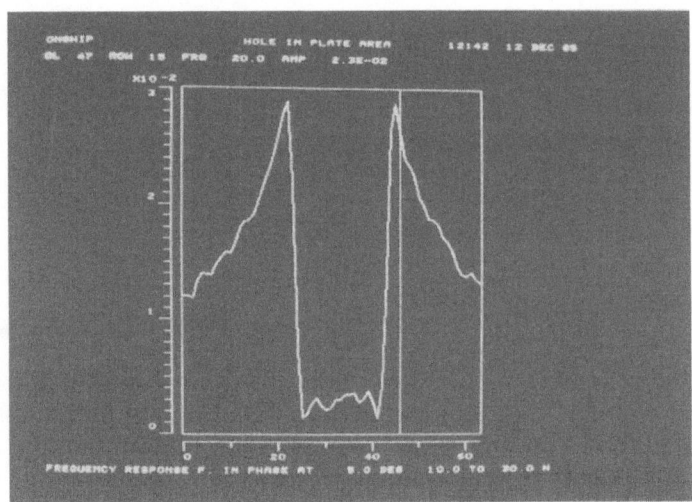

FIGURE 6 — HORIZONTAL LINE PLOT SHOWS STRESS DISTRIBUTION ACROSS HOLE IN
PLATE SPECIMEN

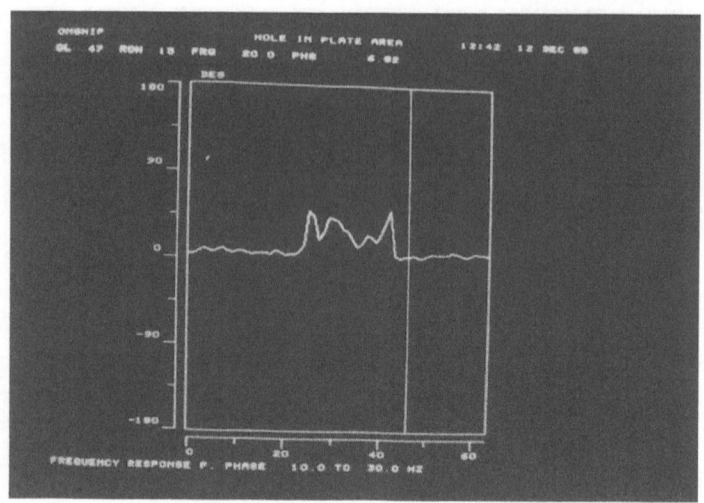

FIGURE 7 - PHASE ALONG A HORIZONTAL LINE THROUGH THE HOLE CENTER

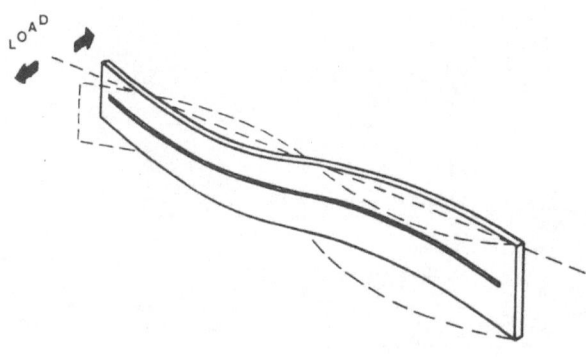

FIGURE 8 - STRAIGHT BEAM UNDER BROAD BAND RANDOM LOADING

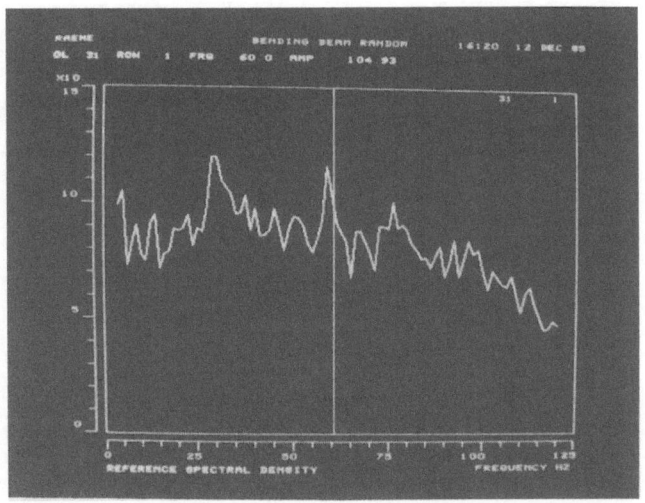

FIGURE 9 - BENDING BEAM REFERENCE SPECTRAL DENSITY

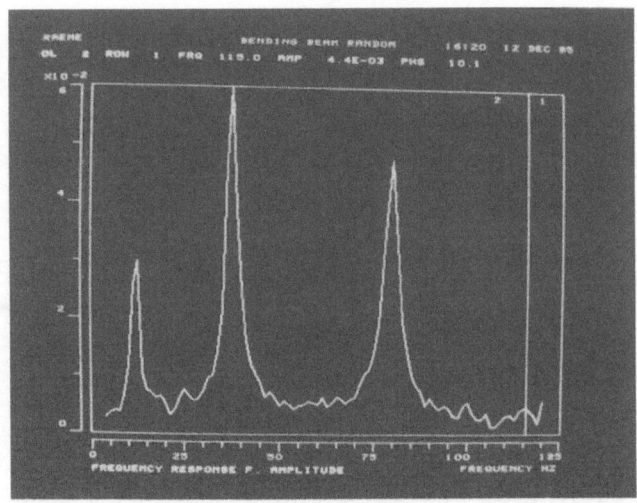

FIGURE 10 - FRF AT ONE POINT ON THE BENDING BEAM VERSUS FREQUENCY

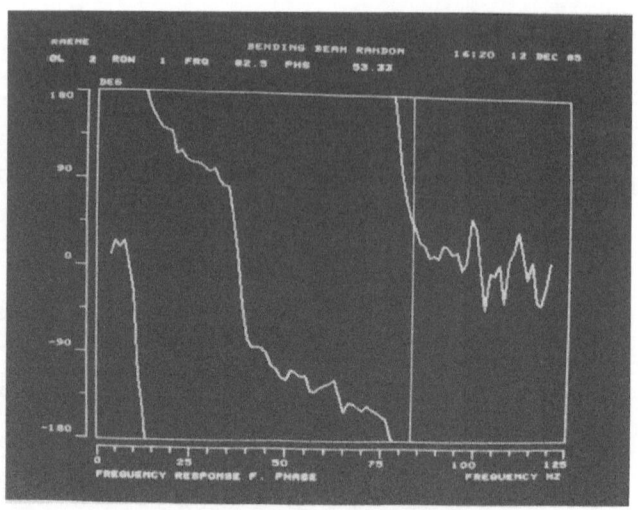

FIGURE 11 - PHASE VERSUS FREQUENCY AT A POINT ON THE BEAM

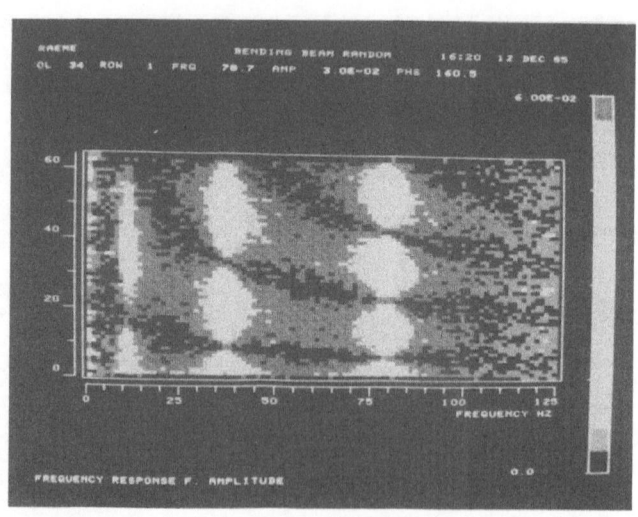

FIGURE 12 - FRF ALONG THE BEAM VERSUS FREQUENCY

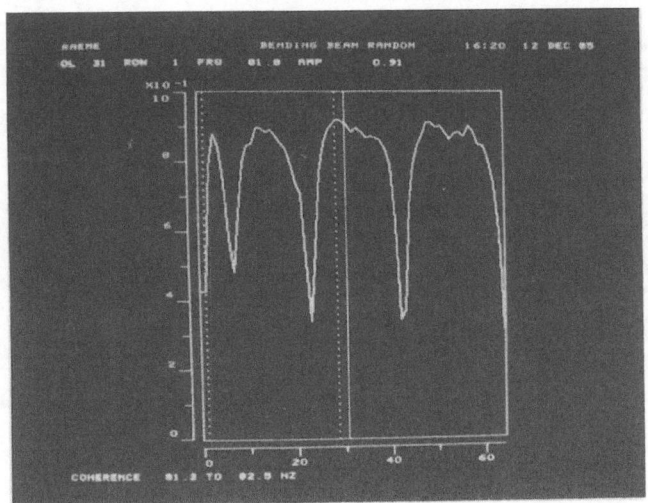

FIGURE 13 - COHERENCE ALONG THE BEAM AT 82 Hz

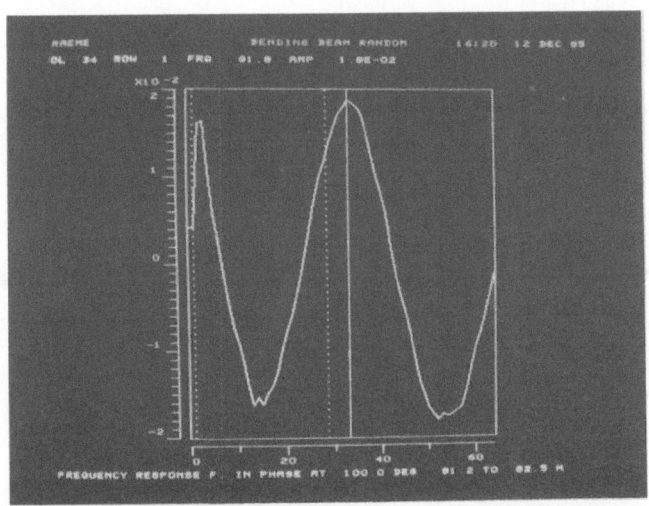

FIGURE 14 - IN-PHASE FRF VERSUS DISTANCE AT 82 Hz

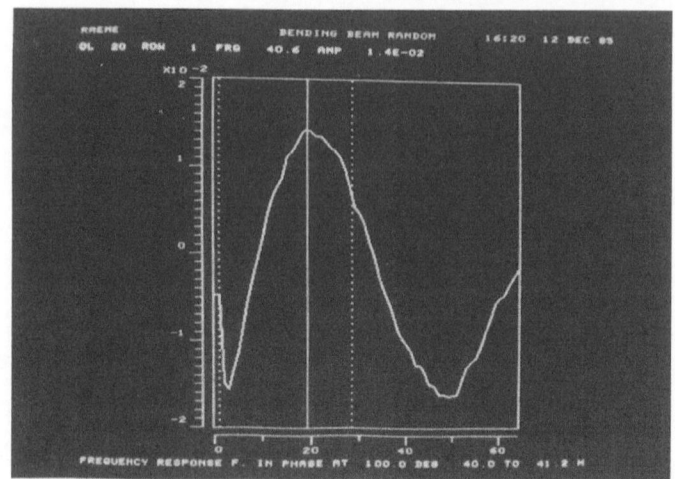

FIGURE 15 - IN-PHASE FRF VERSUS DISTANCE AT 41 Hz

REFERENCES

1. RANDOM DATA: - ANALYSIS AND MEASUREMENT PROCEDURES by Julius S. Bendat and Allan G. Piersol, Published WILEY 1971
2. TECHNICAL REVIEW NO. 1 - 1984 DUAL CHANNEL FFT ANALYSIS PARTS I AND II, Published by Bruel and Kjaer
3. MODAL ANALYSIS SUPPLEMENT SOCIETY FOR EXPERIMENTAL MECHANICS (SEM) EXPERIMENTAL TECHNIQUES - Volume 9 Number 10 - OCTOBER 1985
4. WHYS AND WHEREFORES OF MODAL TESTING, by D.J. Ewins, Imperial College, London, Presented at Soc. of Environ, Eng., MODAL TESTING USING DIGITAL TECHNIQUES - JUNE 6, 1978
5. MODAL SURVEY ACTIVITY VIA F.R.F., by Albert Klosterman and Ray Zimmerman
6. APPLIED TIME SERIES ANALYSIS VOL. I BASIC TECHNIQUES, by Robert K. Otnes - EM Systems Labs, Sunnyvale and Loren Enochson - Gen Rad, John Wiley
7. OBTAINING GOOD RESULTS FROM AN EXPERIMENTAL MODAL SURVEY, by Edward Peterson and Albert Klosterman, Published from Journal of Soc. E.E. March 1978, (This paper covers meaning of coherence function)
8. INNOVATIVE FUNCTIONS FOR TWO CHANNEL FFT ANALYZERS, by Roger Upton, B & K, Denmark, SOUND AND VIBRATION - MARCH 1984
9. PARAMETER ESTIMATION TECHNIQUES FOR MODAL ANALYSIS, By D.L. Brown, R.J. Allemang and Ray Zimmerman - University of Cincinnati, and M. Mergeay, Katholieke - University of Leuven, SAE Congress and Expo. Cobo Hall, Detroit, Feb. 26 - March 2, 1979
10. PROCEEDINGS OF THE INTERNATIONAL MODAL ANALYSIS CONFERENCES, Annual conference since 1983 organized by Union College, Schenectady, NY 12308 USA
11. THERMOELASTIC STRESS ANALYSIS UNDER BROAD BAND RANDOM LOADING, By W.M. Cummings and N. Harwood - National Engineering, Laboratory, Scotland, Proceedings of the 1985 SEM Spring Conference on Experimental Mechanics, pp. 844-850
12. NON-CONTACT STRESS PATTERN ANALYSIS OF STRUCTURES LOADED WITH COMPLEX WAVEFORMS, By D.E. Oliver - Ometron Inc., W.R.S. Webber - Sira Ltd. and J. Gilby - Sira Ltd., Proceedings of the Fourth International Modal Analysis Conference, February, 1986

A NEW EXPERIMENTAL STRESS ANALYSIS TECHNIQUE OF WIDE APPLICATION

P.STANLEY AND W.K.CHAN

SIMON ENGINEERING LABORATORIES, UNIVERSITY OF MANCHESTER, OXFORD ROAD, MANCHESTER M13 9PL, ENGLAND.

1. INTRODUCTION

Since the Haifa conference, which included a presentation (1) on some preliminary thermoelastic stress analysis work, the considerable potential of the SPATE (Stress Pattern Analysis by the measurement of Thermal Emission) technique has been explored and the technique itself has been considerably developed. At present there are some 20 equipments in use in Europe (including the UK), 20 in the USA and Canada, and 7 elsewhere including Australia and Japan. The important role of this major new development in stress analysis and design appraisal studies has now been firmly established, and it is appropriate that its potential should be made widely known.

Recent applications include basic stress evaluations (2), a design assessment of railway coach body panels (3) and dynamic studies with random loading (4). Work on non-metals (5) and orthotropic composites (6) has also been described and crack-tip stress intensity factors have been successfully determined (7) for a variety of loading modes.

In this paper the underlying theory is summarized, the SPATE equipment is described and the main practical aspects of the technique are reviewed. Calibration techniques are outlined and a number of illustrative applications are described. The technique is compared and contrasted with other experimental stress analysis techniques.

2. THERMOELASTIC THEORY

The theory of the technique, based on the work of Lord Kelvin, Biot and Belgen (8), is outlined as follows. It can be shown from classical thermodynamics that, under adiabatic conditions, a change in the stress state in a homogeneous elastic solid gives rise to a temperature change ΔT of the form

$$\Delta T = \frac{T}{\rho \, c_\varepsilon} \sum_{i,j} \frac{\partial \sigma_{ij}}{\partial T} \varepsilon_{ij} \qquad (1)$$

(with i,j = 1, 2 or 3)

where T is the absolute temperature,
c_ε is the specific heat at constant strain,
ρ is the mass density,
σ_{ij} is the stress change tensor, and
ε_{ij} is the strain change tensor.

For an isotropic elastic solid the stress derivatives are readily obtained and, using the established relationship (9) between c_ε and c_p (the specific heat at constant pressure), equation (1) can be expressed as

$$\Delta T = - \frac{\alpha}{\rho \, c_p} T \, (\sigma_1 + \sigma_2 + \sigma_3) \qquad (2)$$

480

where α is the coefficient of linear thermal expansion and σ_1, σ_2 and σ_3 are the principal stress changes.
For an orthotropic solid (6) the equation becomes

$$\Delta T = - \frac{T}{\rho \, c_p} (\alpha_1 \, \sigma_{\alpha 1} + \alpha_2 \, \sigma_{\alpha 2} + \alpha_3 \, \sigma_{\alpha 3}) \qquad (3)$$

where α_1, α_2 and α_3 are the principal coefficients of thermal expansion and

$\sigma_{\alpha 1}$, $\sigma_{\alpha 2}$ and $\sigma_{\alpha 3}$ are the normal stress changes in the directions of the

principal material axes.

   In either case, the temperature change gives rise to an emission of infra-red radiation from the surface of the body, the magnitude of which is proportional to the temperature change (8), and if a linear infra-red detector is used, the relationship between the detector signal S and the stress change sum in an isotropic body (equation (2)) is

$$\sigma_1 + \sigma_2 + \sigma_3 = AS \qquad (4)$$

where A is a calibration factor. (A corresponding expression for an orthotropic material can be readily developed from equation (3).) The proportionality expressed in equation (4) is the basis of the SPATE stress analysis technique.

2. EQUIPMENT AND EXPERIMENTAL PROCEDURE

   The equipment is shown schematically in Fig.1. There are three principal parts, i) the detector/scanner unit, ii) the analogue signal processing unit and iii) the digital control/storage/display unit. The detector is a highly sensitive lead/tin/telluride cell about 100 microns in width, cooled by liquid nitrogen and operating over the 8-14 μm waveband. A germanium lens (f 0.8) focuses incident infra-red radiation onto the detector and a two-mirror raster-type scanning system incorporated in the unit allows a selected part of the surface of a specimen or component to be "scanned". The operator can select a grid of elemental "sampling areas" (minimum dimensions 0.5 mm x 0.5 mm) within the area to be scanned, from each of which in turn radiation is received. He can also select the "sampling time" i.e. the period over which radiation from each sampling area is received. The received signals are digitised, stored in a

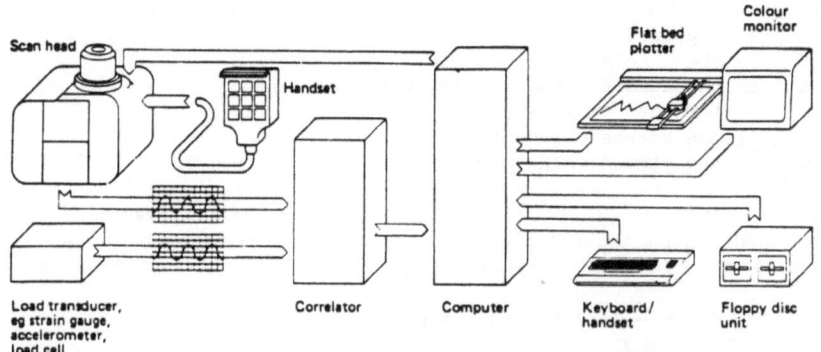

FIG. 1.   BLOCK DIAGRAM OF SPATE EQUIPMENT.

non-volatile 128 k byte bubble memory and simultaneously displayed on a high-resolution video monitor as a 16-colour linearly sub-divided stress contour map over the selected area or as a stress plot along a chosen line within the scanned area. The stored data can be interrogated to provide further stress plots either on the video monitor or in hard copy from a plotter.

The system is designed for use with cyclic loading. Provided that the cyclic frequency is not less than a lower limit (8), such loading readily ensures that adiabatic thermal conditions pertain. The cyclic loading mode also permits correlation of the received signals, in frequency, magnitude and phase, with a reference signal derived from a suitable transducer in the loading system, so that the signal quality is enhanced and a linear response is ensured.

In use, the detector unit, supported on a simple tripod stand, is positioned at a convenient point in front of the object under test and scanning can proceed when the controls have been set. Typically a full detailed scan may take about an hour. The only preparatory work required before testing is to clean the specimen surface and to apply a matt-black surface coating by means of a hand-applied aerosol spray. (This minimum effort contrasts markedly with that required in comparable work using other experimental techniques.)

Important features of the SPATE technique are as follows:

i) The equipment is designed for a frequency range of 0.5 Hz to 20 kHz. For a given material there is a lower frequency limit below which the adiabatic condition no longer holds. The loss of signal with decreasing frequency for mild steel is shown in Fig.2; for this material the lower frequency limit for adiabatic behaviour is approximately 2 Hz. The lower limit for other materials is proportional to the thermal diffusivity of the material (10). Regarding high frequency testing, it is noteworthy that successful applications at a frequency of 26 kHz have been reported (11).

ii) It is not of critical importance that the scanned surface should be normal to the optical axis of the detector unit. Signal variations along four lines (H1 to H4) across the full projected width of a thin cylinder subjected to cyclic internal pressure are illustrated in Fig.3. (A matt-black coating (see above) had been applied). It can be seen there is no significant signal loss until the angle between the surface normal and the optical axis exceeds 55°.

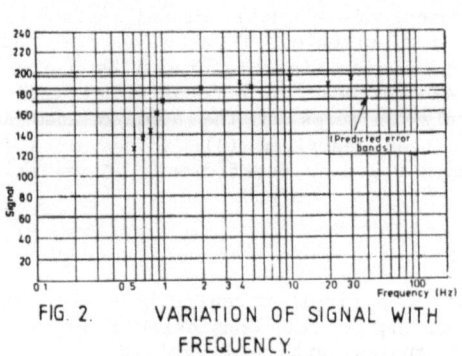

FIG. 2.   VARIATION OF SIGNAL WITH
                     FREQUENCY.

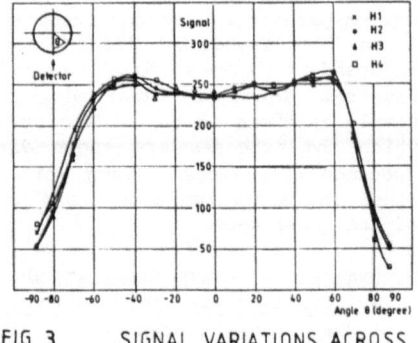

FIG. 3.   SIGNAL VARIATIONS ACROSS
                     CYLINDRICAL VESSEL.

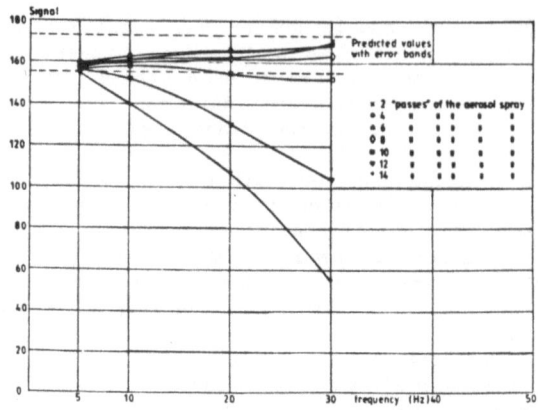

FIG. 4.   EFFECT OF COATING THICKNESS.

iii) The effect of the thickness of the aerosol spray coating on the received signal is insignificant at low frequencies. Results from a simple beam specimen tested at 5, 10, 20 and 30 Hz with surface coatings applied by 2, 4, 6, 8, 10, 12 and 14 "passes" of the RS aerosol spray are shown in Fig.4. With up to 8 passes the signal is independent of the number of passes over the frequency range from 5 to 30 Hz.

iv) The "stand-off" distance between the detector and the body being scanned can vary from about 0.5 m to at least 5 m. With the minimum distance the sampling area is approximately 0.5 mm x 0.5 mm, and these dimensions increase proportionately as the "stand-off" distance is increased. The depth of focus is of the order ± 10 mm with the minimum "stand-off" distance and increases rapidly as this distance is increased.

v) The equipment responds to infra-red flux changes due to a temperature change of the order of 10^{-3} degree Celsius. This corresponds to stress changes of the order of 1 N/mm^2 in steel and 0.4 N/mm^2 in aluminium.

3. CALIBRATION

The constant A in equation (4) is comparable, in a sense, with the material fringe value for photoelastic materials and the gauge factor associated with an electrical resistance strain gauge. Some work on the determination of A in terms of the separate material property values and system parameters incorporated in it has been described (12). Clearly also A can be found by direct comparison of a SPATE signal value with independent stress data obtained for example, from a strain gauge rosette. There are clear advantages, however, in obtaining A by use of a simple specimen (e.g., a 3-point or 4-point beam, or a "Brazilian" disc) in which the stresses are known directly in terms of the applied load and specimen dimensions. Such work has been described (8); values of A for steel and aluminium determined in this way show coefficients of variation of the order 4-6%.

4. EXAMPLES OF SPATE APPLICATIONS

4.1 Cylinder with hole in torsion

The stress concentrating effect of a single circular hole in a thin cylinder subjected to torsional loading is of considerable practical importance. Analytical and numerical studies, with some limited

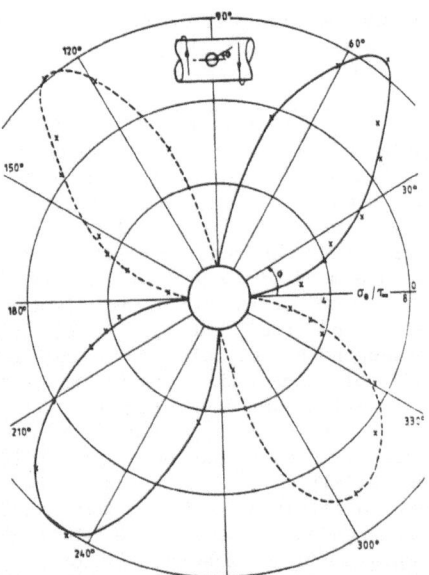

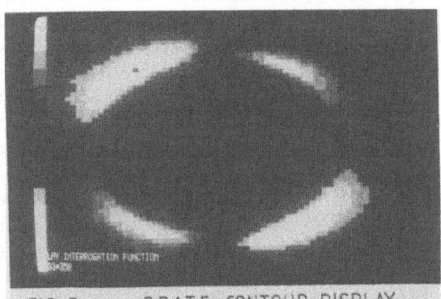

FIG. 5   S PATE CONTOUR DISPLAY
FOR HOLE IN CYLINDER.

FIG. 6.   STRESS INDEX AROUND
HOLE BOUNDARY.

experimental support, have been reported (13,14). The S PATE assessment of the stresses around such a hole (one from a comprehensive test series) is described here. The steel cylinder was 300 mm long, 100 mm mean diameter and 1.63 mm thick; the central circular hole was 30 mm in diameter. Using a specially designed fixture, the cylinder was subjected to cyclic torsional loading (frequency 10 Hz) in a 250 kN Losenhausen hydraulic test machine. The load limits were set to give nominal shear stress limits in the plain cylinder of 27.8 N/mm^2 and 13.9 N/mm^2. The stand-off distance was 1.18 m, giving an adequate depth of focus over the scanned area and a sampling area of 1.2 x 1.2 mm; the sampling time was 3 s. The scanned area was 45 mm long (axially) and 41 mm in projected width (circumferentially) with the hole at the centre. The scanned portion of the cylinder surface was cleaned and spray-coated before testing as described above.

The surface scan (reproduced in "shades of grey") is shown in Fig.5. (The vertical linear scale in this and the other scans reproduced in this paper has been reduced by a factor of 0.67; this reduction does not occur in the production equipments and should be disregarded.) The "diagonal" symmetry of the S PATE signal is immediately evident; it is noteworthy that for a pure shear stress (i.e. in the absence of the hole) the S PATE signal would be zero.

Fig.6 is a polar plot of the elastic stress index (defined as circumferential stress divided by the shear stress in the plain cylinder) around the periphery of the hole. The separate edge values were obtained from transverse signal plots from the stored S PATE data, extrapolating manually to allow for small edge errors (8). Radial stresses are zero around the hole periphery; the circumferential stress follows directly therefore from the S PATE data. A previously determined (8) value of the calibration factor A for steel (0.393 N/mm^2 per unit signal) was used. The mean positive peak stress index was 12.3 (Lekkerkerker's predicted value for this configuration (13) is 13.5) occurring at an angular position of φ = 53°(see inset, Fig.6). A detailed account of this investigation will be presented elsewhere.

4.2 Pressure vessel model

A small (76 mm diameter) thin cylindrical pressure vessel model subjected to cyclic internal pressure (frequency 10 Hz) has been studied. Particular interest attached to the upper end-closure of the vessel which has been "freely-formed" into a "dished" shape (height/diameter ratio 0.145) from an originally flat end by the application of a steadily increasing internal pressure prior to the SPATE test. (The buckling failure mode of very thin torispherical ends is well known (15).)

A SPATE scan, with the viewing direction approximately normal to the centre of the "knuckle" portion of the end, is shown in Fig.7. A corresponding signal plot along a meridion of the vessel (i.e. from the pole of the end, around the "knuckle" and longitudinally along the cylinder) is shown in Fig.8. There is a clear indication of a predominant compressive stress in the end close to the end-cylinder junction and it would seem that the buckling failure mode (15) cannot be dismissed as a possibility for these ends. The profile of the "freely-formed" end has been determined and further analytical work has established a close correspondence between the SPATE data and stress sum results derived from simple membrane theory. This has led to the conclusion that the knuckle bending stresses in a "freely-formed" end are small relative to the membrane stresses; the possibility of defining an optimum form of end of this type will be explored. The effort involved in this model study was notably less than that which would have been required had an alternative experimental technique been used.

FIG. 7.   SPATE CONTOUR DISPLAY
FOR MODEL PRESSURE VESSEL

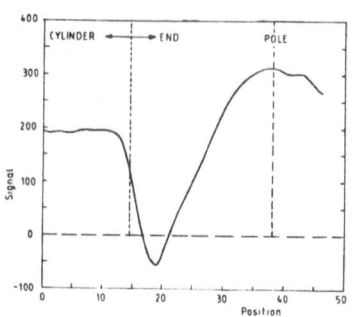

FIG. 8.   TYPICAL SIGNAL PLOT
ALONG PRESSURE VESSEL MERIDION

4.3 Crack-tip studies

It has been established (7) that there is considerable scope for SPATE applications in the study of crack-tip stress fields and, in particular, in the determination of stress intensity factors. A brief account of a K_I determination is given here to illustrate this scope. The specimen used was a flat steel plate, 350 mm long, 300 mm wide and 1.6 mm thick, containing a central circular hole 3 mm in diameter and subjected to a uniformly distributed applied stress at each end; under controlled cyclic loading straight fatigue cracks had been grown symmetrically from the two

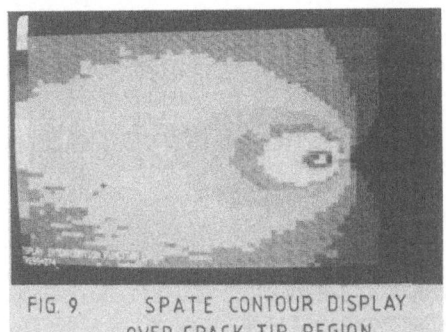

FIG. 9. SPATE CONTOUR DISPLAY
OVER CRACK–TIP REGION

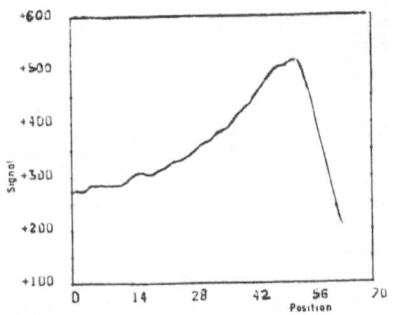

FIG. 10. TYPICAL SIGNAL PLOT
ALONG LINE PARALLEL TO CRACK.

peak stress positions at the hole edge. The cracked specimen was then sprayed with the matt-black coating and subjected to a smaller cyclic load (frequency 10 Hz) under which the cracks did not propagate, and a SPATE scan was obtained over one crack-tip region. The scan - see Fig.9 - showed ample detail with a clearly defined signal concentration at the crack-tip. It is readily shown (2) for this mode I loading that over the part of the stress field where the simple Westergaard equations apply, the SPATE signal S is related to the stress intensity factor K_I by the equation

$$K_I = \frac{AS \sqrt{2 \pi r}}{2 \cos (\Theta /2)} \tag{5}$$

where r, Θ are the conventional polar coordinates with origin at the crack tip and A is the calibration factor. Based on this equation several methods have been developed for obtaining K_I from the SPATE data. For example, it can be shown (7) that the maximum signal value S_{max} along any line parallel to the crack (and distance y from the crack) occurs at $\Theta = 60°$. Fig.10 shows a typical signal plot along such a line. It can also be shown that $(1/S_{max})^2$ varies linearly with y and that a good value of K_I can be derived from the slope of the best line through the $(1/S_{max})^2$ versus y data. This plot is shown in Fig.11. There are clear indications

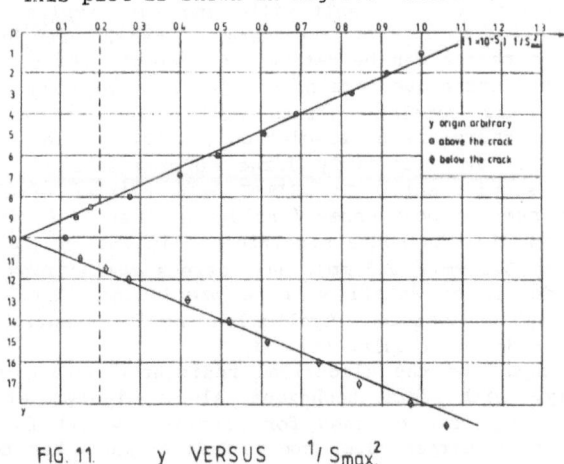

FIG. 11. y VERSUS $^1/S_{max.}^2$

that equation (5) is valid only over a limited range. The K_I values derived from the linear portions of the graph were 585 and 568 $N/mm^{-3/2}$. An independent value (16) for this configuration was obtained as 590 $N/mm^{-3/2}$. Clearly the SPATE technique has considerable potential in the fracture mechanics area.

Related work on K_{II} and mixed mode loading, on the determination of crack-tip velocities and the determination of the Paris law coefficient and index (C and m) has also been completed.

5. DISCUSSION

There is no single simple basis upon which the principal techniques for experimental stress analysis can be usefully compared. Each technique has its own distinctive characteristics which, depending on the circumstances, may be advantageous or otherwise, and generalisations may be misleading. Nevertheless, it is worthwhile to review the principal features of the SPATE technique in relation to those of two widely used alternative experimental techniques - photoelasticity and electrical resistance strain gauges.

Notable features of the SPATE technique are i) there is no direct physical or mechanical contact with the object under test, ii) the preparatory effort (particularly in terms of time) is minimal, iii) it provides a "field" coverage of the scanned area, iv) the actual component or structure can be studied directly, and v) the stress data becomes available practically instantaneously as the scanning proceeds and is stored for subsequent interrogation. By contrast, two- and three-dimensional photoelastic work is based on the use of models, the manufacture, loading and analysis of which require a considerable commitment of time and effort. (It is noted that fringe pattern analysis can be greatly facilitated by the use of one of several alternative forms of automatic polariscope.) This approach also provides a "field" coverage but the derived stress data may contain systematic errors due to atypical Poisson's ratio values and relatively large strains in the loaded models. Strain gauges can usually be used on the component itself but they provide only a discrete coverage of a stress field and prior work (e.g. the use of brittle lacquer) may be required; in any event the preparatory work can be very considerable.

The full scope of the photoelastic approach (including the different forms of loading that can be dealt with) and the simplicity of the basic equipment are noteworthy. In the general case, with some additional effort, separate stresses can be determined from the photoelastic data, and the "frozen-stress" technique remains unique in permitting, in principle, a determination of the complete stress distribution throughout a three-dimensional body. There are no resolution limitations in photoelastic work but the effects of the relatively large deformations (e.g. in crack-tip studies) may be significant. The SPATE data relates essentially to the principal stress sum; in most cases (including those of a "free" edge and a point of symmetry) this is not a significant limitation. With the present resolution limits (0.5 mm x 0.5 mm), only stress evaluations calling for an unusually fine degree of detail will be precluded. The scanned surface must be (optically) accessible. Cyclic loading is essential but the cycle need not be sinusoidal or regular (4).

The "gauge length" of the electrical resistance strain gauge is finite and consequently, with this technique also, highly detailed work is precluded. The gauge can be used for practically all forms of loading, over a wide range of temperature, and once the gauge has been bonded and

wired-in accessibility is not a problem. There are environmental
limitations and reinforcement effects may be significant in some cases.
The supporting technology for use with these gauges is very well developed.

Sound judgement is called for in assessing the relative roles of the
techniques mentioned here. The SPATE technique is not yet fully developed;
there may well be important developments in other fields. Nevertheless, it
can be concluded that the SPATE technique offers an important new element
of choice in experimental stress investigations and related work.

ACKNOWLEDGEMENTS

This work has been supported by a cooperative research grant from the
Science and Engineering Research Council, in association with SIRA Ltd.

REFERENCES
1. Cox, L.J., Holborn, P.E., Oliver, D.E. and Webber, J.M.B., "Stress
   analysis of complex structures using the thermo-elastic effect", Proc.
   VIIth Int. Conf. on Exptl. Stress Anal., Haifa, 1982, 538-544.
2. Stanley, P. and Chan, W.K., "'SPATE' studies of the stress
   distributions in steel plates and rings under in-plane loading", Proc.
   SEM Spring Conf. on Exptl. Mechs., Las Vegas, Nevada, 1985, 747-757.
3. Archer, R. and Razdan, D., "Stress analysis of railway coach body
   panels by means of the SPATE technique", ibid, 740-746.
4. Cummings, W.H. and Harwood, N., "Thermoelastic stress analysis under
   broad-band random loading", ibid, 844-850.
5. Harwood, N. and Cummings, W.M., "Applications of thermoelastic stress
   analysis", Strain, 1986, 22, 7-12.
6. Stanley, P. and Chan, W.K., "Stress studies in composite cylinders
   based on measurement of infra-red emissions due to cyclic loading", To
   be presented at Int. Symp. on Mechanics of Polymer Composites, Prague,
   April, 1986.
7. Stanley, P. and Chan, W.K., "The determination of stress intensity
   factors and crack-tip velocities from thermoelastic infra-red
   emissions", To be presented at the Int. Conf. on Fatigue of Eng. Mats.
   and Structures, Sheffield, Sept., 1986.
8. Stanley, P. and Chan, W.K., "Quantitative stress analysis by means of
   the thermoelastic effect", J. Strain Anal., 1985, 20, 129-137.
9. Rogers, G.F.C. and Mayhew, Y.R., Engineering Thermodynamics Work and
   Heat Transfer, 1967 (Longmans, London).
10. Stanley, P. and Chan, W.K., "The thermoelastic effect for stress
    analysis" Report No.2, University of Manchester, Dept. of Eng.,
    Research Report, Nov., 1983.
11. Oliver, D.E. Private communication.
12. Oliver, D.E. and Webber, J.M.B., "Absolute calibration of the SPATE
    technique for non-contacting stress measurement", Proc. Vth Int. Cong.
    on Exptl. Mechs., Montreal, 1984, 539-546.
13. Lekkerkerker, J.G. "Stress concentration around circular holes in
    cylindrical shells", Proc. Eleventh Int. Conf. Appl. Mechs., Munich,
    1964, 283-288.
14. Bull, J.W., "Stresses around large circular holes in uniform circular
    cylindrical shells", J. Strain Anal., 1982, 17, 9-12.
15. Stanley, P. and Campbell, T.D., "Very thin torispherical pressure
    vessel ends under internal pressure : strains, deformations and
    buckling behaviour", J. Strain Anal., 1981, 16, 187-203.
16. Rooke, D.J. and Cartwright, D.J., Compendium of stress intensity
    factors, 1967 (HMSO, The Hillingdon Press, Uxbridge).

USE OF THE THERMOELASTIC TECHNIQUE IN EXPERIMENTAL STRESS ANALYSIS

N Harwood and W M Cummings

National Engineering Laboratory, East Kilbride, Glasgow, UK

S U M M A R Y
    The practical exploitation of the thermoelastic effect by the
development of a dynamic stress measurement system has extended the range
of techniques currently available to the engineer or designer who requires
full-field experimental stress analysis data. This paper describes the
application of the thermoelastic technique to several engineering
structures which were examined in the National Engineering Laboratory's
structural testing facility. The development of a system for processing
thermoelastic data from structures under random loading is also discussed.

1.   INTRODUCTION
    Recent advances in infra-red technology have led to the development of
the SPATE system[1] for full-field experimental stress analysis. This
instrument has been used in a three-year research project at the National
Engineering Laboratory with the aim of extending the applications of
thermoelastic stress analysis to a wide range of engineering structures
made from a variety of materials.
    The equipment is based on a scanning radiometer which measures the
very slight temperature changes produced on the surface of a structure
which is undergoing cyclic loading. The thermodynamic relationship[2]
between these temperature changes and the corresponding stresses produced
by the applied load signal (ie the thermoelastic effect) is used to
determine a full-field surface stress distribution. A colour-coded contour
map, showing the distribution of the sum of the principal stresses, is
produced as a real-time raster scan in the form of square pixels in a range
of 16 colours which indicate the amplitude and phase polarity of the stress
at a given point on the surface of a structure.
    The evaluation programme at NEL[3-5] has demonstrated the system's
great potential for extending the present capabilities of experimental
stress analysis.
2.   CALIBRATION
    Calibration investigations have been performed on several commonly
used, simple test specimens manufactured from a wide range of metallic and
non-metallic structural materials. The testpieces used for the calibration
investigations are:

a   a beam in four-point bending,
b   a disc loaded in compression, and
c   an axially loaded plate.

    Calibration using such test specimens[6] is performed by comparing the
amplitudes of the thermoelastic response signal with the theoretical stress

solution for given positions on the specimen. This allows an average stress value to be allocated to each of the 16 colours in an area scan measured on a structure with the same material properties as those of the calibration specimen. At NEL the preferred calibration testpiece is the uniformly stressed plate (Fig. 1), since its simplicity makes the calibration procedure very straightforward.

Calibration may either be performed by manufacturing one of the above calibration specimens from a material identical to that of the component to be examined, or more generally, by bonding a strain gauge rosette to the component. Once the relationship between stress/strain and the corresponding thermoelastic response has been determined for a given spot on the component, the same calibration factor will be applicable to the entire surface area provided that the detector remains in focus and that there is no significant variation in material properties throughout the areas to be scanned. If appropriate physical data are available for the component material, they may be used to determine a theoretical solution[1] which may then be compared with the measured calibration factor.

A fourth type of simple specimen which is widely used for a rapid assessment of equipment performance on structural materials is the hole-in-plate testpiece. Software has been written at NEL to display the theoretical solution for the sum of the principal stresses distribution around a hole in an axially loaded plate. This theoretical stress pattern may either be displayed in colour-coded or isometric (Fig. 2) form, thus enabling a direct comparison of theoretical and experimental results to be made.

3. APPLICATION TO ENGINEERING STRUCTURES

At NEL thermoelastic stress analysis has been used to examine a range of complex-geometry structures and several examples are described below.

As part of a study into the fatigue performance of wire ropes intended for offshore applications, an experimental stress analysis was carried out on a short section of 70 mm diameter 6 × 41 wire rope close to the fixed end in a test rig. A wire rope is manufactured by spinning a collection of wires into strands, several of which are then wound together to form a single rope. A SPATE area scan revealed distinctly the surface stress distribution in the strands and individual wires in that section of the rope. The success of the thermoelastic technique for stress measurement from the surface of wire rope has aroused much interest within the industry, since knowledge of the way in which stress is distributed in such components is very limited, and other full-field techniques are unsuitable for this very important engineering application. Despite the fact that the thermoelastic stress patterns are only a surface measurement, broken wires appear as unstressed areas when they emerge at the surface. Thus, the thermoelastic technique shows potential for NDT applications in such components.

A large cast steel X-node (Fig. 3), undergoing fatigue testing in relation to offshore applications, was also examined. The amplitude and phase polarity of the stress distribution both on the outside of the structure and around the bore of the chord were revealed clearly in the thermoelastic data. The overall stress distribution measured over one side of the entire structure can be seen in Fig. 4. Four strain gauges and associated lead wires are clearly visible around the saddle point on the outside surface of the X-node. Such tests have indicated the technical and economic advantages of thermoelastic full-field stress measurements on large-scale structures compared with the highly labour-intensive nature of the corresponding strain gauge techniques. Thus for many full-scale test applications thermoelastic stress analysis is viable economically despite the high capital cost of the equipment.

As part of a bio-engineering research programme a short investigation was conducted on hip prostheses. Thermoelastic stress analysis was used to determine the stress distribution in several different designs of prosthesis in a simulated 'worst' loading condition. The 'worst' situation represents a totally unsupported prosthesis due to a combination of bone resorption and cement failure. The thermoelastic data indicated that the maximum stress occurred at the prosthesis/cement interface, later confirmed as the point of failure in subsequent fatigue tests.

The principal weld in a large tubular, welded T-joint was examined as part of an extensive testing programme investigating the fatigue of offshore structures. An incipient crack was detected in the weld, and the effects of progressive crack growth were monitored by a series of consecutive area scans. A second area of the T-joint was investigated in order to determine whether the stress distribution associated with internal welded stiffeners could be detected on the smooth outer surface of the chord. The surface effect of the hidden stiffeners was clearly visible in the thermoelastic patterns.

Thermoelastic stress analysis was used in conjunction with a brittle lacquer/strain gauge experimental stress analysis of a lorry chassis cross-member in order to refine a finite element model of the component. The stress patterns measured on swaged and flat end-caps were compared in order to ascertain the effectiveness of the stiffened configuration in reducing the stress concentrations around the bolt heads. Five puddle welds which joined the base of the end-caps to the cross-members were also examined. The data revealed that the swaging was not particularly effective and that a single puddle weld was carrying virtually all of the applied load.

A preliminary fracture mechanics study has been carried out on a plain flat bar specimen with a double edge notch. The resolution of the SPATE instrument appears to be adequate for such detailed work. The results from these tests are being compared with a theoretical solution for the same specimen in order to identify any limitations in the current equipment for this important application.

Thermoelastic stress analysis has been carried out at NEL on a wide range of engineering materials, including several steels, Dural, brass, copper, Nimonic 90, zirconium, titanium, a ceramic (Hylox 961), Perspex, Tufnol, brick, concrete, fresh bone, and several fibre-reinforced plastics and woods. The investigations showed that most non-metallic materials have high surface emissivities, and thus that no pretest paint spraying was necessary. In these cases no surface preparation other than cleaning and removal of any loose debris was required. Experimental data indicate that thermoelastic stress analysis may be applied to non-homogeneous or anisotropic materials, although calibration is not as straightforward as in the case of standard components.

4. COMPLEX LOADING DEVELOPMENTS

A major limitation of the SPATE 8000 system is that it can only be used on structures subjected to uniform cyclic loading. The drive signal which is usually chosen is a sinusoidal waveform, since noise reduction is more efficient if the energy is concentrated at a single frequency. The stress patterns described previously in this report are quasi-static, ie they are identical to the sum of the principal stress distributions which would be produced by a static load of the same amplitude as that of the sinusoidal load waveform. Thermoelastic stress analysis has proved to be highly effective in a quasi-static role, but using the system only in this way does not exploit the potential of the technique for genuine dynamic applications. Structures operating in dynamic service conditions are usually subjected to variable-amplitude loading. The main scientific effort on the thermoelastic stress analysis development project at NEL has been concentrated on research into, and development of, methods to extend the range of the SPATE 8000 system to service loading conditions, and thus greatly enlarge the applications area of the thermoelastic technique beyond its present limitation to laboratory environments. The SPATE equipment has been linked to a GenRad 2515 transportable, computer-aided test system provided with hardware and a software language[7] specifically intended for rapid, sophisticated signal-processing applications. At NEL software has been developed which enables thermoelastic signals from structures undergoing random loading to be analysed and stress patterns to be generated, plotted, and stored in a SPATE-compatible format. Use of the SPATE 8000 equipment in conjunction with the GenRad 2515 system means that random signals up to a bandwidth of 25 kHz may be analysed and that thermoelastic data may be interfaced directly with a powerful commercial modal analysis package.

In addition to extending the range of thermoelastic stress analysis to variable-amplitude loading conditions, the use of random drive signals enables modal behaviour to be investigated much more rapidly than is possible under sinusoidal loading, since multiple excitation greatly reduces the required data acquisition time.

A transfer function showing the gain and phase relationship between a 60-120 Hz white noise loading signal and the corresponding thermoelastic response from a stiffened plate is displayed in Fig. 5. Two resonances are clearly evident in the form of peaks and phase changes in the transfer function. When the real and imaginary coefficients of this transfer function around the first structural mode were plotted in the Argand plane, the circular graph shown in Fig. 6 was produced, thus indicating that a standard modal analysis procedure (ie Nyquist plotting) can be used to extract the modal contribution from variable-amplitude thermoelastic signals. A modal analysis using a fixed force gauge and a roving accelerometer showed that these first two structural resonances were torsion and flapping modes of the plate. The thermoelastic stress patterns measured at each of these natural frequencies were consistent with the displacement data obtained from the experimental modal analysis.

The NEL random signal analysis software allows the operator to select either the conventional form (ie H1) of the transfer functions or the inverse form (ie H2). It has been shown[8] that for measurement of very sharp resonances the inverse form is usually more accurate. Under modal conditions either peak magnitude or quadrature[9] frequency response may be used to determine the stress level. Quadrature response gives better isolation from adjacent coupled modes than is afforded by the peak magnitude technique. A more sophisticated technique for the separation of closely coupled modes involves the curve fitting of Nyquist plots such as the one described previously. However, this technique tends to diverge when the circles are ill-defined, thus leading to poor results under purely automatic operation. The data acquisition and analysis software has been developed to incorporate a zoom facility for cases in which a fine resolution over a narrow bandwidth is required. An overlap processing capability is also available to compensate for the 40 per cent energy attenuation produced by the window function which must be applied to impose a quasi-periodicity on random time frames before they are transformed to the frequency domain.

Calibration under random-loading conditions is somewhat more complicated than for the standard sinusoidal case. Stresses are determined in terms of the r.m.s. amplitude since this is statistically constant for a stationary[10] signal. In the case of white noise excitation the calibration procedure is fairly straightforward, following a similar form to that of the sinusoidal technique detailed earlier, except that values are computed in the frequency rather than time domain. Calibration under more complex loading conditions requires further computation, since the transfer function is insensitive to the precise form of the forcing function. In a service load condition where no force signal is available, it would be advantageous to use as a reference the summed outputs from two strain gauges in a T-configuration, since this gives a signal which, like the thermoelastic response, is invariant[11]. Thus the transfer functions would be scalar rather than vector arrays.

In order to assess the effectiveness of the frequency-domain analysis software in measuring stress under variable-amplitude loading, two components with known stress distributions were examined under random (1-40 Hz white noise) loading conditions. It was known that no resonances would be excited in the components over this bandwidth, and thus the frequency domain data could be averaged over the excitation bandwidth. Load cell and thermoelastic response signals were acquired simultaneously on the 2515 system which has a programmable analogue output channel capable of producing a trigger pulse to move the detector scanning mirrors synchronously with the time window required for data acquisition of the random signals from the component. The transfer functions between the load reference signal and the corresponding thermoelastic responses were both bandwidth and ensemble averaged to produce an array of scalar values proportional to the sum of the principal r.m.s. stress amplitudes at each measurement position over the surface of the component.

Since the 2515 system has a monochrome screen, techniques have had to be developed to display stress patterns without using a colour code to indicate the stress level. A monochrome display showing the distribution of the sum of the principal stresses around a hole in a flat bar subjected to axial random loading is shown in Fig. 7. The linear dimension of each pixel box is proportional to the r.m.s. stress at that point. The phase polarity of the stress relative to the load waveform is indicated by the sign symbol inside the box. An alternative monochrome display technique is shown in Fig. 8. This display takes the form of a quasi-isometric stress profile map around a hole-in-plate after mathematical smoothing has been applied to the raw random data. This type of display is more suited to simple flat components rather than to complex-geometry structures. As part of a study to improve the fatigue performance of large chain links for offshore applications, a 480 mm steel chain link was examined under random loading conditions with a data acquisition time of 1 s/pixel. A monochrome representation of the stress pattern is displayed in Fig. 9; the stress concentrations around the internal corners of the link can be seen clearly in the density of the pattern.

Although the 2515 system is only supplied with a monochrome screen as a standard display facility, the computer is compatible with a range of Tektronix colour screens, provided that suitable drive software is written. Several algorithms exploiting the colour facility have been developed. The stored transfer function data may be displayed and copied in the conventional colour-coded stress pattern format when the transportable hardware is connected to our colour-display workstation.

494

5. CONCLUDING REMARKS

Over the past three years the number of users of thermoelastic stress analysis equipment has steadily expanded to include companies in North America, Europe, Japan and Australia, as well as Britain. Their experience has demonstrated that the thermoelastic technique is a very rapid and convenient means of performing full-field experimental stress analysis on engineering structures.

The technique is applicable to a wide range of structural materials, and the ease of interpretation and the non-contacting form of measurement are great advantages compared with other methods. The ability to stand-off at a sufficient distance to examine large areas of a structure, and then zoom in to the regions of interest has also been found to be a particularly useful facility. Thermoelastic stress analysis has great potential in engineering design, particularly for fatigue conditions, for which a knowledge of the location of stress concentrations is essential, and where failures in service can be catastrophic. Nevertheless, since the thermoelastic output is insensitive to shear stresses, care must be taken in using thermoelastic stress analysis equipment to determine the location of highly stressed areas. The technique should be considered to be complementary to other methods, such as strain gauges, rather than a replacement for them.

Random-loading investigations at NEL have indicated that the thermoelastic technique has great potential as a unique method of measuring full-field surface stress on structures operating under dynamic loading conditions in their normal service environment, including resonant behaviour. The technology complements existing computer-based analysis methods by providing rapid experimental feedback to validate finite element models, thus reducing the lead time in the design cycle. The use of such technology as a design aid for troubleshooting in service and for the verification of computer models is an invaluable new technique which, if exploited fully, will enhance product quality in many sectors of industry.

ACKNOWLEDGEMENTS

This paper is published by permission of the Director, National Engineering Laboratory, Department of Trade and Industry. It is Crown copyright.

The thermoelastic equipment used to produce the stress patterns described in this paper is manufactured by Ometron Ltd, Chislehurst, Kent.

REFERENCES

1   MOUNTAIN, D. S. and WEBBER, J. M. B. Stress pattern analysis by thermal emission (SPATE). Proc. Soc. Photo-opt. Instrum. Engrs., 164, October 1978.

2   BIOT, M. A. Thermoelasticity and irreversible thermodynamics, J. Appl. Phys. 27(3), 241-242, 1956.

3   CUMMINGS, W. M. Thermoelastic stress analysis. Engineering, 132, February 1982.

4   CUMMINGS, W. M. and HARWOOD, N. Design analysis of engineering structures using SPATE. First Scottish Design Engineering Conference, Glasgow, April 1985.

5   HARWOOD, N. and CUMMINGS, W. M. Applications of thermoelastic stress analysis. Strain 22(1), 7-12, February 1986.

6 STANLEY, P. and CHAN, W. K. Quantitative stress analysis by means of the thermoelastic effect. J. of Strain Analysis, 20(3), 129-137, July 1985.

7 CUMMINGS, W. M. and HARWOOD, N. Thermoelastic stress analysis under broad-band random loading. SEM Conf. on Exptl. Mechs., Las Vegas, 740-746, June 1985.

8 UPTON, R. Innovative functions for two-channel FFT analysers. Sound and Vibration, March 1984.

9 BROWN, D. L. et al. Parameter estimation techniques for modal analysis. Society of Automotive Engineers, Paper No 790221, 1979.

10 BENDAT, J. S. and PIERSOL, A. G. Random data: Analysis and measurement procedures. John Wiley, 1971.

11 TIMOSHENKO, S. P. and GOODIER, J. N. Theory of elasticity. McGraw-Hill, 1970.

496

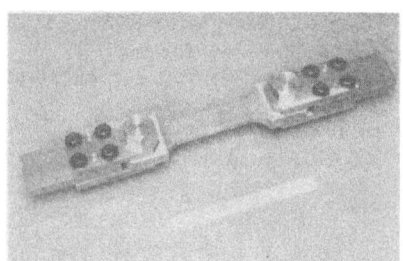

FIG 1   Calibration test piece

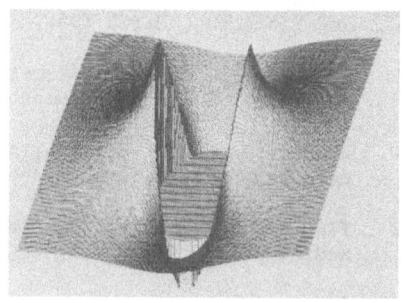

FIG 2   Isometric representation
of theoretical stress pattern

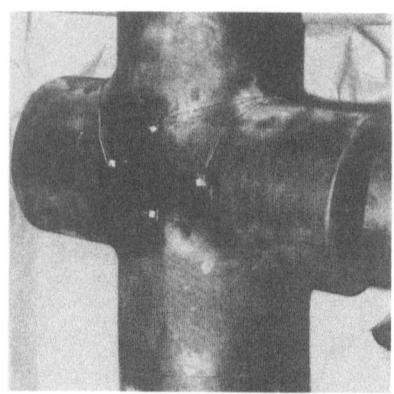

FIG 3   Cast X-node for
offshore application

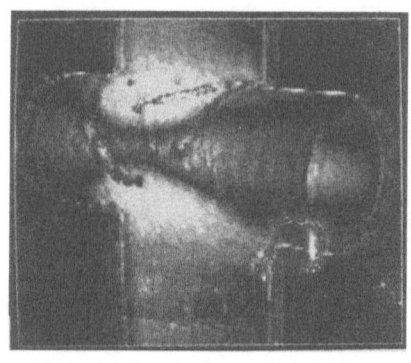

FIG 4   Stress pattern
from X-node

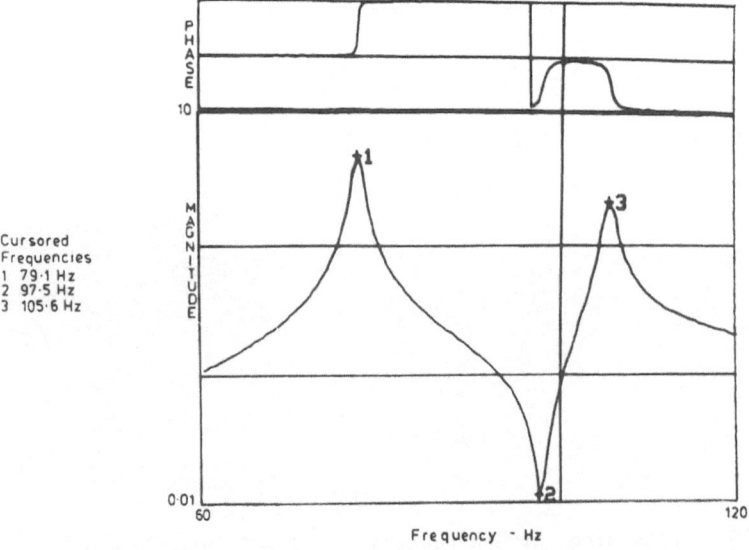

FIG 5   SPATE transfer function acquired under random excitation

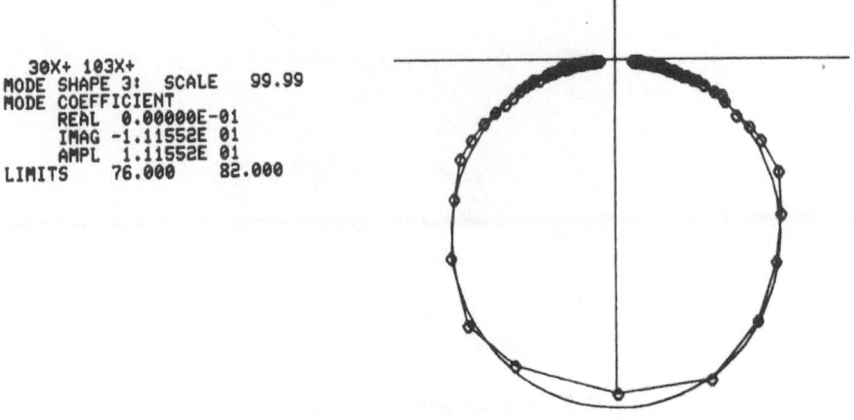

FIG 6   SPATE Nyquist plot

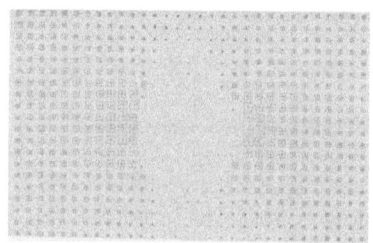

FIG 7 Monochrome stress pattern
from random excitation

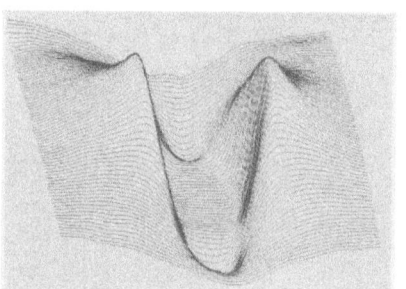

FIG 8 Smoothed isometric stress
profile map from random excitation

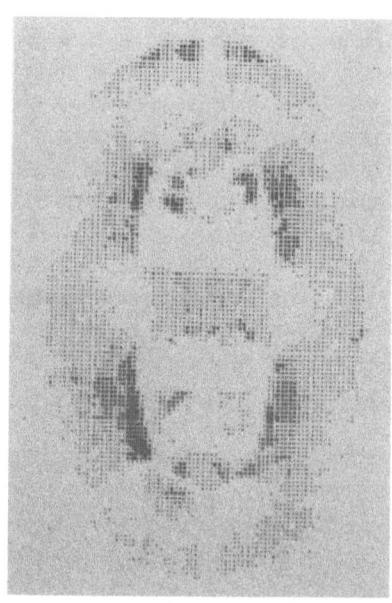

FIG 9 Stress pattern from chain
link under random excitation

OPPORTUNITIES AND LIMITATIONS OF THERMOELASTIC STRESS ANALYSIS

W. KIZLER

Daimler-Benz AG, Abt. V1PS, Mercedesstraße,
7000 Stuttgart 60, West-Germany

SUMMARY

Aside from the appealing advantages of thermoelastic stress analysis, it has characteristics, which do not predestine it to all jobs of a stress analyst. In this presentation, possible users are shown some types of testing applications for which thermoelastic stress analysis is useful and some for which it is overstressed. We think that the knowledge and instruments of a good stress analyst are essential for successful application of SPATE 8000.

1. INTRODUCTION

For the past few years the most enticing way to analyse stresses has been thermoelastic stress analysis with SPATE 8000.
The primary features of SPATE 8000 are:
- Quick, non-contact measurement of the sum of the principal stresses of structures under dynamic loading
- Spatial resolution as fine as 0,5 mm
- Minimal surface preparation

Of coures, thermoelastic stress analysis enters into competition with brittel coating, photoelasticity, strain gauges and the Finite Element Method (FEM). For the engineer it is a question of which system is the best one to do the actual job. The best system may be the fatest, the most sensitive, the cheapest, the one with most complete information, as determined by the engineer's needs. However the features of thermoelastic stress analysis do not make it equally well suited for all types of component optimization. These features are discussed using actual measurements as examples.

2. UNSUITABLE APPLICATIONS

2.1 Reason: Dynamic Loading

Naturally most of the failures in automobile construction and comparable disciplines are not due to static but rather dynamic loading. A static load on a component in service can also be applied cyclically for testing. An exception, however, is forces resulting from fittings.

Example:
The criterion for fracture was to be determined on a cracked crankcase (prototype). In addition to the effect of the combustion pressure, stress analysis showed a high assembly force, which is very dangerous for the prestress-sensitive aluminium material (Fig. 1). It was easier to reduce the effects of the prestress load than the combustion pressure. The prerequisite for this was to determine the problematic value of the assembly force. This would not be possible with thermoelastic stress analysis. Photoelasticity and strain gauges were used for optimization.

500

2.2 Reason: Test Duration

Measurement time is a function of desired quality, fieldlength, and load frequency. The system can be transported in a car and then set up ready for testing by 2 people within a half-hour. However the stress pattern is not available immediately after switching on the camera; it must be composed dot for dot. The diagram (Fig. 2) shows the measurement time with three vibrations per test point. Below 5 Hz the measuring time becomes quite long. The user must then limit the field size.

Example

A rear axle carrier is supported on rubber mountings, of a welded sheet metal construction (Fig. 3). The carrier must be examined from 6 sides.

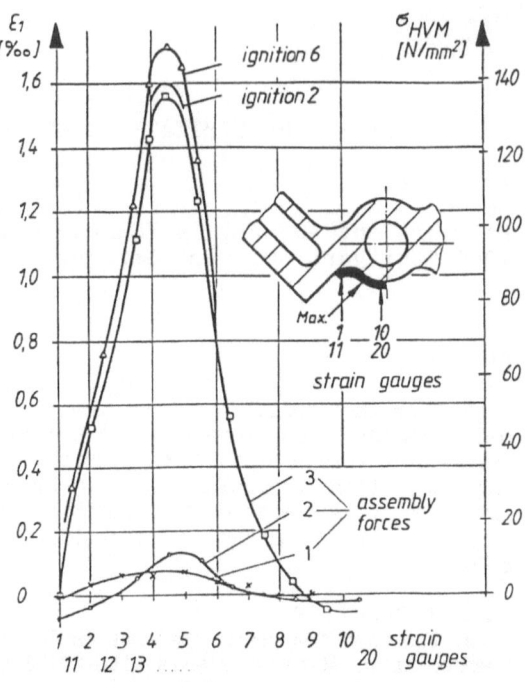

FIGURE 1
Crank case, strain gauges measurement of prestress and service stresses

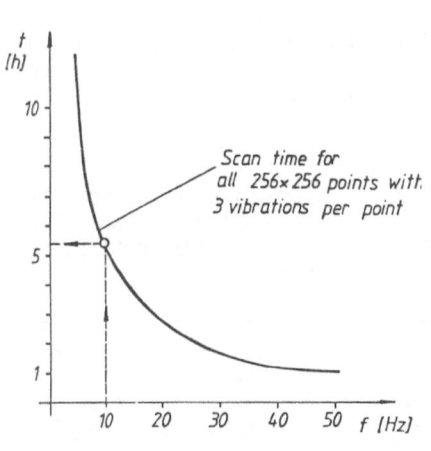

FIGURE 2: Test duration

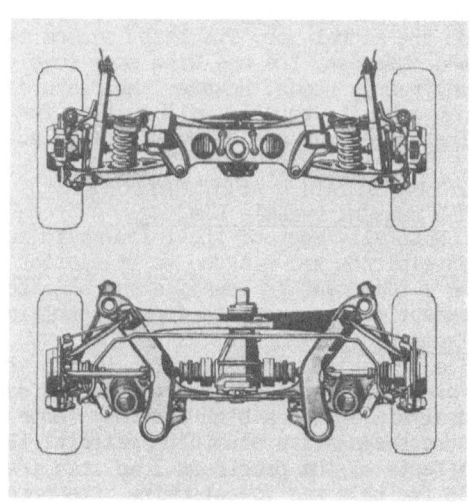

FIGURE 3: Rear axle carrier

The forces to be measured are:
1. Vertical wheel force
2. Cornering force, left
3. Cornering force, right
4. Braking force, service brake, forwards
5. Braking force, service brake, backwards
6. Acceleration from dead stop against parking brake
7. Drive shaft torque, forwards
8. Drive shaft torque, backwards

This results in 8 x 6 = 48 individual measurements. As the carrier is held in rubber mountings, time compression does not appear practical, because the characteristics of the rubber mountings changes. A test point with a diameter of 2 mm is to large for welded seams. The width of the carrier (1 m), therefore, can not be covered with one measurement. In this case the quick set-up time is offset by the long testing time, combined with the demand for a very extensive test machine (e.g. road simulator with 9 channels). Moreover the violent rigid body motion of the component under loading resulting from the rubber mountings would lead to "unsharp" stress pictures.

2.3 Reason: Measurement Information

The information from thermo-elastic stress analysis alone is inadequate: the sole obtainable signal - the sum of the principal stresses - is of no value for determining fatigue life of a solid. The difference between the sum of principal stresses and Henky v. Mises equivalent stresses for varied strain ratios shows this (Fig. 4). According to our experience casted parts have an strain ratio $\varepsilon_2/\varepsilon_1$ between -1 and 0,2, thin walled components represent the entire spectrum from -1 to +1. A complicated structure must never brake at the point where the sum of principal stresses is at its maximum! The crack can start at every other point. A point not delivering a signal can either not in fact be subject to any stress or a shear stress of any magnitude may be present. In this case calibration of the readings is pointless.

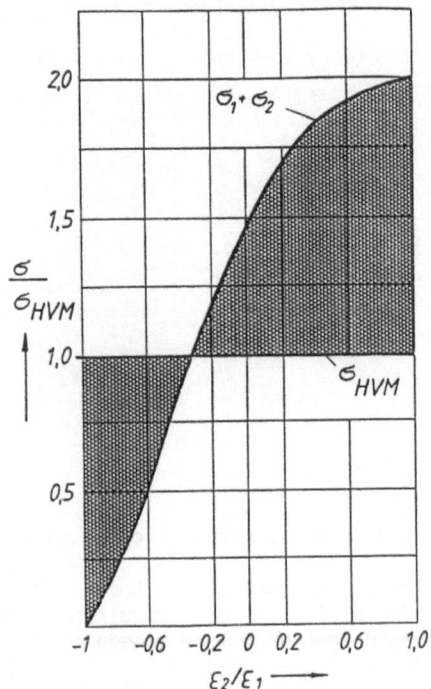

FIGURE 4: SPATE-signal $\sigma_1 + \sigma_2$ at constant Henky v. Mises stesses by varied strain ratio

3. SUITABLE APPLICATIONS

The use of thermoelastic stress analysis is not suitable for field measurements without additional information. Such additional information can be obtained with a second method (e.g. FE calculation). It is, however, better when the information is obtained from the component itself. This thermoelastic stress analysis is useful for the following without using a second procedures:
- Ribbed structures (including framework)
- Edge stress patterns
- loading problems around threaded fittings

3.1 Ribbed Components

The rib forces the stress in its geometrical direction. Moreover the strain ratio of the rib is clearly $\varepsilon_2/\varepsilon_1 = -\nu$. The maximum stress is always present on the upper side of the rib, and when constructed unfavorably, on the end section of the rib as well.

3.1.1 Example: Differential Case Cover

The stress pattern allows recognition of the structure without marking points. It can be seen immediately that one rib is not subject to stress in the case illustrated (Fig. 5).

 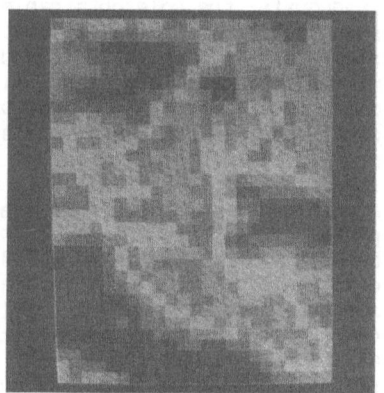

FIGURE 5: Differential case cover

The measurement itself took only 10 min. at a frequency of 20 Hz. A test with photoelasticity requires a considerably longer time.

Photoelasticity Time Schedule
1st & 2nd day Coating
3rd day   Measurement, evaluation
SPATE 8000 Time Schedule
1st day   Entire test

3.1.2 Example: Wheels

Aluminium wheels measured with SPATE 8000 (1) and photoelasticity (Fig. 6). In this case thermoelastic stress analysis is a very quick method for initial selection of component variants, which would otherwise have to be compared using fatigue strength tests. Please note, however, the limitation due to simplified loading without tires. The stress distribution in the wheel depends on the load transfer to the wheel via the tire. The load cannot be transferred fast enough dynamically by the tire when the wheel is standing still (long paths). Dynamic loading is only possible in the laboratory on a rotating bending test machine. However in this case the rim is clamped rigidly in place and therefore

FIGURE 6: Loading of wheel for measurement by Photoelastic Coatings

other stress conditions are present in comparison to the load is transferred via the tire. The stress distribution during one wheel rotation must either be measured statically in the laboratory (e.g. with photoelasticity), dynamically in laboratory with a complex biaxial wheel testing machine (2) or while driving with strain gauges.

3.1.3 Example: Sheet Metal Housing with Rib

Catalyser with rib, measured with engine running, excited to vibrations of the 2nd order of engine speed. For full field stress maps at high temperature SPATE 8000 is without competition. At a temperature of 500 °C measurement with photoelasticity is not possible. Testing with SPATE is possible, because the catalytic converter housing is designed with a rib. The measuring time is very short: 1 - 2 min. at frequencies from 150 - 200 Hz. A problem is presented by the fact that the temperature of the catalytic converter varies. However, since only one calibration value can be selected, the range of the maximum is distorted. The figure 7 shows clearly the stress concentration on the rib with the engine on the overrun, temperature 80 °C.

The resonant frequencies can be determined most simply and accurately with acceleration sensors. The trigger signal was obtained by conversion of the needle pulses from a tachometer via a ratio tuner to a sine signal of the 2nd order of the engine speed. In this manner, components determined to be subject to high vibration via stroboscope tests - this includes, in addition to the exhaust system, generally component mounts, supports and lines - can be tested for stress concentrations on the same day with SPATE.

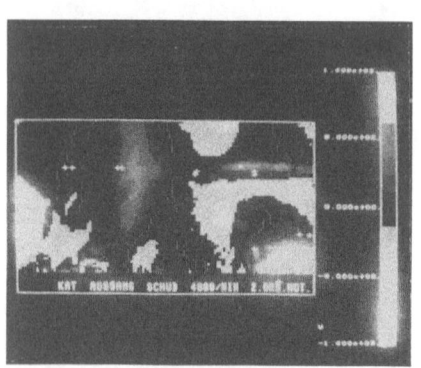 

FIGURE 7: Catalyser, engine running

3.2 Edge Stresses and Load Transfer

Flexplate (membrane), axially loaded, with crack initiation under the threaded fittings. No other experimental technique allows measurement as close to the start of the crack as SPATE 8000. Twenty min. were sufficient to produce the picture at a load frequency of 30 Hz. Compare with FE calculation (Fig. 8).

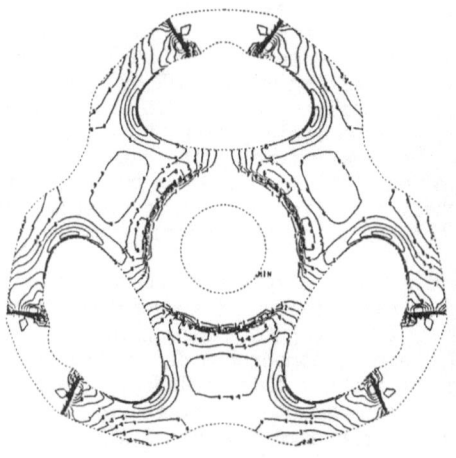 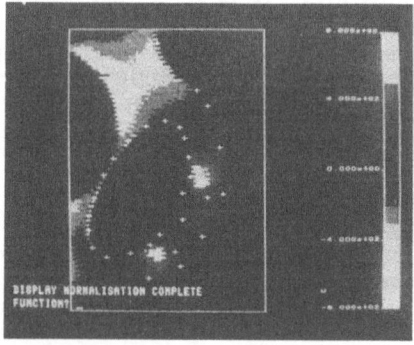

FIGURE 8: Flexplate, axially loaded, SPATE 8000 and FEM

Edge stresses themselves were indicated without error with SPATE, because $\sigma_2 = 0$ at the edge and the edge itself is not notched (3).

4. OUTLOOK

Thermoelastic stress analysis is still new and only approximately 45 units are presently in use worldwide. It cannot replace any of the other experimental procedures; it is one of many methods and has its specific advantages and disadvantages. If thermoelastic stress analysis is to be used successfully, its natural limitations must not be exceeded. Since the handicap of the sum of the principal stresses cannot be overcome, applications for component optimization are limited, in our opinion, to:

- Ribbed components
- Edge stresses (but only if stress gradients are not too steep)
- Very small parts or locations
- Measurements on components, whose characteristics are already known through other analysis methods.

In these cases, all of the advantages of this method can be utilized: the speed and the comparatively easy measurement at high temperature. The alternative, of remeasuring all uncertain locations with strain gauges, does not appear to be very effective: All points with little signal would have to be checked with strain gauges. However, this would degrade SPATE to a very expensive pretest. Moreover brittel coating is better suited as a preliminary test for strain gauge measurement, as it indicates the direction of principal strain. The thermoelastic stress analysis with SPATE 8000 allows very easy measurement; however, the results can be complex to the point of uninterpreteability. For success the user must be a highly experienced stress analyst.

REFERENCES

(1) S. Angerer und C. Kolitsch
Thermoelastische Spannungsanalyse bei der Erprobung von Kfz-Bauteilen,
(Thermoelastic stress analysis by testing of automotive parts),
VDI Bericht 552, Berlin 1985
(2) V. Grubisic
Evaluation of fatigue strength of automotive wheels, SAE International Congress, Detroit, 1983
(3) M. Nishimura et al
Measurement of stress concentration in structures
First international conference Stress Analysis by Thermoelastic Techniques, London 1984

SOME PRACTICAL LIMITS TO THE APPLICABILITY OF THE THERMO-ELASTIC EFFECT

J. McKELVIE

UNIVERSITY OF STRATHCLYDE

1. INTRODUCTION

Thermo-elasticity is the phenomenon that heat is produced, (or in some materials "lost"), when a body undergoes elastic compression, and vice-versa when it undergoes extension. The effect has been appreciated since the time of Kelvin, and, in the past few decades various workers, (e.g., [1], [2]) have studied it for the purpose of applying it for experimental stress analysis. The work of Belgen [2] is particularly substantial and relevant and Biot [3] gives a thermodynamic analysis.

In recent years a radiometric method of considerable sophistication has been developed, (see for example [4]), and a variety of applications have been reported, (see for example [5], [6], [7]).

Thermoelastic methods involve the deduction of the state of stress and strain at a point by assessing the heat produced, - the temperature change being the measured quantity. Consequently, it is generally a requirement that all of the heat produced goes into raising the local temperature, (rather than "leaking away"); it might, in principle, be possible to make allowance for leakage, but this approach has never been adopted. In order to achieve the necesssary degree of adiabaticity, it has been commonplace to cycle the loading conditions. A principal concern of this work is the effect of heat leakage on the temperature change.

Where the temperature change is measured radiometrically, the measured quantity is actually radiant energy, and this can be related to surface temperature only if the value of the local surface emissivity is known, either explicitly, or implicitly by calibration in the locality of a strain gauge. To make measurements away from the strain gauge it is necessary either to know that the emmisivity is reasonably constant across the surface or to know its variation.

The normal method is to ensure constancy of emissivity by coating the surface with a thin layer of matt black paint [8], and this present work is also concerned with the relationship between on the one hand the temperature change on the surface of the object (i.e., below the paint), and, on the other, the temperature change actually experienced on the paint surface.

Some additional factors, - concerning the assessment of sub-surface stresses, the variation of thermo-elastic contant, and strain gauge heating, - are also considered.

2. GENERAL CONSIDERATIONS REGARDING COATINGS
2.1 The need for a coating

As Belgen [2] pointed out, there are in fact three good reasons for the use of a high emissivity coating:-
i) To achieve constancy of emissivity: it appears to be the nature of

things that low emissivity surfaces are liable to have large relative variations, whereas the high emissivities tend to be constant within a few percent. (See for example Ref [10]).

ii) To maximise the radiant energy: the temperature changes are very small, and high emissivity enhances the radiometer signal.

iii) To avoid reflected heat: to the extent that the surface is reflective, radiation from sources of heat in the general locality (e.g., operators), may appear as emission from the point under examination; such reflected radiation may well, due to the movements of the test object as the load is cycled, vary at the load cycle frequency, and thus be highly correlated with the thermoelastic signal; surface of high emissivity will, additionally, have high absorptivity, thus alleviating this problem. Ref [8] indicates that some non-metallic surfaces may be adequately emissive without coating but the problem of spurious reflected radiation has not been investigated except in a qualitative manner by Belgen [2].

2.2 The effects of an inert coating

A coating which is thermoelastically inert will attenuate the temperature change by two different and distinct mechanisms:-

i) Simple thermal lag: a coating has a resistance and a capacitance, and even with no appreciable heat loss to the environment there will be a temperature drop across it under cyclic conditions:

ii) Thermal "drag-down": since no heat is being generated in the coating, its temperature can only rise by the extraction of heat from the substrate; if the latter has not a high enough conductivity then heat will not flow into the substrate surface layers quickly enough to maintain the temperature change adequately close to its value when uncoated.

Belgen [2] analysed and investigated the problem of thermal lag, but the "drag-down" effect has not previously been recognised. Both effects could intuitively be expected to be minimised by operating at low frequencies and vice-versa.

2.3 Modifications due to a thermoelastic coating

If the coating has its own significant thermo-elasticity, then the two effects above will be modified.

This work does not deal in detail with the thermo-elastic coating; most of the likely coatings are plastics-based, and typically will have a relatively low thermo-elastic effect when compared in particular to metallic substrates, (on an equal strain basis).

3. ANALYSIS OF HEAT-LEAKAGE EFFECTS

3.1 General

We are not concerned at this point with coating effects. We consider only the effects of departure from adiabaticity. The considerations are separated into external and internal effects, the latter being treated from, firstly, the general point of view of attenuation of spatial frequencies, and secondly from consideration of the particular case of bending.

3.2 External Effects

The external effect is the tendency for the surface to exchange heat with the environment by radiation and convection; such heat exchange will obviously be in the sense whereby temperature changes of the surface are attenuated.

Belgen [2] gives an analysis: a plate is exposed on two sides, and has heat generated within it; the heat loss at the surface affects the temperature of the plate uniformly through its thickness. He finds that, above 1 Hz, significant attenuation due to this mechanism occurs only for thin (1 mm) or very thin (0.1 mm) plates under, respectively, "severe" or "moderate" forced convection.

However, Belgen's analysis did not take into account the need for heat to flow to the surface from the inside to maintain the surface temperature. This simplification has only very recently been appreciated and it is hoped to present a more complete analysis in due course. For the purpose of the analyses which follow, the heat flux to the environment is taken to be effectively zero.

3.3 Internal Effects

3.3.1 Attenuation of spatial frequencies.

The ability of any system to elucidate the detail of a stress or strain distribution can be expressed through the fidelity which it retains in measuring the amplitude of the Fourier components of that distribution as the spatial frequency increases.

We now consider the attenuation of the amplitude of the temperature fluctuation in a surface due to a distribution of stress sum which is sinusoidal in space and in time.

The general equation for the temperature θ at a point (x,y) at time t in a solid in which heat is being produced at a rate $A(x,y,z,t)$ per unit volume, is

$$k\nabla^2\theta = \rho c \frac{\partial \theta}{\partial t} - A(x,y,z,t) \qquad \dots\dots\dots\dots(1)$$

The usual expression of the thermoelastic effect is in terms of a linear relationship between the adiabatic temperature rise and the stress sum [4] (or the strain sum [3]). For a surface element, we have

$$\theta_a = K_\sigma(\sigma_1 + \sigma_2) = K_\epsilon(\epsilon_1 + \epsilon_2 + \epsilon_3)$$

Thus, for a sinusoidally varying stress sum at a point, the rate of heat production must also be sinusoidal,

$$A(t) = A_0\cos(\omega t) \qquad \dots\dots\dots(2)$$

with corresponding temperature rise

$$\theta_a = \frac{A_0}{\omega\rho c}\sin(\omega t) \qquad \dots\dots\dots(3)$$

and for such a stress sum distributed sinusoidally across the surface, the rate of heat production will be similarly sinusoidally distributed

$$A(x,t) = A_0\cos(\omega t)\cos(ax)$$

or, in two dimensions,

$$A(x,y,t) = A_0\cos(\omega t)\cos ax \cos by \qquad \dots\dots\dots(4)$$

Considering firstly the one-dimensional case; Carslaw & Jaeger [9a] give a solution[*] for a plate of width 4l, and infinitely long in the y-direction, with no heat flow across the sides, and a constant heat production rate $A_c(x)$,

$$\theta_c(x,t) = \frac{4l}{\pi^2 k}\sum_{n=1}^{\infty}\frac{1}{n^2}\left[1 - e^{\frac{-\kappa\pi^2 n^2 t}{4l^2}}\right]\cos\{\frac{n\pi x}{2l}\}\int_{-2l}^{2l}A_c(\xi)\cos\{\frac{n\pi\xi}{2l}\}d\xi \quad \dots(5)$$

For convenience of analysis we will consider spatial frequencies which are a multiple of 4l, and which maximise at x=0

$$A(x) = A_{0c}\cos(\frac{a\pi x}{2l}) \qquad a = 1,2,3\dots \qquad \dots\dots\dots(6)$$

---

[*](The integral limits of Ref.[9a] are incorrect.

510

This gives the solution as

$$\theta_c(x,t) = \frac{4l^2 A_{0C}}{k\pi^2 a^2}\left[1 - e^{-\frac{\kappa\pi^2 a^2 t}{4l^2}}\right]\cos\{\frac{a\pi x}{2l}\} \quad \ldots(7)$$

For A_c varying in time (viz. $A_0\cos(\omega t)$) we use Duhamel's Theorem [9b], which gives

$$\theta(x,t) = \int_0^t \frac{A_0\cos(\omega\lambda + \epsilon)}{\rho C}\left[e^{-\frac{\kappa\pi^2 a^2 (t-\lambda)}{4l^2}}\right]\cos\{\frac{a\pi x}{2l}\}d\lambda \quad \ldots(8)$$

$$= \frac{A_0}{\rho C}\cos\{\frac{a\pi x}{2l}\}\left[\sqrt{\frac{1}{K^2 + \omega^2}}\sin(\omega t + \gamma) - \frac{K^2}{\sqrt{K^2 + \omega^2}}e^{-Kt}\right] \quad \ldots(9)$$

where

$$K = \frac{\kappa\pi^2 a^2}{4l^2} \quad \text{and} \quad \gamma = \tan^{-1}\frac{K}{\omega}$$

The exponential term disappears for large t, leaving a spatial sinusoid oscillating sinusoidally in time, but phase-shifted in time from the oscillation of the principal stress sum, (Equation (2)).

The ratio of the amplitude of the non-adiabatic to the adiabatic temperature is seen to be

$$\frac{\theta(x,t)}{\theta_{ad}} = \frac{\omega}{\sqrt{\omega^2 + K^2}} \quad \ldots(10)$$

Equation (10) allows us to judge the fidelity for any specified material and frequencies in space and time.

In general, the stress will be a function of both x and y; it is readily confirmed that for an insulated square plate of side 4l and

$$A(x,y,t) = A_0\cos\{\frac{a\pi x}{2l}\}\cos\{\frac{b\pi y}{2l}\}\sin(\omega t)$$

the solution of equation (1) is analogous to (9), except that K is modified to

$$K_{xy} = \frac{\kappa(a^2 + b^2)\pi^2}{4l^2} \quad \ldots(11)$$

This results in a greater attenuation, since the heat is no longer restricted to leaking in only one direction.

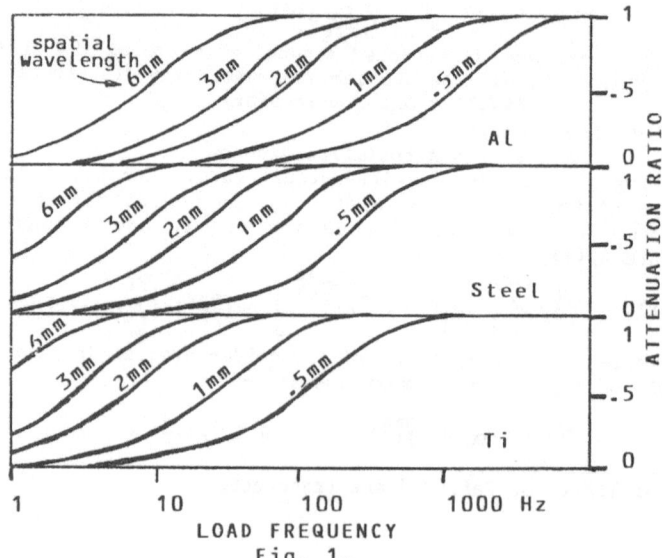

Fig. 1.

Figure 1 shows the effect for three materials. (Throughout this work, materials properties are taken from Ref.[10] unless otherwise indicated.) For the three-dimensional case, we find, mutatis mutandis,

$$K_{xyz} = \frac{\varkappa(a^2 + b^2 + c^2)\pi^2}{4l^2} \qquad ..(12)$$

For a distribution where we have the same spatial frequency occurring in two (or three) directions, the attenuation corresponds to that for a one-dimensional distribution having $\sqrt{2}$ (or $\sqrt{3}$) times the spatial frequency.

It is to be noted that, strictly speaking, these attenuations do not of themselves represent an intrinsic limitation. For if the three-dimensional Fourier transform of the spatial distribution - as measured - can be established, then appropriate matched amplification can, in principle, be applied to obtain the adiabatic distribution.

3.3.2 Effect of bending.

The case of out-of-plane bending is a particularly important one involving the z-direction. If a plate is subject to out-of-plane bending, then heat will tend to leak from the side under compressive stress to that under tension.

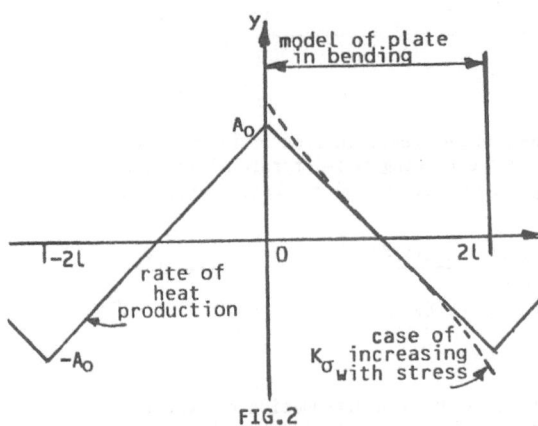

FIG.2

This time, we consider the plate to have thickness 2l in the z-direction and bent in the y-z plane, with no stress in the x-direction. The stress sum is proportional to z.

The mathematical model is equation (5) in which the rate of heat production has the saw-tooth form shown in Figure 2; it will be clear that there is no heat flow across not only the sides but also the mid-section of the model, so that the region $0 < z < 2l$ represents the plate in bending.

The solution of equation (5) is, in these circumstances, for the steady state,

$$\theta(z,t) = \frac{8A_0}{\pi^2\rho c} \sum_{n=0}^{\infty} \frac{1}{(2n+1)^2} \cos\{\frac{(2n+1)\pi z}{2l}\} \frac{\sin(\omega t + \gamma_n)}{\sqrt{K_{ri}^2 + \omega^2}} \quad ..(13)$$

This is the Fourier series for a saw-tooth but with each of the spatial components attenuated according to equation (10) and phase-shifted appropriately in time (equation (9)).

Since the addition of phase-shifted sinusoids of the same frequency results in another sinusoid, we see that for at any point z, the temperature is sinusoidal in time; we may therefore measure the amplitude as ($\sqrt{2}$.rms).

With appropriate integration and laborious algebra it can be shown that this gives,for z=0, the ratio of the non-adiabatic and the adiabatic peak values to be, on the plate surface,

$$\frac{\theta}{\theta_{ad}} = \frac{8}{\pi^2} \sqrt{\left[\sum_{n=0}^{\infty} \frac{\omega}{(K^2 + \omega^2)(2n+1)^2}\right]^2 + \frac{K}{\omega^2}\left[\sum_{n=0}^{\infty} \frac{K\omega}{(K^2 + \omega^2)(2n=1)^2}\right]^2} \quad ..(14)$$

512

This result in agreement with that developed by Belgen [2].* It is illustrated for several materials in Figure 3.

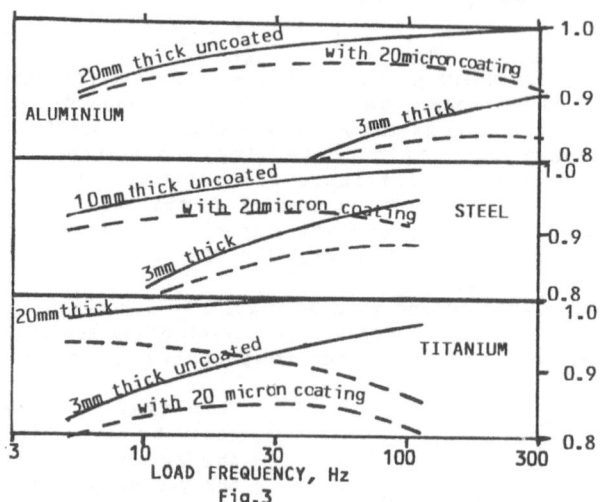

LOAD FREQUENCY, Hz

Fig.3

If we have bi-axial bending, then again, there is less restriction on the heat leakage. We may presume the corresponding two-dimensional Fourier series to be modified analogously in equation (11) to incorporate an attenuation term

$$K_{nm} = \frac{x\pi^2[(2n+1)^2 + (2m+1)^2]}{4l^2} \qquad \ldots\ldots(15)$$

but this case is not analysedin detail.

4. ANALYSIS OF COATING EFFECTS
4.1 Thermal lag

Belgen [2] uses an approximate model where the attenuation through the coating is described as a simple first-order exponential lag. Carslaw & Jaeger [9c] give an exact description of the temperature distribution through an infinitely long slab of thickness 2l (in the y-direction), with a temperature oscillation $\theta_i.sin\omega t$ on the two surfaces. Since there is no heat flow across the mid-section of such a slab (x=0), we can model the coating as the region 0<x<l. For x=0, the relationship between oscillation amplitudes on the outer and inner surfaces of the coating is

$$\frac{\theta_o}{\theta_i} = \left| \frac{cosh\mu x(1+i)}{cosh\mu l(1+i)} \right| \qquad where \quad \mu = \left[\frac{\omega}{2x} \right]^{0.5} \ldots\ldots(16)$$

(This gives a weaker attenuation than Belgen's formula)

A difficulty at this point is in the choice of coating properties and thickness for the analysis. A recommended coating [8] is Radio Spares Matt Black Heat Radiator. Little information is available on this paint, but it is apparently a carbon-filled acrylic-based formulation.

Harwood & Cummings [8] estimate a value of 1×10^{-6} m^2/s for the thermal diffusivity; Belgen [2] measured the thermal diffusivity of a (*Belgen's derivation is based upon the erroneous Ref [9a], but in the particular case in question it nevertheless yields the correct result).

35% graphite 65% acrylic as 0.5 x 10^{-6} m^2/s.

The matter of thickness is also problematical. Belgen [2] found that 7.5 microns was required to ensure opacity to the infrared that his measurement accuracy was estimated as 1.25 microns. The author has studied coatings of the recommended paint in low-power microscopy:- paint was sprayed onto glass, and the requirement for opacity to visible light was found to vary from 5 microns to close on 10 microns. (Opacity to infra-red may require thicker coating).

Tests were carried out to assess the variation in paint thickness, by spraying 100 x 150 mm glass lying vertically against a light grey background; the criterion was that opacity be ensured. The glass was then cut into 12 pieces and viewed edgewise. Coating thickness was in the range 15-25 microns over much of the surface; the minimum observed was 10 microns (with some transparency) along one edge, and a significant area near the bottom was in the range 35 to 40 microns. A 60 mm diameter white porcelain cylinder, (a drinking cup), was sprayed to the same criterion, and broken and examined. Thickness varied from 5 microns to 20 microns. (It was not possible to test for opacity).

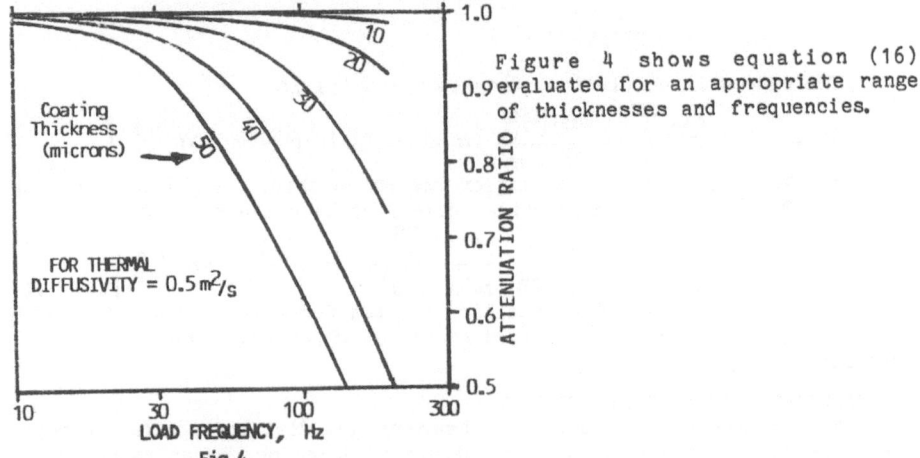

Figure 4 shows equation (16) evaluated for an appropriate range of thicknesses and frequencies.

Fig.4

4.2 Thermal "drag-down"

The effect of the coating in dragging down the temperature of the substrate surface is modelled as follows:

The substrate is considered to have heat produced within its thickness according to the stress distribution. In addition, there is considered to exist on the surface a thin layer, also of substrate material, in which there is no heat production. This inert layer is the model for the coating, and its thickness h_s is adjusted in the ratio of the volumetric specific heat capacities, $$h_S = (h_c\rho_c c_c) \div (\rho_s c_s)$$

We therefore have an inert layer of substrate of the same heat capacity as the coating; the actual substrate thickness is 2(1-h_s) so that the total thickness of the bar is 2l, as before.

The coating is taken to have a heat capacity of 0.25 (Carbon = 0.2, Acrylic = 0.3), and a density of 1500 Kg/m^3; thus for most metals the coating is modelled as a thinner layer of substrate.

Inspection indicates that if for more conductive materials we calculate the attenuation on the paint surface, and, for the less conductive, we calculate it at the paint/substrate interface then we shall, if anything, underestimate the attenuation. Also as a first approximation to find the combined effects of lag and drag-down, we might multiply the thermal lag ratio by the thermal drag-down ratio.

The above argument allows us to dispense with the need to construct a model consisting of two different materials.

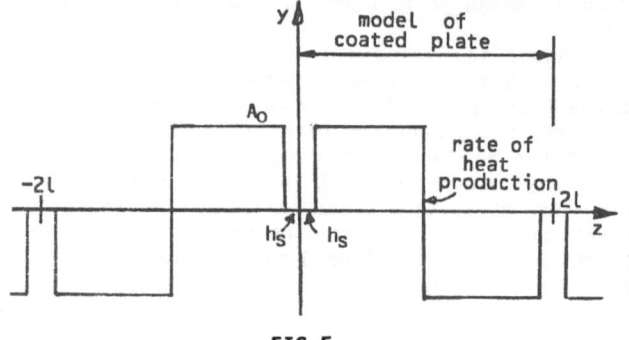

The model is shown in Figure 5. Provided that thick sections are considered, then the effect of the temperature change at the mid-section has no significant effect at the surface.

FIG.5

The appropriate solution, from equation (5) is

$$\theta(z,t) = \frac{4A_0}{\pi\rho C} \sum_{n=1}^{\infty} \frac{\cos\{\frac{n\pi z}{2l}\}}{n\sqrt{K_h^2 + \omega^2}} \sin(\omega t + \gamma_n)\left[\sin\frac{n\pi}{2} - \sin\frac{n\pi h}{2l}\right] \quad \ldots(17)$$

Ignoring this time the effect of the phase shifts (which actually make very little difference in practical cases) we have the ratio for any z,

$$\frac{\theta(z,t)}{\theta_{oad}} \doteq \frac{4}{\pi} \sum_{n=0}^{\infty} \frac{\omega\cos\{\frac{n\pi z}{2l}\}}{n\sqrt{K_h^2 + \omega^2}} \left[\sin\frac{n\pi}{2} - \sin\frac{n\pi h}{2l}\right] \quad \ldots\ldots(18)$$

We evaluate at z=0 for metallic, and z=h for non-metallic, substrate.

The results of this analysis are combined with those of the next section.

5. COMBINED DRAG-DOWN AND BENDING

The nature of the adiabatic temperature distribution in the bending case renders it particularly susceptible to drag-down by an inert coating on the tensile and compressive surfaces.

These combined effects were modelled as above, except that the rate of heat production rate in the substrate was again set proportional to z. Figure 6 shows the model.

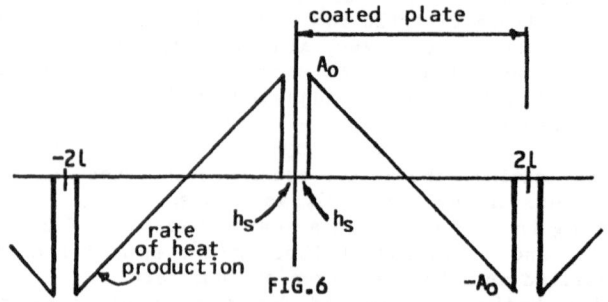

FIG.6

Again using Equation (5), we obtain for this distribution,

$$\frac{\theta(z,t)}{\theta_{oad}} = \frac{4}{\pi} \sum_{n=0} \frac{\omega \cos\left(\frac{n\pi z}{2l}\right)}{n \sqrt{K_n^2 + \omega^2}} \left[\frac{2l}{n\pi} \sin\frac{n\pi(l-h)}{2l} - (l-h)\cos\frac{n\pi(l-h)}{2l} \right] \sin\frac{n\pi}{2} \quad ..(19)$$

The outcome is shown on Figure 3, the dotted lines being the values with drag-down taken into account.

In fact it is found that the drag-down effect for 20 cm thick substrate from equation (18) is a good approximation for the case represented by equation (17), which is therefore not presented separately.

6. POSSIBILITIES OF ASSESSMENT OF SUB-SURFACE STRESS DISTRIBUTION.

Considering equation (13), we see, as indicated earlier, that the summation at any point z results in a sinusoidal oscillation with a phase-shift in time. Thus, if we consider the surface (z=0) of the plate in bending, we ought, by observing the phase-shift, to be able to assess whether bending is taking place. Knowing the geometry and the material properties, it is possible, in principle, to make quantitative estimates of the stress distribution through the thickness.

The effect is not analysed in detail in this work, but what can be said is that the phase change to be measured will increase with lower frequency, whereas phase changes due to coatings will reduce; clearly this is helpful.

However, since there are phase-lags associated also with in-plane stress variations, we would be restricted to cases where there is no in-plane variation of significance. The alternative would be to carry out three-dimensional spatial frequency analysis.

7. VARIATION OF THERMO-ELASTIC CONSTANT

Belgen [2] measured the thermoelastic constant for various metals as a function of the stress level itself. He found for steel, aluminium, and titanium, a linear increase of thermoelastic constant with stress level, (of the order of 10% change over the stress range zero to yield).

Such an increase would cause the adiabatic temperature distribution to have a weakly parabolic shape through the thickness of a bent specimen, intersecting the specimen edge at a steeper angle (e.g., dotted line in Figure 2). Such a shape would exacerbate the tendency for the edge adiabatic temperature to be attenuated. (Similarly, the attenuations shown in Figure 1 would be greater). This aspect has not been analysed in detail in this work.

Of course, such a non-linearity may of itself introduce error, unless it is calibrated into the system.

8. HEATING OF SUBSTRATE BY REFERENCE STRAIN GAUGES

It is known (e.g., [11]) that strain gauge measurements are affected by the conductivity of the substrate. With substrates of low conductance it is found that errors arise due to the strain gauge temperature rising. By the same token, heat is being conducted away from the gauge when the substrate is conductive. With an oscillating strain, we must presume that there will be some synchronous oscillation of temperature close to the gauge. A full analysis is not given here, but the following is helpful:-

Presuming that all the power generated is conducted into the substrate, we may write, for constant supply voltage,

$$W = \frac{V^2}{R}$$

$$\text{or,} \quad \Delta W = \frac{-V^2}{R}\left[\frac{\Delta R}{R}\right] = \frac{-V^2 F\epsilon}{R} \quad ......(20)$$

The heat absorbed by the substrate during a half cycle of sinusoidal strain $\epsilon_0 \sin 2\pi f t$ is

$$\Delta E = \int_0^{1/2f} \Delta \dot{W} dt = \int_0^{f/2} \frac{-V^2 F}{R} \epsilon_0 \sin 2\, ft.dt = \frac{V^2 F \epsilon_0}{\pi f R} \text{ Joules} \quad \ldots\ldots(21)$$

If we consider the heat generated to be absorbed, effectively, within a hemisphere of radius r, then we have the average temperature rise of the whole hemisphere is given by

$$\Delta T_m = \frac{.36 V^2 F \epsilon_0}{\pi^2 \rho\, CR f r^3} \quad \text{per 1000 micro-strain.}$$

Evaluating for Aluminium and Titanium (the low heat-capacity materials) we find, with the standard 120Ω gauge, with F=2, and V=1, and taking r=2 mm,

$$\Delta T_m = \frac{.1 \times 10^{-3}}{f} \, ^\circ C \qquad \text{(in both cases.)}$$

Even allowing for the undoubted non-linearity of the temperature distribution and for the use of several gauges (to evaluate $\sigma_1 + \sigma_2$), it does appear as though the effect at a distance in excess of 1 mm should be negligible for frequencies above 1c/s. At particularly low frequencies, however, this effect may have to be considered as the temperature rise will correlate perfectly with the thermoelastic signal, and therefore will be a source of error. Furthermore, if we lack the knowledge that any area has a sufficiently uniform strain, we will want to make the calibrating measurement very close to the gauge; since the temperature distribution will be distinctly exponential, a very close proximity to the gauge could give an error due to gauge heating.

9. CONCLUSIONS

1.   Analysis has been carried out on the attenuation of thermo-elastic signals due to heat leakage. Considerable attenuations can occur as the frequency of load oscillation reduces. Methods to determine the frequency required to maintain acceptable attenuation levels, for any spatial frequency and/or bending geometry, are presented for the case of constant thermoelastic constant. For materials whose thermoelastic constant increases with stress, the frequencies necessary will be somewhat higher than calculated.

2.   Analyses of thermal lag through the coating and of thermal drag-down of the substrate by the coating, have been presented. It is seen that attenuation from both of these effects increases with increasing frequency and increasing thickness of coating. As a first approximation, the combined attenuation ratio may be calculated as the product of the individual ratios. The heat leakage effect and the coating effects act in contrary senses as the frequency varies, such that for many common combinations of materials and geometry there may be no excitation frequency at which the possibility of significant attenuation is absent. Similarly, there is a limit to the detail which may reliably be obtained from a specimen.

3.   A crude analysis indicates that errors are possible due to the heating effect of the reference strain gauges in their own vicinity. This effect would, however, appear to be of little significance except at particularly low frequency or if the calibration were done very close to the gauge. High-resistance gauges will reduce the problem. A more exact analysis

would be of interest.

4. It would greatly assist in resolving the above difficulties if accurate values of the physical properties of the coating material were available. This, together with a method of measuring coating thickness would be most helpful.

5. The method has potential for providing information about the sub-surface stress distribution by measurement of the phase of the temperature oscillation.

10. REFERENCES

1. Dillon OW Jnr., Taucher TR: The Experimental Technique for Observing the Temperatures due to the Coupled Thermoelastic Effect. Int. Jnl. Solids Structures, 2, (3), 385-391, (1966).

2. Belgen MH: Infrared Radio/metric Stress Instrumentation Application Range Study. NASA Report CR-1067 (1967).

3. Biot MA: Thermoelasticity and Irreversible Thermodynamics. Jnl. App. Phys. 27 (3), 49-53 (1967).

4. Mountain DS, Webber JMB: Stress Pattern Analysis by Thermal Emission (SPATE). Proc. Soc. Photo-Opt. Engrs, Vol 164, 189-196, (1978).

5. Stanley P, Chan WK: Quantitative Stress Analysis by Means of the Thermoelastic Effect. Proc. V. Inst. Congress on Exp. Mech, 547-554, Montreal. Soc. Exp. Stress Analysis (1984).

6. Sarihan V, Oliver D, Russell SS: Thermoelastic Stress Analysis of an Automobile Engine Connecting Rod. Proc. 1985, Soc. Exp. Mech. Annual Spring Conf. 729-739, Las Vagas (1985).

7. Archer R, Razdan D: Stress Analysis of Railway Body Panels by Means of the SPATE Technique. Proc. 1985, Soc. Exp. Mech. Annual Spring Conf. 740-746, Las Vagas (1985).

8. Harwood N, Cummings WN: Calibration and Qualitative Assessment of the SPATE Measurement System. Report DE/1/85, National Engineering Laboratory, (1985).

9. Carslaw HS, Jaeger JC: Condition of Heat in Solids; 2nd Edition, Oxford University Press, 1954.
   (a) Article 3.14, Eqn 10. (b) Article 1.14, (c) Article 3.6,

10. Perry RH, Editor: Chemical Engineers' Handbook, 4th Edition, McGraw-Hill Book Co. (1963).

11. Little EG: Effects of Self Heating when using a Continuous Bridge Voltage for Strain-gauging Epoxy Models. Strain 18, (4), 131-135, (1982).

11. NOMENCLATURE

$A(\ )$ general rate of heat production per unit volume (J/m^3s)

A_c constant rate of heat production pre unit volume (J/m^3s)

A_o amplitude of sinusoidal rate of heat production per unit volume (J/m^3s)

a, b, c spatial frequency parameters in x,y,z -directions,
                               (dimensionless)

C specific heat J/Kg^oK

F strain gauge factor (dimensionless)

f frequency of oscillation (1/s)

$h, (h_c, h_s)$ thickness of layer (of coating and substrate)   (m).

i $\sqrt{-1}$

$K, K_n, K_{nm}, K_x, K_{xy}, K_{xyz}$ - defined in text (1/s)

K_σ	Thermoelastic constant, stress based ($^\circ K$ m^2/N)
K_ϵ	Thermoelastic constant, strain based ($^\circ K$)
k	thermal conductivity (J/m s $^\circ K$)
l	length parameter; various local interpretations, (m)
n	series counter
m	series counter
R	strain gauge resistance (Ω)
r	radius (m)
V	strain gauge voltage (V)
x, y, z	direction coordinates (m)
γ	defined in text (dimensionless)
∇	Laplace operator
Δ	incremental operator
ϵ_1 ϵ_2 ϵ_3	principal strains
θ	temperature (subscripted as explained locally)$^\circ K$
$\varkappa$	thermal diffusivity = $\frac{k}{\rho c}$ (m^2/s)
μ	defined in text (1/m)
ρ	density (Kg/m^3)
σ_1 σ_2	principal stresses (N/m^2)
ξ	intermediate parameter
ω	angular velocity = $2\pi f$ (1/s)
E	energy (J)
W	power (J/s)

PRESSURE-SENSITIVE PHOTOELASTIC SHEETS

A. S. REDNER

STRAINOPTIC TECHNOLOGIES, INC.
Norristown, PA 19404, U.S.A.

1. INTRODUCTION
    Measuring of the pressure distribution between two contacting bodies
is a difficult and challenging problem. The tools that are available
to researchers are based on a variety of concepts:
    –Micro-encapsulated spheres.
    A sheet containing a layer of microencapsulated color-forming
material (Fuji Film), is pressed against another sheet, containing the
color developer. The measured pressure ruptures the film, releasing
the color-forming material and producing stains in the sheet containing
the developer. The color density is indicating the measured pressure.
This technique (3).(4) can be applied to a large variety of industrial
processes. While the simplicity of this approach is very appealing,
only the maximum pressure reached during a cycle can be assessed, and
results are qualitative.
    –Combination imprinter – photoelastic polyurethane sheet
    Arcan and Brull developed a pressure-pad consisting of a layer of
small balls pressed against a photoelastic sheet (2). As a result of
the stress field created by each ball, a set of circular fringes is
observed. The diameter of the first order fringe increases as the
pressure increases.
    The Arcan-Brull concept is quantitative, but severely limited.
The points of information are distant, separated by 12mm or more, and
the response is highly non-linear. The retrieval of information requires
a microscope examination of circles, representing hours of labor for
each photograph obtained.
    The imprinter concept was applied in a foot-print platform, but
its practical use for diagnostic purposes was severely limited, because
of the information retrieval cost.
    –Photoplastic photoelastic sheet
    Arcan and Zandman developed "Memory" sheets, using a photoplastic
material, to analyze dental occlusion problems. The method was based
on the ability of a photoplastic material to retain the photoelastic
pattern after the pressure is removed. The patient was required to
maintain the "bite" for approximately 10 seconds after which time the
photoplastic sheet was removed for examination.
    The method found only limited interest, because of the inherent
limitations of this concept: The photoplastic sheet responds to the
difference of principal strains in its plane, rather than pressure.
In addition, material properties are limiting the minimum and maximum
response.
    Only the maximum pressure attained during the cycle can be estimated
if a relation could be devised correlating the birefringence to the
pressure.

Wieringa, H (ed), Experimental Stress Analysis.
© *1986. Martinus Nijhoff Publishers, Dordrecht.*

Some additional research was conducted more recently (5) (7) on this system, and the method was abandoned, due to the difficulties in finding such a correlation. (8)
  –Other techniques
  More recently, an interesting method was proposed by Betts et al, (6) using the variation of critical light reflection angles. The method appears interesting and potentially useful; so far, efforts to develop a quantitative tool based on this concept were not successful. Of all the methods discussed above only the Fuji Film is commercially produced and marketed.

2. NEW METHOD: PRESSURE–SENSITIVE SHEET

A photoelastic sheet responds to stresses applied in its plane, xy. A pressure P, or a stress σ_z perpendicular to its plane, will not produce any birefringence, since the strains in the plane of the sheet are equal in all directions, and their differences $\varepsilon_x-\varepsilon_y$ are zero.
  The new concept proposed here renders the photoelastic sheet responsive to pressure.
  To accomplish this objective, two equivalent methods could be used:
a) An anisotropic photoelastic material could be used. As result of a mechanical or optical anisotropy, $\varepsilon_x \neq \varepsilon_y$ thus creating photoelastic response to pressure.
b) A geometrically anisotropic sheet can be designed whereby the two mutually perpendicular strains are controlled using a geometry of grooves to enhance unidirectional strain response (1).
  In the research work described here, the approach (b) above was selected. An example of geometry that can be easily evaluated is a grooved sheet:

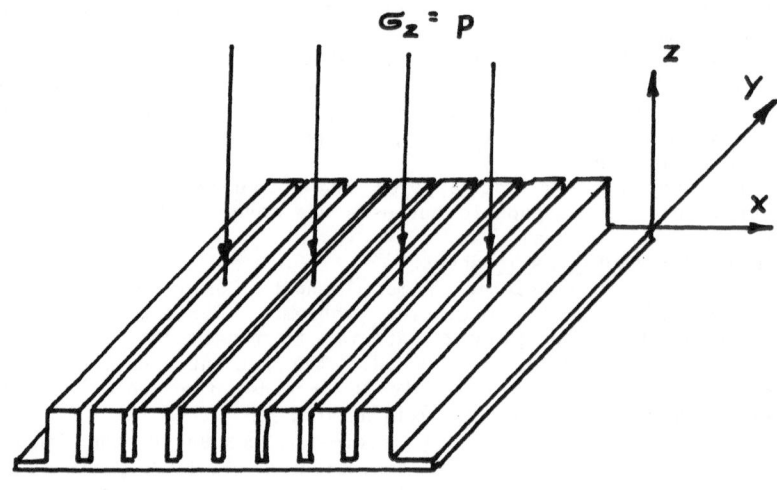

FIGURE 1

In this design, the stiffness in x and y directions is substantially different: The strain ε_y due to the applied pressure is negligible when compared to ε_x. The photoelastic response to the pressure p can be evaluated approximately as follows: (See Figure 1).

$$\varepsilon_y = 0 \text{ (negligible)}$$

$$\tag{1}$$

$$\sigma_z = p$$

and $\sigma_x = 0$ (a free Boundary)

It follows from the above that:

$$\varepsilon_x = -\frac{\nu}{E} (\sigma_z + \sigma_y)$$

$$\tag{2}$$

and $\sigma_y = \nu\sigma_z$

The observed fringe order N is proportional to the difference of principal strains in the plane perpendicular to the interrogating ray

$$N = \frac{1}{f} (\varepsilon_x - \varepsilon_y)$$

$$\tag{3}$$

combining (1) (2) and (3) above yields :

$$N = \frac{\nu(1 + \nu)}{fE} \cdot \sigma_z = \frac{\nu(1 + \nu)}{fE} \times P$$

where N is the measured photoelastic response ,and

f   is the strain fringe value
E, ν are the elastic constant of material used .

Since the stress σ_z represents the measured pressure p, we have:
N = (material constant) x (pressure)
The above equation proves that the grooved sheet response is indeed directly proportional to the pressure applied in the direction perpendicular to the sheet plane.
Making the grooves sufficiently narrow, one obtains a quasi-continuous photoelastic pattern, composed of closely spaced lines of information.
The information provides a full-field topography of the contact or foot-print pressure, and is measurable quantitatively at any desired point. The thickness and material selection determines the output. The overall sensitivity can be easily adjusted to provide a simple visual interpretation, using a color vs. pressure chart.

The information that is displayed in real time, can be stored using photographic techniques or recorded on videotape for static or dynamic interpretation. A totally automated readout (9) can be easily incorporated.

3. SYSTEM CONFIGURATION

The proposed system requires a nearly-normal light incidence access to the observed sheet. A typical arrangement is shown on the Figure 2 below:

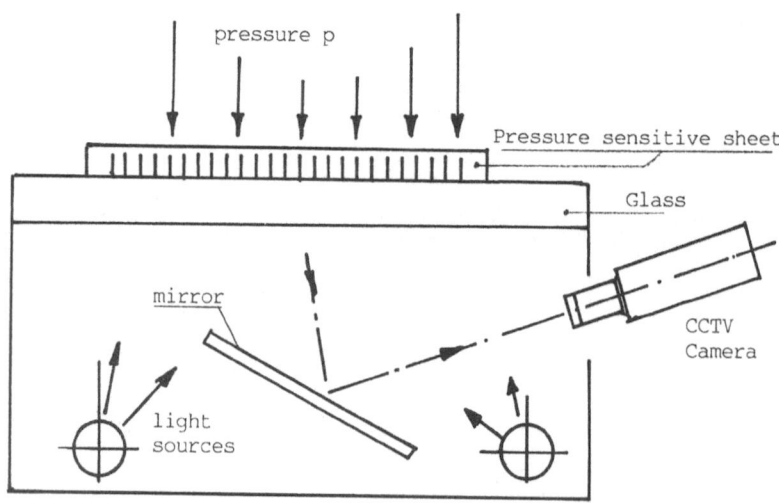

FIGURE 2

The polarizing elements can be placed over the source(s) and the camera, (2a) or directly underneath the observed sheet.

The groove geometry design allows some freedom in modifying the response linearity, mostly when dealing with soft, highly deformable polymers. The research on this topic is now in progress. The range of measured pressure is not limited and depends on the choice of photoelastic material.

4. CONCLUSIONS

A totally new type of pressure-sensitive device is now available to the experimentalists and practitioners in the medical field and industry. The ability to see the pressure distribution directly will not only provide the desired quantitative tool for diagnosis of foot disorders, but also help the shoe designer to positively and quantitatively assess the merits of a new design. It will permit the automotive tire designer to obtain the quantitative information on the tire/road interface pressure, permit the piping industry to develop criteria on flange pressures and required bolt spacings. The method is also applicable to analyze cylinder crowns, gear pressure profiles and a large variety of pressure distribution problems.

REFERENCES
1. O'Regan, R., "New Method for Determining Strain on the Surface of a Body with Photoelastic Coatings", Experimental Mechanics, 5,8,241-246, 1965.
2. Arcan, M. and Brull, M.A., "A Fundamental Characteristic of the Human Body and Foot, the Foot-Ground Pressure Pattern", J. Biomechanics 9, 453-457, 1976.
3. "Fujifilm Prescale - Pressure Detecting Sheet", Technical Bulletin Fuji Photo Film company, Tokyo, Japan (1980).
4. Schopf, H.J., and Karg, E., "Ermittlung von Pressungsuer-teilungen an Kentakt-und Dichtflachen" Messen und Pruffen, 6, pp. 388-395, 1980.
5. Arcan, M. and Zandman, F., "Mechanics of Contact and Memorized Birefringence", C.R. Seances Academie des Sciences, Series B, Jan. 1980, Paris.
6. Betts, R.P., Austin, I.G., Duckworth, T., Crocker, S. and Moore, S., "Critical Light Reflection at a Plastic-Glass Interface and its Application to Foot Pressure Measurements", J. Med. Eng. Tech. 4, 136-142, 1980.
7. Arcan, M. and Zandman, F., "A Method for InVivo Quantitative Occlusal Strain and Stress Analysis", J. Biomechanics, Vol 17,2, pp. 67-79, 1984.
8. Redner, A.S. and Lorek, C., "Evaluation of Contact Stresses from Photocclusion Records", Research Report (internal), May 1982.
9. Redner, A.S., "Photoelastic Measurements by Means of Computer-Assisted Spectral contents Analysis", Experimental Mechanics, 25,2, pp. 148-153, 1985.

THE BOUNDARY ELEMENT METHOD, AN ACCELERATING CALCULATION PROCESS USED IN EXPERIMENTAL STRESS ANALYSIS

H. MARWITZ, W. KIZLER and K. ENSLIN

Daimler-Benz AG, Abt. V1PS, Mercedesstraße,
7000 Stuttgart 60, West-Germany

SUMMARY

The user-oriented computer program DBETSY makes the Boundary Element Method (BEM) a valuable enrichment to the experimental procedures for hybrid stress analysis (1,2). It can be used by experimenters just as effectively as the proven steps for intermediate optimization: plane photoelasticity analysis, table values or analytic methods.
The advantages of integrated experimental calculation with BEM for stress analysis are shown using an axially symmetrical automobile component.

1. INTRODUCTION

Increasingly, stress analysis requires shorter response times in practical industrial applications. In addition to the qualities: accuracy, reliability, completeness of the information and efficiency of the method, speed is becoming more important.
In the development phase, stress analysis uses the model technique. In addition to the experimental procedures such as 3-dimensional photoelasticity - which is characterized by high reliability and acceptance as well as high time requirements - calculatory methods are available, which offer advantages particularly in terms of the manifold application possibilities and the diversity of their models as well as their facilities for representation and rapidity. In each case the responsible Research & Development department decides which method is to be used according to their experience with the procedures and personnel.
Many efforts have been made (3) to retain the advantages of the experimental methods while eliminating their disadvantages through the use of numerical methods. Nevertheless the hybrid method is still not standard.
The reasons for this to date are the high demands encountered in handling the required tools, the high expenditure, as well as the established organizational structures and patterns of thinking found everywhere.
For several years now, the Boundary Element Method has been installed at Daimler-Benz with the program system DBETSY. The program exhibits new perspectives opening the way to a "one-man hybrid procedure". The following descriptions provide a general view of the possibilities offered by this method as well as its usefulness for experimental stress analysis.

2. THE TOOLS FOR THE HYBRID METHOD

In the design phase of a component, when a concrete form exists only on paper, models are used. For stress analysis these models may be of a physical nature (e.g. three-dimensional photoelastic stress models) as well as of an abstract nature (e.g. in the form of numerical models). Both types of models fulfill the same purpose.

If the two model techniques are compared, in order to identify their strengths and weaknesses, the discussion can be limited primarily to two criteria:
- Reliability of the information obtained
- Time required for results

Photoelastic stress analysis offers optimum prerequisites for reliability of the information obtained:
- Equality of the total volume with uniformly equal linear-elastic material characteristics
- Automatically correct force transfer for contact problems (the stiffness of the microstructure automatically assures proper force transfer due to its natural deformation) and for surface and volume loads (internal pressure and centrifugal forces)
- Physical production of the models allow the shapes to be simulated accurately rights up to smallest details in terms of exact shape as well as stiffness through the selection of suitable materials
- Proper simulation at a justifiable expense of even complicated loading characteristics by providing the model to be studied with modelled attachment elements until the interfaces become clear. Improper loads are generally indicated by fracture or implausible stress distribution.
- Representations of the interesting points exact in terms of location and size. The frozen model preserves the stressed state. Even the complete frozen model signals extreme loads in polarized light, which can then be analyzed specifically by sectioning and slicing.

However these clear advantages are offset by the considerable time required (Fig. 1): evaluation is only possible after completion of the master form, manufacture of the casting mould, casting, hardening, machining, design and construction of the load applying device, stress-freezing, sectioning and slicing.

For certain problems intermediate optimization can be accomplished

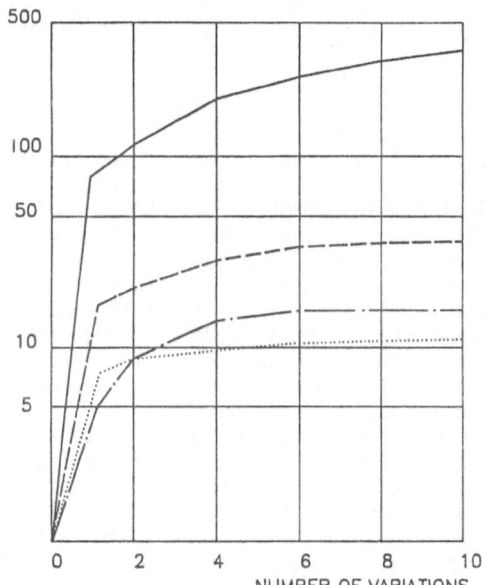

CONSUMTION OF TIME FOR THE OPTIMATION OF A FLANGE [h]

NUMBER OF VARIATIONS

3D–PHOTOELASTICITY—— FINITE ELEMENTS————
2D–PHOTOELASTICITY—·— BOUNDARY ELEMENTS······

FIGURE 1

very quickly (Fig. 1) and elegantly with the plane photoelastic stress analysis using a tracer milling machine. However, this possibility is not universally applicable.

The numerical procedures, until now primarily the Finite Element Method and recently the Boundary Element Method as well, offer the advantage of considerably reducing the time required for a solution, particularly axially symmetrical problems. With DBETSY the work required for discretization is significantly reduced by definition of the surface only. Moreover DBETSY is installed at Daimler-Benz in such a manner, that interactive operation is possible from every terminal and the software is prepared so that the familiarization time remains within acceptable limits. Today an experimenter can therefore work with and analyze his physical and his numerical model from his own laboratory.

This eliminates obstacles presented for the experimenter by FEM. This method is reserved for the design & calculation department due to the extensive investments for hardware and software, the great personnel expenses and the high degree of specialization of the programs. They were therefore used too infrequently as a computation method for hybrid stress analysis, because the experimenter is limited by the capacity and priority of the calculation department for processing his problem and can therefore not continually utilize the advantage of the quick reaction time.

The significant disadvantage of all computation methods, however, is the reliability of the results: any model is only as good as the engineer's ingenuity. The more abstract the starting basis for the shape of the model, the greater the corresponding demands. The most difficult situation is presented for the 'model designer', whose elements are defined only as geometric areas with a certain range for displacement.
- In dividing up the model the designer already assumes an expected result. If he does not select the structure at the true points of the local maxima in such a manner that these stresses can also be measured here, the result will be erroneous qualitatively and quantitatively.
- In the attempt to limit personnel expenditures and computation time, simplifications are made. This affects the shape, the material characteristics and, to a significant degree, the force transfer.
- No self-controlling indication of errors neither in the properties of materials nor in the quantity and quality of the loads.
- Complicated contact and load transfer problems depend highly on the skill of the designer, particularly in the case of load transfer via rubber.
- The influential stiffness at the boundary conditions must either be estimated or determined experimentally.

Starting from this situation primary information is worked out in practice through redundancy of computations and experiments. The hybrid method is therefore the logical consequence.

3. NUMERICAL METHOD
3.1 Physical Principle of BEM
BEM is based on the fact that the linear-elastic behavior in an area is determined by description of singular boundary effects. In comparison to the finite area problem formulation with the FE methods, description of the problem using the boundary problem formulation is reduced by one dimension. In this manner the degrees of freedom are reduced, which, among other things, facilitates numerical solution and for the user means that it is only necessary to set up and enter the geometrical data for the surface of the component. For two-dimensional problems this means that the surface description is reduced to enter the boundary curve.

The boundary curve is allocated source points, at which weighted, singular force solutions are applied. Each such singular force results in a boundary stress vector and a displacement vector at another point. On the basis of this familiar solution a boundary integral equation is derived according to the principle of virtual work (4), in which the real boundary stress and displacement values are included in the integrals as stresses for this solution. This integral equation forms the basic equation for BEM in linear elastostatics and directly links the boundary stresses and boundary deformations with one another in the formulation used here.

3.2 BETSY - DBETSY

The Technical University of Munich has developed the program package BETSY (Boundary Element code for Thermoelastic SYstems) for use of BEM in linear elasticity problems (4). Subdivided into individual routines BETSY presently solves the following problems (Fig. 2):

Stress and deformation for

- Plane and axisymmetrical geometry with symmetrical and asymmetrical loads
- Spatial geometry with spatial loads

One part of the program realizes the substructure technique (mentally cutting the component into a number of subsections) as well as processes the thermoelastic loading via stationary temperature fields. In addition volume force loads such as the force of gravity and centrifugal forces are permissible.

BETSY was expanded by Daimler-Benz to DBETSY (5). As long as the applications are limited to plane and axisymmetrical components, its use is extremely simple thanks to the convenient pre- and postprocessors.

A more precise time comparison between NASTRAN and BETSY is described in (6) using a plane model (Fig. 3). The great time advantage of BEM of 10 : 1 results primarily from the simpler and quicker dis-

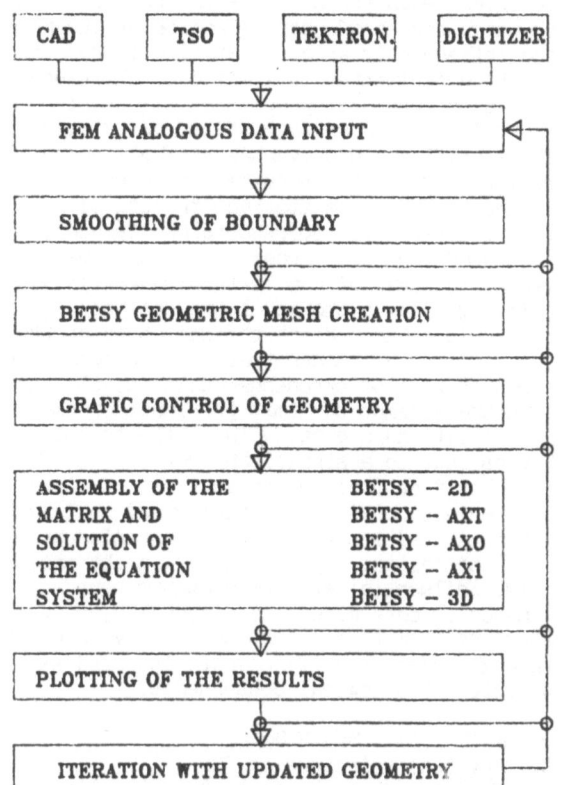

FIGURE 2

cretization with BEM in comparison to FEM. With stress concentration problems the automatic mesh generation possible with FEM presents problems: good results can be obtained only when the user develops the most favorable network corresponding to the expected force flux. DBETSY as well as BETSY are ideally suited supplemental aids even for the experimenter with experience of numerical methods as he can check the results with his own methods.

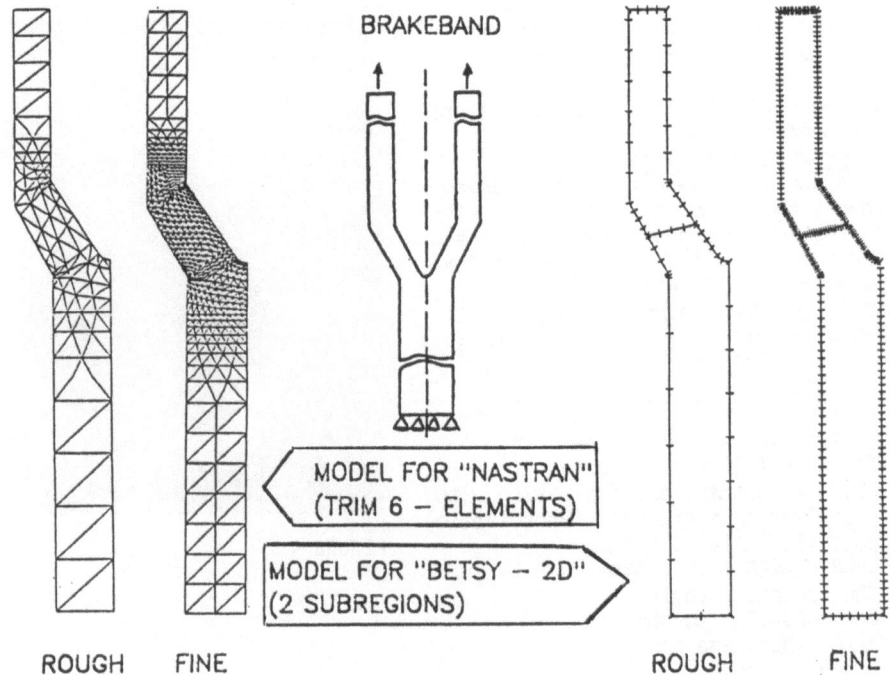

FIGURE 3

3.3 Tips for Using DBETSY

BEM is particularly suitable for compact components (small ratio between component surface and volume). In such cases the calculation accuracy is the highest and the time advantage greatest in comparison to FEM. The stresses and deformations on the surface are the focal point of calculation; stresses on the inside of the component can also be determined using the substructure technique.

In the case of torsional loads the system does not react sensitively to different fine meshes. However, for bending loads the convergence must be checked under all circumstances: the resulting stress values depend to a very great extent on the fineness of the mesh, particularly for elongated components. While a number of elements are usually required anyway in notches for precise description of the stress characteristics, particular attention must be given to relatively fine discretization for bending loads beyond the notch position as well, in order to also assure proper transfer of the bending load up to the notch. In terms of the time required for the mesh generation fine subdivision does not take any longer than coarse subdivision (maximum 200 nodal points). Only the CPU-time is longer due to the increasing number of degrees of freedom. However, this can be counteracted by dividing the component up into subsections. When this is done, the separation lines must not be too close to the notch points, as otherwise the resulting stress values could be affected locally. In such cases another computation without subsections is advantageous for parts with unknown characteristics.

4. PHOTOELASTICITY

In the continuing attempt to make 3-dimensional photoelastic stress analysis competitive in terms of the time requirement, two processing steps have been shortened:

4.1 Grinding the Slices

The use of a plane grinding machine reduces the grinding time by 70 % with considerably improved thickness consistency of the slice (+/- 0.015 mm for 150 mm section length). 1/2 hour is sufficient for finishing a work plate (250 x 150 mm^2) complete with slices ready for evaluation.

4.2 Oblique Incidence

An new device for oblique incidence with two plano-convex lenses (Fig. 4) offers the following advantages:
- Slice adjustment and evaluation with polarization microscope (Photolastic) possible while sitting.
- Lens design eliminates the incidence angle corresponding to the slice thickness and notch radius.

The other steps required for this method from production of the mould to the frozen model correspond to the standard and further shortening is hardly possible.

5. USE OF HYBRID METHOD

5.1 Purpose

The test sequence is described for a drive flange for a passenger car differential (Fig. 5). The purpose of the study was to optimize nine different versions in terms of strength and weight and, if possible, reduce the number (Fig. 6).

FIGURE 4

FIGURE 4

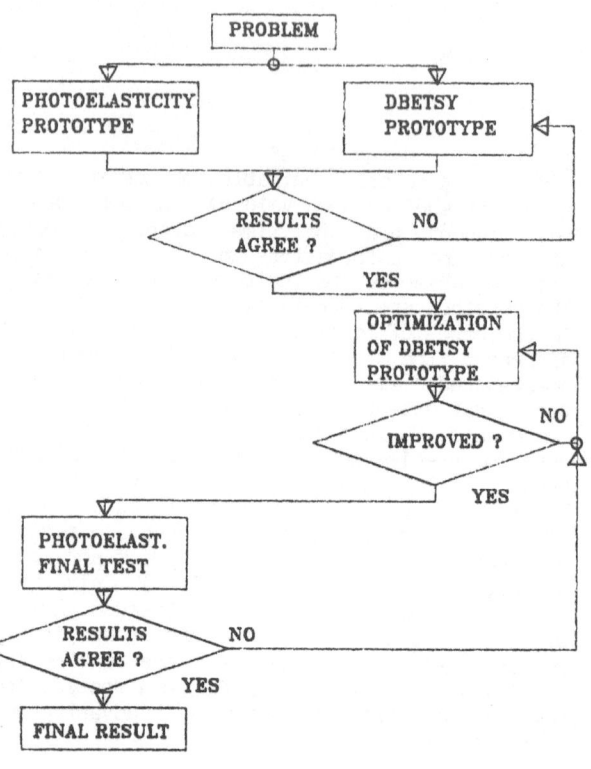

FIGURE 5

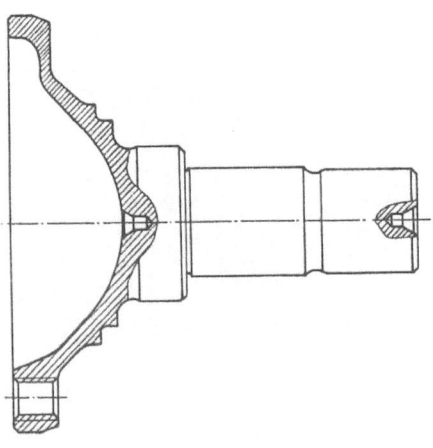
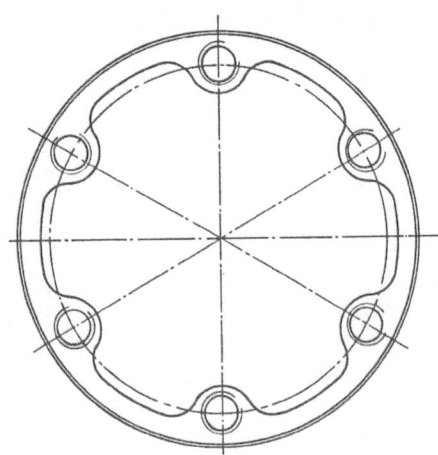

FIGURE 6

5.2 Experiment Plan

Since load transfers can generally
be simulated more realistically in an
experiment (due to the real existence
of the actual stiffness) and errors
in the experiment are automatically
indicated as implausible results,
basic experimentation is a criterion
for creation of the theoretical
models. Ten workdays must be scheduled
for a 3-dimensional photoelasticity
analysis experiment for the first
flange, while initial computation for
the flange with DBETSY, including CPU
time and result checking, requires
approximately 4 h. Modifications of
the radii can be accomplished with
DBETSY in 10 min. (Fig.1).

It is obvious that photoelastic experiments must be limited to that
which is indispensable (Fig. 7).
When DBETSY is used the number of
versions in secondary.

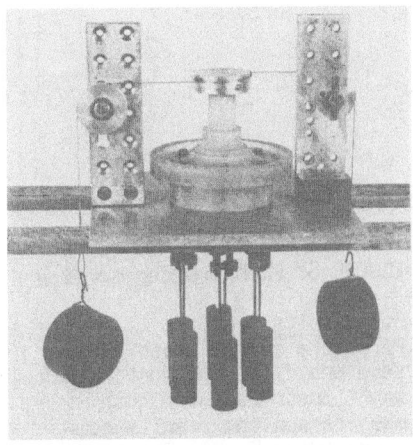

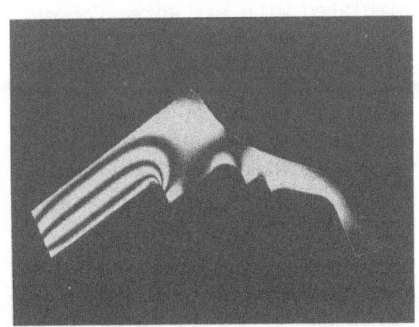

FIGURE 7

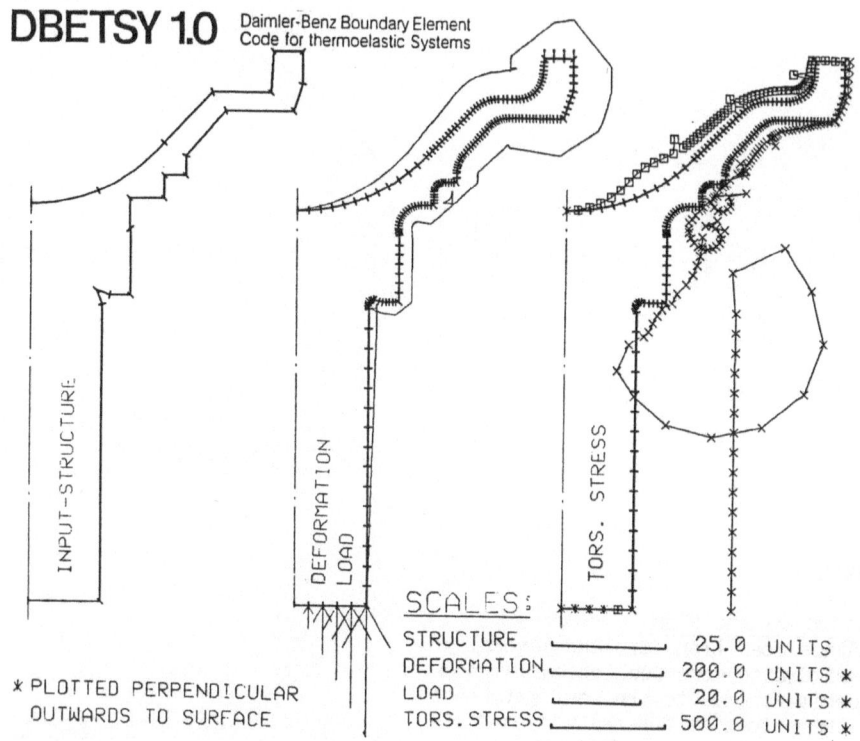

DBETSY 1.0 Daimler-Benz Boundary Element Code for thermoelastic Systems

INPUT-STRUCTURE

DEFORMATION LOAD

TORS. STRESS

* PLOTTED PERPENDICULAR
OUTWARDS TO SURFACE

SCALES:

STRUCTURE		25.0	UNITS
DEFORMATION		200.0	UNITS *
LOAD		20.0	UNITS *
TORS. STRESS		500.0	UNITS *

FIGURE 8: Flange under torsion

5.3 Results

First the actual state of the versions differing most highly was illustrated with both methods and the results compared (Fig. 8, 9). The load was distributed symmetrically for the calculated model and simulated with six screw forces for the photoelastic model. Nevertheless the coincidence was good.

The notch radius with the highest load (relief grove on end of shaft) is still in the hardened zone of the spline and can therefore withstand a higher permanent load than the radii in the area of the cup.

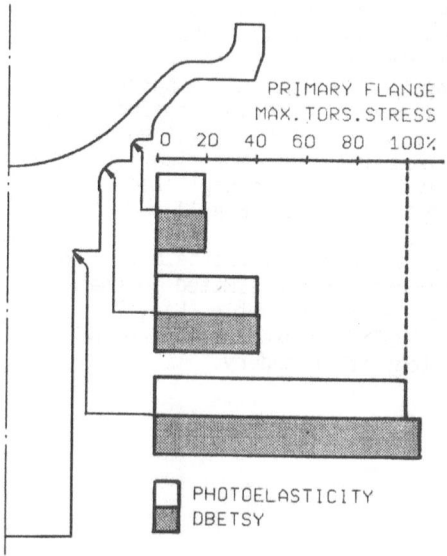

PRIMARY FLANGE
MAX. TORS. STRESS

0 20 40 60 80 100%

☐ PHOTOELASTICITY
▨ DBETSY

FIGURE 9

The following versions were
studied with DBETSY:

- Tolerance limits at the
  highly loaded radius
- Thinner walls on the low
  loaded cup with two
  versions
- Drawing the inner contour
  more deeply in the area
  of the shaft

Varying the radii as well
as making the cup wall
thinner changed the design
to such a small extent that
a final test is not required. A final test with a
photoelatic model was made
for a reduced weight
version. This showed good
correspondence to the
calculation (Fig. 10).

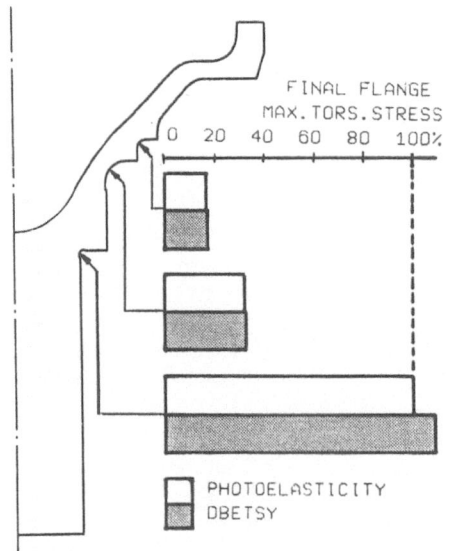

FIGURE 10

REFERENCES

1. H.-J. Schöpf, Trends der Spannungs- und Verformungsanalyse im
   Automobilbau, VDI-Bericht 514, 1984

2. E. Daffner, Untersuchung rotationssymmetrischer Probleme unter
   ökonomischer Verwendung experimenteller und rechnerischer Verfahren,
   GESA-Symposium 1979, Braunschweig

3. Laermann, K. H., Entwicklung und Bedeutung der Modellstatik in Theorie
   und Praxis, VI. Internationale Konferenz Experimentelle Spannungsanalyse, München, FRG, 18. - 22.09.1978

4. Forschungsberichte der Forschungsvereinigung
   Verbrennungskraftmachinen e. V., Frankfurt 1982
   Heft 310 - 1, 2, 3 Abschlußberichte - theoretische Grundlagen von BETSY

5. Möhrmann, W., "DBETSY - die Boundary Element Methode in der
   industriellen Berechnung",
   VDI-Tagung "Berechnung im Automobilbau", Fellbach 10/84,
   VDI-Bericht 527, 1984

6. Radaj, D., Möhrmann, W., Schilberth, G.,
   Economy and convergence of notch stress analysis using boundary and
   finite element methods. Int. J. Num. Meth. Eng., Vol. 20 (1984),
   p. 565/572.

COMPUTER-AIDED MEASUREMENT OF RELATIVE RETARDATIONS IN PLANE PHOTOELASTI-
CITY

F.W. HECKER and B. MORCHE

MECHANIK-ZENTRUM, TECHNISCHE UNIVERSITÄT BRAUNSCHWEIG
D 3300 BRAUNSCHWEIG, FEDERAL REPUBLIC OF GERMANY

1. INTRODUCTION

Computer-aided methods for evaluation of photoelastic patterns use video
technique and digital image processing. They are based on localization of
fringe centers [1, 2, 3, 4]. Neighbourhood operations are needed to reduce
the influence of nonuniform illumination, inhomogeneous optical components
and models, etc. Fractional orders of the relative retardation and of the
isoclinic parameter at points between the fringes are computed by spline-
functions, if the components of the plane stress state have to be derived
from photoelastic data. In the following a method is proposed, which enables the complete extraction of photoelastic information at local picture
elements(pixel) from series of related images of the same stress state.
For this purpose the well-established phase-shifting technique (see e.g.
[5, 6]) was modified to meet the special requirements of photoelastic
patterns.

2. BASIC CONCEPT
2.1 Phase-shifting interferometry

The intensity of an interference pattern depends sinusoidally on the relative phase δ between the two interfering waves. Considering absolute phase
shifts Δ, introduced into the reference wave with respect to a zero configuration, the local intensity is given by

$$I = I_m + I_a \cos(\delta + \Delta) , \tag{1}$$

where I_m denotes the local mean illumination. I_a is the local amplitude of
the intensity component, which alternates sinusoidally when the optical
path of the reference wave is linearly shifted. The local extreme values
of the intensity, I_ℓ and I_u, and the local constrast K, are related to the
quantities of Eq. (1) by

$$I_\ell = I_m - I_a, \ I_u = I_m + I_a, \ K = I_a/I_m . \tag{2}$$

The basic idea of phase-shifting, introduced by Carré [7], is simple: to
change the absolute phase of the reference wave by equal steps Δ and to
measure the local intensity I after each phase step. At least three readings are necessary for evaluation of the three unknown values I_m, I_a and δ.
Typical steps are $\pi/2$, [7], and $2\pi/3$, [5, 6]. The main advantage of the
technique is, that the interference phase δ can be calculated at any point
in the interferogram, independently of local mean illumination and local
fringe contrast.

This method is sometimes called phase-stepping, while the name phase-
shifting is reserved to averaging procedures, where the reference phase is

Wieringa, H (ed), Experimental Stress Analysis.
© *1986. Martinus Nijhoff Publishers, Dordrecht.*

continuously shifted in time and three or four "integrated buckets" are recorded [8].

2.2 Phase measurement in photoelasticity

In plane photoelasticity, two data fields are to be determined: relative phase δ and isoclinic parameter α. The relative phases δ of the two refracted waves, transmitted through the specimen, are related to the fringe order m and the principal-stress difference $(\sigma_I - \sigma_{II})$ by

$$\delta = 2\pi\, m = b(\sigma_I - \sigma_{II})/S , \tag{3}$$

where b denotes the thickness of the specimen. The photoelastic coefficient S depends on the wavelength of the monochromatic light source. α denotes the inclination of the principal directions with respect to the chosen coordinate system, Fig. 1.

A standard diffused-light polariscope consists of a circular polarizer CP, a quarter-wave plate Q, and a plane analyzer A, Fig. 1, where the axes of Q and A can be adjusted independently of each other, e.g. for compensation. For the most general set-up, the intensity of the light emerging from the analyzer is given by [9]

$$I = I_m + I_a\left[\sin2(\psi-\varphi)\ \cos\delta + \cos2(\psi-\varphi)\ \sin2(\varphi-\alpha)\ \sin\delta\right], \tag{4}$$

when left circularly polarized monochromatic light is used. Examination of Eq. (4) indicates, that sinusoidal manipulation of the local intensity can be accomplished by rotation of the quarter-wave plate Q and of the plane analyzer A. The well-known compensation methods and the dynamic ac-phase detection techniques (see [10]) utilize this phenomenon. For convenience, two special systems are investigated. Using the circular-plane polariscope, [11], which is realized by $\varphi = \psi$, Fig. 1, one arrives at the reduced relationship

$$I = I_m + I_a \cdot \sin\delta \cdot \sin2(\psi - \alpha) . \tag{5}$$

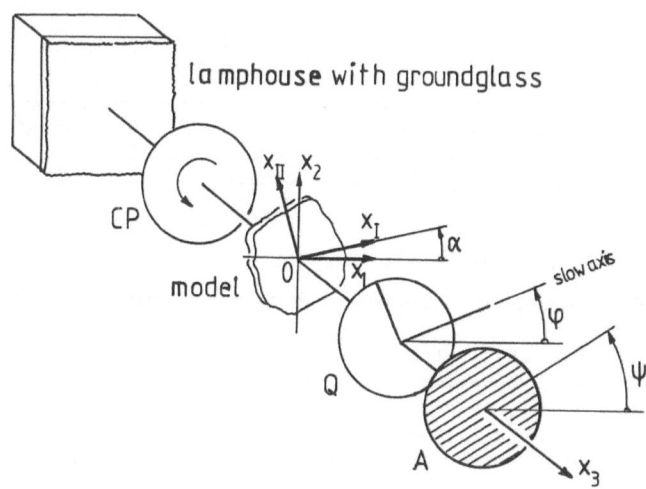

FIGURE 1. Diffused-light polariscope with circular polarizer, quarter-wave plate and plane analyzer.

Three different values of ψ are sufficient for the determination of the absolute value $(I_a \cdot |\sin\delta|)$. Setting $\psi = 0$, $\pi/3$ and $-\pi/3$ yields the relations

$$I_1 = I_m + I_a \sin\delta \; (\qquad\qquad -2\sin 2\alpha)/2 \; , \tag{6}$$

$$I_2 = I_m + I_a \sin\delta \; (+ \sqrt{3} \cos 2\alpha + \sin 2\alpha)/2 \; , \tag{7}$$

$$I_3 = I_m + I_a \sin\delta \; (- \sqrt{3} \cos 2\alpha + \sin 2\alpha)/2 \; , \tag{8}$$

respectively. Two equations, which do not contain I_m, are derived:

$$I_a \sin\delta \cdot \sin 2a = (I_2 + I_3 - 2I_1)/3 \tag{9}$$

$$I_a \sin\delta \cdot \cos 2\alpha = (I_2 - I_3)/\sqrt{3} \; , \tag{10}$$

and after squaring and adding up one obtains

$$I_a \; |\sin\delta| \; = \; \sqrt{2\left[(I_1 - I_2)^2 + (I_2 - I_3)^2 + (I_3 - I_1)^2 \right]/3} \; . \tag{11}$$

A second set of equations is gained by means of the circular polariscope with $\varphi = \psi - \pi/4$ and $\varphi = \psi + \pi/4$, respectively, where the simple expressions

$$I_4 = I_m + I_a \cos\delta \; , \tag{12}$$

$$I_5 = I_m - I_a \cos\delta \; , \tag{13}$$

hold. Eqs. (12) and (13) yield

$$I_a \cos\delta = (I_4 - I_5)/2 \; , \tag{14}$$

and finally the raw phase data can be calculated by

$$\tilde{\delta} = \arctan \frac{2\sqrt{2}\sqrt{(I_1 - I_2)^2 + (I_2 - I_3)^2 + (I_3 - I_1)^2}}{3(I_4 - I_5)} \; . \tag{15}$$

An example is given below.

2.3 Measurement of the isoclinic parameter in photoelasticity

Information about the isoclinic parameter α is preferably collected in a white-light polariscope with crossed plane polarizers. The light intensity of the photoelastic pattern may be found by summing up the effects of discretized elements of the continuous light source spectrum. The resultant irradiance at a point of the sensing device may be presented as

$$\bar{I} = \bar{I}_\ell + \bar{I}_a \cdot \bar{g} \cdot \left[1 - \cos 4 \, (\psi - \alpha) \right] \; , \tag{16}$$

where ψ denotes the angular position of the second plane polarizer. $\bar{g}$ is a positive-definite function of the local principal-stress difference, which vanishes at singular points $(\sigma_I - \sigma_{II} = 0)$ only, see e.g. [3]. By selecting $\psi = 0$, $\pi/6$ and $-\pi/6$, respectively, the corresponding irradiances read

$$\bar{I}_1 = \bar{I}_\ell + \bar{I}_a \cdot \bar{g} \cdot \left[1 + (\qquad\qquad - 2 \cos 4\alpha)/2 \right] \; , \tag{17}$$

$$\bar{I}_2 = \bar{I}_\ell + \bar{I}_a \cdot \bar{g} \cdot \left[1 + (\; \sqrt{3} \sin 4\alpha + \cos 4\alpha)/2 \right] \; , \tag{18}$$

$$\bar{I}_3 = \bar{I}_\ell + \bar{I}_a \cdot \bar{g} \cdot \left[1 + (-\sqrt{3} \sin 4\alpha + \cos 4\alpha)/2 \right] \; , \tag{19}$$

and the raw data of the isoclinic parameter are computed by

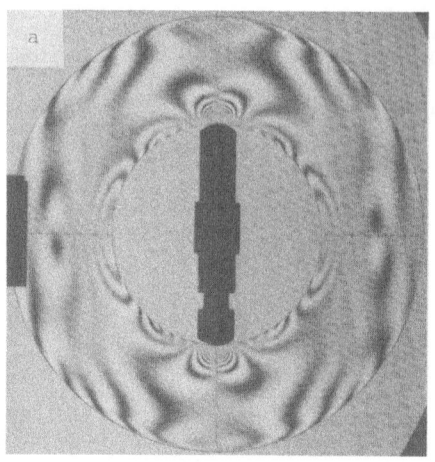

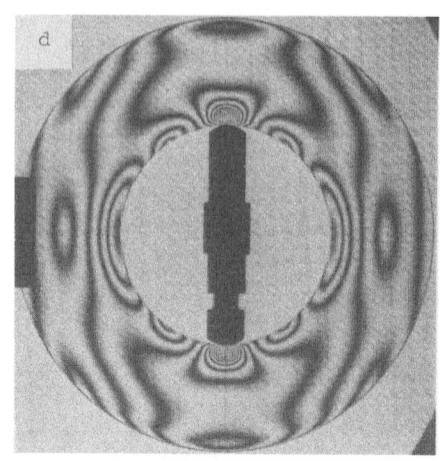

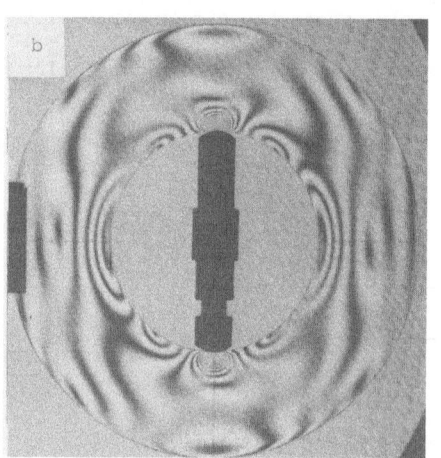

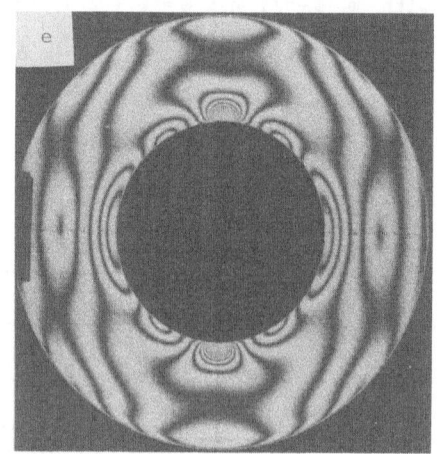

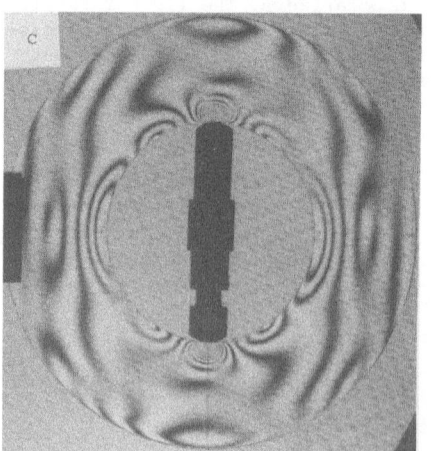

FIGURE 2. Photoelastic patterns of a circular ring in diametral expansion. a) to c) circular-plane polariscope with $\varphi = \psi = 0°$, 60° and -60°, respectively. d) and e) circular polariscope with light-field and dark-field fringes, respectively.

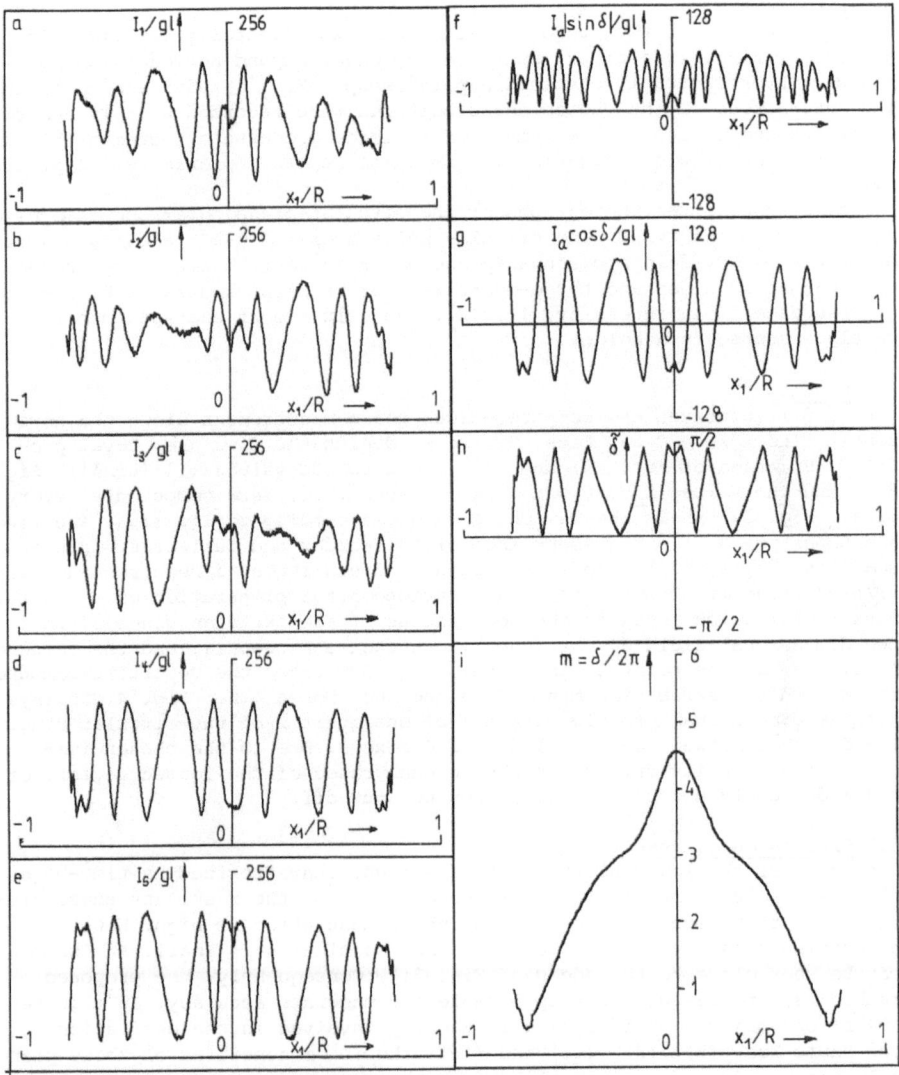

FIGURE 3. Phase measurement along the horizontal line $x_2/R = 0.63$ of the circular ring under investigation. a) to e) digitized grey levels of the intensities I_1 to I_5, compare Figs. 2a) to e). f) to i) evaluated distributions $I_a|\sin\delta|$, $I_a\cos\delta$, $\tilde{\delta}$ and $m = \delta/2\pi$, respectively.

$$\alpha = \frac{1}{4} \arctan \frac{(\bar{I}_2 - \bar{I}_3)}{\sqrt{3}(\bar{I}_2 + \bar{I}_3 - 2\bar{I}_1)} . \tag{20}$$

3. ILLUSTRATIVE EXAMPLE OF DIGITAL PHASE MEASUREMENT

3.1 Experiment

The capability of the proposed method was experimentally verified with a circular ring, made of polyester resin, under expanding load. The specimen was investigated in a standard polariscope, Fig. 1, with sodium light. The intensities of the photoelastic patterns were recorded by a TV camera, digitized and stored in the memory of an image processing system of 512 x 512 pixels and 256 grey levels. The image scale was approximately 2 pixels/mm.

Fig. 2 shows three photographs of the circular-plane polariscope ($\varphi = \psi = 0$, $\pi/3$ and $-\pi/3$) and two of the circular polariscope ($\varphi = \psi - \pi/4$ and $\varphi = \psi + \pi/4$). A pattern observed in a circular-plane polariscope is built up by fragments of one-quarter-order and three-quarter-order fringes, separated by isoclinic fringes of mean intensity [11, 12]. The left rim of the specimen is partly obscured by a holder.

3.2 Results

Fig. 3 illustrates the computer-aided phase measurement along the horizontal line $x_2/R = 0.63$. Figs. 3a) to e) depict the five grey level profiles. By means of Eqs. (9) and (10) the absolute values $(I_a \cdot |\sin\delta|)$, Fig. 3f), and the values $I_a \cos\delta$, Fig. 3g), respectively, are computed at every pixel. Fig. 3h) shows the raw phase data $\tilde{\delta}$ according to Eq. (15). The lower and upper peaks of $\tilde{\delta}$ represent loci of half-order and full-order fringes, see Figs. 2d) and 3d). Proper adjusting and scaling of δ requires additional processing of $\tilde{\delta}$. This can be automated after preparation of a special data field, which contains the coordinates of the skeleton lines of the two-dimensional field $\tilde{\delta}$, obtained by a standard procedure, and the value of δ at least at one point of the model. Fig. 3i) shows the resultant computed fringe order distribution $m = \delta/2\pi$ along the chosen line. Fig. 4 displays a three-dimensional profile $m(x_1, x_2)$ of one quarter of the circular ring. The distance between contour lines is 4 pixels. Due to the chosen image scale, high density fringes in the neighbourhood of the loaded regions of the model could not be resolved; they wer cut off.

3.3 Discussion of results

Despite of the low quality of the specimen, characterized by time-edge effects, included bubbles and scratched surfaces, the resultant phase profile is rather smooth as long as the phase modulation is high. However, discontinuities may occur, when the phase modulation is small. An example can be seen close to the edges of Fig. 3i). Consequently, the proposed method is, at present, only applicable for moderate accuracy. In this respect, one major simplification is tacitly involved in the derivation of the basic Eqs. (5), 12), (13) and (16): the inhomogeneities of those plane polarizers and quarter-wave plates, which are rotated for the manipulation of the photoelastic patterns, are neglected. In order to estimate quantitatively the change of the mean intensity I_m by certain rotation of optical components, the fields I_m were measured for $\varphi = \psi = 0$ and $\pi/3$, respectively, when no model was present ($\delta = 0$), compare Eqs. (5) and (7). The relative deviation may easily reach ± 10 %, Fig. 5. It appears that inhomogeneities of rotated optical components are especially critical at points where the fringe density as well as the modulatin of I is low, the latter occuring

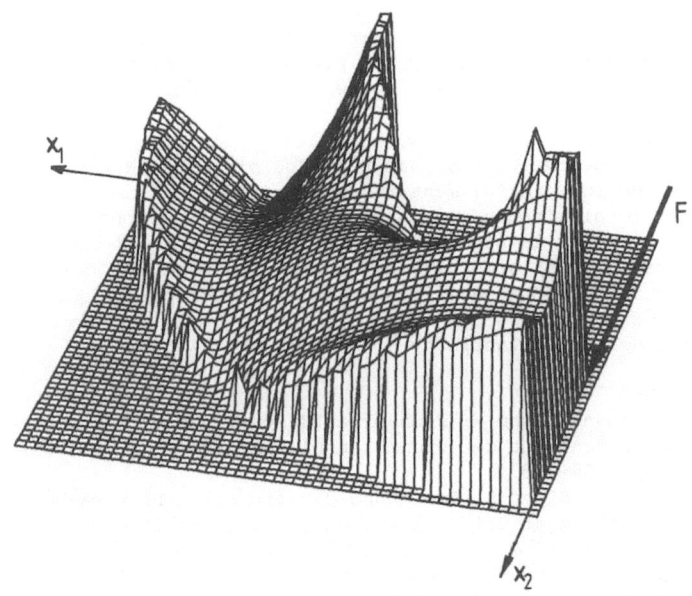

FIGURE 4. Three-dimensional plot m (x_1, x_2) of one quarter of the circular ring in diametral expansion. Grid distance of this presentation is 4 pixels.

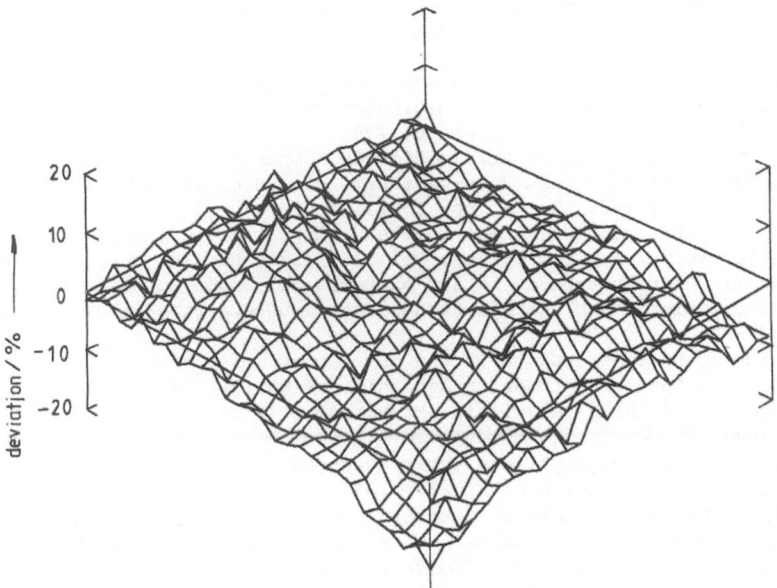

FIGURE 5. Relative change of the digitized mean intensity distribution $I_m(x_1, x_2)$ of the circular-plane polariscope after a common 60°-rotation of the analyzing components. Grid distance of this presentation is 16 pixels.

for $\tan 2\alpha \approx \pm\sqrt{3}/2$ and $\sin 2\alpha \approx 0$, compare Eqs. (6) to (8). Other factors may also influence the accuracy of the present state of the proposed method, e.g. noise and nonlinearities of the video system.

4. CONCLUSIONS

It has been demonstrated, that the idea of phase-shifting can be sucessfully applied for digital measurement of fractional orders of isochromatic fringes at local picture elements. Suitable manipulation of photoelastic intensity patterns is accomplished by rotating quarter-wave plates and/or plane polarizers. At present, the accuracy is moderate, because it does not yet consider the inhomogeneities of the rotated optical components. Improvement of the proposed method may contribute to further development of computer-aided photoelasticity.

4. REFERENCES

1. Müller RK and Gaupp M: Rechnergestützte Auswertung spannungsoptischer Modellversuche mit der digitalen Bildverarbeitung. VDI-Berichte Nr.514, 1984, pp. 165-168.
2. Rasche N and Simon B: Anwendung der digitalen Bilverarbeitung in der Spannungsoptik am Beispiel des Schubspannungsdifferenzenverfahrens. VDI-Berichte Nr. 480, 1983, pp. 63-68.
3. Perzborn V and Laermann K-H: Einsatz der digitalen Bildverarbeitung zur Auswertung von Interferenzstreifenmustern in der Spannungsoptik. VDI-Berichte Nr. 552, 1985, pp. 1-13.
4. Hapel K-H: Digitale Bildverarbeitung spannungsoptischer Isochromatenbilder. Forschung im Ingenieurwesen, Vol. 51, 1985, Nr. 6, pp. 193-199.
5. Dändliker R, Thalmann R and Willemin J-F: Fringe interpolation by two-reference-beam holographic interferometry: reducing sensitivity to hologram misalignment. Optics Communications, Vol. 42, 1982, Nr. 5, pp. 301-306.
6. Dörband B: Die 3-Interferogramm-Methode zur automatischen Streifenauswertung in rechnergesteuerten digitalen Zweistrahlinterferometern. Optik, Vol. 60, No. 2, 1982, pp 161-174.
7. Carré P: Installation et utilisation du comparateur photoélectrique et interférentiel du Bureau Internationale des Poids et Mesures. Metrologia, Vol. 2, 1966, No. 1, pp. 13-23.
8. Cheng Y-Y and Wyant JC: Phase shifter calibration in phase-shifting interferometry. Applied Optics. Vol. 24, 1985, Nr. 18, pp. 3049-3052.
9. Hudec M: Détermination du retard photoélastique par compensation [in Serbo-Croatian]. Publikacije techničkog faculteta v. Sajarevu, 1959, pp. 21-28.
10. Lagarde A: On some aspects in the development of photoelastic measurements. In: Optical Methods in Mechanics of Solids, edited by A. Lagarde. Aalphen an den Rijn, Rockfield, Maryland: Sijthoff & Nordhoff 1981, pp. 1-40.
11. Hecker FW: Spannungsoptische Untersuchungen von Scheibenmodellen im Halbtonfeld. VDI-Berichte Nr. 297, 1977, pp. 29-36.
12. Oheix P: Méthode globale de détermination de l'ordre de frange en valeur algébrique d'un modèle photoélastique bidimensionel et visualisation des franges de quart d'ordre. C. R. Acad. Sc. Paris, t. 279, No. 15, Série B- 28, 1974, pp. 361-363.

CHECKING THE SHEAR-DIFFERENCE METHOD IN CASE OF PHOTO-ELASTIC COATING

F. THAMM,   L. BORBÁS
                Technical University Budapest /Hungary/

Dept. for Engineering   Dept. of Machine Elements
     Mechanics

Photoelastic coating is used mainly to obtain qualitative
results as an aid to strain gauge technique. Separation of
the principal stresses is usually avoided because of doubts
of its accuracy, lacking reliable initial values for the
integration, because the state of stress in the coating
ceases to be two-dimensional in the neighbourhood of the
circumference of the coat. As for some special investigations the necessity of a thorough evaluation of such measurements were needed, investigations were involved for accuracy-
problems in connection with integration-methods using the
equilibrum-equations.

To get a first impres-
sion of the distortion
of the state of stress
at the circumference of
the coating, the theore-
tical solution of
G. Vörös [1] for the
case of a sheet of finite
dimensions bonded upon an
infinite half-space stretched homogenously in the
x-direction by the am-
ount ε_{xo} was used. His
formulas are shown in
Fig.1. /a misprint in his
publication corrected/.
The distribution of the
stress components along
the x-axis of Fig.1. is
shown in Fig.2. As only
the stress-component σ_x
can be detected photo-
elastically, the error
of the measurement can
be approximated by the
deviation of σ_x from
$\sigma_0 = E_L \varepsilon_{xo}$.

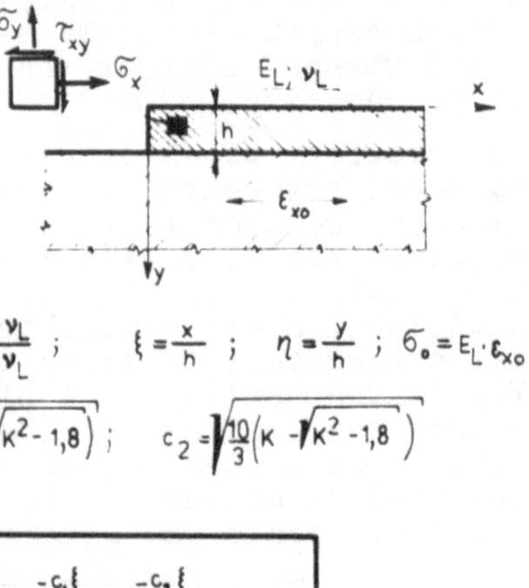

$$K = \frac{2+\nu_L}{1-\nu_L} \quad ; \quad \xi = \frac{x}{h} \quad ; \quad \eta = \frac{y}{h} \quad ; \quad \sigma_0 = E_L \cdot \varepsilon_{xo}$$

$$c_1 = \sqrt{\frac{10}{3}\left(K + \sqrt{K^2 - 1,8}\right)} \quad ; \quad c_2 = \sqrt{\frac{10}{3}\left(K - \sqrt{K^2 - 1,8}\right)}$$

$$\frac{\sigma_x}{\sigma_0} = \frac{c_2 e^{-c_1 \xi} - c_1 e^{-c_2 \xi}}{c_1 - c_2} + 1$$

$$\frac{\sigma_y}{\sigma_0} = 2\sqrt{5} \cdot \eta^2 \cdot \frac{c_1 e^{-c_1 \xi} - c_2 e^{-c_2 \xi}}{c_1 - c_2}$$

$$\frac{\tau_{xy}}{\sigma_0} = 2\sqrt{5} \cdot \eta \cdot \frac{e^{-c_1 \xi} - e^{-c_2 \xi}}{c_1 - c_2}$$

. . . .(1)

Fig.1. Formulas of G.Vörös [1] for
the stress distribution in an elastic layer upon an elastic
half-space stretched homogeniously along the x-axis.
E_L: Young's modulus, ν_L Poissons ratio of the layer.

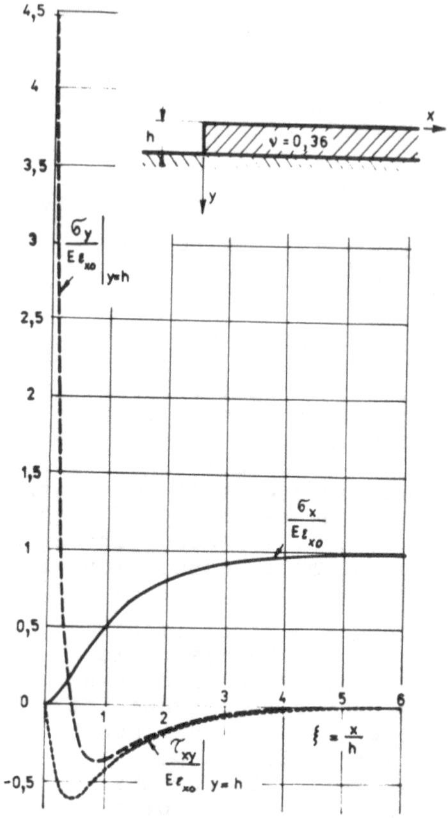

Fig.2. The stress distribution in an elastic layer bonded on a half-space stretched homogenously in the x-direction based on the formulas of eqs. (1).

The error in case of a stretch in the x-direction is plotted in curve A of Fig.3.

In the case of a stretch of the half-space parallel to the edge of the coating, the error is caused by the deviation of Poissons ratios of half-space and coating resulting in the curve B in Fig.3. for a Poissons ratio of $\nu_L = 0,36$ for the coating and $\nu_S = 0,3$ for the half-space.

As case A in Fig.3. cannot be used as initial value for a stress-separation procedure, case B indicates a fairly good measurement result beyond the distance 2h from the edge of the coating.

The accuracy of an actual measurement procedure was evaluated by an experiment on a specimen for which the theoretical solution of the stress distribution was available. the arrangement is shown in

Fig.4. The bending of a plate in two mutually perpendicular planes was produced by loading a square plate in its corners with concentrated forces of opposite sign as shown in Fig.5. The bending moments in this case are evenly distributed over the whole plate and have the value

$$m_x = -m_y = m_0 = \frac{F}{2} \tag{2}$$

The actual shape of the specimen investigated, shown in Fig. 6. differed from that in Fig.5. Therefore the even distribution of the bending moments was realized only at the central part of the plate, the size of which could be evaluated by inspecting the fringe pattern of the photoelastic coating prior to the drilling of the central hole.

The stress distribution in an infinite plate in pure uniaxial bending around a central hole is given by Goodier [2]. With the notations of Fig.5. the distribution of bending moments is given by

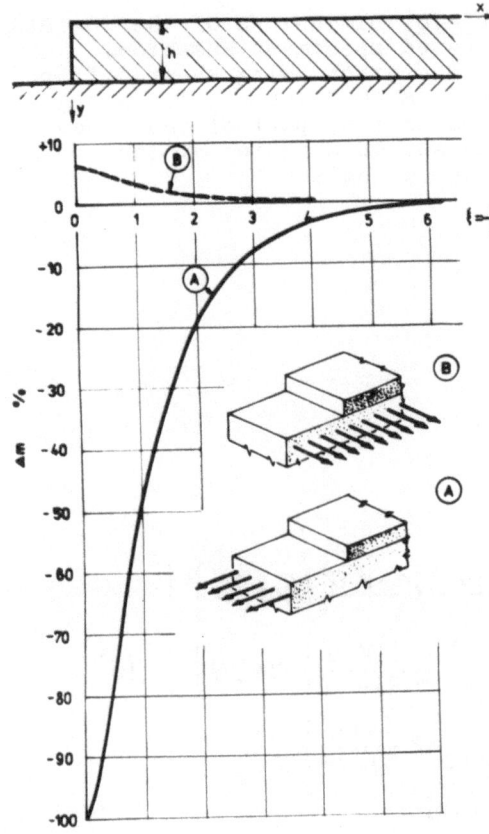

Fig.3. The deviation of the fringe-order of the photo-elastic layer from the strain-difference in the specimen under the coating basing on the formulas [1]. Curve A: strainig the specimen perpendicular to the edge of the coating; curve B: strainig the specimen parallel to the edge of the coating.

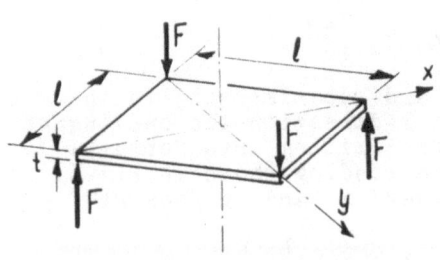

Fig.4. Specimen for producing homogenous biaxial pure bending in a plate of uniform thickness.

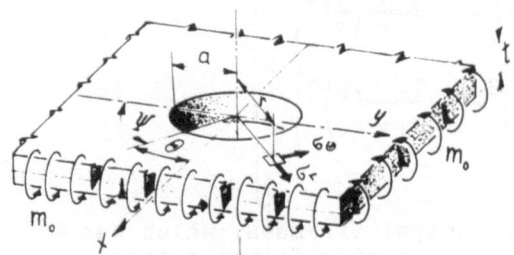

Fig.5. Circular hole in an infinite plate in homogenous biaxial bending. Dimensions and notations.

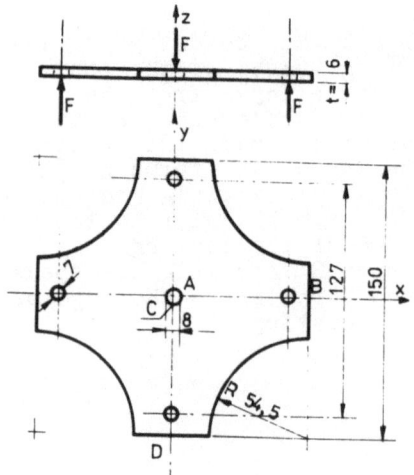

Fig.6. Dimensions of the investigated specimen. Numerical
integration using shear-difference method was applied along
the initial part of the lines
AB and CD starting at the
central hole

$$m_r = \frac{m_o}{2}\left[(1+\nu)\left(1-\frac{a^2}{r^2}\right)-(1-\nu)\left\{1-\frac{4\nu}{3+\nu}\cdot\frac{a^2}{r^2}-3\frac{1-\nu}{3+\nu}\cdot\frac{a^4}{r^4}\right\}\cos 2\theta\right]$$

$$m_\theta = \frac{m_o}{2}\left[(1+\nu)\left(1+\frac{a^2}{r^2}\right)+(1-\nu)\left\{1+\frac{4}{3+\nu}\cdot\frac{a^2}{r^2}-3\frac{1-\nu}{3+\nu}\cdot\frac{a^4}{r^4}\right\}\cos 2\theta\right] \quad \Big\} \quad (3)$$

$$m_{r\theta} = \frac{m_o}{2}(1-\nu)\left\{1-2\frac{3-\nu}{3+\nu}\frac{a^2}{r^2}+3\frac{1-\nu}{3+\nu}\frac{a^4}{r^4}\right\}\sin 2\theta$$

with ν denoting the Poissons ratio.
The stress components at the upper surface of the plate of
the thickness   t   are given by

$$\sigma_r = \frac{6m_r}{t^2} \; ; \quad \sigma_\theta = \frac{6m_\theta}{t^2} \; ; \qquad \tau_{r\theta} = \frac{6m_{r\theta}}{t^2} \qquad (4)$$

Inserting eq. (2) into eq. (3) and superimposing to the distribution of the bending in the x-direction the bending of
opposite sign in the y-direction, writing into formulas (3)
the angle ψ instead of θ , the loading shown in Fig.5. is
obtained. This yields for the axes   x   and   y /cos 2θ = ±1;
sin 2θ = 0/

$$\sigma_r = \mp \frac{3F}{t^2}(1-\nu)\left[1-\frac{4\nu}{3+\nu}\left(\frac{a}{r}\right)^2-3\cdot\frac{1-\nu}{3+\nu}\left(\frac{a}{r}\right)^4\right]$$

$$\sigma_\theta = \pm \frac{3F}{t^2}(1-\nu)\left[1+\frac{4}{3+\nu}\left(\frac{a}{r}\right)^2-3\cdot\frac{1-\nu}{3+\nu}\left(\frac{a}{r}\right)^4\right] \quad \Big\} \quad (5)$$

$$\tau_{r\theta} = 0$$

And the difference of the principal stresses, which can
directly compared with the isochromatic fringe pattern

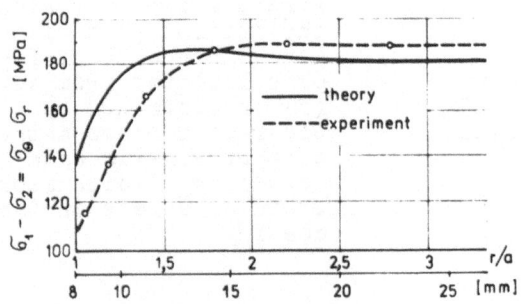

Fig.7. The plot of the difference of principal stresses /measured and calculated/ along the lines AB and CD of Fig. 6.

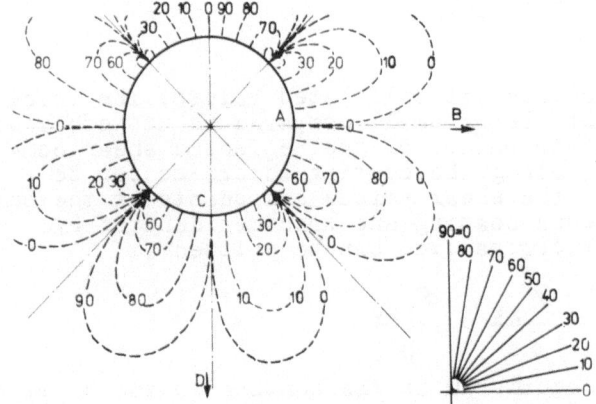

Fig.8. The plot of isoclinics of the photoelastic image of the coating around the central hole of the specimen of Fig.6.

$$\sigma_1 - \sigma_2 = \pm \frac{3F}{t^2}(1-\nu)\left[2 - 4\frac{1-\nu}{3+\nu}\left(\frac{a}{r}\right)^2 - 6\frac{1-\nu}{3+\nu}\left(\frac{a}{r}\right)^4\right] \qquad (6)$$

The measurement was carried out on an aluminium plate with the dimensions shown in Fig.6. having a 2mm thick photoelastic coating of Hungarian origin. Reflection polariscope type 030 of Vishay was used, the fringe orders were measured by Babinet compensator.

The distribution of measured and calculated values of principal stress-difference is shown in Fig.7., a plot of isoclinics is presented in Fig.8. Shear-difference method was applied along the initial part of the lines AB and CD of Fig.6. starting at points A resp. C. Calculated and measured values of the stress components itself are shown in Fig.9. As the stress distribution along lines AB and CD differed from each other only in sign, Figs. 7. and 9. show the combination of the results along both paths of integration.

As shown previously, stress distribution ceases to be plane in the neighbourhood of the edge of the photoelastic coating, in this case at the circumference of the central hole. This effect was neglected, and the integration was carried out on the usual way of shear difference method, thus producing the

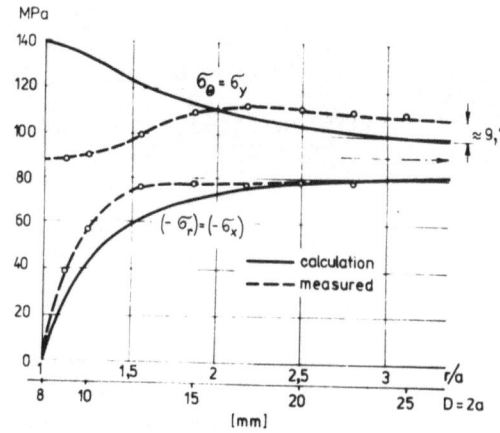

MPa

Fig.9. The plot of the stress components along the initial part of the lines AB and CD of Fig. 6. calculated from eq. (3) and obtained by shear-difference method of photoelastic data neglecting 3-dimensional state of stress in the coating.

deviation between measured and calculated values. The thickness h of the coating being small compared to its other dimensions, a linear distribution of the neglected shear components τ_{xz} and τ_{yz} along the coating thickness can be assumed. The values of the shear stress components on the surface of the bond between coating and specimen denoted by τ_{xzo} and τ_{yzo} their derivatives can be written

$$\frac{\partial \tau_{xz}}{\partial z} = \frac{\tau_{xzo}}{h} \quad ; \quad \frac{\partial \tau_{yz}}{\partial z} = \frac{\tau_{yzo}}{h}$$

and the equations of equilibrum in the coating in the 3-dimensional case will get following form

$$\left. \begin{aligned} \frac{\partial \sigma_x}{\partial x} + \frac{\partial \tau_{xy}}{\partial y} + \frac{\tau_{xzo}}{h} &= 0 \\ \frac{\partial \tau_{xy}}{\partial x} + \frac{\partial \sigma_y}{\partial y} + \frac{\tau_{yzo}}{h} &= 0 \end{aligned} \right\} \qquad (7)$$

Based on the theoretical an measured values of the stress components eqs.(7) permit the evaluation of τ_{xzo} and τ_{yzo}. Assumed, that the measurement error in τ_{xy} can be neglected

$$\left(\frac{\partial \tau_{xy}}{\partial y} \right)_{experimental} = \left(\frac{\partial \tau_{xy}}{\partial y} \right)_{theory}$$

The comparison of the equilibrum equation of the state of plane stress with eq.(7) applied to finite steps of integration Δx resp. Δy yields following equations

$$\left. \begin{aligned} \tau_{xzo} &= \frac{h}{\Delta x} \left[(\Delta \sigma_x)_{exp} - (\Delta \sigma_x)_{theory} \right] \\ \tau_{yzo} &= \frac{h}{\Delta y} \left[(\Delta \sigma_y)_{exp} - (\Delta \sigma_y)_{theory} \right] \end{aligned} \right\} \qquad (8)$$

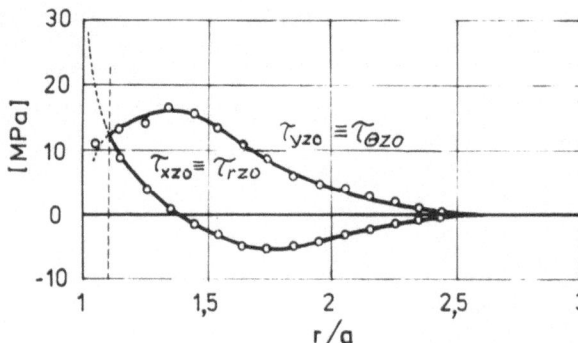

Fig.10. The distribution of shear stresses at the bonding surface between the photoelastic coating and the specimen computed from eqs. (8) based on the comparison of measured and calculated values.

The distribution of $\tau_{xzo} \equiv \tau_{rzo}$ and $\tau_{yzo} \equiv \tau_{\theta zo}$ along the lines AB and CD of integration is shown in Fig.10. The dotted line on the figure marks the limit, where formulas (8) begin to be uncertain. As far as these results permit a more generalized conclusion, they show, that conventional shear-difference method may be applied to the photoelastic coating method with an accuracy sufficient for many technical applications, provided, that the measurement procedure is performed carefully using pointwise compensation and skilled bonding technique.

REFERENCES

[1] VÖRÖS,G.M.: Thin elastic layers; cemented joints, coats. Periodica Polytechnica. Mechanical Engineering. 18/1974/ pp. 113-121.

[2] GOODIER,J.N.: Influence of circular and elliptical holes on transverse flexure if elastic plates. Phil.Mag. 22/1936 pp.69-80.

F.THAMM.   L.BORBÁS.
Technical University
H - 1521 Budapest /Hungary/

ON A NONLINEAR THEORY OF PHOTOELASTICITY

K.-H. LAERMANN

DR.-ING., O.PROF. F. BAUSTATIK, BUGH WUPPERTAL, FB 11 - BAUTECHNIK

As yet, in photoelasticity it has always been supposed Hooke's theory of elasticity to be valid and consequently linear relations between bire-fringence effects and stresses to be existing [1]. However, in areas of high stress concentration, e.g. in the vicinity of crack tips, notches and inclusions, and with respect to some of the mainly used photoelastic model materials, considerable uncertainties may result. Therefore the following nonlinear-elastic stress - strain relations will be introduced. On the supposition that the strains are still small, such relations may be formu-lated according to Kauderer [2] for an isothermal state:

$$\varepsilon_{ij} = \left[\frac{1}{3K} \varkappa(s) - \frac{1}{2G} g(\tau_o^2) \right] s\, \delta_{ij} + \frac{1}{2G} g(\tau_o^2)\, \sigma_{ij}, \tag{1}$$

where $\varkappa(s)$ denotes a compression function, formulated as a potential series

$$\varkappa(s) = 1 + \sum_{\nu=1}^{n} \frac{\varkappa_{\nu}}{(3K)^{\nu}}\, s^{\nu}, \tag{2}$$

and similarly, $g(\tau_o^2)$ denotes a shear function

$$g(\tau_o^2) = 1 + \sum_{\nu=1}^{n} \frac{g_{2\nu}}{(2G)^{2\nu}}\, \tau_o^{2\nu}. \tag{3}$$

With K and G, the initial values of the compression modulus and the shear modulus respectively, the coefficients $\varkappa_{\nu}$ and $g_{2\nu}$ describe the nonlinear material response. These values are to be determined by material testing procudures, such as tensile tests and shear tests. Furthermore, in eq.s (2) and (3), s denotes the mean tension, and τ_o the reference stress. According to Neumann [3] (see also Coker/Filon [4], Mindlin [5]), in amorphous material the birefringence effect is assumed to be linear, de-pending on the mechanically induced strains. And of course there are some materials for which the validity of this assumption has been proved experi-mentally.
With n_o, the refraction index of the unstrained material, n_{ij}, the refrac-tion tensor, and the strain-optical coefficients d_1 and d_2, the relation between birefringence and strain holds

$$n_{ij}^{-2} - n_o^{-2}\delta_{ij} = d_1 e_{ij} + \frac{1}{3} d_2\, e\, \delta_{ij}. \tag{4}$$

Wieringa, H (ed), Experimental Stress Analysis.
© *1986. Martinus Nijhoff Publishers, Dordrecht.*

In eq. (4), e denotes the volume change and e_{ij} the strain deviation. Together with eq. (1), a nonlinear relation between the stress tensor and the refractive tensor will be obtained:

$$n_{ij}^{-2} - n_o^{-2} \delta_{ij} = A_1 s \delta_{ij} + A_2 \sigma_{ij} .$$ (5)

For abbreviation it has been introduced

$$A_1 = d_2 \frac{\kappa(s)}{3K} - d_1 \frac{g(z_o^2)}{2G} ; \quad A_2 = d_1 \frac{g(z_o^2)}{2G} .$$

In the following, a plane stress state may be considered in plane (x_1, x_2); the direction of the incident light ray may be parallel to the x_3-axis. Assuming a rectilinear light path, the component n_{33} of the refraction tensor is unessential to the further consideration, and because of $\sigma_{13} = \sigma_{23} = 0$, the components n_{13} and n_{23} are also equal zero. Thus, the refraction tensor is reduced to $n_{\alpha\beta}$, $\alpha,\beta \in [1,2]$. The principal axes ψ_N of the index ellipse with reference to the x_1-axis are given by the proper vectors of $n_{\alpha\beta}^{-2}$:

$$\tan 2\psi_N = 2 n_{12}^{-2} [n_{11}^{-2} - n_{22}^{-2}]^{-1} .$$ (6)

It can be proved easily, that the principal directions of the index ellipse coincide with the principal axes of the stress tensor. The eigen values of $n_{\alpha\beta}^{-2}$ are the principal values n_α^{-2}; they are given by

$$n_1^{-2} = n_o^{-2} + (A_1 + \tfrac{3}{2} A_2) s + \tfrac{1}{2} A_2 [(\sigma_{11} - \sigma_{22}) \cos 2\psi_N + 2\sigma_{12} \sin 2\psi_N] ,$$

$$n_2^{-2} = n_o^{-2} + (A_1 + \tfrac{3}{2} A_2) s - \tfrac{1}{2} A_2 [(\sigma_{11} - \sigma_{22}) \cos 2\psi_N + 2\sigma_{12} \sin 2\psi_N] .$$ (7)

Because of the weak birefringence, from eq.s (7) follows

$$n_1 - n_2 = -\tfrac{1}{2} n_o^3 A_2 [(\sigma_{11} - \sigma_{22}) \cos 2\psi_N + 2\sigma_{12} \sin 2\psi_N] ,$$ (8)

where n_α denote the refraction indices.
The birefringence is defined as the difference of the refraction indices n_α

$$\Delta := n_1 - n_2 ;$$ (9)

and by the relation

$$\Delta = \frac{\lambda}{d} \delta ,$$ (10)

related to the wave length of the used monochromatic light, the thickness d of the model and the order of birefringence δ. The principal axes of the stress state, i.e. the angles $\psi_S = \psi_N$, are determined in linearly polarized light. With A_2 and the shear function according to eq. (3), eq. (8) yields

$$\left[1+\sum_{\nu=1}^{n} g_{2\nu} \frac{1}{(2G)^{2\nu}} \tau_o^{2\nu}\right] \sigma_{12} = -\frac{2G}{d_1} \frac{1}{n_o^3} \Delta \sin 2\psi_N = \hat{S} \frac{\delta}{d} \sin 2\psi_N \qquad (11)$$

with the stress-optical coefficient $\hat{S}$.
Eliminating the stress component σ_{22} in τ_o^2, the reference stress in case of a two-dimensional stress state is given by

$$\tau_o^2 = \frac{2}{9} \left[\sigma_{11}^2 - 2\sigma_{11}\sigma_{12} \tan^{-1}2\psi_N + (3+4\tan^{-2}2\psi_N)\sigma_{12}^2\right]; \qquad (12)$$

furthermore, the stress component σ_{11} will be expressed by the equilibrium condition

$$\sigma_{11} = C'(x_2) - \int \sigma_{12,2} \, d\bar{x}_1 . \qquad (13)$$

For integration, a predictor - corrector - procedure based on the method of Heun [6] will be derived. In discrete points (j,k), the predictor is determined according to the recurrent formula of Euler-Cauchy (Fig. 1a)

$$\mathcal{J}_{(j,k)}^{(o)} = \mathcal{J}_{(j-1,k)}^{(o)} - \left[\sigma_{12}^{(o)}(j-1,k+1) - \sigma_{12}^{(o)}(j-1,k-1)\right] \frac{\Delta x_1}{2\Delta x_2} \qquad (14)$$

with the initial value

$$\mathcal{J}_{(0,k)}^{(o)} = \sigma_{11}(0,k) = (\sigma_{11})_o . \qquad (15)$$

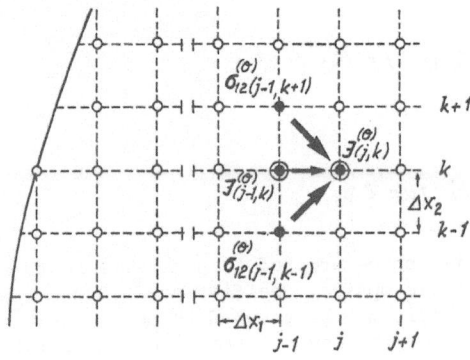

FIGURE 1.a)

Eq. (14) yields a first approximate value of the stress component σ_{11}

$$\sigma_{11}^{(o)}(j,k) \equiv \mathcal{J}^{(o)}(j,k) . \tag{16}$$

The result is introduced into eq. (12) and this equation then into eq. (11). This leads to an algebraic equation of grade 2n+1 in σ_{12}, the solution of which according to the "Pegasus"-method [7] yields a first approximation of the shear stress $\sigma_{12}^{(o)}(j,k)$. This approach is now improved by the so-called corrector in an iterative procedure (Fig. 1b)

$$\mathcal{J}^{(\mu+1)}(j,k) = \mathcal{J}^{(\mu+1)}(j-1,k) \div \left[\sigma_{12}^{(\mu+1)}(j-1,k+1) + \sigma_{12}^{(\mu)}(j,k+1) - \sigma_{12}^{(\mu+1)}(j-1,k-1) - \sigma_{12}^{(\mu)}(j,k-1) \right] \frac{\Delta x_1}{4\Delta x_2} \tag{17}$$

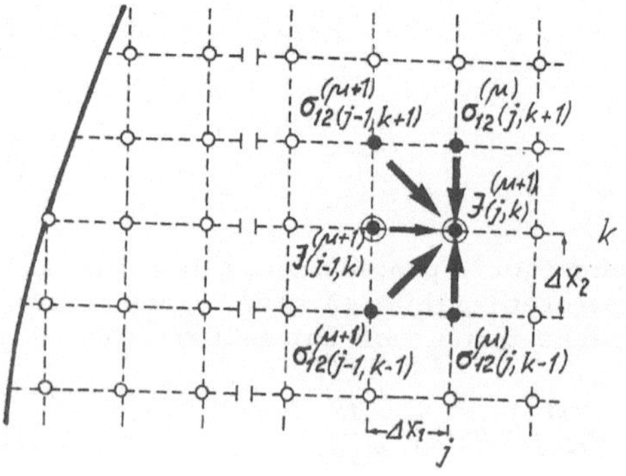

FIGURE 1.b)

Introducing the results of eq. (17) into eq. (11) yields improved values of the shear stress component $\sigma_{12}^{(\mu+1)}(j,k)$. If the intervals of the discrete grid in the direction of integration are sufficiently small, generally one to two iterative steps may only be necessary.
Finally, with the results

$$\sigma_{11}(j,k) = \sigma_{11}^{(m+1)}(j,k) \quad ; \quad \sigma_{12}(j,k) = \sigma_{12}^{(m+1)}(j,k)$$

the component $\sigma_{22}(j,k)$ is given as well

$$\sigma_{22}(j,k) = \sigma_{11}^{(m+1)}(j,k) - 2\,\sigma_{12}^{(m+1)}(j,k) \cdot \tan^{-1} 2\,\gamma_s . \tag{18}$$

The stress – strain relation will be determined in a tensile test as well as Poisson's ratio ν and the stress-optical coefficient $\hat{S}$. Because of less changes of ν over σ, Poisson's ratio may be assumed to be constant. The stress – strain relation of some of the investigated model material can be described approximately by

$$\varepsilon_{11} = \frac{1}{E} \left[1 + f(\sigma_{11}) \right] \sigma_{11} , \tag{19}$$

with E, the initial value of Young's modulus. If the response of the material is considered to be homogeneous, f must be an even function such that

$$\varepsilon_{11} = \frac{1}{E}\left[1 + a_2\,\sigma_{11}^2\right]\sigma_{11}\,. \tag{20}$$

The material coefficient a_2 will be taken from the results of the tensile tests.
In a one-dimensional stress state then the compression function κ holds

$$\kappa = 1 + a_2\,\sigma_{11}^2 \tag{21}$$

and the shear function as well

$$g = 1 + a_2\,\sigma_{11}^2\,. \tag{22}$$

Consequently, the first two terms in the potential series (eq. (3)) are considered only; then g_2 holds

$$g_2 = 18\,G^2\,a_2\,. \tag{23}$$

In case of pure bending, the strain and the order of birefringence δ must be linear functions of z. However, in experiments with the model material VP 1527 (Fig. 2)

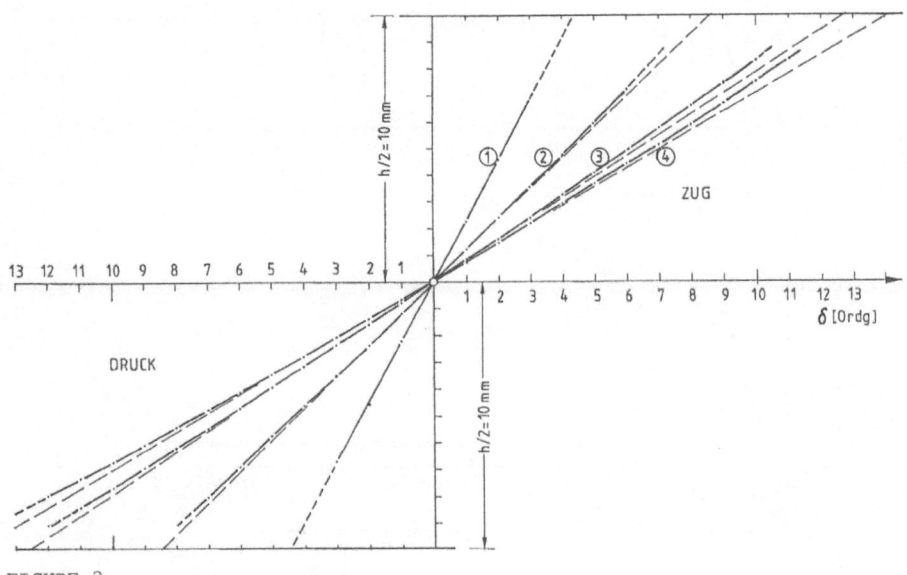

FIGURE 2

nonlinear and inhomogeneous effects have been observed. These effects are caused by lateral contraction:

$$\delta_i = \frac{Ed_o}{S}\left(1-\nu\,\varepsilon_i\right)\varepsilon_i \ . \tag{24}$$

The relation between strain and fringe order runs

$$\varepsilon_i = -\frac{1}{2\nu}\left(1-\sqrt{1+4\nu\,\frac{S}{Ed_o}\,\delta_i}\right) . \tag{25}$$

It should be pointed out, that in areas of high stress concentration lateral contraction must be taken into consideration. The boundary values of strain have been taken from experiments for different values of the applied bending moments.

Having determined the function of strain over the cross-section according to eq. (25), the stress $\sigma_{11}(z)$ will be calculated from the inverse relation of eq. (20)

$$\sigma_{11}^3 + \frac{1}{a_2}\,\sigma_{11} - \frac{1}{a_2}\,E\varepsilon_{11} = 0 \ . \tag{26}$$

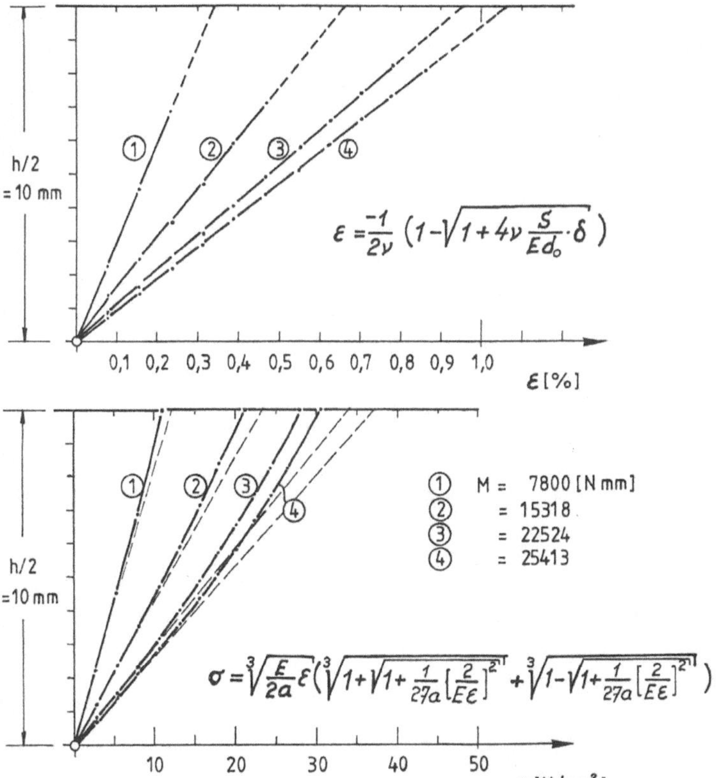

FIGURE 3

Fig. 3 may demonstrate the influence of nonlinear response on the distribution of stress over the cross-section of the beam under pure bending.

REFERENCES

1. Brewster D: On the Production of Cristalline Structure in Cristallised Powders by Compression and Traction. Trans. Royal Soc., Edinburgh, 20, 1853.
2. Kauderer H: Nichtlineare Mechanik. Springer-Verlag, Berlin/Göttingen/Heidelberg, 1958.
3. Neumann, FE: Die Gesetze der Doppelbrechung des Lichtes in comprimierten oder ungleichförmig erwärmten unkristallinen Körpern. Abh. Königl. Akad. d. Wissenschaften zu Berlin, 8. Nov. 1841.
4. Coker EG; Filon LNG: A Treatise on Photoelasticity. 2nd. Edit., Edit. H.T. Jessop, University Press, Cambridge, 1957.
5. Mindlin, RD: A Mathematical Theory of Photo-Viscoelasticity. J. Appl. Physics, Vol. 20, 1949.
6. Engeln-Müllges G; Reutter F: Numerische Mathematik für Ingenieure, 4. Aufl. Wissenschaftsverlag, Bibliograph. Inst., Mannheim/Wien/Zürich, 1985.
7. Dowell M; Jarratt P: The "Pegasus"-Method for Computing the Root of an Equation. BIT 11, 1971.

STRAIN DISTRIBUTION DURING HOT ROLLING
OF STRIP BY PHOTOPLASTIC SIMULATION

H.A.GOMIDE
Federal University of Uberlândia
Uberlândia, Minas Gerais
Brazil

C.P.BURGER
Iowa State University
Ames, Iowa 50011
USA

1. INTRODUCTION

Hot rolling of billets into sheets is an old forming process. Yet, the causes of many of the defects in the finished, that clearly originate   dur ing rolling, are not understood. Those defects in the finished product caus e the metal industry significant additional costs each year. Gusminsky and Ellis [1] reported that 10 to 15 percent of the total width of rolled strip is recirculated because of craking. Nussbaum [2] pointed out that as   much as 30 percent of rolled or extruded metals may be scrap. The defects that occur in the hot-formed metal are a consequence of various parameters   not taken into account in process design. A good theoretical analysis   is not available to predict satisfactorily the way in which metals flow during hot forming.

The best attempt, so far, to provide a general solution is the free body equilibrium approach developed by Orowan [3] . Unfortunately, this approach requires numerical integrations to be perfomed at various stages   in   the analysis. Orowan's experimental technique consisted of placing a   grid   of one material on top of another, usually along a plane of symmetry.   After the composite is rolled and sectioned, the deformed grid pattern is used to determined the strains in the two directions of the original grid. Averbach [4] employed this technique and radiographed an embedded grid of lead in a cast tin bar. After deformation, he extracted data directly from the radio graph on two vertical sections along the direction of rolling. This allowed him to compute the longitudinal and vertical strains.

Most of the rolling theories in use today are restricted to predicting pa rameters of load and torque required in a rolling mill [5,6] .   Relatively successful analyses are also available for predicting lateral spread when billets with a width-to-thickness ratio larger than 6:1 are rolled [7].When the ratio is less than 6, investigators had no choice but to resort to   em pirical formulas with limited ranges of application. The most   significant of these are based on experimental results conducted on small   slabs   by Sparling [8] , El-Kalay and Sparling [9] , Helmi and Alexander [10],   and Beese [11] .

Few analytical solutions exist in the literature dealing with   the side spread in rolling. Hill [12] proposed a general method of analysis for met al working problems. Recently, attempts to apply Hill's analysis to   roll ing were made by Lahoti and Kobayashi [13] and Oh and Kobayashi [14] .

To date, no finite-element solution are available for general three-dimen sional flow. Rao and Kumar [15] have analyzed the cold rolling of strip us ing finite-element techniques, but the process is one of plane strain   and neglects deformation of the roll. Coulomb friction is assumed. Dawson and Thompson [16] and Dawson [17] used finite-element analysis methods that per mit integration of the deformation rate tensor or deformation gradient tens or to obtain values of strain in a material as it flows through a   steady-

560

state Eurelian reference frame.

Slipline field solution for rolling is another area that has occupied the attention of several researchers. Alexander [18] was the first to propose a complete slipline field solution for the plain-strain hot rolling problem . Alternative and less exacting slipline solution followed [19,20] . The re view by Sparling of literature on the rolling process outlines important trends and major advances in practice [21] .

In a different approach to the problem several experimental attempts have been made to evaluate the three-dimensional strain distribution in hot-form ing processes using photoplastic techniques [22-27] . The approach is based on the permanent birefringence that remains locked into a photoplastic mate rial after large plastic deformation. From this,predictions of the three-dimensional strain field in hot-worked metal may be made. Burger,Oyinlola and Scott [23] used polyester resins to model hot rolling by the methods of photomecanics. They used a material developed by Morris and Riley [25] and rolled a number of billets of different dimensions in an attempt to simulate the hot-rolling process. Their results demonstrated that photoplasticity can be used to find the whole field strain distribution in billets.Further work in this direction was reported by Burger, El-Hout and Gomide [26] . They ap plied photoplastic technique to evaluated the three-dimensional strain fi eld of a rolled billet with a particular geometry.

Gomide and Burger [24] determined the three-dimensional strain distributi on in upset rings and describes how an experimental method using photopla stic simulation can be used, as a viable method for studying the monuniforn strain in general three-dimensional problems. They also [27] , using a small polyester billets with width-to-thickness ratio less than 6 , as is common in hot rolling of ingots into billets, shown the influence of the geometry of the billets on the formation of a double bulge and provide gen eral information about the internal flow pattern in the billets.This paper reports continuing progress in this modeling approach. The way in which the three principal strain progressively develop during passage of the billets through the rolls are presented, from entro into to exit from the rolls for the condition of sticking friction.

2. EXPERIMENTAL TECHNIQUE

The experimental technique was performed by rolling small billets of lam inac polyester resin [27] through an electric-powered rolling mill which was completely enclosed by an insulated hot-air chamber that maintained constant temperature, pre-set for the test. The air was heated by a hair dryer and continually circulated by means of a small blower. The electrical powered roll drive system used a two-way, two-speed motor connected throuth a variable transformer and linked to a speed reducer gear box which trans mits the power to a large gear by a link chain. The large gear engaged the drive gears on the rolls. Using this system, different roll speed from 0,5 to 1,5 in/min could be used. A transparent and removable window made from clear plexiglass was placed on the insulated box in front of the rolls to better align and observe the models during a test. The system was left to heat for 3 hours at a temperature of ~169°F(~76°C). The test temperature was monitored with two thermo-couples adjacent to the upper and lower rolls.The models to be tested were preheated in a separate oven. Figure 1 shows sche matically the system used to test the samples.

In order to get the proper roughness of the rool surfaces, a technique similar to that described in Reference [27] was used. A particular strip geometry, W = 0.8 in. (20.32mm) and H = 0,4 in. (10.16mm) was rolled in such a way that it did not pass completely through the rolls. After it was

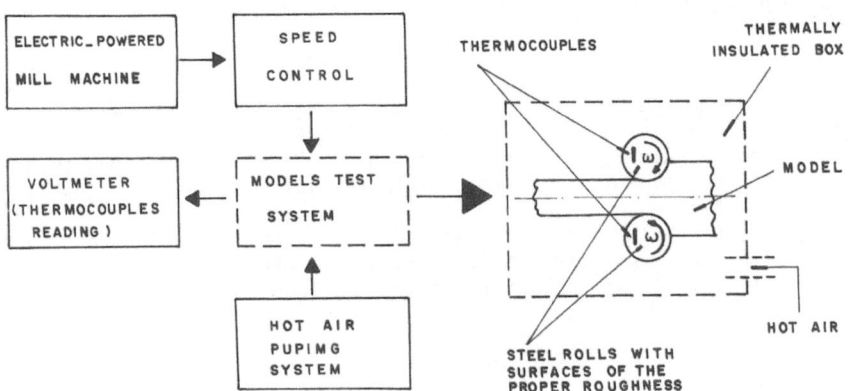

FIGURE 1 - Schematic view of the test system

rolled 2/3 of its total length, the rotation of the rolls were reversed such that the final strip was only partially deformed. The specimens, as shown on Figure 2, had a length of 4 in. (101,60mm). This was long enough to obtain uniform deformation.

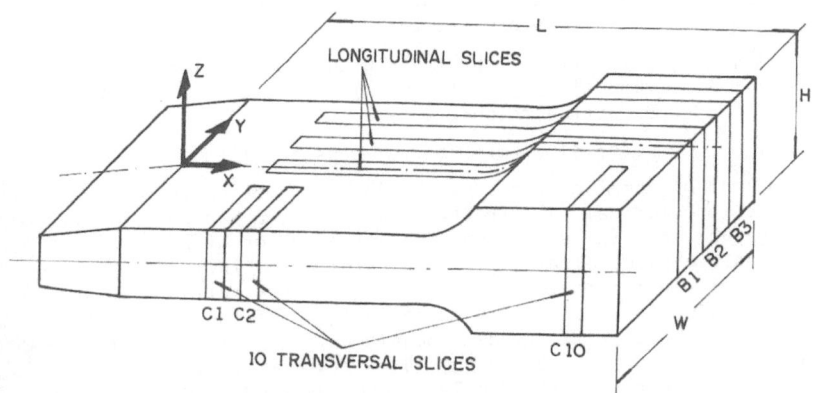

FIGURE 2 - Sketch showing the slices taken from the model and coordinate system.

After deformation the specimens were taken from the heated environment as soon after rolling as possible so that the maximum amount of deformation remained fixed in the models [24] . To facilitate entrance into the rolls during rolling, one end of the strips was tapered down one-sixth of the total length. The rolls were set for a nominal deformation of 20%. The roll velocity was adjusted to 1 in./min. This corresponded to strain rate of about 0,01 sec^{-1} for the chosen nominal deformation and a roll diameter of 2,75 in. The strain rate was computed by estimating the total time for a particular point on a strip to pass through the rolls if there was no back slipping.

3. EXPERIMENTAL RESULTS

The main objective of this paper was to evaluate the strain distribution of slabs when the strip drives through the rolls. In other words, the results that will be presented simulate the instantaneous strain distribution at a particular position of the strip when it is between the rolls,as shown schematically on Figure 1. After removal from the rolls the partially deformed strip was sliced to yield three longitudinal slices and ten transverse slices. Figure 2 presents a sketch of the strip showing how the slices were taken from the model and the coordinate system used to evaluate the strain distribution.

The theory of photoplasticity then indicates that the algebraic difference between the larger and smaller in-plane strain at any point on the slices is related to the respective fringe patterns by the following expressions:

$$\varepsilon_x - \varepsilon_z = \frac{N_y f_\varepsilon}{t_y} \qquad \text{for } \varepsilon_x > \varepsilon_z \qquad (1)$$

$$\varepsilon_y - \varepsilon_z = \frac{N_x f_\varepsilon}{t_x} \qquad \text{for } \varepsilon_y > \varepsilon_z \qquad (2)$$

where

N_x, N_y = the fringe order at the point of interest on the transverse and longitudinal slices,respectively. The slices are viewed in x and y directions.

t_x, t_y = thickness of the transverse and longitudinal slices, respectively. Thickness is measured in the x or y direction,respectively.

f_ε = material-strain optical constant [24].

In this kind of forming process the major deformation is a large extension in the longitudinal (x) direction, which is caused by a large compression in the vertical (z) direction. For large ratios of W/H (>20), it is common to assume plane strain, i.e., no deformation in the transverse (y) direction. For smaller ratios of W/H,especially for ratios less than 6, there is noticeable side flow (which causes bulging of the sides).The bulging may be barrel-shaped or it may be a double bulge. It is, therefore,reaonable to assume that $\varepsilon_x > \varepsilon_y > 0$ and $\varepsilon_z < 0$ everywhere in the rolled strip[27].

A conventional parallel field polariscope was used to determine the fringe orders. Tardy compensation [28] was used at each of the intersecting points to obtains the exact relative retardation along selected lines at intervals of 0.04 in. (1 mm). Since the slices were very thin, the fringe orders were low. Under these conditions, white light correctly used provides better resolution than the usual monochromatic system.

To evaluate the strain distribution, a third independent equation involving the three strain components was obtained by assuming constant volume under conditions of large (finite) deformation. Thus,

$$\varepsilon_1 + \varepsilon_2 + \varepsilon_3 + \varepsilon_1\varepsilon_2 + \varepsilon_2\varepsilon_3 + \varepsilon_1\varepsilon_2\varepsilon_3 = 0 \qquad (3)$$

It is important to point out that the third-order strain term in this equation was small enough to be neglected. The maximum error introduced by neglecting these terms was always less than 4%.

The surface width (W) of the strip that was in contact with the rolls was measured before and after rolling. There was no increase. This confirms that the condition of sticking friction was achieved.

The photoplastic fringe patterns (isochromatic) in the longitudinal slice which was taken from the center of the models (slice B1) are shown on Figure 3a, while Figure 3b reveals the isoclinics in the same slice, i.e., in the vertical (z) and horizontal (x) directions. Figure 3c and 3d shown the iso chromatic and isoclinics, respectively on a set of half transverse slices (C2, C4, C6, C8, C9 and C10).

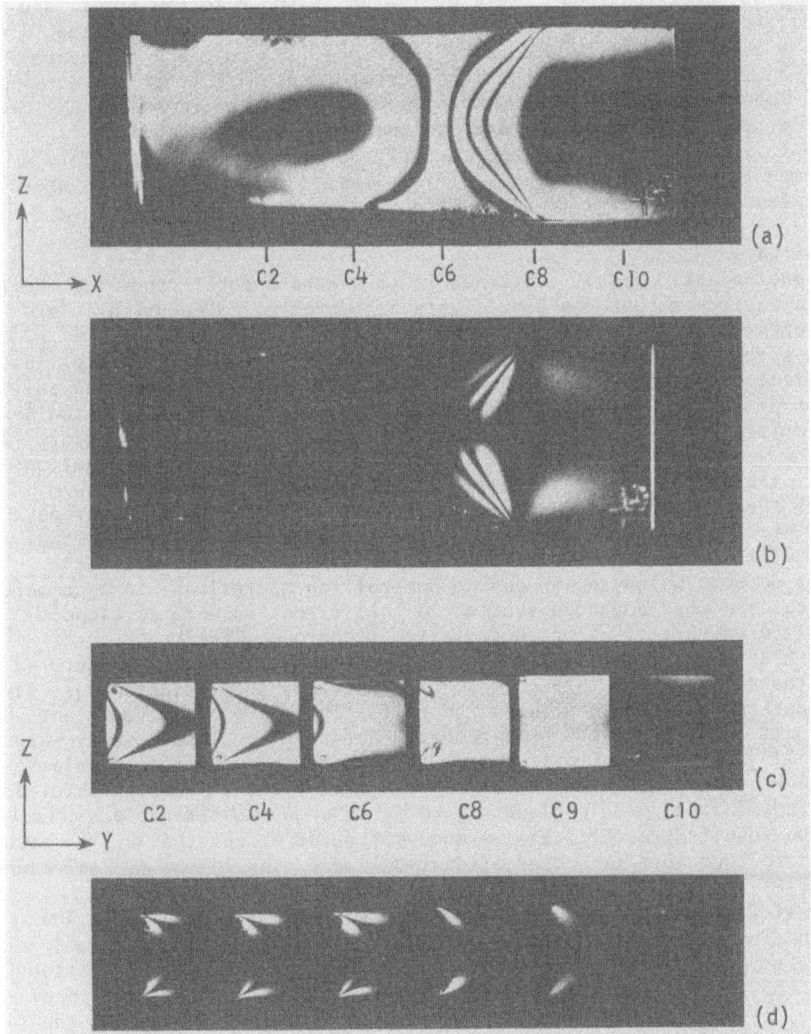

FIGURE 3 - Photographs of photoplastic data from the rolled strip.

From Figure 3b and 3d it can be observed that the isoclinic parameter 0° covers most of the area of the slices, i.e., $\sigma_x = \sigma_1, \sigma_y = \sigma_2$ and $\sigma_z = \sigma_3$ in

all blach regions. Only small area had isoclinics other than zero. The iso
clinics never exceede 15°. It was, therefore, assumed that $\varepsilon_x = \varepsilon_1$, $\varepsilon_y = \varepsilon_2$, and $\varepsilon_z = \varepsilon_3$ over the region studied. Then, Eqs. (1), (2), and (3) can be
solved to obtain the strain distributions.

With the photoplastic data taken from both sets of slices, the strain dis
tribution was obtained for three diferent planes in the y direction and
three different z positions in each plane. Figure 4a to Figure 4c shows the
complete three-dimensional strain distribution. Each figure is for a diffe
rent longitudinal slice (B slice). The positions of the lines (line A)where
the strains distributions were evaluated is shown to the right and the
initial and final positions of the curvature are indicated by dashed lines.

It was used the same procedure of reference [27] to check the overall tech
nique and to confirm that the higher order terms in Eq. (3) are of sec
ondary importance. In this way, the measured reduction in height ΔH^*, retai
ned after relaxation, was compared to the value obtained when ε_z is inte
grated to yield $w = \int \varepsilon_z \, dz$, and the difference was about 3%. This was
performed in the region that has already passed through the rolls where ε_z
is constant with respect to x-direction.

DISCUSSION

The photoplastic results obtained in this experimental study for the ratio
W/H < 6 explore an unknown area. Current theoretical predictions are in
error when the billets have width-to-thickness ratios less than 6 [7],yet
narrow slabs are common in hot rolling of industrial ingots into billets.Fi
nite element approaches [17] and slip line solutions [18] are unrealistic
because of the assumption of plain strain no longer hold. It is well-know
that for small ratios W/H (less than 6) a noticeable side flow occurs [27].
Over the years a number of attempts have been made to produce equations that
predict the extent of the increase in width [8,10,11] for ratios W/H < 6.
Most of these empirical equations are able to predict the side spread but
do not provide any information about the internal flow pattern in the bil
lets. Until now no theoretical or experimental approach is available to
predict defects which appear during hot rolling operations in commercial
practice. The most advanced studies in this direction were developed at
Iowa State University using photoplastic technique [22-27] .

The three-dimensional strain distribution presented here on Figure 4a to
4c as the strip drives through the rolls provides the instantaneous strain
distribution in a region between the rolls. This is another important set
of results that has never been reported before. The closest study was re
ported [17] using a finite elements approach to solve similar problems un
der the assumption of plain strain. The finite element results obtained in
that study for the middle-plane of the strip shows considerable similarities
with the results presented on the abouve figures except that ε_1 is assumed
to be zero. Much greater deformation (40%) were used in the finite element
analysis.

Several interesting feature can be observed on Figures 4a to 4c.The strain
distribution ε_x assumes positive value all along the roll-direction(x-direc
tion). It is relatively constant in the region that has already passed throu
gh the rolls. This corresponds, on the figures, to the area left of the
dasked lines at the biginning of the curvature of the top edge of the slice.
For the region which corresponds to the curvature on the strip,i.e.,the area
between the dasked lines, the ε_x values decrease fairly smoothly to near
zero for the positions A1 and A2 on the slices B1 to B3. This is not
true for the position A3 where ε_x does not have such a rapid decrease in
the region between the rolls. ε_x is far from zero at the entrance to the
rolls. Position A3 in close to the contact surface and for this reason the

565

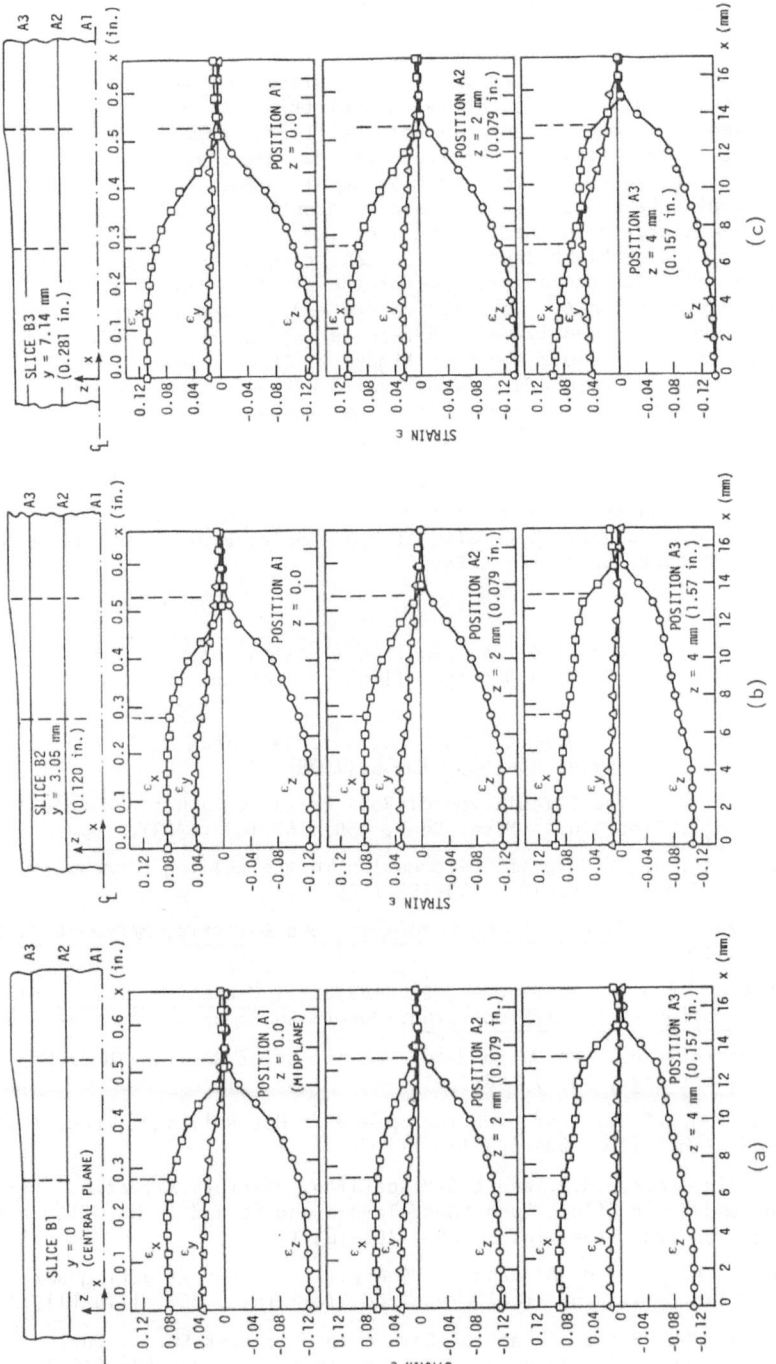

Figure 4 - Variations in three-dimensional strains in the direction of rolling.

strains are affected very strougly by the friction between the rolls and the strip. This trend can also be observed on the fringe distributions shown on Figure 3a. In the region of the strip free from deformation, i.e., .the region on the right of the dasked lines, the figures show that the ε_x strain distributions have small values equal to or close to zero. The only excep tion is again the position A3. The strain distribution ε_z has the same fea ture observed for the ε_x values, except that it has negative values all along the roll-direction. The magnitude of the values for both ε_x and ε_z appear to increase with y, i.e., from slice B1 to B3.

The ε_y strains have positive values all through to model. Their values are small and decrease in the z-direction, i.e., from A1 to A3 on slices B1 and B2. This is probably because of the constraints imposed by the friction between the roll surface and strip. For the slice B3 (close to the edge) the strain distribution ε_y increases in the z-directions. This was also observed on the strips analyzed on Reference [27] where the values for ε_y are gratter than the values shown on Figure 4. That is because the particu ular strip presented here had a larger ratio W/H = 2. The lateral flow constraint is then more noticeable, and the problem is slightly closer to plane strain than for the thicker billets.

ACKNOWLEDGMENTS

The research reported here was conducted with support from Iowa State University - USA, Federal University of Uberlândia, Brazil and the Brazili an National Research Council - CNPq.

REFERENCES

[01] - Gusminsky.G. and F. Ellis, "An Investigation into the Influence of Edge Shape on Craking During Rolling", J. of the Inst. of Metals,95, 33-37 (1967)

[02] - Nussbaum, A.T., "Remelt Shop Systems for Aluminum Mill Scrap Recy cling", Light Metal Age, 33, 10-15 (1975).

[03] - Orowan, E., "The Calculation of Roll Pressure in Hot and Cold Flat Rolling", Proc. Inst. Mech. Eng., 150, 140-167 (1943).

[04] - Averbach, B.L., "Plastic Deformation in the Rolling Processes", J. Metals Trans. ASME, 188, 150-155 (1950).

[05] - Avitzur, B., Metal Forming: Processes and Analysis, McGraw-Hill,New York (1968).

[06] - Thomson, E.G., C.T. Yang and S. Kobayashi, Mechanics of Plastic Deformation in Metal Processing, Macmillan Company, New York(1965).

[07] - El-Waziri,A.H., "An Up-to-Date Examination of Rolling Theory", Iron and Steel Eng., 40, 73-80(1363).

[08] - Sparling,L.G.M., "Formulas for Spreed in Hot Rolling", Proc. Inst . Mech. Eng., 175 , 604-609 (1961).

[09] - El-Kalay, A.K.E.H.A. and L.G.M.Sparling, "Factors Affecting Frictio nal and Their Effect Upon Load, Torque and Spread in Hot Flat Roll ing", J. Iron Steel Inst., 206, 162-163 (1968).

[10] - Helmi, A. and J.M. Alexander, "Geometric Factors Affecting Spread in Hot Flat Rolling of Steel", J. Iron Steel Ins., 206, 1110-1117(1968)

[11] - Beese,J.G.,"Ratio of Lateral Strain to Thickness Strain During Hot Rolling of Steel Slabs",J.Iron Steel Inst.,210, 433-436 (1972).

[12] - Hill, R.A. "General Method of Analysis for Metal-Working Processes", J.Mech.Phys.Solids, 11, 305-326 (1963).

[13] - Lahoti,G.O. and S. Kobayashi, "On Hill's General Method of Analysis for Metal-Working Processes", J. Mech. Sci., 16, 521-540 (1974).

[14] - Oh,S.I. and S. Kobayashi, "An approximate Method for a Three - Dimensional Analysis of Rolling", J.Mech. Sci., 17, 293-305 (1975).

[15] - Rao, S.S. and A. Kumar, "Finite Element Method in Cold Strip Rolling" Int. J. Mech. Tool Des. Res., 17, 159-165 (1977).

[16] - Dawson, P.R. and E.G. Thompson, "Finite Element Analysis of Steady State Elasto-Visco-Plastic Flow by the Initial Stress-Rate Method" . Int. J. Numerical Methods Eng., 12, 47-57 (1978).

[17] - Dawson,P.R., "Viscoplastic Finite Element Analysis of Steady - State Forming Processes Including Strain History and Stress Flux Dependence", American Soc. Mech. Eng., 28, 55-66 (1978).

[18] - Alexander,J.M. "A Slip Line Field for the Hot Rolling Process",Proc. Inst. Mech. Eng., 169, 1021-1028 (1955).

[19] - Crane,F.A. and J.M. Alexander, "Slip Line Fields and Deformation in Hot Rolling Strip", J. Inst. Met., 96, 289-300(1968).

[20] - Firbank, T.C. and P.R. Lancaster, "A Suggestion Slip Line Field for Cold Rolling with Slipping Friction", Int. J. Mech. Sci., 7,847-852, (1965).

[21] - Sparling,L.G.M., "Hot and Cold Rolling", Int. Met. Rev., 22,303-313, (1977).

[22] - Burger,C.P., "Nonlinear Photomechanics", Exp. Mech., 20 (11),381-389 (1980)

[23] - Burger,C.P., A.K. Oynlola and T.E. Scott, "Full Field Strain Distribution in Hot Rolled Billet by Simulation", Met. Eng. Q.,Trans.ASME, 26-29 (1976).

[24] - Gomide,H.A. and C.P. Burger, "Three-Dimensional Strain Distributions in Upset Rings by Photoplastic Simulation", Exp. Mech, 21 (10), 361 370 (1981).

[25] - Morris,D.H. and W.F. Riley, "A Photomechanics Material for Elasto Plastic Strees Analyses", Exp. Mech, 12 (10) 448-452 (1972).

[26] - Burger,C.P., J.N.El-Hount and H.A.Gomide, "Three-dimensional Strain Field for a Hot Rolled Billet by Photoplastic Simulation", Proc. 3rd Int'l Conf. Mech. Behavior of Mat'l, University of Cambridge, Cambridge, England (1979).

[27] - Burger, C.P. and H.A.Gomide, "Three-Dimensional Strain in Rolled Slabs by Photoplastic Simulation", Exp. Mech.,22, 441-447(1982).

[28] - Dally,J.W. and W.F. Riley, Experimental Stress Analysis, McGraw-Hill New York (1978).

ON VIBRATION AND SHOCK ISOLATION OF SENSITIVE GOODS BY VISCOELASTIC PACKAGING MATERIALS

T. Plitt[+], H. Weber[+], A. Geißler[*], K.-F. Ziegahn[*]

[+] University of Karlsruhe, [*] ICT Pfinztal-Berghausen

1. INTRODUCTION

To investigate transportions induced accelerations, loads, and stresses on sensitive goods which are protected against vibrations and shocks by viscoelastic packaging materials it is necessary to measure acceleration spectra of the transportation environment. Though these spectra are of a stochastic nature, there are typical frequencies for different means of transportation (truck, train, aeroplane etc.) which show periodic behaviour. Therefore it is useful to study harmonic vibrations of systems consisting of a good and a certain configuration of packaging material. Fig. 1 shows an example for such a system for which the input vibrations are produced by the loading platform of a truck. The analysis of the vibrations of the good, the loads or stresses on the good which may not exceed a critical value can be done theoretically or experimentally. The different methods dependent on the aim of the analysis are indicated in fig. 2.
In the case of a theoretical solution the dynamic mechanical behaviour, e.g. the stress-strain-behaviour of the packaging material, must be known. Then with methods of structural mechanics either stresses and strains in the packaging material and on the surface of the good may be calculated as a function of time and space (continuum mechanics problem), or a spring-dashpot-model of the system having one or more degrees of freedom may be derived (system identification problem). With such a model the answer of the system to different types of inputs (vibrations, shock) follow from the theory of rigid body dynamics.

The experimental investigation of the problem yields no major difficulties, if only accelerations of the good are of interest. For this purpose shock and vibration tests with the system under investigation have to be run. The determination of the spring-dashpot-model of the system, however, or the evaluation of the time dependent strains and stresses in the packaging material or on the surface of the good may become rather involved. To solve the second problem, different methods of experimental mechanics, e.g. moiré or photoviscoelasticity, are applicable. When using the method of photoviscoelasticity the dynamic optical behaviour, e.g. the birefringence-stress-or the birefringence-strain-behaviour of the packaging material must be studied.

In this paper we will present the methods of measuring acceleration spectra and some typcial results for transportation
by truck. Then mathematical models for the mechanical and
optical behaviour of nonlinear viscoelastic packaging materials under large prestrains and small superimposed vibrations are explained. For Polyvinylchloride (PVC) the dynamical
material behaviour is investigated. The testing setup and the
method of experimentation as well as some results are reported. To demonstrate the influence of material nonlinearities,
resonsance curves for a rigid good enclosed in a packaging
arrrangement of PVC are measured.

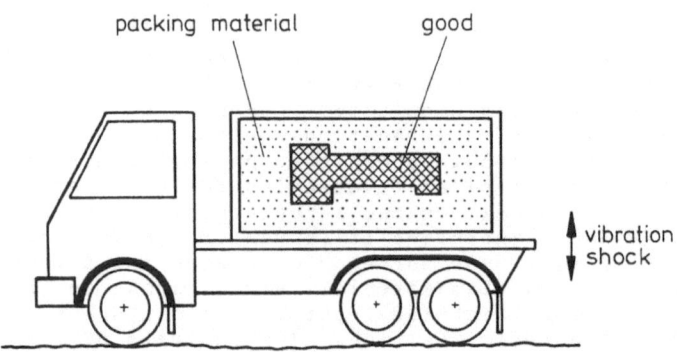

Fig. 1: Good enclosed by packaging material

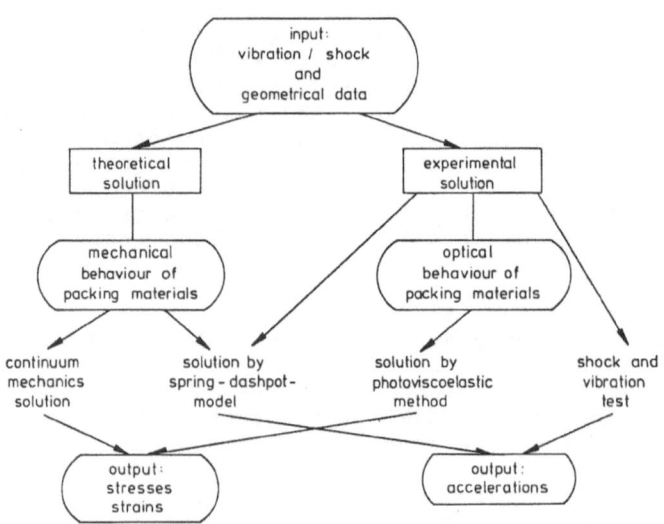

Fig: 2: Methods for analyzing stresses, strains, and
        accelerations subject to transportation induced
        vibrations and shocks

2. ACCELERATION SPECTRA OF A TRUCK

Acceleration spectra of vertical truck vibration can be obtained by measuring the acceleration versus time and computing
the acceleration versus frequency spectra. To compare different types of transportation vehicles one has agreed to define
the loading platform as an interface between truck and cargo
and to measure vertical accelerations at this platform /1,2/.

It is usual to collect vibration data on a tape recorder during a longer period of driving on different types of roads
and then, after having finished measurements, to compute
Fourier-Transformation of the signal. In this paper frequency
analysis is made just during the measuring drive in-situ by a
mobile frequency analyzer (FFT). This offers the advantage to
observe the averaging process of the individual computed spectra and to interpret the measure in relation to the circumstances, for example velocity, surface roughness, measuring
point within the vehicle and way of driving.

It was found that two main resonance frequencies occur independent of all these parameters varying only with the intensity of acceleration not with the frequency. These main resonances can be explained as the basic resonance of the whole
vehicle body and the basic resonance of the wheel/axle unit.
This means that, though vibrations on a vehicle platform are
normally described as a broadband random signal, there are
superposed two or more typical frequencies which clear surpass
the rest of the acceleration level of the averaged spectra. A
typical result of those frequency - analyzed and averaged
measurements on the platform of a Volkswagen -cargovan is
shown in fig. 3.

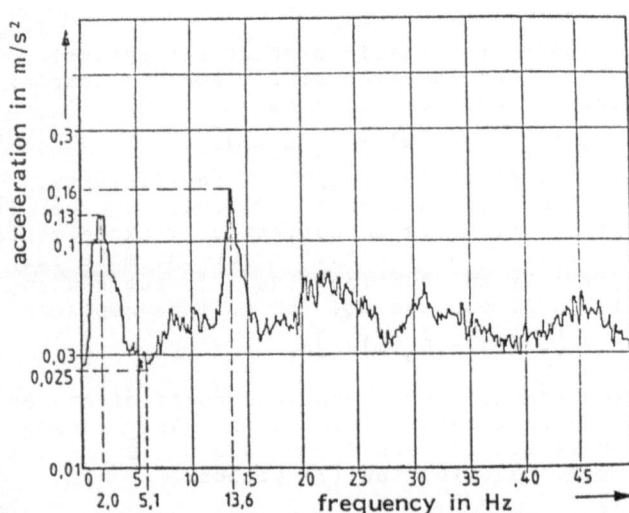

Fig. 3: Typical acceleration spectra of a truck

3. UNIAXIAL BEHAVIOUR OF VISCOELASTIC PACKAGING MATERIALS

3.1 Mathematical model of material behaviour

If we restrict our considerations to uniaxial harmonic vibrations of a nonlinear viscoelastic specimen and assume an input of the form

$$\varepsilon(t) = \varepsilon_m + \varepsilon_0 e^{i\omega t} \tag{1}$$

where ε_m is a prestrain, $\varepsilon_0 \ll \varepsilon_m$ the amplitude of vibration, and ω the frequency we get for a quadratic approximation of the mechanical material response in the stationary state /3/

$$\sigma(t) = \sigma_m + \tilde{\sigma}_0 e^{i\omega t}. \tag{2}$$

σ_m is the stationary part of the stress output. It depends nonlinearly on ε_m. $\tilde{\sigma}_0$ is a complex stress amplitude depending on ω, ε_m and ε_0

$$\tilde{\sigma}_0 = \sigma_0 e^{i\delta} = [\tilde{E}(\omega) + \tilde{H}(\omega)\varepsilon_m]\varepsilon_0. \tag{3}$$

Here δ is the phase difference between stress and strain. $\tilde{E}$ and $\tilde{H}$ are complex dynamic moduli for the linear and nonlinear contribution to the stress output, respectively.

For the optical response to the input (1) we get

$$\Delta n(t) = \Delta n_m + \Delta \tilde{n}_0 e^{i\omega t}. \tag{4}$$

Δn is the birefringence which depends on the fringe order N, the wavelength λ of the light and the thickness d of the specimen by

$$\Delta n = N\lambda/d. \tag{5}$$

In eq. (4) Δn_m represents the static part of the optical response. It is a nonlinear function of ε_m. For the complex amplitude $\Delta \tilde{n}_0$ a similiar eq. to (3) holds

$$\Delta \tilde{n}_0 = \Delta n_0 e^{i\gamma_\varepsilon} = [\tilde{C}_\varepsilon(\omega) + \tilde{D}_\varepsilon(\omega)\varepsilon_m]\varepsilon_0. \tag{6}$$

$\tilde{C}_\varepsilon$ and $\tilde{D}_\varepsilon$ are called complex strain-optical coefficients. γ_ε is the phase difference between strain and birefringence. If the dynamic optical behaviour is investigated by experiments with a stress input instead of a strain input the material response is also given by an equation like (4). But now Δn_m depends nonlinearly on stress and eq. (6) must be replaced by

$$\Delta \tilde{n}_0 = \Delta n_0 e^{i\gamma_\sigma} = [\tilde{C}_\sigma(\omega) + \tilde{D}_\sigma(\omega)\sigma_m]\sigma_0. \tag{7}$$

$\tilde{C}_\sigma$ and $\tilde{D}_\sigma$ are called complex stress-optical coefficients and γ_σ is the phase difference between stress and birefringence. There is a simple relation between $\tilde{C}_\sigma$, $\tilde{D}_\sigma$ and $\tilde{C}_\varepsilon$, $\tilde{D}_\varepsilon$ /4/. A generalization of eqs. (3), (6) and (7) yields

$$\tilde{\sigma}_0 = \tilde{E}^*(\omega, \varepsilon_m)\varepsilon_0 \tag{8}$$

$$\Delta \tilde{n}_0 = \tilde{C}_\varepsilon^*(\omega, \varepsilon_m)\varepsilon_0 = \tilde{C}_\sigma^*(\omega, \sigma_m)\sigma_0 \tag{9}$$

$\bar{E}^*$ being now a generalized complex modulus and $\tilde{C}_\varepsilon^*$, $\tilde{C}_\sigma^*$ being generalized optical coefficients.

3.2 Experimental set-up

To analyse the nonlinear dynamical behaviour of candidate packaging materials, e.g. to determine the complex moduli and the complex optical coefficients, the testing set-up shown in fig. 4 was used. It consists of an electronically controlled dynamic shaker for frequencies up to 1 kHz, displacement am- plitudes up to 50 mm, and a maximum force amplitude of 0,5 kN. The optical measurements are performed by means of tardy com- pensation using a strobe light as light source which is trig- gered by the control signal of the shaker to establish a stan- ding picture of the isochromatic field in a test specimen. A phase shifter produces phase differences between the shaker signal and the trigger event up to one period. It also serves to indicate the instantaneous values of force or displacement at different phase positions.

Legend:

1 shaker
2 loading frame
3 power supply
4 control unit
5 amplifiers
6 phase shifter for trigger signal
7 oscilloscope
8 photomultiplier power supply
9 tension specimen

Fig. 4: Testing set-up

To avoid individual heating of the specimens during dynamic
excitaton a controlled heating chamber with transparent walls
of PMMA was built. With this device all experiments could be
done at the same raised temperature corresponding to the
highest heating of the specimens occuring in tests without
temperature control.

3.3 Results

For a typical packaging material, e.g. a soft Polyvinylchloride (PVC) the investigation of the dynamical material behaviour yields the following results. The dependence of the
absolute value of the generalized complex modulus $\bar{E}^*$ on frequency is given in fig. 5a. Fig. 5b shows the frequency dependence of the damping coefficient tan δ and fig. 5c demonstrates the influence of the prestrain ε_m on $\bar{E}^*$.

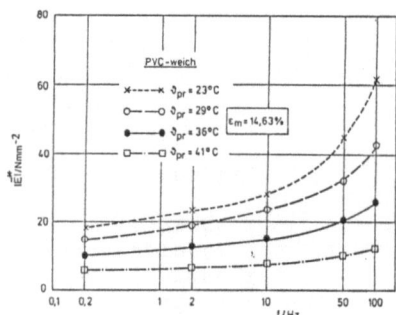

Fig. 5a: Dependence of $|\bar{E}^*|$
on frequency

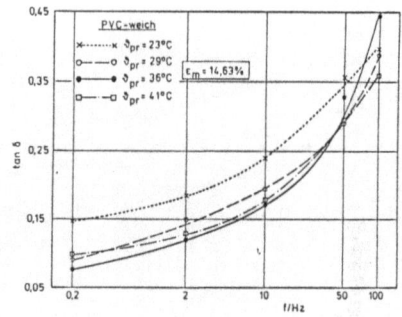

Fig. 5b: Dependence of tan δ
on frequency

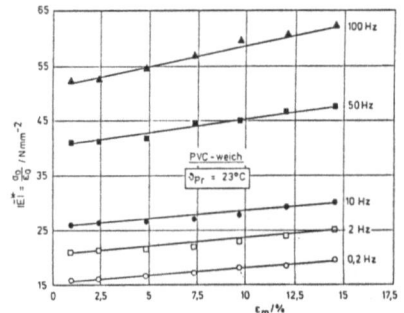

Fig. 5c: Dependence of $|\bar{E}^*|$
on prestain

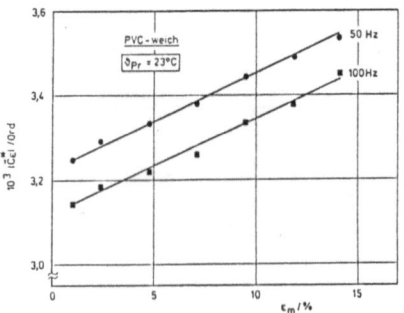

Fig 5d: Dependence of $|\tilde{C}_\varepsilon^*|$
on prestrain

As $\bar{E}^*$ is a linear function of ε_m eq. (3) is a good approximation for the nonlinear dynamic material behaviour of PVC. Similiar to $\bar{E}^*$ the absolute value of the complex strain-optical coefficient $\bar{C}_\varepsilon^*$ is also linearly dependent on ε_m as fig. 5d indicates.

4. MODEL TESTING

4.1 Resonance curves

We consider goods which are enclosed in the packaging similar to the example shown in fig. 1. During transportation these goods are often stacked on one another. This results in different prestrains in the material of the indiviual packaging. As the dynamic behaviour of packaging materials like PVC depends on prestrain we suppose that the resonance frequencies of the system depend on prestrain, too.

To demonstrate this tests were conducted with the model of a package shown in fig.6. It consists of a rigid mass simulating the good and a plate of PVC surrounding the mass and representing the packaging arrangement. By a frame the PVC-plate may be strained to a certain degree. If this system is excited harmonically we measure by an accelerometer fixed to the mass the resonance frequencies shown in fig. 7. They confirm the predicted behaviour of the system.

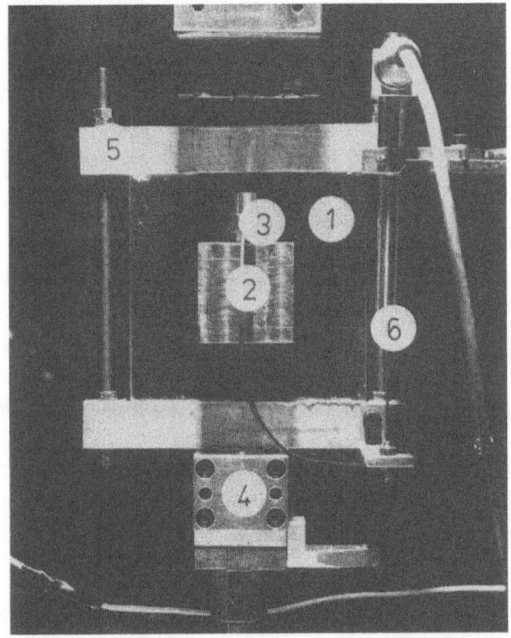

Legend:
1 PVC-plate (packaging)
2 mass (good)
3 accelerometer
4 shaft of the shaker
5 loading frame
6 displacement trans-
  ducer

Fig. 6: Model test of a good enclosed by packaging material

4.2   Photoviscoelasticity

Sometimes it is not sufficient to know the acceleration of a
good due to a harmonic excitation. Beyond the resonance behaviour of a good the stressing of it may also be of interest.
As fig. 2 indicates this may be done in a photoviscoelastic
experiment, e.g. by measuring the time varying isochromatic
and isoclinic fields in the PVC plate of the model investigated in chapter 4.1. An example of an isochromatic field at a
certain time is given in fig. 8. The determination of the
stress from the fringe orders is simply performed at points
located at the free surface of the plate where the stress
state is uniaxial using eq. (7). At all other points we must
refer to the two dimensional form of eq. (7). in combination
with one of the usual stress separation techniques of photoelasticity, e.g. shear difference method and oblique incidence.

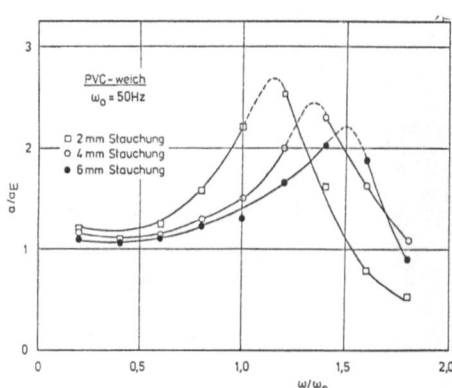

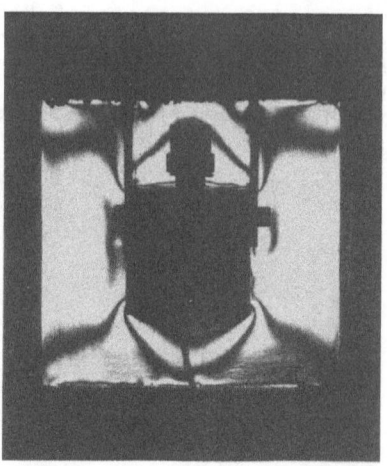

Fig 7: Resonance curves for the
      model test

Fig. 8: Isochromatic field
       in the packaging
       material

LITERATURE

/1/ Meßtechnik bei Transportbeanspruchungen.
Schriftreihe Transportkette, Band 35, Verkehr-Verlag
J. Fischer, Frankfurt/Pfinztal 1982

/2/ K.-F. Ziegahn
Wechselwirkungen zwischen Transportgut, Transportmittel
und Verpackung als Grundlage logistischer Planung
FhG-Berichte 3/4-84, S. 53-57, Hg.: Fraunhofer-Gesell-
schaft e.V. München, Pfinztal 1984

/3/ T. Plitt
Zum mechanischen und photomechanischen Verhalten von
vorbelasteten viskoelastischen Bauteilen unter dynamischer
Beanspruchung
Doctoral Thesis, Univ. of Karlsruhe, 1985

/4/ H. Weber, T. Plitt, A. Geißler
Betrachtungen viskoelastischer Materialien im Hinblick auf
die Ermittlung von Transportbeanspruchungen
to be published in VFI, Baltz Verlag München

LABORATORY TESTS ON ENCAPSULATED HIGH TEMPERATURE STRAIN
GAGES SG 425 FOR MEASUREMENTS UP TO 530°C

PETER HOFSTÖTTER

TÜV RHEINLAND, INSTITUT FOR MATERIAL TESTING,
COLOGNE, GERMANY

1. INTRODUCTION

In-service measurements of strain at temperatures up to 315°C
have been performed for years mainly in nuclear power plants
using strain gages SG 125/SG 128. All these measurements
helped to solve many problems for example to determine the
actual stresses during service or to supervise components to
show no dangerous vibrations. Before using this technique in
conventional power stations too, that means at temperatures up
to 530°C, strain gages SG 425 had to be tested for their suitability. This was done in numerous laboratory tests in the Institute for Materials Testing of TÜV Rheinland in Cologne on
behalf of the German technical inspection agencies and two
great German electric power generating companies (RWE and
PREAG). Testing took three years and was completed in January
1986 [1].

2. BACKGROUND

2.1. Strain gages

Strain measurements at temperatures up to 530°C can be performed using capacitive and resistance strain gages. In the
reported research program only resistance gages SG 425 of
EATON were tested. These gages, principally shown in Fig. 1,
consist of a PtW measuring wire with an active length of
17.3 mm which is pressed into a tube by means of MgO-powder.
This tube is connected with a flange which will be spotwelded
on the surface of the part to be measured. Outside the active
gage the thickness of the PtW-wires is increased by gold coating. These four wires, coming from the active and the dummy
circuit are TIG welded to copper wires within a junction box
and the copper wires, passing through a tube, will be connected with the measuring equipment. So the gage and the connecting cable are totally encapsulated in tubes and do not require more protection against dust, water etc. At the end of
the high temperature cable the circuit will be completed to a
full bridge using two resistances; the connection to the measuring equipment will be made in a six wires circuit. If a gage
of this type is heated in an oven up to 530°C without any connection to a part of material it will show an apparent strain.
If the gage is mounted on a part of material this apparent
strain is influenced by the coefficient of thermal expansion

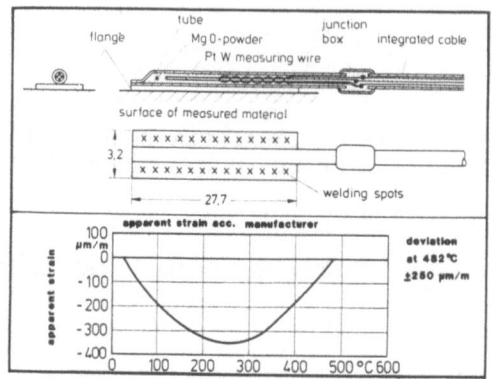

FIGURE   1:   Encapsulated high-temperature strain
gage SG 425

of the material. So the apparent strain of the gage depends on
the gage itself and on the material measured. This apparent
strain which is shown in Fig. 1 according to the manufacturer's data must be known to perform an accurate strain measurement. The manufacturer states that the deviation of the
apparent strain is ± 250 μm/m at 482°C. This deviation is
too high for a measurement to be accurate. So before taking
measurements the apparent strain of each gage has to be
determined on the material on which it will be used.

2.2. Research program
The aim of the research work was to develop a method to calibrate the gages before use, that means to measure their apparent strain, to control influences on this apparent strain, to
control the gage factor and to test the long time behaviour of
the gages.
It was planned first to test 10 gages SG 425. After having
obtained positive results 5 gages MG 425 with an active length
of only 8.6 mm were to be tested. Long time tests were planned
to be done at 530°C over a period of 1000 hours. During the
tests the results required some changes to be made: it became
necessary to test more gages and gages of another type and to
perform the long time tests at more temperature stages and
over longer periods. So 17 gages SG 425 with an austenitic
case - the standard version - and 10 gages SG 425 with a
case of Inconel were tested. Long time tests were performed
at 310, 340, 436, 494, 530 and 600°C over periods of up to
5000 hours.

3. APPARENT STRAIN
3.1. Shape
A typical shape of the apparent strain of gages SG 425 is
shown in Fig. 2, where both gages are spotwelded on Nimonic.
Gage "a" is a standard version with case and flange fabricated from an austenitic material. The first cycle shows a
broad hysteresis and a remaining strain, the second cycle a
broad hysteresis too. The next cycles were identical with
the second one.

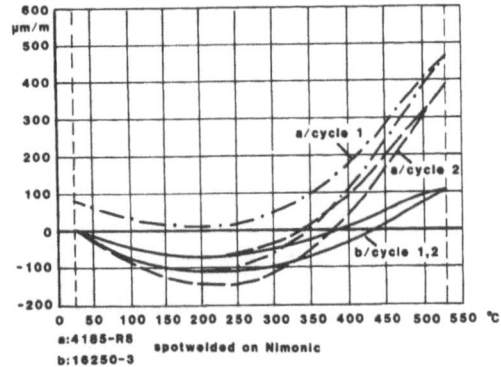

FIGURE 2: Apparent strain of SG 425

The difference between the first and second cycle requires accurate measurements to be taken at the second cycle. The shape of gage "b", a gage with case and flange fabricated from Inconel, a material with an thermal expansion coefficient similar to that of a ferritic material, is more satisfactory with the first and second cycle being identical, and hysteresis and gradient being smaller. The reason why is that there are no forces between the gage and the ground material if the thermal coefficient of expansion is the same. This is true for Inconel and ferritic material and so the manufacturer was asked [2] to offer standard gages for ferritic material fabricated from Inconel and standard gages for austenitic material from austenite as done so far.

3.2. Influence of heated cable
During the research work the gages were tested with a length of 300 mm of cable within the oven. This length is realistic for measurements on pipes, vessels etc. to lead the cable out of the heated zone through the insulation. If length of cable heated is different, the apparent strain will change as shown in Fig. 3 for 530°C. The influence of the heated length of cable was found to be -100 µm/m per 1 m cable at 530°C with a deviation of ± 30 µm/m. It is recomended to calibrate the gage with the length of cable that will be heated during the measurement. In this case no correction is necessary.

3.3. Influence of temperature compensation resistance
The manufacturer promises to adapt the apparent strain of a gage to the coefficient of thermal expansion of any material. This is done using the temperature compensation resistance RTC. With one gage the influence of this resistance on the apparent strain was found to be -300 µm/m per 1 Ω at 530°C. If each gage of a test is calibrated on the material on which it will be used no adaption to the material is necessary except for using gages with austenitic flange for austenitic material and with flange of Inconel for ferritic material. This method is recommended.

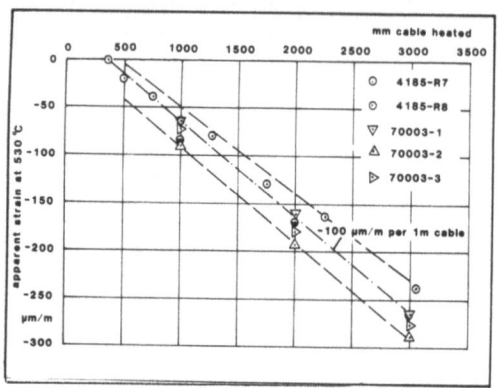

FIGURE   3:   Change of apparent strain of SG 425
with length of heated cable

4. CALIBRATION

The term "Calibration" as used in this paper means the measurement of the apparent strain of a gage on the material on
which it will be mounted.

4.1. Method of calibration

There are two methods of calibration of high temperature
strain gages used in W. Germany: twice spotwelding [3] and
clamping of gages [4], [5].

In this research work the methods according to [3] and [4]
were tested for comparison, but the method described in [5]
was applied as TÜV has gained extensive experience in practical measurements with this method. The clamping device used
is described in detail in [5].

First tests with this clamping device and standard gages SG
425 with a flange of austenitic material showed no sufficient
reproducibility. Each new clamping of a gage lead to a change
in apparent strain and the same effect was observed with the
clamping device described in [4]. The reason was found to be
the difference in the coefficient of thermal expansion beween
the material of the clamping device and the material of the
gage. During heating this difference leads to forces building
up which, with each clamping, cause changes in the gage and in
the apparent strain. To prevent this effect it is necessary to
have gages and material with the same coefficient of thermal
expansion. This is the second reason to ask the manufacturer
for standard gages fabricated from Inconel for measurements
on ferritic materials.

If austenitic gages are to be used on ferritic material, the
only one calibration method is to spotweld the gage at a specimen to get the apparent strain, to cut it off with a scal-
pel and to spotweld it to the material.

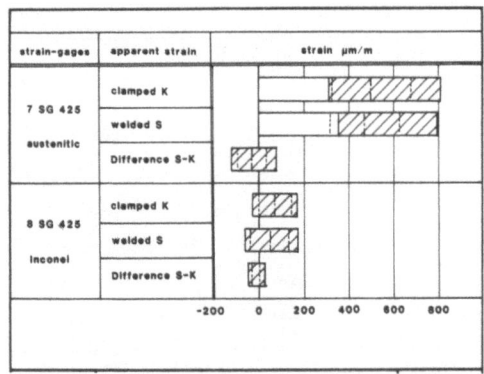

strain-gages	apparent strain	strain µm/m

FIGURE   4:   Apparent strain of clamped and
                spotwelded gages

4.2. Reproducibility

The difference between the apparent strain of the spotwelded
and the clamped gage shows the reproducibility of the method
of calibration. Fig. 4 illustrates this for 7 gages with
austenitic flanges and 8 gages with flanges fabricated from
Inconel. The beams show maximum and minimum values and the
standard deviation of the 7 or 8 values respectively. It can
be seen that both the single values of the apparent strain
of clamped and spotwelded gages and the difference are much
better with the Inconel flange gages than with the austenitic flange gages. The difference of the apparent strain at
530°C showed a standard deviation of 69 µm/m for the 7
austenitic gages and a standard deviation of 25 µm/m for 8
gages fabricated from Inconel. This is the same deviation
we established for gages SG 125/SC 128 at 315°C using our
clamping device.

5. GAGE FACTOR
5.1. Control of gage factor

To control the gage factor a triangle bending beam was used,
according to [6]. As the thickness of this beam was 4 mm, the
nominal strain of ± 1000 µm/m arises from the middle of the
beam to the surface from 0 to 1000 µm/m. So it can easily be
seen that there is a great influence of the height of the
active gage above the surface. Our tests showed that this
method is not sufficient to control the gage factor.

5.2. Change of the gage factor with temperature

The triangle bending beam which could be loaded from outside
the oven was a good tool to test the change of the gage factor
with temperature. Fig. 5 shows the results of 4 gages which
confirm the value stated by the manufacturer to be 3.24 %
per 100 K.

5.3. Change of the gage factor with time

Due to damages of gages during long term tests, which will be
reported later, the change of the gage factor with time could
be tested only with one of the 4 gages at the bending beam

584

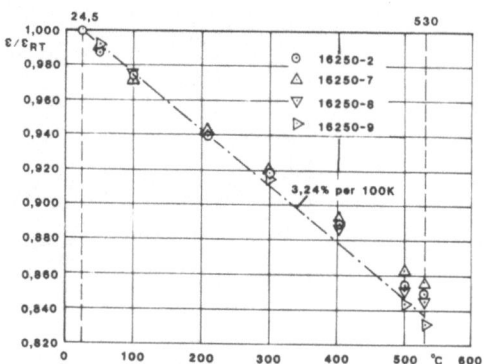

FIGURE   5:   Change of gage factor of SG 425 with temperature

after 3042 hours at 494°C. This gage showed no change of the gage factor with time.

6. DRIFT OF THE GAGES

One of the main issues of the research work was the drift of the gages, that means the zero-shift of the gages, mounted on a specimen of a stable material — in the research work Nimonic was used — without any external load. The following table gives an overview about the drift tests:

test	temperature	testing time	gage	drift rate per day after 1000 h	end of signal	load
	°C	h	No.	µm/md	h	µm/m
	310	516	4185-R7	+ 1.13	–	0
	340	2370	5050- 7	+ 0.11	–	0
			4185-R9	+ 0.32	–	0
			16250- 5	+ 0.14	–	0
	436	+1258	4185-R7	– 1.80	–	0
	530	2932	4185-R8	+ 0.65	–	0
			4287- 4	– 6.90	1272	0
drift		3437	5050-10	– 7.64	1657	0
		3963	16250- 3	– 4.28	1895	0
			16250- 4	– 5.55	1679	0
			16250- 5	– 1.58	3527	0
		2900	70003- 1	– 5.63	–	0
			70003- 2	– 9.70	–	0
			70003- 3	– 3.93	–	0
	600	+4936	5050- 7	+ 3.13	–	0
			4185-R9	+15.18	–	0
			16250- 6	+ 0.74	–	0
creep	494	3042	16250- 2	– 2.81	–	+ 1000
			16250- 7	– 2.01	2400	+ 1000
			16250- 8	– 1.54	2928	– 1000
			16250- 9	– 6.81	2856	– 1000

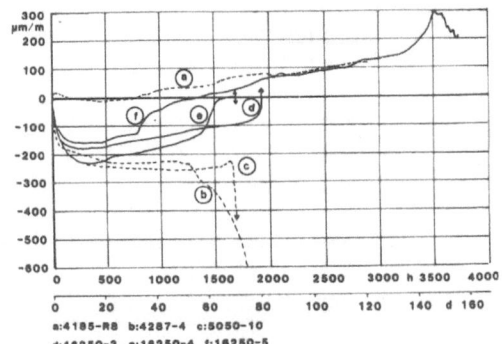

FIGURE   6:   Drift of 6   SG 425 at 530°C

Fig. 6 shows the drift of 6 gages over the time at 530°C.
At the end of this test after 3963 hours 5 out of 6 gages had
failed due to an interruption in the cable, the last one gage
was tested only for 2932 hours. On the contrary, no failure
was found with 3 gages at 600°C over a time of 4936 hours and
with 3 gages with a filled junction box over more than 2900
hours at 530°C.
In Fig. 7 the drift rates in μm/m per day after 1000 hours are
summarized over the temperature. This figure is showing all
values of this research program and also a great many of values found in the literature as well as the drift rates of
NiCr gages SG 125/SG 128. There is a sudden break in the drift
rates of SG 425 from 300°C to 550°C compared with values at
about 600°C.

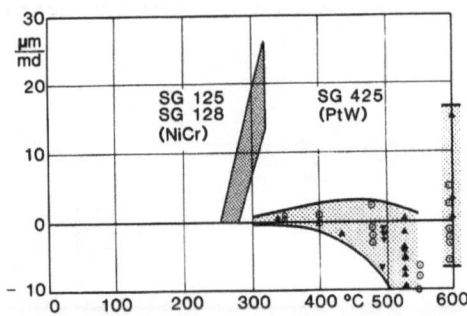

FIGURE   7:   Drift rate of SG 425

7. CREEP OF THE GAGES
Creep of strain gages is the lessening of an imposed strain.
4 gages were tested over a period of 3042 hours at 494°C, 2 of
them loaded with + 1000 μm/m and the other 2 with - 1000 μm/m
at the triangle bending beam.
Fig. 8 shows the strain versus time. The signals do not indicate any influence of the imposed loads; they are fully within
the range of the drift rates in Fig. 7. So it can be stated

586

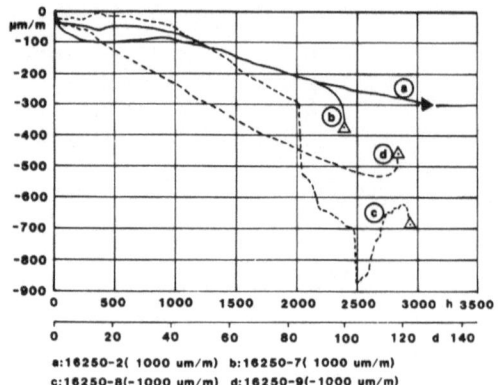

a:16250-2(1000 um/m) b:16250-7(1000 um/m)
c:16250-8(-1000 um/m) d:16250-9(-1000 um/m)

FIGURE   8:   Creep and drift of 4   SG 425
               at 494°C

that no significant creep could be found. 3 out of the 4 gages
tested failed before the end of the testing time of 3042 hours.

8. CHANGE OF APPARENT STRAIN WITH TIME
In a research work on NiCr strain gages SG 125/SG 128 [7] a
change of apparent strain at higher temperatures with time was
found. This effect was observed also at gages SG 425, as shown
in Fig. 9 compared with the results of SG 125/SG 128. Unfortunately due to the failure of many gages during long time
tests, only a few values could be obtained. Strange values
were recorded with 3 gages SG 425 with a filled junction box.
This problem needs more careful investigation.

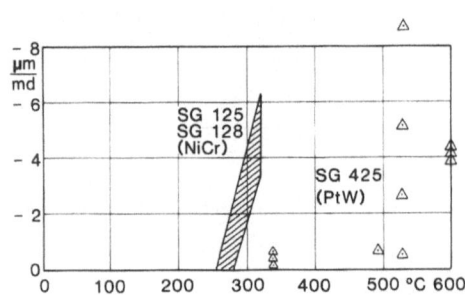

FIGURE   9:   Change of apparent strain of SG 425

9. LIFETIME

The main problem with gages SG 425, encountered in this research program, was lifetime. This can be seen from the table in Chapter 5: 3 out of 4 gages tested at 494°C failed after about 2000 hours. At 530°C 5 out of 6 gages failed after 1200 to 3500 hours. 3 more gages with filled junction box did not show a failure over 3000 hours at 530°C. 3 gages tested at 600°C did not fail after 5000 hours. The failure was found to be due to a separation within the junction box, starting with an increasing resistance up to a full separation of the circuit. Fig. 10 shows an opened junction box of a gage after failure. 4 gold wires extend from below the gage, 3 copper wires from the upper side leading to the integrated cable; both copper and gold wires TIG welded to a sphere. The separation can be seen in Fig. 10 to be located at the transition from the copper wire to the spere of copper and gold. A scanning electron microscope photo in Fig. 11 shows the surface of the separation at the side of the sphere in a higher enlargement. This surface is not like that of a typical crack caused by force or vibration. So it is supposed that the reason of the separation is an effect of diffusion of two materials with different diffusion rates like the Kirkendall-effect. A possibility to prevent this effect and to improve the long time behaviour of the gages might be to use other materials.

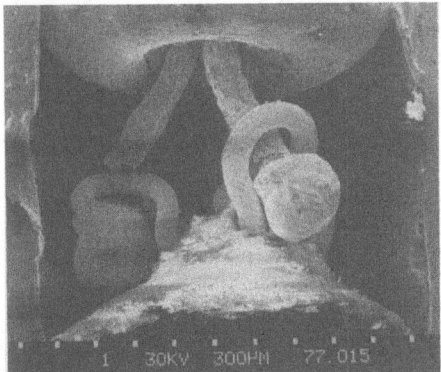 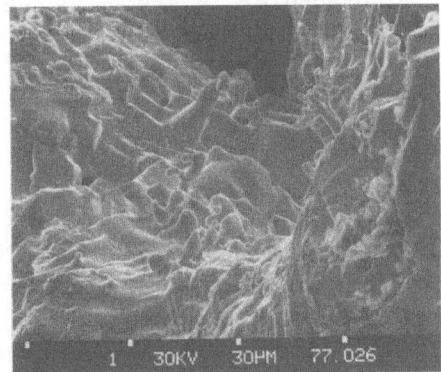

FIGURE 10: Junction box of SG 425 with connection of wires

FIGURE 11: Surface of separation of spere within the junction box

Another problem of the gages is the sensitivity to vibration. As in most cases it is not known whether or not there will be vibration on the gage installed it is necessary to use gages with filled junction box. So the manufacturer is asked to offer standard gages with filled junction boxes.

10. CONCLUSION

In a research work strain gages SG 425 were tested at the Institute for Materials Testing of TÜV Rheinland, Cologne, for use at temperatures up to 530°C, a temperature common in conventional power stations. Short time and long time tests produced the following results:

- Strain gages SG 425 are suitable for strain measurements within a short time.
- It is necessary to calibrate the gages on the material on which they will be used for example by using the clamping device, as described in this paper.
- For measurements on ferritic materials it is necessary to use gages fabricated from Inconel, a material with a coefficient of thermal expansion similar to that of ferritic materials. So in contrast to austenitic standard gages the apparent strain has a better shape and calibration can be done by using a clamping device.
- It is recommended to calibrate the gages with the length of cable needed or to calculate the influence of the heated length of cable.
- At temperatures of 260°C and over this type of gage is much more stable than NiCr gages SG 125/SG 128 are.
- At temperatures of 494 to 530°C many gages failed after 1200 to 3500 hours, probably because of an effect of diffusion between copper and gold within the junction box. Here the manufacturer is asked to use other materials.
- Practical measurements have shown that in most cases, even in static measurement, vibration on the gage cannot be prevented. So the manufacturer is asked to offer standard gages with filled junction box.

REFERENCES

1. Hofstötter P: Erprobung des Einsatzes gekapselter Hochtemperatur-Dehnungsmeßstreifen für Dehnungsmessungen bei Temperaturen bis 530°C, VdTÜV-Forschungsvorhaben F 172
2. Hofstötter P: The Use of encapsulated High Temperature Strain Gages at Temperatures up to 315°C, Experimental Techniques, Aug. 1985, Pages 24-29
3. Böhm W: Hinweise zur Anwendung des 2-fachen Aufschweißens von EATON-Hochtemperatur-Dehnungsmeßstreifen zur Voruntersuchung von DMS-Eigenschaften, VDI-Berichte Nr. 480, 1983 Seite 167-169
4. Amberg C, Czaika N: Zur Vorherbestimmung und Reproduzierbarkeit des Nullpunkttemperaturganges aufschweißbarer DMS bis 320°C bzw. 600°C, VDI-Berichte Nr. 480, 1983 Seite 145-150
5. Hofstötter P: Calibration of High-temperature Strain Gages with the Aid of a Clamping Device, Experimental Mechanics, June 1982, Pages 223-225
6. VDI/VDE-Richtlinien 2635, Dehnungsmeßstreifen mit metallischem Meßgitter, Kenngrößen und Prüfbedingungen, August 1974
7. Hofstötter P: Untersuchung gekapselter Hochtemperatur-Dehnungsmeßstreifen unter Anwendung zweier Kalibrierverfahren, Technische Überwachung 24(1983), Nr. 4(April), Seite 162-165

STRAIN MEASUREMENTS
AT VARIABLE TEMPERATURES UP TO 300 °C
WITHIN A PRESTRESSED CONCRETE PRESSURE VESSEL
BY MEANS OF ENCAPSULATED WELDED STRAIN GAGES

C. Amberg

Bundesanstalt für Materialprüfung, Berlin'

1. INTRODUCTION

This paper represents the 2nd part of the report on a test series with fully encapsulated weldable tube-sized strain gages (s. g.) from Ailtech/USA installed within a prestressed concrete pressure vessel (PCPV) with a hot steel liner /1/ in the Forschungszentrum Seibersdorf in Austria.

The 1st part shows that these s. g. are very well suited for long-term supervision of concrete structures over a period of more than 10 years under temperatures up to 120 °C. To avoid repetitions it is referred to this part regarding information

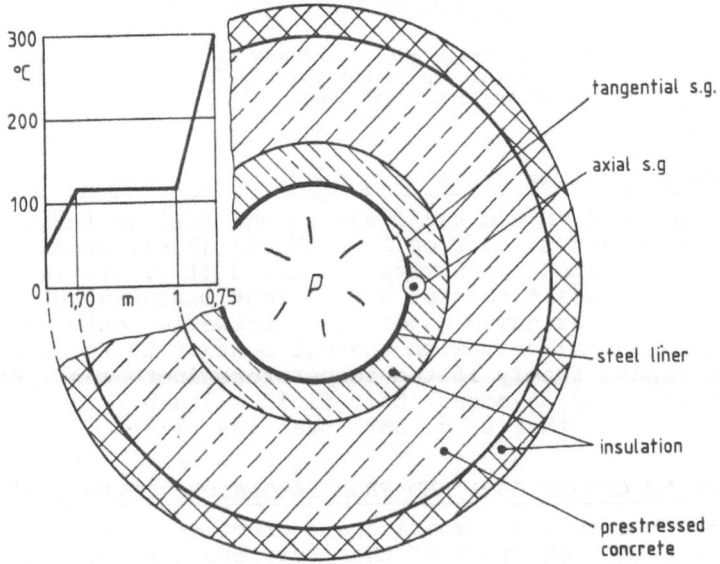

fig. 1   Cross-section of the PCPV in Seibersdorf/Austria
with values of temperatures at nominal operating
conditions

on the construction of the s. g., the measured area in the vessel, the arrangement of the s. g. at the liner and the installation techniques as well as to the covers of the active and dummy measuring points at the liner and within the concrete.

This 2nd part deals with the suitability of these s. g. to determine and to control the expansion restraint of the hot liner during the heating (and pressure) processes up to 300 °C. Correcting curves obtained from laboratory investigations are used and error evaluations are given.

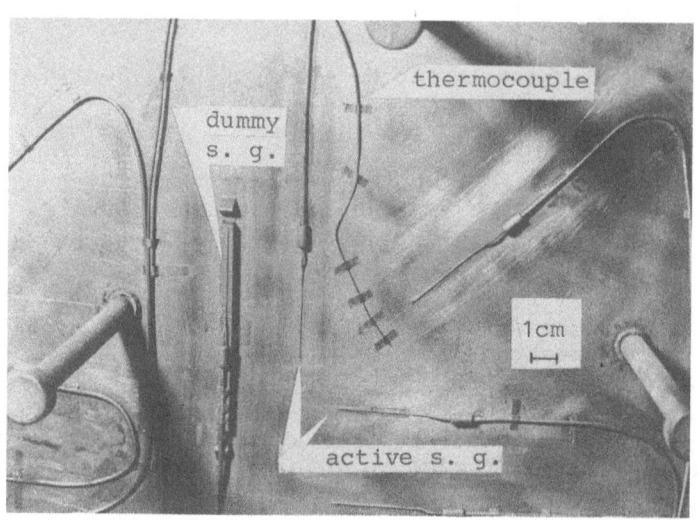

fig. 2 4 active strain gages, a dummy s. g. and a thermocouple within an anchor bolt field at the concrete
side of the hot steel liner

2. 18 ENCAPSULATED WELDED STRAIN GAGES AT THE LINER

At the concrete side of the liner in the PCPV 18 fully encapsulated tube-sized s. g. from Ailtech/USA (3 different s. g. types) and 7 thermocouples were installed (fig. 1). 14 active s. g. have been positioned at 3 different measuring directions; 4 (active) s. g. as dummy gages were spotwelded onto small bars from liner material only with thermal contact to the liner (the bars expand freely if the temperature increases). Fig. 2 is a photograph of an anchor bolt field; within 4 fields located next to each other s. g. have been installed.

3. CORRECTING CURVES OBTAINED FROM LABORATORY INVESTIGATIONS

In order to ensure a temperature correction of the strain measurements the behaviour of the 3 different s. g. types under variable temperatures between 30 and 300 °C was previously determined in the laboratory: The producer's specifications concerning the value and temperature dependence of the sensitivi-

ty (fig. 3a) were verified by means of a few (lost) gage specimens within a high temperature strain transducer calibration
device with a 4-point bending beam.

The apparent strain (fig. 3b, c) on the liner material (at
free thermal expansion) was predetermined for each s. g. individually within clamping devices /3/ manufactured from liner
material (fig. 4) and heated within furnaces.

Subsequent investigations showed that this predetermination
of the apparent strain up to 300 °C is possible within an
error of + 25 µm/m /4, 5/. An additional error is to be considered which occurs due to different heated cable lengths in

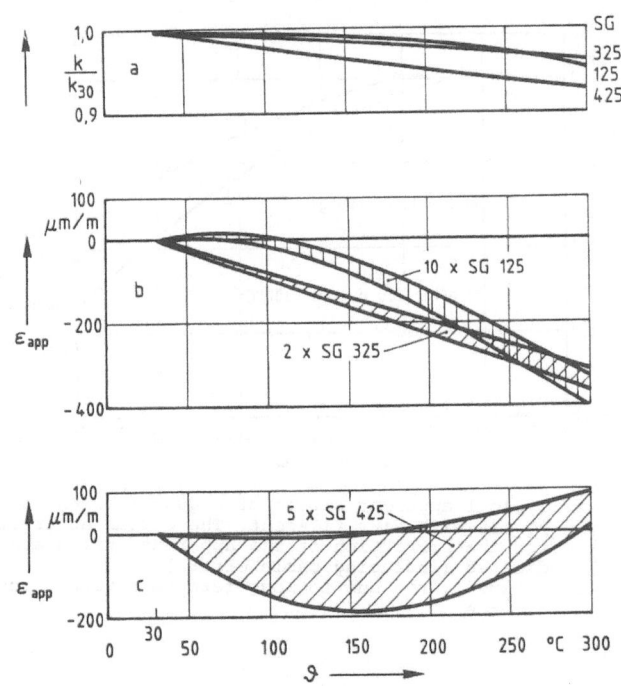

fig. 3 Predetermined correcting curves of the weldable
strain gages: sensitivity change by temperature (a)
and apparent strain (b, c) of the 3 s. g. types
SG 125, SG 325 (both with NiCr sensing filament)
and SG 425 (PtW filament)

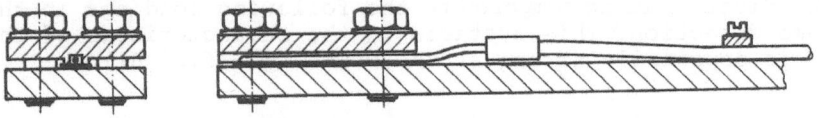

fig. 4 Clamping device for the predetermination of the
apparent strain of the s. g. on liner material

592

the laboratory and later within the vessel. The apparent strain of all 3 s. g. types can be corrected with the evaluated errors in fig. 5a.

A systematic investigation /4/ showed a typical alteration of the apparent strain curve of the quarter bridge s. g. SG 125 during the influence of temperatures above 260 °C caused by metallurgical alterations within the NiCr filament of the s. g. Regarding the vessel at a liner temperature of 300 °C over a time period of (summarized) 250 days an altered curve like that of fig. 5b is to be expected for the s. g. type SG 125. The respective alteration of the 2 halfbridge s. g. types SG 325 (NiCr filaments) and SG 425 (PtW) is negligible.

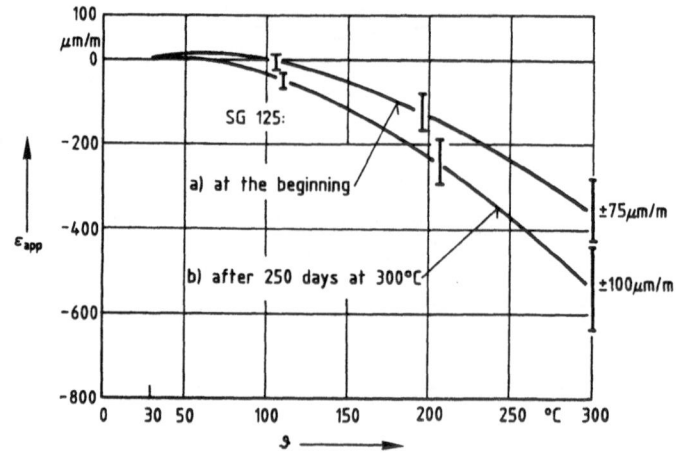

fig. 5 a) Typical apparent strain of a quarterbridge s. g.
           SG 125 with NiCr filament. The estimated errors
           for the correction of the apparent strain are in-
           dicated (including cable errors)
        b) Altered curve due to long-term temperature in-
           fluence

4. HYSTERESIS ERROR DURING THE FIRST LOADING OF THE GAGES

In the heating process of the vessel up to 300 °C liner temperature the liner was compressed up to -2000 µm/m. The NiCr
measuring wires were compressed heavier and the PtW wires less
because of the different thermal coefficients of the wire materials and the liner. However, as our s. g. investigations
show /3/ an essential smaller s.g. reading is to be expected at
the first loading compared to the following loadings in the
same direction. This hysteris in the characteristic of the s.
g. is especially strong at the PtW-s. g. (fig. 6I).

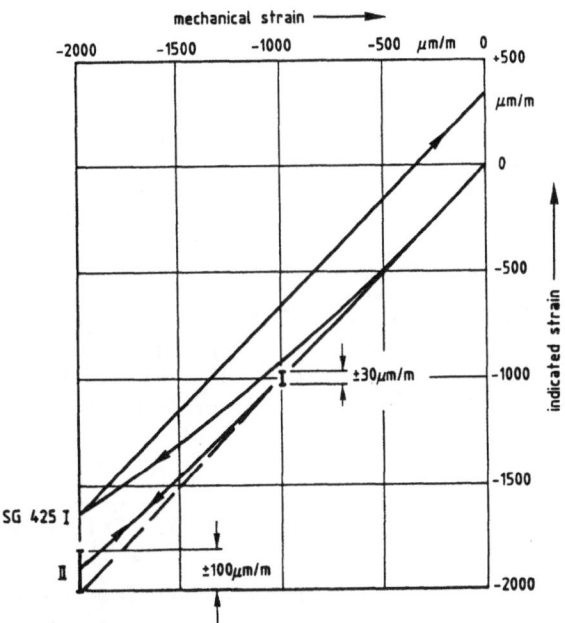

fig. 6 I) Typical characteristic of a PtW-s. g. at the
            first compression
        II) Characteristic with evaluation of errors at the
            following compressions of all 3 types of s. g.

At the following loadings up to about -1000 μm/m the gage
factor and its spread of ± 3 % - both given by the producer -
can be used for the strain computation and error evaluation;
at -2000 μm/m an additional nonlinearity of about 5 % is to be
expected (fig. 6 II).

The s. g. hysteresis can be recognized clearly when comparing
the consecutive vessel heating and cooling processes (cycles)
No. 5 and No. 6A (the values in fig. 7 are already temperature
corrected by using the values of the curves of fig. 3):

During cycle No. 5 the liner was heated to 300 °C for the
first time, previously only up to about 200 °C. At the second
identical loading of the liner during the cycle No. 6A both
the tangential and the axial s. g. indicated larger compressions than during cycle No. 5 (vessel condition 3), and after
cooling down to 200 °C (condition $\overline{2}$) in cycle No. 6A a considerably better agreement with the 200 °C-values (condition 2,
before the heating procedure to 300 °C) is obtained than in
cycle No. 5: in cycle No. 5 the values of the vessel condi-
tions 3, $\overline{2}$, $\overline{1}$ and $\overline{0}$ are influenced by the hysteresis error of
the s. g. occuring during their first loading (fig. 7).

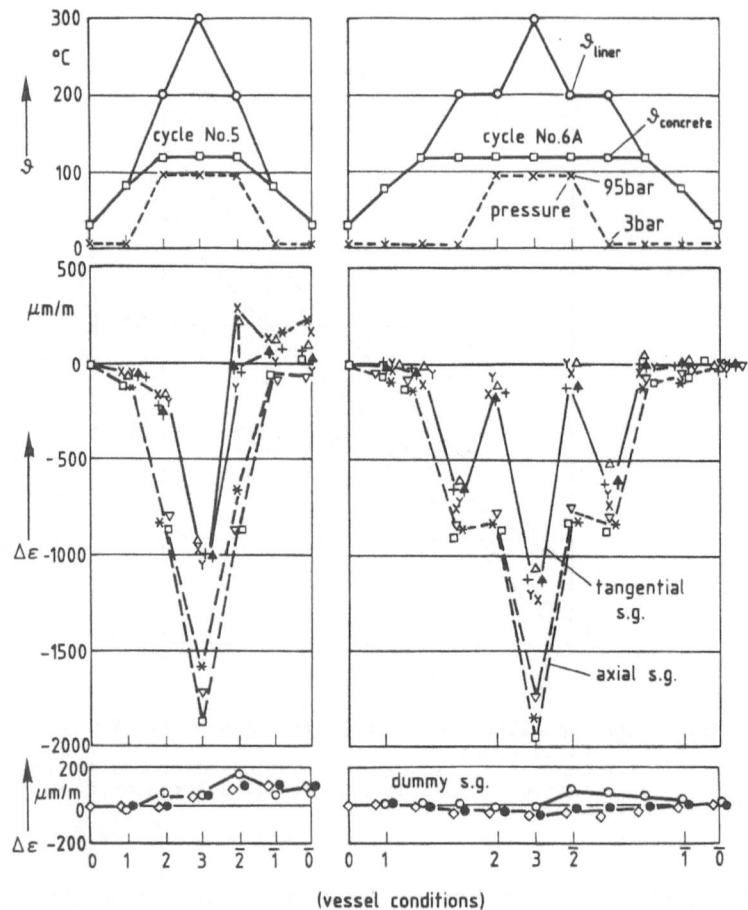

fig. 7 Cycle No. 5: The hysteresis of the s. g. during the
compression at the first heating procedure to 300 °C

Cycle No. 6A: Typical values of the measured liner
strains at different vessel conditions

5. THE OPERATING BEHAVIOUR OF THE LINER

The strain values from the cycle No. 6A in fig. 7 have also
been measured during the subsequent cycles. They represent the
operating behaviour of the measured liner area at the different temperature and pressure conditions of the vessel.

In fig. 8 these values are taken as a function of temperature. Due to the temperature correction by using the values of
the curves of fig. 3 no strain change (during changing temperatures) in fig. 8 indicates that the measured liner part expands freely. Thus, especially the small dummy s. g. values
prove that the correction of the apparent strain by the clamping method was possible with a slight error.

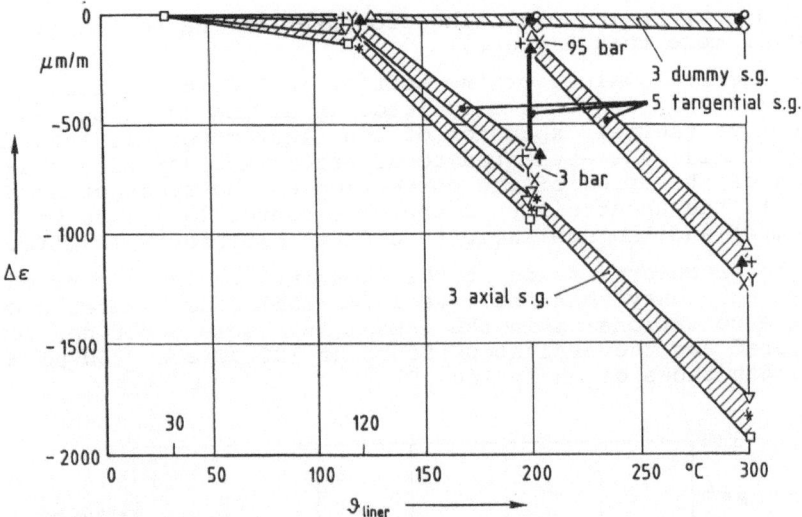

fig. 8   The operating behaviour of the measured liner area
          as a function of the liner temperature and the inner
          vessel pressure; scatter of measured (temperature
          corrected) strains

Fig. 8 also shows that the liner expands nearly freely up to
120 °C (the prestressed concrete (fig. 1) was simultaneously
heated up to 120 °C and then kept at constant temperature).
During the further heating procedure the concrete restrained
the thermal liner expansion, and at 300 °C caused an axial
compression of nearly -2000 μm/m.

The inner vessel pressure was increased from 3 to 95 bar at
200 °C so that the liner did extend in tangential direction remained, however, under compression as intended. Pressure has
only little effect on the axial strain.

6. SUITABILITY OF THE S. G. AND MEASURING ERRORS

Fig. 8 shows the scatter of the temperature corrected values
of 11 s. g.

The readings of one further tangential s. g. were up to 200
μm/m higher than the maximum readings of the other 5 tangential s. g.

The values of 3 s. g. positioned in a direction of 45 ° were
ranging between those of the tangential and the axial s. g. as
was to be expected. They are not given in this paper.

Three of the 18 s. g. at the liner failed already long before
the start of the heating cycles /2/.

The expansion restraint of the liner could be maesured within
the scatter shown in fig. 8 reproducibly at 14 s. g. (disregarded the errors occuring during first loading as explained by

fig. 6 and 7) during 9 heating processes of the vessel over a period of more then 6 years.

From the mean values of the scatter in fig. 8 the real liner strain (fig. 9) is to be calculated by adding the known values of the free thermal expansion of the liner material. Into fig. 9 additionally the estimated total errors are entered, consisting of the error of the correction of the apparent strain (fig. 5), the uncertainty of the gage factor (+ 3 % up to 1000 µm/m) and the nonlinearity of the characteristic (fig. 6).

If the predetermination of the apparent strain is carried out in such a way that the integrated cables are heated under the same conditions as in the vessel, no cable error has to be considered and the evaluated errors at 300 °C are ± 50 µm/m lower than those given in fig. 9.

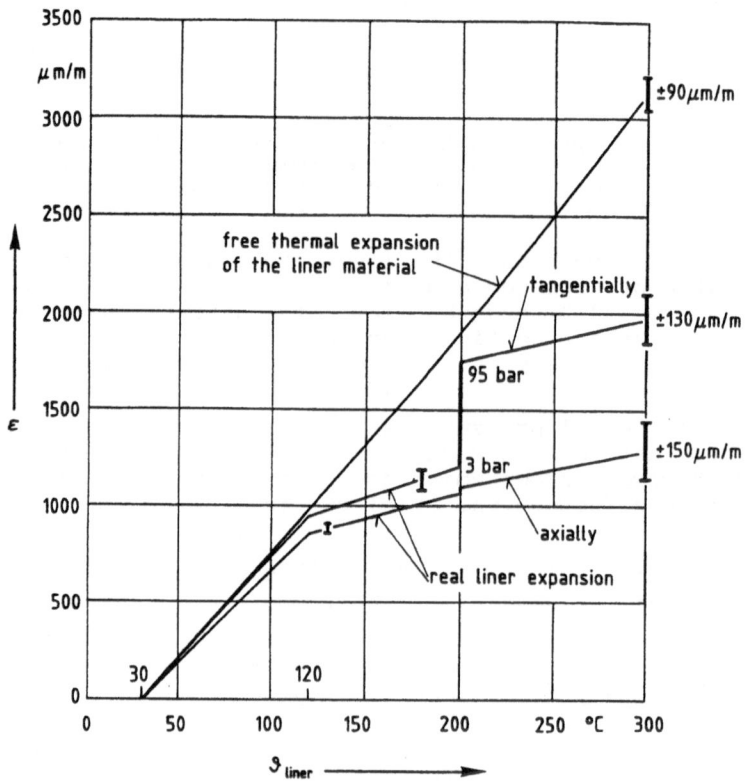

fig. 9   The real tangential and axial strains of the mea-
         sured liner area and real strain of the dummy bars
         from liner material; with evaluation of the measu-
         ring errors at some of the temperatures

By carrying out the corrections given here (especially concerning the apparent strain) and by considering the listed non-
correctable measuring errors all 3 types of investigated s. g.
are well suited for the long-term supervision of the expansion
restraint of the liner within a PCPV between room temperature
and 260 °C. This applies also to all 3 s. g. types up to 300 °C
if one temperature step that is to be measured only needs a
few hours (or one day).

The presentation and assessment of the long-term errors of
these s. g. at constant temperatures (especially above 260 °C)
will be given in a further part of this test report.

ACKNOWLEDGMENTS

The author once again thanks for the very helpfull cooperation, as was already done in the first part /2/, and wants to express his appreciation to
Mr. L. Weißbacher from the Reaktorbau Forschungs- und Baugesellschaft (RFB)
for the technical support in-situ and to Mr. A. Witt from the Österreichi-
sches Forschungszentrum Seibersdorf (ÖFZS) for the coordination of the work
during the vessel heating processes and to Prof. N. Czaika from the Bundesanstalt für Materialprüfung (BAM) Berlin, for the useful discussions on
the assessment of the s. g. behaviour.

REFERENCES

/1/ Zemann, H., Weißbacher, L., Mayer, N. and Amberg, C.:
     In-service Supervision of a Prestressed Concrete Pressure Vessel.
     Int. J. Pres. Ves. + Piping 21 (1985) 121-156 and ÖFZS-Bericht 4308,
     Jan. 1985, pp 1-55.

/2/ Amberg, C., Czaika, N.:
     On long term strain measurements in prestressed concrete structures at
     temperatures up to 120 °C with encapsulated weldable strain gages.
     Prepr. IMEKO '85, April 22-26, 1985, Praha/CSSR, Vol. 3, pp. 78-85.

/3/ Amberg, C., Czaika, N.:
     About Applicability of High Temperature Strain Gauges and Differential
     Transformer Displacement Transducers for the Instrumentation of Pres-
     tressed Concrete Reactor Vessels.
     Prepr. 2nd Int. Conf. SMIRT, Sept. 10-14, 1973, Berlin, Vol. VI - part
     A Suppl., H 4/8.

/4/ Amberg, C., Czaika, N.:
     Zur Vorherbestimmung und Reproduzierbarkeit des Nullpunkttemperatur-
     ganges aufschweißbarer DMS bis 320 bzw. 600 °C.
     VDI-Berichte Nr. 480 (1983), 145-150.

/5/ Hofstötter, P., Weichsel, J.:
     Einsatz einer Klemmvorrichtung zur Kalibrierung von Hochtemperatur-
     Dehnungsmeßstreifen.
     Techn. Überwachung 21 (1980) No. 4, pp. 147-150.

LOSS OF STABILITY OF UNDERGROUND EXCAVATIONS - VIEW ON A POSSIBILITY OF PREDICTION

Z.KŁECZEK, A.ZORYCHTA, A.SZUMIŃSKI, A.ŁUCZAK, Z.IWULSKI
University of Mining and Metallurgy, Cracow, Poland

1. INTRODUCTION

The continuous demand for raw materials causes an increase of the exploitation depth and as a result of this the increase of stress values in the rock massive is observed. In consequence the phenomena of the loss of stability of underground excavations take place very often. Some of them cause disturbances in mining processes and the other /rock bursts/ make the threat for miners life [1]. So the need of conducting underground measurements for prediction such phenomena are evident. There is also another reason which speaks for in-situ measurements. It is caused by their cognitive character. The results of underground measurements may be used as a verification of the theoretical considerations of rock mechanics

2. A METHOD OF THE STRESS MEASUREMENTS

At present a lot of methods of stress measurements are known [1]. These methods divide the measurements of :
- primary stresses in rock massive i.e. the stresses which exist when there are no underground excavations,
- secondary stresses due to mining.

The suggested method permits measuring the both kinds of stresses. The idea of the method consists in setting up in rock massive a stress gauge /Fig.1/ and next in making the destressing hole near by the gauge /Fig.2/.

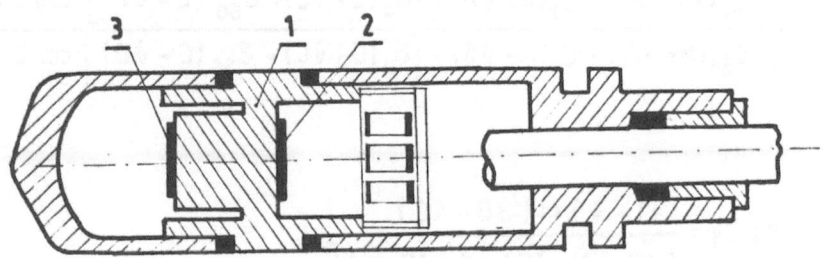

1 - measuring element
2 - active strain gauge
3 - compensative strain gauge

Fig. 1

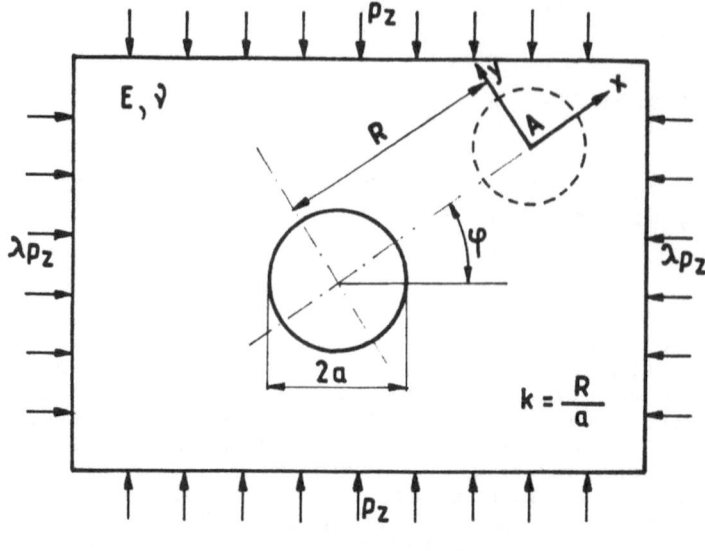

Fig. 2

After making this hole the state of stresses changes and in consequence the change of stresses in the gauge can be measured. The state of stresses /primary or secondary/ is determined by the formulae :

a/. For rectangular rosette /strains $\varepsilon_0, \varepsilon_{45}, \varepsilon_{90}$ are measured/:

$$p_z = \frac{E}{B(1-\nu^2)}\left[(\varepsilon_0 - \nu\varepsilon_{90}) + \frac{1-\nu}{2F}\alpha\left(\frac{B}{\sin 2\varphi} + \frac{C}{tg\,2\varphi}\right)\right]$$

$$\lambda = \frac{\varepsilon_0(A-\nu B) - \varepsilon_{90}(B-\nu A) + \left[\varepsilon_0(D-\nu C) + \varepsilon_{90}(C-\nu D)\right]\cos 2\varphi}{\varepsilon_{90}(B-\nu A) - \varepsilon_0(A-\nu B) + \left[\varepsilon_0(D-\nu C) + \varepsilon_{90}(C-\nu D)\right]\cos 2\varphi}$$

/1/

$$tg\,2\varphi = \frac{(1-\nu)\alpha(BD-AC)}{F\left[\varepsilon_0(A-\nu B) - \varepsilon_{90}(B-\nu A)\right]}$$

where:

$$\alpha = 2\varepsilon_{45} - \varepsilon_0 - \varepsilon_{90}$$

b/. For delta rosette /strains ε_0, ε_{60}, ε_{120} are measured/:

$$p_z = \frac{E}{B(1-\vartheta^2)}\left[\varepsilon_0 - \frac{\vartheta\beta}{3} + \frac{2\sqrt{3}(1-\vartheta)\beta}{3F}\left(\frac{B}{\sin 2\varphi} + \frac{C}{tg 2\varphi}\right)\right]$$

$$\lambda = \frac{\beta(B-\vartheta A) + \varepsilon_0(A-\vartheta B) - \left[\beta(C+\vartheta D) - 3\varepsilon_0(D+\vartheta C)\right]\cos 2\varphi}{\beta(\vartheta A - B) - \varepsilon_0(A-\vartheta B) - \left[\beta(C+\vartheta D) - 3\varepsilon_0(D+\vartheta C)\right]\cos 2\varphi}$$

$$tg\, 2\varphi = \frac{\sqrt{3}\cdot\gamma\,(BD+AC)}{3F\left[(B-\vartheta A)\,\beta + \varepsilon_0(A-\vartheta B)\right]}$$

/2/

$$\beta = \varepsilon_0 - 2\varepsilon_{60} - 2\varepsilon_{120}$$

$$\gamma = \varepsilon_{60} - \varepsilon_{120}$$

$$A = 1 + k^2$$

$$B = 1 - k^2$$

$$C = 1 - 4k^2 + 3k^4$$

$$D = 1 + 3k^4$$

$$F = 1 + 2k^2 - 3k^4$$

$$k = a/R$$

It should be stressed that this method can be used only when the stiffness of the gauge is the same as the stiffnes of rocks. Adopting the solution [3] the necessary condition has a form :

$$\left(\frac{R_0}{R_1}\right)^2 = \frac{\xi(1+\vartheta) - 2\left[1 + \vartheta_g(1-2\vartheta)\right]}{\xi(1+\vartheta) - 2(1+\vartheta_g)}$$

/3/

$$\xi = E_g/E$$

It should be mentioned that the method can be used not only for elastic rocks but also for rocks which have rheological properties.

3. PREDICTION THE LOSS OF STABILITY

Underground observations show us that there are two kinds of a loss of stability of underground excavations: quasistatic loss of stability or dynamic one. The first one causes an increase of loads on underground support and the second one is the reason of rock - bursts phenomena. It can be proved that these processes are caused by fracture of rocks. So let us show the method of forecasting the rock fracture. This method rely on in-situ measurements.

At present a sufficient accuracy for mining purposes gives us the Coulomb-Mohr criterion of fracture

$$p_z \left(1 - \frac{1+\sin \varrho}{1-\sin \varrho} \lambda \right) = R_c \qquad /4/$$

where:

$$p_z > \sqrt{} (1+\lambda) p_z > \lambda p_z$$

ϱ - angle of internal friction
R_c - compressive strength

From this criterion it can be shown [4] that the zone of fractured rocks will occur when:

$$\lambda_s < \frac{1-\sin \varrho}{1+\sin \varrho} \qquad /5/$$

So taking into account /1/ or /2/ and /5/ we see that prediction of the quasi-static loss of stability is possible.

For prediction the dynamic loss of stability the additional date /beside the existence of fractured rocks/ must be taken into consideration. These data are connected with an energy capacity of rocks. If we consider a relation between stresses and strains /Fig.3/ or a relation between energy A_r

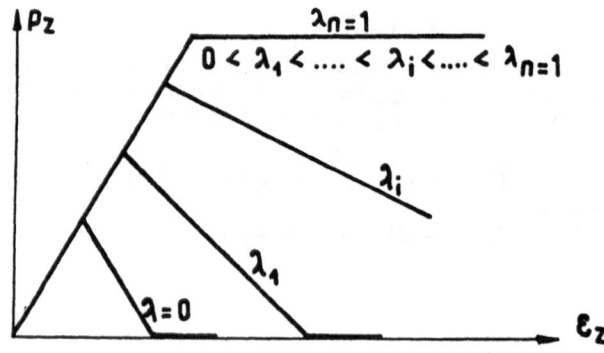

Fig.3

and strains /Fig.4/ we can see that the dynamic fracture will occur when

$$A_s > A_r$$ /6/

where:

A_s - energy supplied to rocks
A_r^s - energy capacity of rocks

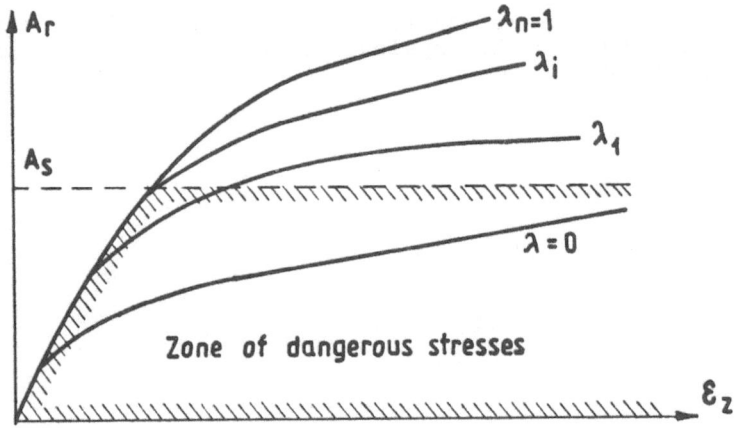

Fig.4

This condition means that the dynamic fracture will take place when simultaneously:

$$\lambda < \lambda_s$$ /7/

$$\lambda < \lambda_t$$

4. CONCLUSIONS

Summing up our consideration it should be stressed that methods for prediction the loss of stability of underground excavations must base on:
- underground measurements of changes of stresses in the rock massive; these measurements determine the supplied energy,
- measurements of energy capacity of rocks; for instance these measurements can be made in a laboratory or in-situ.
The presented in the paper method of measurements concern the first group of factors.

Once again we want to emphasize that for prediction the loss stability phenomena both kinds of measurements must be made simultaneously.

REFERENCES

1. Filcek H.,Kłeczek Z.,Zorychta A.: Opinions and Solutions
   Concerning Rock Bursts in the Coal Mines.Sc.Bull.University
   of Mining and Metallurgy Górnictwo, No 123,1985.
2. Leeman E.R.: The Measurement of Stress in Rock.J.South Afr.
   Inst.Min.Metall.V.8,1964.
3. Zorychta A.: The Influence of External Factors on the
   Stress Changes Determined by the Stressmeters.Quarterly
   Archives of Mining Sciences.Vol.28, No 1,1983 /in Polish/.
4. Kłeczek Z.,Szumiński A.,Zorychta A.: In-situ Stress Measurements versus Forecasting of Rock Bursts.Proc.of the Conf.
   on Applying the Geophysical Methods in Mining, Cracow,1985
   /in Polish/.

SPECIFIC TECHNIQUES IN 2D STRESS ANALYSIS

H.N. TEODORESCU[+], D. CRISTEA[++], E. SOFRON[x], M. POPA[+]

[+]INSTITUT POLITEHNIC, IASI, ROMANIA; [++]UNIVERSITATEA IASI;
[x]INSTITUTUL POLITEHNIC BUCURESTI, ROMANIA

1. INTRODUCTION

We feel that the stress analysis is at an important cross road. We charge ourself to clearly establish some main requirements and directions in this realm. This is a programmatic paper rather than a descriptive one. It is the third in a series of papers, the first ones /1/,/2/ being devoted, as those to follow, to particular aspects. This rather general paper, as well as our global goal, is based and legitimated by a comprehensive analysis of facts and by experiments we have performed last years in order to introduce new methods in stress analysis.

The directions we want to put into evidence are given by the critical examination of the requirements, of the methods for data collection, of the physical effects to be used, and of the specific techniques for data collection and data processing in the bi-dimensional stress analysis (2D), as well as by the comparative examination of the above mentioned tools, with their advantages and drawbacks.

Our programm encompasses the following basical aspects which are to be clarified:

i) what is to be done : a general overview of the problem, namely establishing and defining the domain and its major matters;

ii) by which means to estimate stress : aspects related to physical effects under stress;

iii) by which means to collect data : aspects related to transducers and surface exploration (two intercorrelated problems);

iv) how to process data, and the influence of this stage on the above stages.

Although we plan to contribute in the future to the renew of the stress analysis in all these points, our goal is to produce basical research and not devices devoted to the market.

2. THE STATE OF THE ART

The last ten years yielded important changes in the possibilities offered to the experimentalist in stress analysis, by the advent of the two-dimensional techniques. However, these techniques refer only to processing data. Moreover, it is an obvious discrepancy between the means of processing and those of collecting data, the last ones remaining

essentially zero-dimensional (punctual). It could be believed that such discrepancy is of little importance, as collecting data processes are discontinuous. This point of
view is erroneous, as discrete transducers are to be used
in very large numbers, let say 10,000,for obtaining a
100 x 100 image of the stress in the plane.

Much work is to be devoted to the development of bi-dimensional measurement techniques for stress, some of them
based on classical physical effects, as used in punctual
measurements, and some others not previously used. The only
classical technique being truly 2D is photoelasticity. Much
work is also to be devoted to improvements and adaptation
of existing algorithms for image processing. Techniques
used in measuring other parameters related to surfaces, as
film thickness, surface resistivity etc., are to be taken
into account for possible adaptation to stress measurements (2D).

Some of these aspects are briefly discussed in this
paper.

3. PHYSICAL EFFECTS AND STRESS MEASUREMENT TECHNIQUES
3.1. Physical effects under stress
3.1.1.Generalities. The stress changes the thermodynamical state of materials and following, it is to be expected that all material characteristics are to be changed,
in a well definite manner. However, the variation of the
parameters with respect to the stress is not linear, ex-
cepting the approximation of infinitezimal, linearized
changes. When linearizabile behaviour occurs, no phase
changes are allowed, and no significant dislocations are
to be expected. On the other hand, as obvious, the change
of other thermodynamical parameters, mainly temperature,
will produce similar, indiscernible effects, thus limiting
the accurancy of the stress determination.

The measured effects may be ranged in different classes:
mechanical (compression etc.), electrical and magnetical
(changes in conductivity/resistivity; changes in the electrical permitivity; in magnetical permeability; in electrical polarisation, in magnetisation and in related, second order parameters, as temperature coefficients of the
above, electrical or magnetical loss factors etc.); optical - electromagnetical (refraction index, absorbance
index, specific optical polarization, at different wave-
lengths), a.s.o.

3.1.2.Measurement methods and basical properties. There
are two general methods to determine stress induced changes:

i) direct methods - the measured material is the one
for which deformations are to be surveied, i.e. its own
parameters are determined; it is to be underlined that
these methods are well fitted for 2D stress analysis;

ii) indirect methods - on the stressed material a thin
film of 'sensitive' material is deposited; the stress induced from the substrate (ideally, the same deformation
occurs in both the substrate and the film) is determined

by means of changes in film parameters.

It is not unusual that, when measuring by means of sensitive films, to allow phase changes in the basical material or in the film. It is the case, for example, of the determinations performed via liquid crystal films.

On the other hand, there are two different problems in stress analysis: linear and nonlinear behaviour, which are to be dealt with in specific manners, regarding both physical effects and methodologies.

It could seem that this discussion is trivial, but we do insist because many researchers, too well accustomed with a given technique, are tempted to erroneously extend it to cases where it is ineffective.

Concluding, the matters of stress analysis, as regarding its first step, could be summarized as bellow:

TABLE 1. Stress Measurement Techniques

Dimensionality	Measured object	Material behaviour	Transducer behaviour	Noted
0 D (punctual)	basically indirect	basically linear	linear	
1 D				+
2 D	direct	linear or plastic	linear or unlinear	++
	indirect	linear or plastic	linear or unlinear	+++

+-Generally omitted case, as easily reduced to measurements by punctual transducers.

++-In this case, the transducer is the material itself; the measured parameter may be linearly or unlinearly dependent on the stress, even if the material is in the linear region. See also +++).

+++-The standard use of computers in 2D analysis allows determinations by means of nonlinear physical effects, or nonlinear transducers. This is no more a limitation, as inverse functions are easily computed or read from tables.

The first step in designing 2D measuring methods -and these methods are to be invented, as the domain is just emerging- is to choose the proper definition of the problem, as indicated in the Table 1, and then to search for and find the optimum physical effect on which to base measurements.

In the following paragraphs we exemplify the use of two unusual optical methods.

3.2. Use of the optical emission effect

It is well known that materials submitted to stress generate light. The phenomenon is named triboluminescence and is similar in many respects to the acoustic emission under stress. It was proved that light emission is due to the plastic deformation of the material, and is directly related to the dynamical excitation. Even if the exact explanation of this phenomenon is not yet performed, the effect can be advantageously used to determine equi-deformation areas. We have proposed the use of this effect in stress analysis, demonstrated a feasible experimental arrangement and derived some applications in /2/. Further details will be given in a separate paper, to be delivered to the Int. Conf. on Residual Stresses, Oct. 1986, Garmisch-Partenchirchen (FRG) /3/.

3.3. Use of liquid crystals

Liquid crystals (lc) can be used to simulate the dynamical behaviour of plates under stresses, in a manner similar to the photoelastometric method. The use of the lc based method is profitable when dynamical, plastic properties are to be investigated, or when dislocation behaviour is to be demonstrated. As seen in Figure 1, the method yields clear images under polarized light. Further details can be find in /1/.

Both methods, optical emission and "lc dynamical elastometry" are true 2D techniques, producing images of the stresses fields.

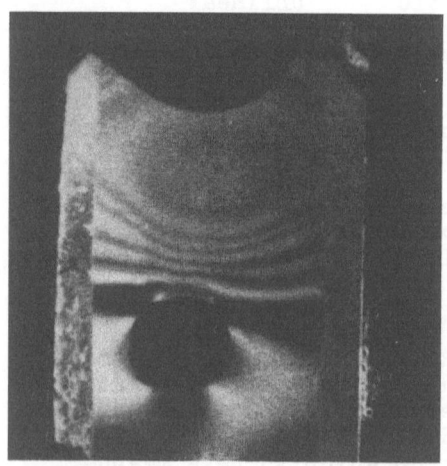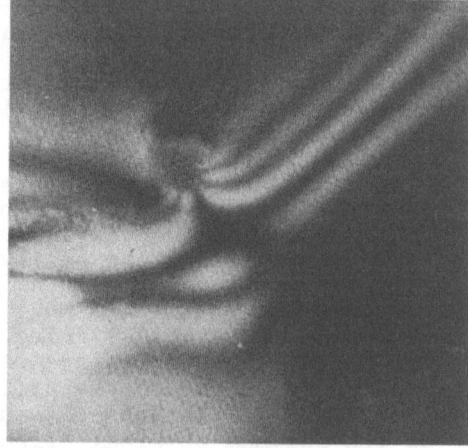

FIGURE 1. Examples of images obtained under dynamical stresses in unhomogeneous structures, by the method of lc dynamical elastometry (polarized light).

3.4. Other effects

Optical interferometric, magnetical and resistive methods, as well as direct measurement of films thickness (mechanical measurement) are extensivelly used in other fields -as electronics or fine mechanics- in order to establish the properties of the surfaces. We have proposed a methodology based on such measurements in order to determine stresses; the measurements are performed on thin films deposited on the surfaces of the measured bodies /2/. The proposed method has two advantages: it is easily adapted to robotic measurements, and make use of techniques already developped; of course, it yields 2D determinations.

4. WHAT THE COMPUTER IS CHANGING?

It is not obvious that the use of computers into integrated stress analysis systems may change basical aspects of the classical techniques, or may lead to completly new techniques. This question and the confusion it creates is obvious in the early attempts to introduce 2D techniques: in fact, they are just simple extensions of the classical, zero-dimensional ones. Following, it is of interest to note some of the characteristics the use of computers involves.

First at all, computers make utilizable physical effects previously not used; in fact, only the photoelastometric effect is trully bi-dimensional. Its use was possible because it is a visible effect, thus allowing determinations by the human specific 2D transducer (the eye), or by TV cameras. It is not surprising that this effect is nowaday the only one (true 2D) to be applied in 2D analysis. However, other transducers may allow the use of a large range of effects, as stated above. It is curious that much work was done to reduce, by corrections, intrinsic 2D effects, resistivity/eddy currents/magnetic permeability etc. dependence on stress, to punctually useable effects, but no important work was devoted up to now to their use in natural 2D measurements. Computers may help by determining local parameters from measurements in the neighbouring area.

Second, the range of the physical effects computers make useable is further extended by the possibility of automatically applying linearization functions to nonlinear effects.

Third, also due to the fast computation capabilities, it is possible to automatically correct errors due to sensors, temperature variations, methodology or variability of the probes.

Fourth, by the same capability, it is possible to make differential measurements, further reducing the errors due to the imperfections of the measured material, to the deposited sensitive film etc. Differentiation is applied to data collected before and after the stress acts.

The improvements, as above mentioned, enlarge significantly the range of methodologies to be expected, in the near future, to emerge in the stress analysis. Today, this is a quasi-virgin realm and much work is to be accomplished, but the results will surely recompense it.

The computer requires the discretization of the surface on which data are collected. This implies advanced circuitry and other means for geometrical measurements. Moreover, the method of collecting data is dramatically changed by the computer, as two dimensional "images" can be obtained not only by true 2D transducers (as Tv cameras, used in photo-elastometry), but also by robotic exploration of the surface, as proposed in /2/.

It is already well known that computers allow the enhancement of data by image processing algorithms. The enhancement can be done interactivelly with the human operator.

Finally, computers are able to storage and manage large amounts of collected data, obtained on long periods, thus enableing recordings of the history of the mechanical changes under stresses. This opens new possibilities in non-linear, time-dependent, slow relaxation processes experiments.

We are not concerned here with computer simulations.

5. COMPUTER ASSISTED EXPLORATION

5.1. Exploration methods

The exploration methods are not to be invented. They can be borrowed from television or from methods used recently in other 2D materials testing techniques (for example, resistivity tests in microelectronics). It seems that circular (spiral) exploration is advantageous for low and medium resolution, due to the ease of driving the measuring head and of data representation.

5.2. Mechanical exploration

This method is derived from techniques used in data storage devices (discs, tapes). Unlike optical bidimensional data collection, as used in Tv cameras, in this case the exploration process is submitted to errors due to the mechanical device. Thermal offtrack and tilt in the picking up arm and imprecision in the explored body (disc) drive are the same as in electronic discs. The precision is further limited by the dynamic friction forces and tempo-rary adhesivness in the mechanical system, and, if contact methods are used, by the forces occuring between the transducer and the measured surface. Imprecision may occur in the driving (electronic) system as well, and is dependent on the resolution of the conversion processes (analog-to-digital, and digital - to-analogue) in the driving and sensing systems. Large surfaces are difficult to be explored with high precision, and the existence of the stressing devices - not encountered in electronic systems for data storage - can strongly influence the positioning precision. Moreover, another difficult problem is risen by the defor-mation of the surface and body; this occurs in an apriori-cally unknown manner, making that differential measurements become erroneous. (Space 'wrapping' could be used, but no such algorithm is available today).

6. PROPOSALS FOR 3D STRESS ANALYSIS

Computers and ideas borrowed from medicine and material testing can lead to a completly new technique in stress analysis: the three-dimensional (3D), or thomographic stress analysis. This technique, whose feasibility we like only to mention here, consists, as indicated by its name, in the determination of stresses in the body volume, without any direct access. This can be performed only by computer and could make use of the propagation of the ultrasound waves in the stressed material, or of techniques derived from photoelastometry. X rays and microwaves could also be used. In the 3D technique, differential measurements and computations are essential.

7. CONCLUSION

We have suggested some main purposes and directions in the development of stress analysis and briefly reported on our researches. It seem obvious that stress analysis methods will much change in the near future.

REFERENCES

1. Teodorescu H.N. et al : Photoelastometry with liquid crystals, VDI Berichte Nr. 439, p. 197, 1982
2. Teodorescu H.N. et al : Use of special physical effects in stress analysis, VDI Berichte, p.279, 1985
3. Teodorescu H.N. et al : The use of the optical emission in the experimental stress analysis, to be published

EXPERIMENTAL STRESS-STRAIN FIELD IN ELASTOMERIC O-RING SEALS

A.STROZZI

INSTITUTE OF APPLIED MECHANICS, BOLOGNA UNIVERSITY, ITALY

1. INTRODUCTION

This paper deals with the mechanical analysis of elastomeric seals having a circular cross section. Examples of applications of such seals encompass mechanical actuators, chemical plants and nuclear power stations.A detailed analysis of the stress-strain field in elastomeric seals is fundamental in estimating the seal performances in static and dynamic applications.

When high integrity containment is required, an assessment of the perform ance and failure modes is recommendable. Avoidance of leakage requires the surface roughness to be infilled by the seal material. This in turn can be achieved by resorting to high contact pressures, which can cause the seal failure according to fracture mechanics modes.

Regarding dynamic applications, the contact pressure profile is a good ind icator of the sealing performances, and its knowledge is required in elasto hydrodynamic studies.

Analytical results concerning the stress-strain field in elastomeric units are difficult to attain for technically relevant situations, owing to the material and geometrical nonlinearities related to high strains. The finite element method may produce an oscillatory stress field, since the limited cubic compressibility of the elastomers is responsible for numerical ill-conditioning.

On the other side, elastomeric seals lend themselves to be investigated experimentally. In paricular, the experimental analysis of rubber seals pr-ofits from the following considerations:

- photoelastic materials are available the physical properties of which are very close to those of the elastomers employed in practical applications;
- the actual stress-strain field can be generally likened to a plane state of strain;
- the seal is subject to finite deformations and to mixed (i.e. imposed par tially on displacements and partially on stresses) boundary conditions.In addition, the contact is usually progressive (i.e. the contact width incr eases with the imposed seal compression).

These facts suggest that the photoelastic technique is particularly suit-able for examining the stress field in elastomeric seals. If plane stress and plane strain models produced the same in-plane stress field, the former state of stress would be preferable from the experimental analysis and the loading device viewpoints. Unfortunately, since the contact is progressive, plane-stress and plane-strain models would produce different results in terms

of stress field and contact width even in the case of small deformations (1).
In addition, the presence of finite deformations totally precludes the possibility of transferring the in-plane stresses obtained for plane stress mod
els to plane strain situations.

The first aim of this paper is to present a device which permits a plane
state of strain to be achieved in a plane model describing the seal   cross
section, without precluding a photoelastic investigation. The   traditional
separation of the photoelastic stresses is lengthy and affected by cumulative errors. The second aim of this paper is to present a stress   separation
method which profits from the high deformability of the elastomeric material
employed and avoids cumulative errors.

The photoelastic analysis is applicable to unpressurized, laterally unres
trained O-Ring seals, for which a variety of analytical, numerical and experimental results already exists. An exhaustive bibliography on this   topic
is given in (2).

The third aim of this paper is to compare the experimental findings to the
existing analytical models, in order to assess the proposed procedure.

Overall, the results are encouraging. Current research is dedicated to the
analysis of laterally restrained O-Ring seals, for which the data available
are not so exhaustive and quite sparse.

The paper is organized as follows. A description of the loading device
and of the stress separation method is followed by a presentation of   the
existing analytical models. The subsequent section is devoted to the presen
tation of the experimental results and to a critical comparison with the av
ailable analytical models.

2. THE EXPERIMENTAL METHOD

A 110 mm-diameter disk representing the O-Ring cross section is machined
out of a polyurethane (Hysol 4485), 12.7 mm-thick sheet, defined by a Young
modulus E of 3.52 MPa, by a Poisson's ratio of 0.489 and by a photoelastic
constant of 0.182 MPa mm/fringe order (3).

The effects of seal axisymmetry on the stress field could be photoelastic
ally investigated by resorting to a pseudo-two dimensional photoelastic met
hod (4). This approach would require a "thermal loading" which involves exp
erimental difficulties, and therefore a more traditional plane strain model
and a loading by compression are employed.

The seal loading is achieved by resorting to the testing machine of Figure
1 (3). The disk is positioned between two parallel, 20-mm thick perspex plates which prevent any increase in the disk initial thickness, thus simulat
ing an authentic situation in which the mean plan seal diameter is not modified by the fractional compression.

The seal model actually undergoes a plane state of strain if no contraction of its thickness occurs as a result of the seal compression. Unfortunat
ely, during previous experiments a gap was signalled between disk   lateral
sides and restraining perspex plates in restricted zones (5). As a consequen
ce, a plane state of strain is not achieved for the whole model,   whereas
limited parts by the disk free boundary are subject to a plane stress situat
ion. This point is discussed further in section 4.

A horizontal, sliding, top loading bar enables various diametral compress ions to be achieved.

The stresses within the model can be classified into internal stresses and boundary stresses. In addition, the boundary stresses can be divided into those concerning the free boundary portion and those referred to the contact region, where their component perpendicular to the contact profile is the contact pressure. The free boundary stresses are investigated by tradition- al photoelastic methods (6). The contact pressure is examined via the "Mul- ler method" (7), which is based upon a compensating pressure concept. A ser ies of narrow holes are drilled through a metal surface in contact with the sealing profile. Then, the holes are progressively and sequentially pressur ized with air until the fluid begins to leak. The corresponding contact pr- essure is approximately equal to the air pressure for which the fluid leaks. To this aim, ten pairs of adjacent, 1.5 mm-diameter holes are drilled thro- ugh the loading bar. The distance between the axes of the adjacent holes is 6 mm, while the pitch between the ten pairs of holes is 10 mm. Of every pa- ir, one of the holes supplies pressurized air to the contact zone under exa mination, while the other acts as a drainage hole. If the pressure profile exhibits a minimum between two pressure maxima, during the scanning of the zone of pressure minimum the "compensating pressure" can build up if the drainage holes are inefficient or not included (3). As a consequence, spur- ious contact pressure values higher than the real ones would be attributed to such zones. The available results (5) show that the pressure profile is of hertzian type, and therefore no pressure minimum exists for the case under examination.

The analysis of the internal stresses is dealt with in the followup. In this paper a stress separation based upon a combination of the photoelastic and Muller techniques is presented. It is applicable to plane strain models made of soft materials and, in comparison to more conventional methods, avo- ids cumulative errors (3). The salient aspects of the proposed stress separ ation method are explored in the sequel. The photoelastic technique permits the difference between the in-plane principal stresses to be computed, since:

$$|\sigma_1 - \sigma_2| = \frac{an}{s} \tag{1}$$

where a is the photoelastic constant, n is the isochromatic fringe order and s is the thickness of the disk. In addition, if the material is assumed to behave as linear elastic, the principal stress σ_z perpendicular to the plane of the seal model can be expressed as a function of the in-plane principal stresses σ_1 and σ_2 :

$$\sigma_z = \nu(\sigma_1 + \sigma_2) \tag{2}$$

where ν is the Poisson's ratio. The stress perpendicular to the plane of the model is evaluated by the Muller method. To this aim, ten pairs of equispac ed holes similar to those in the loading bar are drilled through one of the perspex plates. The separation of the in-plane principal stresses is achiev ed by combining equations (1) and (2). Since the Poisson's ratio for elast- omers approaches $\frac{1}{2}$, one obtains:

$$\sigma_1 = \sigma_z + \frac{an}{2s} \quad ; \quad \sigma_2 = \sigma_z - \frac{an}{2s} \tag{3}$$

where the uncertainty on the sign can be removed by comparison with the existing data. Moreover, if the isocline pattern is known, the in-plane stres
ses σ_x and σ_y, referred to an x-y cartesian frame can be obtained.

Figures 2,3 and 4 show the isocline patterns for seal compressions of 10,
20 and 30 percent respectively. Apparently, this novel stress separation met
hod was first developed at Risley Nuclear Labs, U.K.. Actually, holes were
drilled through the lateral restraining plates in (4) and in (8), but they
were mainly introduced to lower the frictional effects between disk lateral
walls and restraining plates. A repeated pressurization of such holes aimed
at facilitating a disk settlement as free as possible from frictional effec
ts. The proposed approach is the "dual" of that based upon the measurement of
the variation in thickness for plane stress problems (6). An alternative ap
proach for obtaining the individual in-plane principal stresses in seal mod
els is based upon a combination between photoelasticity and the grid method,
which also permits the sum of the principal stresses to be obtained (9).

3. ANALYTICAL MODELS

Analytical models exist concerning stresses and displacements in laterally unrestrained, unpressurized O-Ring seals.Two parts, dealing respectively with displacement-related and to stress-related parameters, comprise th
is section.

3.1. Displacement-related parameters

The salient geometrical parameters describing a compressed O-Ring seal in
clude the deformed shape, the deformed chord diameter, d, and the contact
width, w. The increase in the chord diameter with the seal fractional compr
ession is relevant for determining which groove width avoids any interaction
between compressed seal and groove lateral walls. The seal contact width is
fundamental in elastohydrodynamic studies. Also, w is necessary for determining the total force applied to a seal as the resultant of the contact pres
sures acting along the contact width.

The most rational geometrical model appears to be that presented in (10),
according to which the free boundary of the O-Ring seal cross section is
closely described in terms of an arc having the same radius of the undeform
ed disk and laterally positioned in such a way that the seal undergoes isoch
oric deformations. Obviously, such a model fails by the ends of the contact
between seal and counterface, since it would involve a non differentiable
deformed seal periphery. Since in the real configuration a fillet exists
which connects the arcued seal contour with the contact zone, it can be speculated that such an analytical model would produce values of w higher than
the actual ones. According to the above mentioned model, the deformed chord
diameter, d, as a function of the diametral compression, C, is:

$$\frac{d}{D} = \frac{1}{4(1-C)} \{\pi - 2\sin^{-1}(1-C)\} - \frac{1}{2} \{1 - (1-C)^2\}^{\frac{1}{2}} + 1 \qquad (4)$$

where:

$$C = 1 - \frac{h}{D} \qquad (5)$$

where h is the height of the compressed disk.

Similarly, the ratio between contact width, w, and undeformed chord diameter ,D , is:

$$\frac{w}{D} = \frac{d}{D} + \{1 - (1-C)^2\}^{\frac{1}{2}} - 1 \tag{6}$$

3.2. Stress-related parameters

The salient parameters are the peak contact pressure, p_c , the tensile stress at the shoulder of the contact zone (10), σ_s (index s stands for "shoulder"), and the tensile stress at the seal centre, σ_c (index c stands for "centre"). The peak pressure is relevant for the sealing mechanism,while the tensile stresses can be responsible for the seal cracking.

In pioneering (and fundamental) studies on design directives for flexible seals it was conjectured that the hertzian tehory might not hold true in the case of O-Ring seals, since the ratio between contact width and undeformed seal diameter approaches unity for realistic fractional compressions (see ref. 11, page 69). Quite surprisingly, the photoelastic results of (12) did not depart significantly from Hert'z predictions even for values of the ratio between contact width and undeformed diameter up to 0.3. In addition, specific experimental research on O-Ring seals (13) clarified that the experimental peak contact pressure closely agrees with the hertzian theory up to a seal compression of 25 percent, for which w/D ≈ 0.65, while an empirical correction is needed to extend the hertzian theory up tp a fractional compression of 40 percent , for which w/D ≈ 1. According to (13), the normalised expression of p_c is :

$$\frac{p_c}{E} = \frac{2}{(3\pi)^{\frac{1}{2}}} 2^{\frac{1}{2}} \{1.25 \ c^{\frac{3}{2}} + 50 \ c^6\}^{\frac{1}{2}} \tag{7}$$

In (14) it is speculated that the validity of the hertzian theory up to high strains is due to different errors being compensatory.

A tensile stress occurs by the seal centre in a direction parallel to the upper and lower counterfaces, and it tends to promote a crack which can literally split the seal into two half-disks. The outcome of tensile stresses by the disk centre is exploited to assess the tensile strength in rocks . This method is appropriately called "indirect tensile test" (15). It might be employed to determine the conditions of failure in elastomeric materials.

The normalised expression of the tensile stress at the centre of a disk compressed by two radial forces acting along a diameter is (15) :

$$\sigma_c = \frac{2 \ P}{\pi \ D} \tag{8}$$

where P is the resultant force acting at each of the contact widths. The stress value by the seal centre is scarcely influenced by the extent of the arc over which the distributed load acts (15). An expression for P as a function of the fractional compression is reported in (13). An alternative expression for P can be derived by assuming an elliptical contact pressure profile, where the pressure peak value is given by expression (7) and the contact width is that of equation (6). The latter model has been employed in this paper, since it is more consistent with the other modellings adopted throughout the paper.

Another seal protion is subject to high tensile stresses. These occur at the shoulder of the contact zone, and promote radial cracks (10). It might be surmised that an analytical value of such a tensile stress was provided by the hertzian theory. Unfortunately, while in the case of a deformable sphere compressed against a rigid plane the hertzian theory supplies a non vanishing value of such a tensile stress (which, by the way, vanishes for $\nu = 0.5$), in the case of a deformable cylinder compressed against a rigid plane no stress occurs at the disk shoulder according to the hertzian theory . The lack of a reliable analytical model for the tensile stress by the disk shoulder is discomforting, and this gap should be hopefully filled.

4. RESULTS AND DISCUSSION

The analytical and experimental deformed chord diameters for fractional compressions up to 30 percent are presented in Figure 5. The good agreement between analytical and experimental findings confirms the reliability of the analytical model of section 3.1. Anyway, such a model produces values which are lower than the experimental ones. This can be attributed to the absence of a fillet by the seal shoulder in the analytical model, which is responsible for the analytical deformed chord diameter being underestimated.

The analytical and experimental contact widths are displayed in Figure 6. Again, the approximations inherent in the analytical model are seemingly responsible for the disagreement between analytical and experimental values , as it was speculated in section 3.1.

The peak contact pressure was evaluated via the Müller method. Figure 7 illustrates the analytical and experimental peak contact pressures, which correlate favourably.

Figure 8 displays the tensile stress by the seal centre as a function of the seal diametral compression. The experimental data are lower than the analytical ones. This is partially ascribable to the fact that the analytical tensile stress by the seal centre assumes slightly lower values as the boundary arc along which the contact pressure acts is increased, partially to frictional effects, and partially to the large strains involved. The peak contact pressure is of the order of twice the tensile stress by the seal centre, but the two curves are not proportional.

The tensile stress by the shoulder of the contact zone is displayed in Figure 9. Here, an analytical model is missing. The experimental results are of the order of twice the numerical data of (2). This disagreement cannot derive from the tendency of the compressed disk lateral walls to move apart from the perspex restraining plates. Indeed, the presence of a plane stress zone by the disk free contour in the experimental study instead of the numerical plane strain state should produce a reduced stress by the shoulder of the contact zone. Thus, the discrepancy mentioned above may be imputable to the "weak" character of the numerical method, according to which the concentrated peak describing the tensile stress by the contact end is rounded off. Despite these uncertainties, the experimental value of the tensile stress by the shoulder of the contact is lower than that by the seal centre. This explains why a crack is more prone to initiate by the seal centre than by the end of the contact zone.

Figures 10,11 and 12 present the internal stresses in a 10 percent compressed seal for ten equispaced points (a selection of which labels the tic marks of x-axis) lying along three lines parallel to the counterfaces. Figure 10 is referred to a line 5-mm far from the lower counterface. Figure 11 deals with a line one quarter of the compressed seal height far from the counterf ace. Finally, Figure 12 describes a line equidistant from the two counterf aces. Such lines are displayed in the insets of the corresponding figures . Figures 13,14 and 15 are referred to a 20 percent compressed seal, while Figures 16,17 and 18 deal with a compression of 30 percent. The reported val ues are the mean of three experiments, where the stress separation has been performed according to section 2. The peak stress along a direction perpend icular to the counterfaces and referred to the line which is nearly superimp osed to the contact segment is in good agreement with the peak contact pres sure of Figure 7. Consistently with the analytical findings (15), the peak value of the stress component parallel to the counterfaces is tensile and of the same order of magnitude in Figures 11 and 12, while it changes its sign as the counterface is approached (Figure 10). This tendency is increasingly less evident in the 20 and 30 percent compressed seals, and this is possibly due to the seal deformed shape being subject to considerably high strains. The relative importance of the frictional effects between disk lateral walls and restraining plates can be assessed by computing the scatter of the res ultant of the stress component perpendicular to the counterfaces and acting along the three lines mentioned above. In the case of a 10 percent compress ed seal, the three resultant forces referred to Figures 13, 14 and 15 differ by less than 10 percent. Similar results hold for the other seal compressions.

5. CONCLUSIONS

A novel stress separation method has been applied to the mechanical analy sis of an unpressurized, laterally unrestrained O-Ring seal. The salient fea tures of the stress-strain field have been underlined. The experimental res ults retrieved compare favourably with the available analytical models both for displacement-related and for stress-related parameters, thus confirming the validity of the proposed methodology. The analysis of more complex seal configurations is the object of current research.

REFERENCES

1. Dundurs J, Stippes M, J.Appl.Mech.,pp.965-970, 1970
2. George AF, Strozzi A, Rich JI, Int.Report, Berkeley Nucl. Labs,1985
3. Strozzi A, ASLE Transactions, to appear
4. Fourney ME, Exp. Mech.,pp.19-25,1971
5. Molari PG, VI Int.Conf. Fluid Sealing BHRA,pp.B2/15-31,1973
6. Kuske A, Robertson G.: Photoelastic Stress Analysis, Wiley,1974
7. Muller HK, II Int.Conf. Fluid Sealing BHRA,pp.B2/13-28,1964
8. Forster MJ, J.Appl.Phys.,pp.1104-1106,1955
9. Durelli AJ : Applied Stress Analysis, Prentice-Hall,1967
10.Ebisu T, Yamamoto H, Maekawa H, Odonera A, PATRAM 83,pp.672-679,1983
11.Blok H, Symp. on Lubr. and Wear, Houston,pp.1-151,1963
12.Fessler H, Ollerton E, J.Appl.Phys.,pp.387-393,1957
13.Lindley PB, Jo. IRI,pp.209-213,1967
14.Johnson KL, Proc.Instn.Mech.Engrs,pp.363-378,1982
15.Hondros G., Australian J.Appl.Sci.,pp.243-268,1959

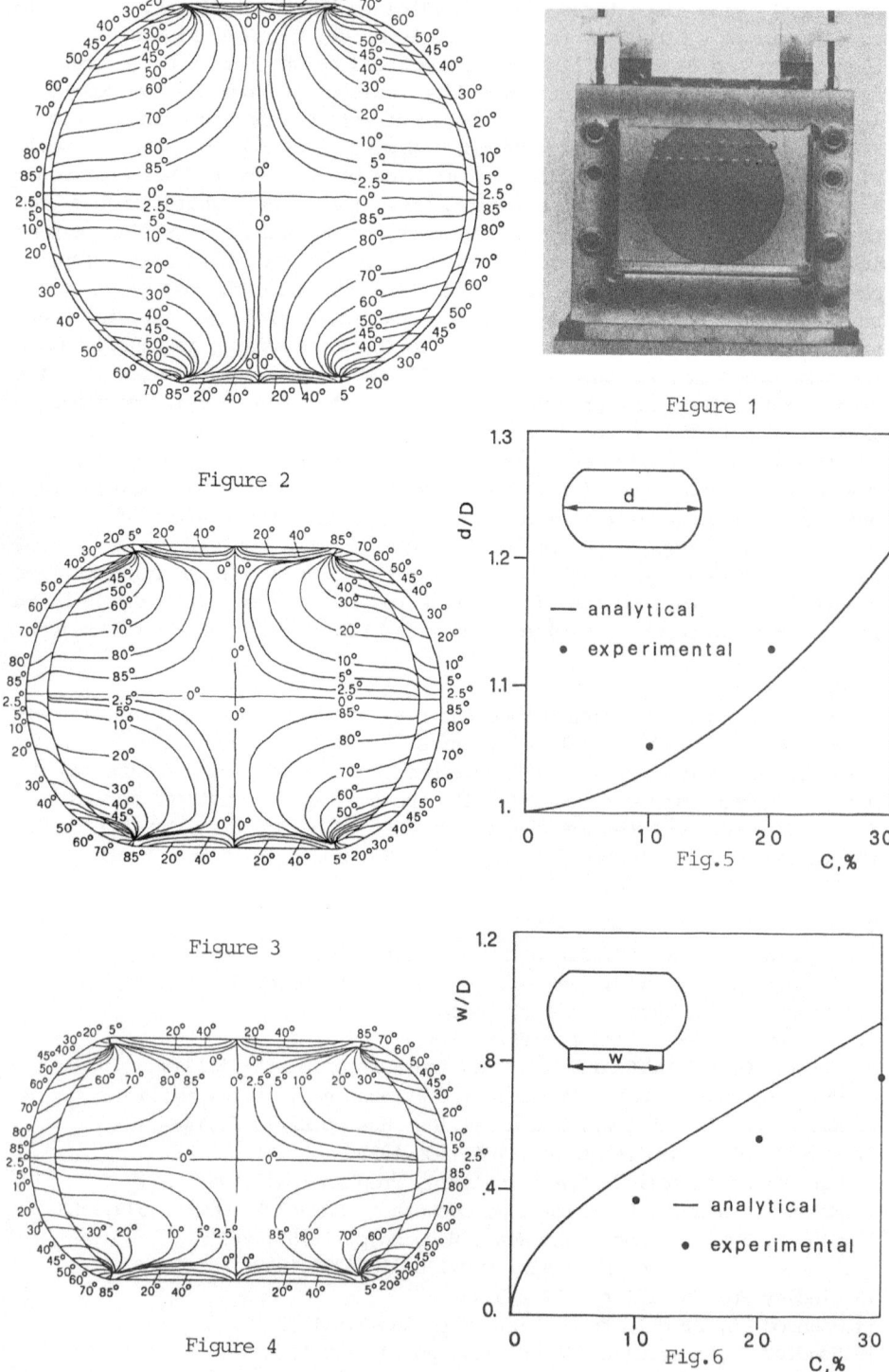

Figure 1

Figure 2

Figure 3

Figure 4

Fig.5

Fig.6

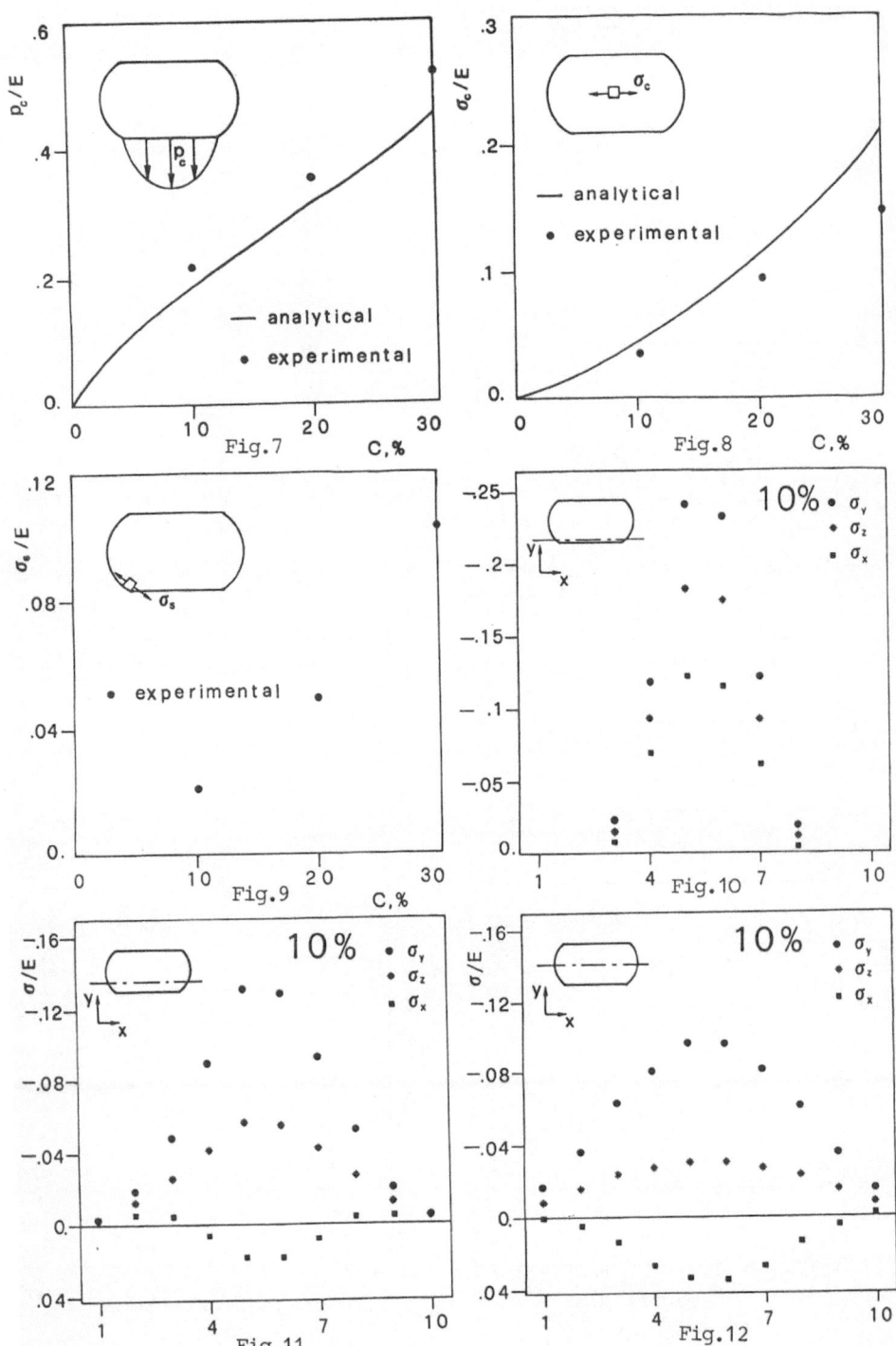

Fig.7

Fig.8

Fig.9

Fig.10

Fig.11

Fig.12

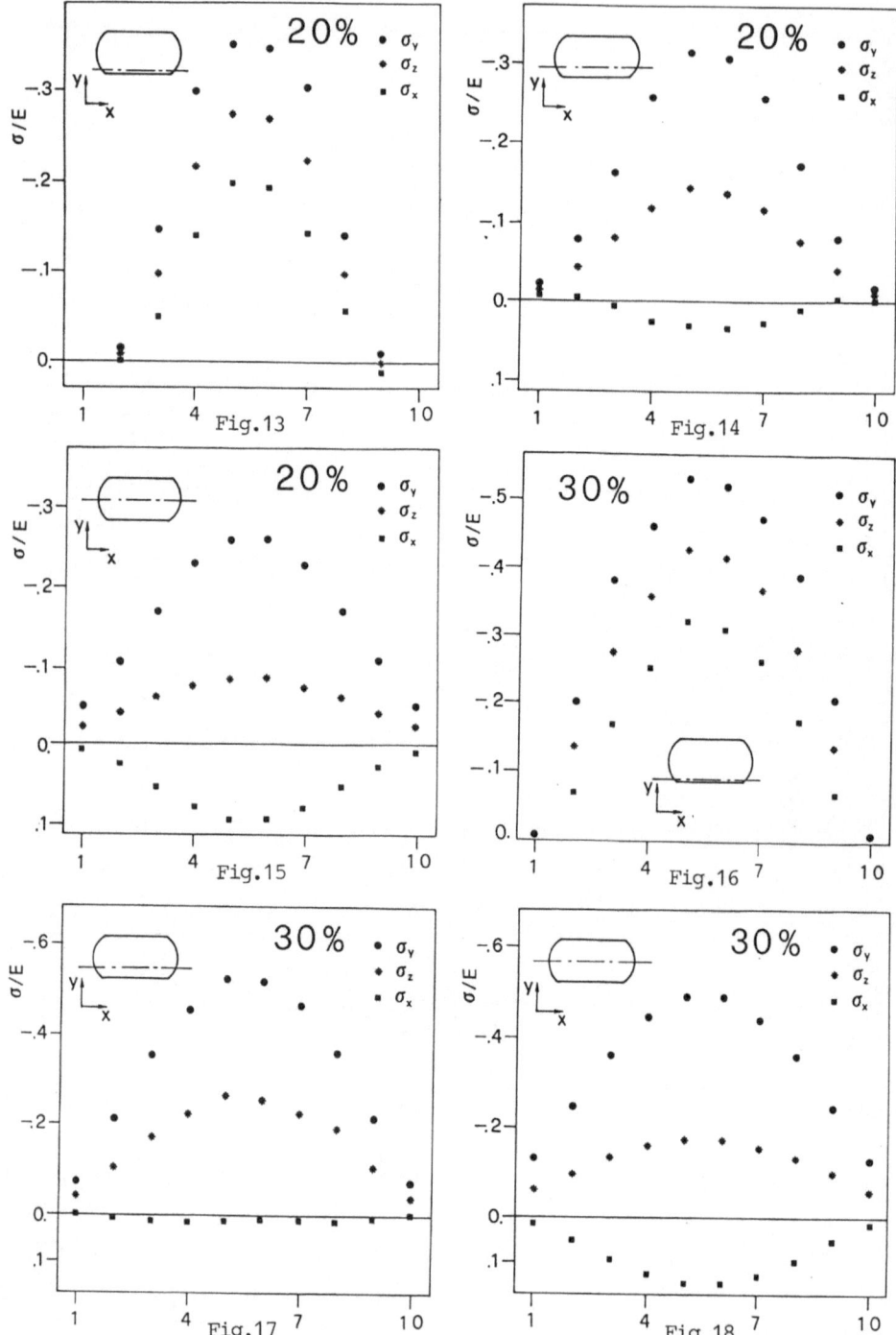

Fig.13

Fig.14

Fig.15

Fig.16

Fig.17

Fig.18

MEASUREMENT OF TECTONIC ROCK STRAIN IN ICELAND

Stefan Keil, Hottinger Baldwin Messtechnik GmbH, Darmstadt
Karlheinz Schäfer, Universität Bayreuth

GEOLOGICAL FRAME WORK

The lithosphere of the earth which comprises the entire crust and a part of the upper mantle consits of 16 major plates that drift horizontally in different directions by 1 cm to 10 cm per year. Thus, there are three different tectonic mechanismes which occur at the plates' boundaries:

1. Divergence at the plates creates fissures filled with upwelling magma from the upper mantle. Since new earthcrust is formed by this process it is defined as a constructive plate boundary.
2. Convergence of the plates occurs at the deep ocean trenches, where heavy oceanic plates are subducted by overriding buoyant plates or at folded mountain belts, where two buyoant plates collide. This kind of plate boundary is called destructive.
3. Transform faults are active at boundaries of plates moving parallel but in opposite directions to each other. No lithospheric material is formed or destroyed at this type of plate boundary which is therefore called conservative.

Since all three types of processes are accompanied by energy release in form of earthquakes, the world-wide distribution of epicentres delineates all the active plates boundaries of which only a minor part appears above sea-level. The Mediterranean-Himalayan fold belt is one extensive stretch of a destructive plate boundary system, the San Andreas Fault and the North Anatolic fault are other plate boundaries of the conservative type which are extending a long distance on land. But of the more than 70 000 km long midoceanic rift system which is the location of the divergence of plates extending from the East Pacific across the Mid-Indian ocean and along the Mid-Atlantic-Ridge, only 350 km reach heights above sea-level. Iceland is the only place where a complete oceanic rift system is directly accesible for geoscientific research. The boundary between the European and American plates corresponds to the fissured axial rift zone that cuts across Iceland from the southwest to the northeast. The nature of the driving mechanism which splits Iceland in two parts and causes the plate movements is not yet well understood. The plates might be driven by push due to the upwelling and injection of magma or by drag force induced by friction between lithosphere and mantle currents or by pull due to subduction at the oceanic trenches. As a result in Iceland an average spreading rate of 2 cm/yr in the E-W-direction has been determined over the past 16 Mio years. We expected that the diverging plate movement would induce forces aligned in the direction of the plate drift and would create regional uniformly horizontal stress as well as corresponding strain in the elastic deformable rocks.

In December 1975 an abrupt rifting episode started in northeast Iceland. Below the Krafla calder a magma chamber was filled with a constant magma flow of 5 m³/s. Consequently the land surface was elevated by a rate of a few millimeters per day for weeks or months until the volume of magma in the chamber was sufficient to initiate fissuring of the elastic crust above the magma chamber and to the north and south of

Fig. 1: A volcanic fissure eruption in the centre of the Krafla caldera occured during the rifting event of September 8-9, 1977

the Krafla calder. At the moment of magma injecting the new formed fissures the land surface collapsed during a period of a few hours to a few days. Until Sept. 4, 1984, a total of 21 pulses or rifting events occured leading to the opening of a new 100 km -long and 7.5 m wide fissure swarm and to 9 volcanic fissure eruptions one of which can be seen in Fig. 1. Geodetic measurements [1; 2] revealed that during each rifting event the new formed fissure swarm subsided whereas the area east and west of the fissures was contracted and uplifted.

PREVIOUS MEASUREMENTS with STRAIN GAGES

In 1976 we started an interdisciplinary geological-experimental mechanics project in order to monitor elastic deformation of the Icelandic crust due to those abrupt rifting events. First we applied the overcoring method to determine in-situ rock deformation by means of strain gage rosettes at 20 sites located within the area of active rifting, between the Krafla fissure swarm and the east coast and across entire South Iceland.
Horizontal principal tensile stress which was E 34°S oriented could only be recorded in a 2.5 km wide NNE-SSW Running zone of active fissuring, whereas outside that zone horizontal principal stresses were always compressive indicating a consistent NW-SW orientation. Only within the western branch of the southern rift zone horizontal principal stresses were compressive and parallel to the strike (N35°E) of the fissures [3] .

We used strain gages also for long-period recordings of strain fluctuation from 1976 to 1980 at 14 of these sites. In 1977 the zone of
NW-SE directed tensile stress had expanded across the entire nothern
axial rift zone.
In summer 1978 a complete systematic reorganisation of the strain components could be observed. The strain ellipses were in a radial arrangement with the axes of horizontal principal compressive strain directed toward the area of intensive rifting in northern Iceland. In 1979
the axes of horizontal principal compressive strain still pointed to
the area of intensive rifting in northern Iceland indicating that radial stresses originating from the Krafla pulsation centre were still
recorded over most of the island.
The strain ellipses of 1980 had rotated by 90° according to those of
1979. The results of strain measurements and the confirmation of the
rifting events encouraged to continue the investigation with a caliper
type displacementtransducer, being safer under field conditions as
strain gages.

PROBLEMS AND SOLUTIONS

The aim of the measurements is to ascertain displacement due to strain
at the surface of horizontal rock exposures over a long period of time.
For this purpose, bench marks are cemented into the rock surface at
selected measuring sites and the changes in distance between the bench
marks measured with a special displacement transducer. The transducer
is not permanently installed at one site, but is transported from site
to site and measurements made at discrete intervals, the transducer
being set on the bench marks.

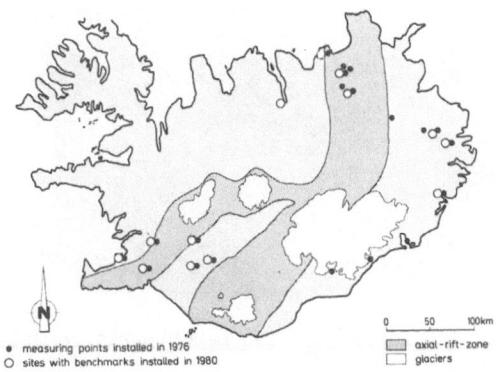

Fig.2: Location of the rift zone across Iceland and sites of
       measurement

The measuring systems used comprises the transducer and a battery-
powered measuring device which outputs a digital display of the signal
from the transducer. The highest resolution possible at the most sensitive setting is where one digit of the display is equivalent to
a displacement between the bench marks of 0.000125 mm.

Measurements at each site in at least three directions are required to ascertain the principal strains, principal directions and, possibly, principal stresses. For this reason, each measuring site has four bench marks in the directions South-North, East-West and South-West/North-East. The nominal distance between the measuring points is 10 cm. From each site we took essays for laboratory investigations to determine the modulus of elasticity, Poisson's ratio, thermal coefficient of expansion and possible anisotropic behaviour of the material. All investigated rocks revealed thermal and elastic anisotropies. Thermally induced stresses were eleminated by referring to the annual mean temperature at each site. Thus, we could calculate tectonic stresses from the strain data. In 1980, 13 sites ware installed for displacement transducer measurement. These sites were located within the area of active rifting between the northern axial rift zone and the east coast, and along the south coast, as shown in Fig. 2.

DISPLACEMENT TRANSDUCER

The displacement transducer used is capable of detecting changes in distance between to bench marks fixed in a surface with high sensitivity. The bench marks are steel balls 1.5 mm in diameter fixed on steel bolts, sunk into the surface of the object to be measured. The sensing feet of the transducer are set on these balls. The measuring principle is based on a soft measuring tongue on which a temperature-compensated strain gage full bridge circuit is rested. The transducer emits an electrical signal proportional to the deflection of the measuring tongue. The measuring tongue can handle displacements of±2.5 mm from centre. The linearity error over the entire measuring range is less than 0.05 %. The device has been so calibrated that a deflection of the sensing tip of 1 mm produces a signal of 1 mV per Volt excitation voltage.

Fig. 3: Displacement transducer and two bench marks cemented into the
         rock surface

The displacement transducer as used has a reference length of 100 mm, so the nominal distance between two bench marks must also be 100 mm. The rod on the transducer spanning 100 mm is of austenitic steel with a linear thermal expansion coefficient of $16*10^{-6}$/K. Fig. 3 shows the displacement transducer and two bench marks cemented into the rock. A plastics thermal insulator has been affixed to the grip to prevent the heat from the user's hand influencing the temperature of the rod (Fig. 3, left).

INSTALLATION OF THE MEASURING SITES

Most of the test sites were installed on the flat rock bottom of wide glacial valleys at a suitable distance from the valley walls. At the other sites it was ensured that the distance of the measuring site from the next mountain or valley was at least twice as great as the difference in altitude between that of the measuring point and the peak of the next mountain or the bottom of the next valley. The weathered encrustations of the rock surface were abraded away and the unweathered rock exposed at the measuring sites.

Each bench mark comprises a stainless steel bolt with a ball at its head upon which the displacement transducer rests. The bolt is cemented into the rock. The dimensions of the bolt can be seen in Fig. 4. Each site has four bench marks in the arrangement shwon in Fig. 5. The template for installing the bolts, a 10 mm thick aluminium plate with inset steel holders, is also shown in Fig. 5.

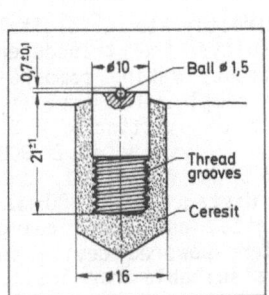

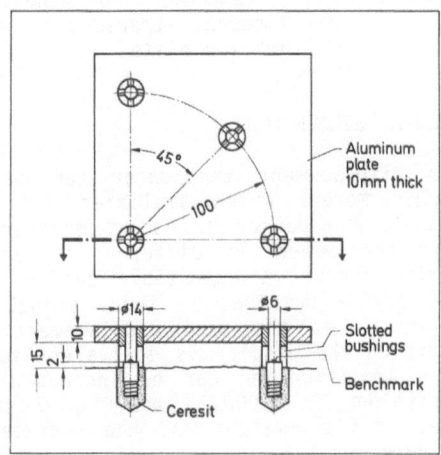

Fig. 4: Bench mark cemented into the rock for setting the displacement transducer

Fig. 5: Template for the installation of the benchmarks at the measuring site

First the template is positioned with the aid of a compass; then, four 6 mm diameter holes are drilled into the rock using the template as a guide. The template is then removed and the holes drilled out to a diameter of 16 mm. The bolts (as shown in Fig. 5) are then wedged into the holders in the base of the template. A retainer in the template ensures that the heads of the bolts are wedged at a uniform depht. Ceresit rapid-hardening cement is inserted into the holes and the bolts inserted using the template. After a few minutes have elapsed, the cement has hardened and the template can be removed. The distance between the bench marks is then 100 mm with a good degree of accuracy. The measuring device described is then used to take an absolute measurement. This procedure is shown in Fig. 6, which also shows the four measuring points at a measuring site.

628

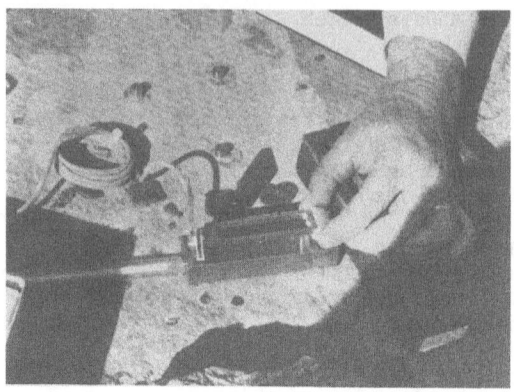

Fig. 6: Zero adjustment of the measuring system by setting the displacement transducer to the reference distance of the calibration plate

DIGITAL STRAIN METER

The displacement transducer can be connected directly to a digital strain meter, which amplifies and digitally displays the transducer signal. The device has a compensatory zero adjustment, making comparative measurement of different distances extremely simple. The amplification factor is adjustable over a wide range. At the setting $k = 2$, the device resolves an input signal of 1 mV/V to a display of 2 000 digits, i.e. the used displacement transducer DD1 is connected, one digit on the display is equivalent to a change in distance of 0.0005mm. This amplification can be increased to 8 000 digits, one digit being equivalent to 0.000125 mm. Fig. 7 shows the battery-powered device in use. Its low weight and handy dimensions make it suitable for insitu measurements.

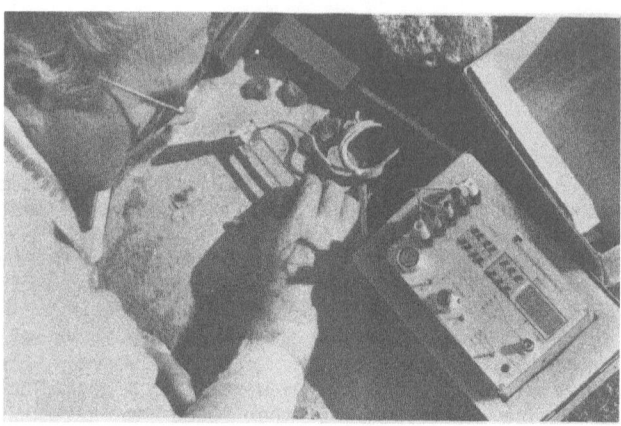

Fig. 7: Measuring distances at a site using the displacement transducer and digital strain meter

TAKING THE MEASUREMENT

The variable to be measured is the distance between two bench marks. This can be simply achieved if a known reference distance is available for on-site calibration of the measuring system. This is available in the form of a calibratiuon plate. The transducer is set on the measuring points of the calibration plate as shown in Fig. 6 and zero compensation carried out on the strain meter. Then the transducer is set on the bench marks in the rock. The value displayed is the difference between the reference value and measured variable. The measured variable can therefore easily be ascertained, depending on sign, from the sum or difference of calibration distance and measured value. The temperature of the calibration plate is measured during zero calibration to allow for thermal expansion. As the thermal coefficient of the plate is known, the actual reference distance can be ascertained with the relevant corrections. As the aim of the measurements is ascertain changes in distance and all zero calibration measurements are taking using the same calibration plate, the amount of the reference distance at reference temperature is of minor importance as it is the basis of all measurements.

RESULTS

In 1981 the most distant sites from the Krafla volcanic centre still revealed the same orientation of the strain ellipses as of 1980, whereas the sites being closer to the Krafla area again indicated a rectan-

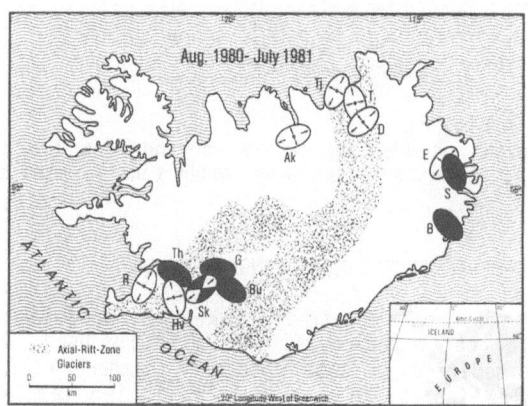

Fig. 8: Dark ellipses reveal compressive stress, the short axis corresponds to the maximum, the long axis to the minimum horizontal compressive stress. Light ellipses depict tensile stress, the short axis corresponds to minimum, the long axis to maximum horizontal tensile stress. The rifting episodes occured during the recording period of August 1980 to July 1981. The site T is located within the Krafla fissure swarm, the sites D and Tj are outside of it. Tensile stress built up from 1980 to 1981 in northeast Iceland, whereas compression and tensile stress was recorded in east and southwest Iceland at the same time.

gular rotation. In summer of 1982 only the sites which are nearer to Krafla area than the sites of the east coast and in the southwest of Iceland revealed again a rotation of the strain ellipses (Fig. 8). There was no rifting event from November 18-23, 1981 to September 4, 1984, but land elevation in the area of the Krafla caldera continued all the time leading to a general expansion of the crust which we recorded at all sites (Fig. 9)
We consider all stress fluctuations determined from the strain measurements to be caused by the pulsation of the Krafla volcanic centre. We intend to continue our measurements of crustal strain in Iceland as long as the rifting episode is active and will proceed even beyond the time when abrupts rifting activity has ceased to record the following consequences of crustal strain behaviour.

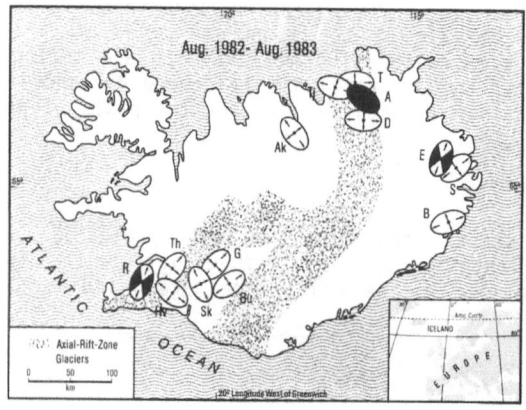

Fig. 9: (Symbols as in Fig. 8)
No rifting event occured during the period from 1982 to 1983. Except site A, all other sites were under tensile stress

PLANNED FURTHER DEVELOPMENTS OF THE MEASURING TECHNIQUE

The good results of measurements already taken encourage us to continue. On the basis of our experiences, the technique used for future measurements will be modified in some points. A hammer drill will be used to drill the holes instead of a normal drill. Two templates will therefore be required for drilling the holes and cementing the bolts.

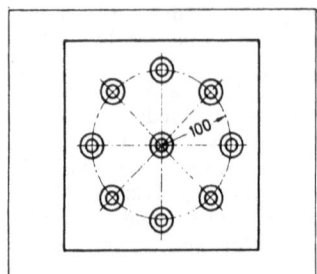

Fig. 10: Proposed arrangement of bench marks for future measurements

The number of measuring points per site will be increased to 9 in the arrangement as shown in Fig. 10. This arrangement permits two measurements to be taken for each direction, which increases their reliability. The distance bar of the transducer will be constructed of Invar steel. Trials with measuring distances greater then 100 mm are in progress.

REFERENCES

[1] Möller, D., B. Ritter, K. Wendt: Geodetic measurement of horizontal deformations in northeast Iceland; Earth Eval. Sci 2 (1982) p. 149-154

[2] Kanngieser, E.: Vertical component of ground deformation in north Iceland; Ann. Geophys. 1 (1983) p. 321 - 328

[3] Schäfer, K., St. Keil: In situ Gesteinsspannungsermittlungen in Island; Messtechnische Briefe 15 (1979) p. 35 - 46

[4] Björnsson, A.: Dynamics of Crustal Rifting in NE Iceland; J. Geophys. Res. 90 (1985) p. 10151 - 10162

[5] Schäfer, K.: Horizontal and Vertical Movements in Icelands; Tectonophysics 29 (1975) p. 223 - 231

INDEX